T0238657

Lecture Notes in Computer Science 8384

Commenced Publication in 1973
Founding and Former Series Editors:
Gerhard Goos, Juris Hartmanis, and Jan van Leeuwen

Editorial Board

David Hutchison
Lancaster University, Lancaster, UK

Takeo Kanade
Carnegie Mellon University, Pittsburgh, PA, USA

Josef Kittler
University of Surrey, Guildford, UK

Jon M. Kleinberg
Cornell University, Ithaca, NY, USA

Alfred Kobsa
University of California, Irvine, CA, USA

Friedemann Mattern
ETH Zurich, Zürich, Switzerland

John C. Mitchell
Stanford University, Stanford, CA, USA

Moni Naor
Weizmann Institute of Science, Rehovot, Israel

Oscar Nierstrasz
University of Bern, Bern, Switzerland

C. Pandu Rangan
Indian Institute of Technology, Madras, India

Bernhard Steffen
TU Dortmund University, Dortmund, Germany

Demetri Terzopoulos
University of California, Los Angeles, CA, USA

Doug Tygar
University of California, Berkeley, CA, USA

Gerhard Weikum
Max Planck Institute for Informatics, Saarbruecken, Germany

For further volumes:
http://www.springer.com/series/7407

Roman Wyrzykowski · Jack Dongarra
Konrad Karczewski · Jerzy Waśniewski (Eds.)

Parallel Processing and Applied Mathematics

10th International Conference, PPAM 2013
Warsaw, Poland, September 8–11, 2013
Revised Selected Papers, Part I

 Springer

Editors
Roman Wyrzykowski
Konrad Karczewski
Institute of Computer and
  Information Science
Czestochowa University of Technology
Czestochowa
Poland

Jerzy Waśniewski
Informatics and Mathematical Modelling
Technical University of Denmark
Kongens Lyngby
Denmark

Jack Dongarra
Department of Computer Science
University of Tennessee
Knoxville, TN
USA

ISSN 0302-9743   ISSN 1611-3349   (electronic)
ISBN 978-3-642-55223-6   ISBN 978-3-642-55224-3   (eBook)
DOI 10.1007/978-3-642-55224-3
Springer Heidelberg New York Dordrecht London

Library of Congress Control Number: 2014937670

LNCS Sublibrary: SL1 – Theoretical Computer Science and General Issues

© Springer-Verlag Berlin Heidelberg 2014
This work is subject to copyright. All rights are reserved by the Publisher, whether the whole or part of the material is concerned, specifically the rights of translation, reprinting, reuse of illustrations, recitation, broadcasting, reproduction on microfilms or in any other physical way, and transmission or information storage and retrieval, electronic adaptation, computer software, or by similar or dissimilar methodology now known or hereafter developed. Exempted from this legal reservation are brief excerpts in connection with reviews or scholarly analysis or material supplied specifically for the purpose of being entered and executed on a computer system, for exclusive use by the purchaser of the work. Duplication of this publication or parts thereof is permitted only under the provisions of the Copyright Law of the Publisher's location, in its current version, and permission for use must always be obtained from Springer. Permissions for use may be obtained through RightsLink at the Copyright Clearance Center. Violations are liable to prosecution under the respective Copyright Law.
The use of general descriptive names, registered names, trademarks, service marks, etc. in this publication does not imply, even in the absence of a specific statement, that such names are exempt from the relevant protective laws and regulations and therefore free for general use.
While the advice and information in this book are believed to be true and accurate at the date of publication, neither the authors nor the editors nor the publisher can accept any legal responsibility for any errors or omissions that may be made. The publisher makes no warranty, express or implied, with respect to the material contained herein.

Printed on acid-free paper

Springer is part of Springer Science+Business Media (www.springer.com)

Preface

This volume comprises the proceedings of the 10th International Conference on Parallel Processing and Applied Mathematics, PPAM 2013, which was held in Warsaw, Poland, September 8–11, 2013. The jubilee PPAM conference was organized by the Department of Computer and Information Science of the Czestochowa University of Technology, under the patronage of the Committee of Informatics of the Polish Academy of Sciences, in cooperation with the Polish-Japanese Institute of Information Technology. The main organizer was Roman Wyrzykowski.

PPAM is a biennial conference. Nine previous events have been held in different places in Poland since 1994. The proceedings of the last six conferences have been published by Springer-Verlag in the *Lecture Notes in Computer Science* series (Nałęczów, 2001, vol. 2328; Częstochowa, 2003, vol. 3019; Poznań, 2005, vol. 3911; Gdańsk, 2007, vol. 4967; Wrocław, 2009, vols. 6067 and 6068; Toruń, 2011, vols. 7203 and 7204).

The PPAM conferences have become an international forum for exchanging ideas between researchers involved in parallel and distributed computing, including theory and applications, as well as applied and computational mathematics. The focus of PPAM 2013 was on models, algorithms, and software tools that facilitate efficient and convenient utilization of modern parallel and distributed computing architectures, as well as on large-scale applications.

This meeting gathered the largest number of participants in the history of PPAM conferences – more than 230 participants from 32 countries. A strict refereeing process resulted in the acceptance of 143 contributed presentations, while approximately 44 % of the submissions were rejected. Regular tracks of the conference covered such important fields of parallel/distributed/cloud computing and applied mathematics as:

- Numerical algorithms and parallel scientific computing
- Parallel non-numerical algorithms
- Tools and environments for parallel/distributed/cloud computing
- Applications of parallel computing
- Applied mathematics, evolutionary computing, and metaheuristics

The plenary and invited talks were presented by:

- Fran Berman from the Rensselaer Polytechnic Institute (USA)
- Ewa Deelman from the University of Southern California (USA)
- Jack Dongarra from the University of Tennessee and Oak Ridge National Laboratory (USA), and University of Manchester (UK)
- Geoffrey Ch. Fox from Indiana University (USA)
- Laura Grigori from Inria (France)
- Fred Gustavson from the IBM T.J. Watson Research Center (USA)
- Georg Hager from the University of Erlangen-Nuremberg (Germany)
- Alexey Lastovetsky from the University College Dublin (Ireland)

- Miron Livny from the University of Wisconsin (USA)
- Piotr Luszczek from the University of Tennessee (USA)
- Rizos Sakellariou from the University of Manchester (UK)
- James Sexton from the IBM T.J. Watson Research Center (USA)
- Leonel Sousa from the Technical University of Lisbon (Portugal)
- Denis Trystram from the Grenoble Institute of Technology (France)
- Jeffrey Vetter from the Oak Ridge National Laboratory and Georgia Institute of Technology (USA)
- Richard W. Vuduc from the Georgia Institute of Technology (USA)
- Robert Wisniewski from Intel (USA)

Important and integral parts of the PPAM 2013 conference were the workshops:

- Minisympsium on GPU Computing organized by José R. Herrero from the Universitat Politecnica de Catalunya (Spain), Enrique S. Quintana-Ortí from the Universidad Jaime I (Spain), and Robert Strzodka from NVIDIA
- Special Session on Multicore Systems organized by Ozcan Ozturk from Bilkent University (Turkey), and Suleyman Tosun from Ankara University (Turkey)
- Workshop on Numerical Algorithms on Hybrid Architectures organized by Przemysław Stpiczyński from the Maria Curie Skłodowska University (Poland), and Jerzy Waśniewski from the Technical University of Denmark
- Workshop on Models, Algorithms and Methodologies for Hierarchical Parallelism in New HPC Systems organized by Giulliano Laccetti and Marco Lapegna from the University of Naples Federico II (Italy), and Raffaele Montella from the University of Naples Parthenope (Italy)
- Workshop on Power and Energy Aspects of Computation organized by Richard W. Vuduc from the Georgia Institute of Technology (USA), Piotr Luszczek from the University of Tennessee (USA), and Leonel Sousa from the Technical University of Lisbon (Portugal)
- Workshop on Scheduling for Parallel Computing, SPC 2013, organized by Maciej Drozdowski from Poznań University of Technology (Poland)
- The 5th Workshop on Language-Based Parallel Programming Models, WLPP 2013, organized by Ami Marowka from the Bar-Ilan University (Israel)
- The 4th Workshop on Performance Evaluation of Parallel Applications on Large-Scale Systems organized by Jan Kwiatkowski from Wrocław University of Technology (Poland)
- Workshop on Parallel Computational Biology, PBC 2013, organized by David A. Bader from the Georgia Institute of Technology (USA), Jarosław Żola from Rutgers University (USA), and Bertil Schmidt from the University of Mainz (Germany)
- Minisymposium on Applications of Parallel Computations in Industry and Engineering organized by Raimondas Čiegis from Vilnius Gediminas Technical University (Lithuania), and Julius Žilinskas from Vilnius University (Lithuania)
- Minisymposium on HPC Applications in Physical Sciences organized by Grzegorz Kamieniarz and Wojciech Florek from A. Mickiewicz University in Poznań (Poland)

- Minisymposium on Applied High-Performance Numerical Algorithms in PDEs organized by Piotr Krzyżanowski and Leszek Marcinkowski from Warsaw University (Poland), and Talal Rahman from Bergen University College (Norway)
- Minisymposium on High-Performance Computing Interval Methods organized by Bartłomiej J. Kubica from Warsaw University of Technology (Poland)
- Workshop on Complex Colective Systems organized by Paweł Topa and Jarosław Wąs from AGH University of Science and Technology in Kraków (Poland)

The PPAM 2013 meeting began with five tutorials:

- Scientific Computing on GPUs, by Dominik Göddeke from the University of Dortmund (Germany), and Robert Strzodka from NVIDIA
- Design and Implementation of Parallel Algorithms for Highly Heterogeneous HPC Platforms, by Alexey Lastovetsky from University College Dublin (Ireland)
- Node Level Performance Engineering, by Georg Hager from the University of Erlangen-Nuremberg (Germany)
- Delivering the OpenCl Performance Promise: Creating and Optimizing OpenCl Applications with the Intel OpenCl SDK, by Maxim Shevtsov from Intel (Russia)
- A History of A Central Result of Linear Algebra and the Role of that Gauss, Cholesky and Others Played in Its Development, by Fred Gustavson from the IBM T.J. Watson Research Center (USA)

The PPAM Best Poster Award is granted to the best poster on display at the PPAM conferences, and was established at PPAM 2009. This award is bestowed by the Program Committee members to the presenting author(s) of the best poster. The selection criteria are based on the scientific content, and on the quality of the poster presentation. The PPAM 2013 winners were Lars Karlsson, and Carl Christian K. Mikkelsen from Umea University, who presented the poster "Improving Perfect Parallelism." The Special Award was bestowed to Lukasz Szustak, and Krzysztof Rojek from the Częstochowa University of Technology, and Pawel Gepner from Intel, who presented the poster "Using Intel Xeon Phi to Accelerate Computation in MPDATA Algorithm."

A new topic was introduced at PPAM 2013: *Power and Energy Aspects of Computation (PEAC)*. Recent advances in computer hardware rendered the issues related to power and energy consumption as the driving metric for the design of computational platforms for years to come. Power-conscious designs, including multicore CPUs and various accelerators, dominate large supercomputing installations as well as large industrial complexes devoted to cloud computing and the big data analytics. At stake are serious financial and environmental impacts, which the large-scale computing community has to now consider and embark on careful re-engineering of software to fit the demanding power caps and tight energy budgets.

The workshop presented research into new ways of addressing these pressing issues of energy preservation, power consumption, and heat dissipation while attaining the best possible performance levels at the scale demanded by modern scientific challenges.

The PEAC Workshop, as well as the conference as a whole, featured a number of invited and contributed talks covering a diverse array of recent advances, including:

- Cache-aware roofline model for monitoring performance and power in connection with application characterization (by L. Sousa et al.)
- Resource scheduling and allocation schemes based on stochastic models (by M. Oxley et al.)
- A comprehensive study of iterative solvers on a large variety of computing platforms including modern CPUs, accelerators, and embedded computers (by Enrique S. Quintana-Ortí et al.)
- Energy and power consumption trends in HPC (by P. Luszczek)
- Sensitivity of graph metrics to missing data and the benefits they have for overall energy consumption (by A. Zakrzewska et al.)
- Cache energy models and their analytical properties in the context of embedded devices (by K. de Vogeleer et al.)
- Predictive models for execution time, energy consumption, and power draw of algorithms (by R. Vuduc)

The organizers are indebted to the PPAM 2013 sponsors, whose support was vital to the success of the conference. The main sponsor was the Intel Corporation. The other sponsors were: IBM Corporation, Hewlett-Packard Company, Rogue Wave Software, and AMD. We thank to all the members of the international Program Committee and additional reviewers for their diligent work in refereeing the submitted papers. Finally, we thank all of the local organizers from the Częstochowa University of Technology, and the Polish-Japanese Institute of Information Technology in Warsaw, who helped us to run the event very smoothly. We are especially indebted to Grażyna Kołakowska, Urszula Kroczewska, Łukasz Kuczyński, Adam Tomaś, and Marcin Woźniak from the Częstochowa University of Technology; and to Jerzy P. Nowacki, Marek Tudruj, Jan Jedliński, and Adam Smyk from the Polish-Japanese Institute of Information Technology.

We hope that this volume will be useful to you. We would like everyone who reads it to feel invited to the next conference, PPAM 2015, which will be held September 6–9, 2015, in Kraków, the old capital of Poland.

January 2014

Roman Wyrzykowski
Jack Dongarra
Konrad Karczewski
Jerzy Waśniewski

Organization

Program Committee

Jan Węglarz	Poznań University of Technology, Poland (Honorary Chair)
Roman Wyrzykowski	Częstochowa University of Technology, Poland (Program Committee Chair)
Ewa Deelman	University of Southern California, USA (Program Committee Vice-Chair)
Francisco Almeida	Universidad de La Laguna, Spain
Pedro Alonso	Universidad Politecnica de Valencia, Spain
Peter Arbenz	ETH, Zurich, Switzerland
Piotr Bała	Nicolaus Copernicus University, Poland
David A. Bader	Georgia Institute of Technology, USA
Michael Bader	TU München, Germany
Włodzimierz Bielecki	West Pomeranian University of Technology, Poland
Paolo Bientinesi	RWTH Aachen, Germany
Radim Blaheta	Institute of Geonics, Czech Academy of Sciences
Jacek Błażewicz	Poznań University of Technology, Poland
Adam Bokota	Częstochowa University of Technology, Poland
Pascal Bouvry	University of Luxembourg
Tadeusz Burczyński	Silesia University of Technology, Poland
Jerzy Brzeziński	Poznań University of Technology, Poland
Marian Bubak	AGH Kraków, Poland, and University of Amsterdam, The Netherlands
Christopher Carothers	Rensselaer Polytechnic Institute, USA
Jesus Carretero	Universidad Carlos III de Madrid, Spain
Raimondas Čiegis	Vilnius Gediminas Technical University, Lithuania
Andrea Clematis	IMATI-CNR, Italy
Jose Cunha	University Nova of Lisbon, Portugal
Zbigniew Czech	Silesia University of Technology, Poland
Jack Dongarra	University of Tennessee and ORNL, USA, and University of Manchester, UK
Maciej Drozdowski	Poznań University of Technology, Poland
Erik Elmroth	Umea University, Sweden
Mariusz Flasiński	Jagiellonian University, Poland
Franz Franchetti	Carnegie Mellon University, USA
Tomas Fryza	Brno University of Technology, Czech Republic
Pawel Gepner	Intel Corporation

Domingo Gimenez	University of Murcia, Spain
Mathieu Giraud	LIFL and Inria, France
Jacek Gondzio	University of Edinburgh, UK
Andrzej Gościński	Deakin University, Australia
Laura Grigori	Inria, France
Adam Grzech	Wroclaw University of Technology, Poland
Inge Gutheil	Forschungszentrum Juelich, Germany
Georg Hager	University of Erlangen-Nuremberg, Germany
José R. Herrero	Universitat Politecnica de Catalunya, Barcelona, Spain
Ladislav Hluchy	Slovak Academy of Sciences, Slovakia
Florin Isaila	Universidad Carlos III de Madrid, Spain
Ondrej Jakl	Institute of Geonics, Czech Academy of Sciences
Emmanuel Jeannot	Inria, France
Bo Kågström	Umea University, Sweden
Alexey Kalinov	Cadence Design System, Russia
Aneta Karaivanova	Bulgarian Academy of Sciences, Sofia
Eleni Karatza	Aristotle University of Thessaloniki, Greece
Ayse Kiper	Middle East Technical University, Turkey
Jacek Kitowski	Institute of Computer Science, AGH, Poland
Jozef Korbicz	University of Zielona Góra, Poland
Stanislaw Kozielski	Silesia University of Technology, Poland
Dieter Kranzlmueller	Ludwig Maximillian University, Munich, and Leibniz Supercomputing Centre, Germany
Henryk Krawczyk	Gdańsk University of Technology, Poland
Piotr Krzyżanowski	University of Warsaw, Poland
Mirosław Kurkowski	Częstochowa University of Technology, Poland
Krzysztof Kurowski	PSNC, Poznań, Poland
Jan Kwiatkowski	Wrocław University of Technology, Poland
Jakub Kurzak	University of Tennessee, USA
Giulliano Laccetti	University of Naples Federico II, Italy
Marco Lapegna	University of Naples Federico II, Italy
Alexey Lastovetsky	University College Dublin, Ireland
Joao Lourenco	University Nova of Lisbon, Portugal
Hatem Ltaief	KAUST, Saudi Arabia
Emilio Luque	Universitat Autonoma de Barcelona, Spain
Vyacheslav I. Maksimov	Ural Branch, Russian Academy of Sciences
Victor E. Malyshkin	Siberian Branch, Russian Academy of Sciences
Pierre Manneback	University of Mons, Belgium
Tomas Margalef	Universitat Autonoma de Barcelona, Spain
Svetozar Margenov	Bulgarian Academy of Sciences, Sofia
Ami Marowka	Bar-Ilan University, Israel
Norbert Meyer	PSNC, Poznań, Poland
Jarek Nabrzyski	University of Notre Dame, USA
Raymond Namyst	University of Bordeaux and Inria, France
Maya G. Neytcheva	Uppsala University, Sweden

Gabriel Oksa	Slovak Academy of Sciences, Bratislava
Ozcan Ozturk	Bilkent University, Turkey
Tomasz Olas	Częstochowa University of Technology, Poland
Marcin Paprzycki	IBS PAN and SWPS, Warsaw, Poland
Dana Petcu	West University of Timisoara, Romania
Enrique S. Quintana-Ortí	Universidad Jaime I, Spain
Jean-Marc Pierson	Paul Sabatier University, France
Thomas Rauber	University of Bayreuth, Germany
Paul Renaud-Goud	Inria, France
Jacek Rokicki	Warsaw University of Technology, Poland
Gudula Runger	Chemnitz University of Technology, Germany
Leszek Rutkowski	Częstochowa University of Technology, Poland
Robert Schaefer	Institute of Computer Science, AGH, Poland
Olaf Schenk	Università della Svizzera Italiana, Switzerland
Stanislav Sedukhin	University of Aizu, Japan
Franciszek Seredyński	Cardinal Stefan Wyszyński University in Warsaw, Poland
Happy Sithole	Centre for High Performance Computing, South Africa
Jurij Silc	Jozef Stefan Institute, Slovenia
Karolj Skala	Ruder Boskovic Institute, Croatia
Peter M.A. Sloot	University of Amsterdam, The Netherlands
Leonel Sousa	Technical University of Lisbon, Portugal
Radek Stompor	Université Paris Diderot and CNRS, France
Przemysław Stpiczyński	Maria Curie Skłodowska University, Poland
Maciej Stroiński	PSNC, Poznań, Poland
Ireneusz Szcześniak	Częstochowa University of Technology, Poland
Boleslaw Szymanski	Rensselaer Polytechnic Institute, USA
Domenico Talia	University of Calabria, Italy
Christian Terboven	RWTH Aachen, Germany
Andrei Tchernykh	CICESE Research Center, Ensenada, Mexico
Suleyman Tosun	Ankara University, Turkey
Roman Trobec	Jozef Stefan Institute, Slovenia
Denis Trystram	Grenoble Institute of Technology, France
Marek Tudruj	Polish Academy of Sciences and Polish-Japanese Institute of Information Technology, Warsaw, Poland
Bora Uçar	Ecole Normale Superieure de Lyon, France
Marian Vajtersic	Salzburg University, Austria
Jerzy Waśniewski	Technical University of Denmark
Bogdan Wiszniewski	Gdańsk University of Technology, Poland
Andrzej Wyszogrodzki	IMGW, Warsaw, Poland
Ramin Yahyapour	University of Göttingen/GWDG, Germany
Jianping Zhu	Cleveland State University, USA
Julius Žilinskas	Vilnius University, Lithuania
Jarosław Żola	Rutgers University, USA

Contents – Part I

Parallel Non-Numerical Algorithms

Tools and Environments for Parallel/Distributed/Cloud Computing

Minisymposium on GPU Computing

Special Session on Multicore Systems

Workshop on Numerical Algorithms on Hybrid Architectures

Workshop on Models, Algorithms, and Methodologies for Hierarchical Parallelism in New HPC Systems

Workshop on Power and Energy Aspects of Computation

Contents – Part II

The 4th Workshop on Performance Evaluation of Parallel Applications on Large-Scale Systems

Workshop on Parallel Computational Biology (PBC 2013)

Minisymposium on Applications of Parallel Computation in Industry and Engineering

Minisymposium on HPC Applications in Physical Sciences

Minisymposium on Applied High Performance Numerical Algorithms in PDEs

Minisymposium on High Performance Computing Interval Methods

Workshop on Complex Collective Systems

Algebra and Geometry Combined Explains How the Mind Does Math

Fred G. Gustavson[1,2(✉)]

[1] IBM T.J. Watson Research Center, Ossining, USA
[2] Umeå University, Umeå, Sweden
fg2935@gmail.com

Abstract. This paper updates my talk on Cache Blocking for Dense Linear Algorithms since 1985 given at PPAM 11; see [11]. We again apply Dimension Theory to matrices in the Fortran and C programming languages. New Data Structures (**NDS**) for matrices are given. We use the GCD algorithm to transpose a n by m matrix A in CMO order, standard layout, in-place. Algebra and Geometry are used to make this idea concrete and practical; it is the reason for title of our paper: make a picture of any matrix by the GCD algorithm to convert it into direct sum of square submatrices. The picture is Geometry and the GCD algorithm is Algebra. Also, the in-place transposition of the GKK and TT algorithms will be compared. Finally, the importance of using negative integers will be used to give new results about subtraction and finding primitive roots which also make a priori in-place transpose more efficient.

Keywords: New data structures (**NDS**) · In-place matrix transposition · Cache blocking · Dimension theory · Negative integers

1 Introduction

This paper updates research about **NDS** that were given at PPAM 11 [11] and so we just overview [11] to make this paper cogent and self-contained. We claim that partitioning matrices into disjoint submatices using **NDS** is very important for DLA (**D**ense **L**inear **A**lgebra). The standard matrix data layouts, column and row major order, (CMO and RMO) are **very unlikely** to change. Hence, submatrices must be reformatted so that moving them between memory and caches allow them to benefit from the principles underlying cache blocking. In this regard, the concluding summary of [11] said: We indicated [3,4] that DLAFA (**F**actorization **A**lgorithms) are mainly MM (**M**atrix **M**ultiply) algorithms. The standard API, **A** **P**rogramming **I**nterface, for matrices and Level-3 BLAS use arrays; see page 739 of [2]. All standard array layouts are *one* dimensional. It is *impossible* to maintain locality of reference in a matrix or any higher than 1-D object using a 1-D layout; see [19]. MM requires row and column operations and thus requires MT (**M**atrix **T**ransformation) to NDS. Our results on in-place MT show that performance suffers greatly when one uses a 1-D layout. Using NDS

R. Wyrzykowski et al. (Eds.): PPAM 2013, Part I, LNCS 8384, pp. 1–11, 2014.
DOI: 10.1007/978-3-642-55224-3_1, © Springer-Verlag Berlin Heidelberg 2014

for matrices "approximates" a 2-D layout; thus, one can dramatically improve in-place MT performance as well as DLAFA performance. Our message is that DLAFA are mostly MM. MM requires MT and both require NDS. Thus, DLAFA can and do perform well on multicore but only if one uses NDS.

An application of the Algorithms and Architecture Approach [4] describes a "fundamental principle" of Linear Algebra called the "Principle of Linear Superposition". We use it to describe the factorization algorithms of DLA of a matrix A in terms of its sub matrices A_{ij} instead of its elements a_{ij}. These submatrices must be laid out optimally on a given platform to ensure automatic cache blocking! The LAPACK and ScaLAPACK libraries were also based on this fundamental principle. However, both of these libraries use standard data layouts for matrices and the Level-3 BLAS to gain their performance on all platforms. This decision worked until the introduction of **M**ulti-**C**ore, MC. Peak performance for DLA factorization only occurs if all matrix operands are used multiple times when they enter an L1 cache or core. This ensures that the initial cost of bringing an operand into cache is then amortized by the ratio of $O(n^3)$ arithmetic to $O(n^2)$ elements or **nb** flops[1] per matrix element a_{ij}. Multiple reuse of all operands *only* occurs if all matrix operands map well into the L1 caches. For MC processors, an "L1 cache" is the data area of a core. For MC it is critical to get submatrices to the cores as fast as possible. The standard programming interface, called API, that hold matrices A that the BLAS and DLA libraries use is the 2-D array of the Fortran and C programming languages. For this API, submatrices A_{ij} of A are *not* stored contiguously. Thus it is *impossible* to move A_{ij} to and from the memory hierarchy from and to the various cores in a fast or optimal manner! This problem is corrected by using NDS to hold these submatrices A_{ij}. By using dimension theory [19] we shall indicate why this is true.

For MC the disproportion between multiple CPU processing and memory speed has become much higher. On a negative side, the API for BLAS-3 hurts performance; it requires repeated matrix data reformatting from its API to NDS. A new "BLAS-3" concept has emerged; it is to use NDS in concert with "BLAS-3" kernels [1,4,16,17]. For MC, the broad idea of "cache blocking" is mandatory as matrix elements must be fed to SPE's or GPU's as fast as possible. Also important is the arrangement in memory of the submatrices A_{ij} of A that are to be processed. This then defines "cache blocking" on MC processors for DLA.

We describe three matrix layouts. First, we assume that the matrices are stored in **R**ectangular **B**lock (RB) format. RB format stores a M by N matrix A as contiguous rectangular submatrices A_{ij} of size MB by NB. Square Block (SB) format is a special case of RB format when the rectangle is a square. A SB of order NB is also a contiguous 1-D array of size NB^2 and for most cache designs a contiguous array whose size fits into the cache is mapped from its place in memory into the cache by the *identity* mapping. RB format has a number of other advantages. A major one is that it naturally partitions a matrix to be a matrix of sub-matrices. This allows one to view matrix transposition of a M by

---

[1] **nb** is the order of a square submatrix A_{ij} of A that enters a core.

N matrix A where $M = m$MB and $N = n$NB as a **block transposition** of a much smaller m by n block matrix A. However, usually M and N are *not* multiples of MB and NB. So, RB format as we define it here, would pad the rows and columns of A so that M and N become multiples of some blocking factors MB and NB. We think padding is an **essential condition** for this type of "cache blocking".

The second format for storing matrices is the standard 2-D array format of the Fortran and C programming languages. The third format is defined by the GCD in-place transpose algorithm [6]. The two key ideas are "the GCD algorithm is used to represent any CMO matrix A as a direct sum of square submatices A_i" and "knowing any square matrix can be transposed in-place".

Section 2 gives a discussion of Dimension Theory. It describes the "random nature" of the memory layout of the standard API and thus shows why Fortran and C arrays *cannot* be truly multi-dimensional. Section 3 describes the features of In-Place Transformations between standard full layouts of matrices and the RB or square block SB formats of NDS. The key algorithm is "vector transpose": it demonstrate a novel form of "cache blocking" where CMO A is reorganized to be very efficient for DLA algorithms. Section 4 compares the in-place transposition algorithms of TT [13, 20] and GKK [10]. Section 5 describes GCD transpose and Sect. 6 describes new results based on using negative integers.

2   Dimension Theory and Its Relation to Standard CM and RM Arrays of Fortran and C

Fortran and C use 1-D layouts for their multi-dimensional arrays. Most library software for DLA use the Fortran and C API for their matrices which are clearly 2-D. The Fundamental Theorem of Dimension Theory states that it is *impossible* to preserve closeness of all points p in a neighborhood $\mathcal{N}$ of a D dimensional object when one uses a $d < $ D dimensional coordinate system to describe the object; see pages 106 to 120 of [19]. We are concerned with a submatrix A_{ij} which is representing a $\mathcal{N}$ of a matrix A that is a 2-D object. The result says that it is *impossible* to lay out matrix A in a 1-D fashion and maintain closeness of all of the elements of A_{ij}. The result warns us about moving A_{ij} to and from cache as cache represents a $\mathcal{N}$ of 2-D object A. Computer scientists use the phrase "preserve data locality" when data is mapped from memory into a cache and we note that when data A_{ij} is contiguous in computer memory then its mapping into cache is by the *identity* mapping. Clearly, this is the **fastest way to move data** and it also **preserves data locality** in cache.

2.1   Submatrices A_{ij} of A in Fortran and C and Their Generalization

Let A have m rows and n columns with LDA $\geq m$. In Fortran the columns of A are stored stride one and the row elements are stored LDA elements apart. This is a 1-D layout. A^T has n rows and m columns with LDAT $\geq n$. Its rows are stored stride one and its columns are laid out LDAT elements apart. Again, this is a 1-D layout. Actually A and A^T are the same object and this is how we "view" A or A^T. Clearly both A and A^T contain the same information. Now, everything we just said about A and A^T applies equally well to every submatrix A_{ij} of A and its transpose A_{ij}^T. However, copies of submatrices are usually made during the processing of DLA algorithms when A is in standard layout. To avoid this we generalize the layout of A to be RB format where each scalar $a_{i,j}$ element of standard format becomes a rectangular or square submatrix $A(\texttt{I : I + MB} - 1, \texttt{J : J + NB} - 1)$ of size MB rows and NB columns. All submatrices are contiguous, meaning LDA = MB. Simple and non-simple layouts of $A(\texttt{I : I + MB} - 1, \texttt{J : J + NB} - 1)$ are used; see Sect. 2.1 of [5] and [7] for the meaning of non-simple format. Today, these non-simple formats are called rb formats standing for **r**egister **b**lock formats. The last block rows and columns of RB A are called left-over blocks. These A_{IJ} blocks reside in MB*NB locations of array storage even though they require less storage to store. It is very important to pad these left-over blocks; otherwise the theory behind in-place fast data movement of RB A breaks down somewhat or becomes less efficient. With these changes one can transpose or transform RB A submatrices both in-place and out-of-place.

2.2   Tutorial on the Essence of Dimension

We now return to our description of Dimension Theory. This is a deep mathematical subject and we think it sheds light on the subject of cache blocking. Before going further let us study 2-D domains; e.g. a matrix or a City Map. The concept to emerge is "closeness of *all* points in a neighborhood" of an arbitrary domain point p. How does one describe an arbitrary neighborhood? Here is the key question: when does a labeling of a domain satisfy the neighborhood property of closeness? This notion can be made mathematically precise and correct and so we can define dimension in a satisfactory manner.

Before answering we back up and try other labelings of domain points. Let us try natural Numbers: $1, 2, 3, \ldots$ Fortran and C use this labeling to lay out matrices A; e.g., in Fortran scalar element a_{ij} is located in computer memory at word location $\texttt{i + (j - 1)*LDA}$ past the beginning word of A. Notice that some neighboring a_{ij} elements are widely separated with this single labeling; e.g., in Fortran CM format row elements are widely separated in memory by LDA elements. Is this true for all single labelings of 2-D objects? The answer is yes. Next we need to measure distance. We give a metric for a neighborhood that uses two coordinates. We use a one norm: let $p = (u, v)$ and $q = (x, y)$ be two points. Then $\texttt{norm}(p, q) = |u - x| + |v - y|$.

We are now ready to give the mathematical essence of dimension. Indexing with single numbers, or simple enumeration is applicable only to those cases where the objects have the character of a sequence. Simple, single indexing must obey the neighborhood property and these objects are labeled one dimensional. Now consider maps, matrices, etc. and note they *cannot* be labeled by a simple sequential ordering as the "the neighborhood property will be violated" (we have said this was so above). However, two simple sequences suffice. The use of the one norm shows us why visually. Now we discuss some prior history about dimension and its resolution. There was an "erroneous prior notion" that a rectangle had more points than a line; and that a solid had more points than a rectangle! Later, Cantor's theory of infinities asserted that "All domains have the same number of points"! Thus this "erroneous prior notion" needed to be corrected. However, a difficult problem remained: "is it possible to label a domain with two different labelings that both obey the neighborhood principle of a higher to lower labeling"? The Fundamental Theorem of Dimension Theory says the answer is **no**! In 1913 L. E. J. Brouwer stated and proved this theorem which we now phrase in a slightly different manner. "It is *not* possible to label a domain with two different labelings that both obey the neighborhood principle".

3 Converting Standard Format to RB Format In-Place via Vector Transposition

In [8] we demonstrated in-place transposition for matrices stored in CM format. In terms of speed they improved the existing state-of-the-art slightly. However, they were very slow compared to out-of-place transpose algorithms. There is no way to overcome this problem if one insists on storing A in a 1-D layout. Let M $\times$ N matrix A be laid out in CM format. In Fortran we have A(0:M-1,0:N-1). The element a_{ij} is stored at offset $k = i + j$M or $A[k]$ and the in-place permutation $P(k)$ has $a_{ij} = A[k]$ ending up at offset $\bar{k} = i$N$ + j$ or at $A[\bar{k}]$. Note that $P(k)$ also equals kN mod q where $q =$ MN $- 1$ and so $P(k)$ is governed by modular arithmetic. Now modular arithmetic with parameters N and q also define different pseudo-random number generators; see [15, Section 3.2.1.3]. It follows that in-place transpose algorithms in Fortran and C will exhibit a random memory access pattern and thus have very poor performance: each memory access will likely miss in each level of the cache hierarchy. Cache miss penalties are huge (in the hundreds of cycles) for MC processors. This number theory above and dimension theory of Sect. 2, explaining poor performance, are conjectured to be related.

3.1 Dense Linear Algebra Algorithms for MC Use RB or SB Format

We need a fast way to transform A, in a standard CM or RM format, to be in RB format [9,10,14]. The idea is to move contiguous lines of data. The VIPX algorithm of [9] maps in-place CM A residing in a Fortran array A(0:mMB-1,0:nNB-1) to be RB A also residing in A. A column swath is an mMB by NB subarray of A; we call it B(0:mMB-1,0:NB-1). The LDA's of both these array are LDA $= m$MB. Under the column swath vector mapping CM A becomes m size MB by NB RB's. Algorithm VIPX is embarrassingly parallel: it is applied n times on the n column swaths of A to produce RB A. Section 3.2 describes the VIPX algorithm. CM A and RB format A will occupy array A with LDA $= m$MB where $m = \lceil M/\text{MB} \rceil$; see Sect. 2.1 where this layout padding was mentioned as being important. In [10] we improved the speed of the VIPX algorithm by using a number theory algorithm to find, a priori, the exact nature of the vector P mapping.

3.2 The VIPX Vector Transpose Algorithm

We overview how one gets from standard CM format to RB format. Recall, from Sect. 3.1, array B holds an m by NB submatrix C whose elements are column vectors of length MB. Let C^T also occupy array B. C^T is a size NB by m matrix of vectors of length MB. To see what is going on we give a small example. Let $m = 4$, MB=2 and NB=3. Then C, as a scalar matrix of size mMB $= 8$ by NB $= 3$, is also a $m = 4$ by NB $= 3$ matrix of vectors of length MB $= 2$ residing in array B. Originally, the mNB $= 12$ vectors of C are stored in CM format in array B. After in-place transposition these mnB $= 12$ vectors of C are stored in RM format in array B. This means the original mnB $= 12$ vectors now occupy array B in the permuted order $\mathcal{M} = 0, 4, 8, 1, 5, 9, 2, 6, 10, 3, 7, 11$. $\mathcal{M}$ is the list form of the permutation vector transpose mapping of CMO C to RMO C or CMO C^T. Also, C^T is **identical to** $m = 4$ RB's of size MB by NB concatenated together.

For array A holding matrix A we do n parallel B $\rightarrow$ B^T operations for each of the n concatenated subarrays B that make up the array A. After completion of these n parallel computation steps we have transformed CM matrix A in array A in-place to become the same matrix A but now A is represented in RB format in array A. Thus, after vector in-place transpose we have "cache blocked" matrix A! In conclusion, we hope the reader now clearly sees at a deeper level why NDS significantly improves MC DLA algorithm performance. The transformation of A in standard format to RB format by in-place *vector* transposition was orders of magnitude faster than ordinary scalar in-place methods.

4   A Comparison of GKK and TT Transpose Algorithms

Both these algorithms are claimed to be in-place. TT uses $\min(m, n)$ extra storage and has an operation count that is $\leq 24mn$ swaps plus $O(mn)$ index operations. GKK uses a compact representation of the number of cycles, NC used by the transposition or permutation mapping $\mathcal{M}$ of CMO A to RMO A. Its operations count is $mn + $ NC moves plus the same number of index computations Also, there is the apriori computation of the cycle structure of $\mathcal{M}$. The GKK algorithm has good parallel implementations and is organized to use vector and cache blocking type operations as much as possible. Except for [13] there has been no implementation of the TT algorithm. The paper [20] does not mention parallelism. It does mention operations on "contiguous entries that could decrease cache misses" on page 380. Also, on the bottom of page 383 and the top of page 384 there is explicit mention of moving contiguous entries of A. Otherwise, there is no further mention of contiguous and or cache considerations.

4.1 Storage Amounts for GKK and TT Algorithms

The TT algorithm requires $\min(m, n)$ storage· locations. The GKK algorithm generates a compact representation of the cycle structure of $\mathcal{M}$. Figure 2 of [10] studied random matrices whose sizes m, n were between 2 and 100. 45 % of these matrices had less than 9 cycles in their $\mathcal{M}$. This was a surprising result for us. However, the storage cost to represent the cycle structure of any $\mathcal{M}$ was very tiny. In Sect. 5 of [10] square matrices were considered and a order 19 example was discussed as having 190 cycles! Not mentioned was the storage cost of this structure which was a sparse matrix type row pointer and column index structure to represent these 190 cycles. Only 4 row pointers and 12 column indices were needed. This compares favorably with the TT amount or 19 locations. The TT papers discusses matrices with $n = m^q$ whose size is $|A| = m^{q+1}$. These matrices have a maximum cycle length of size $q + 1$. They represent a generalization of square matrices where $q = 1$. Such matrices represent a set worst cases for GKK type algorithms. To give some idea what GKK does for this case we let $m = 10$ and choose $q = 1, 2, 3, 4, 5$. The GKK algorithm found the number of cycles to be $55, 340, 2530, 20008, 166870$. However, the cycle representation storage for these five matrices is $15, 46, 65, 289, 93$. These latter values compare favorably with the constant TT value of $m = 10$.

4.2 Vector Transpose and $n \times m$ A Matrices Where $n = km^q$

Both the TT and vector transpose, see Sect. 3, algorithms consider matrices A where $n = km^q$. For TT, $k < m$. For vector transpose, $k > 1$ is arbitrary so q can be one. In Sect. 3 these types of matrices give good performance. A user can set the LDA of A so this condition always holds; see Sect. 1.2 of [10]. In Sect. 3 of [10] we discuss the case where m and n are not multiples of blocking factors m_b and n_b. The TT authors handle this related issue by using their Lemmas 3 and 4. We close this section by noting that the cycle structure of A can vary greatly when $k > 1$ versus $k = 1$ in the TT algorithm; we give two examples where $k = 1, 2$: $q = 2, m = 3$ and $q = 1, m = 29$. In the first example, we get 8 cycles of length 3 and 3 singleton cycles versus 1 cycle of length 52 and 2 singleton cycles; in the second example, we get 406 cycles of length 3 and 29 singleton cycles versus single cycles of length 1640 and 40 and 2 singleton cycles.

4.3 A Clarifying Example with $n = 673$ and $m = 384$

The GKK algorithm can be used as subroutine of the TT algorithm: instead of using TT just get the apriori amount of storage used by GKK. For this problem it is 21 integers which gives a total of 30 cycle leaders. However, going further with TT, A becomes a vertical partition of submatrices A_1 of order 384 and a 289×384 A_2. At this point, TT just needs to find A_2^T in-place. Again, the GKK storage amount for finding A_2^T is 12 integers giving a total of 70 cycles. Now going still further with TT it naturally partitions A_2 as order 289 $A2_1$ and 289×95 $A2_2$ submatrices. Wanted for finding A_2^T are $A2_1^T$ and $A2_2^T$. Now again checking GKK storage of $A2_2^T$ one finds 13 integers for a total of 171 cycles. At this point our implementation of TT makes a recursive call on $A2_2$ to get a vertical partition of $A2_2$ into submatrix 285×95 $A3$ and 4×95 $A4$. Now $|A4| = 4 * 95 \leq m$ so $A4$ can be transposed out-of-place. Since $A3$ can be also transposed in-place, it follows that this modification of the TT algorithm is complete. In summary, we have shown that the very fast apriori feature of GKK can be used by the TT algorithm. If its compact representation of the number of cycles is too high; i.e., $\geq m$ then one can continue with the TT algorithm.

5   A GCD Transpose Algorithm

In [6] the GCD Transpose algorithm did *not* represent CMO A as a direct sum, or partitioning, of square matrices. We use Vector Transposition of Sect. 3 and Lemma 4 of TT [13,20] to get a direct sum. Doing so makes this algorithm more efficient; e.g. it is easily made parallel. Let CMO A be a $m = r_0$ by $n = r_1$, with $m > n$, rectangular matrix. Assume that $g = \gcd(m,n) = 1$. Otherwise, $m = Mg$ and $n = Ng$ and A is an M by N matrix whose elements are order g SB's and $\gcd(M,N) = 1$; i.e, the above GCD Transpose algorithm works for all A. Note that this GCD algorithm has $r_{i-1} = q_i r_i + r_{i+1}$ for $i = 1,\ldots,k$ and $r_k = 1$. The GCD Transpose algorithm starts with $A = A0_2$ of size r_0 by r_1 and produces submatrix $A2_2$ of size r_2 by r_3 with the rest of A as square submatrices.

1. partition A vertically into $q_1 r_1 \times r_1$ A_1 and $r_2 \times r_1$ A_2 using Lemma 4 of [13]
2. fully process all of A_1 (by transposing it) as follows:
   (a) vector transpose A_1 into q_1 SB's of order r_1 if $q_1 > 1$; see Sect. 3
   (b) in-place transpose q_1 SB's of order r_1.
3. process the $r_2 \times q_2 r_2$ $A2_1$ part of A_2 leaving alone its $r_2 \times r_3$ $A2_2$ part. Note that $A2_1$ is partitioned into q_2 SB's. So, in-place transpose these q_2 SB's.

After this initial GCD Transpose step submatrix $A2_2$ needs further processing; however, the remainder of A has been partitioned into SB's. Thus, apply the above Algorithm again on submatrix $A2_2$ producing $r_4 \times r_5$ CMO $A4_2$ which needs further processing and now the remainder of A has been partitioned into SB's. So, by the principle of infinite descent, this process must terminate.

We use the example, in [6], of $n = 673 \times m = 384$ A to clarify the above GCD transpose algorithm. We have $r_0 = 673$ and $r_1 = 384$ and steps $1,2$ of the GCD gives $q_{1,2} = 1,2$ and $r_{2,3} = 289,95$. $q_1 r_1 \times r_1$ A_1 and $r_2 \times r_1$ A_2 vertically partitions A in step 1. Step 2a is a no-operation and square matrix A_1 is transposed in Step 2b. A_2 is naturally partitioned as $A2_1$ and $A2_2$ during Step 1. In Step 3, square matrix $A2_1$ is transposed leaving matrix $r_2 \times r_3$ $A2_2$ as input for the next stage of the algorithm. Now steps $3,4$ of GCD give $q_{3,4} = 3,23$ and $r_{4,5} = 4,3$. Then $q_3 r_3 \times r_3$ A_3 and $r_4 \times r_3$ A_4 vertically partitions $A2_2$ in step 1. Step 2a does vector transpose on a $q_3 \times r_3$ "vector matrix" of $VL = r_3$. One gets q_3 contiguous (partititioned) SB's of order r_3. All q_3 SB's are transposed in Step 2b. A_4 is naturally partitioned as $A4_1$ and $A4_2$ during Step 1. $A4_1$ is q_4 contiguous SB's of order r_4! In Step 3, all q_3 SB's of $A4_1$ are transposed in-place.

$A4_2$ is input for the next stage of the algorithm. Next steps $5, 6$ of GCD give $q_{5,6} = 1, 3$ and $r_{6,7} = 1, 0$. Thus, step 1 give the vertical partition A_5 as a SB of order 3 and A_6 as 1 by 3. Step 2b is done and the algorithm ends.

6   The Power of Negative Integers

After negative integers are defined subtraction becomes $A - B \equiv A + (-B)$ where, for clarity, it is important to distinguish between between a negative sign and the subtraction operator. In fact, subtraction "disappears" and it can be replaced by ten's complement addition, tc, as Lagrange [18] noted to French teachers of grades K-12 in 1795. This idea does *not* eliminate carrying. However, **a form of duality exists between addition/subtraction and carrying/borrowing: when carry occurs in addition replace it by non-borrow subtraction.** For example, $8 + 7 = 10 + (7 - 2) = 15$. **When borrowing occurs in subtraction replace it by non-carry addition** [12]; e.g, $13 - 4 = 3 + 6$. In these examples $2 = \text{tc}(8)$ and $6 = \text{tc}(4)$. **This duality result leads to learning simpler addition and subtraction facts by omitting the harder facts** [12].

By using negative digits in a base 10 representation of an integer one *can* obtain the correct solution of any subtraction problem by just "doing" digit by digit subtraction. Consider $613 - 204 = 4\,1\text{-}1$ done by $6 - 2 = 4$ in the hundreds position $1 - 0 = 1$ in the tens position and $3 - 4 = -1$ in the ones position. In general, **any subtraction problem "breaks up" into only two types of subproblems : 1 digit no borrow and $k > 1$ digit borrow subproblems** [12]. Now, $-1, k - 2\,9's, t$ where $t = 10$ is **a k digit representation of zero** as $10^{k-1} = 9 \times 10^{k-2} + \cdots + 90 + 10$. **This representation of zero added to the answer of *any* $k > 1$ digit borrow subproblem yields the standard answer of this subproblem** [12]. Take $3207 - 1228 = 2\,0\text{-}2\text{-}1$. Clearly $2000 - 21 = 1979$ is the correct answer and one sees that $2 - 1 = 1$, $9 - 0 = 9$, $9 - 2 = 7$ and $10 - 1 = 9$ gives $k = 4$ digit by digit subtractions yielding 1979.

6.1   Finding Primitive Roots Using Smaller Integers

Doing arithmetic with smaller integers is preferable than using larger positive integers. Gauss noticed that he could simplify his original proof of the Law of Quadratic Reciprocity by using both positive and negative integers to represent the integers modulo an odd prime p, i.e., $\pm 1, \ldots, (p - 1)/2$ instead of $1, \ldots p - 1$. He discovered Gauss's Lemma using this idea. Now using the same idea, we show how to find smaller primitive roots for odd primes of the form $p = 4k - 1$. Take $p = 7$. The positive primitive roots of 7 are $3, 5$. However, using $\pm 1, \pm 2, \pm 3$ these roots become $-2, 3$. A more striking example is $p = 191$. The smallest positive primitive root is 19; using integers -2 is a primitive root. A new result is that one of $\pm g^{(p-1)/2} \neq 1$; g is a potential primitive root for p. Thus the GKK computation of [10] becomes more efficient as it can now more quickly find primitive roots for prime powers .

References

1. Buttari, A., Langou, J., Kurzak, J., Dongarra, J.: A class of parallel tiled linear algorithms for MC architectures. Parallel Comput. **35**(1), 38–53 (2009)
2. Gustavson, F.G.: Recursion leads to automatic variable blocking for dense linear-algebra algorithms. IBM J. R. & D. **41**(6), 737–755 (1997)
3. Gustavson, F.G.: New generalized data structures for matrices lead to a variety of high-performance algorithms. In: Boisvert, R.F., Tang, P.T.P. (eds.) Proceedings of the IFIP WG 2.5 Working Group on The Architecture of Scientific Software, Ottawa, Canada, pp. 211–234. Kluwer Academic Publishers, Boston, October 2–4 2000
4. Gustavson, F.G.: High performance linear algebra algs. using new generalized data structures for matrices. IBM J. R. & D. **47**(1), 31–55 (2003)
5. Gustavson, F.G.: New generalized data structures for matrices lead to a variety of high performance dense linear algebra algorithms. In: Dongarra, J., Madsen, K., Waśniewski, J. (eds.) PARA 2004. LNCS, vol. 3732, pp. 11–20. Springer, Heidelberg (2006)
6. Gustavson, F.G., Gunnels, J.A.: Method and structure for cache aware transposition via rectangular subsections. U.S. Patent US20060161607 A1, Application No. 11/035,953, submitted 14 January 2005, published 20 July 2006
7. Gustavson, F.G., Gunnels, J.A., Sexton, J.C.: Minimal data copy for dense linear algebra factorization. In: Kågström, B., Elmroth, E., Dongarra, J., Waśniewski, J. (eds.) PARA 2006. LNCS, vol. 4699, pp. 540–549. Springer, Heidelberg (2007)
8. Gustavson, F.G., Swirszcz, T.: In-place transposition of rectangular matrices. In: Kågström, B., Elmroth, E., Dongarra, J., Waśniewski, J. (eds.) PARA 2006. LNCS, vol. 4699, pp. 560–569. Springer, Heidelberg (2007)
9. Gustavson, F.G.: The relevance of new data structure approaches for dense linear algebra in the new multicore/manycore environments. IBM Research report RC24599, also, to appear in PARA'08 proceeding, 10 p. (2008)
10. Gustavson, F.G., Karlsson, L., Kågström, B.: Parallel and cache-efficient in-place matrix storage format conversion. ACM TOMS 38(3), Article 17, 1–32 (2012)
11. Gustavson, F.G.: Cache blocking for linear algebra algorithms. In: Wyrzykowski, R., Dongarra, J., Karczewski, K., Waśniewski, J. (eds.) PPAM 2011, Part I. LNCS, vol. 7203, pp. 122–132. Springer, Heidelberg (2012)
12. Gustavson. F.G.: A subtraction algorithm based on adding C to both A and B. Power Point Presentation, fg2935@gmail.com, 50 slides, 28 October 2013
13. Gustavson, F.G., Walker, D.W.: Algorithms for in-place matrix transposition. In: Wyrzykowski, R., Dongarra, J., Karczewski, K., Waśniewski, J. (eds.) PPAM 2013, Part II. LNCS, vol. 8385, pp. 105–117. Springer, Heidelberg (2014)
14. Karlsson, L.: Blocked in-place transposition with application to storage format conversion. Technical report UMINF 09.01. Department of Computing Science, Umeå University, Umeå, Sweden. January 2009. ISSN 0348–0542
15. Kunth, D.: The Art of Computer Programming, 3rd edn., vol. 1, 2 & 3. Addison-Wesley, Reading (1998)
16. Kurzak, J., Buttari, A., Dongarra, J.: Solving systems of linear equations on the Cell processor using Cholesky factorization. IEEE Trans. Parallel Distrib. Syst. **19**(9), 1175–1186 (2008)
17. Kurzak, J., Dongarra, J.: Implementation of mixed precision in solving mixed precision of linear equations on the Cell processor: Research Articles. Concurr. Comput.: Pract. Exper. **19**(10), 1371–1385 (2007)

18. Lagrange, J.L.: Lectures On Elementary Mathematics, 156 p. Dover Publications, New York (2008)
19. Tietze, H.: Three Dimensions-Higher Dimensions. Famous Problems of Mathematics, pp. 106–120. Graylock Press, Rochester (1965)
20. Tretyakov, A.A., Tyrtyshnikov, E.E.: Optimal in-place transposition of rectangular matrices. J. Complex. **25**, 377–384 (2009)

Numerical Algorithms and Parallel Scientific Computing

Exploiting Data Sparsity in Parallel Matrix Powers Computations

Nicholas Knight$^{(\boxtimes)}$, Erin Carson, and James Demmel

University of California, Berkeley, USA
{knight,ecc2z,demmel}@cs.berkeley.edu

Abstract. We derive a new parallel communication-avoiding matrix powers algorithm for matrices of the form $A = D + USV^H$, where D is sparse and USV^H has low rank and is possibly dense. We demonstrate that, with respect to the cost of computing k sparse matrix-vector multiplications, our algorithm asymptotically reduces the parallel latency by a factor of $O(k)$ for small additional bandwidth and computation costs. Using problems from real-world applications, our performance model predicts up to 13× speedups on petascale machines.

Keywords: Communication-avoiding · Matrix powers · Graph cover · Hierarchical matrices · Parallel algorithms

1 Introduction

The runtime of an algorithm can be modeled as a function of *computation* cost, proportional to the number of arithmetic operations, and *communication* cost, proportional to the amount of data movement. On modern computers, the time to move one word of data is much greater than the time to complete one arithmetic operation. Technology trends indicate that the performance gap between communication and computation will only widen in future computers, resulting in a paradigm shift in the design of high-performance algorithms: to achieve efficiency, one must focus on *communication-avoiding* approaches.

We consider a simplified machine model, where a parallel machine consists of p processors, each able to perform arithmetic operations on their M words of *local memory*. Processors communicate point-to-point *messages* of $n \leq M$ contiguous words, taking $\alpha + \beta n$ seconds on both sender and receiver, over a completely connected network (no contention), and each processor can send or receive at most one message at a time. For simplicity, we do not model overlapping communication and computation. Given an algorithm's *latency cost*, number of messages sent, *bandwidth cost*, number of words moved, and *arithmetic (flop) cost*, the number of arithmetic operations performed, we estimate the runtime T (along the critical path) on a parallel machine with latency α, reciprocal bandwidth β, and arithmetic (flop) rate γ as

$$T = (\#\text{messages} \cdot \alpha) + (\#\text{words moved} \cdot \beta) + (\#\text{flops} \cdot \gamma). \tag{1}$$

R. Wyrzykowski et al. (Eds.): PPAM 2013, Part I, LNCS 8384, pp. 15–25, 2014.
DOI: 10.1007/978-3-642-55224-3_2, © Springer-Verlag Berlin Heidelberg 2014

Computing k repeated sparse matrix-vector multiplications (SpMVs), or, a *matrix powers* computation, with $A \in \mathbb{C}^{n \times n}$ and $x \in \mathbb{C}^{n \times q}$, where typically $q \ll n$, can be written as

$$K_{k+1}(A, x, \{p_j\}_{j=0}^k) := [x^{(0)}, \ldots, x^{(k)}] := [p_0(A)x, p_1(A)x, \ldots, p_k(A)x], \quad (2)$$

where p_j is a degree-j polynomial. Due to a small ratio of arithmetic operations to data movement, the performance of this computation is bound by communication on modern computers. Matrix powers computations constitute a core kernel in a variety of applications, including steepest descent algorithms and Krylov subspace methods for linear systems and eigenvalue problems, including the power method to compute PageRank.

Previous efforts have produced parallel communication-avoiding matrix powers algorithms to compute (2) that achieve an $O(k)$ reduction in parallel latency cost versus computing k repeated SpMVs for a set number of iterations [4, 11]. This improvement is only possible if A is *well partitioned* (to be defined in Sect. 1.1). Although such advances show promising speedups for many problems, the requirement that A is well partitioned often excludes matrices with dense components, even if those components have low rank (*data sparsity*). In this work, we derive a new parallel communication-avoiding matrix powers algorithm for matrices of the form $A = D + USV^H$, where D is well partitioned and USV^H may not be well partitioned but has low rank. (Recall $x^H = \bar{x}^T$ denotes the Hermitian transpose of x.) There are many practical situations where such structures arise, including power-law graph analysis and circuit simulation. Hierarchical ($\mathcal{H}$-) matrices (e.g., [1]), common preconditioners for Krylov subspace methods, also have this form. Our primary motivation is enabling preconditioned communication-avoiding Krylov subspace methods, where the preconditioned system has hierarchical semiseparable (HSS) structure. There is a wealth of literature related to communication-avoiding Krylov subspace methods; we direct the reader to the thesis of Hoemmen for an overview [5, Sects. 1.5 and 1.6].

With respect to the cost of computing k SpMVs, our algorithm asymptotically reduces parallel latency by a factor of $O(k)$ with only small additional bandwidth and computational costs. Using a detailed complexity analysis for an example HSS matrix, our model predicts up to 13× speedups over the standard algorithm on petascale machines. Our approach is based on the application of a blocking covers technique [9] to communication-avoiding matrix powers algorithms [4,10]. We briefly review these works below.

1.1   The Blocking Covers Technique

Hong and Kung [6] prove a lower bound on data movement for a sequential matrix powers computation on a regular mesh. Given directed graph $G = (V, E)$ representing nonzeros of A, vertex $v \in V$, and constant $\tau \geq 0$, let the τ-*neighborhood* of v, $N^{(\tau)}(v)$, be the set of vertices in V such that $u \in N^{(\tau)}(v)$ implies there is a path of length at most τ from u to v; a τ-*neighborhood-cover* of G is a sequence of subgraphs, $\mathcal{G} = \{G_i = (V_i, E_i)\}_{i=1}^k$, such that $\forall v \in V$,

$\exists G_i \in \mathcal{G}$ for which $N^{(\tau)}(v) \subseteq V_i$ [9]. If G has a τ-neighborhood cover with $O(|E|/M)$ subgraphs, each with $O(M)$ edges where M is the size of the primary memory, Hong and Kung's method reduces data movement by a factor of τ over computing (2) column-wise. A matrix that meets these constraints is also frequently called *well partitioned* [4] (we use this terminology for the parallel case as well).

Certain graphs with low diameter (e.g., multigrid graphs) may not have τ-neighborhood covers that satisfy these memory constraints. Leiserson et al. overcome this restriction by "removing" a set $B \subseteq V$ of *blocker* vertices, chosen such that the remaining graph $V - B$ is well partitioned [9]. Let the τ-neighborhood with respect to B be defined as $N_B^{(\tau)}(v) = \{u \in V : \exists \text{ path } u \to u_1 \to \cdots \to u_t \to v, \text{ where } u_i \in V - B \text{ for } i \in \{1,\ldots,t < \tau\}\}$. Then a (τ, r, M)-blocking cover of G is a pair $(\mathcal{G}, \mathcal{B})$, where $\mathcal{B} = \{B_i\}_{i=1}^k$ is a sequence of subsets of V such that: (1) $\forall i \in \{1,\ldots,k\}, M/2 \leq |E_i| \leq M$, (2) $\forall i \in \{1,\ldots,k\}, |B_i| \leq r$, (3) $\sum_{i=1}^k |E_i| = O(|E|)$, and (4) $\forall v \in V, \exists G_i \in \mathcal{G}$ such that $N_{B_i}^{(\tau)}(v) \subseteq V_i$ [9]. Leiserson et al. present a 4 phase sequential matrix powers algorithm that reduces the data movement by a factor of τ over the standard method if the graph of A has a (τ, r, M)-blocking cover that meets certain criteria. Our parallel algorithm is based on a similar approach. Our work generalizes the blocking covers approach [9], both to the parallel case and to a larger class of data-sparse matrix representations.

1.2   Parallel Matrix Powers Algorithms

Parallel variants of matrix powers, for both structured and general sparse matrices, are presented in the thesis of Mohiyuddin [10], which summarizes and elaborates upon previous work and implementations [4,11]. We review two of these parallel matrix powers algorithms, referred to as PA0, the naïve algorithm for computing (2) via k SpMV operations, and PA1, a communication-avoiding variant. We assume the polynomials $\{p_l\}_{l=0}^k$ in (2) satisfy a recurrence,

$$p_0(z) := 1, \qquad p_{j+1}(z) = \left(z p_j(z) - \sum_{i=0}^j h_{i,j} p_i(z) \right) / h_{j+1,j}, \qquad (3)$$

whose coefficients we store in an upper Hessenberg matrix

$$H_k := \begin{bmatrix} h_{0,0} & h_{0,1} & \cdots & h_{0,k-1} \\ h_{1,0} & h_{1,1} & \cdots & h_{1,k-1} \\ 0 & h_{2,1} & \ddots & h_{2,k-1} \\ \vdots & \ddots & \ddots & \vdots \\ 0 & 0 & \cdots & h_{k,k-1} \end{bmatrix}. \qquad (4)$$

Let $\mathrm{nz}(A) = \{(i,j) : A_{ij} \text{ treated as nonzero}\}$ represent the edges in the directed graph of A, and let $A_{\mathcal{I}}$ indicate the submatrix of A consisting of rows $i \in \mathcal{I}$. For simplicity, we ignore cancellation, i.e., we assume $\mathrm{nz}(p_j(A)) \subseteq \mathrm{nz}(p_{j+1}(A))$ and every entry of $x^{(j)}$ is treated as nonzero for all $j \geq 0$.

We construct a directed graph $\mathcal{G} = (\mathcal{V}, \mathcal{E})$ representing the dependencies in computing $x^{(j)} := p_j(A)x$ for every $0 \leq j \leq k$. First, denoting row i of $x^{(j)}$ by $x_i^{(j)}$, we define the $n(k+1)$ vertices $\mathcal{V} := \{x_i^{(j)} : 1 \leq i \leq n, 0 \leq j \leq k\}$. The edge set $\mathcal{E}$ consists of k copies of nz(A), between each adjacent pair of the $k+1$ levels $\mathcal{V}^{(j)} := \{x_i^{(j)} : 1 \leq i \leq n\}$, unioned with the edges due to the polynomial recurrence, i.e.,

$$\mathcal{E} := \left\{ \left(x_{i_1}^{(j+1)}, x_{i_2}^{(j)}\right) : \begin{smallmatrix} 0 \leq j < k, \\ (i_1,i_2) \in \text{nz}(A) \end{smallmatrix} \right\} \cup \left\{ \left(x_i^{(j+d')}, x_i^{(j)}\right) : \begin{smallmatrix} 1 \leq d' \leq d, \\ 0 \leq j \leq k-d', \\ 1 \leq i \leq n \end{smallmatrix} \right\} \quad (5)$$

where H_k has d nonzero superdiagonals (including main diagonal).

Now we partition $\mathcal{V}$ 'rowwise,' that is, each $x_i^{(j)}$ is assigned a processor *affinity* $m \in \{0, \ldots, p-1\}$, for $0 \leq j \leq k$. Let $\mathcal{V}_m$ and $\mathcal{V}_m^{(j)}$ restrict $\mathcal{V}$ and $\mathcal{V}^{(j)}$ to their elements with affinity m. Let $\mathcal{R}(\mathcal{S})$ denote the *reachability set* of $\mathcal{S} \subseteq \mathcal{V}$, i.e., the set $\mathcal{S}$ and vertices reachable from $\mathcal{S}$ via paths in $\mathcal{G}$; then, as with $\mathcal{V}$, we define the subsets $\mathcal{R}^{(j)}$, $\mathcal{R}_m$, and $\mathcal{R}_m^{(j)}$ of $\mathcal{R}$.

At the end of the computation, processor m has computed/stored the entries $\mathcal{V}_m$. Thus, for PA0, processor m must own $A_{\{i:x_i^{(j)} \in \mathcal{V}_m\}}$ and $\mathcal{V}_m^{(0)}$, and for PA1, processor m must own $A_{\{i:x_i^{(1)} \in \mathcal{R}(\mathcal{V}_m)\}}$ and $\mathcal{R}^{(0)}(\mathcal{V}_m)$. We assume that the rows of A are distributed to processors offline, while the source vector $x^{(0)}$ must be distributed at runtime (online).

With this notation, we present the parallel matrix powers algorithms PA0 (Algorithm 1) and PA1 (Algorithm 2), as pseudocode for processor m. The advantage of PA1 over PA0 is that it may send fewer messages between processors: whereas PA0 requires k rounds of messages, PA1 requires only one. If the number of other processors with whom processor m must communicate is within a constant factor for both algorithms, PA1 obtains a $\Theta(k)$-fold latency savings. In general, however, PA1 incurs greater bandwidth, arithmetic, and storage costs, as processors may perform redundant computations to avoid communication. Furthermore, in practice, PA1 requires additional data structures to encode the reachability sets; we assume these data structures are populated offline in a preprocessing phase.

We refer the reader to the complexity analysis in Tables 2.3 and 2.4, performance modeling in Sect. 2.6, and performance results in Sects. 2.10.3 and 2.11.3 of Mohiyuddin's thesis [10], which demonstrate that this optimization can lead to speedups in practice. For example, for a 9-point stencil on a $n^{1/2}$-by-$n^{1/2}$ mesh with p processors, assuming $k \ll (n/p)^{1/2}$ and the monomial basis $(p_j(z) = z^j)$, the number of arithmetic operations grows by a factor $1 + 2k(p/n)^{1/2}$, the number of messages decreases by a factor of k, and the number of words moved grows by a factor of $1 + k(p/n)^{1/2}$ [10]. Therefore, since the additional costs are lower order terms, we expect PA1 to give $\Theta(k)$ speedup when performance is latency-bound. Our performance modeling has confirmed this results [7].

Algorithm 1. PA0. Code for proc. m.	**Algorithm 2.** PA1. Code for proc. m.
1: **for** $j = 1, \ldots, k$ **do**	1: **for** all procs. $\ell \neq m$ **do**
2: **for** all procs. $\ell \neq m$ **do**	2: Send $x_i^{(0)} \in \mathcal{R}_m^{(0)}(\mathcal{V}_\ell^{(k)})$ to proc. ℓ.
3: Send $x_i^{(j-1)} \in \mathcal{R}_m^{(j-1)}(\mathcal{V}_\ell^{(j)})$ to proc. ℓ.	3: Recv. $x_i^{(0)} \in \mathcal{R}_\ell^{(0)}(\mathcal{V}_m^{(k)})$ from proc. ℓ.
4: Recv. $x_i^{(j-1)} \in \mathcal{R}_\ell^{(j-1)}(\mathcal{V}_m^{(j)})$ from proc. ℓ.	4: **end for**
5: **end for**	5: **for** $j = 1, \ldots, k$ **do**
6: Compute $x_i^{(j)} \in \mathcal{V}_m^{(j)}$ via (3).	6: Compute $x_i^{(j)} \in \mathcal{R}^{(j)}(\mathcal{V}_m)$ via (3).
7: **end for**	7: **end for**

2 Derivation of Parallel Blocking Covers

Recall that, given matrices $A \in \mathbb{C}^{n \times n}$ and $x \in \mathbb{C}^{n \times q}$, and $k \in \mathbb{N}$, our task is to compute (2). If A is not well partitioned, PA0 must communicate at every step, but now the cost of PA1 may be much worse: when $k > 1$, every processor needs all rows of A and $x^{(0)}$; there is no parallelism in computing all but the last SpMV. (Note when $k = 1$, PA1 degenerates to PA0.)

If, however, A can be split in the form $D + USV^H$, where D is well partitioned and USV^H has low rank, we can use a generalization of the blocking covers approach [9] to recover parallelism. In this case, D has a good cover and US can be applied locally, but the application of V^H incurs global communication. Thus, the application of V^H will correspond to the blocker vertices in our algorithm, PA1-BC, which we now derive.

First, we recursively partition $H_k := \begin{bmatrix} H_{k-1} & h^{(k-1)} \\ 0_{1,k-1} & h_{k,k-1} \end{bmatrix}$ with $H_1 := [h_{0,0}, h_{1,0}]^T$,

so $h^{(0)}, \ldots, h^{(k-1)}$ forms the upper triangle of H_k; substituting $z := A =: D + USV^H$, the recurrence for $x^{(j)} = p_j(A)x^{(0)}$ is

$$x^{(j+1)} = \left(Dx^{(j)} - [x^{(0)}, \ldots, x^{(j)}](h^{(j)} \otimes I_{q,q}) + USV^H x^{(j)} \right) / h_{j+1,j}. \quad (6)$$

We exploit the following identity, established by induction [7], to avoid performing $V^H \cdot x^{(j)}$ explicitly.

Lemma 1. *Given the additive splitting* $z = z_1 + z_2$, *(3) can be rewritten as*

$$p_j(z) = p_j(z_1) + \sum_{i=1}^{j} p_{i-1}^{j-i+1}(z_1) z_2 p_{j-i}(z) / h_{j-i+1,j-i} \quad (7)$$

for $j \geq 0$, *where* $p_j^i(z)$ *is a degree-j polynomial related to* $p_j(z)$ *by reindexing the coefficients* $h_{l,j} := h_{l+i,j+i}$ *in (3).*

Now substitute $z := A = D + USV^H =: z_1 + z_2$ in (7), premultiply by SV^H, and postmultiply by $x^{(0)}$, to obtain

$$SV^H x^{(j)} = S \left(V^H p_j(D)x^{(0)} + \sum_{i=1}^{j} V^H p_{i-1}^{j-i+1}(D) U \frac{SV^H x^{(j-i)}}{h_{j-i+1,j-i}} \right). \quad (8)$$

Let $W_i := V^H p_i(D) U$ for $0 \le i \le k-2$, $y_i := V^H p_i(D) x$ for $0 \le i \le k-1$, and $b_j := S V^H x^{(j)}$ for $0 \le i \le k-1$. We can write p_i^j in terms of $p_i = p_i^0$ via the following result, established by induction [7].

Lemma 2. *There exist coefficient vectors $w_i^j \in \mathbb{C}^{i+1}$ satisfying*

$$[W_0, \ldots, W_i](w_i^j \otimes I_{r,r}) = V^H p_i^j(D) U \qquad (9)$$

for $0 \le i \le k-2, 1 \le j \le k-i-1$, that can be computed by $w_0^j := 1$ and

$$w_{l+1}^j := \left(H_{l+1} w_l^j - \left[\begin{bmatrix} w_0^j \\ 0_{l,1} \end{bmatrix}, \begin{bmatrix} w_1^j \\ 0_{l-1,1} \end{bmatrix}, \ldots, \begin{bmatrix} w_l^j \\ 0 \end{bmatrix} \right] h_{\{j,\ldots,j+l\}}^{(j)} \right) / h_{j+l+1,j+l}. \qquad (10)$$

Using this result, we write (8) as

$$b_j = S \left(y_j + [W_0, \ldots, W_{j-1}] \cdot \sum_{i=1}^j \left(\begin{bmatrix} w_{i-1}^{j-i+1} \\ 0_{j-i,1} \end{bmatrix} \otimes I_{r,r} \right) \frac{b_{j-i}}{h_{j-i+1,j-i}} \right); \qquad (11)$$

however, in case H_k is Toeplitz, the summation simplifies to $\left[\frac{b_{j-1}^T}{h_{j,j-1}}, \ldots, \frac{b_0^T}{h_{1,0}} \right]^T$, so we need not compute $\{w_i^j\}$.

Ultimately we must evaluate (6), substituting b_j for $S V^H x^{(j)}$. This can be accomplished by applying PA1 to the following recurrence for $p_j(z,c)$, where $c := \{c_0, \ldots, c_{j-1}, \ldots\} := \{U b_0, \ldots, U b_{j-1}, \ldots\}$:

$$p_0(z,c) := 1, \quad p_{j+1}(z,c) := \left(z p_j(z) - \sum_{i=0}^j h_{i,j} p_i(z) + c_j \right) / h_{j+1,j}. \qquad (12)$$

Given the notation established, we construct PA1-BC (Algorithm 3). In terms of the graph of D, $\mathcal{G} = (\mathcal{V}, \mathcal{E})$, processor m must own

$$D_{\{i : x_i^{(1)} \in \mathcal{R}(\mathcal{V}_m)\}}, \ U_{\{i : x_i^{(1)} \in \mathcal{R}(\mathcal{V}_m)\}}, \ V_{\{i : x_i^{(j)} \in \mathcal{V}_m\}}, \text{ and } \mathcal{R}^{(0)}(\mathcal{V}_m), \qquad (13)$$

in order to compute the entries $x_i^{(j)} \in \mathcal{V}_m$. In exact arithmetic, PA1-BC returns the same output as PA0 and PA1. However, by exploiting the splitting $A = D + U S V^H$, PA1-BC may avoid communication when A is not well partitioned. Communication occurs in calls to PA1 (Lines 1 and 4), as well as in Allreduce collectives (Lines 2 and 5). As computations in Lines 1, 2, and 3 do not depend on the input $x^{(0)}$, they need only be computed once per matrix $A = D + U S V^H$, thus we assume their cost is incurred offline.

For the familiar reader, the sequential blocking covers algorithm [9] is a special case of a sequential execution of Algorithm 3, using the monomial basis, where $U = [e_i : i \in \mathcal{I}]$ and $S V^H = A_{\mathcal{I}}$, where e_i is the i-th column of the identity and $\mathcal{I} \subseteq \{1, \ldots, n\}$ are the indices of the blocker vertices. In Algorithm 3, Lines $\{1,2,3\}$, $\{4,5\}$, 6, and 7 correspond to the 4 phases of the sequential blocking covers algorithm, respectively [9]. In the next section, we demonstrate the benefit of our approach on a motivating example, matrix powers with HSS matrix A.

Algorithm 3. PA1-BC. Code for proc. m.

1: Compute local rows of $K_{k-1}(D, U, H_{k-1})$ with PA1, premultiply by local columns of V^H.
2: Compute $[W_0, \ldots, W_{k-2}]$ by an Allreduce.
3: Compute w_i^j for $0 \leq i \leq k-2$ and $1 \leq j \leq k-i-1$, via (10).
4: Compute local rows of $K_k(D, x^{(0)}, H_k)$ with PA1, premultiply by local columns of V^H.
5: Compute $[y_0, \ldots, y_{k-1}]$ by an Allreduce.
6: Compute $[b_0, \ldots, b_{k-1}]$ by (11).
7: Compute local rows of $[x^{(0)}, \ldots, x^{(k)}]$ with PA1, modified for (12).

3 Hierarchical Semiseparable Matrix Example

Hierarchical ($\mathcal{H}$-) matrices are amenable to the splitting $A = D + UV^H$, where D is block diagonal and UV^H represents the off-diagonal blocks. Naturally, U and V are quite sparse and it is important to exploit this sparsity in practice. In the special case of HSS matrices, many columns of U and V are linearly dependent, and we can exploit the matrix S in the splitting USV^H to write U and V as block diagonal matrices. We review the HSS notation and the algorithm for computing $v = Ax$ given by Chandrasekaran et al. [3, Sects. 2 and 3]. For any $0 \leq L \leq \lfloor \lg n \rfloor$, where $\lg = \log_2$, we can write A hierarchically as a perfect binary tree of depth L by recursively defining its diagonal blocks as $A =: D_{0;1}$ and

$$D_{\ell-1;i} =: \begin{bmatrix} D_{\ell;2i-1} & U_{\ell;2i-1}B_{\ell;2i-1,2i}V_{\ell;2i}^H \\ U_{\ell;2i}B_{\ell;2i,2i-1}V_{\ell;2i-1}^H & D_{\ell;2i} \end{bmatrix} \quad (14)$$

for $1 \leq \ell \leq L$, $1 \leq i \leq 2^{\ell-1}$, where $U_{0;1}, V_{0;1} := []$, and for $\ell \geq 2$,

$$U_{\ell-1;i} =: \begin{bmatrix} U_{\ell;2i-1}R_{\ell;2i-1} \\ U_{\ell;2i}R_{\ell;2i} \end{bmatrix}, \quad V_{\ell-1;i} =: \begin{bmatrix} V_{\ell;2i-1}W_{\ell;2i-1} \\ V_{\ell;2i}W_{\ell;2i} \end{bmatrix}; \quad (15)$$

the subscript expression $\ell; i$ denotes vertex i of the 2^ℓ vertices at level ℓ.

The action of A on a matrix x, i.e., $v := Ax$, satisfies $v_{0;1} = D_{0;1}x_{0;1}$, and for $1 \leq \ell \leq L$, $1 \leq i \leq 2^\ell$, satisfies $v_{\ell;i} = D_{\ell;i}x_{\ell;i} + U_{\ell;i}f_{\ell;i}$, with $f_{1;1} = B_{1;1;2}g_{1;2}$, $f_{1;2} = B_{1;2;1}g_{1;1}$, and, for $1 \leq \ell \leq L-1$, $1 \leq i \leq 2^\ell$,

$$f_{\ell+1;2i-1} = \begin{bmatrix} R_{\ell+1;2i-1}^T \\ B_{\ell+1;2i-1,2i}^T \end{bmatrix}^T \begin{bmatrix} f_{\ell;i} \\ g_{\ell+1;2i} \end{bmatrix}, \quad f_{\ell+1;2i} = \begin{bmatrix} R_{\ell+1;2i}^T \\ B_{\ell+1;2i,2i-1}^T \end{bmatrix}^T \begin{bmatrix} f_{\ell;i} \\ g_{\ell+1;2i-1} \end{bmatrix}, \quad (16)$$

where, for $1 \leq \ell \leq L-1$, $1 \leq i \leq 2^\ell$, $g_{\ell;i} = \begin{bmatrix} W_{\ell+1;2i-1} \\ W_{\ell+1;2i} \end{bmatrix}^H \begin{bmatrix} g_{\ell+1;2i-1} \\ g_{\ell+1;2i} \end{bmatrix}$, and $g_{L;i} = V_{L;i}^H x_{L;i}$ for $1 \leq i \leq 2^L$. For any HSS level ℓ, we assemble the block diagonal matrices

$$U_\ell := \bigoplus_{i=1}^{2^\ell} U_{\ell;i}, \qquad V_\ell := \bigoplus_{i=1}^{2^\ell} V_{\ell;i}, \qquad D_\ell := \bigoplus_{i=1}^{2^\ell} D_{\ell;i}, \quad (17)$$

denoted here as direct sums of their diagonal blocks. We also define matrices S_ℓ, representing the recurrences for $f_{\ell;i}$ and $g_{\ell;i}$, satisfying $v = Ax =: D_\ell x + U_\ell S_\ell V_\ell^H x$. We now discuss parallelizing the computation $v = Ax$, to generalize PA0 and PA1 to HSS matrices.

3.1   PA0 for HSS Matrices

We first discuss how to modify PA0 for HSS A, exploiting the $v = Ax$ recurrences for each $1 \leq j \leq k$; we call the resulting algorithm PA0-HSS. For brevity, we defer a detailed description [7]. PA0-HSS can be seen as an HSS specialization of known approaches for distributed-memory $\mathcal{H}$-matrix-vector multiplication [8].

We assume the HSS representation of A has perfect binary tree structure to some level $L > 2$, and there are $p \geq 4$ processors with p a power of 2. For each processor $m \in \{0, 1, \ldots, p-1\}$, let L_m denote the smallest level $\ell \geq 1$ such that $p/2^\ell$ divides m. We also define the intermediate level $1 < L_p := \lg(p) \leq L$ of the HSS tree; each $L_m \geq L_p$, and equality is attained when m is odd.

First, on the upsweep, each processor locally computes $V_{L_p}^H x$ (its subtree, rooted at level $L_p = \lg(p)$) and then performs L_p steps of parallel reduction, until there are two processors active, and then a downsweep until level L_p, at which point each processor is active, owns $S_{L_p} V_{L_p}^H x$, and recurses into its local subtree to finally compute its rows of $v = D_L x + U_L S_L V_L x$. More precisely, we assign processor m the computations $f_{\ell;i}$ and $g_{\ell;i}$ for

$$\left\{ \ell, i : \begin{array}{c} L \geq \ell \geq L_p \\ 2^\ell m/p+1 \leq i \leq 2^\ell (m+1)/p \end{array} \right\} \text{ and for } \left\{ \ell, i : \begin{array}{c} L_p-1 \geq \ell \geq L_m \\ i = 2^\ell m/p+1 \end{array} \right\} \quad (18)$$

and D_L, U_L, and V_L are distributed contiguously block rowwise, so processor m stores blocks $D_{L_p;m+1}$, $U_{L_p;m+1}$, and $V_{L_p;m+1}$. The $R_{\ell;i}$, $W_{\ell;i}$, and $B_{\ell;i}$ matrices are distributed so that they are available for the computations in the upsweep/downsweep; memory requirements are listed in Table 1.

3.2   PA1 for HSS Matrices

The block-diagonal structure of D_ℓ, U_ℓ, and V_ℓ in (17) suggests an efficient parallel implementation of PA1-BC, which we present as PA1-HSS (Algorithm 4). However, now each processor must perform the *entire* upsweep/downsweep between levels 1 and L_p locally. The additional cost shows up in our complexity analysis (see Table 1) as a factor of p, compared to a factor of $\lg(p)$ in PA0-HSS; we also illustrate this tradeoff in Sect. 4.

We assume the same data layout as PA0-HSS: each processor owns a diagonal block of D_{L_p}, U_{L_p}, and V_{L_p}, but only stores the smaller blocks of level L. We assume each processor is able to apply S_{L_p}. We rewrite (6) for the local rows, and exploit the block diagonal structure of D_{L_p} and U_{L_p}, to write

$$x_{L_p;m+1}^{(j+1)} = \left(D_{L_p;m+1} x_{L_p;m+1}^{(j)} - [x_{L_p;m+1}^{(0)}, \ldots, x_{L_p;m+1}^{(j)}](h^{(j)} \otimes I_{q,q}) \right.$$
$$\left. + U_{L_p;m+1}(b_j)_{\{mr+1,\ldots,(m+1)r\}} \right) / h_{j+1,j}. \quad (19)$$

Each processor locally computes all rows of $b_j = S_{L_p} V_{L_p}^H x^{(j)} = S_{L_p} \cdot z$, where z is the maximal parenthesized term in (11), using the HSS recurrences:

$$V_{L_p}^H x^{(j)} = z =: \left[g_{L_p,1}^T \cdots g_{L_p,p}^T \right]^T \mapsto \left[f_{L_p,1}^T \cdots f_{L_p,p}^T \right]^T := b_j = S_{L_p} V_{L_p}^H x^{(j)}. \quad (20)$$

The rest of PA1-HSS is similar to PA1-BC, except Allgather operations replace Allreduce operations in Lines 1 and 5 to exploit block structure of V^H.

Algorithm 4. PA1-HSS (Blocking Covers). Code for proc. m.

1: Compute $K_{k-1}(D_{Lp;m+1}, U_{Lp;m+1}, H_{k-1})$, premultiply by $V_{Lp;m+1}^H$.
2: Compute $[W_0, \ldots, W_{k-2}]$ by an Allgather.
3: Compute w_i^j for $0 \leq i \leq k-2$, and $1 \leq j \leq k-i-1$, via (10).
4: Compute $K_k(D_{Lp;m+1}, x_{Lp;m+1}^{(0)}, H_k)$, premultiply by $V_{Lp;m+1}^H$.
5: Compute $[y_0, \ldots, y_{k-1}]$ by an Allgather.
6: Compute $[b_0, \ldots, b_{k-1}]$ by (11), where $S = S_{L_p}$ is applied by (20).
7: Compute local rows of $[x^{(0)}, \ldots, x^{(k)}]$ according to (19).

3.3 Complexity Analysis

We gave a detailed complexity analysis of PA0-HSS and PA1-HSS in [7]; we summarize the asymptotics (i.e., ignoring constant factors) in Table 1. We assume A is given in HSS form, as described above, where all block matrices are dense. For simplicity, we assume n and *HSS-rank* r are powers of 2 and leaf level $L = \lg(n/r)$. Note that one could use faster Allgather algorithms (e.g., [2]) for PA1-HSS to eliminate the factor of $\lg p$ in the number of words moved.

4 Performance Model

We model speedups of PA1-HSS over PA0-HSS on two machine models used by Mohiyuddin [10] – 'Peta,' an 8100 processor petascale machine, and 'Grid,' 125 terascale machines connected via the Internet. Peta has a flop rate $\gamma = 2 \cdot 10^{-11}$ s/flop, latency $\alpha = 10^{-5}$ s/message, and bandwidth $\beta = 2 \cdot 10^{-9}$ s/word, and Grid has flop rate $\gamma = 10^{-12}$ s/flop, latency $\alpha = 10^{-1}$ s/message, and bandwidth $\beta = 25 \cdot 10^{-9}$ s/word. Complexity counts used can be found in [7].

Speedups of PA1-HSS over k invocations of PA0-HSS, for both Peta and Grid, are shown in Fig. 1. We used parameters from the parallel HSS performance tests of Wang et al. [13], where $p = (4, 16, 64, 256, 1024, 4096)$, $n = (2.5, 5, 10, 20, 40, 80) \cdot 10^3$, $r = (5, 5, 5, 5, 6, 7)$. Note that for Grid we only use the first 3 triples (p_i, n_i, r_i) since $p_{max} = 125$. We assume a three-term recurrence (H_k is tridiagonal), as these suffice in practice to obtain well-conditioned polynomial bases, even for large k [12].

On Peta, we see $O(k)$ speedups for smaller p and k, but as these quantities increase, the expected speedup drops. This is due to the extra multiplicative factor of p in the bandwidth cost and the extra additive factor of $k^3 qp$ in the

Table 1. Asymptotic complexity of PA0-HSS and PA1-HSS, ignoring constant factors. 'Offline' refers to Lines 1–3 and 'Online' refers to Lines 4–7 of PA1-HSS.

Algorithm		Flops	Words moved	Messages	Memory
PA0-HSS		$kqrn/p + kqr^2 \lg p$	$kqr \lg p$	$k \lg p$	$(kq+r)n/p + r^2 \lg p$
PA1-HSS	(offline)	$kr^2n/p + k^3$	$kr^2p \lg p$	$\lg p$	$(kq+r)n/p + k(q+r)rp$
	(online)	$kqrn/p + k(k+r)^2qp$	$kqrp \lg p$	$\lg p$	

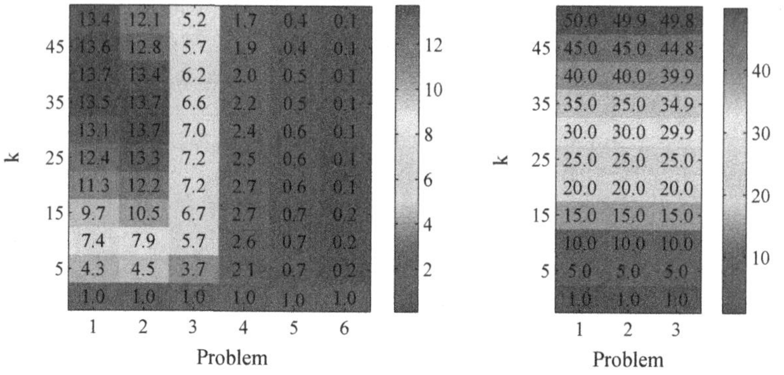

Fig. 1. Predicted PA1-HSS speedups on Peta (left) and Grid (right). Note that p and n increase with problem number on x-axis.

flop cost of PA1-HSS. Since the relative latency cost is lower on Peta, the effect of the extra terms becomes apparent for large k and p. On Grid, PA0-HSS is extremely latency bound, so a $\Theta(k)$-fold reduction in latency results in a $\Theta(k)\times$ faster algorithm. This is the best we can expect. Note that many details are abstracted in these models, which are meant to give a rough idea of asymptotic behavior. Realizing such speedups in practice remains future work.

5   Future Work and Conclusions

In this work, we derive a new parallel communication-avoiding matrix powers algorithm for $A = D + USV^H$, where D is well partitioned and USV^H has low rank but A may not be well partitioned. This allows speedups for a larger class of problems than previous algorithms [4,10], which require well-partitioned A. Our approach exploits low-rank properties of dense blocks, asymptotically reducing parallel latency cost. We demonstrate the generality of our parallel blocking covers technique by applying it to matrices with hierarchical structure. Performance models predict up to $13\times$ speedups on petascale machines and up to $3k$ speedups on extremely latency-bound machines, despite tradeoffs in arithmetic and bandwidth cost. Future work includes a high-performance parallel implementation of our algorithm to verify predicted speedups, as well as integration into preconditioned communication-avoiding Krylov solvers.

Acknowledgments. We acknowledge support from the US DOE (grants DE-SC000 4938, DE-SC0005136, DE-SC0010200, DE-SC0008700 and AC02-05CH11231) and DARPA (grant HR0011-12-2-0016), as well as contributions from Intel, Oracle, and MathWorks.

References

1. Bebendorf, M.: A means to efficiently solve elliptic boundary value problems. In: Bart, T., Griebel, M., Keyes, D., Nieminen, R., Roose, D., Schlick, T. (eds.) Hierarchical Matrices. LNCS, vol. 63, pp. 49–98. Springer, Heidelberg (2008)
2. Chan, E., Heimlich, M., Purkayastha, A., Van De Geijn, R.: Collective communication: theory, practice, and experience. Concurrency Comput.: Pract. Exper. **19**, 1749–1783 (2007)
3. Chandrasekaran, S., Dewilde, P., Gu, M., Lyons, W., Pals, T.: A fast solver for HSS representations via sparse matrices. SIAM J. Matrix Anal. Appl. **29**, 67–81 (2006)
4. Demmel, J., Hoemmen, M., Mohiyuddin, M., Yelick, K.: Avoiding communication in computing Krylov subspaces. Technical report UCB/EECS-2007-123, University of California-Berkeley (2007)
5. Hoemmen, M.: Communication-avoiding Krylov subspace methods. Ph.D. thesis, University of California-Berkeley (2010)
6. Hong, J., Kung, H.: I/O complexity: the red-blue pebble game. In: Proceedings of the 13th ACM Symposium on Theory of Computing, pp. 326–333. ACM, New York (1981)
7. Knight, N., Carson, E., Demmel, J.: Exploiting data sparsity in parallel matrix powers computations. Technical report UCB/EECS-2013-47, University of California-Berkeley (2013)
8. Kriemann, R.: Parallele Algorithmen für $\mathcal{H}$-Matrizen. Ph.D. thesis, Christian-Albrechts-Universität zu Kiel (2005)
9. Leiserson, C., Rao, S., Toledo, S.: Efficient out-of-core algorithms for linear relaxation using blocking covers. J. Comput. Syst. Sci. Int. **54**, 332–344 (1997)
10. Mohiyuddin, M.: Tuning hardware and software for multiprocessors. Ph.D. thesis, University of California-Berkeley (2012)
11. Mohiyuddin, M., Hoemmen, M., Demmel, J., Yelick, K.: Minimizing communication in sparse matrix solvers. In: Proceedings of the Conference on High Performance Computing Networking, Storage, and Analysis, pp. 36:1–36:12. ACM, New York (2009)
12. Philippe, B., Reichel, L.: On the generation of Krylov subspace bases. Appl. Numer. Math. **62**, 1171–1186 (2012)
13. Wang, S., Li, X., Xia, J., Situ, Y., de Hoop, M.: Efficient scalable algorithms for hierarchically semiseparable matrices. SIAM J. Sci. Comput. (2012, under review)

Performance of Dense Eigensolvers on BlueGene/Q

Inge Gutheil[1]($\boxtimes$), Jan Felix Münchhalfen[2], and Johannes Grotendorst[1]

[1] Institute for Advanced Simulation, Jülich Supercomputing Centre,
Forschungszentrum Jülich GmbH, 52425 Jülich, Germany
i.gutheil@fz-juelich.de
http://www.fz-juelich.de/ias/jsc
[2] IT Center of RWTH Aachen University,
Seffenter Weg 23, 52704 Aachen, Germany
muenchhalfen@rz.rwth-aachen.de
http://www.rz.rwth-aachen.de

Abstract. Many scientific applications require the computation of about 10–30 % of the eigenvalues and eigenvectors of large dense symmetric or complex hermitian matrices. In this paper we will present performance evaluation results of the eigensolvers of the three libraries Elemental, ELPA, and ScaLAPACK on the BlueGene/Q architecture. All libraries include solvers for the computation of only a part of the spectrum. The most time-consuming part of the eigensolver is the reduction of the full eigenproblem to a tridiagonal one. Whereas Elemental and ScaLAPACK only offer routines to directly reduce the full matrix to a tridiagonal one, which only allows the use of BLAS 2 matrix-vector operations and needs a lot of communication, ELPA also offers a two-step reduction routine, first transforming the full matrix to banded form and thereafter to tridiagonal form. This two-step reduction shortens the reduction time significantly but at the cost of a higher complexity of the back transformation step. We will show up to which part of the eigenspectrum the use of the two-step reduction pays off.

Keywords: Eigenvalue and eigenvector computation · Elemental · ELPA · ScaLAPACK · BlueGene/Q

1 Introduction

Many scientific applications in the fields of materials science and quantum chemistry require the computation of eigenvalues and eigenvectors of dense real symmetric or complex hermitian matrices. For example, in DFT (Density Functional Theory) calculations on modern supercomputers [1], typical sizes of those matrices are about 50000×50000 and 10–30 % of the eigenvalues and eigenvectors have to be computed. As the computation of eigenvalues and eigenvectors is of complexity N^3 where N denotes the problem size, these computations require

R. Wyrzykowski et al. (Eds.): PPAM 2013, Part I, LNCS 8384, pp. 26–35, 2014.
DOI: 10.1007/978-3-642-55224-3_3, © Springer-Verlag Berlin Heidelberg 2014

highly parallel algorithms to speed up the computation on modern computers with thousands of cores and the possibility to start even more threads than cores.

The oldest parallel library for dense linear algebra which is still in use in many applications is ScaLAPACK [2]. It is a pure MPI library requiring a two-dimensional block cyclically distributed matrix as input. In the last years two new libraries have been developed, Elemental [3], a complete C++ framework for dense linear algebra and ELPA [4] which is an add-on to ScaLA-PACK for the solution of real symmetric and complex hermitian eigenproblems. Both offer the choice to compute all or some of the eigenvalues and eigenvectors.

Whereas ScaLAPACK and Elemental only offer the one-step reduction of a full matrix to tridiagonal form, ELPA also includes a routine for two-step reduction, first to banded and then to tridiagonal form. This reduction allows the use of highly optimized BLAS 3 matrix-matrix operations for the first step, thus reducing significantly the execution time for the reduction, which still is the bottleneck of all dense eigensolvers. The gain in the reduction phase however goes with a loss in the back transformation phase, as the back transformation now is also done in two steps. The complexity of the back transformation depends on the number of eigenvectors to be computed, thus the gain in the reduction phase will pay off, if only a part of the eigenvectors are to be computed.

2   The BlueGene/Q Architecture

The BlueGene/Q Architecture is built from IBM PowerPC A2 compute cards, each card with 16 cores at 1.6 GHz and with 16 GB DDR3 Memory. Each core has two 4-way SIMD units which each can deliver 4 results per cycle. 32 compute cards form a node card, meaning each node card consists of 512 cores and has a total amount of 512 GB of main memory. 16 node cards form a midplane, two midplanes together with one or two I/O drawers form one rack. Thus each rack consists of 16384 cores and has a peak performance of 0.2 Petaflops. Each core allows 4-way hyperthreading, so up to 64 MPI processes or for example 16 MPI processes with up to 4 threads (OpenMP or pthreads) per MPI-process can be started on each compute card.

The BlueGene/Q called JUQUEEN [5] installed at Jülich Supercomputing Centre in May 2012 now consists of 28 racks (7 rows 4 racks), which means that the full machine has 28672 compute cards (458752 cores) and thus a peak performance of 5.9 Petaflops.

3   Parallel Libraries Investigated

The first library for the solution of dense symmetric eigenvalue problems we studied is ScaLAPACK. We used release 2.0.1, the current release 2.0.2 has not changed in the eigensolver part. It is mainly written in FORTRAN 77 with a few C-routines hidden from the user. An add-on to ScaLAPACK for eigenvalue problems is ELPA (EigensoLver for Petaflop Applications), written in Fortran 95. Both libraries use the same block-cyclic two-dimensional data distribution. The user has to distribute the matrix according to that distribution and calls the routines with the distributed matrix. ScaLAPACK only offers a pure MPI version whereas in the development version of ELPA an OpenMP/MPI hybrid version is available. ELPA does not use the BLACS for the computational routines but only creates two MPI sub-communicators per process, one for the process row and one for the process column of each MPI process. We used the ELPA development version of November 2012.

A new library for linear algebra with dense matrices is the Elemental C++ framework which uses a two-dimensional data distribution with block size 1. The version we used was 0.75. Elemental also offers a hybrid parallel version.

3.1 Routines Tested in the Libraries

All routines in all libraries under investigation follow the same three steps for the computation of eigenvalues and eigenvectors of a real dense symmetric matrix: Reduction to tridiagonal form via orthogonal transformations, solution of the tridiagonal eigenvalue problem, back transformation of the eigenvectors of the tridiagonal matrix to those of the original matrix.

ScaLAPACK 2.0.1 offers four different routines for the solution of the dense symmetric eigenvalue problem. They differ in the reduction routine used and, more important, in the way the tridiagonal eigenvalue problem is solved. PDSYEV and PDSYEVD use the original version PDSYTRD of the reduction to tridiagonal form whereas PDSYEVX and PDSYEVR use a reduction called PDSYNTRD which is usually faster than the original version due to some improvements in communication costs. This reduction always uses a square processor grid for the reduction phase. Even though data has to be redistributed in cases where the original process grid is not square, we found out that this routine is faster than the old one [6]. PDSYEV uses the QR algorithm for the computation of eigenvalues and eigenvectors of the tridiagonal matrix. This method is rather slow although it can be parallelized well. It delivers all eigenvalues and eigenvectors and the eigenvectors are orthogonal to working precision. We did not investigate that routine because it is too slow.

PDSYEVX is the most often used eigensolver routine in ScaLAPACK as it was for a long time the only one allowing to compute only a part of the eigenspectrum. It uses bisection and inverse iteration for the computation of the eigenvalues and eigenvectors of the symmetric tridiagonal matrix. This method can be easily parallelized and works well if the eigenvalues are well separated. The input parameter ORFAC can be used to decide when eigenvalues are treated as clustered and thus the eigenvectors have to be re-orthogonalized. The re-orthogonalization of eigenvectors belonging to clustered eigenvalues is not parallelized and thus it can be very expensive both in terms of compute time and in terms of memory requirements. For our measurements we set ORFAC $=$ $1.0 * 10^{-4}$.

PDSYEVD uses the divide-and-conquer-method for the tridiagonal eigenproblem. It was up to version 1.8 the fastest routine if all eigenvalues and eigenvectors had to be computed. There is no version of this routine allowing to compute only a part of the spectrum.

PDSYEVR is new in release 2.0.1 and uses the MRRR [7,8] algorithm for the tridiagonal eigenproblem. This routine allows to compute only a part of the eigenspectrum and is supposed to be faster than bisection and inverse iteration if eigenvalues are clustered.

The **ELPA** library consists of two routines which differ in the way the full eigenproblem is reduced to tridiagonal form and as a result of that in the back transformation of eigenvectors. The routine ELPA1 uses the well-known one-step reduction to tridiagonal form, similar to the old ScaLAPACK reduction routine, ELPA2 uses a two-step reduction first to band and then to tridiagonal form and a two-step back transformation [9,10]. Both routines use a modification of the divide-and-conquer method for the solution of the tridiagonal eigenvalue problem which allows to compute only a part of the eigenspectrum.

In the library **Elemental** there is only one routine for the solution of the symmetric or hermitian eigenvalue problem, HermitianEig. For the reduction to tridiagonal form there are two choices, one for general rectangular processor grids and one that uses the largest square processor grid that fits to the number of processors given. As with the ScaLAPACK reduction routines, the routine using a square processor grid is faster than the other even if the original grid is not square. This has been shown in [11]. For the solution of the tridiagonal eigenproblem the MRRR algorithm is used in the implementation PMRRR by M. Petschow [12]. In contrast to the other libraries the distribution block size does not determine the algorithmic block size thus allowing to use different algorithmic block sizes for different computation steps. We did not investigate this possibility.

4   Measurements Done

As most applications on JUQUEEN should use at least one midplane with 512 compute cards and thus 8192 cores we tried to measure the performance of the eigensolvers on a complete midplane and on one rack. Most of those measurements were done in the pre-production phase of JUQUEEN, later on we only had limited resources for tests. The up to 4-way multi-threading allows to start up to 64 processes per compute card. We therefore tried to use one, two, or four MPI processes per core to see, whether multithreading pays even with pure MPI. In the pre-production phase of JUQUEEN we measured execution times for the computation of the full eigenspectrum on one node card, one midplane, and one rack with matrix sizes ranging from 5000 to 60000. Later on measurements on one node card were done with matrix sizes from 6000 to 50000 by steps of 4000.

For ScaLAPACK and ELPA block sizes were in the beginning chosen to be 32. In private communication with Thomas Auckenthaler, one of the ELPA authors, we learned that for ELPA smaller blocks should be better and so we used a block size of 16 for ELPA [13] for the measurements of this article. Elemental had not yet been ported to JUQUEEN in the pre-production phase, thus all measurements were done with limited resources. We chose the default algorithmic block size of 128 which was seen to be optimal on BlueGene/P and on a preliminary BlueGene/Q hardware [11].

Only the ScaLAPACK measurements of the pre-production phase are still used in this presentation. For ELPA we repeated the measurements with the development version of November 2012 and block size 16. This version contains a new routine for the two-step back transformation, which is optimized for the BlueGene/Q vector instructions. The new optimized ELPA library turned out to be much faster than the older version.

ELPA and Elemental both contain hybrid versions, but due to compiler problems we could not compare these versions to the pure MPI implementation.

For all measurements we used a test program that constructs a dense matrix by transforming a diagonal matrix with known eigenvalues by a random Householder matrix and compares the computed eigenvalues to the given ones. We always included tests for the correctness of the results and the orthogonality of the computed eigenvectors, but the timing was only done for the routine investigated. All measurements of the partial spectrum were done only on one node card.

5   Scaling Results up to One Rack of JUQUEEN

Figure 1 shows the results of the fastest routine for the full eigenspectrum of each library on a node card of JUQUEEN using 16 (left) and 32 (right) MPI processes per compute card. We measured the performance for matrices of sizes between 1000 and 50000. Due to our checking of results and other additional memory consumptions PDSYEVD could only be measured up to $N = 34000$, PDSYEVR up to $N = 29000$ and PDSYEVX with ORFAC $= 1.0 * E^{-4}$ up to $N = 42000$. All ScaLAPACK routines investigated performed similarly, there is not much difference between 16 and 32 MPI processes per compute card except for PDSYEVR, which for matrices of size up to 20000 is significantly slower with 32 processes per compute card than with 16 processes. Overall for matrices of sizes up to 25000 the fastest ScaLAPACK routine was PDSYEVX, followed by PDSYEVD for matrices smaller than 6000 and PDSYEVR for matrices larger than 6000.

On a node card computing all eigenvalues and -vectors ELPA1 is faster than ELPA2. Overall the performance of ELPA is better than ScaLAPACK's and Elemental's performance. All routines show a small speedup if 32 MPI processes per compute card are used.

On a midplane we could see that due to a bug in the communication that we had already seen on BlueGene/P with 1024 MPI processes (see [11]) PDSYEVR

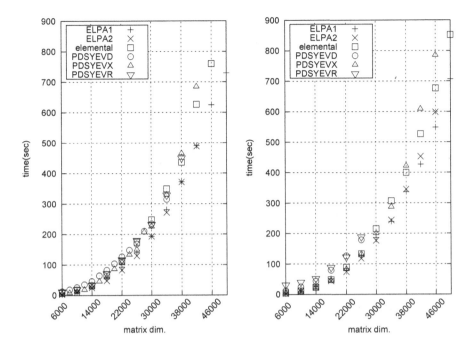

Fig. 1. Comparison of all routines of each library on a node card using 16 (left) and 32 (right) MPI processes per compute card.

took more than 900 s for the smallest matrix of size $N = 6000$ for 16 MPI processes per compute card and thus could not be shown in the figures. We did not further investigate PDSYEVR for the computation of the full eigenspectrum.

Overall the ScaLAPACK routines could no longer compete with the new libraries ELPA and Elemental, and PDSYEVX showed a very high variation in execution times for large matrices as can be seen from Fig. 2. ScaLAPACK and Elemental still showed a small speedup with 32 MPI processes per compute card, whereas both ELPA routines became slower. On one rack ELPA2 again was the fastest routine, ELPA1 and Elemental almost equal and ScaLAPACK slower than the new routines.

For a matrix of size $N = 50000$ Elemental showed a speedup for one midplane compared to a node card of about 8, ELPA1 of about 9, ELPA2 even 10. It is the fastest routine on a midplane. The speedup of the ScaLAPACK routines could not be measured as we could not run the programs with $N = 50000$. For one rack there was a speedup compared to the fastest run on a midplane for all routines except PDSYEVX, Elemental even got a speedup of more than two compared to the run on a midplane with 16 MPI processes per compute card. As the speedup is only slightly more than two, we think that it is due to the fact that we did only one measurement per matrix size.

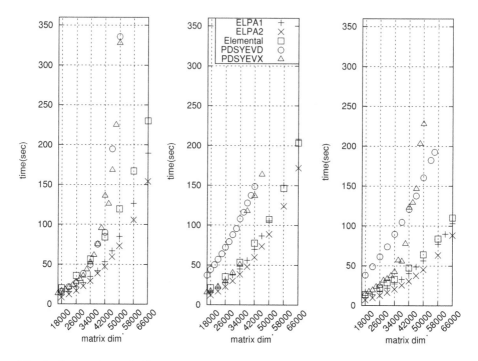

Fig. 2. Comparison of all routines of each library on a midplane left: using 16 MPI processes per compute card, middle: using 32 MPI processes per compute card and right: on one rack using 16 MPI processes per compute card.

All new routines scale up to one rack of BlueGene/Q, for Elemental using 32 MPI processes per compute card is faster, for ELPA using only 16 processes per compute card is faster. Thus we are indeed waiting for a hybrid parallelization to see whether that allows to explore hyperthreading.

6   Results for Different Parts of the Spectrum

Figure 3 shows that even for the computation of the full eigenspectrum the routine with the two-step reduction is not much slower than the one with the one-step reduction on a node card. The parts of the time for reduction and back transformation almost change their roles. Thus it can be expected that if less eigenvectors have to be transformed the time for the routine with the two-step reduction will become much smaller.

As most applications require only a part of the eigenspectrum we also compared the performance of the different libraries and routines for the computation of 5 % and 45 % of the eigenvalues and eigenvectors. From Fig. 4 it can be seen that for the computation of 45 % of the spectrum ELPA2 is almost twice as fast as the other routines, for 5 % even more than three times as fast.

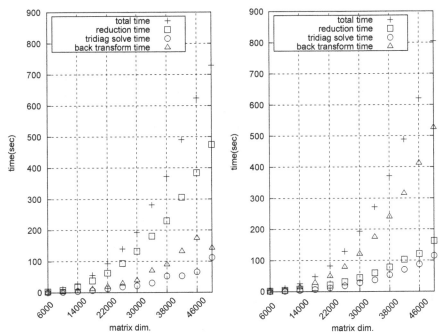

Fig. 3. Comparison of ELPA1 (left) and ELPA2 (right) on one node card using 16 MPI processes per compute card, computation of the full eigenspectrum

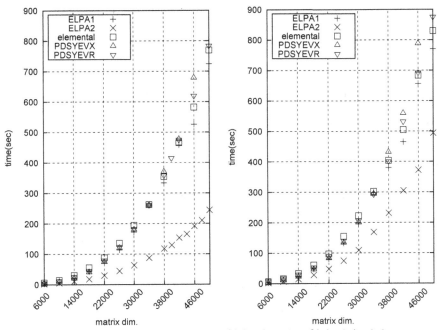

Fig. 4. Comparison of all routines if only 5 % (left) and 45 % (right) of the spectrum is computed, one node card using 16 MPI processes per compute card

7   Conclusions

For the computation of all eigenvalues and eigenvectors of a large full real symmetric matrix the old ScaLAPACK routine PDSYEVX has big problems with clustered eigenvalues. The new ELPA library solves this problem by using the divide-and-conquer algorithm for the computation of the eigenvalues of the tridiagonal matrix. On a BlueGene/Q both ELPA implementations are faster than ScaLAPACK's routine PDSYEVD. A better implementation of the BLACS directly based on the basic communication layer or a vendor optimized ScaLA-PACK could perhaps speed up ScaLAPACK. Without an optimized version of ScaLAPACK and the overhead of BLACS above MPI we recommend not to use ScaLAPACK for the computation of eigenvalues and eigenvectors of dense real symmetric matrices on a BlueGene/Q.

An alternative to ELPA is the new library Elemental which uses the PMRRR algorithm and is very flexible because the algorithmic block size can be chosen without changing the data layout. Also, filling of the C++ class DistMatrix can be easier than distributing a matrix in the block-cyclic two-dimensional way ScaLAPACK and ELPA require.

For the computation of only a part of the spectrum (even more than 50 %) the ELPA2 routine performs much better than all the other routines. It is in fact the only one that is significantly faster if only a part of the spectrum is needed compared to the full spectrum. This is mainly because for all other routines the reduction phase is so dominant, that gains in the other parts of the computation have almost no influence on the execution time. This means that for computations where only some of the eigenvalues and -vectors have to be computed, ELPA2 should be used if possible, especially, if the application already was written to use ScaLAPACK and thus already has the matrix in the way it is also needed for ELPA.

Acknowledgements. The authors thank Jack Poulson, the author of the Elemental library and the ELPA team, especially Thomas Auckenthaler, for their immediate responses to problem reports.

References

1. FLEUR: The Jülich FLAPW code family. Website (May 2013). http://www.flapw. de
2. Choi, J., Demmel, J., Dhillon, I., Dongarra, J., Ostrouchov, S., Petitet, A., Stanley, K., Walker, D., Whaley, R.: Scalapack: a portable linear algebra library for distributed memory computers-design issues and performance. Comput. Phys. Commun. **97**(1–2), 1–15 (1996)
3. Poulson, J., Marker, B., van de Geijn, R.A., Hammond, J.R., Romero, N.A.: Elemental: a new framework for distributed memory dense matrix computations. ACM Trans. Math. Softw. **39**(2), 13:1–13:24 (2013)
4. ELPA: Eigenvalue Solvers for Petaflop Applications home page. Website (May 2013). http://elpa.rzg.mpg.de

5. FZJ-JSC: IBM Blue Gene/Q - JUQUEEN home page. Website (May 2013). http://www.fz-juelich.de/ias/jsc/juqueen
6. Gutheil, I.: Performance evaluation of scalapack eigensolver routines on two hpc systems. In: 6th International Workshop on Parallel Matrix Algorithms and Applications (PMAA'10) (2010). http://juser.fz-juelich.de/record/10376
7. Dhillon, I., Parlett, B., Vömel, C.: The design and implementation of the MRRR algorithm. ACM Trans. Math. Softw. (TOMS) **32**(4), 533–560 (2006)
8. Dhillon, I.: A new $O(n^2)$ algorithm for the symmetric tridiagonal eigenvalue eigenvector problem. Ph.D. thesis, University of California, Berkeley (1997)
9. Auckenthaler, T., Blum, V., Bungartz, H.J., Huckle, T., Johanni, R., Krämer, L., Lang, B., Lederer, H., Willems, P.: Parallel solution of partial symmetric eigenvalue problems from electronic structure calculations. Parallel Comput. **37**(12), 783–794 (2011)
10. Auckenthaler, T., Bungartz, H.J., Huckle, T., Krämer, L., Lang, B., Willems, P.: Developing algorithms and software for the parallel solution of the symmetric eigenvalue problem. J. Comput. Sci. **2**(3), 272–278 (2011)
11. Gutheil, I., Berg, T., Grotendorst, J.: Performance analysis of parallel eigensolvers of two libraries on BlueGene/p. J. Math. Syst. Sci. **2**(4), 231–236 (2012)
12. Petschow, M., Peise, E., Bientinesi, P.: High-performance solvers for dense hermitian eigenproblems. SIAM J. Sci. Comput. (SISC) 35(1), C1–C22 (2013). arXiv:1205.2107v2[cs.MS]
13. Münchhalfen, J.: Performance analysis and comparison of parallel eigensolvers on blue gene architectures. Berichte des Forschungszentrums Jülich (4359) 65 p. (2013). http://juser.fz-juelich.de/record/128657

Experiences with a Lanczos Eigensolver in High-Precision Arithmetic

Alexander Alperovich[1], Alex Druinsky[2(✉)], and Sivan Toledo[2]

[1] Microsoft, Haifa, Israel
[2] Tel Aviv University, Tel Aviv, Israel
alexdrui@post.tau.ac.il

Abstract. We investigate the behavior of the Lanczos process when it is used to find all the eigenvalues of large sparse symmetric matrices. We study the convergence of classical Lanczos (i.e., without reorthogonalization) to the point where there is a cluster of Ritz values around each eigenvalue of the input matrix A. At that point, convergence to all the eigenvalues can be ascertained if A has no multiple eigenvalues. To eliminate multiple eigenvalues, we disperse them by adding to A a random matrix with a small norm; using high-precision arithmetic, we can perturb the eigenvalues and still produce accurate double-precision results. Our experiments indicate that the speed with which Ritz clusters form depends on the local density of eigenvalues *and on the unit roundoff*, which implies that we can accelerate convergence by using high-precision arithmetic in computations involving the Lanczos iterates.

Keywords: Lanczos · Mixed-precision arithmetic · Ritz clusters

1 Introduction

Existing software libraries offer us several choices when we wish to compute the eigenvalues of a sparse matrix. One option is to use a dense eigensolver, such as one of those implemented in LAPACK [1], which compute all of the eigenvalues at a cost of $\Theta(n^3)$ arithmetic operations. This is a high cost for a sparse matrix. Krylov-subspace eigensolvers, such as ARPACK [16], take advantage of sparsity, but they only offer to compute small, user-selected subsets of the spectrum. Other algorithms compute the whole spectrum, but they require the matrix to have a special sparsity structure, such as the MRRR method for symmetric tridiagonal matrices [7]. In this paper we explore the possibility of accomplishing both goals simultaneously: computing all of the eigenvalues and also taking advantage of sparsity, even when there is no specific sparsity structure.

The Lanczos process is a long-established and well-known eigensolver [15] (see also [10,17,23,24,29]). It takes as input an n-by-n Hermitian matrix A and produces a sequence of matrices $T^{(m)}$ and $Q^{(m)}$ such that

$$AQ^{(m)} = Q^{(m)}T^{(m)} + r^{(m)}e_m^* ,$$

R. Wyrzykowski et al. (Eds.): PPAM 2013, Part I, LNCS 8384, pp. 36–46, 2014.
DOI: 10.1007/978-3-642-55224-3_4, © Springer-Verlag Berlin Heidelberg 2014

where $Q^{(m)}$ is an n-by-m orthonormal matrix, $T^{(m)}$ is an m-by-m tridiagonal matrix, e_m is the last unit vector of dimension m, and $r^{(m)}$ is some n-vector. The sequences $Q^{(m)}$ and $T^{(m)}$ are nested: each iteration of the Lanczos process adds one column to Q and a row and a column to T. The process is a short-recurrence Krylov-subspace iteration; in each iteration, the algorithm multiplies one vector by A and performs a small number of vector operations on vectors of size n.

In exact arithmetic, the residual vector $r^{(m)}$ vanishes after at most k iterations, where k is the number of distinct eigenvalues of A. When $r^{(m)}$ vanishes, $T^{(m)}$ is an orthonormal projection of A onto the column space of Q, and therefore every eigenvalue of $T^{(m)}$ is an eigenvalue of A. For all the starting vectors except for a set of measure 0, $r^{(m)}$ vanishes after exactly k iterations and all the eigenvalues of A appear in the spectrum of $T^{(k)}$.

Practitioners quickly discovered that the behavior of Lanczos in floating-point arithmetic differs significantly from that predicted by the theoretical results. In particular, the columns of Q quickly lose orthogonality, and r never vanishes in practice. Researchers mostly explored two families of techniques for addressing this difficulty. One set of techniques attempts to prevent the loss of orthogonality in Q. This can be done using a full orthogonalization process or using selective orthogonalization and related techniques [11, 20, 25, 26]. The other set of techniques [2, 31] attempts to extract useful spectral information from the process after a relatively small number of iterations; this rarely results in the identification of all the eigenvalues, but it can result in useful approximations to a subset of the eigenvalues that are important in a given application (e.g., the smallest ones). These families of techniques are not mutually exclusive; many Lanczos codes use both.

However, around 30 years ago a group of researchers explored the use of Lanczos without sophisticated orthogonalization for finding all the eigenvalues of A [3, 9, 19]; we refer to such methods as *classical Lanczos* methods. This line of research was based on a deep numerical analysis of the Lanczos process that ultimately showed that in floating point, *the eigenvalues of T eventually approximate all the eigenvalues of A* [8]. (This fact was recognized years before it was actually proved; see, for example, [3]). These researchers produced two Lanczos codes, both in the 1980s. To the best of our knowledge none of the Lanczos codes that have been published since 1985 have been *classical* Lanczos. Even though development of new codes has slowed down, there has been intense ongoing theoretical interest in classical Lanczos, resulting in a large body of results (see [17] and the numerous references therein). Experimental studies have also been published [13].

Our goal in this paper is to present the challenges that are involved in developing a classical Lanczos eigensolver, to describe our ideas for addressing these challenges, and to explore the feasibility of these ideas. A major tool in our toolkit is high-precision arithmetic, which our experiments show can be used to tackle hard matrices with tight clusters of eigenvalues and also to devise a reliable termination criterion for the algorithm. Our computational cost of using high-precision arithmetic is negligible because we need it only for forming the matrix T and not for computing its eigenvalues.

2   Background and Methodology

A key issue in Lanczos solvers, including ours, is deciding which Ritz value (eigenvalue of $T^{(m)}$) is an approximate eigenvalue of A. A growing body of work suggests that non-trivial clusters of Ritz values are only found very close to eigenvalues of A (see Wülling [32,33], Knizhnerman [14, Theorem 2], Strakoš and Greenbaum [30], and Greenbaum [12]). That is, if we find two or more eigenvalues of $T^{(m)}$ that are very close to each other, they normally indicate the location of an eigenvalue of A; we call such Ritz values *doubly-converged*. This phenomenon was already known to Cullum and Willoughby [3] and to Parlett and Reid [19], but back then there were no provable bounds on the location of eigenvalues relative to non-trivial Ritz clusters. We write that Ritz clusters *normally* indicate eigenvalues because all the results in the literature are conditioned on properties of the spectrum of A and/or $T^{(m)}$, which might not hold. However, exceptions seem very rare and some conditions are easily tested (in particular, conditions that only involve Ritz values).

If all the eigenvalues of A are simple, we can stop Lanczos once we have n distinct Ritz clusters (doubly-converged eigenvalues). If there are multiple eigenvalues, we need another strategy. The codes of Cullum and Willoughby [3] and Parlett and Reid [19] used heuristics to decide when to stop. These heuristics sometimes cause the algorithms to fail to find all the eigenvalues; these failures are sometimes silent and sometimes explicit (reported to the user).

In order to address this problem, we use a conceptually simple solution that we call *dispersion*. Instead of running Lanczos on A itself, we will run it on $A + P$, where P is a random symmetric matrix (from some appropriate distribution) with a small norm $\|P\|_2 \le \delta$. We choose P so that it is cheap to apply to vectors; this results in Lanczos iterations that are about as cheap as those performed on A alone. The perturbation P perturbs the eigenvalues, but only by δ or less. Hopefully, $A + P$ has no multiple eigenvalues; multiple eigenvalues of A are transformed into clusters of close but distinct eigenvalues of $A + P$. The choice of P determines how close the eigenvalues of $A + P$ are; we do not have a complete theory that guarantees good separation with high probability, but experiments have shown that dispersion works well. We omit these experiments from this paper, and focus instead on the convergence for a given operator (which the reader can take to be $A + P$).

The size of the perturbation δ and machine precision $\epsilon_{\text{machine}}$ must be tailored according to the accuracy ϵ required by the user, using high-precision arithmetic to reduce $\epsilon_{\text{machine}}$ if necessary. The relation $\epsilon > \delta > \epsilon_{\text{machine}}$ must hold with sufficient safety margins so that the perturbation can simultaneously separate multiple eigenvalues and preserve the required accuracy.

An alternative approach to obtaining the required accuracy is to use a first-order correction. Here, we consider A as a matrix that we obtain from $A + P$ by adding the perturbation $-P$. To first order, the eigenvalues of A are equal to $\mu_i - v_i^T P v_i$, where μ_i and v_i are eigenpairs of $A + P$ for $i = 1, 2, \ldots, n$; for details, see [29, pp. 45–48]. Computing the correction $v_i^T P v_i$ requires that we compute the eigenvectors v_i, which is accomplished by multiplying the n-by-m matrix of iterates Q by the eigenvectors of T. This costs $\Theta(mn)$ arithmetic operations for each v_i and $\Theta(mn^2)$ overall. Because $m > n$, this is at least as expensive as the $\Theta(n^3)$ cost of computing the eigenvalues of A directly using a dense eigensolver. This cost is too high for sparse matrices and therefore we do not use first-order corrections in this paper.

From a broader perspective, matrices with multiple eigenvalues are a sort of singularity, a set of measure zero that causes algorithmic difficulty. The idea of perturbing problems in order to avoid such singularities is well-established in existing research [4, 6, 28].

To compute the eigenvalues of T in our experiments we used the LAPACK subroutine DSTEMR and the MPACK subroutine RSTEQR. Both of these subroutines are symmetric tridiagonal eigensolvers; DSTEMR implements the MRRR algorithm and RSTEQR the implicit QL or QR methods. The MPACK library, to which RSTEQR belongs, is a collection of multiple-precision versions of BLAS and LAPACK subroutines [18]. Although we only needed the eigenvalues of T to double-precision accuracy, in some of our experiments we used higher precision, which is why we used MPACK.

3   Convergence on Real-World Matrices

Figure 1 explores the behavior of high-precision Lanczos on a large set of real-world matrices. We ran our code on all 133 symmetric matrices of dimension 2500 or less from the University of Florida Sparse Matrix Collection [5] (we only used matrices with numerical values; we omitted sparsity-pattern-only matrices). We did not attempt to disperse multiple eigenvalues; instead, we compared the eigenvalues computed by our code to those computed by LAPACK and counted how many agree to within $10^{-9}\|A\|$ or better.[1] This tells us when our code fails to find isolated eigenvalues or entire clusters, but not whether it converges to all the eigenvalues in a tight cluster.

The results show that Lanczos can compute all the eigenvalues of most of the matrices after $16n$ iterations: of over 60 % of the matrices with 64-bit arithmetic, and of more than 70 % with 128- and 256-bit arithmetics. After only $4n$ or $8n$ iterations, Lanczos can still compute all the eigenvalues of many matrices. As we perform more iterations, Lanczos tends to resolve more eigenvalues *in sparse areas of the spectrum*. The remaining non-converged eigenvalues tend to be shadowed by their neighbors.

4   The Effects of Clusters of Eigenvalues

As we saw in Sect. 3, the key to fast convergence of the Lanczos iteration is dealing with clusters of eigenvalues. One of the problems caused by clustering has been discovered by Parlett et al. [21], who showed that a Ritz value can become

---

[1] We used the LAPACK unsymmetric eigensolver DGEEV to compute the eigenvalues. Although the symmetric subroutine DSYEV is more efficient, we did not use it here.

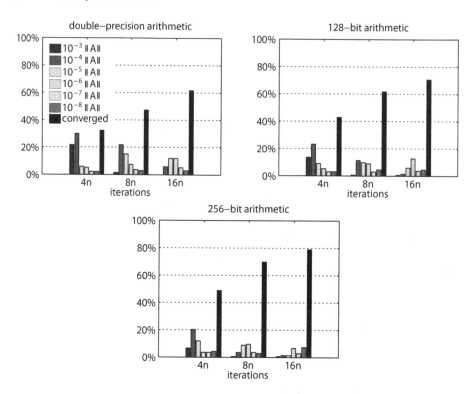

Fig. 1. Convergence behavior on a set of 133 real-world matrices. The graphs show the percentage of matrices that have converged to all the eigenvalues after $4n$, $8n$, and $16n$ iterations (in brown). The code did not attempt to find the multiplicity of each eigenvalue. The graphs also show, for matrices that have not converged, the greatest distance from a non-converged eigenvalue to its nearest neighbor. The graphs show results for 64-, 128- and 256-bit arithmetic, clockwise from top left. The number of bits refers to the accuracy of the arithmetic with which T was computed; its eigenvalues were always computed in 64-bit arithmetic (Color figure online).

fixed between two nearby eigenvalues and hold there for a number of iterations before finally migrating towards one of the eigenvalues (see also [22,27]). In this section we study this phenomenon, called *misconvergence*, and we also show that the problems caused by clustering go beyond misconvergence.

In our experiments we found that small clusters do not substantially affect convergence outside the cluster. When we ran Lanczos on a synthetic matrix whose eigenvalues were evenly spaced in the interval $[-1, 1]$, and then added a small cluster of $0.05n$ eigenvalues, we found that adding the cluster caused no visible artifacts on the plot of the Ritz values produced by the iteration. However, we found severe misconvergence within the cluster. Figure 2 shows that a Ritz value that shows up in a cluster tends to wander around near and between eigenvalues and then typically settles for a long time *in-between* eigenvalues. As more Ritz values show up, a misconverged eigenvalue tends to shift closer to an

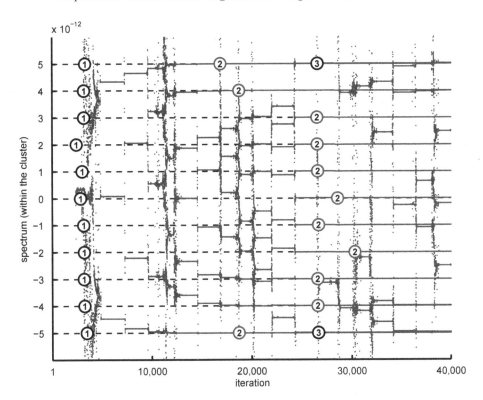

Fig. 2. Evolution of Ritz values near a cluster. The cluster consists of 11 eigenvalues 10^{-12} apart, represented by black horizontal dashed lines. Blue dots represent Ritz values. A circled numeral k shows the first time that there are k Ritz values near an eigenvalue. Red circles around the numeral 2 show where double convergence first occurs. Blue lines are formed by converged or misconverged Ritz values (Color figure online).

eigenvalue, until it actually converges. If we inspect the eigenvalue at 5×10^{-12}, for example (the topmost one), we see a misconverged Ritz value that shifts between 3 or 4 stable locations before converging.

Additional Ritz values appear periodically near an eigenvalue. The most important effect of clusters is on this periodicity. In a cluster, the periodicity is longer; a Ritz value appears near a specific eigenvalue less often than near non-clustered eigenvalues. This is shown in the left plot of Fig. 3. This phenomenon causes Lanczos to converge more slowly in the presence of clusters. If we examine the raw density of Ritz values, ignoring the distribution of eigenvalues, we see that the cluster attracts more Ritz values than intervals of the same size elsewhere in the spectrum. This is shown in the right plot of Fig. 3. This increased attraction is not sufficient, however, to compensate for the larger number of eigenvalues in the interval, so convergence to all eigenvalues is still adversely affected by the cluster.

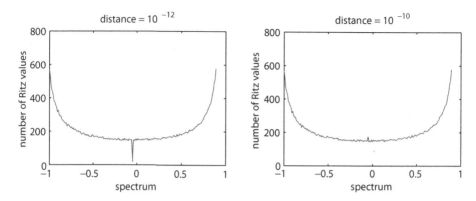

Fig. 3. The number of Ritz values within a distance of 10^{-12} (left) and 10^{-10} (right) from each eigenvalue after $200n$ iterations. The matrix has order $n = 200$ and has a cluster of 10 eigenvalues positioned in the center of the spectrum at regular distances of 10^{-12}. There are fewer Ritz values near each eigenvalue within the cluster than elsewhere, yet there are more Ritz values in the cluster area than if there was a single eigenvalue there.

5   The Effects of High-Precision Arithmetic on the Lanczos Process

The distribution of Ritz values within a cluster is typically *not* uniform, just like within the spectrum as a whole. When eigenvalues in the cluster are evenly distributed, more Ritz values appear at the edges of the cluster than near its center, as shown in Fig. 4. Even when there are *on average* 2 or 3 Ritz values per eigenvalue in the cluster, we may be very far from convergence, because there are not enough Ritz values near eigenvalues in the center of the cluster.

Increasing the precision of the floating-point arithmetic also increases the attractive power of clusters upon Ritz values, as shown in Fig. 5. As we increase the precision, the number of Ritz values in a cluster increases, speeding up the convergence. This phenomenon was already observed by Edwards et al. [9], but it does not appear that Lanczos codes have used this insight.

As the size of an eigenvalue cluster grows, its effect on convergence becomes devastating, even in high precision. Figure 6 shows that as the size of a cluster grows, the number of Ritz values in it increases, but not nearly fast enough to obtain convergence on all eigenvalues. When the cluster is small, say containing 10 eigenvalues, there are more than 3 Ritz values per eigenvalue after $20n$ iterations, even in 64-bit arithmetic (and more Ritz values in higher precision). When the cluster contains 100 eigenvalues, there is not even a single Ritz value per eigenvalue after $20n$ iterations; we cannot expect convergence in that many iterations. Things get much worse as the cluster size continues to grow.

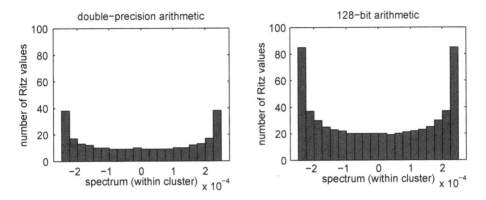

Fig. 4. A histogram of the Ritz values within a cluster of 5000 eigenvalues that are spaced 10^{-7} apart after $20n = 20 \cdot 7000$ Lanczos iterations. Apart from the cluster, the spectrum contains 2000 eigenvalues spaced evenly between -1 and 1. The histogram on the left shows the results in 64-bit arithmetic and the results on the right in 128-bit.

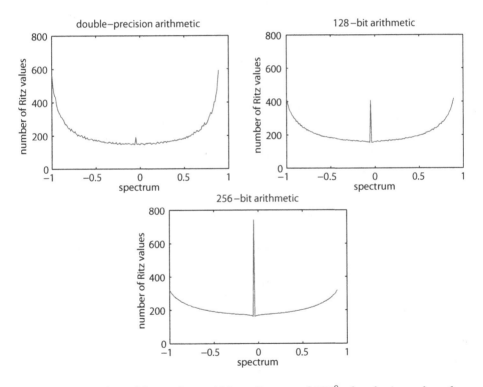

Fig. 5. The number of Ritz values within a distance of 10^{-9} of each eigenvalue after $200n$ iterations. The matrix has order $n = 200$ and has a cluster of 10 eigenvalues positioned in the center of the spectrum at regular distances of 10^{-11}. The iteration is carried out using different levels of floating-point precision: IEEE-754 double precision (64 bits; top left), 128-bit floating point (top right) and 256-bit floating point (bottom).

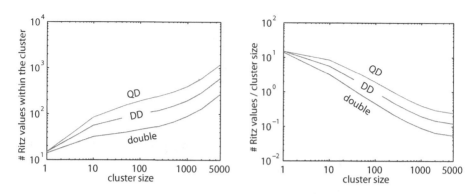

Fig. 6. The number of Ritz values within a cluster of eigenvalues that are spaced 10^{-7} apart after $20n$ Lanczos iterations. Apart from the cluster, the spectrum contains 2000 eigenvalues spaced evenly between -1 and 1. In both graphs, the X axis shows the number of eigenvalues in the cluster, ranging from 1 to 5000. (The dimension of the matrices therefore ranged from 2000 to 7000). On the left, the Y axis shows the number of Ritz values in the cluster. On the right, the Y axis shows the same number, but divided by the size of the cluster. Both graphs show the results of computations in 64-bit arithmetic (double), 128-bit (DD), and 256-bit (QD).

6   Conclusions

Our experiments suggest several conclusions. The results indicate that Lanczos can find *all* the eigenvalues of many real-world matrices. When it fails, convergence is impaired by the existence of dense areas in the spectrum, which is manifested by misconvergence, and more importantly by relatively low density of Ritz values in such dense areas. High-precision arithmetic helps to neutralize the effect of clustering, but the level of precision must be commensurate with the severity of clustering. How can we find the correct level of precision? One strategy is to start iterating in double precision and then repeatedly increase the accuracy after completing each sequence of n iterative steps. At the limit we have infinite accuracy and n additional steps are enough, although we expect a moderate level of accuracy to be sufficient for most matrices. We do not know how effective this strategy is in practice; this question is left for future work.

The slowdown in convergence due to clustering may make Lanczos impractical for problem matrices unless measures are taken to address this issue. Initial experimentation on small matrices suggests that randomized dispersion is effective when the spectrum contains clusters but they are not too large, but ineffective when clusters are very large (say an eigenvalue of multiplicity 3000 in a matrix of dimension 10000).

These technical conclusions lead us to two higher-level observations. First, classical Lanczos may be the only practical way of finding all the eigenvalues for some matrices. If the $\Theta(n^2)$ space required for dense methods is not available, and if shift-invert operations are too expensive (e.g., matrices for which there is no sparse factorization), and if the spectrum contains only mild clustering, then classical Lanczos may be the method of choice. This motivates further development of Lanczos codes and techniques.

We are used to resolving details at the scale of ϵ using floating-point arithmetic with unit roundoff near ϵ ($\approx 10^{-16}$ for 64-bit arithmetic). For symmetric eigensolvers, resolving eigenvalues at this scale does not mean finding 16 significant digits per eigenvalue; it merely means finding 16 digits relative to the scale of the largest one. For ill-conditioned matrices, resolving eigenvalues to that scale may not be excessive at all. Our second high-level observation is that in Lanczos, resolving eigenvalues at the scale of ϵ may require arithmetic with significantly smaller unit roundoff, perhaps 10^{-32} or 10^{-64}, or even less. More efficient implementations of high-precision floating-point arithmetic will enable computational scientists to resolve details that are currently beyond reach, like the eigenvalues of matrices with highly-clustered spectra.

Acknowledgments. We thank the referees for their valuable comments. The idea of using first-order corrections that we discuss in Sect. 2 was proposed by one of the referees.

This research was supported in part by grant 1045/09 from the Israel Science Foundation (founded by the Israel Academy of Sciences and Humanities), and by grant 2010231 from the US–Israel Binational Science Foundation.

The first author was at Tel Aviv University while conducting this research.

References

1. Anderson, E., Bai, Z., Bischof, C., Blackford, S., Demmel, J., Dongarra, J., du Croz, J., Greenbaum, A., Hammarling, S., McKenney, A., Sorensen, D.: LAPACK Users' Guide. SIAM, Philadelphia (1999)
2. Calvetti, D., Reichel, L., Sorensen, D.C.: An implicitly restarted Lanczos method for large symmetric eigenvalue problems. Electron. Trans. Numer. Anal. **2**, 1–21 (1994)
3. Cullum, J.K., Willoughby, R.A.: Lanczos Algorithms for Large Symmetric Eigenvalue Computations: Vol. 1 Theory. Birkhäuser, Basel (1985)
4. Davies, E.B.: Approximate diagonalization. SIAM J. Matrix Anal. Appl. **29**, 1051–1064 (2007)
5. Davis, T.A., Hu, Y.: The University of Florida sparse matrix collection. ACM Trans. Math. Softw. **38**, 1:1–1:25 (2011)
6. Dhillon, I.S., Parlett, B.N., Vömel, C.: Glued matrices and the MRRR algorithm. SIAM J. Sci. Comput. **27**, 496–510 (2005)
7. Dhillon, I.S., Parlett, B.N., Vömel, C.: The design and implementation of the MRRR algorithm. ACM Trans. Math. Softw. **32**, 533–560 (2006)
8. Druskin, V.L., Knizhnerman, L.A.: Error bounds in the simple Lanczos procedure for computing functions of symmetric matrices and eigenvalues. USSR Comput. Math. Math. Phys. **31**(7), 20–30 (1991)
9. Edwards, J.T., Licciardello, D.C., Thouless, D.J.: Use of the Lanczos method for finding complete sets of eigenvalues of large sparse symmetric matrices. J. Inst. Math. Appl. **23**, 277–283 (1979)
10. Golub, G.H., van Loan, C.F.: Matrix Computations, 3rd edn. Johns Hopkins University Press, Baltimore (1996)

11. Grcar, J.F.: Analyses of the Lanczos algorithm and of the approximation problem in Richardson's method. Ph.D. thesis, University of Illinois at Urbana-Champaign (1981)
12. Greenbaum, A.: Behavior of slightly perturbed Lanczos and conjugate-gradient recurrences. Linear Algebra Appl. **113**, 7–63 (1989)
13. Kalkreuter, T.: Study of Cullum's and Willoughby's Lanczos method for Wilson fermions. Comput. Phys. Commun. **95**, 1–16 (1996)
14. Knizhnerman, L.A.: The quality of approximations to a well-isolated eigenvalue, and the arrangement of "Ritz numbers" in a simple Lanczos process. Comput. Math. Math. Phys. **35**(10), 1175–1187 (1995)
15. Lanczos, C.: An iteration method for the solution of the eigenvalue problem of linear differential and integral operators. J. Res. Nat. Bur. Stand. **45**(4), 255–282 (1950)
16. Lehoucq, R.B., Sorensen, D.C., Yang, C.: ARPACK Users' Guide. SIAM, Philadelphia (1997)
17. Meurant, G.: The Lanczos and Conjugate Gradient Algorithms: From Theory to Finite Precision Computations. SIAM, Philadelphia (2006)
18. Nakata, M.: The MPACK: multiple precision arithmetic BLAS and LAPACK. http://mplapack.sourceforge.net/ (2010)
19. Parlett, B.N., Reid, J.K.: Tracking the progress of the Lanczos algorithm for large symmetric eigenproblems. IMA J. Numer. Anal. **1**, 135–155 (1981)
20. Parlett, B.N., Scott, D.S.: The Lanczos algorithm with selective orthogonalization. Math. Comp. **33**, 217–238 (1979)
21. Parlett, B.N., Simon, H., Stringer, L.M.: On estimating the largest eigenvalue with the Lanczos algorithm. Math. Comp. **38**, 153–165 (1982)
22. Parlett, B.: Misconvergence in the Lanczos algorithm. In: Cox, M., Hammarling, S. (eds.) Reliable Numerical Computation, pp. 7–24. Clarendon Press, Oxford (1990)
23. Parlett, B.N.: The Symmetric Eigenvalue Problem. Prentice-Hall, Englewood Cliffs (1980)
24. Saad, Y.: Numerical Methods for Large Eigenvalue Problems, 2nd edn. SIAM, Philadelphia (2011)
25. Simon, H.D.: Analysis of the symmetric Lanczos algorithm with reorthogonalization methods. Linear Algebra Appl. **61**, 101–131 (1984)
26. Simon, H.D.: The Lanczos algorithm with partial reorthogonalization. Math. Comp. **42**, 115–142 (1984)
27. van der Sluis, A., van der Vorst, H.A.: The convergence behavior of Ritz values in the presence of close eigenvalues. Linear Algebra Appl. **88–89**, 651–694 (1987)
28. Spielman, D.A., Teng, S.H.: Smoothed analysis: an attempt to explain the behavior of algorithms in practice. Commun. ACM **52**, 76–84 (2009)
29. Stewart, G.W.: Matrix Algorithms, Volume 2: Eigensystems. SIAM, Philadelphia (2001)
30. Strakoš, Z., Greenbaum, A.: Open questions in the convergence analysis of the Lanczos process for the real symmetric eigenvalue problem. IMA Preprint 934, University of Minnesota (1992)
31. Wu, K., Simon, H.: Thick-restart Lanczos method for large symmetric eigenvalue problems. SIAM J. Matrix Anal. Appl. **22**, 602–616 (2000)
32. Wülling, W.: The stabilization of weights in the Lanczos and conjugate gradient method. BIT Numer. Math. **45**, 395–414 (2005)
33. Wülling, W.: On stabilization and convergence of clustered Ritz values in the Lanczos method. SIAM J. Matrix Anal. Appl. **27**, 891–908 (2006)

Adaptive Load Balancing for Massively Parallel Multi-Level Monte Carlo Solvers

Jonas Šukys[(✉)]

ETH Zürich, Zürich, Switzerland
jonas.sukys@sam.math.ethz.ch

Abstract. The Multi-Level Monte Carlo algorithm was shown to be a robust solver for uncertainty quantification in the solutions of multi-dimensional systems of stochastic conservation laws. For random fluxes or random initial data with large variances, the time step of the explicit time stepping scheme becomes random due to the *random* CFL stability restriction. Such *sample path dependent* complexity of the underlying deterministic solver renders our *static* load balancing of the MLMC algorithm very inefficient. We introduce an *adaptive* load balancing procedure based on two key ingredients: (1) pre-computation of the time step size for *each* draw of random inputs (realization), (2) distribution of the samples using the greedy algorithm to "workers" with heterogeneous speeds of execution. Numerical experiments showing strong scaling are presented.

Keywords: Uncertainty quantification · Conservation laws · Multi-level Monte Carlo · FVM · Load balancing · Greedy algorithms · Linear scaling

1 Introduction

A number of problems in physics and engineering are modeled in terms of systems of conservation laws, defined on the d-dimensional physical domains:

$$\begin{cases} \mathbf{U}_t(\mathbf{x}, t) + \mathrm{div}(\mathbf{F}(\mathbf{U})) = \mathbf{S}(\mathbf{x}, \mathbf{U}), \\ \mathbf{U}(\mathbf{x}, 0) = \mathbf{U}_0(\mathbf{x}), \end{cases} \quad \forall (\mathbf{x}, t) \in \mathbb{R}^d \times \mathbb{R}_+. \tag{1}$$

Here, $\mathbf{U} : \mathbb{R}^d \to \mathbb{R}^m$ denotes the vector of conserved variables, $\mathbf{F} : \mathbb{R}^m \times \mathbb{R}^m \to \mathbb{R}^{m \times d}$ is the collection of directional flux vectors and $\mathbf{S} : \mathbb{R}^d \times \mathbb{R}^m \to \mathbb{R}^m$ is the source term. The partial differential equation is augmented with initial data $\mathbf{U}_0$.

Examples for conservation laws include the shallow water equations of oceanography, the Euler equations of gas dynamics, the Magnetohydrodynamics (MHD) equations of plasma physics, the wave equation and others.

As the equations are non-linear, analytic solution formulas are only available in very special situations. Consequently, numerical schemes such as finite volume methods [4] are required for the study of systems of conservation laws.

R. Wyrzykowski et al. (Eds.): PPAM 2013, Part I, LNCS 8384, pp. 47–56, 2014.
DOI: 10.1007/978-3-642-55224-3_5, © Springer-Verlag Berlin Heidelberg 2014

Existing numerical methods for approximating (1) require initial data $\mathbf{U}_0$, source $\mathbf{S}$ and flux function $\mathbf{F}$ as input. However, in most practical situations, it is not possible to measure *some* of these inputs precisely; for wave equation, material coefficients are often uncertain due to the scarcity of seismic measurements, whereas fluid dynamics and surface wave propagation in shallow water equations often lacks precision in initial data and source terms. Such uncertainty in inputs propagates to the solution, leading to the *stochastic* system of conservation laws:

$$\begin{cases} \mathbf{U}(\mathbf{x}, t, \omega)_t + \mathrm{div}(\mathbf{F}(\mathbf{U}, \omega)) = \mathbf{S}(\mathbf{x}, \omega), \\ \mathbf{U}(\mathbf{x}, 0, \omega) = \mathbf{U}_0(\mathbf{x}, \omega), \end{cases} \quad \mathbf{x} \in \mathbb{R}^d, \ t > 0, \ \forall \omega \in \Omega. \quad (2)$$

where $(\Omega, \mathcal{F}, \mathbb{P})$ is a complete probability space, the initial data $\mathbf{U}_0$ and the source term $\mathbf{S}$ are random fields [6,7], and the flux $\mathbf{F}$ is a Ω-uniformly Lipschitz random function [6]. The solution is also realized as a random field; its statistical moments (e.g. expectation $\mathbb{E}[\mathbf{U}]$ and variance $\mathbb{V}[\mathbf{U}]$) are the quantities of interest. Numerical methods for approximation of (2) include the stochastic Galerkin, stochastic collocation (see references in [6]) and stochastic Finite Volume [5]. Currently these methods are not able to handle large number of uncertainty sources, are intrusive and hard to parallelize. Alternatively, an estimate of $\mathbb{E}[\mathbf{U}]$ can be obtained by the Monte Carlo finite volume method (MC-FVM) [6], i.e. by computing the sample mean (ensemble average) of M solutions $\mathbf{U}_{\mathcal{T}}^{i,n}$, each of them approximated using FVM method [4] on a mesh $\mathcal{T}$ with mesh width Δx:

$$E_M[\mathbf{U}_{\mathcal{T}}^n] := \frac{1}{M} \sum_{i=1}^{M} \mathbf{U}_{\mathcal{T}}^{i,n}, \quad M \in \mathbb{N}. \quad (3)$$

MC-FVM estimate $E_M[\mathbf{U}_{\mathcal{T}}^n]$ was proven to converge with the rate $-s/(d+1+2s)$, where s denotes the convergence rate of the FVM solver [6–8]. Slow convergence makes MC-FVM method computationally unfeasible if high accuracy is needed.

The *multi-level* Monte Carlo finite volume method (MLMC-FVM) was recently proposed in [6,7]. The key idea behind MLMC-FVM is to simultaneously draw MC samples on a hierarchy of nested grids. There are four main steps:

1. **Nested meshes:** Consider *nested* triangulations $\{\mathcal{T}_\ell\}_{\ell=0}^{\infty}$ of the spatial domain with corresponding mesh widths Δx_ℓ that satisfy $\Delta x_\ell = \mathcal{O}(2^{-\ell}\Delta x_0)$. An example of such hierarchy with the first 3 levels is provided in Fig. 1.
2. **Sample:** For each level of resolution $\ell \in \mathbb{N}_0$, we draw M_ℓ independent identically distributed (i.i.d) samples $\{\mathbf{U}_{0,\ell}^i, \mathbf{S}_\ell^i, \mathbf{F}_\ell^i\}$ with $i = 1, 2, \ldots, M_\ell$ from the random fields $\mathbf{U}_0, \mathbf{S}, \mathbf{F}$ and approximate $\mathbf{U}_{0,\ell}^i$ and $\mathbf{S}_\ell^i$ by cell averages.
3. **Solve:** For each resolution level ℓ and each realization $\{\mathbf{U}_{0,\ell}^i, \mathbf{S}_{0,\ell}^i, \mathbf{F}_\ell^i\}$, the underlying balance law (1) is solved by the finite volume method [4] with mesh width Δx_ℓ; denote solutions by $\mathbf{U}_{\mathcal{T}_\ell}^{i,n}$ at the time t^n and mesh level ℓ.
4. **Estimate solution statistics:** Fix the highest level $L \in \mathbb{N}_0$. Denoting the MC estimator defined in (3) for the level ℓ by E_{M_ℓ}, we approximate $\mathbb{E}[\mathbf{U}]$ by

$$E^L[\mathbf{U}(\cdot, t^n)] := \sum_{\ell=0}^{L} E_{M_\ell}[\mathbf{U}_{\mathcal{T}_\ell}^n - \mathbf{U}_{\mathcal{T}_{\ell-1}}^n], \quad (4)$$

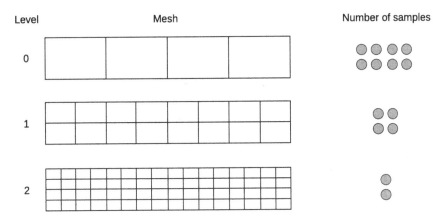

| Level | Mesh | Number of samples |

Fig. 1. Example of the first three levels ($L = 2$) of the hierarchy of nested grids for the two dimensional case. Example for the number of samples M_ℓ is provided according to (5) for $s = 1/2$ with the number of samples on the finest mesh level set to $M_L = 2$.

In order to equilibrate statistical and spatio-temporal discretization errors in (4), the following number of samples on each mesh level ℓ is needed [7,8]:

$$M_\ell = M_L 2^{2(L-\ell)s}, \quad M_L \in \mathbb{N}. \tag{5}$$

Notice that most of MC samples are computed on the coarsest mesh level $\ell = 0$, and only a small fixed number M_L of samples is needed on the finest mesh $\ell = L$, see Fig. 1. The error vs. work estimate for MLMC is given by [6,7],

$$\text{error} \lesssim (\text{Work})^{-s/(d+1)} \log(\text{Work}). \tag{6}$$

Estimate (6) shows that MLMC is superior to MC; in particular, at the relative error level of 1 %, MLMC-FVM was two orders of magnitude faster [6–8].

MLMC-FVM is *non-intrusive* as any standard FVM solver can be used in step 3. Furthermore, MLMC-FVM is amenable to *efficient parallelization*, which is the main topic of this paper. In Sect. 2 the limits of the *static* load balancing are discussed as a motivation for a novel *adaptive* load balancing, which is introduced in Sect. 4. Parallel scaling and efficiency analysis is provided in Sect. 5.

2 Scalable Parallel Implementation of MLMC-FVM

We use 3 levels of parallelization: across mesh levels, across MC samples and using domain decomposition (DDM) for FVM solver, see example in Fig. 2.

In [9] all required ingredients for parallelization were introduced and analyzed: parallel robust pseudo random number generation (WELL512a RNG was used), numerically stable parallel "online" variance computation algorithms, domain decomposition method within each FVM solver and *static* load balancing, which distributes (at compile-time) computational work of multiple concurrent solve steps *evenly* among the available cores using the a-priori

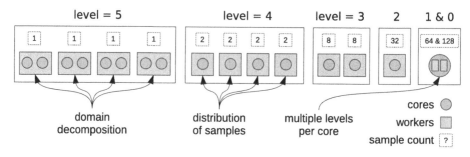

Fig. 2. Parallelization over mesh levels, MC samples and using domain decomposition.

estimates on the computational work of any sample on a given mesh resolution $\mathcal{T}$. Static load balancing appeared to be very efficient for stochastic systems of conservation laws (2) with *deterministic fluxes* $\mathbf{F}$ and stochastic initial data $\mathbf{U}_0$ with *small* variance $\mathbb{V}[\mathbf{U}_0]$, i.e. $\mathbb{V}[\mathbf{U}_0] \ll \mathbb{E}[\mathbf{U}_0]$. In particular, we have verified strong scaling (fixed problem size while increasing the number of cores) of *static* load balancing in the parallel (using MPI [10]) code ALSVID-UQ [1] up to 40 000 cores [9].

The goal of this paper is to design a novel *adaptive* load balancing (at runtime) which would be efficient for a much broader range of stochastic systems of conservation laws (2), for instance, where the flux function $\mathbf{F}$ is *random*.

3   Problem Setting and Estimates for Computational Work

Static load balancing [9] uses two assumptions on the computational work $\mathrm{Work}_{\mathcal{T}}$ needed to solve (1) for a given *sample* of random input data on a given mesh $\mathcal{T}$:

1. *accurate* relative (w.r.t. another mesh $\mathcal{T}'$) estimates for $\mathrm{Work}_{\mathcal{T}}$ are available
2. for a *fixed* mesh $\mathcal{T}$, estimate $\mathrm{Work}_{\mathcal{T}}$ is almost the *same* for *all* realizations

3.1   Estimates for the Computational Work of the FVM Solver

Assuming *a homogeneous computing environment* (all cores have identical CPUs, RAM, and equal bandwidth/latency to all other cores), accurate computational work estimates were derived in [9]. For a given mesh $\mathcal{T}$ with mesh width Δx and total number of cells $N = \#\mathcal{T}$, the required computational work for one *time step* (numerical flux approximations) of *one* sample was computed to be

$$\mathrm{Work}_{\mathcal{T}}^{\mathrm{step}} = \mathrm{Work}^{\mathrm{step}}(\Delta x) = \mathcal{O}(N) = K\Delta x^{-d}, \tag{7}$$

where constant K depends on FVM that is used, but does *not* depend on mesh width Δx. In most explicit FVM schemes [4], lower order terms $\mathcal{O}(\Delta x^{-d+1})$ in (7) are negligible, even on a very coarse mesh. To ensure the stability of the FVM scheme, a CFL condition [4] is imposed on the time step size $\Delta t := t^{n+1} - t^n$,

$$\Delta t = \frac{C_{\text{CFL}}}{\lambda} \Delta x, \quad 0 < C_{\text{CFL}} \leq 1, \quad \lambda > 0, \tag{8}$$

where the so-called CFL number C_{CFL} does not depend on Δx and λ is the absolute value of the maximal wave speed [4]. Hence, the computational work $\text{Work}_{\mathcal{T}}^{\text{det}}$ for *one* complete *deterministic* solve using the FVM method on the triangulation $\mathcal{T}$ with mesh width Δx is given by multiplying the work for one step (7) by the total number of time steps Δt^{-1} for the time horizon $T > 0$,

$$\text{Work}_{\mathcal{T}}^{\text{det}} = \text{Work}_{\mathcal{T}}^{\text{step}} \cdot \frac{T}{\Delta t} = K \Delta x^{-d} \lambda \frac{T}{C_{\text{CFL}} \Delta x} = \frac{KT}{C_{\text{CFL}}} \lambda \Delta x^{-(d+1)}. \tag{9}$$

For *deterministic fluxes* $\mathbf{F}$ and stochastic initial data $\mathbf{U}_0$ with *small* variance $\mathbb{V}[\mathbf{U}_0]$, the maximum wave speed λ does *not* vary significantly among all MC samples and hence the second property in Sect. 3 holds. For *random* flux $\mathbf{F}$, however, λ can *strongly depend* on the particular realization of $\mathbf{F}$. As an example, we consider acoustic wave equation in random heterogeneous medium.

3.2 Acoustic Wave Equation in Random Medium

Acoustic wave equation in random heterogeneous d-dimensional domain $\mathbf{D} \subset \mathbb{R}^d$ can be written in a form of a *linear system* of *first order* conservation laws [8],

$$\begin{cases} p_t(\mathbf{x}, t, \omega) - \nabla \cdot (c(\mathbf{x}, \omega) \mathbf{u}(\mathbf{x}, t, \omega)) = 0, \\ \mathbf{u}_t(\mathbf{x}, \omega) - \nabla p(\mathbf{x}, \omega) = 0, \end{cases} \quad \mathbf{x} \in \mathbf{D}, \ t > 0, \ \omega \in \Omega, \tag{10}$$

with *deterministic* initial data $p(\mathbf{x}, 0) \in C^\infty(\mathbf{D})$, $\mathbf{u}_0(\mathbf{x}, 0) \in (C^\infty(\mathbf{D}))^d$ and *random* coefficient $c \in L^0(\Omega, L^\infty(\mathbf{D}))$ with $\mathbb{P}[c(\mathbf{x}, \omega) > 0, \forall \mathbf{x} \in \mathbf{D}] = 1$. As the system (10) is *linear* and random coefficient c is independent of t, the maximum wave speed λ does *not* depend on $\mathbf{u}_0, p_0$ or t, but *explicitly depends* [8] on c,

$$\lambda(\omega) = \max_{\mathbf{x} \in \mathbf{D}} \sqrt{c(\mathbf{x}, \omega)}. \tag{11}$$

Depending on c, the variance of $\lambda(\omega)$ can be very large. As an example, consider domain $\mathbf{D} = [0, 3]^2$ and the wave speed c given by its Karhunen-Loève expansion,

$$\log c(\mathbf{x}, \omega) = \log \bar{c}(\mathbf{x}) + \sum_{\mathbf{m} \in \mathbb{N}_0^2 \setminus \{0\}}^{\infty} \sqrt{\alpha_{\mathbf{m}}} \Psi_{\mathbf{m}}(\mathbf{x}) Y_{\mathbf{m}}(\omega), \tag{12}$$

with eigenvalues $\alpha_{\mathbf{m}}$, eigenfunctions $\Psi_{\mathbf{m}}(\mathbf{x})$, and the mean field $\bar{c}(\mathbf{x})$ set to

$$\alpha_{\mathbf{m}} = |\mathbf{m}_1 + \mathbf{m}_2|^{-2.5}, \quad \Psi_{\mathbf{m}}(\mathbf{x}) = \sin(\mathbf{m}_1 \pi \mathbf{x}_2) \sin(\mathbf{m}_2 \pi \mathbf{x}_1), \quad \bar{c}(\mathbf{x}) \equiv 0.1,$$

and with *independent standard normal* random variables $Y_m \sim \mathcal{N}[0, 1]$. Then, $c, c^{-1} \notin L^\infty(\Omega, L^\infty(\mathbf{D}))$, i.e. there is *positive* probability such that $\lambda(\omega)$ attains any *arbitrary large* or *arbitrary small* value. Nevertheless, according to Proposition 1 and Theorems 2 and 5 in [8], solutions to (10) with (12) are well-defined, have *finite* statistical moments and can be approximated by the MLMC-FVM.

For such class of problems, the computational work (9) required for one sample (draw) on a given mesh $\mathcal{T}$ is a random variable, proportional to $\lambda(\omega)$,

$$\text{Work}_{\mathcal{T}}^{\text{rand}}(\omega) = \frac{KT}{C_{\text{CFL}}} \lambda(\omega) \Delta x^{-(d+1)}. \tag{13}$$

Proceeding with our analysis, we consider the *expected* computational work,

$$\mathbb{E}[\text{Work}_{\mathcal{T}}] = \mathbb{E}[\text{Work}_{\mathcal{T}}^{\text{rand}}(\omega)] = \frac{KT}{C_{\text{CFL}}} \mathbb{E}[\lambda(\omega)] \Delta x^{-(d+1)}, \tag{14}$$

which is *finite*, as long as the expected value of maximal wave speed λ is finite. The direct consequence of this is that the *static* load balancing from [9], at least *on average*, is expected to scale. Furthermore, in [8], MLMC-FVM algorithm is analyzed in the case of (13) and the resulting complexity of error vs. *expected* amount of computational work is proven to be analogous to (6),

$$\text{error} \lesssim (\mathbb{E}[\text{Work}])^{-s/(d+1)} \log(\mathbb{E}[\text{Work}]). \tag{15}$$

Note, that *all* systems (linear and non-linear) of conservation laws exhibit analogous phenomenon if the flux function $\mathbf{F}$ is *random* with large variance, see [6] for more examples. Another class of problems is with *non-linear* fluxes (can be deterministic), where λ *depends* not only on $\mathbf{F}$ but also on $\mathbf{U}$ (hence also on $\mathbf{U}_0$) and potentially has large variance if $\mathbf{U}_0$ has large variance.

3.3   Limits of the Static Load Balancing

Despite the estimates (14) of *average* work, the *efficiency* of the balancing is expected to drop significantly for a *single* run of the MLMC-FVM due to the non-uniform λ_ℓ^i. This can be clearly seen in Fig. 3, where the scaling analysis of *static* load balancing in MLMC-FVM for the wave equation (10) with material coefficient given by (12) is performed. As expected, the algorithm *scales* linearly with the number of cores, but the efficiency is consistently low, where:

$$\text{efficiency} := 1 - \frac{(\text{total clock time of all MPI routines and idling})}{(\#\text{cores}) \times (\text{wall clock time})}. \tag{16}$$

Labels "MLMC" and "MLMC2" indicate $s = 1/2$ and $s = 1$ in (5), respectively. The runtime of all simulations is measured by `MPI_Wtime()` routine [10].

Simulations were executed on Cray XE6 (see [11]) with 1496 AMD Interlagos 2×16-core 64-bit CPUs (2.1 GHz), 32 GB DDR3 memory per node, 10.4 GB/s Gemini 3D torus interconnect with a theoretical peak performance of 402 TFlops.

To improve the efficiency of load balancing, *sample-dependent* computational work estimates (13) need to be taken into account. To this end, we introduce an *adaptive* load balancing, where samples are distributed during run-time, after computing $\lambda(\omega)$ for *each* required realization, but *before* actually starting the FVM time stepping of *any* sample (hence, it is *not* a dynamic balancing).

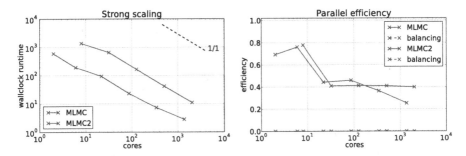

Fig. 3. Inefficient strong scaling of *static* load balancing in case of *random* maximum wave speeds $\lambda(\omega)$ with large relative variance $\mathbb{V}[\lambda]/\mathbb{E}[\lambda]$ resulting from (12).

4 Adaptive Load Balancing

We assume to have a "pool" $\mathcal{G}$ of cores (processing units), consisting of groups $\mathcal{G}_m$ (of arbitrary size) of cores indexed by "multi level" $m = L, L-1, \ldots, m_0 \geq 0$, which are themselves divided into *equal* groups $\mathcal{G}_m^s$ of cores indexed by "sampler" $s = 1, \ldots, P_m$. The number of cores in a given sampler $\mathcal{G}_m^s$ is independent on s and denoted by D_m. An example of such pool with $L = 5$, $m_0 = 1$, $\{P_m\} = \{1, 1, 2, 4, 8\}$, $\{D_m\} = \{1, 1, 1, 1, 2\}$ is depicted in Fig. 2. We assume, that *any* of the MC samples from *any* mesh level ℓ can be efficiently computed on *any* sampler $\mathcal{G}_m^s$ in the pool, in serial or by using domain decomposition if $D_m > 1$. By efficient computation we assume strong scaling of the domain decomposition.

4.1 Computation and Distribution of Loads

Define Load_ℓ^i to be the normalized (constants are neglected) required computation time for the i-th *difference* of samples between mesh levels ℓ and $\ell - 1$,

$$\mathrm{Load}_\ell^i = \lambda_\ell^i \left(\Delta x_\ell^{-(d+1)} + \Delta x_{\ell-1}^{-(d+1)} \right), \qquad \ell = 0, \ldots, L, \quad i = 1, \ldots, M_\ell, \quad (17)$$

where all λ_ℓ^i are computed *in parallel* on all P_L samplers $\mathcal{G}_L^1, \ldots, \mathcal{G}_L^{P_L}$, each consisting of D_L cores, and then broadcast to *every* core. Computations of λ_ℓ^i do *not* need time stepping and hence are cheaper by a factor $\mathcal{O}(\Delta x_\ell)$ compared to the full FVM; required global communication is also small, of order $\mathcal{O}(2^{2Ls}|\mathcal{G}|)$.

The goal of the load balancing is to distribute all samples with required computational time Load_ℓ^i to samplers $\mathcal{G}_m^s$. Greedy algorithm for *identical* samplers has been analyzed in [3] and was proven to be a 4/3-approximation, i.e. the makespan (maximum run-time among all workers) is at most 4/3 times larger than the *optimal* (minimal) makespan. If loads are *not* ordered, then greedy algorithm is only a 2-approximation [3]. Here we present a generalization of the greedy algorithm for samplers with *heterogeneous speed* of execution. The main idea of the algorithm is the *recursive* assignment of the *largest* available Load_ℓ^i to the sampler $\mathcal{G}_m^s$ for which the total run-time $\mathcal{R}_m^s$ *including* Load_ℓ^i is *minimized*.

The pseudo code of the *adaptive* load balancing is provided as Algorithm 1, where the notation $\text{Load}_\ell^i \in \mathcal{G}_m^s$ means that i-th difference of samples between mesh resolution levels ℓ and $\ell - 1$ is assigned to be computed on sampler $\mathcal{G}_m^s$.

Algorithm 1. Greedy load balancing (with non-identical speeds of execution)

$\mathcal{L} = \{\text{Load}_\ell^i : \ell = 0, \ldots, L, i = 1, \ldots, M_\ell\}$
while $\mathcal{L} \neq \varnothing$ **do**
    $\text{Load}_\ell^i = \max \mathcal{L}$
    $\mathcal{G}_m^s = \arg\min\limits_{\mathcal{G}_m^s} \left(\mathcal{R}(\mathcal{G}_m^s) + \text{Load}_\ell^i / D_m \right), \quad \mathcal{R}(\mathcal{G}_m^s) = \sum \{\text{Load}/D_m : \text{Load} \in \mathcal{G}_m^s\}$
    $\mathcal{G}_m^s = \mathcal{G}_m^s \cup \text{Load}_\ell^i$
    $\mathcal{L} = \mathcal{L} \backslash \text{Load}_\ell^i$
end while

Note, that if samplers have *identical* speeds of execution, i.e. D_m are all equal, then the above Algorithm 1 reduces to the standard greedy algorithm.

If loads are *not* ordered (replace "$\max \mathcal{L}$" by "*any* load from $\mathcal{L}$"), then Algorithm 1 is only a $(1 + D_{\max}/D_{\min})$-approximation (analogous proof as in [3]). Hence, if samplers $\mathcal{G}_m^s$ have very *heterogeneous* speeds of execution $1/D_m$, Algorithm 1 may provide a *much* longer makespan, compared to the optimal. However, if we assume that loads are *ordered* and are as heterogeneous as samplers,

$$\frac{\text{Load}_{\max}}{\text{Load}_{\min}} := \frac{\max_{\ell,i} \text{Load}_\ell^i}{\min_{\ell,i} \text{Load}_\ell^i} \geq \frac{D_{\max}}{D_{\min}}, \tag{18}$$

then Algorithm 1 is a 2-approximation. We present this result as a theorem.

Theorem 1. *If* (18) *holds and the last load of the bottle-neck sampler is bounded by* $(D_{\min}/D_{\max}) \cdot \text{Load}_{\max}$, *then Algorithm 1 is a 2-approximation.*

Proof. Let $R(\mathcal{G}^*)$ be the run-time of the bottle-neck sampler $\mathcal{G}^*$ and Load^* be the last sample assigned to $\mathcal{G}^*$. Then, according to distribution procedure,

$$R(\mathcal{G}^*) \leq R(\mathcal{G}_m^s) + \text{Load}^*/D_m, \quad \forall m = m_0, \ldots, L, \quad s = 1, \ldots, P_m.$$

Summing the above inequality over all samplers $\mathcal{G}_m^s$, we obtain a bound

$$R(\mathcal{G}^*) - \frac{1}{\#\{\mathcal{G}_m^s\}} \sum_{m,s} \frac{\text{Load}^*}{D_m} \leq \frac{1}{\#\{\mathcal{G}_m^s\}} \sum_{m,s} R(\mathcal{G}_m^s) \leq \mathcal{R}^o,$$

where $\mathcal{R}^o$ is the optimal timespan, which is certainly *not* smaller than the average of all runtimes $R(\mathcal{G}_m^s)$. Next, we use (18) and $\text{Load}^* \leq \text{Load}_{\max} D_{\min}/D_{\max}$,

$$\frac{1}{\#\{\mathcal{G}_m^s\}} \sum_{m,s} \frac{\text{Load}^*}{D_m} \leq \frac{1}{\#\{\mathcal{G}_m^s\}} \sum_{m,s} \frac{D_{\min}}{D_{\max}} \frac{\text{Load}_{\max}}{D_{\min}} \leq \frac{\text{Load}_{\max}}{D_{\max}} \leq \mathcal{R}^o.$$

Combining both bounds, the desired inequality $R(\mathcal{G}^*) \leq 2\mathcal{R}^o$ is obtained. $\square$

In case of MLMC-FVM, the assumption (18) is often satisfied, since loads Load_ℓ^i scale asymptotically as $\text{Work}_{\mathcal{T}_\ell} = \mathcal{O}(2^{(d+1)\ell})$ due to (13), and the speeds of execution D_m using domain decomposition scale only as $\#\mathcal{T}_m$, i.e. $D_m = \mathcal{O}(2^{dm})$.

4.2   Implementation Remarks

Once the loads have been distributed to samplers $\mathcal{G}_m^s$, the parallel execution of FVM solves and the final assembly of the MLMC-FVM estimator remained analogous as in [9], i.e. Message Passing Interface (MPI) was chosen, making heavy use of the appropriate local MPI intra-communicators [10]. The *new* part for the *adaptive* balancing is the parallel computation (and broadcast) of the maximum wave speeds λ_ℓ^i, which is problem-specific. For the wave equation (10), λ_ℓ^i were computed by computing random coefficients c_ℓ^i and then using (11). Note, that in the computation of loads (as well as the samples themselves), the parallelization configuration (D_m, P_m) might need to be *adapted* to the required *mesh level* ℓ: due to memory limitations, samples on fine meshes use larger $D_{m,\ell}$ and fewer samplers $P_{m,\ell}$, and due to inefficiency of DDM, samples on coarse meshes use smaller $D_{m,\ell}$ and more samplers $P_{m,\ell}$, keeping $P_{m,\ell} \cdot D_{m,\ell} = P_m \cdot D_m$.

5   Efficiency and Linear Scaling in Numerical Simulations

To ensure a fair comparison, the *adaptive* load balancing algorithm was tested on the same problem as the *static* load balancing, see (10) in Subsect. 3.3.

In Fig. 4 we verify *strong scaling* of our implementation. We observed scaling to be maintained for up to almost 40 000 cores at high efficiency. Simulations were executed on the same Cray XE6 (see [11]) architecture as in Subsect. 3.3. We believe that our parallelization algorithm will scale linearly for a much larger number of cores if the problem size is increased. The computational complexity of the *adaptive* balancing (computation of loads and distribution of samples) is $\mathcal{O}(2^{L \cdot \max(2s,d)} + |\mathcal{G}|2^{2Ls})$, where both terms are dominated by full simulation.

For *non-linear* fluxes and random initial condition, knowledge of the input data (at $t = 0$) might *not* be sufficient to accurately estimate $\lambda(\omega)$. In such cases the performance of the *adaptive* load balancing might be sub-optimal; *dynamic* load balancing could be used, possibly introducing a large overhead [2].

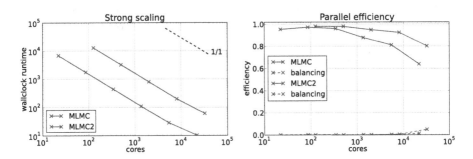

Fig. 4. Strong scaling of *adaptive* load balancing up to 40 000 cores. The efficiency is nearly optimal and is *much* better if compared to the *static* load balancing in Fig. 3.

6   Conclusion

MLMC-FVM algorithm is superior to standard MC algorithms for uncertainty quantification in hyperbolic conservation laws, and yet, as most sampling algorithms, it still scales linearly w.r.t. number of uncertainty sources. Due to its non-intrusiveness, MLMC-FVM was efficiently parallelized for multi-core architectures. For systems with *deterministic* fluxes, *static* load balancing was already available [9], which was shown to scale strongly and weakly on the high performance cluster [11] in multiple space dimensions. For *linear* systems with *random* fluxes, *adaptive* load balancing was introduced, which maintains the same scaling properties, but, by design, is applicable to a much wider class of problems.

Acknowledgments. This work is performed under ETH interdisciplinary research grant CH1-03 10-1 and CSCS production project grant ID S366.

References

1. ALSVID-UQ, v3.0. http://www.sam.math.ethz.ch/alsvid-uq
2. Dandamudi, S.P.: Sensitivity evaluation of dynamic load sharing in distributed systems. IEEE Concurrency **6**(3), 62–72 (1998)
3. Graham, R.L.: Bounds on multiprocessing timing anomalies. SIAM J. Appl. Math. **17**(2), 416–429 (1969)
4. LeVeque, R.A.: Numerical Solution of Hyperbolic Conservation Laws. Cambridge University Press, Cambridge (2005)
5. Mishra, S., Risebro, N.H., Schwab, C., Tokareva, S.: Multi-level Monte Carlo finite volume methods for scalar conservation laws with random flux. Technical Report, SAM, 2012. http://www.sam.math.ethz.ch/reports/2012/35
6. Mishra, S., Schwab, C., Šukys, J.: Multi-level monte carlo finite volume methods for uncertainty quantification in nonlinear systems of balance laws. In: Bijl, H., Lucor, D., Mishra, S., Schwab, C. (eds.) Uncertainty Quantification in Computational Fluid Dynamics. Lecture Notes in Computational Science and Engineering, vol. 92, pp. 225–294. Springer, Zurich (2013)
7. Mishra, S., Schwab, C.: Sparse tensor multi-level monte carlo finite volume methods for hyperbolic conservation laws with random initial data. Math. Comp. **280**(81), 1979–2018 (2012)
8. Šukys, J., Schwab, C., Mishra, S.: Multi-level Monte Carlo finite difference and finite volume methods for stochastic linear hyperbolic systems. In: Dick, J., et al. (eds.) Monte Carlo and Quasi-Monte Carlo Methods 2012, vol. 65, pp. 649–666. Springer, Berlin (2013) http://www.sam.math.ethz.ch/reports/2012/19
9. Šukys, J., Mishra, S., Schwab, C.: Static load balancing for multi-level Monte Carlo finite volume solvers. In: Wyrzykowski, R., Dongarra, J., Karczewski, K., Waśniewski, J. (eds.) PPAM 2011, Part I. LNCS, vol. 7203, pp. 245–254. Springer, Heidelberg (2012)
10. MPI: A Message-Passing Interface Standard. Version 2.2, 2009. http://www.mpi-forum.org/docs/mpi-2.2/mpi22-report.pdf
11. Rosa, Swiss National Supercomputing Center (CSCS), Lugano, www.cscs.ch

Parallel One–Sided Jacobi SVD Algorithm with Variable Blocking Factor

Martin Bečka and Gabriel Okša[✉]

Institute of Mathematics, Slovak Academy of Sciences, Bratislava, Slovak Republic
{Martin.Becka,Gabriel.Oksa}@savba.sk

Abstract. Parallel one-sided block-Jacobi algorithm for the matrix singular value decomposition (SVD) requires an efficient computation of symmetric Gram matrices, their eigenvalue decompositions (EVDs) and an update of matrix columns and right singular vectors by matrix multiplication. In our recent parallel implementation with p processors and blocking factor $\ell = 2p$, these tasks are computed serially in each processor in a given parallel iteration step because each processor contains exactly two block columns of an input matrix A. However, as shown in our previous work, with increasing p (hence, with increasing blocking factor) the number of parallel iteration steps needed for the convergence of the whole algorithm increases linearly but faster than proportionally to p, so that it is hard to achieve a good speedup. We propose to break the tight relation $\ell = 2p$ and to use a small blocking factor $\ell = p/k$ for some integer k that divides p, ℓ even. The algorithm then works with pairs of logical block columns that are distributed among processors so that all computations inside a parallel iteration step are themselves parallel. We discuss the optimal data distribution for parallel subproblems in the one-sided block-Jacobi algorithm and analyze its computational and communication complexity. Experimental results with full matrices of order 8192 show that our new algorithm with a small blocking factor is well scalable and can be 2–3 times faster than the ScaLAPACK procedure `PDGESVD`.

Keywords: Singular value decomposition · Serial one–sided Jacobi algorithm · Parallel one–sided block–Jacobi algorithm

1 Introduction

The one-sided block-Jacobi SVD algorithm is suited for the SVD computation of a general, dense, complex matrix A of order $m \times n$, $m \geq n$. However, we will restrict ourselves to real matrices with obvious modifications for the complex case.

We start with the block-column partitioning of A in the form

$$A = [A_1, A_2, \ldots, A_\ell],$$

where the width of A_i is n_i, $1 \leq i \leq \ell$, so that $n_1 + n_2 + \cdots + n_\ell = n$.

R. Wyrzykowski et al. (Eds.): PPAM 2013, Part I, LNCS 8384, pp. 57–66, 2014.
DOI: 10.1007/978-3-642-55224-3_6, © Springer-Verlag Berlin Heidelberg 2014

The serial algorithm can be written as an iterative process:

$$A^{(0)} = A, \quad V^{(0)} = I_n,$$
$$A^{(r+1)} = A^{(r)}U^{(r)}, \quad V^{(r+1)} = V^{(r)}U^{(r)}, \quad r \geq 0. \tag{1}$$

Here the $n \times n$ orthogonal matrix $U^{(r)}$ is the so-called *block rotation*, and the purpose of matrix multiplication $A^{(r)}U^{(r)}$ in (1) is to mutually orthogonalize individual columns between block columns i and j of $A^{(r)}$. During the iterative process (1), the block column pairs (i, j) are mutually orthogonalized according some *ordering*, which defines the algorithm's strategy. The most common cyclic strategies are the *row-cyclic* one and the *column-cyclic* one, where the orderings are given row-wise and column-wise, respectively, with regard to the upper triangle of A. The first $\ell(\ell - 1)/2$ iterations constitute the first *sweep*. When the first sweep is completed, the pairs (i, j) are repeated during the second sweep, and so on, up to the convergence of the entire algorithm.

Notice that in (1) only the matrix of right singular vectors $V^{(r)}$ is iteratively computed by orthogonal updates. If the process ends at iteration t, say, then $A^{(t)}$ has mutually highly orthogonal columns. Their norms are the singular values of A, and the normalized columns (with unit 2-norm) constitute the matrix of left singular vectors.

Alternatively [7], the right singular vectors collected in V can be computed *a posteriori* from the matrix equation

$$AV = A^{(t)}. \tag{2}$$

This strategy spares the orthogonal updates of right singular vectors in each parallel iteration step, but the original matrix A is needed.

A parallel version of the one-sided block-Jacobi SVD algorithm, implemented on p processors, mutually orthogonalizes $\ell/2$ pairs of block columns in each parallel iteration step. The orthogonalization of block columns A_i and A_j is implicitly equivalent to the diagonalization of the auxiliary symmetric Gram matrix $G = (A_i, A_j)^T(A, A_j)$, which can be computed by its eigenvalue decomposition (EVD). This leads to the natural matrix data distribution with the blocking factor $\ell = 2p$ where each processor contains two full block columns of width $n/2p$. Consequently, the computations of the Gram matrix, the EVD and the update of block columns are *local* with respect to processors.

To proceed in parallel computation, some parallel ordering is required that defines p independent pairs of block columns of A which are simultaneously mutually orthogonalized in a given parallel iteration step by computing p eigenvalue decompositions of p Gram matrices G. Up to now, some cyclic (static) parallel ordering has been used [1]. In [5], three new variants of so-called *dynamic* ordering were proposed and tested that take into account the actual degree of the mutual perpendicularity of any pair of block columns based on the estimation of *principal angles*. Moreover, the design of global and local convergence criteria was also analyzed in some detail. These new ideas led to the substantial decrease of the number of parallel iteration steps needed for the convergence of whole algorithm for a wide class of random matrices with various distributions of singular values.

The basic variant of the above algorithm can be augmented by pre-processing and post-processing steps similarly to the two-sided block-Jacobi SVD algorithm; see [4,9]. Pre-processing consists of pivoted or un-pivoted QR factorization of matrix A, followed by the un-pivoted LQ factorization of R-factor. The Jacobi process is then applied to the lower triangular factor L. Notice that in this case the *a posteriori* computation of right singular vectors in (2), which is implemented in both variants of our Jacobi algorithm (see below), is very simple and fast: the system matrix is lower triangular so that the appropriate columns of V are computed locally using $O(n)$ flops. In the post-processing step, the computed left and right singular vectors are multiplied by orthogonal matrices that arise in QR and LQ factorization. This type of pre-processing helps to concentrate the Frobenius norm of Gram matrices near their main diagonal so that less iterations are needed for inner EVDs of 2×2 block subproblems.

In this paper we break the tight relationship $\ell = 2p$ and wish to use a small blocking factor $\ell = p/k$ for some natural number k that divides p, ℓ even. Each processor contains one *physical* block column of size $m \times n/p$, but the Jacobi algorithm will use *logical* block columns, whereby one logical column consists of k physical block columns, i.e., it is shared by k processors. This data distribution requires to parallelize the computation of Gram matrices, the EVD, the update of block columns and the computation of right singular vectors. Next section contains the description of steps needed for such a parallelization.

2 Parallel Computation in a Parallel Iteration Step

As mentioned above, an input matrix $A \in \mathbb{R}^{m \times n}$ is distributed among p processors with the processor grid $1 \times p$ so that each processor contains one physical block column of size $m \times n/p$. For large p and fixed n, one can have many blocks with a small width. In [2] it was shown that the number of parallel iteration steps needed for the convergence increases linearly but faster than proportionally to ℓ. Hence, to decrease a parallel execution time of the whole algorithm, it can be advantageous to work with a small blocking factor ℓ. This is possible by using logical block columns that consist of k physical block columns each, where the "logical" blocking factor is $\ell = p/k$.

One subtask in the parallel one-sided Jacobi algorithm needs two logical block columns, i.e., it requires a context of type 1 (CTXT1) with the processor grid $1 \times 2k$. Note that there are exactly $\ell/2$ subproblems in each parallel iteration step. Inside each CTXT1, the cyclic data distribution is given by the processor grid, i.e., the row block size $mb = m$ and the column block size $nb = n/p$. At the beginning of algorithm, all CTXT1s are constructed using subsequent block columns of matrix A.

Next, one has to compute the Gram matrix inside each CTXT1, which can be done using the ScaLAPACK procedure PDSYRK. However, it is well known that the most efficient data distribution for parallel matrix-matrix products of type

$Y^T Y \in \mathbb{R}^{2k \times 2k}$ as well as for the EVD is the two-dimensional one of type $r \times c = 2k$ with $r \approx c \approx \sqrt{2k}$; see [6]. Therefore, each CTXT1 is re-defined by using the ScaLAPACK procedure PDGEMR2D into a context of type 2 (CTXT2), where the corresponding processors are arranged into a grid $r \times c$. The corresponding matrix Y consisting of two logical block columns is cyclically re-distributed onto this processor grid.

The CTXT2 is then used for a parallel EVD computation of the Gram matrix and for a parallel update of block columns of A inside each subtask. Since Gram matrices are symmetric, their EVDs are computed using the ScaLAPACK procedure PDSYEV. Updates of block columns can be computed using the procedure PDGEMM. Data are then re-distributed to the original CTXT1 because this context is more suited for the parallel computation of weights in the re-ordering step.

In [5], three variants of dynamic ordering for the one-sided Jacobi algorithm were designed, implemented and tested, and variant 3 has been recommended for general use. Let $e \equiv (1, 1, \ldots, 1)^T \in \mathbb{R}^{k \times 1}$, and for each logical column block A_i define its *representative vector*,

$$c_i \equiv \frac{A_i\, e}{\|e\|}, \quad 1 \leq i \leq \ell. \tag{3}$$

The choice of e ensures a uniform participation of all k one-dimensional subspaces, which constitute $\mathrm{span}(A_i)$, in the definition of c_i. For any pair of logical block columns (A_i, A_j), the weight w_{ij} describes the mutual position of the whole subspace $\mathrm{span}(A_i)$ with respect to the representative vector c_j. Hence,

$$w_{ij} \equiv \|A_i^T c_j\| = \frac{\|A_i^T A_j e\|}{\|e\|}. \tag{4}$$

It can be expected that this weight will estimate the mutual position of two subspaces $\mathrm{span}(A_i)$ and $\mathrm{span}(A_j)$ quite precisely because the orientation of c_j with respect to the *whole* orthonormal basis of $\mathrm{span}(A_i)$ is taken into account. Then the dynamic ordering chooses $\ell/2$ pairs of logical block columns that are mutually inclined mostly (i.e., they have the largest values of w_{ij}), and these pairs will be orthogonalized in next parallel iteration step. More details can be found in [5].

In our implementation, the weight computation begins with computing the representative vector for each logical block column, which can be easily done by summing all matrix column vectors within each block column communicator. Also the computation of $\ell(\ell - 2)/2$ weights in (4) is done in parallel and in a perfectly balanced way. At the end, all processors contain all weights and can compute the re-ordering in parallel. Note that choosing $\ell/2$ logical block columns for the subsequent orthogonalization does not require any explicit Send/Receive operation because only contexts are generated.

3   Computational and Communication Complexity

Supposing a perfect parallelization of each procedure, the number of flops for individual subtasks (GRAM, EVD, MM = matrix-matrix multiplications,

WC = weight computation) in one parallel iteration step are summarized in Table 1 for a square matrix of order n.

Table 1. Comparison of computational complexity per one iteration step

task	$\ell = 2p$	$\ell = p/k$
GRAM	$n^3/(2p^2)$	$n^3/(\ell p)$
EVD	$c_1 n^3/p^3$	$4c_1 n^3/(\ell^2 p)$
MM	$2c_2 n^3/p^2$	$4c_2 n^3/(\ell p)$
WC	n^2	$n^2\ell/(2p)$

The values c_1 and c_2 are small constants; see [2]. Given p processors, it can be seen that for fixed blocking factor $\ell = 2p$ the computation of weights is not parallelized at all, which is in contrast to the case of $\ell = p/k$. Since for $\ell = p/k$ one has $2 \leq \ell \leq p$, the computational complexity of GRAM, EVD and MM is worse than for fixed $\ell = 2p$.

Table 2 shows the comparison of an overall computational work needed in one sweep defined as a sequence of ℓ parallel iteration steps.

Table 2. Comparison of computational work per one sweep

task	$\ell = 2p$	$\ell = p/k$
GRAM	n^3/p	n^3/p
EVD	$2c_1 n^3/p^2$	$4c_1 n^3/(\ell p)$
MM	$4c_2 n^3/p$	$4c_2 n^3/p$
WC	$n^2\ell$	$n^2\ell^2/(2p)$

Whereas the computational work for GRAM and MM is equal in both cases, the computation of EVD is cheaper in the case of fixed blocking factor. For fixed p, this difference can be compensated by less parallel iteration steps needed for the convergence of the whole Jacobi algorithm as can be expected for small blocking factors $\ell = p/k$; see [2,3]. On the other hand, the computation of weights for the dynamic ordering is much cheaper for $\ell = p/k$, especially for small values of ℓ.

With respect to the communication, each ScaLAPACK routine has its own communication complexity that is not "visible" to the user. Here we can analyze the WC, which turns out to be the most important part of the overall communication complexity. In the case of a cyclically distributed matrix, a representative vector is computed using MPI_REDUCE($\ell/p, n$) (first parameter is a number of processors, the second one is data size). Next, the representative vectors are gathered in one matrix using MPI_ALLGATHER(ℓ, n) and broadcast by MPI_BCAST($\ell/p, n\ell$). Inside each logical block, the weights are computed using

MPI_REDUCE($\ell/p, \ell/2$) and gathered by MPI_ALLGATHER($\ell, 3\ell/2$). As discussed in next section, the communication complexity of the WC and its efficient implementation is crucial for the performance of the Jacobi algorithm with $\ell = 2p$.

4   Numerical Experiments

The parallel one-sided block-Jacobi SVD algorithm with the un-pivoted QRLQ pre-processing step (see [4]) was implemented on the Doppler Cluster at the University of Salzburg, Austria. The Doppler Cluster consists of 32 nodes where each node has 16 or 64 cores of type Opteron Series 6200, 2.2 GHz, and with 2–8 GB RAM per core. The parallel system is equipped with a variety of GPGPU accelerator hardware and uses the QDR Infiniband / Mellanox interconnection network.

We used up to six nodes in a stand-alone mode of computation, so that measured execution times are quite reliable.

All computations were performed using the IEEE standard double precision floating point arithmetic with the machine precision $\epsilon_M \approx 2.22 \times 10^{-16}$. The number of processors p was $p = 16$, 32, 64 and 128 for random (normally distributed), real, square matrices of order $n = 8192$. The matrix condition number κ was 10^1 (well conditioned matrix). Singular values were also distributed normally.

Various blocking factors of form $\ell = p/k$ were used, where k is a natural number. For $\ell = 2$, the smallest blocking factor possible, one computes the EVD of the Gram matrix $A^T A$ in exactly one parallel iteration step and the parallel execution time T_p is the minimal one, about three times less than for other blocking factors. However, this approach is not recommended in general, because the computation of the *whole* Gram matrix squares the singular values (SVs) of A which can cause serious numerical problems in the case of very small SVs. Hence, the cases with $2 < \ell \leq p$ are more suitable from the numerical point of view and only these blocking factors are discussed in detail in the following.

The local EVDs for $\ell = 2p$ were computed by using the LAPACK procedure DGESVJ (see [7,8]) that implements the serial one-sided Jacobi SVD method and ensures the high relative accuracy also inside the 2×2 block subproblems. Since Gram matrices are symmetric and positive semidefinite, their EVD is identical with SVD. However, the procedure DGESVJ does not exploit the matrix symmetry to decrease the number of arithmetic operations so that it is not optimal with respect to the speed of computation and memory. For other blocking factors, the ScaLAPACK procedure PDSYEV was used. Note that PDSYEV is based on matrix tri-diagonalization so that the relative accuracy of computed eigenvalues can be lost. However, the nowadays ScaLAPACK library does not contain the parallel Jacobi procedure for the EVD of symmetric matrices.

For all processor grids, the block cyclic matrix distribution with $mb = nb = 50$ was used. In all cases, the global stopping criterion was

$$\max_{i,j} w_{ij} < n \, \epsilon_M. \tag{5}$$

Locally, two block columns were not mutually orthogonalized if

$$w_{ij} < n \, \epsilon_{\mathrm{M}}. \tag{6}$$

The comparison of the scalability of our new algorithm with the variant using $\ell = 2p$ and the ScaLAPACK procedure PDGESVD with respect to the total parallel execution time T_{p} is depicted in Fig. 1.

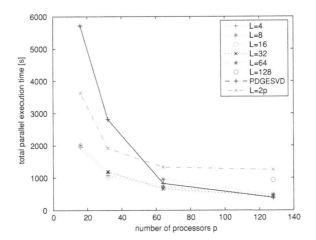

Fig. 1. Scalability for $n = 8192$ and $\kappa = 10^1$

For almost all blocking factors ℓ, our new algorithm is well scalable with only one exception for $\ell = p = 128$. For smaller number of processors ($p \leq 64$), our new variant is about 2–3 times *faster* than the ScaLAPACK procedure PDGESVD; otherwise the both algorithms are practically identical w.r.t. T_{p}. However, our older variant with $\ell = 2p$ is not scalable for a larger number of processors ($p > 32$). To get insight into its behavior, it is necessary to analyze the performance of individual parallel tasks w.r.t. a number of processors. Such profiling is depicted in Fig. 2.

As can be immediately seen, all computations are well scalable with exception of the WC, which does not scale at all (here GM denotes the Gram matrix computation and V the *a posteriori* computation of right singular vectors). This experimental fact is in accordance with the analysis of the communication complexity performed at the end of previous section. Using $\ell = 2p$, many global communication steps are in fact independent of p or *increase* with p. We suspect that the communication complexity can be decreased by a reformulation of the WC so that the procedures of type PBLAS 2 or PBLAS 3 will be used (the current implementation uses PBLAS 1).

This communication bottleneck was eliminated to a large extent in our new variant with a variable blocking factor ℓ; see Fig. 3 for chosen $\ell = 16$. The good scalability is in accordance with the theoretical communication complexity, since

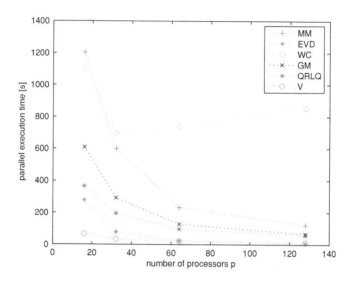

Fig. 2. Profiling for $\ell = 2p$ with respect to p

using a small, constant ℓ means that global communication steps of the WC scale as p^{-1} or are independent of p. Consequently, it only requires about 5–10 percent of T_{p} except for $p = 128$. Clearly, using $p = 128$ processors for matrices of order $n = 8192$ seems to be too much.

Regarding the accuracy of computation, we regularly computed three "quality indices" Q_1, Q_2 and Q_3 that describe the relative error (with respect to the Frobenius norm) in the orthogonality of left and right singular vectors, and in the matrix assembled from the computed SVD triple, respectively:

$$Q_1 = \frac{\|U^T U - I\|_{\mathrm{F}}}{\sqrt{n}}, \quad Q_2 = \frac{\|V^T V - I\|_{\mathrm{F}}}{\sqrt{n}}, \quad Q_3 = \frac{\|A - U\Sigma V^T\|_{\mathrm{F}}}{\|A\|_{\mathrm{F}}}.$$

For example, using $p = \ell = 64$, the achieved accuracy was $Q_1 \approx 5 \times 10^{-11}$, $Q_2 \approx 1 \times 10^{-13}$ and $Q_3 \approx 9 \times 10^{-14}$. The relatively low level of orthogonality of U is connected to an inaccurate computation of Gram matrices, especially of their off-diagonal blocks, in later parallel iteration steps (when the block columns are mutually "almost orthogonal"), and to the inherently low relative accuracy of the ScaLAPACK procedure PDSYEV based on the tri-diagonalization of symmetric matrices. In theory, the orthogonality of U could be improved using stronger global and local convergence criteria in the right-hand side of (5) and (6). We tested various approaches, e.g., using the value n/ℓ instead of n, but then the ScaLAPACK procedure PDSYEV did not converge for p, $\ell \geq 64$. This problem can be circumvented by avoiding the work with Gram matrices and computing the inner SVD subproblems using parallel version of some Jacobi procedure. However, at the moment, no such procedure is available in the ScaLAPACK library.

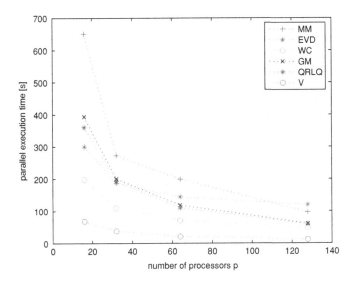

Fig. 3. Profiling for $\ell = 16$ with respect to p

On the other hand, the level of orthogonality of the *a posteriori* computed right singular vectors is excellent. This is a consequence of remarkable numerical properties of the lower triangular factor L to which the Jacobi method is applied after the preprocessing QRLQ step. As discussed in [7] in detail, when written as $L = Y_1 D_1 = D_2 Y_2$, where D_1 and D_2 are diagonal scaling matrices containing the Euclidean norms of columns and rows of L, respectively, both matrices $Y_i, i = 1, 2$, are well-conditioned. Well-conditioning of Y_1 means a high relative accuracy of computed singular values and left singular vectors, whereas well-conditioning of Y_2 leads to a very accurate solution of the matrix system (2), i.e., to a high relative accuracy of the *a posteriori* computed right singular vectors. Moreover, in the case of $\ell = 2p$, the same is true for the serial Jacobi SVD procedure DGESVJ in the computation of inner EVDs of symmetric Gram matrices, since any such computation starts with a (pivoted) local QRLQ pre-processing step.

5   Conclusions

Traditionally, the Jacobi SVD/EVD method (both its one- and two-sided variants) is considered to be the slowest one. However, recent progress in the serial one-sided Jacobi algorithm (see [7,8]) has shown that its speed is now comparable to that of algorithms, which are based on a matrix bi- or tri-diagonalization. Moreover, the Jacobi method computes *all* singular values and vectors with high relative accuracy for a large class of matrices, which is not generally true for other methods. Now we have shown that the same is true for the parallel one-sided block Jacobi algorithm with a variable (small) blocking factor, so that its performance is comparable to or significantly better than that of the ScaLAPACK procedure PDGESVD. The basic "building blocks" for a successful implementation are: (i) matrix pre-processing by the parallel QRLQ step, (ii) new parallel dynamic ordering of subproblems and its efficient parallelization, and (iii) *a posteriori* computation of right singular vectors.

To exploit fully the superb numerical properties of the parallel one-sided block Jacobi method, including the inherently high level of orthogonality of computed left singular vectors, it is desirable to avoid completely the computation of Gram matrices in inner SVD subproblems. Our idea is to use some sort of "recursive" one-sided Jacobi algorithm in the future, because other parallel SVD procedures, based on a matrix bidiagonalization, do not consistently converge down to a small multiple of machine precision (say, $10\epsilon_M$).

Acknowledgment. Authors were supported by the VEGA grant no. 2/0003/11.

References

1. Bečka, M., Vajteršic, M.: Block-Jacobi SVD algorithms for distributed memory systems: II. Meshes Parallel Algorithms Appl. **14**, 37–56 (1999)
2. Bečka, M., Okša, G., Vajteršic, M.: Dynamic ordering for a parallel block-Jacobi SVD algorithm. Parallel Comput. **28**, 243–262 (2002)
3. Bečka, M., Okša, G.: On variable blocking factor in a parallel dynamic block-Jacobi SVD algorithm. Parallel Comput. **28**, 1153–1174 (2003)
4. Bečka, M., Okša, G., Vajteršic, M., Grigori, L.: On iterative QR pre-processing in the parallel block-Jacobi SVD algorithm. Parallel Comput. **36**, 297–307 (2010)
5. Bečka, M., Okša, G., Vajteršic, M.: New dynamic orderings for the parallel one-sided block-Jacobi SVD algorithm. Parallel Process. Lett. (2013) (Sent for publication)
6. Blackford, L.S., et al.: ScaLAPACK Users' Guide, 1st edn. SIAM, Philadelphia (1997)
7. Drmač, Z., Veselić, K.: New fast and accurate Jacobi SVD algorithm: I. SIAM J. Matrix Anal. Appl. **29**, 1322–1342 (2007)
8. Drmač, Z., Veselić, K.: New fast and accurate Jacobi SVD algorithm: II. SIAM J. Matrix Anal. Appl. **29**, 1343–1362 (2007)
9. Okša, G., Vajteršic, M.: Efficient preprocessing in the parallel block-Jacobi SVD algorithm. Parallel Comput. **31**, 166–176 (2005)

An Identity Parareal Method
for Temporal Parallel Computations

Toshiya Takami[1]([⊠]) and Daiki Fukudome[2]

[1] Research Institute for Information Technology, Kyushu University,
6-10-1 Hakozaki, Higashi-ku, Fukuoka 812-8581, Japan
takami@cc.kyushu-u.ac.jp
[2] Graduate School of Information Science and Electrical Engineering,
Kyushu University, 744 Motooka, Nishi-ku, Fukuoka 819-0395, Japan

Abstract. A new simplified definition of time-domain parallelism is introduced for explicit time evolution calculations, and is implemented on parallel machines with bucket-brigade type communications. By the use of an identity operator instead of introducing an approximate solver, a recurrence formula for the parareal-in-time algorithm is much simplified. In spite of such a simple definition, it is applicable to many of explicit time-evolution calculations. In addition, this approach overcomes several drawbacks known in the original parareal-in-time method. In order to implement this algorithm on parallel machines, a parallel bucket-brigade interface is introduced, which reduces programming and tuning costs for complicated space-time parallel programs.

Keywords: Parareal-in-time · Bucket-brigade communication · Strong scaling · Massively parallel machine · Scientific computing

1 Introduction

The spatial domain decomposition technique is widely used in various scientific computing problems with neighborhood collective communications [1] included in the MPI 3.0 Standard [2] to resolve spatial dependencies between adjacent regions. However, when the number of independent components in the spatial direction is limited, it becomes difficult to carry out effective executions on massively parallel computers. Then, another direction for the domain decomposition, e.g., the time axis, should be considered in order to achieve strong scaling even in relatively small problems.

One of the famous methods to realize the time-domain decomposition is the "parareal-in-time" algorithm [3], while various approaches have been done in this field [4]. More than one decade from the first paper by J. Lions, et al. [3], a large number of articles have been published in the fields of applied mathematics, physics, chemistry, parallel computing, etc. The parareal method is actually applied to various dependent calculations from linear iterations [5] to large-scale

R. Wyrzykowski et al. (Eds.): PPAM 2013, Part I, LNCS 8384, pp. 67–75, 2014.
DOI: 10.1007/978-3-642-55224-3_7, © Springer-Verlag Berlin Heidelberg 2014

scientific time-evolution problems [6], and can be used to obtain further acceleration over saturation in the spatial decomposition.

On the other hand, it is known that there are several drawbacks in this algorithm: (1) definition of a tailor-made approximate solver to the original one is necessary, and its performance affects convergence property and speed-up ratio; (2) its parallel implementation is complicated when we introduce space and time parallelism, simultaneously. In the present work, these shortcomings are resolved through the use of an identity operator and a parallel bucket-brigade communication interface. This article is organized as follows. A new definition of the simplified parareal-in-time method is given in Sect. 2, and a new interface for the parallel bucket-brigade communication is introduced in Sect. 3. In Sect. 4, we present results of performance measurements by the space-time parallel code developed on this interface.

2   The Identity Parareal Method

Suppose that $x_k = x(t_k)$ is a dynamical variable defined by an explicit iterator $x_{k+1} = F_k(x_k)$. In this case, sequential execution is required with respect to k, since x_{k+1} depends directly on the previous result x_k. When $F(x)$ is an a-th order solver, it allows the truncation error of the order δt^{a+1} for the discrete time representation $\delta t = t_{k+1} - t_k$,

$$x(t_{k+1}) - x(t_k) = F_k(x_k) - x_k = \frac{\partial x}{\partial t}\delta t + \cdots + O(\delta t^{a+1}) \tag{1}$$

In the original parareal-in-time algorithm, with an approximate solver $G_k(x_k)$, an approximate sequence $\{x_k^{(r)}\}$ is calculated by the recurrence relation,

$$x_{k+1}^{(r+1)} = G_k(x_k^{(r+1)}) + F_k(x_k^{(r)}) - G_k(x_k^{(r)}), \tag{2}$$

where $\{x_k^{(r)}\}$ converges to the exact sequence, $\{x_k\}$, in a sufficiently large r. The key to an efficient implementation is the definition of $G(x)$. It must be a good approximation of $F(x)$ in order to achieve fast convergence, and is also expected to be much faster than $F(x)$ for sufficient speed-up in parallel computing. Since these requirements are conflicting each other and are strongly dependent on the property of the original solver $F(x)$, it is difficult to give a general strategy to introduce $G(x)$. Thus, many analyses have been done on applicability of the parareal-in-time algorithm to various scientific problems.

Our approach is somewhat different. We introduce an identity transformation as the approximate solver $G(x)$. Then, the recurrence formula for the parareal iteration is simplified to

$$x_{k+1}^{(r+1)} = x_k^{(r+1)} + F_k(x_k^{(r)}) - x_k^{(r)}, \tag{3}$$

which is the simplest definition of the parareal-in-time method. We call this implementation of the parareal-in-time method "identity parareal" (iParareal).

In continuous problems, this procedure converges for sufficiently small δt and sufficiently large r since the leading term of the time evolution is in proportion to δt (see Eq. (1)). The problem is how effectively we can construct the time-parallel scheme with a finite δt and r.

2.1 Convergence Analysis

Convergence properties and the remaining errors by the parareal-in-time algorithm have already been analyzed in case that those integrators, $F_k(x)$ and $G_k(x)$, are linear functions [5]. This analysis is also valid for the "iParareal" method since $G(x)$ is the simplest linear function, i.e., an identity.

When F_k is a linear operator, the exact state x_k at k-th time step is calculated by ordered operations of F_k to the initial value x_0,

$$x_k = F_k F_{k-1} \cdots F_1 x_0 = [I + (F_k - I)] \cdots [I + (F_1 - I)]x_0. \tag{4}$$

The last expression can be expanded as

$$x_k = \left[I + \sum_{j=1}^{k}(F_j - I) + \sum_{j'>j''}(F_{j'} - I)(F_{j''} - I) + \cdots \right] x_0. \tag{5}$$

An approximate sequence $\{x_k^{(r)}\}$ by the iParareal method is defined by ignoring higher order terms in the right-hand-side, i.e., this is r-th perturbation with respect to the operator $F - I$. Then, the remaining error after r-th iteration for k-th time step is estimated by

$$\frac{\left| x_k - x_k^{(r)} \right|}{|x_0|} \approx \frac{k!}{(k-r)!r!}\left[\rho(F - I) \right]^r \approx \sqrt{\frac{k}{2\pi(k-r)r}}\left[\frac{ek}{r}\rho(F - I) \right]^r \tag{6}$$

where $\rho(A)$ represents the spectral radius of the operator A.

On the other hand, the truncation error by the original integrator F is $k\delta t^{a+1}$ at k-th time step. Thus, the errors by the iParareal, Eq. (6), should be compared to this value.

2.2 Applications in Scientific Computing

There are various types of scientific calculations to be parallelized by the space-time domain decomposition. In this work, we analyze an applicability of the iParareal method to particle dynamics described by quantum and classical mechanics. Its applicability to the other types of calculations such as fluid dynamics [7–9], plasma physics [10], classical wave propagation [11,12], etc., should be analyzed elsewhere.

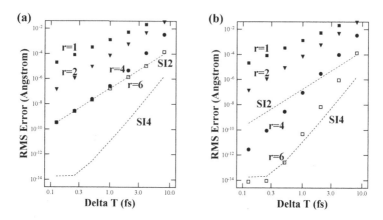

Fig. 1. Root mean square (RMS) errors of atom coordinates after 16 steps by iParareal iterations $r = 1$ (closed square), 2 (triangle), 4 (circle), and 6 (open square) compared with the standard 2nd-order symplectic integrator (SI2) and 4th-order SI (SI4). The iParareal method is applied to (a) 2nd order SI, and (b) 4th order SI.

Quantum Mechanics. As many studies on the related systems [13] have been done, the standard parareal method is applied to time evolutions in quantum mechanics,

$$|\psi(t + \delta t)\rangle = \exp\left[\frac{\delta t}{i\hbar} H(t)\right] |\psi(t)\rangle. \tag{7}$$

The Hamiltonian, $H(t)$, contains time-dependent external fields in optimal control problems [13]. Even if $H(t)$ is time dependent, i.e., the system is not isolated, we often represent the time evolution by unitary transformations. Then, we can decompose the unitary operator into the form,

$$\exp\left[\frac{\delta t}{i\hbar} H(t)\right] = I + \left(\exp\left[\frac{\delta t}{i\hbar} H(t)\right] - I\right). \tag{8}$$

Thus, the time-evolution operator is approximated by an identity. This type of successive multiplications of a unitary matrix has already been analyzed [5], which guarantees that the iParareal implementation stably converges to the exact result.

Classical Mechanics. Molecular dynamics (MD) simulations are widely used to investigate to represent dynamic processes of microscopic systems, where an extremely large number of time steps are calculated. For a large scale calculation with more than 10^6 atoms, spatial domain decomposition is effectively used with various approximate theories to reduce computational costs such as the particle mesh Ewald summation, the fast multipole method, etc. For relatively small systems, however, the effective use of the spatial parallel computation is limited. Then, the time-domain parallelism for MD [14,15] is desired.

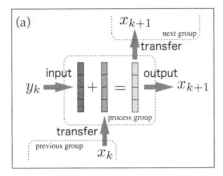 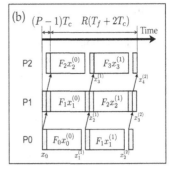

Fig. 2. (a) Schematic picture of the bucket-brigade communications; (b) a parallel configuration of the iParareal method with $P = 3$ and $R = 2$

The usual time evolution in MD is implemented by a second-order symplectic integrator (SI) [16], which is traditionally known as the Velocity Verlet algorithm. Here, we use an Ar cluster with 249 atoms as an example to demonstrate convergence properties of the iParareal method. In Fig. 1(a) and (b), errors of the iParareal implementation by the second and forth order SIs are shown for the number of the iParareal iterations $r = 1$, 2, 4, and 6. It is realized that these errors are bounded by ones of the original solvers imposed as dashed curves, 2nd order SI (SI2) and 4th order SI (SI4). Thus, the iParareal method converges when we use the sufficient number of iterations corresponding to the order of the original solver.

3   Bucket-Brigade Implementation of iParareal

There are several ways to implement the parareal-in-time method [17,18]. The original algorithm consists of sequential and parallel parts, where the total performance heavily depends on the sequential calculations distributed over multiple resources. The iParareal implementation in Sect. 2 is defined by substituting an identity operation for the coarse solver in the sequential part. Although this new definition much simplified the sequential part, its performance is still sensitive to configurations of communications. In this section, we define a new parallel bucket-brigade interface to simplify the programming of the space-time parallel code.

3.1   Bucket-Brigade Communication Interface

At first, we divide the iParareal iteration Eq. (3) into two equations,

$$y_k^{(r+1)} = F_k(x_k^{(r)}) - x_k^{(r)}, \tag{9}$$

$$x_{k+1}^{(r+1)} = x_k^{(r+1)} + y_k^{(r+1)}. \tag{10}$$

While the first relation, Eq. (9), is calculated within each parallel resource independently, the second part, Eq. (10), should be implemented with communications between adjacent resources. Note that it consists of communications with neighbor processes but is globally dependent over the whole resources since Eq. (10) should be executed sequentially in the order of k.

In order to represent the dependent part (10), we introduce a parallel bucket-brigade communication interface. Consider the case of P-parallel resources in the time domain and Q-parallel execution for the local calculations, Eq. (9). Then, the dependent part, Eq. (10), consists of P-stage pipeline communications. When computational data are divided into Q processes, Q parallel pipelines are executed as Q-line parallel bucket brigades.

The interface for this communication pattern is,

```
int BB_Reduce(void* x, void* y, int n, Data type, Op op,
    int src, int dst, int nrank, Comm comm);
```

where the data x with n elements is received from the group src, and the result of op(x,y) is transferred to the group dst, where each group has nrank $(= Q)$ processes. In the present implementation, we assume that each parallel resource has the same number of ranks (nrank).

3.2   Speed-Up Ratio

As shown in Fig. 2(b), we obtain $K = P + R - 1$ time steps by R iterations of the iParareal method with P parallel resources in time direction. Then, an expected speed-up ratio is given by

$$S_{\text{iParareal}}(P, R, T_f, T_c) = \frac{(P + R - 1)T_f}{(P - 1)T_c + R(T_f + 2T_c)}, \tag{11}$$

where T_f is the computational time of Eq. (9), and $(P - 1)T_c$ represents extra costs for the bucket-brigade communications (Fig. 2(a)) which includes costs for calculations of Eq. (10) and communication between resources.

4   Performance Measurement

We show several results of measurements on a Sandy Bridge cluster connected by the InfiniBand FDR. We have 16 cores in each node, and a flat MPI configuration is used for parallel computing. In the iParareal implementation, the dependent part, Eq. (10), is written by the bucket-brigade communication. As an example, parallel speed-ups in quantum time-evolutions is measured, where wave functions are represented by complex vectors with 2048 elements and the time-step computation, Eq. (7), is implemented by a matrix-vector multiplication.

In Table 1, we show measured values and speed-ups in several configurations of parallel resources, where P represents the number of resources used in the time direction, Q is the number of processes for the single time evolution, Eq. (7), and

Table 1. Measured speed-up ratios by the iParareal method

$P \times Q$	Resources nodes	proc.	Parameters P	Q	R	K	Time (ms) T_Q	$T_{P \times Q}$	Temporal speed-up
4×8	2	32	4	8	4	7	3.48	14.43	1.69
8×8	4	64	8	8	4	11	3.48	15.15	2.53
16×8	8	128	16	8	4	19	3.48	16.11	4.10
4×16	4	64	4	16	4	7	1.95	8.87	1.54
8×16	8	128	8	16	4	11	1.95	10.05	2.13
16×16	16	256	16	16	4	19	1.95	11.71	3.16

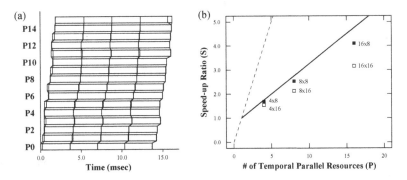

Fig. 3. (a) Timeline of an iParallel execution, (b) speed-up by the iParareal method compared with the theoretical limit of $R = 4$ and the linear speed-up (dashed line).

R is the number of iterations of the iParareal algorithm. The total number of time-steps K is given by $P + R - 1$, and the total number of processes is $P \times Q$. T_Q is a duration of Eq. (9) executed by Q processes, and $T_{P \times Q}$ is a total time by $P \times Q$ processes. The speed-up ratio in the time parallel method defined by

$$[\text{Speed-up Ratio}] \equiv \frac{T_Q \times K}{T_{P \times Q}}. \tag{12}$$

is shown in the last column of the same table.

Figure 3(a) is a visualized example of a timeline of the case 16×8 in Table 1. Wide rectangles represent the part of local calculations and thin ones represent communications including unexpected waiting time. In Fig. 3(b), measured speed-up ratios in Table 1 are shown with a theoretical limit of the speedup,

$$S_{\text{ideal}}(P, R) = \frac{P + R - 1}{R}, \tag{13}$$

which is obtained by ignoring the costs T_c for the bucket-brigade communication and computations in Eq. (11).

5 Summary

In this contribution, we proposed a new implementation of the parareal-in-time method, where an identity operation is used as the approximate solver $G(x)$ in the original algorithm Eq. (2). Although the convergence of the parareal iteration becomes slower than the case with an efficient approximate solver $G(x)$, it was shown that this implementation is applicable to explicit time evolution problems.

We also introduced a new communication interface called a parallel bucket-brigade pattern, which is used in the simple implementation of the parareal method. Parallel performance was measured on a cluster machine and several visualized examples were shown. While the current code is written by blocking communications in MPI, further improvements will be done with the non-blocking implementation or neighbor collective operations [1].

Acknowledgments. This work is supported by JST, CREST.

References

1. Hoefler, T., Jumsdaine, A., Rehm, W.: Implementation and performance analysis of non-blocking collective operations for MPI. In: Proceedings of SC07, IEEE Computer Society (2007)
2. MPI 3.0 released at September 21, 2012: http://www.mpi-forum.org/
3. Lions, J.-L., Maday, Y., Turinici, G.: A "parareal" in time discretization of PDE's. C. R. Acad. Sci. Ser. I Math. **332**, 661–668 (2001)
4. Gander, M.J., Vandewalle, S.: Analysis of the parareal time-parallel time-integration method. SIAM J. Sci. Comput. **29**, 556–578 (2007)
5. Takami, T., Nishida, A.: Parareal acceleration of matrix multiplication. Adv. Parallel Comput. **22**, 437–444 (2012)
6. Speck, R., Ruprecht, D., Krause, R., Emmett, M., Minion, M., Winkel, M., Gibbon, P.: A massively space-time parallel n-body solver. In: Proceedings of SC12, IEEE Computer Society (2012)
7. Farhat, C., Chandesris, M.: Time-decomposed parallel time-integrators: theory and feasibility studies for fluid, structure, and fluid, structure applications. Int. J. Numer. Meth. Eng. **58**, 1397–1434 (2003)
8. Fischer, P.F., Hecht, F., Maday, Y.: A parareal in time semi-implicit approximation of the navier-stokes equations. In: Barth, T.J., Griebel, M., Keyes, D.E., Nieminen, R.M., Roose, D., Schlick, T., Kornhuber, R., Hoppe, R., Périaux, J., Pironneau, O., Widlund, O., Xu, J. (eds.) Domain Decomposition Methods in Science and Engineering. LNCSE, vol. 40, pp. 433–440. Springer, Heidelberg (2005)
9. Bal, G.: On the convergence and the stability of the parareal algorithm to solve partial differential equations. In: Barth, T.J., Griebel, M., Keyes, D.E., Nieminen, R.M., Roose, D., Schlick, T., Kornhuber, R., Hoppe, R., Périaux, J., Pironneau, O., Widlund, O., Xu, J. (eds.) Domain Decomposition Methods in Science and Engineering. LNCSE, vol. 40, pp. 425–432. Springer, Heidelberg (2005)
10. Samaddar, D., Newman, D.E., Sánchez, R.: Parallelization in time of numerical simulations of fully-developed plasma turbulence using the parareal algorithm. J. Comput. Phys. **229**, 6558–6573 (2010)

11. Duarte, M., Massot, M., Descombes, S.: Parareal operator splitting techniques for multi-scale reaction waves: numerical analysis and strategies. ESAIM: Math. Model. Numer. Anal. **45**, 825–852 (2011)
12. Ruprecht, D., Krause, R.: Explicit parallel-in-time integration of a linear acoustic-advection system. Comput. Fluids **59**, 72–83 (2012)
13. Maday, Y., Turinici, G.: Parallel in time algorithms for quantum control: parareal time discretization scheme. Int. J. Quantum Chem. **93**, 223–228 (2003)
14. Baffico, L., Bernard, S., Maday, Y., Turinici, G., Zérah, G.: Parallel-in-time molecular-dynamics simulations. Phys. Rev. E **66**, 057701 (2002)
15. Srinivasan, A., Chandra, N.: Latency tolerance through parallelization of time in scientific applications. Parallel Comput. **31**, 777–796 (2005)
16. Bal, G., Wu, Q.: Symplectic parareal. In: Langer, U., Discacciati, M., Keyes, D.E., Widlund, O.B., Zulehner, W. (eds.) Domain Decomposition Methods in Science and Engnieering XVII. LNCSE, vol. 60, pp. 401–408. Springer, Heidelberg (2008)
17. Aubanel, E.: Scheduling of tasks in the parareal algorithm. Parallel Comput. **37**, 172–182 (2011)
18. Elwasif, W.R., Foley, S.S., Bernholdt, D.E., Berry, L.A., Samaddar, D., Newman, D.E., Sanchez, R.: A dependency-driven formulation of parareal: parallel-in-time solution of PDEs as a many-task application. In: Proceedings of MTAGS'11, pp. 15–24 (2011)

Improving Perfect Parallelism

Lars Karlsson[(✉)], Carl Christian Kjelgaard Mikkelsen, and Bo Kågström

Department of Computing Science and HPC2N, Umeå University, Umeå, Sweden
{larsk,spock,bokg}@cs.umu.se

Abstract. We reconsider the familiar problem of executing a perfectly parallel workload consisting of N independent tasks on a parallel computer with $P \ll N$ processors. We show that there are memory-bound problems for which the runtime can be reduced by the forced parallelization of individual tasks across a small number of cores. Specific examples include solving differential equations, performing sparse matrix–vector multiplications, and sorting integer keys.

Keywords: Perfectly parallel problem · Resource contention · Forced parallelization

1 Introduction

Super-linear speedups are frequently observed when the number of cores is increased to the point where the problem at hand fits in the union of the caches [2]. A related—but different—effect is observed when multiple cores share a cache but execute independent tasks. In such cases, the *effective* cache capacity per task is increased by the forced parallelization of each individual task. To be specific, if P cores share a cache of capacity C, then the effective cache capacity per task is only C/P when the cores execute independent tasks. If, on the other hand, they all collaborate to solve a single task, then the effective cache capacity is restored to C.

A *perfectly parallel workload* is one which has already been decomposed into a large number of independent tasks. Perfect parallelism can be found in many applications, often when exploring a parameter space or a large database. On a parallel computer with P identical and independent processors, any perfectly parallel workload should obtain close to linear (perfect) speedup. The only real issue is whether the load can be balanced or not. Here the key assumption is that the processors are *independent* in the sense that they do not interfere with each other by competing for shared resources. However, this assumption is violated on a multicore processor where the cores typically share one or more levels of cache as well as a memory bus.

The aim of this paper is to stress and provide experimental support for the following conclusion: If resources are shared between cores, then it is frequently possible to improve the execution time of a perfectly parallel workload by the forced parallelization of each task over a small number of cores. We refer to

R. Wyrzykowski et al. (Eds.): PPAM 2013, Part I, LNCS 8384, pp. 76–85, 2014.
DOI: 10.1007/978-3-642-55224-3_8, © Springer-Verlag Berlin Heidelberg 2014

this as a *hybrid (parallel) scheme* for executing perfectly parallel workloads in contrast to a *perfectly parallel scheme* where each task remains sequential. We want to emphasize that cache sharing is most detrimental for memory-bound applications. There are compute-bound applications for which the performance is almost completely insensitive to the cache size.

The rest of the paper is organized as follows. In Sect. 2, we describe the hardware, Intel/AMD, used in our experiments. In Sect. 3, we introduce a number of perfectly parallel workloads with varying characteristics in order to explore the applicability of such hybrid schemes. Simultaneously, we present and analyze the experimental results. Hybrid schemes are not universally applicable, but likely candidates can be identified using the guidelines laid down in Sect. 4.

2   Test Systems

We have performed tests on two contemporary multicore systems with different architectures. The *Triolith* system is based on Intel Xeon E5-2660 processors with a base frequency of 2.2 GHz. The eight cores of a processor share a 20 MiB L3 cache and a memory bus. Each core has a private 256 KiB L2 cache. The *Abisko* system is based on AMD Opteron 6238 processors with a base frequency of 2.6 GHz. The twelve cores of a processor are partitioned into two NUMA domains consisting of six cores each. The six cores in a domain share a 6 MiB L3 cache and a memory bus. There are three 2 MiB L2 caches per domain and each cache is shared by a pair of cores.

In our experiments, we restrict ourselves to one NUMA domain, since cores from different domains typically share little or no resources and can often be viewed as independent. In particular, we use eight cores per processor on Triolith and six cores per processor on Abisko.

3   Examples

We provide four examples of perfectly parallel workloads with varying characteristics in order to shed light on the benefits and limitations of hybrid schemes.

The *copy workload* is an idealized workload in which memory is accessed with unit stride and each task can be parallelized without overhead. It puts high pressure on the shared bandwidth resources and thereby exposes negative effects of sharing caches and memory buses. Conversely, it emphasizes how the forced parallelization of each task alleviates some of the competition for shared resources.

The *power workload* is a more realistic example consisting of repeated sparse matrix–vector multiplications in the context of the power method for computing a dominant eigenvector. The sparse matrix data is accessed contiguously, while there is random access to the elements of the vector being multiplied. The forced parallelization of the power method requires two synchronizations per iteration: One for the matrix–vector multiplication and one for the normalization. It also requires communication of the normalized vector.

The *Spike workload* has unit stride accesses and memory is updated in-place. The forced parallelization overhead in terms of redundant work is non-negligible. Each core needs to synchronize with its nearest neighbors and the volume of communication is minimal.

The *radix sort workload* is an example with quasi-random access to memory, strict synchronization requirements, and large communication volume. This workload accesses memory in an inherently data-dependent way, which negates the possibility of prefetching, and a large communication volume puts pressure on the cache coherence protocol.

In the following, we use the term *effective memory bandwidth* to refer to the number of bytes accessed at the source code level divided by the execution time. We only count accesses to data structures whose sizes depend on the size of the problem. This provides a useful tool to interpret results for a particular workload and scheduling scheme but one should not use these figures to compare different workloads and scheduling schemes. Traditionally, the term *parallel speedup* refers to the ratio of the execution times of a sequential run and a parallel run. Since the aim of this paper is to compare the performance of different parallelization strategies, we instead emphasize the *relative improvement* of one strategy over another, again by computing a ratio of execution times.

3.1   Memory Copy

In this workload, each task consists of copying data back and forth a fixed number of times between two contiguous vectors. The individual tasks can be parallelized without overhead. For a vector length of n, iteration count k, and execution time T, the working set size is $8n$ bytes (32-bit integer elements) and the effective memory bandwidth is $8kn/T$ bytes per second. In all experiments presented here, $k = 100$.

Figure 1 (left) illustrates the effective memory bandwidths of the copy workload on Abisko (top) and Triolith (bottom) for various hybrid schemes as a function of working set size. The curves are labeled by $p \times q$, where p is the number of thread groups and q is the number of threads per forcibly parallelized task. In particular, sequential execution is denoted by 1×1 and the perfectly parallel scheme is denoted by $p \times 1$.

First consider the sequential execution on Abisko (top left) labeled 1×1. The parallel speedup, that is, the ratio of the execution time for the 6×1 perfectly parallel scheme and the 1×1 sequential scheme ranges from 1.24 to 4.18, rather than the expected value of 6. This indicates that a severe competition for resources is taking place. There is a marked drop near the L2 cache capacity (2 MiB) and a second drop near the sum of the L2 and L3 cache capacities $(2 + 6 = 8\,\text{MiB})$[1]. Next consider the perfectly parallel scheme labeled 6×1. There are again two marked drops, but this time the first drop occurs already near the 1 MiB line since the L2 cache is shared by two cores. Similarly, but more dramatically, the second drop occurs already near the 2 MiB line as the

---

[1] The L3 cache on Abisko is a non-inclusive victim cache, hence the addition.

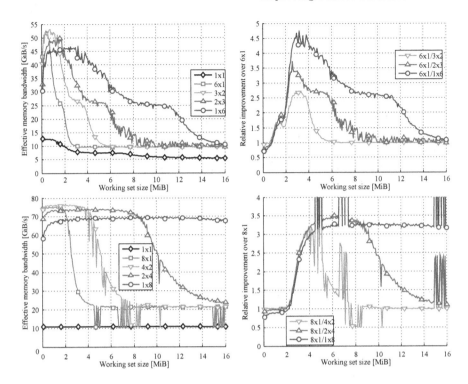

Fig. 1. Results for the *copy* workload. Left: Effective memory bandwidths. Right: Speedups relative to the perfectly parallel configuration. Top: The Abisko system. Bottom: The Triolith system.

combined effective capacities of the L2 and L3 caches is only $2/2 + 6/6 = 2\,\text{MiB}$. Finally, consider the opposite configuration 1×6, where six cores work together on one task. In addition to using the full capacity of the L3 cache, each task now also has access to an *aggregate* L2 capacity of $3 \cdot 2 = 6\,\text{MiB}$, which correlates somewhat with the first drop, and the combined capacities of the L2 and L3 caches are $12\,\text{MiB}$, which correlates well with the second drop.

The results obtained on Triolith (bottom left) are similar except that there is only one drop and the point where the working set size spills over into the L3 cache is not observed. The drops correlate with the $20/8 = 2.5$, $20/4 = 5$, $20/2 = 10$, and $20/1 = 20\,\text{MiB}$ effective L3 cache capacities per task, although the latter is not included in the figure. At this point we have no explanation for the distinct spikes in the graphs.

Figure 1 (right) illustrates the relative improvement of the various hybrid schemes over the perfectly parallel scheme. A relative improvement greater than unity implies that the hybrid scheme is faster than the perfectly parallel scheme. On Abisko the relative improvement peaks at 4.5 and on Triolith at 3.5.

3.2   The Power Method

In this workload, we apply the power method, $w_{k+1} = Ax_k$ and $x_{k+1} = \frac{w_{k+1}}{\|w_{k+1}\|_2}$, to approximate the dominant eigenvector of every member of a large set of sparse matrices. The matrices have a random sparsity pattern which is generated with predetermined lower (1) and upper (13) bounds on the number of nonzero entries per row. The total number of nonzeros is $7n$, where n is the order of the matrix.

The forced parallelization of a single instance of the power method is achieved by splitting the matrix into block rows and applying the matrix–vector multiplications in parallel. The matrix is stored in compressed sparse row format.

The working set size of the power method includes storage for one sparse matrix and two vectors. The vectors require $8n$ bytes each (64-bit floating point elements). To store the matrix, the row pointer array requires $4(n + 1)$ bytes, the column index array requires $4 \cdot 7n$ bytes, and the value array requires $8 \cdot 7n$ bytes. Hence, the working set size is $104n + 4$ bytes. For each matrix–vector multiplication $y \leftarrow Ax$, there is one element access of x for each nonzero in A, the vector y is accessed once, and A is accessed once. Hence, the effective memory bandwidth for k iterations of the power method is $k(152n + 4)/T$, where T is the execution time[2]. In all our experiments, $k = 20$. The small cost of normalization is not included in the calculation of the effective bandwidth.

Figure 2 illustrates the effective bandwidths (left) and speedups (right) on Abisko (top) and Triolith (bottom). Notice that the speed deterioration is much more gradual than in the copy workload on both systems. On Triolith, the initial drastic drops correlate nicely with the working set size spilling over the L3 cache capacity into main memory. However, a second, slower, drop appears with the following possible explanation. As the working set size increases, so does the probability that the required elements of x belong to different pages of memory with a consequent increase in the effective memory latency due to TLB misses. The results on Abisko are noteworthy. The curves are intertwined with a drop after 2 MiB and again a second, slower, drop. In contrast to Triolith, the optimal number of threads per task depends non-trivially on the working set size. The maximum relative improvement on Abisko is 1.7 and on Triolith it is 2.4.

3.3   The Spike Algorithm

This workload consists of computing functions $\phi_j(x) = u^{(j)}(x, L)$, where each function $u^{(j)}$ is given implicitly as the solution of a heat equation

$$u_t^{(j)} = a_j(x) \cdot u_{xx}^{(j)}(x, t), \quad 0 < x < 1, \quad 0 < t < L,$$

with a known initial condition and homogenous boundary conditions. This is an example of a perfectly parallel parameter study. The goal could be to determine a material, i.e., a function a, such that a specific distribution of the heat is

---

[2] Note that the effective memory bandwidth is a tool used to illustrate the time measurements and does not reflect the memory bandwidth that is actually consumed at the hardware level.

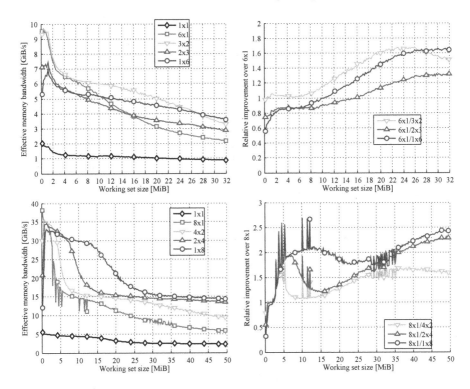

Fig. 2. Results for the *power* workload. Left: Effective memory bandwidths. Right: Speedups relative to the perfectly parallel configuration. Top: The Abisko system. Bottom: The Triolith system.

obtained at time $t = L$. The individual equations are solved using the implicit Euler method, a finite difference method which hinges on the solution of a sequence of tridiagonal linear systems that are strictly diagonally dominant by rows. Let n denote their dimension. The linear systems are solved sequentially using Gaussian elimination without pivoting. After the initial factorization, the solution is advanced in time by solving a sequence of systems using the factored matrix. Each sequential solve (forward and backward substitution) requires only $5n$ arithmetic operations while accessing $7n$ matrix and vector elements.

For the forced parallellization we have used a new variant of the truncated Spike algorithm [6]. Normally, the Spike algorithm requires $10n$ arithmetic operations for the solve phase and accesses $14n$ elements. However, it is possible to reduce the number of accesses to $9n$. The details will be presented in a future paper. Here, we concentrate on the effects on the perfectly parallel workload. Since the solve phase is memory bound, it is essential to reduce the memory footprint rather than the number of arithmetic operations.

The working set sizes and effective memory bandwidths are different for the sequential and parallel algorithms. The working set size is $4 \cdot 8n = 32n$ bytes (64-bit floating point elements) in the sequential case and $5 \cdot 8n = 40n$ bytes in the parallel case, an increase of 25 %. The effective memory bandwidth for k time steps and execution time T is $7 \cdot 8nk/T = 56nk/T$ bytes per second in the sequential case and $9 \cdot 8nk/T = 72nk/T$ bytes per second in the parallel case. In all our experiments, $k = 100$.

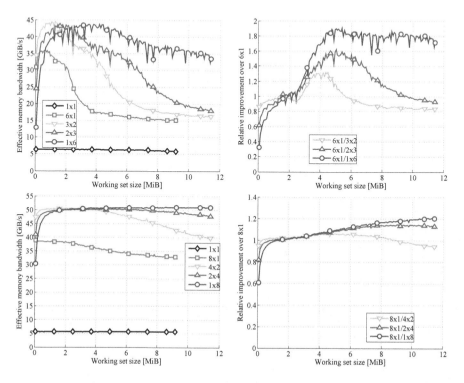

Fig. 3. Results for the *Spike* workload. Left: Effective memory bandwidths. Right: Speedups relative to the perfectly parallel configuration. Top: The Abisko system. Bottom: The Triolith system.

Since the sequential and parallel algorithms have different working set sizes and bandwidths, one must be cautious when comparing the effective bandwidths of a sequential (1×1 and 8×1) and a hybrid (4×2, 2×4, and 1×8) scheme. In the right side of Fig. 3, the relative improvements are plotted against the working set size for the parallel runs.

As demonstrated in Fig. 3, the behaviour of the Spike workload is similar to the copy workload except that the distinction between L2 and L3 on Abisko is much less pronounced. On Triolith, there is only a marginal slow-down when surpassing the effective L3 capacity in the perfectly parallel 8×1 scheme. The maximum relative improvement on Abisko is approximately 1.8 while on Triolith the relative improvement is a more modest 1.2. It should be noted that these improvements are obtained despite the fact that the parallel algorithm moves more data and must communicate and synchronize.

3.4   Radix Sort

In this workload, n 32-bit keys are sorted using a radix-16 sort. More specifically, each key is viewed as an eight digit number in base 16 and the numbers are sorted one hexadecimal digit at a time in phases starting with the least significant digit. Each phase consists of three steps: First compute the frequency of each digit to determine the size of each bucket, then find the starting position of each bucket via a prefix sum of 16 integers, and finally map the elements into their buckets.

Our parallel algorithm mirrors the sequential one and partitions the input array into blocks. The counting step is perfectly parallel. The prefix sum is performed by one thread with barrier synchronization before and after. In the permutation step, each thread is responsible for the permutation of its own block of the input array and hence writes up to 16 contiguous blocks in the output array. The permutation step is followed by a third barrier.

The working set size is $2 \cdot 4n = 8n$ bytes since there are two arrays of length n involved in each phase. The effective memory bandwidth is $8 \cdot 3 \cdot 4n/T = 96n/T$ bytes per second since in each of the eight phases the input array is read *twice* and the output is written once.

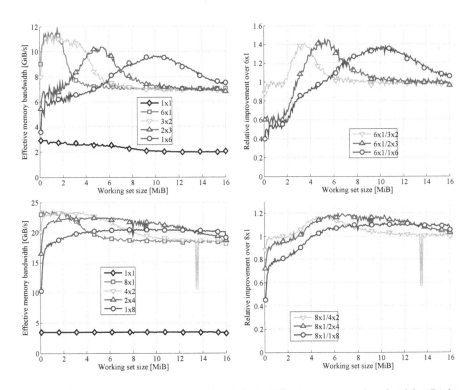

Fig. 4. Results for the *radix sort* workload. Left: Effective memory bandwidths. Right: Speedups relative to the perfectly parallel configuration. Top: The Abisko system. Bottom: Th Triolith system.

Figure 4 displays the effective memory bandwidths and relative improvements of the radix sort workload. One of its important characteristics is that the threads read and write data written by other threads in the preceding phase. If the entire working set fits in the aggregate L2 cache, then pressure is put on the cache coherence protocol to transfer data between sibling L2 caches. As the working set size increases, data is necessarily evicted from the L2 caches to a shared resource. Later accesses to these data do not generate any cache coherence traffic. The observations in the left half of Fig. 4 (left) suggest that communicating via the cache coherence protocol is far more costly than accessing a shared higher level resource. This correlates with the low performance observed for small working set sizes on both systems (2×3 and 1×6 on Abisko and all hybrid schemes on Triolith) and also correlates with the high performance of the 3×2 hybrid scheme on Abisko since each pair of cooperating threads share the *same* L2 cache.

4   Guidelines

We propose the following guidelines to help users identify workloads that are likely to benefit from forced parallelization: (i) establish that resource sharing is an issue by running the perfectly parallel workload both sequentially and in parallel. Sub-linear speedup indicates competition for shared resources; (ii) ensure that the tasks exhibit data reuse since otherwise caching has no effect; (iii) verify that the tasks can be parallelized with little parallel overhead as a large overhead negates any gains obtained from the parallelization.

After the guidelines have been used to identify a likely candidate, the user can start to analyze the independent tasks in detail using, e.g., the roofline model [8] and profiling tools that measure hardware performance counters [1]. Scheduling tasks with the cache hierarchy in mind [4,5] has proven to be effective on multicore-based systems and is one of many techniques that can be used when forcibly parallelizing the tasks.

5   Conclusion

We have reconsidered the familiar problem of executing a single perfectly parallel workload consisting of independent tasks on many multicore processors. We argued that the presence of shared caches in multicores can have detrimental effects on the performance. We have shown that a number of memory-bound workloads can significantly benefit from the forced parallelization of individual tasks across a few cores. If the applications are sufficiently compute-bound, then it is unlikely that forced parallelization will yield any benefits.

The benefits of forced parallelization vary, both quantitatively and qualitatively, between different systems and workload characteristics, making it hard to predict if a workload will benefit from a hybrid approach. We provided guidelines that help identify likely candidates.

Based on this work, we conclude that there is reason to reconsider the parallelization strategy employed for perfectly parallel workloads. The relative improvements we have observed over the perfectly parallel scheme are modest but it is worth remembering that the absolute savings across a large number of processors is still significant. Moreover, we observe that the development of efficient multicore algorithms for relatively small problems remains relevant.

In the Grid scheduling community, people have considered the related problem of scheduling multiple workloads with different resource usage characteristics (e.g., memory-bound and compute-bound workloads) in attempts to mitigate resource contention [9].

It is possible to view the hybrid schemes as another example of the benefits of M-tasks as used by Rauber and Rünger [7]. Recently, we successfully used forced parallelization of small and otherwise insignificant tasks in order to improve the scalability of en eigenvalue solver [3].

The source code used for the experiments is available upon request.

Acknowledgements. Financial support by the Swedish Research Council grant VR A0581501 and eSSENCE, a strategic collaborative eScience programme. This research was conducted using the resources of HPC2N and NSC.

References

1. Grant, R.E., Afsahi, A.: A comprehensive analysis of OpenMP applications on dual-core Intel Xeon SMPs. In: IPDPS, pp. 1–8 (2007)
2. Gustafson, J.L.: Fixed time, tiered memory, and superlinear speedup. In: Proceedings of the Fifth Distributed Memory Computing Conference (DMCC), pp. 1255–1260 (1990)
3. Karlsson, L., Kågström, B., Wadbro, E.: Fine-grained bulge-chasing kernels for strongly scalable parallel QR algorithms. Parallel Comput. (2013, accepted)
4. Muddukrishna, A., Podobas, A., Brorsson, M., Vlassov, V.: Task scheduling on manycore processors with home caches. In: Caragiannis, J., et al. (eds.) Euro-Par Workshops 2012. LNCS, vol. 7640, pp. 357–367. Springer, Heidelberg (2013)
5. Olivier, S.L., Porterfield, A.K., Wheeler, K.B., Prins, J.F.: Scheduling task parallelism on multi-socket multicore systems. In: Proceedings of ROSS'11, pp. 49–56. ACM, New York (2011)
6. Polizzi, E., Sameh, A.H.: A parallel hybrid banded system solver: the SPIKE algorithm. Parallel Comput. **32**(2), 177–194 (2006)
7. Rauber, T., Rünger, G.: M-task-programming for heterogeneous systems and grid environments. In: IPDPS (2005)
8. Williams, S.W.: Auto-tuning performance on multicore computers. Ph.D. thesis, EECS Department, University of California, Berkeley (2008)
9. Zhuravlev, S., Blagodurov, S., Fedorova, A.: Addressing shared resource contention in multicore processors via scheduling. SIGARCH Comput. Archit. News **38**(1), 129–142 (2010)

Methods for High-Throughput Computation of Elementary Functions

Marat Dukhan$^{(\boxtimes)}$ and Richard Vuduc

School of Computational Science and Engineering,
College of Computing, Georgia Institute of Technology,
266 Ferst Drive NW, Atlanta, GA 30332, USA
mdukhan3@gatech.edu

Abstract. Computing elementary functions on large arrays is an essential part of many machine learning and signal processing algorithms. Since the introduction of floating-point computations in mainstream processors, table lookups, division, square root, and piecewise approximations were essential components of elementary functions implementations. However, we suggest that these operations can not deliver high throughput on modern processors, and argue that algorithms which rely only on multiplication, addition, and integer operations would achieve higher performance. We propose 4 design principles for high-throughput elementary functions and suggest how to apply them to implementation of log, exp, sin, and tan functions. We evaluate the performance and accuracy of the new algorithms on three recent x86 microarchitectures and demonstrate that they compare favorably to previously published research and vendor-optimized libraries.

Keywords: Elementary functions · SIMD · Fused multiply-add

1 Introduction

In this paper we discuss the methods for high-throughput computation of elementary functions, namely exponent, logarithm, sine, and tangent. Due to extensive use of elementary functions in mathematical models, they are widespread in a number of application areas. Examples include non-parametric statistics [18], information theory measures entropy and Kullback–Leibler divergence, and random number generation for normal [2], Student-t [1], logistic [15], exponential [15], and Cauchy [9] distributions.

This paper brings three contributions to the study of high-throughput elementary functions:

1. **Design principles for portable high-throughput elementary functions** (Sect. 3): We argue why the methods traditionally used for elementary functions might be suboptimal for a high-throughput implementation, and propose four design principles to improve performance of high-throughput elementary functions on modern processors.

R. Wyrzykowski et al. (Eds.): PPAM 2013, Part I, LNCS 8384, pp. 86–95, 2014.
DOI: 10.1007/978-3-642-55224-3_9, © Springer-Verlag Berlin Heidelberg 2014

2. **Elementary functions design using only hardware-efficient operations** (Sects. 4, 5): We show that it is possible to implement elementary functions without violating the four design principles.
3. **Experimental evidence on the advantages of the four design principles** (Sects. 5, 6): We verify that following the proposed design principles leads to higher-throughput elementary functions by comparing our implementation to 15 other implementations on 3 microarchitectures.

2   Background

Introduction of the IEEE754 floating-point standard and wide spread of hardware floating-point implementations boosted research on elementary functions, and by the first years of the 1990s the high-level design of elementary functions was settled. A modern elementary function implementation has 3 steps [11,12]:

1. **Range reduction** where the argument of the function with potentially infinite domain is transformed into a reduced argument from a finite, and usually small, range where the function can be efficiently approximated.
2. **Approximation** where the value of elementary function is calculated using a combination of polynomial, piecewise-polynomial, and rational approximation, table lookups, and other transformations.
3. **Reconstruction** where the value of the function at its initial argument is reconstructed from its computed value on reduced argument.

Several methods, introduced between mid-80s to mid-90s, combined both good accuracy and good performance, and became classical in this field: Gal's algorithm for accurate table-based approximations [8], P.T.P. Tang's table-based algorithms [17], and ATA method by Wong and Goto [19]. These methods used tables of pre-computed values to save precious floating-point operations at the time when floating-point performance of processors was poor.

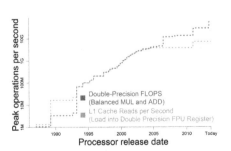

Fig. 1. Evolution of performance of Intel x86 desktop processors

However, as the floating-point capabilities of processors evolved, the advantages of table-based method started to diminish. Newer processors brought pipelined floating-point units, capable of starting an FP operation every cycle, SIMD extensions, and Fused Multiply-Add (FMA) instructions. Figure 1 suggests that while the throughput of FP operations increased by several times during the last decades, performance of table lookups did not match this rate of improvements. This shift in hardware capabilities created an opportunity to design high-throughput elementary functions without table lookups. An earlier attempt to design table-free functions is presented in SLEEF library [16]. The functions presented in this paper differ from SLEEF in five aspects: we refrain from using divisions and square roots, we use branch instructions to handle special cases, we hide the latency in polynomial evaluations by computing a function on a batch of elements, we use the state-of-the-art polynomial approximation method [3] which produces substantially more accurate polynomials than Taylor series in [16], and we prefer very high-degree polynomial approximations with minimal transformations of the input arguments while [16] favors lower-degree polynomials with special function-specific argument transformations to improve accuracy.

3   Design Principles

In this section we present the four principles for high-throughput design of elementary functions: avoid table lookups, use multiplication and addition instead of division and square root, avoid unpredictable branches, and process elements in batches to hide latency.

Table 1. Latency of GATHER operation, emulated via multiple simple instructions, cycles per instruction

Operation Processor	2x DP GATHER	4x DP GATHER
Piledriver	18	23
Nehalem	12.3	–
Ivy Bridge	9	14

Table-based methods became popular in elementary functions implementations because they allowed to design highly accurate elementary function without using extended precision floating-point formats. However, we suggest that they could be a poor building block for a high-throughput implementation. First, as demonstrated by Fig. 1 floating-point operations deliver several times higher throughput than table lookups. Secondly, the latency of SIMD table lookup operations in prohibitively high. Most SIMD instruction sets, such as SSE, AVX, and NEON, do not support GATHER operation, which loads data from memory using indices from SIMD register, and it needs to be emulated with multiple instructions. Intel Haswell processors support GATHER instructions in hardware, but its does not improve the latency. Table 1 indicates that GATHER latencies are too high to be hidden with software pipelining or out-of-order execution. Thus, our first design principle is to **avoid table lookups in high-throughput elementary functions**.

Even on modern hardware not all floating-point operations are equally efficient. While multiplication and addition operations are usually pipelined and can execute every cycle, floating-point division and square root occupy the execution units for many cycles, and a new operation can start only after the

Table 2. Peak single-thread throughput of floating-point operations per cycle [6]

Operation Processor	DP ADD	DP MUL	DP MAC	DP DIV	DP SQRT
Piledriver	4	4	4	4/18	4/29
Nehalem	2	2	2	2/22	2/32
Sandy Bridge	4	4	4	4/44	4/43

previous has finished. Table 2 illustrates this fact, and suggests the second design principle: **division and square root operations should be avoided in elementary functions**.

Piecewise approximations were important at the time when floating-point operations required dozens of cycles to execute. This approach is still used in LibM libraries, but doesn't perform well on modern architectures: branch instructions which depend on the input data tend to be unpredictable for processors while the cost of branch misprediction in the last 20 years followed an exponential trend and increased from 4 DP FLOPs in the original Pentium (1993) to 240 in Intel Haswell microarchitecture (2013). Therefore, we argue that **algorithms for elementary functions should avoid unpredictable branches**, and in particular piecewise approximations. However, we consider acceptable to use branches for detection of special cases (such as infinite or NaN inputs) because special inputs are rare and thus predictable.

Finally, modern processors are optimized for high throughput as opposed to low latency. Often they can start two floating-point instructions per cycle, but it takes several cycles for each instruction to complete. To maximize the efficiency of floating-point units it is important to have multiple input elements, so that each cycle processor can start computing new elements while previously started instructions complete in background. Thus, we suggest that **high-throughput vector elementary functions should process elements in batches to hide the latency of floating-point instructions** as the last design principle.

4   Building Blocks for Elementary Functions

The analyses in Sect. 3 suggest that elementary functions should be composed from addition and multiplication only. Previous studies in floating-point computations suggest several function approximations which satisfy this requirement.

4.1   Polynomial Approximation

One of the oldest methods to evaluate a function is to approximate is with a polynomial $f(x) = \sum_{i=0}^{n} c_i x^i$ and evaluate the polynomial at the point where we want to evaluate the function. The computationally efficient method to evaluate the polynomial is the Horner scheme, which can be illustrated with the formula

$$\sum_{i=0}^{n} c_i x^i = c_0 + x\left(c_1 + x\left(c_2 + \ldots + x(c_{n-1} + x \cdot c_n)\ldots\right)\right)$$

Computing the value of the polynomial starts with the value $y \leftarrow c_n$, and on each step $k = 1 \ldots n$ the value is updated as

$$y \leftarrow c_{n-k} + x \cdot y$$

After n steps y will have the value of the polynomial evaluated at point x. Each step requires one addition and one multiplication which can be combined into an FMA operation.

Finding a good approximating polynomial for a given function is not trivial. In this paper we use the recently proposed method [3] for constructing optimal polynomial approximations with floating-point coefficients.

4.2 Newton-Raphson Iterations

Newton-Raphson iterations [10,12] represent a high-throughput alternative to hardware division operation. A Newton-Raphson iteration begins with an approximation r_n for a reciprocal of x and produces an improved, more accurate, approximation r_{n+1}. Each iteration approximately doubles the number of correct bits in the approximation. If FMA instruction is available, a Newton-Raphson iteration for reciprocal will compute $\varepsilon_n = 1 - r_n \cdot x$ and $r_{n+1} = r_n + \varepsilon_n \cdot r_n$ and will converge to the true reciprocal with 0.5 ULP accuracy [10] for almost all inputs. On processors without FMA instruction, one can use a simpler Newton-Raphson iteration for a reciprocal: $r_{n+1} = r_n \cdot (2 - r_n \cdot x)$, but due to roundoff errors it will converge only to about 1.5 ULP accuracy. The number of iterations to convergence depends on the accuracy of the initial approximation r_0. SSE extension for x86 ISA provides a fast RCPPS instruction to compute an initial approximation to reciprocal. If the initial value is computed with RCPPS instruction, Newton-Raphson iterations converge in $2 - 3$ iterations, depending on the accuracy of RCPPS instruction, which is implementation-dependent. Newton-Raphson iterations were also suggested for computing square root and reciprocal square root [10,12].

5   Implementation of High-Throughput LibM

Our design of elementary functions uses two code paths for each function: the fast path correctly computes the function value for all but special cases, and the special path correctly computes all cases, but executes more instructions and thus has lower throughput. To check for special values we examine only the high 32 bits of the input double-precision elements. Our algorithms rely on availability of the following operations in the machine instruction set:

1. Floating-point addition and multiplication. Fused multiply-add support is not required, but is beneficial for accuracy.
2. Simple integer and logical SIMD instructions which operate on the same registers as floating-point instructions.

Below we describe the algorithms we developed for computing vector mathematical functions, and demonstrate the measured accuracy and performance of the functions compared to other implementations. The labels GNU and Intel correspond to GNU LibM and Intel LibM correspondingly. AMD and AMD/V represents the scalar and vector version of AMD LibM 3.1, SLEEF, SLEEF/S, and SLEEF/A — scalar, SSE, and AVX versions of the SLEEF 2.80 library. MKL/HA and MKL/LA correspond to high-accuracy and low-accuracy versions in Intel MKL 11.1.0. IPPvm/HA and IPPvm/LA denote high-accuracy and low-accuracy versions in Intel IPP 8.0.1. FDLibM and Cephes labels mark results for FDLibM 5.3 and Cephes 2.7 libraries. The CRLibM label represents results for the latest (April 11, 2011) version of CRLibM library and CRLibM/x87 marks its special version which internally uses double-extended instructions of x87 FPU [5]. In cases where several implementations from the same vendor demonstrate similar accuracy, we present only results with better performance. We present results for three processors with different microarchitectures: Intel Core i7 950 with Nehalem microarchitecture, Intel Xeon E5-2603 on Sandy Bridge microarchitecture, and Piledriver-based AMD FX-6300. FDLibM, CRLibM, Cephes, and SLEEF libraries were compiled with gcc-4.8.1 with all performance optimizations (-O3) and architecture-specific optimizations for each processor (-march=native).

5.1 Log Function

For the `log` function we decompose the argument into two parts: $\frac{\sqrt{2}}{2} \leq t < \sqrt{2}$ and integer s such that $\log x = s \cdot \log 2 + \log t$. Then we approximate $\log t$ on $\left[\frac{\sqrt{2}}{2}, \sqrt{2}\right)$ with a 20-degree polynomial in $(t-1)$: $\log t = (t-1) + \sum_{i=2}^{20} c_i \cdot (t-1)^i$. Finally, we reconstruct and reconstruct the value of $\log x$ as $\log t + s \cdot l2_{low} + s \cdot l2_{high}$ where $l2_{high}$ is a floating-point representation of $\log 2$ with the 11 least significant bits zeroed out, and $l2_{low}$ is a floating-point representation of $(\log 2 - l2_{high})$ (Fig. 2).

Fig. 2. Performance and accuracy of `log` function implementations on $[0.1, 10000.0]$.

5.2 Exp Function

For the range reduction in the `exp` function denote $n = \lfloor x \cdot \log_2 e \rceil$, $t = x - n \cdot \log(2)$ then $-\frac{\log(2)}{2} \leq t \leq \frac{\log(2)}{2}$, and $e^x = 2^n \cdot e^t$. To save accuracy in computing the reduced argument t we use the Cody-Waite range reduction [4] (Fig. 3):

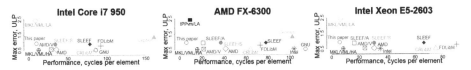

Fig. 3. Performance and accuracy of `exp` function implementations on $[-707.0, 707.0]$.

$t = x - n \cdot l2_{hi} - n \cdot l2_{low}$, where $l2_{hi}$ is the floating-point representation of log 2 with 11 zeroes in the lowest bits, and $l2_{low}$ is the floating-point representation of $(\log 2 - l2_{hi})$. We approximate e^t on $[-\frac{\log(2)}{2}, \leq \frac{\log(2)}{2}]$ by an 11-degree polynomial and reconstruct the final value of exp x as

$$e^x = 2^n + 2^n \cdot \left(t + \sum_{i=2}^{11} c_i \cdot t^i \right).$$

5.3 Trigonometric Functions

For trigonometric functions we perform range reduction over $\frac{\pi}{2}$ using Cody-Waite range reduction algorithm [4] (Fig. 4). First, we compute $n = \lfloor x \cdot \frac{2}{\pi} \rfloor$, and then we compute the reduced argument $t = x - n \cdot \frac{\pi}{2}$ as

$$t = x - n \cdot pio2_{hi} - n \cdot pio2_{me} - n \cdot pio2_{lo}$$

Fig. 4. Performance and accuracy of `sin` function implementations on $[-10000.0, 10000.0]$.

where $pio2_{hi} + pio2_{me} + pio2_{lo}$ together approximate $\frac{\pi}{2}$, the last 15 bits of the $pio2_{hi}$ and $pio2_{me}$ are zeroed out, and the second subtraction is performed in double-double arithmetics. The range reduction is reasonably accurate only if n has at most 15 significant bits. Larger arguments are considered special cases because table-based Payne-Hanek algorithm [14] is required to reduce them. After the reduced arguments t and n are computed, the final values of `sin`, `cos` and `tan` can be expressed in terms of $\sin t$ or $\cos t$ as suggested in [13]:

$n \bmod 4$	$\sin x$	$\cos x$	$\tan x$
0	$\sin t$	$\cos t$	$\sin t / \cos t$
1	$\cos t$	$-\sin t$	$-\cos t / \sin t$
2	$-\sin t$	$-\cos t$	$\sin t / \cos t$
3	$-\cos t$	$\sin t$	$-\cos t / \sin t$

The range for t is symmetric relative to zero $-\frac{\pi}{4} \leq t \leq \frac{\pi}{4}$, which contributes to efficiency of approximations. We approximate $\sin t$ on $[-\frac{\pi}{4}, \frac{\pi}{4}]$ by a 13-degree polynomial in $(t, t^3, t^5, \ldots, t^{13})$ and $\cos t$ by a 14-degree polynomial in $(1, t^2, t^4, \ldots, t^{14})$.

Fig. 5. Performance and accuracy of **tan** function implementations on [−1000.0, 1000.0].

In our implementation of **tan** we conditionally swap the values for sine and cosine of t depending on the value of n, so only one division is needed. We take advantage of the FMA capability on Piledriver and replace division with multiplication by the reciprocal, with an additional correction with two FMA operations. The reciprocal is computed using Newton-Raphson iterations with FMA. The measured accuracy of the **tan** function is same as if division was done via the hardware instruction (Fig. 5).

5.4 Architecture-Specific Optimizations

We optimized the vector mathematical functions described above for three x86 microarchitectures. The functions were implemented in assembly, targeting different instruction sets: SSE4.2 on Nehalem, AVX on Sandy Bridge, and AVX with FMA4 on Piledriver.

All functions compute elements in batches to hide the latency. For non-trigonometric functions the batch size is 8 SIMD registers for Intel processors and 5 AVX registers for AMD Piledriver. Trigonometric functions process elements in batches of 5 SIMD registers on all processors.

6 Performance and Accuracy Evaluation

We measured performance and accuracy on the intervals suggested by Intel VML library reference. The plots in Sect. 5 specify the intervals for each elementary function. We used two different probability distributions to generate input data. For accuracy test we used a probability distribution which assigned equal probability to each floating-point number on the test interval. We choose this probability distribution because it maximizes coverage and is able to generate small floating-point numbers. Since the density of floating-point numbers exponentially increases near zero, it might generate small numbers more often, than in typical use. Thus, for performance tests we used a conventional uniform distribution over the same intervals. Accuracy and performance of all libraries were measured on exactly the same input data.

The accuracy of elementary functions was tested on one million random points, and compared against the 160-bit accurate values computed with MPFR library [7]. The errors are reported in units-in-the-last-place (ULP). ULP is the minimum difference between two nearby floating-point numbers; it can be thought of as a change in the last bit of a result. It is a standard measure or errors in floating-point computation: the theoretical minimum on the maximum error for elementary functions is $\frac{1}{2}$ ULP, and most LibM libraries target the maximum error below 1 ULP. In the experiments our elementary functions demonstrated errors below 2 ULP for non-trigonometric functions and below 3 ULP for trigonometric functions. We suggest that this accuracy level is sufficient for practical use and matches industry standards: Intel C++ and FORTRAN compilers default to 4 ULP accuracy for vectorization of elementary functions.

We measured performance on 2000 random points and recorded the best timings among 1000 measurements to filter out the effect of cache and TLB misses and system interrupts. To minimize systematic error, we disabled Turbo Boost, Hyper Threading, and power-saving capabilities of processors for the length of the test.

7   Conclusion

In this paper we considered the effect the changes in hardware may have on design of high-throughput elementary functions. We analysed which methods for elementary functions are suitable for a high-throughput implementation, and evaluated the hardware-friendly implementation on 3 different hardware platforms. Our results suggest that hardware-friendly methods can successfully compete with traditional elementary function designs in terms of performance, and provide enough accuracy for practical use.

We aim to make the mathematical functions designed for this project available to researchers in applied areas through Yeppp! library (www.yeppp.info). The functions considered in this paper are released in Yeppp! version 1.0.0.

To facilitate further research in this area we released the performance and accuracy measurement tool used in this research as open source software.[1]

Acknowledgements. This work was supported in part by the National Science Foundation (NSF) under NSF CAREER award number 0953100 and the U.S. Dept. of Energy (DOE), Office of Science, Advanced Scientific Computing Research under award DE-FC02-10ER26006/DE-SC0004915. Any opinions, findings and conclusions or recommendations expressed in this material are those of the authors and do not necessarily reflect those of NSF or DOE.

References

1. Bailey, R.: Polar generation of random variates with the t-distribution. Math. Comput. **62**(206), 779–782 (1994)
2. Box, G., Muller, M.: A note on the generation of random normal deviates. Ann. Math. Stat. **29**(2), 610–611 (1958)
3. Brisebarre, N., Chevillard, S.: Efficient polynomial L^∞-approximations. In: 18th IEEE Symposium on Computer Arithmetic, 2007. ARITH'07. pp. 169–176. IEEE (2007)
4. Cody, W., Waite, W.: Software Manual for the Elementary Functions. Prentice-Hall, New Jersey (1980)

---

[1] Repository is hosted on bitbucket.org/MDukhan/hysteria.

5. de Dinechin, F., Defour, D., Lauter, C., et al.: Fast correct rounding of elementary functions in double precision using double-extended arithmetic (2004)

6. Fog, A.: Instruction tables: lists of instruction latencies, throughputs and micro-operation breakdowns for Intel. AMD and VIA CPUs, Technical report (2012)

7. Fousse, L., Hanrot, G., Lefèvre, V., Pélissier, P., Zimmermann, P.: MPFR: a multi-pleprecision binary floating-point library with correct rounding. ACM Trans. Math. Softw. (TOMS) **33**(2), 13 (2007)

8. Gal, S.: An accurate elementary mathematical library for the ieee floating point standard. ACM Trans. Math. Softw. (TOMS) **17**(1), 26–45 (1991)

9. Gentle, J.E.: Random Number Generation and Monte Carlo Methods. Springer, New York (2003)

10. Markstein, P.: IA-64 and Elementary Functions: Speed and Precision. Prentice Hall, New Jersey (2000)

11. Muller, J.-M.: Elementary Functions: Algorithms and Implementation. Birkhauser, Boston (1997)

12. Muller, J.-M., Brisebarre, N., de Dinechin, F., Jeannerod, C.-P., Lefevre, V., Melquiond, G., Revol, N., Stehle, D., Torres, S., Muller, J.-M., Brisebarre, N., Dinechin, F., Jeannerod, C.-P., Lefevre, V., Melquiond, G., Revol, N., Stehle, D., Torres, S.: Handbook of Floating-Point Arithmetic. Birkhauser, Boston (2010)

13. Ng, K.C.: Argument reduction for huge arguments: Good to the last bit (1992)

14. Payne, M., Hanek, R.: Radian reduction for trigonometric functions. ACM SIGNUM Newsl. **18**(1), 19–24 (1983)

15. Press, W., Teukolsky, S., Vetterling, W., Flannery, B.: Numerical Recipes: The Art of Scientific Computing. Cambridge University Press, Cambridge (2007)

16. Shibata, N.: Efficient evaluation methods of elementary functions suitable for SIMD computation. Comput. Sci.-Res. Dev. **25**(1), 25–32 (2010)

17. Tang, P.: Table-lookup algorithms for elementary functions and their error analysis. In: Proceedings of the 10th IEEE Symposium on Computer Arithmetic, 1991. pp. 232–236. IEEE (1991)

18. Williams, C.K.I., Rasmussen, C.E.: Gaussian Processes for Machine Learning, MIT Press, Cambridge (2006)

19. Wong, W.-F., Goto, E.: Fast evaluation of the elementary functions in single precision. IEEE Trans. Comput. **44**(3), 453–457 (1995)

Engineering Nonlinear
Pseudorandom Number Generators

Samuel Neves[(✉)] and Filipe Araujo

CISUC, Department of Informatics Engineering,
University of Coimbra, Coimbra, Portugal
`sneves@dei.uc.pt`, `filipius@uc.pt`

Abstract. In the era of multi and many-core processors, computer simulations increasingly require parallel, small and fast pseudorandom number generation. Although linear generators lend themselves to a simpler evaluation that ensures favorable properties like guaranteed period, they may adversely affect the result of simulations or be quite large. Conversely, nonlinear generators may provide apparently random sequences, but are either very slow or difficult to analyze regarding their period. This is the case of our previous functions, Tyche and Tyche-i. Despite being among the fastest in their class and having average periods of 2^{127}, they may contain smaller cycles of arbitrary size. To overcome this limitation, in this paper we explore different forms of counters impacting either the state or the speed of the generator. We also introduce two number-theoretic generators that use 2×127 bits for periods of 2^{116} and 2^{125} and low to moderate computational costs. We experimentally demonstrate the efficiency of our new generators and observe that they exchange speed for period guarantees in a tradeoff that seems widespread in state-of-the-art random number generators.

Keywords: Nonlinear pseudorandom generator · Elliptic curves · Elliptic curve linear congruential generator

1 Introduction

Pseudorandom number generators (PRNGs) attempt to generate sequences that have similar properties to that of genuinely random sequences, while simultaneously being deterministic and reproducible. Most commonly-used generators are based on number-theoretic constructions: linear congruential generators (LCGs) work over the ring of integers modulo m, linear feedback shift registers from the ring of polynomials modulo p, and so on. Working in such mathematical structures makes it easier to reason about period, equidistribution, and other desirable properties of PRNGs. However, they are often *linear*, which adversely affects real-world simulation results, and speed, due to the required arithmetic.

Nonlinear generators can be divided in two main camps. On one hand, we have the number-theoretic generators, which rely on number theory to demonstrate their properties, such as period and distribution. On the other hand,

R. Wyrzykowski et al. (Eds.): PPAM 2013, Part I, LNCS 8384, pp. 96–105, 2014.
DOI: 10.1007/978-3-642-55224-3_10, © Springer-Verlag Berlin Heidelberg 2014

there are the cryptography-based generators, which rely on the avalanche effect of cryptographic primitives to mask any relationship from one state to the next.

Recently, non-number-theoretic, nonlinear generators have been proposed with some success [20,24]. These often try to "mix" the bits of the PRNG's state as best as possible across iterations, thus obtaining quality random sequences. Despite passing statistical tests, these generators tend to rely on heuristic assumptions, making it impossible to guarantee their period or distribution properties.

Some generators, most notably the Mersenne Twister [17], use very large states to thwart some of the drawbacks of their linearity. The SIMD-oriented version of the Mersenne Twister [23] is indeed quite faster, but still uses a state of similar size as the original MT, requiring hundreds of 128-bit words. Smaller-state generators, like Xorshift [16], have small state but struggle to provide adequate statistical quality [22].

Meanwhile, computer architecture is shifting. It is becoming harder to put more transistors within the same chip area, and manufacturers are often forced to choose between memory and execution units. For now, general-purpose chips, such as x86-64 and ARM, tend to favor fast memory (cache), while special-purpose chips (e.g., GPUs) lean toward more computational power. The current trend appears to be towards more cores, and therefore one can expect less memory per running thread. Future PRNGs should therefore save memory and share no state between threads, to avoid contention. By decreasing order of relevance, modern PRNGs should have the following properties:

High Quality: Any proposed generator must pass stringent statistical tests, such as TestU01's "Big Crush" battery [13].

Large Period: While opinions vary about the minimum acceptable period, we aim for a minimum of 2^{128}.

Small State: The size of the state should not be significantly larger than the binary logarithm of its period, which we assume to be roughly 2^{128}.

Fast: The generator must be as quick as possible, because it is often in the critical path of simulations.

Linearity enables design of fast generators with provable properties, such as period, statistical distribution, and small state [14,26]. However, linear congruential generators often possess a lattice structure that can skew the results of a simulation [9,15]. To run away from linear generators' drawbacks, we add another requirement to a good generator: *nonlinearity*. This means that a generator should not be representable as an affine transformation in $\mathbb{F}_2$, $\mathbb{Z}_n$, or any such ring. Since provable properties such as period often affect the speed of PRNGs, in this paper we propose a number of different algorithms, including elliptic curve generators, to explore the speed/period tradeoff. We believe that our generators provide excellent state-of-the-art speeds for their properties.

Our contribution in this paper is twofold: firstly, we present two nonlinear number-theoretic generators based on elliptic curves over a prime field. Secondly, we introduce two new tweaks to the Tyche generator [20], representing two different tradeoffs between period guarantee and speed. We analyze and discuss the results in Sect. 3.

2   Small Nonlinear Generators

Generators based on number-theoretic structures are among the most common in the literature, and include the linear congruential generator, the linear feedback shift register and its many variants, the inversive congruential generator, Blum-Blum-Shub [5], and others. Generally, generators in this category that are linear are reasonably fast, while nonlinear ones tend to underperform and are less commonly used in real applications.

The most general construction for a generator is of the form

$$S_i = f(S_{i-1})$$
$$x_i = g(S_i)$$

The functions f and g are known as the *transition* and *output* functions. This construction allows for much freedom in the design process, and encompasses the vast majority of the existing generators. For example, in the common linear congruential generator [14], f is the $S_i = aS_{i-1} + b \bmod m$ recurrence, while g is usually a truncation of the current value. In this case, the computational effort is skewed almost completely towards f. Conversely, a generator created from, e.g., a block cipher E_K (e.g., AES-128) in counter mode [8] has $S_i = S_{i-1} + 1$, and $x_i = E_K(S_i)$. This case is computationally skewed towards the output function.

There are advantages and disadvantages on each of these extremes. It is easier to find a simple function that achieves full period—for example $f(x) = x + 1$— and use the output function to produce random-looking outputs, than it is to find a random-looking transition function that also has large guaranteed period. However, the latter case is often computationally better than the former; simple functions often do not have good statistical properties, and the output function has to do a large amount of work to achieve enough bit diffusion.

To achieve full period in the least possible space, the transition function must be invertible, i.e., a permutation. Non-invertible functions have an expected period of $2^{n/2}$ for n bits of state ; such functions therefore need $2n$ bits to achieve 2^n period, which breaks our space-efficiency requirement. On the other hand, there are not many restrictions on the output function.

2.1   Elliptic Curves

Elliptic curves have found many uses in cryptography [10,18], and are the current leading candidate for public-key key-exchanges and digital signatures. The set of solutions (x, y) to the Weierstrass equation over some field F

$$E(F) : y^2 + a_1xy + a_3y = x^3 + a_2x^2 + a_4x + a_6, \quad x, y, a_i \in F \qquad (1)$$

together with the "point at infinity" P_∞ and an point addition operation $+$ form an Abelian group. The order of this group, $\#E$, follows the famous Hasse bound:

$$\#F + 1 - 2\sqrt{\#F} < \#E(F) < \#F + 1 + 2\sqrt{\#F} \qquad (2)$$

These two facts allow one to define a linear congruential-style generator of the form

$$P_i = G + P_{i-1} \tag{3}$$

which will then have an order equal to that of the generator G. Apart from having guaranteed period, such generators are also known to have desirable statistical properties [6,11], have small state (i.e., one point), and can skip ahead quickly by noticing that

$$P_n = nG + P_0 \tag{4}$$

where nG means repeated addition and can be done in $O(\log n)$ additions using standard techniques. For a generator of this kind to be fast, two things are required: (i) fast arithmetic over F; (ii) fast point addition.

Despite all elliptic curves being representable in affine coordinates satisfying Eq. 1, several other representations exist that present certain performance tradeoffs: Jacobian projective coordinates [7], Montgomery curves [19], Jacobi quartics [7], Hessian curves [7], Edwards curves [4], Twisted Edwards curves [2], and others[1]. Of these, there are two main representations that are of interest to us, when it comes to cheap point addition: Montgomery curves and twisted Edwards curves.

Algorithm 2.1. MONTGOMERYDOUBLE(X, Z)

---

comment: K is Constant for a given curve

$K \leftarrow (a_2 + 2)/4$
$A \leftarrow (X + Z)^2$
$B \leftarrow (X - Z)^2$
$C \leftarrow A - B$
return $(AB, C(B + KC))$

Twisted Edwards curves have recently been used to break speed records in elliptic curve scalar multiplication [3]. However, these curves require rather large point representations (4 field elements) to achieve the best speeds, and even then have *relatively* high addition and doubling costs. In comparison, Montgomery curves only require two field elements but do not support full addition, instead requiring *differential addition* to perform scalar multiplication. Point doubling of Montgomery curves, listed in Algorithm 2.1, has the best operation count, at 4 multiplications, 4 additions, and 1 multiplication by a constant. This fact motivates our new generator mode, vaguely resembling the Blum-Blum-Shub [5] generator:

$$P_i = P_{i-1} + P_{i-1} = 2P_{i-1}. \tag{5}$$

---

[1] Most known efficient formulas for various curves and point representations are found in the Explicit-Formulas Database: http://hyperelliptic.org/EFD/index.html

This new recurrence has two advantages: it only requires point doubling, and requires less fixed constants by removing the need for G. This recurrence computes the sequence $2P, 4P, \ldots, 2^i P$ and is therefore also possible to skip ahead by computing

$$P_n = (2^n \bmod \operatorname{ord}(P_0))P_0 \tag{6}$$

and the period of this sequence is given by $\operatorname{ord}(2) \bmod \operatorname{ord}(P_0)$. The order of this generator is thus dependent not only on the number of points in the elliptic curve, but also on the order of 2 in the ring of integers modulo the order of the initial point.

2.2   The M127 Generator

Our first concrete proposal of a generator that achieves near 2^{128} period consists of the following Montgomery curve over the integers modulo the Mersenne prime $2^{127} - 1$:

$$y^2 = x^3 + 131074x^2 + x. \tag{7}$$

This curve has order $4 \cdot p_1$, $p_1 = 42535295865117307934202406649106774733^2$. The constant 131074 was chosen so that $32769 = (131074 + 2)/4$ has Hamming weight 2 and multiplication can be performed via one shift and one addition: $32769x = x \ll 15x$.

Points are represented in the traditional Montgomery curve fashion (X, Z). This representation only contains information about the X coordinate of a given point; the Y coordinate is ignored, and thus points are in reality an equivalence class of (X, Y) and $(X, -Y)$. The (X, Z) representation avoids costly inversions by storing X as a fraction, i.e., $X = X/Z$.

The underlying field, $\mathbb{F}_{2^{127}-1}$, was chosen to minimize the cost of modular reduction. Modular reduction by $2^n - 1$ is known to be achievable by the divisionless expression

$$x \bmod (2^n - 1) \equiv (x \bmod 2^n + \lfloor x/2^n \rfloor) \bmod (2^n - 1). \tag{8}$$

To select a starting point from a (say) 128-bit seed s, one can compute $P_0 = s(2, 1)$. Note that $(2, 1)$ has order p_1, and thus there is no chance that an unlucky seed will get stuck in a small order point forever. The only seed that must be avoided is, of course, p_1 itself, since this would result in the point-at-infinity as the starting point.

This generator is appropriate for architectures where integer multiplication is fast. There are 4 $\mathbb{F}_{2^{127}-1}$ multiplications per iteration, each of which requires 4 $64 \times 64 \to 128$-bit multiplications. While there is plenty of exploitable parallelism in both field and curve arithmetic to attenuate the effect of high-latency multiplication instructions, it may not be enough.

---

[2] The order of Montgomery curves is always a multiple of 4.

2.3   The M31x4 Generator

The curve from the previous section relied on fast arithmetic over the under-lying field. While this can be reasonably expected from large general purpose processors, it is often the case that smaller or specialized processors are unable to perform multiple-precision arithmetic very quickly. Furthermore, small inte-ger multiplication has quadratic complexity, and for CPUs with small register sizes that complexity grows quickly. For this reason we propose the following generator, which only requires 32-bit arithmetic.

This generator uses not one, but 4 Montgomery curves in parallel over the Mersenne prime $2^{31} - 1$:

$$y^2 = x^3 + v_i x^2 + x \tag{9}$$

where $v = \{904572996, 1467357171, 1043599384, 1244578513\}$. The 4 curves have orders of respectively $4 \cdot 536871259$, $4 \cdot 536872363$, $4 \cdot 536872907$, and $4 \cdot 536873203$. At the end of each iteration, the generator outputs the *combination* of the x-coordinates of the points:

$$g(x_0, x_1, x_2, x_3) = x_0 \oplus (x_1 \lll 7) \oplus (x_2 \lll 11) \oplus (x_3 \lll 29) \tag{10}$$

where $\oplus$ means XOR and $\lll$ means rotation towards the most significant bits.

The 4 v_i parameters were chosen to maximize the order of 2 in the respective fields. This makes the overall period of this generator $536871258 \cdot 536872362 \cdot 536872906 \cdot 536873202 \approx 2^{116}$.

This generator allows many different implementation approaches, and is suitable for large CPUs (where it can be implemented using general purpose instruction or SIMD) and GPUs alike.

2.4   Tweaking Tyche with a Counter

The Tyche generator [20] is a generator based on the ChaCha core permuta-tion [1], which works in a mode similar to OFB mode in block ciphers. We refer to [20] for the complete description of Tyche. While it shows great performance across many architectures, due to its use of simple 32-bit instructions, it has several drawbacks:

No Provable Period: Treating the core permutation MIX as a random permu-tation allows us to estimate the expected period of a sequence to be roughly 2^{127}. However, this says nothing about the actual cycle structure of Tyche, and unlikely as it may be, there may be some hidden pitfalls in this generator.

No Random Access: While Tyche provides some higher level parallelism sup-port by defining different stream starting points, it is impossible to jump ahead inside a single stream. This may be inconvenient in some situations.

In this section we propose a tweak to Tyche [20], named Tyche-CTR-R, to change the mode of operation of Tyche. Once the initial state is set up, the least significant 64 bits are used as a counter incremented by the odd constant

5871781008561895865^3, while the most significant 64 bits remain constant, serving as identifier (a *nonce*) of the current stream. Then, this state is processed R times by the MIX function, and the least significant word is returned. Algorithm 2.2 describes Tyche-CTR-R.

Algorithm 2.2. TYCHE-CTR-$R(a, b, c, d)$

$(b, a) \leftarrow (a + 2^{32}b) + 5871781008561895865$
$(a', b', c', d') \leftarrow (a, b, c, d)$
for $i \leftarrow 0$ **to** R
  do $(a', b', c', d') \leftarrow MIX(a', b', c', d')$
return (a')

Our experiments have suggested that 5 rounds, i.e., Tyche-CTR-5, are sufficient to achieve enough diffusion to pass known statistical tests. It is easy to see that the period of Tyche-CTR-R is 2^{64} and it is easy to jump ahead arbitrarily within a stream, by adding an appropriate multiple of the constant used in Algorithm 2.2. This also implies the generator is massively parallelizable. One should notice that Tyche-CTR-5 provides 2^{64} distinct streams with a guaranteed period of 2^{64}, and still enables further tweaks to the lengths of the counter and nonce.

2.5 Tyche as a Counter-Dependent Generator

The tweak presented in the previous section was fairly aggressive: instead of one MIX call per iteration, we now require $R \geq 5$ calls to achieve the same effect. This is a massive slowdown, even though it does enable some desirable properties, and the higher latency may be hidden by computing several values in parallel.

We propose in this section another tradeoff: a 2^{32} guaranteed minimum period, 2^{159} average period, and 160 bits of state. The approach we follow is known as *counter-dependent* generators [25], and is pictured in Algorithm 2.3.

Algorithm 2.3. TYCHE-CD-32(a, b, c, d, e)

$e \leftarrow e + (e^2 \vee 5) \bmod 2^{32}$
$(a, b, c, d) \leftarrow MIX(a, b, c, d)$
return $(b + e)$

---
^{3} Counters that add an odd constant different from 1 are often known as Weyl generators.

Table 1. Timing and period information of Sect. 2 generators, a 128-bit LCG, and a 128-bit EC-RNG.

Generator	State bits	Cycles/Iteration	Average period	Min period	Jump ahead
LCG-128	127	32	$2^{127} - 1$	$2^{127} - 1$	Yes
EC-LCG	3×127	238	$\approx 2^{127}$	$\approx 2^{127}$	Yes
XORWOW [21]	192	7	$2^{192} - 2^{32}$	$2^{192} - 2^{32}$	Yes
M127	2×127	96	$\approx 2^{125}$	$\approx 2^{125}$	Yes
M31x4	2×127	38	$\approx 2^{116}$	$\approx 2^{116}$	Yes
Tyche [20]	128	12	$\approx 2^{127}$	1	No
Tyche-i [20]	128	6	$\approx 2^{127}$	1	No
Tyche-CD-32	160	12	$\approx 2^{159}$	2^{32}	No
Tyche-CTR-5	128	44	2^{64}	2^{64}	Yes

Since this tweak does not enable random stream access, we opted to use the T-function $x + (x^2 \vee 5) \pmod{2^n}$, proven by Klimov and Shamir to be invertible and single-cycled [12]. This function is executed in parallel with MIX, and is not expected to significantly slow down the generator. Additionally, the greater complexity of this function provides some more diffusion than a simpler counter.

3   Results and Discussion

The generators described in Sect. 2 have passed the TestU01 [13] "BigCrush" battery of tests. Therefore, to evaluate their performance, we include timings for a 128-bit LCG with modulus $2^{127} - 1$ and multiplier 43, and an EC-LCG using projective coordinates over a Weierstrass curve of prime order over the same prime as M127, $2^{127} - 1$. Additionally, we also compare the original Tyche algorithm against the variants proposed in Sects. 2.4 and 2.5.

Table 1 shows the number of cycles per iteration of the aforementioned generators. The timings were obtained on an Intel Core-i7 2630QM "Sandy Bridge" processor, with Turbo Boost and hyperthreading disabled.

It is apparent from Table 1 that M127, despite much optimization effort, is quite far from the performance of a similar-period LCG. Nevertheless, it is more than twice as fast as an EC-LCG of similar period, most of this stemming from the faster Montgomery doubling operation. The speed of the M31x4 generator is on the same order of magnitude as the LCG, due to the lack of large integer arithmetic in favor of parallel small elliptic curves and a simple combiner.

One pattern that emerges is that the number-theoretic generators do not seem to have significant advantages over the counter-based generator. The counter-based generator Tyche-CTR-5 not only is faster than the nonlinear number-theoretic generators, but also contains every feature found in the latter. Note that if storage space is not an issue, it is possible to, e.g., run 4 or 8 parallel instances of the Tyche-CTR in SIMD registers, resulting in a significant speedup. This would be impossible in recursive generators.

Finally the Tyche-CD-32 generator strikes a balance between features and speed. It is as fast as Tyche, hiding the extra instructions in between Tyche's critical path, guarantees a reasonable minimum period of 2^{32}, and does not require vectorization tricks to be extremely quick.

Ultimately, this is the tradeoff generators seem to make: linear generators can be fast, small, and provably periodic, but suffer in statistical properties; nonlinear generators can be fast and small, but their period will not be provable and are purely sequential; nonlinear number theoretic generators will have provable period, but will suffer in speed due to the heavy arithmetic.

Future work involves investigating additional number-theoretic constructs to obtain faster generators; perhaps elliptic curves are not the optimal choice, despite their popularity in cryptography.

Acknowledgments. This work has been supported by the project CMU-PT/ RNQ/0015/2009, TRONE — Trustworthy and Resilient Operations in a Network Environment.

References

1. Bernstein, D.J.: ChaCha, a variant of Salsa20. In: Workshop Record of SASC 2008: The State of the Art of Stream Ciphers, January 2008
2. Bernstein, D.J., Birkner, P., Joye, M., Lange, T., Peters, Ch.: Twisted edwards curves. In: Vaudenay, S. (ed.) AFRICACRYPT 2008. LNCS, vol. 5023, pp. 389–405. Springer, Heidelberg (2008)
3. Bernstein, D.J., Lange, T.: Analysis and optimization of elliptic-curve single-scalar multiplication. IACR Cryptology ePrint Archive 2007, 455 (2007)
4. Bernstein, D.J., Lange, T.: Faster addition and doubling on elliptic curves. In: Kurosawa, K. (ed.) ASIACRYPT 2007. LNCS, vol. 4833, pp. 29–50. Springer, Heidelberg (2007)
5. Blum, L., Blum, M., Shub, M.: A simple unpredictable pseudo-random number generator. SIAM J. Comput. **15**(2), 364–383 (1986)
6. Chen, Z., Gomez, D., Pirsic, G.: On lattice profile of the elliptic curve linear congruential generators. Period. Math. Hung. **68**, 1–12 (2012)
7. Chudnovsky, D.V., Chudnovsky, G.V.: Sequences of numbers generated by addition in formal groups and new primality and factorization tests. Adv. Appl. Math. **7**(4), 385–434 (1986)
8. Dworkin, M.: Recommendation for block cipher modes of operation: methods and techniques. Special Publication 800–38A, National Institute of Standards and Technology, Gaithersburg, MD 20899–8930 (2001)
9. Ferrenberg, A.M., Landau, D.P., Wong, Y.J.: Monte Carlo simulations: hidden errors from "good" random number generators. Phys. Rev. Lett. **69**, 3382–3384 (1992)
10. Lenstra, H.W. Jr.: Elliptic curves and number-theoretic algorithms. In: Gleason, A.M. (ed.) Proceedings of the International Congress of Mathematicians, vol. 1, pp. 99–120. American Mathematical Society, Providence (1987)
11. Hess, F., Shparlinski, I.E.: On the linear complexity and multidimensional distribution of congruential generators over elliptic curves. Des. Codes Crypt. **35**(1), 111–117 (2005). http://dx.doi.org/10.1007/s10623-003-6153-0
12. Klimov, A., Shamir, A.: Cryptographic applications of T-Functions. In: Matsui, M., Zuccherato, R.J. (eds.) SAC 2003. LNCS, vol. 3006, pp. 248–261. Springer, Heidelberg (2004)

13. L'Ecuyer, P., Simard, R.: TestU01: A C library for empirical testing of random number generators. ACM Trans. Math. Softw. **33**(4), 22 (2007)
14. Lehmer, D.: Mathematical methods in large-scale computing units. In: Proceedings of the 2nd Symposium on Large-Scale Digital Calculating Machinery, pp. 141–146. Harvard University Press, Cambridge, Massachusetts (1949)
15. Marsaglia, G.: Random numbers fall mainly in the planes. PNAS **61**(1), 25–28 (1968). http://dx.doi.org/10.1073/pnas.61.1.25
16. Marsaglia, G.: Xorshift RNGs. J. Stat. Softw. **8**(14), 1–6 (2003)
17. Matsumoto, M., Nishimura, T.: Mersenne twister: a 623-dimensionally equidistributed uniform pseudo-random number generator. ACM Trans. Model. Comput. Simul. **8**(1), 3–30 (1998)
18. Miller, V.S.: Use of elliptic curves in cryptography. In: Williams, H.C. (ed.) CRYPTO 1985. LNCS, vol. 218, pp. 417–426. Springer, Heidelberg (1986)
19. Montgomery, P.L.: Speeding the Pollard and elliptic curve methods of factorization. Math. Comput. **48**, 243–264 (1987)
20. Neves, S., Araujo, F.: Fast and small nonlinear pseudorandom number generators for computer simulation. In: Dongarra, J., Karczewski, K., Waśniewski, J., Wyrzykowski, R. (eds.) PPAM 2011, Part I. LNCS, vol. 7203, pp. 92–101. Springer, Heidelberg (2012)
21. NVIDIA Corporation: CURAND Library (July 2013), http://docs.nvidia.com/cuda/curand/
22. Panneton, F., L'ecuyer, P.: On the Xorshift random number generators. ACM Trans. Model. Comput. Simul. **15**(4), 346–361 (2005)
23. Saito, M., Matsumoto, M.: SIMD-oriented fast Mersenne Twister: a 128-bit pseudorandom number generator. In: Keller, A., Heinrich, S., Niederreiter, H. (eds.) Monte Carlo and Quasi-Monte Carlo Methods 2006, pp. 607–622. Springer, Berlin (2008)
24. Salmon, J.K., Moraes, M.A., Dror, R.O., Shaw, D.E.: Parallel random numbers: as easy as 1, 2, 3. In: Lathrop, S., Costa, J., Kramer, W. (eds.) SC, p. 16. ACM (2011). http://doi.acm.org/10.1145/2063384.2063405
25. Shamir, A., Tsaban, B.: Guaranteeing the diversity of number generators. Inf. Comput. **171**(2), 350–363 (2002)
26. Tausworthe, R.C.: Random numbers generated by linear recurrence modulo two. Math. Comput. **19**, 201–209 (1965)

Extending the Generalized Fermat Prime Number Search Beyond One Million Digits Using GPUs

Iain Bethune[1]([⊠]) and Michael Goetz[2]

[1] EPCC, The University of Edinburgh, James Clerk Maxwell Building,
The King's Buildings, Mayfield Road, Edinburgh EH9 3JZ, UK
ibethune@epcc.ed.ac.uk
http://www.epcc.ed.ac.uk/~ibethune
[2] PrimeGrid, Ardsley, New York, USA
mgoetz@primegrid.com
http://www.primegrid.com

Abstract. Great strides have been made in recent years in the search for ever larger prime Generalized Fermat Numbers (GFN). We briefly review the history of the GFN prime search, and describe new implementations of the 'Genefer' software (now available as open source) using CUDA and optimised CPU assembler which have underpinned this unprecedented progress. The results of the ongoing search are used to extend Gallot and Dubner's published tables comparing the theoretical predictions with actual distributions of primes, and we report on recent discoveries of GFN primes with over one million digits.

Keywords: Generalized Fermat Numbers · Primality testing · Volunteer computing · Computational mathematics · GPU computing · CUDA

1 Background

Computational number theory and in particular the search for large prime numbers has grown steadily in popularity over the last two decades. Led by projects like the Great Internet Mersenne Prime Search (GIMPS), tens of thousands of volunteers now contribute computer time in support of projects such as "Seventeen or Bust" - attempting to solve the Sierpiński Problem [9] - and searches for primes of particular types including Proth ($k \cdot 2^n + 1, k < 2^n$), Riesel ($k \cdot 2^n - 1, k < 2^n$), Cullen ($n \cdot 2^n + 1$) and Woodall ($n \cdot 2^n - 1$) primes. Many of these prime searches are coordinated by the PrimeGrid [10] project, which uses the Berkeley Open Infrastructure for Network Computing (BOINC) [1] to allow client computers to download, process, and return work units consisting of primality tests or sieving.

The Generalized Fermat Numbers (GFN) are defined as having the form $F_{b,n} = b^{2^n} + 1$. Starting in 2000 Yves Gallot led a very active and well-organised

R. Wyrzykowski et al. (Eds.): PPAM 2013, Part I, LNCS 8384, pp. 106–113, 2014.
DOI: 10.1007/978-3-642-55224-3_11, © Springer-Verlag Berlin Heidelberg 2014

distributed search for GFN primes using his 'Proth' and 'Genefer' programs. Many GFN primes were found with over 100,000 digits and preliminary results were published in a seminal paper by Gallot and Dubner [5] in 2002. However, by 2004 the project drew to a gradual conclusion with the exception of a few individual searchers.

In 2009 PrimeGrid restarted the GFN search beginning from where the previous effort left off, searching $n \geq 15$. Due in part to increased CPU power, a very large user base and improvements to Gallot's software, exceptional progress has been made to date, which we report hereafter.

2   Software for GFN Searching

2.1   PRP Testing

During the early years of the GFN search Gallot's original C program 'Genefer' was used to perform probable primality (PRP) tests on GFNs. An overview of the implementation of the PRP test employed is described by Gallot and Dubner [5] and details of FFT multiplication modulo Fermat numbers are given by Crandall and Fagin [4]. The program was later modified by Gallot and David Underbakke, rewriting the critical numerical routines (FFT and modular reduction) using Intel assembly language. One variant 'Genefer80' made use of the Intel x87 instruction set, which allows use of the extended 80-bit precision of the x87 Floating Point Unit compared to the standard 64-bit 'double precision' of the x86 FPU. By taking care to ensure all intermediate values are stored at this higher precision, much larger values of b can be tested for a given n before encountering round-off errors in the conversion from floating-point back into integer representation. Although slightly slower than the C implementation, the ability to test larger b values has been invaluable as the search for $n \leq 16$ has now passed the b limit of 'Genefer' (see Table 2 for the current search limits). Similarly, 'Genefx64' uses the SSE2 vector instruction set, allowing modern CPUs to compute the FFT at nearly twice the speed of 'Genefer' with similar accuracy. Since all Intel 64-bit processors support SSE2, the original C implementation is now essentially obsolete, only used by the few remaining 32-bit processors participating in the search. The speeds and b limits of each of these variants are compared in Table 1.

When PrimeGrid restarted the GFN prime search in 2009, the Genefer applications were extended with a checkpoint/restart capability and integrated with Mark Rodenkirch's PRPNet software which coordinated the distribution of PRP tests to client computers and the recording and reporting of results. Initially, the 'Genefer80' and 'Genefx64' applications were only available for the MS Windows platform, and testing began for $n = 15, 16, 18, 19$ ($n = 17$ continues to be searched independently by participants in the original GFN prime search).

The authors' contributions to the development of these programs began with the porting of the 'Genefer80' and 'Genefx64' assembly codes from Intel-syntax

Table 1. b limits and performance (ms per multiplication) of Genefer variants for selected n. Tests performed on Intel Core 2 Quad 2.4 GHz running Window 7 Pro 64 bit, with an Nvidia GTX460 1350 MHz (Driver 285.86).

| | Genefer80 | | Genefer | | Genefx64 | | GeneferCUDA | |
n	b limit	t (ms)	b limit	t (ms)	b limit	t (ms)	b limit	t (ms)
15	67,210,000	2.34	1,630,000	1.67	1,575,000	0.912	1,840,000	0.212
17	45,450,000	11.2	1,095,000	7.54	1,060,000	4.05	1,270,000	0.601
19	30,020,000	57.4	695,000	35.3	735,000	19.3	815,000	1.98
21	20,250,000	277	490,000	175	515,000	102	580,000	8.23
22	-	-	-	-	-	-	480,000	16.5

to AT&T/GNU syntax, allowing these to be compiled using the GNU GCC Compiler and made available for Mac OS X and Linux platforms. At the same time, an initial port of 'Genefer' was developed by Shoichiro Yamada using Nvidia's 'Compute Unified Device Architecture' (CUDA) programming model, and subsequently optimised and extended by the authors. For a comprehensive overview of CUDA and Graphics Processing Units (GPUs), we refer the reader to Nickolls *et al.* [8]. For our purposes it suffices to say that many modern computers contains GPUs providing performance of 100 to 1000 GFLOPS (billion floating point operations per second), compared with around 10 GFLOPS from a typical CPU core. The FFT operation in 'GeneferCUDA' is performed by Nvidia's CUFFT library, and in order to minimise the cost of repeatedly transferring data to and from the GPU, the remaining steps in the calculation loop have been ported to CUDA kernels able to run on the GPU. As shown in Table 1 this results in significant speedups (4.3x faster than Genefx64 for $n = 15$, and 9.7x for $n = 19$). More importantly, however, the advent of 'GeneferCUDA' has allowed larger values of n to be tackled that would take prohibitively long on a CPU. For example, a typical test at $n = 22$ that takes around a week on a GPU would take over 3 months on a CPU! Testing of GFN for $n = 22$ has already begun, and results of the search so far are reported in Sect. 3. The introduction of the CUDA code in the $n = 19$ search has increased the rate of progress so much that it we have also been able to start searching the $n = 20$ range.

Most recently, in early 2012, the authors added support for BOINC directly in our code, allowing the GFN prime search to be offered via the PrimeGrid BOINC project rather than requiring participants to install the PRPNet client. All the 'Genefer' variants have been unified into a single program, allowing a single consistent interface independent of the actual calculation method employed. In addition, this will make the development of any additional FFT implementations much easier, and will facilitate future maintainability of the software. Finally, we have made our programs freely available in both source and binary forms from https://www.assembla.com/spaces/genefer, which we believe is a significant contribution to the community.

Table 2. Contiguous search limits and largest known primes for each n.

n	b limit (Sep 2013)	Largest prime	Date	Decimal digits
15	6,961,316	$1554729^{32768} + 1$	Jul 2011	235,657
16	3,196,780	$1950221^{65536} + 1$	Jan 2005	477,763
17	1,166,000	$1372930^{131072} + 1$	Sep 2003	804,474
18	1,024,466	$773620^{262144} + 1$	Feb 2012	1,528,413
19	750,244	$475856^{524288} + 1$	Aug 2012	2,976,663
20	201,460	-	-	-
22	10,428	-	-	-

2.2 Sieving

Despite the excellent performance obtained with recent versions of 'Genefer', in common with other prime searches to efficiently search a large range of candidates (here the b values to be tested for each n in $F_{b,n}$) we employ a sieve to remove candidates which have 'small' prime factors. The sieving algorithm used was developed by Phil Carmody [3]. Deciding exactly when to stop sieving - the *depth* of the sieve - is a function of the relative speed at which the sieving program can find factors compared to the rate at which the primality testing program can test the remaining candidates. Initially, we carried out sieving using the 'AthGFNSv' program developed by Underbakke, Gallot and Carmody. However, in May 2012 a CUDA sieving program 'GFNSvCUDA' was implemented by Anand Nair, which was dramatically faster than the existing CPU sieve. For example, at $n = 19$, several years of sieving on CPUs had reached a depth of 3070P (i.e. trial factors up to 3.07×10^{15} had been checked). Within the first 6 months of sieving on GPUs, a depth of 19100P has been reached (including a re-check of the original 3070P), and the sieving effort stopped as it is now more efficient to PRP test the remaining candidates directly.

3 Distribution of Large GFN Primes

To date, PrimeGrid is actively searching $15 \leq n \leq 22$, with the exception of $n = 17$ which is reserved by independent searchers. The $n = 21$ case is still in the process of sieving, but good progress has been made in primality testing the other n, which we summarise in Table 2. Note that for $n = 15, 16, 17$ the largest known GFN prime is significantly beyond the current b reported. This represents the fact that while every b below the reported values is known to have been tested, individual searchers have tested small ranges far in advance of the current organised search limit.

In their 2002 paper [5] Gallot and Dubner presented a method for calculating the expected number of GFN primes for each n up to a particular limit of b. They showed excellent agreement between the predicted and the actual numbers of primes found for $n \leq 12, b \leq 10^6$ and $n = 13, 14, b \leq 10^4$ based on the then current search limits. We have calculated the expected numbers of GFN primes

Table 3. Comparison of predicted and actual number of GFN primes for $13 \leq n \leq 22$ up to current search limits

2^n	$b \leq 10^5$			$b \leq 10^6$			b	Search limit		
	Est.	Act.	Err.	Est.	Act.	Err.		Est.	Act.	Err.
8192	10	3	-2.2	81	74	-0.8	13,000,000	764	730	-1.2
16384	5	1	-1.7	38	33	-0.9	4,560,000	156	137	-1.5
32768	2	1	-0.5	14	16	0.6	6,961,000	84	91	0.8
65536	2	1	-0.5	13	14	0.2	3,196,000	35	38	0.5
131072	1	1	0.2	7	5	-0.6	1,166,000	8	7	-0.4
262144	0	2	2.2	4	7	1.5	1,024,000	4	7	1.5
524288	0	1	1.6	2	-	-	750,000	2	4	2.0
1048576	0	-	-	1	-	-	201,460	0	0	0.0
$\vdots$	$\vdots$			$\vdots$			$\vdots$			$\vdots$
4194304	0	-	-	0	-	-	10,428	0	0	0.0

for each n up to our new search limits using Gallot's method and compared with the actual numbers of primes found to date in Table 3. For ease of comparison with Gallot and Dubner's tables, we also report the difference between estimated and actual numbers of primes in terms of standard deviations. In addition to PrimeGrid's database, the Largest Known Primes Database [2] was used to provide data for smaller b and n values.

We observe that while most of the findings are broadly in line with the predicted values (indeed, over 50 % of the errors are less than one standard deviation), there appear to be significant excesses of GFN primes for $n = 18, 19$, particularly for small b. Unfortunately, with the current b limits, the number of primes is too low to assess the probability that the predicted distribution of primes is correct via the Chi Squared Test. Nevertheless, it is still possible to check the validity of the prediction, since if Gallot's expression for the number of GFN primes for given b, n was too small then we should see that more candidates remain after sieving than expected.

Dubner and Keller [6] showed that a given prime $p = k \cdot 2^{n+1} + 1$ divides $F_{b,n}$ with probability $2^n/p$ (averaged over all b). Thus if we sieve R GFNs with all potential divisors $p < p_{max}$, the number of expected candidates is

$$\prod_{p < p_{max}} \left(1 - \frac{2^n}{p}\right) \cdot R, \quad p \equiv 1 \bmod 2^{n+1} \tag{1}$$

Applying Mertens' 3rd theorem [7] we have

$$\prod_{p < p_{max}} \left(1 - \frac{2^n}{p}\right) = \frac{2C_n}{e^\gamma \log(p_{max})} \tag{2}$$

Table 4. Expected and actual candidates remaining after sieving to a depth of p_{max}

n	p_{max}	Candidates remaining	
		Expected	Actual
18	$2.510 \cdot 10^{18}$	17,228,044	17,300,322
19	$1.855 \cdot 10^{19}$	16,577,985	16,546,522
20	$1.985 \cdot 10^{19}$	18,321,722	18,342,741
21	$1.935 \cdot 10^{19}$	20,355,000	20,378,158
22	$2.120 \cdot 10^{19}$	21,953,527	21,952,320

Table 5. GFN mega-primes found by PrimeGrid

GFN	Digits	Finder	Date	Software
$475856^{524288} + 1$	2,976,633	Masashi Kumagai	Aug 2012	GeneferCUDA
$356926^{524288} + 1$	2,911,151	Tim McArdle	Jul 2012	Genefx64
$341112^{524288} + 1$	2,900,832	Peyton Hayslette	Jun 2012	GeneferCUDA
$75898^{524288} + 1$	2,558,647	Michael Goetz	Nov 2011	GeneferCUDA
$773620^{262144} + 1$	1,543,643	Senji Yamashita	Apr 2012	GeneferCUDA
$676754^{262144} + 1$	1,528,413	Carlos Loureiro	Feb 2012	GeneferCUDA
$525094^{262144} + 1$	1,499,526	David Tomecko	Jan 2012	GeneferCUDA
$361658^{262144} + 1$	1,457,075	Michel Johnson	Nov 2011	GeneferCUDA
$145310^{262144} + 1$	1,353,265	Ricky L Hubbard	Feb 2011	Genefx64
$40734^{262144} + 1$	1,208,473	Senji Yamashita	Mar 2011	Genefx64
$9 \cdot 2^{3497442} + 1*$	1,052,836	Heinz Ming	Oct 2012	LLR
$81 \cdot 2^{3352924} + 1*$	1,009,333	Michał Gasewicz	Jan 2012	LLR

where

$$C_n = \prod_{p \neq 2} \frac{(1 - \frac{a_n(p)}{p})}{(1 - \frac{1}{p})}, \quad a_n(p) = \begin{cases} 2^n & if\ p \equiv 1 \bmod 2^{n+1}, \\ 0 & otherwise. \end{cases} \qquad (3)$$

So sieving the GFNs $F_{b,n}, b \in [2, B_{max}]$ we expect the number of candidates remaining to be

$$e^{-\gamma} C_n B_{max} / \log(p_{max}) \qquad (4)$$

As shown in Table 4, we find excellent agreement between the expected and actual number of candidates remaining after sieving. As a result, we assert that the excess of primes for $n = 18, 19$ is no more than a statistical anomaly. Further searching at these n, as well as $n = 20, 21, 22$ for which we currently have little data, will be needed to confirm or refute this.

4   GFN Mega-Primes

As a result of the aforementioned extensions to the 'Genefer' program and wide participation in the search since it was made available through the BOINC platform we have made rapid progress to high b values, particularly for $n \geq 18$ where the CUDA implementation has been used. Consequently we have discovered a number of GFN mega-primes (primes with over 1 million decimal digits), and they are listed in Table 5. Note that the two primes marked with an asterisk were found by PrimeGrid's Proth prime search, rather than the GFN search, but since they can be expressed as GFNs with $n = 1$ they are included for completeness. Prior to our search efforts, only one GFN mega-prime was known - $24518^{262144} + 1$, with 1,150,678 digits - found in March 2008 by Stephen Scott, searching independently.

5   Continuing the Search

The results reported above are only a snapshot in time from an ongoing, popular prime search project. We intend to continue the search for large GFN primes for all $n \geq 15$, including $n = 21$ which is currently unsearched. Of particular interest to many participants is the search at $n = 22$, where the current GFNs being tested have decimal lengths of over 17.1 million digits, close to the size of the largest known prime $2^{57885161} - 1$ (17.4 million digits). The b limit of 'GeneferCUDA' for $n = 22$ corresponds to GFNs of 23 million digits, meaning that a prime found during this search has a chance of becoming the largest known prime of *any* kind, a position that has been held solely by Mersenne primes since the discovery of M_{756839} in 1992.

In order to support the ongoing search we will continue to develop 'Genefer' to take advantage of the latest computing hardware. In particular versions able to take advantage of other non-Nvidia GPU hardware (for example using the OpenCL library) and Intel's Advanced Vector Extensions (AVX) may prove invaluable in the search for a new world record GFN prime.

Acknowledgments. The first author acknowledges the support of NAIS, the Centre for Numerical Algorithms and Intelligent Software (EPSRC grant EP/G036136/1).

We also wish to thank several people who have contributed to the GFN prime search. First, we thank Yves Gallot for popularising the search, developing the initial Genefer code upon which the entire project is based, and also for useful discussions concerning the purported excess of primes at large n (see Sect. 3). Second, we thank David Underbakke, Mark Rodenkirch, Ken Brazier, Shoichiro Yamada, Ronald Schneider and Anand Nair, who have all contributed to the ongoing development of the PRP testing and sieving software. Third, thanks go to the PrimeGrid team Rytis Slatkevicius, Lennart Vogel and John Blazek without whom the search would not have reached such a wide audience. Finally, we are grateful to all the 'crunchers' who have dedicated their computer resources and made possible the ongoing success of the search.

References

1. Anderson, D.: BOINC: a system for public-resource computing and storage. In: Proceedings of the 5th IEEE/ACM International Workshop on Grid Computing, GRID '04, pp. 4–10. IEEE Computer Society, Washington. http://dx.doi.org/10.1109/GRID.2004.14 (2004)
2. Caldwell, C.: The prime pages - the largest known primes database. http://primes.utm.edu
3. Carmody, P.: GFN filters. http://fatphil.org/maths/GFN/maths.html
4. Crandall, R., Fagin, B.: Discrete weighted transforms and large-integer arithmetic. Math. Comp. **62**, 305–324 (1994)
5. Dubner, H., Gallot, Y.: Distribution of generalized Fermat prime numbers. Math. Comp. **71**, 825–832 (2002)
6. Dubner, H., Keller, W.: Factors of generalized Fermat numbers. Math. Comp. **64**, 397–405 (1995)
7. Mertens, F.: Ein beitrag zur analytischen zahlentheorie. J. reine angew. Math **78**, 46–62 (1874)
8. Nickolls, J., Buck, I., Garland, M., Skadron, K.: Scalable parallel programming with CUDA. Queue **6**(2), 40–53 (2008). http://doi.acm.org/10.1145.1365490.1365500
9. Sierpiński, W.: Sur un problème concernant les nombres k. 2n + 1. Elem. Math. **115**, 73–74 (1960)
10. Slatkevicius, R., Vogel, L., Blazek, J.: PrimeGrid website. http://www.primegrid.com

Iterative Solution of Singular Systems with Applications

Radim Blaheta, Ondřej Jakl$^{(\boxtimes)}$, and Jiří Starý

IT4Innovations Department, Institute of Geonics AS CR, Ostrava, Czech Republic
{blaheta,jakl,stary}@ugn.cas.cz

Abstract. This paper deals with efficient solution of singular symmetric positive semidefinite problems. Our motivation arises from the need to solve special problems of geotechnics, e.g. to perform upscaling analysis of geocomposites. In that and other applications we have to solve boundary problems with pure Neumann boundary conditions. We show that the stabilized PCG method with various preconditioners is a good choice for systems resulting from the numerical solution of Neumann problems, or more generally problems with a known small dimensional null space.

We make use of this scenario to compare parallel implementations of the corresponding solvers, namely implementations in the in-house finite element software GEM and implementations employing components of the general Trilinos framework. The studies show that the solvers based on GEM are highly competitive with its recognized counterpart.

Keywords: Singular system · Symmetric positive semidefinite problem · Stabilized preconditioned conjugate gradient method · GEM software · Trilinos

1 Introduction

This contribution concerns the iterative solution of singular systems which arise in many applications. Let us mention the following

- solution of PDE problems with pure Neumann boundary conditions (which is our main aim), see [7]. Such problems have a specific role in numerical upscaling, see [6],
- solution of Neumann type subproblems in domain decomposition techniques as FETI, Neumann-Neumann, BDDC methods, see [12,15],
- analysis of Markov chain problems, computation of stochastic vector, see e.g. [14],
- computer tomography and inverse problems.

2 Iterative Solution of Singular Symmetric Semidefinite Systems

Let us focus on iterative solution of linear systems of the form

$$Au = b, \tag{1}$$

R. Wyrzykowski et al. (Eds.): PPAM 2013, Part I, LNCS 8384, pp. 114–123, 2014.
DOI: 10.1007/978-3-642-55224-3_12, © Springer-Verlag Berlin Heidelberg 2014

where A is a singular, symmetric, positive semidefinite $n \times n$ matrix, $b \in R^n$. For $u, v \in R^n$ denote $\langle u, v \rangle = u^T v$ and $\|u\|$ the Euclidean inner product and norm. Due to symmetry of A, the range $R(A)$ and the null space $N(A)$ are mutually orthogonal with respect to the Euclidean inner product and the vectors $u \in R^n$ can be uniquely decomposed as

$$u = u_N + u_R, \text{ where } u_N \in N(A) \text{and } u_R \in R(A).$$

Let $b = b_N + b_R$, $N(A) \neq \{0\}$, then the system (1) has infinitely many generalized (least squares) solutions u,

$$\|Au - b\| = \min\{\|Av - b\|, \ v \in R^n\} \tag{2}$$

among which there is a unique least squares solution u^* with the minimal Euclidean norm. Note that $u^* = A^+ b$, where A^+ is the Moore-Penrose pseudoinverse of A, see [8, 12]. If $b \in R(A)$, i.e. the system (1) is consistent, then the generalized solutions are standard solutions of (1).

Let us assume that (1) is solved iteratively with denoting the i-th iteration u^i,

$$u^i \in u^0 + K_i(A, r^0) = u^0 + \text{span}\{r^0, Ar^0, \ldots, A^{i-1}r^0\}, \quad r^0 = b - Au^0, \tag{3}$$

where $K_i(A, r^0) = \text{span}\{r^0, Ar^0, \ldots, A^{i-1}r^0\}$ is a Krylov space. Then

$$u^i = u^0 + q_{i-1}(A)r^0, \quad \text{where } q_{i-1} \text{ is a polynomial of order } \leq i - 1. \tag{4}$$

The convergence can be investigated through behaviour of $e^i = u^i - u^*$. If $e^i \to 0$ then the iterations converge to the minimal least squares solution u^*. If $e^i \to w$, where $w \in N(A)$, then the iterations converge to a (generalized) solution of A.

From (4), it follows that

$$e^i = u^0 - u^* + q_{i-1}(A)(b_N + Au^* - Au^0) = p_i(A)e^0 + q_{i-1}(0)b_N, \tag{5}$$

where $p_i(\lambda) = 1 - \lambda q_{i-1}(\lambda)$.

If $e^0 = e_N^0 + e_R^0$ then $p_i(A)e_N^0 = u_N^0$ and $p_i(A)e_R^0$ depends on values $p_i(\lambda)$ on $\lambda \in \sigma(A) \setminus \{0\}$. The second term is zero for consistent problems, but otherwise can be convergent if $q_{i-1}(0) = -p_i'(0) \neq 0$.

The simplest Richardson's iteration $u^{i+1} = u^i + \omega A(b - u^i)$ fulfill (3), (4), (5) with

$$p_i(\lambda) = (1 - \omega\lambda)^i, \quad p_i(0) = 1, \quad q_{i-1}(0) = -p_i'(0) = (i + 1)\omega.$$

Thus, the method converges ($e^0 \to u_N^0$) for the consistent problems, but diverges (the second terms gradually dominates) for the inconsistent case ($b_N \neq 0$).

To get convergence even for inconsistent case, the method needs a modification. For example, we can use extrapolation of Richardson's iterations [13]. For

$$\bar{u}^{i+1} = u^{i+1} - (i + 1)(u^{i+1} - u^i),$$

we get

$$\begin{aligned}
\bar{u}^{i+1} - u^* &= u^{i+1} - u^* - (i+1)(u^{i+1} - u^i) \\
&= p_{i+1}(A)e^0 + (i+1)\omega(b_N + A(u^* - u^i)) \\
&= p_{i+1}(A)e^0 + (i+1)\omega A e^i \\
&= p_{i+1}(A)e^0 + (i+1)\omega A(p_i(A)e^0 + i\omega b_N)) \\
&= p_{i+1}(A)e^0 + (i+1)\omega(p_i(A)Ae^0)).
\end{aligned}$$

This extrapolated method converges since $p_i(\lambda) \le q^i$ for all $\lambda \in \sigma(A) \setminus \{0\}$, where $q < 1$ for a suitable ω.

It means that there are ways how to damp the divergence of the null space component of the iterations. On the other hand, this divergence in the null space component may not cause a problem in case that we are interested only in quantities, which do not depend on the null space component, like gradients, fluxes, strains and stresses.

A similar analysis can be done for other iterative methods applied to singular systems, see e.g. [8]. For the conjugate gradient (CG) method, the convergence can be proven in the consistent case, see e.g. [1]. But the inconsistence influences both $N(A)$ and $R(A)$ components of the iterations, see [7,11] and below.

3   Solution of Neumann Problems

The solution of boundary value problems with pure Neumann boundary conditions arises in different applications, see the other sections. If the solution of the continuous Neumann problem exists, then global balance (consistency) conditions like (7) are satisfied. On the contrary, these conditions guarantee the existence of the (not unique) solution. For example ([7]), for the Neumann problem,

$$-\mathrm{div}(\nabla u) = f \text{ in } \Omega \quad \text{and} \quad \nabla u \cdot n = g \text{ in } \partial\Omega \tag{6}$$

the solution exits if and only if

$$\int_\Omega f \, dx + \int_{\partial\Omega} g \, dx = 0. \tag{7}$$

In the case (6), (7), if u is a solution, then $u + v$ is a solution for all $v \in \mathcal{N} = \mathrm{span}\{1\}$, where 1 is a constant function in Ω. A finite element discretization then should provide a consistent singular linear system (1) with the nullspace $N(A) = \mathcal{N}_h$ provided by discretization of $\mathcal{N}$. However, the computer arithmetic and numerical integration errors may cause that the FEM system is inconsistent and/or $N(A) \ne \mathcal{N}_h$.

Problems with inconsistency and singularity can be treated by using a priori knowledge about $\mathcal{N}$ and $\mathcal{N}_h$. For example, we are able to regularize the problem by fixing some degrees of freedom and solving the problem $R_{dof} A R_{dof}^T u = R_{dof} b$ instead of (1). Here, R_{dof} is the restriction operator omitting the fixed DOF's.

Such a technique is frequently used in engineering community, but without a special care the modified system matrix $R_{dof} A R_{dof}^T$ can be very ill-conditioned which is a serious drawback for the iterative solution.

Using the knowledge of $\mathcal{N}$, other techniques use the projection $P : R^n \to \mathcal{R}_h$, where $\mathcal{R}_h$ is the orthogonal complement of $\mathcal{N}_h$. The projector can be constructed as $P = I - V(V^T V)^{-1} V^T$, where V is a matrix, whose columns create a basis of $\mathcal{N}_h$. Such projector can be applied within any iterative method, including PCG (Fig. 1). In $PCGstab1$ algorithm, the projection P is used to project the right hand side vectors or all residuals during the PCG iterative process. In $PCGstab2$, the projection P is applied twice per iteration to project both residuals and computed iterations. $PCGstab2$ is equivalent to the replacement of A by PAP which also makes the system matrix singular. The fully stabilized $PCGstab2$ was introduced e.g. in [9]. Note that $g = G(r)$ denotes the action of preconditioner, which can be also nonlinear (variable, flexible), see e.g. [3].

given u^0
compute
$r^0 = P_a(b - Au^0)$, $g^0 = P_b G(r^0)$, $v^0 = g^0$
for $i = 0, 1, \ldots$ until convergence do

$w^i = P_c A P_d v^i$
$\alpha_i = \langle r^i, g^i \rangle / \langle w^i, v^i \rangle$
$u^{i+1} = u^i + \alpha_i v^i$
$r^{i+1} = P_a(r^i - \alpha_i w^i)$
$g^{i+1} = P_b G(r^{i+1})$
$\beta_{i+1} = \langle g^{i+1}, r^{i+1} \rangle / \langle g^i, r^i \rangle$
$v^{i+1} = g^{i+1} + \beta_{i+1} v^i$
end

1. Standard PCG:
$$P_a = P_b = P_c = P_d = I$$
2. $PCGstab1$:
$$P_a = P$$
$$P_b = P_c = P_d = I$$
3. $PCGstab2$:
$$P_a = P_b = P$$
$$P_d = P_c = I$$
or equivalently
$$P_a = P_b = I$$
$$P_c = P_d = P$$

Fig. 1. Stabilizations in the PCG algorithm.

Note that the application of the standard PCG to inconsistent systems is problematic from two reasons. The inconsistent part of the right hand side enters the $N(A)$-part of the iterations and can make them divergent, but the inconsistent part also enters the formulas for α and β and spoils the $R(A)$-part of the iterations, see [5,11].

4   Application in Upscaling

The elastic response of a representative volume Ω is characterized by homogenized elasticity C or compliance S tensors ($S = C^{-1}$). The compliance tensor can be determined from the relation

$$S \langle \sigma \rangle = S \sigma_0 = \langle \varepsilon \rangle, \tag{8}$$

where $\langle\sigma\rangle$ and $\langle\varepsilon\rangle$ are volume averaged stresses and strains computed from the Neumann problem

$$-\mathrm{div}(\sigma) = 0, \quad \sigma = C_m\varepsilon, \quad \varepsilon = (\nabla u + (\nabla u)^T)/2 \quad \text{in } \Omega, \tag{9}$$

$$\sigma n = \sigma_0 n \quad \text{on } \partial\Omega. \tag{10}$$

Above, σ and ε denote stress and strain in the microstructure, C_m is the variable local elasticity tensor, u and n denote the displacement and the unit normal, respectively. The use of Neumann boundary conditions allows us to get a lower bound for the upscaled elasticity tensor [6].

In the analysis of geocomposites (see [6]), the domain Ω is a cube with a relatively complicated microstructure. The FEM mesh is constructed on the basis of CT scans. Consequently using the GEM software [4], the domain is discretized by linear tetrahedral finite elements. The arising singular system is then solved by stabilized *PCGstab1* method, which we have implemented in two different frameworks and run using various preconditioners:

GEM-DD is a solver fully implemented in GEM software. It uses one-level additive Schwarz domain decomposition preconditioner with subdomain problems solved approximately by application of displacement decomposition incomplete factorization described in [2]. The resulting preconditioner is symmetric positive definite.

GEM-DD-CG solver differs in preconditioning, which is a two-level Schwarz domain decomposition arising from the previous GEM-DD by additive involvement of a coarse problem correction. The coarse problem is created by a regular algebraic aggregation of mesh nodes. The aggregation factors (numbers of the original nodes to be merged in into a single coarse mesh node in each coordinate direction) may be different in each coordinate direction and we have three DOF per aggregation. In this case, the coarse problem is singular with a smaller null space containing only the rigid shifts. The resulting preconditioner is again symmetric positive definite. In our implementation, the coarse problem is solved only approximately by inner (not stabilized) CG method with a lower solution accuracy - relative residual accuracy $\varepsilon_0 \leq 10^{-1}$. Thus, we use a variable type of preconditioner.

Trisol ILU is solver based on the Trilinos framework and the linear system is imported from GEM. The preconditioner is similar to GEM-DD, i.e. one-level Schwarz with the minimal overlap and working on the same subdomains as in GEM-DD are used. The subproblems are replaced by ILU without displacement decomposition, using a drop tolerance and a fill limit.

Trisol ML-DD is also Trilinos based (in fact implemented in the same executable as Trisol ILU), but the corresponding command line parameters invoke multilevel-level V-cycle preconditioner exploiting smoothed aggregations with aggressive coarsening, see [10]. Six DOF translational plus rotational are used per aggregation. ILU is applied as smoother at the finest level, other smoothing is realised by symmetrized Gauss-Seidel. The coarsest problem is solved by a direct solver.

4.1   Benchmarks

The starting point of our upscaling analysis are the geocomposite samples of $75 \times 75 \times 75\,\text{mm}$ in size. They represent a very complicated microstructure arising from injection of polyurethane resin into coal environment, with the aim e.g. to reinforce coal pillars during mining. The samples are scanned by an X-ray computer tomograph (CT) which produces a set of images (of the same resolution) corresponding to cuts through the sample. Depending on the resolution and number of cuts, we obtain rectangular and structured voxel representation of various sizes. From the values provided by the CT scanner for each voxel we derive material distribution in the sample. This material identification is one of the trickiest points in the modelling procedure. For example, empty space in the structure such as cavities or air bubbles can lead to a kind of singularity, when some voxel regions weekly hang in the void space sharing only one corner or edge with the surroundings. From the data gained, numerical models are created, employing standard linear tetrahedral finite elements.

For the subsequent demonstrations we chose benchmark models with parameters specified in the following Table 1.

Table 1. Benchmarks representing microstructures of two geocomposite samples. Denotation, applied discretization meshes and sizes of resulting linear systems.

Benchmark	Discretization	Size in DOF	Data size
GEOC-1	$232 \times 232 \times 38$	6 135 936	1.0 GB
GEOC-2s	$257 \times 257 \times 257$	50 923 779	8.5 GB
GEOC-2l	$257 \times 257 \times 1025$	203 100 675	33.5 GB

Note that the GEOC1 scan was obtained by courtesy of the Kumamoto University GeoX CT Center, Japan, while the larger GEOC2 scan was performed with a new CT scanner of the Institute of Geonics AS CR, Czech Republic. In fact, in this case the CT scan was transformed to two meshes that differ only in the number of nodes in the $\mathcal{Z}$ direction (GEOC-2s(mall),GEOC-2l(arge) benchmarks).

4.2   Basic Performance Comparison

The smaller GEOC1 benchmark with about 6 million DOF served well to compare the GEM and Trisol parallel PCG implementations on the solution of the Neumann problem in elasticity. The stopping criterion was the same for all numerical tests and defined as $\|r\|/\|b\| \le \varepsilon = 10^{-5}$. For DD-CG, a coarse problem with aggregation factors $6 \times 6 \times 3$ and 67 200 DOF was applied.

The tests were performed on a 64-core NUMA multiprocessor called Enna:

– eight octa-core Intel Xeon E7–8837/2.66 GHz processors
– 256 GB of DDR2 RAM
– CentOS 6.3, Intel Cluster Studio XE 2013, Trilinos 11.4.1

The following Table 2 summarizes the main runtime characteristics of the GEOC1 benchmark obtained on the Enna multiprocessor.

Table 2. Iteration counts ($\#$ It), wall-clock time (in seconds) for the initial preparation of preconditioner (T_{prep}) and the solution time itself (T_{iter}) are provided for various numbers of subdomains ($\#$ Sd; always corresponding to the number of employed processing units).

	GEM DD			DD-CG			Trisol ILU			ML-DD		
$\#$ Sd	$\#$ It	T_{prep}	T_{iter}	$\#$ It	T_{prep}	T_{iter}	$\#$ It	T_{prep}	T_{iter}	$\#$ It	T_{prep}	T_{iter}
1	–			–			345	666.0	1943.7	–		
2	293	0.3	325.3	137	15.3	182.9	472	492.2	1442.6	53	1187.8	734.9
4	302	0.2	187.3	124	16.2	88.6	463	561.1	714.6	61	750.1	386.0
8	300	0.1	127.6	115	16.1	56.5	441	470.7	372.6	74	614.8	202.2
16	350	0.1	149.4	116	15.1	54.3	387	422.9	219.0	93	905.2	127.1

Those results indicate that the stabilization of PCG algorithm has the expected positive effect on the convergence, because without stabilization the solution converges up to a low residual tolerance about $\varepsilon = 10^{-2}$ and then starts to oscillate, see [5] for details.

In the iterative phase timed by T_{iter}, GEM was up to four times faster than Trisol in our tests, but Trisol showed better scalability. The involvement of the grid problem speeds up the solution by factor of $1.5 - 3$ in both cases.

Table 2 under time T_{prep} lists also the overhead needed for the initial preparation of the preconditioner (incomplete factorization on subdomains, creation of a coarse problem, etc.). While in the case of GEM solvers T_{prep} is small in comparison with the time required by the iterations T_{iter}, it makes a significant part of the Trisol solution and very often dominates over T_{iter}. This makes GEM's solution faster by an order of magnitude.

A bit surprising decrease of the number of iterations reported in Tables 2, 3, 4, especially for DD-CG, can be explained by the fact that smaller subdomain problems are solved more accurately in our implementation.

4.3   Cluster Computations

Recently we gained access to a new supercomputer called Anselm, run by the Czech National Supercomputing Center IT4Innovations. This multicomputer (cluster architecture) comprises 207 compute nodes, among which we employed those equipped with:

- two octa-core Intel E5–2665/2.4 GHz processors
- 64 GB of memory and 500 GB of local disk capacity
- Infiniband QDR interconnection, fully non-blocking, fat-tree
- Bullx Linux OS (Red Hat family), Intel Parallel Studio XE 2013

This platform allowed us to extend the scalability studies, in which we made use of the larger GEOC2 benchmark in both variants, cf. Table 1. Table 3 shows the timings of GEM solvers (without and with coarse grid problem applied) obtained for GEOC2s, i.e. a problem of more than 50 million DOF, where the performance up to 64 processing elements on Enna and up to 128 processing elements on Anselm could be compared. The stopping criterion was again $\|r\|/\|b\| \leq \varepsilon = 10^{-5}$ and the DD-CG solver made use of a coarse problem with aggregation factors $9 \times 9 \times 9$ (81 000 DOF).

Table 3. Timings of the GEOC2s benchmark achieved by the GEM solvers on the multiprocessor Enna and cluster Anselm: Iteration counts (# It), wall-clock time (in seconds) of the solution (T_{iter}) and the corresponding performance ratio Anselm/Enna (A/E) are provided for up to 128 subdomains (# Sd).

| | Enna | | | | Anselm | | | |
| | DD | | DD-CG | | DD | | DD-CG | |
# Sd	# It	T_{iter}	# It	T_{iter}	T_{iter}	A/E	T_{iter}	A/E
2	914	8461.2	437	3523.1	5644.2	0.67	2785.4	0.79
4	1129	4973.3	428	1923.6	3526.2	0.59	1383.4	0.72
8	1421	2942.5	416	922.9	2422.6	0.82	725.7	0.79
16	1655	1994.6	376	415.8	1325.8	0.64	348.7	0.84
32	1847	1923.5	329	348.3	798.3	0.42	194.8	0.56
64	2149	3074.9	295	505.9	620.8	0.20	117.6	0.23
128					515.7		107.1	

For greater number of subdomains systems with distributed memory scale better, because multiprocessors in general suffer from the memory-processor bandwidth contention: On Enna the scalability fades out at about 32 processing elements, but the turning point on Anselm is around 128 elements.

In absolute figures, we were able to solve the benchmark 3–4 times faster on Anselm than on Enna. The advantage of Anselm is to be derived partially from the fact that its newer Intel Sandy Bridge CPU architecture as such outperforms Enna's Westmere one, in our application by 20–40 %, what can be estimated from the tests using up to 8 cores, when the processors work in similar conditions.

In Table 3 we can also once again observe the benefit of the DD-CG approach: The added coarse grid problem speeds up the original one-level domain decomposition by a factor of 2.5–5.

4.4   The Largest Benchmark

The execution of the largest benchmark GEOC2l (about 200 million DOF) was very demanding on resources, so the number of performed tests was rather low. Based on the experience from GEOC2s we focused on the impact of the coarse grid size. The main results are summarized in Table 4.

Table 4. Timings of the GEOC2l benchmark achieved by the GEM-DD-CG solver on the multiprocessor Enna and cluster Anselm: Iteration counts(# It) and wall-clock time (in seconds) for the solution time (T_{iter}) are provided now for different sizes of CG problem involved in computations and for various numbers of subdomains (# Sd).

| | Enna | | | | | | Anselm | |
| | DD- $9 \times 9 \times 9$ | | DD-9 $\times 9 \times 18$ | | DD- $9 \times 9 \times 27$ | | DD- $9 \times 9 \times 27$ | |
# Sd	# It	T_{iter}	# It	T_{iter}	# It	T_{iter}	# It	T_{iter}
4	751	13719.0	858	15737.6	997	18518.4	997	12671.4
8	690	6237.7	800	6960.8	917	8062.9	917	5803.9
16	585	2717.4	674	4010.6	777	4815.6	777	2576.6
32	585	2483.6	622	2923.8	708	3452.5	708	1157.5
64					627	3637.0	627	558.8
128							652	358.5
256							631	299.6
512							649	333.5

Most experiments were carried out on Enna, which is more convenient and accessible for such purposes. Table 4 demonstrates the impact of the coarse grid size on the time of the solution. We can observe that the "standard" aggregation factor $729 = 9 \times 9 \times 9$ (providing a coarse grid of 310 500 DOF) leads to shorter solution times than the "more aggressive" aggregations $1458 = 9 \times 9 \times 18$ (156 600 DOF) and $2187 = 9 \times 9 \times 27$ (105 300 DOF), which naturally have higher number of iterations. However, the coarse problem is solved in parallel with the subdomain problems and with increasing number of (smaller) subdomains this static task develops to a bottleneck when not matching the decreasing amount of computation on the subdomains any more. We could confirm this observation on Anselm, where the best time in the Table 4 (299.6 s with 256 processing elements and aggregation $9 \times 9 \times 27$) was surpassed by an experiment with the coarser aggregation $15 \times 15 \times 31$: The overall best GEOC2l solution time of 249.8 s was achieved after 910 iterations on # Sd=512 subdomains.

5   Conclusions

The first aim of this contribution was to demonstrate efficient parallel solution of singular symmetric positive semidefinite problems. On several examples we could examine that the stabilized preconditioned conjugate gradient approach is a good choice for systems arising from the numerical solution of Neumann problems, or more generally problems with a known small-dimensional null space. We could also observe the great value of the coarse grid correction for the convergence.

The second objective was a comparative case study of software that implements this approach. We compared specialized solvers developed from scratch for the finite element package GEM and solvers implemented from building blocks of the Trilinos library to achieve the same functionality. We compared their performance on practical problems of increasing size and on different parallel architectures and showed that on a modern cluster the iterations to solve a singular system of more than 200 million DOF take about four minutes.

Acknowledgement. This work was supported by the European Regional Development Fund in the IT4Innovations Centre of Excellence project (CZ.1.05/1.1.00/02.0070). We would like to thank Dr. Erhan Turan from Mechran, Istanbul for fruitful discussions and extended assistance in implementing methods in Trilinos.

References

1. Áxelsson, O.: Iterative Solution Methods. Cambridge University Press, New York (1994)
2. Blaheta, R.: Displacement decomposition - incomplete factorization preconditioning techniques for linear elasticity problems. Numer. Linear Algebra Appl. **1**, 107–128 (1994)
3. Blaheta, R.: GPCG - generalized preconditioned CG method and its use with nonlinear and non-symmetric displacement decomposition preconditioners. Numer. Linear Algebra Appl. **9**, 527–550 (2002)
4. Blaheta, R., Jakl, O., Kohut, R., Starý, J.: GEM – a platform for advanced mathematical geosimulations. In: Wyrzykowski, R., Dongarra, J., Karczewski, K., Wasniewski, J. (eds.) PPAM 2009, Part I. LNCS, vol. 6067, pp. 266–275. Springer, Heidelberg (2010)
5. Blaheta, R., Jakl, O., Starý, J., Turan, E.: Parallel solvers for numerical upscaling. In: Manninen, P., Öster, P. (eds.) PARA 2012. LNCS, vol. 7782, pp. 375–386. Springer, Heidelberg (2013)
6. Blaheta, R., Kohut, R., Kolcun, A., Souček, K., Staš, L.: Micromechanics of geocomposites: CT images and FEM simulations. In: Kwasniewski, M., Lydzba, D. (eds.) Rock Mechanics for Resources, Energy and Environment, pp. 399–404. Taylor & Francis Group, London (2013). ISBN 978-1-138-00080-3
7. Bochev, P., Lehoucq, R.B.: On the finite element solution of the pure Neumann problem. SIAM Rev. **47**, 50–66 (2005)
8. Eiermann, M., Marek, I., Niethammer, W.: On the solution of singular linear systems of algebraic equations by semiiterative methods. Numer. Math. **53**, 265–283 (1988)
9. Farhat, C., Mandel, J., Roux, F.X.: Optimal convergence properties of the FETI domain decomposition method. Comput. Methods Appl. Mech. Eng. **115**, 367–388 (1994)
10. Gee, M.W., Siefert, C.M., Hu, J.J., Tuminaro, R.S., Sala, M.G.: ML 5.0 smoothed aggregation user's guide. Report SAND2006-2649, Sandia National Laboratories (2006)
11. Kaasschieter, E.F.: Preconditioned conjugate gradients for solving singular systems. J. Comput. Appl. Math. **24**, 265–275 (1988)
12. Kučera, R., Kozubek, T., Markopoulos, A., Machalová, J.: On the Moore-Penrose inverse in solving saddle-point systems with singular diagonal blocks. Numer. Linear Algebra Appl. **19**, 677–699 (2012)
13. Marchuk, G.I.: Methods of numerical mathematics. Springer, New York (1982). Czech transl. Academia (1987)
14. Marek, I., Szyld, D.B.: Algebraic Schwarz methods for the numerical solution of Markov chains. Linear Algebra Appl. **386**, 67–81 (2004)
15. Toselli, A., Widdlund, O.: Domain Decomposition Methods - Algorithms and Theory. Springer, Heidelberg (2005)

Statistical Estimates for the Conditioning of Linear Least Squares Problems

Marc Baboulin[1]([✉]), Serge Gratton[2], Rémi Lacroix[1], and Alan J. Laub[3]

[1] Université Paris-Sud and Inria, Orsay, France
{marc.baboulin,remi.lacroix}@inria.fr
[2] ENSEEIHT and CERFACS, Toulouse, France
serge.gratton@enseeiht.fr
[3] University of California Los Angeles, Los Angeles, USA
laub@ats.ucla.edu

Abstract. In this paper we are interested in computing linear least squares (LLS) condition numbers to measure the numerical sensitivity of an LLS solution to perturbations in data. We propose a statistical estimate for the normwise condition number of an LLS solution where perturbations on data are mesured using the Frobenius norm for matrices and the Euclidean norm for vectors. We also explain how condition numbers for the components of an LLS solution can be computed. We present numerical experiments that compare the statistical condition estimates with their corresponding exact values.

Keywords: Linear least squares · Condition number · Statistical condition estimation · Componentwise conditioning

1 Introduction

We consider the overdetermined linear least squares (LLS) problem

$$\min_{x \in \mathbb{R}^n} \|Ax - b\|_2, \tag{1}$$

with $A \in \mathbb{R}^{m \times n}, m \geq n$ and $b \in \mathbb{R}^m$. We assume throughout this paper that A has full column rank and as a result, Eq. (1) has a unique solution $x = A^+ b$ where A^+ is the Moore-Penrose pseudoinverse of the matrix A, expressed by $A^+ = (A^T A)^{-1} A^T$. We can find for instance in [7,13,19] a comprehensive survey of the methods that can be used for solving efficiently and accurately LLS problems.

The condition number is a measure of the sensitivity of a mapping to perturbations. It was initially defined in [23] as the maximum amplification factor between a small perturbation in the data and the resulting change in the problem solution. Namely, if the solution x of a given problem can be expressed as a function $g(y)$ of a data y, then if g is differentiable (which is the case for many

R. Wyrzykowski et al. (Eds.): PPAM 2013, Part I, LNCS 8384, pp. 124–133, 2014.
DOI: 10.1007/978-3-642-55224-3_13, © Springer-Verlag Berlin Heidelberg 2014

linear algebra problems), the absolute condition number of g at y can be defined as (see e.g. [12])

$$\kappa(y) = \max_{z \neq 0} \frac{\|g'(y).z\|}{\|z\|}. \tag{2}$$

From this definition, $\kappa(y)$ is a quantity that, for a given perturbation size on the data y, allows us to predict *to first order* the perturbation size on the solution x. Associated with a backward error [26], condition numbers are useful to assess the numerical quality of a computed solution. Indeed numerical algorithms are always subject to errors although their sensitivity to errors may vary. These errors can have various origins like for instance data uncertainty due to instrumental measurements or rounding and truncation errors inherent to finite precision arithmetic.

LLS can be very sensitive to perturbations in data and it is crucial to be able to assess the quality of the solution in practical applications [4]. It was shown in [14] that the 2-norm condition number cond(A) of the matrix A plays a significant role in LLS sensitivity analysis. It was later proved in [25] that the sensitivity of LLS problems is proportional to cond(A) when the residual vector is small and to cond(A)^2 otherwise. Then [12] provided a closed formula for the condition number of LLS problems, using the Frobenius norm to measure the perturbations of A. Since then many results on normwise LLS condition numbers have been published (see e.g. [2,7,11,15,16]).

It was observed in [18] that normwise condition numbers can lead to a loss of information since they consolidate all sensitivity information into a single number. Indeed in some cases this sensitivity can vary significantly among the different solution components (some examples for LLS are presented in [2,21]). To overcome this issue, it was proposed the notion of "componentwise" condition numbers or condition numbers for the solution components [9]. Note that this approach must be distinguished from the componentwise metric also applied to LLS for instance in [5,10]. This approach was generalized by the notion of *partial* or *subspace* condition numbers which corresponds to conditioning of $L^T x$ with $L \in \mathbb{R}^{n \times k}, k \leq n$, proposed for instance in [2,6] for least squares and total least squares, or [8] for linear systems. The motivation for computing the conditioning of $L^T x$ can be found for instance in [2,3] for normwise LLS condition numbers.

Even though condition numbers provide interesting information about the quality of the computed solution, they are expected to be calculated in an acceptable time compared to the cost for the solution itself. Computing the exact (subspace or not) condition number requires $\mathcal{O}(n^3)$ flops when the LLS solution x has been already computed (e.g., using a QR factorization) and can be reused to compute the conditioning [2,3]. This cost is affordable when compared to the cost for solving the problem ($\mathcal{O}(2mn^2)$ flops when $m \gg n$). However statistical estimates can reduce this cost to $\mathcal{O}(n^2)$ [17,20]. The theoretical quality of the statistical estimates can be formally measured by the probability to give an estimate in a certain range around the exact value. In this paper we summarize closed formulas for the condition numbers of the LLS solution and of its components, and we propose practical algorithms to compute statistical estimates of these quantities. In particular we derive a new expression for the statistical estimate of the conditioning of x. We also present numerical experiments to compare LLS conditioning with the corresponding statistical estimates.

Notations. The notation $\|\cdot\|_2$ applied to a matrix (resp. a vector) refers to the spectral norm (resp. the Euclidean norm) and $\|\cdot\|_F$ denotes the Frobenius norm of a matrix. The matrix I is the identity matrix and e_i is the ith canonical vector. The uniform continuous distribution between a and b is abbreviated $\mathcal{U}(a, b)$ and the normal distribution of mean μ and variance σ^2 is abbreviated $\mathcal{N}(\mu, \sigma^2)$. cond($A$) denotes the 2-norm condition number of a matrix A, defined as cond(A) $= \|A\|_2 \|A^+\|_2$. The notation $|\cdot|$ applied to a matrix or a vector holds componentwise.

2   Condition Estimation for Linear Least Squares

In Sect. 2.1 we are concerned in calculating the condition number of the LLS solution x and in Sect. 2.2 we compute or estimate the conditioning of the components of x. We suppose that the LLS problem has already been solved using a QR factorization (the normal equations method is also possible but the condition number is then proportional to cond(A)^2 [7, p. 49]). Then the solution x, the residual $r = b - Ax$, and the factor $R \in \mathbb{R}^{n \times n}$ of the QR factorization of A are readily available (we recall that the Cholesky factor of the normal equations is, in exact arithmetic, equal to R up to some signs). We also make the assumption that both A and b can be perturbed, these perturbations being measured using the weighted product norm $\|(\Delta A, \Delta b)\|_F = \sqrt{\|\Delta A\|_F^2 + \|\Delta b\|_2^2}$ where ΔA and Δb are absolute perturbations of A and b. In addition to providing us with simplified formulas, this product norm has the advantage, mentioned in [15], to be appropriate for estimating the forward error obtained when the LLS problem is solved via normal equations.

2.1   Conditioning of the Least Squares Solution

Exact formula. We can obtain from [3] a closed formula for the absolute condition number of the LLS solution as

$$\kappa_{LS} = \|R^{-1}\|_2 \left(\|R^{-1}\|_2^2 \|r\|_2^2 + \|x\|_2^2 + 1\right)^{\frac{1}{2}}, \tag{3}$$

where x, r and R are exact quantities.

This equation requires mainly to compute the minimum singular value of the matrix A (or R), which can be done using iterative procedures like the inverse power iteration on R, or more expensively with the full SVD of R ($\mathcal{O}(n^3)$ flops). Note that $\|R^{-T}\|_2$ can be approximated by other matrix norms (see [19, p. 293]).

Statistical estimate. Similarly to [8] for linear systems, we can estimate the condition number of the LLS solution using the method called *small-sample theory* [20] that provides statistical condition estimates for matrix functions.

Let us denote by $x(A, b)$ the expression of x as a function of the data A and b. Since A has full rank n, $x(A, b)$ is continuously F-differentiable in a neighborhood of (A, b). If $x'(A, b)$ is the derivative of this function, then $x'(A, b).(\Delta A, \Delta b)$ denotes the image of $(\Delta A, \Delta b)$ by the linear function $x'(A, b)$. By Taylor's theorem, the forward error Δx on the solution $x(A, b)$ can be expressed as

$$\Delta x = x'(A, b).(\Delta A, \Delta b) + \mathcal{O}(\|(\Delta A, \Delta b)\|_F^2). \tag{4}$$

Following the definition given in Eq. (2), the condition number of x corresponds to the operator norm of $x'(A, b)$, which is a bound to first order on the sensitivity of x at (A, b) and we have

$$\|\Delta x\|_2 \le \kappa_{LS} \|(\Delta A, \Delta b)\|_F .$$

We now use [20] to estimate $\|\Delta x\|_2$ by

$$\xi(q) = \frac{\omega_q}{\omega_n} \sqrt{|z_1^T \Delta x|^2 + \cdots + |z_q^T \Delta x|^2}, \tag{5}$$

where $z_1, \cdots, z_q$ are random orthogonal vectors selected uniformly and randomly from the unit sphere in n dimensions, and ω_q is the Wallis factor defined by

$$\omega_1 = 1,$$
$$\omega_q = \frac{1 \cdot 3 \cdot 5 \cdots (q-2)}{2 \cdot 4 \cdot 6 \cdots (q-1)} \quad \text{for } q \text{ odd},$$
$$\omega_q = \frac{2}{\pi} \frac{2 \cdot 4 \cdot 6 \cdots (q-2)}{1 \cdot 3 \cdot 5 \cdots (q-1)} \quad \text{for } q \text{ even}.$$

ω_q can be approximated by $\sqrt{\frac{2}{\pi(q-\frac{1}{2})}}$.

It comes from [20] that if for instance we have $q = 2$, then the probability that $\xi(q)$ lies within a factor α of $\|\Delta x\|_2$ is

$$Pr(\frac{\|\Delta x\|_2}{\alpha} \le \xi(q) \le \alpha \|\Delta x\|_2) \approx 1 - \frac{\pi}{4\alpha^2}. \tag{6}$$

For $\alpha = 10$, we obtain a probability of 99.2 %.

For each $i \in \{1, \cdots, q\}$, using Eq. (2) we have the first-order bound

$$|z_i^T \Delta x| \le \kappa_i \|(\Delta A, \Delta b)\|_F , \tag{7}$$

where κ_i denotes the condition number of the function $z_i^T x(A, b)$. Then using (5) and (7) we get

$$\xi(q) \le \frac{\omega_q}{\omega_n} \left(\sum_{i=1}^q \kappa_i^2 \right)^{\frac{1}{2}} \|(\Delta A, \Delta b)\|_F .$$

$\xi(q)$ being an estimate of $\|\Delta x\|_2$, we will use the quantity $\bar{\kappa}_{LS}$ defined by

$$\bar{\kappa}_{LS} = \frac{\omega_q}{\omega_n} \left(\sum_{i=1}^q \kappa_i^2 \right)^{\frac{1}{2}} \tag{8}$$

as an estimate for κ_{LS}.

We point out that $\bar{\kappa}_{LS}$ is a scalar quantity that must be distinguished from the estimate given in [21] which is a vector. Indeed the small-sample theory is used here to derive an estimate of the condition number of x whereas it is used in [21] to derive estimates of the condition numbers of the components of x (see Sect. 2.2). Now we can derive Algorithm 1 that computes $\bar{\kappa}_{LS}$ as expressed in Eq. (8) and using the condition numbers of $z_i^T x$. The vectors $z_1, \cdots, z_q$ are obtained for instance via a QR factorization of a random matrix $Z \in \mathbb{R}^{n \times q}$. The condition number of $z_i^T x$ can be computed using the expression given in [3] as

$$\kappa_i = \left(\|R^{-1}R^{-T}z_i\|_2^2 \|r\|_2^2 + \|R^{-T}z_i\|_2^2 (\|x\|_2^2 + 1) \right)^{\frac{1}{2}}. \tag{9}$$

The accuracy of the estimate can be tweaked by modifying the number q of considered random samples. The computation of $\bar{\kappa}_{LS}$ requires computing the QR factorization of an $n \times q$ matrix for $\mathcal{O}(nq^2)$ flops. It also involves solving q times two $n \times n$ triangular linear systems, each triangular system being solved in $\mathcal{O}(n^2)$ flops. The resulting computational cost is $\mathcal{O}(2qn^2)$ flops (if $n \gg q$).

Algorithm 1. Statistical condition estimation for linear least squares solution

---

Require: $q \geq 1$, the number of samples
    Generate q vectors $z_1, z_2, ..., z_q \in \mathbb{R}^n$ with entries in $\mathcal{U}(0, 1)$
    Orthonormalize the vectors z_i using a QR factorization
    for $j = 1$ to q **do**
        Compute $\kappa_j = \left(\|R^{-1}R^{-T}z_j\|_2^2 \|r\|_2^2 + \|R^{-T}z_i\|_2^2 (\|x\|_2^2 + 1) \right)^{\frac{1}{2}}$
    end for
    Compute $\bar{\kappa}_{LS} = \frac{\omega_q}{\omega_n} \sqrt{\sum_{j=1}^q \kappa_j^2}$ with $\omega_q = \sqrt{\frac{2}{\pi(q-\frac{1}{2})}}$

2.2   Componentwise Condition Estimates

In this section, we focus on calculating the condition number for each component of the LLS solution x. The first one is based on the results from [3] and enables us to compute the exact value of the condition numbers for the ith component of x. The other is a statistical estimate from [21].

Exact formula. By considering in Eq. (9) the special case where $z_i = e_i$, we can express in Eq. (10) the condition number of the component $x_i = e_i^T x$ and then calculate a vector $\kappa_{CW} \in \mathbb{R}^n$ with components κ_i being the exact condition number for the ith component expressed by

$$\kappa_i = \left(\|R^{-1}R^{-T}e_i\|_2^2 \|r\|_2^2 + \|R^{-T}e_i\|_2^2 (\|x\|_2^2 + 1) \right)^{\frac{1}{2}}. \tag{10}$$

The computation of one κ_i requires two triangular solves ($R^T y = e_i$ and $Rz = y$) corresponding to $2n^2$ flops. When we want to compute all κ_i, it is more efficient to solve $RY = I$ and then compute YY^T, which requires about $2n^3/3$ flops.

Statistical condition estimate. We can find in [21] three different algorithms to compute statistical componentwise condition estimation for LLS problems. Algorithm 2 corresponds to the algorithm that uses unstructured perturbations and it can be compared with the exact value given in Eq. (10). Algorithm 2 computes a vector $\bar{\kappa}_{CW} = (\bar{\kappa}_1, \cdots, \bar{\kappa}_n)^T$ containing the statistical estimate of each κ_i. Depending on the needed accuracy for the statistical estimation, the number of random perturbations $q \geq 1$ applied to the input data in Algorithm 2 can be adjusted. This algorithm involves two $n \times n$ triangular solves with q right-hand sides, which requires about qn^2 flops.

Algorithm 2. Componentwise statistical condition estimate for linear least squares

---

Require: $q \geq 1$, the number of perturbations of input data
   for $j = 1$ to q **do**
      Generate $S_j \in \mathbb{R}^{n \times n}$, $g_j \in \mathbb{R}^n$ and $h_j \in \mathbb{R}^n$ with entries in $\mathcal{N}(0, 1)$
      Compute $u_j = R^{-1}(g_j - S_j x + \|Ax - b\|_2 R^{-T} h_j)$
   end for
   Let $p = m(n+1)$ and compute vector $\bar{\kappa}_{CW} = \frac{\sum_{i=1}^q |u_j|}{q \omega_p \sqrt{p}}$ with $\omega_q = \sqrt{\frac{2}{\pi(q - \frac{1}{2})}}$

---

3   Numerical Experiments

In the following experiments, random LLS problems are generated using the method given in [22] for generating LLS test problems with known solution x and residual norm. Random problems are obtained using the quantities m, n, ρ, l such that $A \in \mathbb{R}^{m \times n}$, $\|r\|_2 = \rho$ and $\text{cond}(A) = n^l$. The matrix A is generated using

$$A = Y \begin{pmatrix} D \\ 0 \end{pmatrix} Z^T, \; Y = I - 2yy^T, \; Z = I - 2zz^T$$

where $y \in \mathbb{R}^m$ and $z \in \mathbb{R}^n$ are random unit vectors and $D = n^{-l} diag(n^l, (n - 1)^l, (n - 2)^l, \cdots, 1)$. We have $x = (1, 2^2, ..., n^2)^T$, the residual vector is given by $r = Y \begin{pmatrix} 0 \\ v \end{pmatrix}$ where $v \in \mathbb{R}^{m-n}$ is a random vector of norm ρ and the right-hand side is given by $b = Y \begin{pmatrix} DZx \\ v \end{pmatrix}$. In Sect. 3.1, we will consider LLS problems of size $m \times n$ with $m = 9984$ and $n = 2496$. All the experiments were performed using the library LAPACK 3.2 [1] from Netlib.

3.1   Accuracy of Statistical Estimates

Conditioning of LLS Solution. In this section we compare the statistical estimate $\bar{\kappa}_{LS}$ obtained via Algorithm 1 with the exact condition number κ_{LS} computed using Eq. (3). In our experiments, the statistical estimate is computed

Table 1. Ratio between statistical and exact condition numbers ($q = 2$)

cond(A)	n^0	$n^{\frac{1}{2}}$	n^1	$n^{\frac{3}{2}}$	n^2	$n^{\frac{5}{2}}$	n^3
$\|r\|_2 = 10^{-10}$	57.68	3.32	1.46	1.19	1.10	1.03	1.07
$\|r\|_2 = 10^{-5}$	57.68	3.33	1.45	1.18	1.07	1.09	1.05
$\|r\|_2 = 1$	57.68	3.36	1.45	1.19	1.19	1.05	1.15
$\|r\|_2 = 10^5$	57.68	3.33	1.24	1.04	1.05	1.05	1.02
$\|r\|_2 = 10^{10}$	57.68	1.44	1.07	1.09	1.00	1.01	1.07

using two samples ($q = 2$). For seven different values for cond(A) = n^l (l ranging from 0 to 3, $n = 2496$) and several values of $\|r\|_2$, we report in Table 1 the ratio $\bar{\kappa}_{LS}/\kappa_{LS}$, which is the average of the ratios obtained for 100 random problems.

The results in Table 1 show the relevance of the statistical estimate presented in Sect. 2.1. For $l \geq \frac{1}{2}$ the averaged estimated values never differ from the exact value by more than one order of magnitude. We observe that when l tends to 0 (i.e., cond(A) gets close to 1) the estimate becomes less accurate. This can be explained by the fact that the statistical estimate $\bar{\kappa}_{LS}$ is based on evaluating the Frobenius norm of the Jacobian matrix [17]. Actually some additional experiments showed that $\bar{\kappa}_{LS}/\kappa_{LS}$ evolves exactly like $\|R^{-1}\|_F^2 / \|R^{-1}\|_2^2$. In this particular LLS problem we have

$$\|R^{-1}\|_F^2 / \|R^{-1}\|_2^2 = \left(1 + (n/(n-1))^{2l} + (n/(n-2))^{2l} + \cdots + n^{2l}\right)/n^{2l}$$

$$= \sum_{k=1}^{n} \frac{1}{k^{2l}}.$$

Then when l tends towards 0, $\|R^{-1}\|_F / \|R^{-1}\|_2 \sim \sqrt{n}$, whereas this ratio gets closer to 1 when l increases. This is consistent with the well-known inequality $1 \leq \|R^{-1}\|_F / \|R^{-1}\|_2 \leq \sqrt{n}$. Note that the accuracy of the statistical estimate does not vary with the residual norm.

Componentwise Condition Estimation. Figure 1 depicts the conditioning for all LLS solution components, computed as $\kappa_i/|x_i|$ where κ_i is obtained using Eq. (10). Figure 1(a) and (b) correspond to random LLS problems with respectively cond(A) = $2.5 \cdot 10^3$ and cond(A) = $2.5 \cdot 10^9$. These figures show the interest of the componentwise approach since the sensitivity to perturbations of each solution component varies significantly (from 10^2 to 10^8 for cond(A) = $2.5 \cdot 10^3$, and from 10^7 to 10^{16} for cond(A) = $2.5 \cdot 10^9$). The normalized condition number of the solution computed using Eq. (3) is $\kappa_{LS}/\|x\|_2 = 2.5 \cdot 10^3$ for cond(A) = $2.5 \cdot 10^3$ and $\kappa_{LS}/\|x\|_2 = 4.5 \cdot 10^{10}$ for cond(A) = $2.5 \cdot 10^9$, which in both cases greatly overestimates or underestimates the conditioning of some components. Note that the LLS sensitivity is here well measured by cond(A) since $\|r\|_2$ is small compared to $\|A\|_2$ and $\|x\|_2$, as expected from [25] (otherwise it would be measured by cond(A)^2).

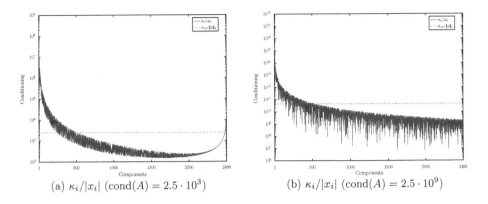

(a) $\kappa_i/|x_i|$ $(\text{cond}(A) = 2.5 \cdot 10^3)$ (b) $\kappa_i/|x_i|$ $(\text{cond}(A) = 2.5 \cdot 10^9)$

Fig. 1. Componentwise condition numbers of LLS (problem size 9984×2496)

In Fig. 2 we represent for each solution component, the ratio between the statistical condition estimate computed via Algorithm 2, considering two samples $(q = 2)$, and the exact value computed using Eq. (10). The ratio is computed as an average on 100 random problems. We observe that this ratio is lower than 1.2 for the case $\text{cond}(A) = 2.5 \cdot 10^3$ (Fig. 2(a)) and close to 1 for the case $\text{cond}(A) = 2.5 \cdot 10^9$ (Fig. 2(b)), which also confirms that, similarly to $\overline{\kappa}_{LS}$ in Sect. 3.1, the statistical condition estimate is more accurate for larger values of $\text{cond}(A)$.

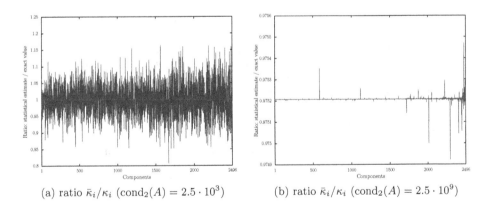

(a) ratio $\overline{\kappa}_i/\kappa_i$ $(\text{cond}_2(A) = 2.5 \cdot 10^3)$ (b) ratio $\overline{\kappa}_i/\kappa_i$ $(\text{cond}_2(A) = 2.5 \cdot 10^9)$

Fig. 2. Comparison between componentwise exact and statistical condition numbers

4 Conclusion

We illustrated how condition numbers of a full column rank LLS problem can be easily computed using exact formulas or statistical estimates at an affordable flop count. Numerical experiments on random LLS problems showed that the statistical estimates provide good accuracy by using only 2 random orthogonal vectors. Subsequently to this work, new routines will be proposed in the public domain libraries LAPACK and MAGMA [24] to compute exact values and statistical estimates for LLS conditioning.

References

1. Anderson, E., Bai, Z., Bischof, C., Blackford, S., Demmel, J., Dongarra, J., Croz, J.D., Greenbaum, A., Hammarling, S., McKenney, A., Sorensen, D.: LAPACK Users' Guide, 3rd edn. SIAM, Philadelphia (1999)
2. Arioli, M., Baboulin, M., Gratton, S.: A partial condition number for linear least squares problems. SIAM J. Matrix Anal. Appl. **29**(2), 413–433 (2007)
3. Baboulin, M., Dongarra, J., Gratton, S., Langou, J.: Computing the conditioning of the components of a linear least squares solution. Numer. Linear Algebra Appl. **16**(7), 517–533 (2009)
4. Baboulin, M., Giraud, L., Gratton, S., Langou, J.: Parallel tools for solving incremental dense least squares problems: application to space geodesy. J. Algorithms Comput. Technol. **3**(1), 117–133 (2009)
5. Baboulin, M., Gratton, S.: Using dual techniques to derive componentwise and mixed condition numbers for a linear functional of a linear least squares solution. BIT **49**(1), 3–19 (2009)
6. Baboulin, M., Gratton, S.: A contribution to the conditioning of the total least squares problem. SIAM J. Matrix Anal. Appl. **32**(3), 685–699 (2011)
7. Björck, Å.: Numerical Methods for Least Squares Problems. SIAM, Philadelphia (1996)
8. Cao, Y., Petzold, L.: A subspace error estimate for linear systems. SIAM J. Matrix Anal. Appl. **24**, 787–801 (2003)
9. Chandrasekaran, S., Ipsen, I.C.F.: On the sensitivity of solution components in linear systems of equations. Numer. Linear Algebra Appl. **2**, 271–286 (1995)
10. Cucker, F., Diao, H., Wei, Y.: On mixed and componentwise condition numbers for Moore-Penrose inverse and linear least squares problems. Math. Comput. **76**(258), 947–963 (2007)
11. Eldén, L.: Perturbation theory for the least squares problem with linear equality constraints. SIAM J. Numer. Anal. **17**, 338–350 (1980)
12. Geurts, A.J.: A contribution to the theory of condition. Numer. Math. **39**, 85–96 (1982)
13. Golub, G.H., Van Loan, C.F.: Matrix Computations, 3rd edn. The Johns Hopkins University Press, Baltimore (1996)
14. Golub, G.H., Wilkinson, J.H.: Note on the iterative refinement of least squares solution. Numer. Math. **9**(2), 139–148 (1966)
15. Gratton, S.: On the condition number of linear least squares problems in a weighted Frobenius norm. BIT **36**, 523–530 (1996)
16. Grcar, J.F.: Adjoint formulas for condition numbers applied to linear and indefinite least squares. Lawrence Berkeley National Laboratory Technical Report, LBNL-55221 (2004)
17. Gudmundsson, T., Kenney, C.S., Laub, A.J.: Small-sample statistical estimates for matrix norms. SIAM J. Matrix Anal. Appl. **16**(3), 776–792 (1995)

18. Higham. N.J.: A survey of componentwise perturbation theory in numerical linear algebra. In: Gautschi, W. (ed.) Mathematics of Computation 1943–1993: A Half Century of Computational Mathematics, Proceedings of Symposia in Applied Mathematics, vol. 48, pp. 49–77. American Mathematical Society, Providence (1994)
19. NJ, Higham: Accuracy and Stability of Numerical Algorithms. SIAM, Philadelphia (2002)
20. Kenney, C.S., Laub, A.J.: Small-sample statistical condition estimates for general matrix functions. SIAM J. Sci. Comput. **15**(1), 36–61 (1994)
21. Kenney, C.S., Laub, A.J., Reese, M.S.: Statistical condition estimation for linear least squares. SIAM J. Matrix Anal. Appl. **19**(4), 906–923 (1998)
22. Paige, C.C., Saunders, M.A.: LSQR: an algorithm for sparse linear equations and sparse least squares. ACM Trans. Math. Softw. **8**(1), 43–71 (1982)
23. Rice, J.: A theory of condition. SIAM J. Numer. Anal. **3**, 287–310 (1966)
24. Tomov, S., Dongarra, J., Baboulin, M.: Towards dense linear algebra for hybrid GPU accelerated manycore systems. Parallel Comput. **36**(5&6), 232–240 (2010)
25. Wedin, P.-Å.: Perturbation theory for pseudo-inverses. BIT **13**, 217–232 (1973)
26. Wilkinson, J.H.: Rounding Errors in Algebraic Processes. Her Majesty's Stationery Office, London (1963)

Numerical Treatment of a Cross-Diffusion Model of Biofilm Exposure to Antimicrobials

Kazi Rahman and Hermann J. Eberl[(✉)]

University of Guelph, Guelph, ON N1G2W1, Canada
{krahman,heberl}@uoguelph.ca

Abstract. We present a numerical method for a highly nonlinear PDE model of biofilm response to antibiotics with three nonlinear diffusion effects: (i) porous medium degeneracy, (ii) super-diffusion singularity, (iii) nonlinear cross-diffusion. The scheme is based on a Finite Volume discretization in space and semi-implicit, non-local time integration. The resulting discretized system is implemented in Fortran and parallelized with OpenMP. The numerical method is validated in a simulation study.

Keywords: Biofilm · Cross-diffusion · Numerical method

1 Introduction

Bacterial biofilms are microbial layers on biotic and abiotic surfaces. Bacteria attach to the surface and start producing extracellular polymeric substances (EPS) in which they are themselves embedded and which protect them against washout and antimicrobials. This protection makes biofilm eradication difficult which causes major problems in the treatment of bacterial infection and in disinfection of medical and industrial surfaces [2]. The mathematical model of biofilm exposure to biocides in our study is a highly nonlinear diffusion reaction system, based on the prototype single-species biofilm model of [5]. In [6] this model was extended to a model of biofilm response to biocides. This is extended in [14] by including cross-diffusion effects describing mixing and separation of active and inactive biomass in more detail. The model shows several non-standard diffusion effects that make numerical treatment difficult: (i) porous medium degeneracy, (ii) super-diffusion singularity, (iii) nonlinear cross-diffusion. We propose a numerical method based on a Finite Volume discretization in space and semi-implicit, non-local time integration. This method is a cross-diffusion extension of the method studied in [12,13] for models of the type [6], and is easy to parallelize.

2 Mathematical Model

The biofilm model is formulated in terms of the dependent variables volume fractions occupied by active biomass X and by inert biomass Y, and the concentrations of nutrient C and biocide B. It reads

R. Wyrzykowski et al. (Eds.): PPAM 2013, Part I, LNCS 8384, pp. 134–144, 2014.
DOI: 10.1007/978-3-642-55224-3_14, © Springer-Verlag Berlin Heidelberg 2014

$$X_t = \nabla \left(D_{11}(X,Y)\nabla X + D_{12}(X,Y)\nabla Y \right) + \mu \frac{CX}{\kappa_1 + C} - \xi_1 X - \xi_2 \frac{BX}{\kappa_2 + B}, \quad (1)$$

$$Y_t = \nabla \left(D_{21}(X,Y)\nabla X + D_{22}(X,Y)\nabla Y \right) + \xi_2 \frac{BX}{\kappa_2 + B}, \quad (2)$$

$$C_t = \nabla \left(D_C (X+Y)\nabla C \right) - \frac{\mu X^\infty}{\Upsilon_1} \frac{CX}{\kappa_1 + C}, \quad (3)$$

$$B_t = \nabla \left(D_B (X+Y)\nabla B \right) - \frac{\xi_2 X^\infty}{\Upsilon_2} \frac{BX}{\kappa_2 + B}. \quad (4)$$

All parameters, are positive; see Table 1 for their definition. Growth of active biomass is due to nutrient uptake. Biocide is degraded during inactivation of active biomass. Spatial expansion of the biofilm is described by the interaction of three nonlinear diffusion effects: (i) a porous medium degeneracy that ensures that the biofilm does not expand as long as there is locally space available to accommodate newly produced biomass, (ii) a super-diffusion singularity that ensures that the maximum biomass density is obeyed, and (iii) cross-diffusion that describes the mixing of both biomass fractions. Following [14], the biomass diffusion coefficients are with $M := X + Y$ given by

$$\begin{cases} D_{11}(X,Y) = \Phi(M) + X\Psi(M), & D_{12}(M) = X\Psi(M), \\ D_{21}(X,Y) = Y\Psi(M), & D_{22}(X,Y) = \Phi(M) + Y\Psi(M), \end{cases} \quad (5)$$

where the functions Φ and Ψ are defined using the density-dependent diffusion coefficient $D(M)$ of a single species biofilm model through

$$D(M) = \delta \frac{M^4}{(1-M)^4} = \Phi(M) + M\Psi(M) \quad (6)$$

$$\Phi(M) = \left(1 - \int_0^M D(m)dm \right) \frac{\int_0^M D(m)dm}{M}. \quad (7)$$

The integral can be expressed in terms of elementary functions. The biofilm phase proper is the region where $M(t) > 0$, the aqueous phase is the region where $X(t) = Y(t) = 0$. Both regions are connected by a moving interface. Diffusion of substrates B, C is slower in the biofilm than in the aqueous phase. Linear interpolation between a fully compressed biofilm with $M = 1$ and the aqueous phase with $M = 0$ gives

$$D_{B,C}(M) = D_{B,C}^0 \cdot (1 + M(\rho_{C,B} - 1)). \quad (8)$$

For the biomass fractions we pose homogeneous Neumann conditions on $\partial\Omega$. For the dissolved substrates we pose Dirichlet conditions on some part of the boundary, from where substrates and biocides are added, and homogeneous Neumann conditions everywhere else. The initial data have compact support: Active biomass is located in small pockets/colonies somewhere at the boundary, everywhere else $X = 0$ initially. We shall assume that initially no inert biomass is in the system $Y = 0$. The initial data for the substrates are $C = C_0 > 0$ and $B = B_0 \geq 0$.

The special form of the biomass cross-diffusivities, in particular $D_{12}(0,\cdot) = D_{21}(\cdot,0) = 0$ maintains non-negativity of X and Y. Adding (1) and (2) we find that $M = X + Y$ is bounded from above by the solution of the prototype single species biofilm model [5], implying with [7] that the solution of the model might come close to its upper bound but never actually attains the super-diffusion singularity, i.e. $X + Y < 1$, at least as long as the interface is separated from the boundary somewhere. A finite speed of interface propagation follows with standard results on degenerate parabolic equations [17]. Nevertheless, regions with $X + Y = 0$ and $X + Y \approx 1$ can be in very close proximity. While the solutions are continuous, the biomass density gradients can blow up at the interface [9].

3 Numerical Method

Each of the three non-Fickian diffusion effects (i), (ii), (iii) mentioned in the Introduction has its own numerical challenges. To deal with (i) and (ii), a semi-implicit numerical method has been developed in [4,12,13], which we extend here to include (iii). A key property of the underlying method is that after semi-implicit discretization it requires the repeated solution of linear systems with at least weakly diagonally dominant M-matrices. Direct discretization of cross-diffusion terms spoils this property. Instead, in the semi-implicit framework used here, the cross-diffusion terms are treated as advection terms. We use upwinding for its discretization, which preserves the M-matrix properties [8,11]. Beside it being only first order, the main disadvantage commonly associated with this method is that it can induce strong artificial numerical diffusion. In our model, where the transition from the biofilm to the aqueous phase is described by a steep interface, interface smearing due to numerical diffusion is an important concern. However, due to the specific form of the diffusion coefficients, we have close to the interface $M \ll 1$, thus $D_{ij} \approx \delta M^4$, and therefore we can hope that these effects are small.

Equations (1), (2) can be written as convection-diffusion-reaction equations

$$X_t = \nabla\left(D_{11}(X,Y)\nabla X - w_X X\right) + R_1(C,B)X, \tag{9}$$
$$Y_t = \nabla\left(D_{22}(X,Y)\nabla Y - w_Y Y\right) + R_2(X), \tag{10}$$

with "cross-diffusion velocities"

$$w_X := -\Psi(X + Y)\nabla Y, \quad w_Y := -\Psi(X + Y)\nabla X.$$

We first discretize in time, using a non-local (in time) representation of non-linearities as suggested in [3,10] for ODE problems, then in space. We use the shorthand notation for dependent variables and nonlinearities,

$$X^k := X(t^k,\cdot), \quad w_X^k := -\Psi(X^k + Y^k)\nabla Y^k, \quad R_1^k := R_1(C^k,B^k), \quad etc.$$

Following [4,12,13] we obtain the semi-implicit time discretization of (9), (10)

$$\frac{X^{k+1} - X^k}{\tau^k} = \nabla \left(D_{11}(X^k, Y^k) \nabla X^{k+1} - w_X^k X^{k+1} \right) + R_1^k X^{k+1} \qquad (11)$$

$$\frac{Y^{k+1} - Y^k}{\tau^k} = \nabla \left(D_{22}(X^k, Y^k) \nabla Y^{k+1} - w_Y^k Y^{k+1} \right) + R_2^k \qquad (12)$$

where $\tau^k := t^{k+1} - t^k$. For the spatial discretization, we use a Finite Volume scheme on a regular grid of size $n \times m$ for the rectangular domain $L \times H$. The dependent variables are evaluated in the center of the grid cells and the diffusive and convective fluxes at the grid cell edges. Thus, for $i = 1, ..., n$, $j = 1, ..., m$, $k = 0, 1, 2, ...$ with $\Delta x := L/n = H/m.$,

$$X_{i,j}^k \approx X \left(t^k, (i - 1/2)\, \Delta x, (j - 1/2)\, \Delta x \right).$$

For the self-diffusive fluxes in (9) we use, as in [4,12,13], the standard second order approximation. For the flux between cells (i, j) and $(i + 1, j)$ we have

$$J_{i+1/2,j}^{k+1} = \frac{1}{2} \left(D(X_{i+1,j}^k, Y_{i+1,j}^k) + D(X_{i,j}^k, Y_{i,j}^k) \right) \frac{X_{i+1,j}^{k+1} - X_{i,j}^{k+1}}{\Delta x},$$

and accordingly for the fluxes across the remaining edges of cell (i, j)

$$J_{i-1/2,j}^{k+1} = \frac{1}{2} \left(D(X_{i-1,j}^k, Y_{i-1,j}^k) + D(X_{i,j}^k, Y_{i,j}^k) \right) \frac{X_{i,j}^{k+1} - X_{i-1,j}^{k+1}}{\Delta x},$$

$$J_{i,j+1/2}^{k+1} = \frac{1}{2} \left(D(X_{i,j+1}^k, Y_{i,j+1}^k) + D(X_{i,j}^k, Y_{i,j}^k) \right) \frac{X_{i,j+1}^{k+1} - X_{i,j}^{k+1}}{\Delta x},$$

$$J_{i,j-1/2}^{k+1} = \frac{1}{2} \left(D(X_{i,j-1}^k, Y_{i,j-1}^k) + D(X_{i,j}^k, Y_{i,j}^k) \right) \frac{X_{i,j}^{k+1} - X_{i,j-1}^{k+1}}{\Delta x}.$$

New in the present model are the cross-diffusion terms, which we represented in (9) as convective terms. At the edge between cells (i, j) and $(i+1, j)$ the velocity component u in x-direction of the velocity vector w_X is calculated as

$$u_{i+1/2,j}^k = \frac{1}{2} \left(\Psi(X_{i+1,j}^k + Y_{i+1,j}^k) + \Psi(X_{i,j}^k + Y_{i,j}^k) \right) \frac{Y_{i+1,j}^k - Y_{i,j}^k}{\Delta x}$$

and accordingly at the edge between cells $(i - 1, j)$ and (i, j)

$$u_{i-1/2,j}^k = \frac{1}{2} \left(\Psi(X_{i-1,j}^k + Y_{i-1,j}^k) + \Psi(X_{i,j}^k + Y_{i,j}^k) \right) \frac{Y_{i,j}^k - Y_{i-1,j}^k}{\Delta x}.$$

At the edges between the cells (i, j) and $(i, j + 1)$ and (i, j) and $(i, j - 1)$, the velocity components in y-direction of w_X are

$$v_{i,j+1/2}^k = \frac{1}{2} \left(\Psi(X_{i,j+1}^k + Y_{i,j+1}^k) + \Psi(X_{i,j}^k + Y_{i,j}^k) \right) \frac{Y_{i,j+1}^k - Y_{i,j}^k}{\Delta x},$$

$$v_{i,j-1/2}^k = \frac{1}{2} \left(\Psi(X_{i,j-1}^k + Y_{i,j-1}^k) + \Psi(X_{i,j}^k + Y_{i,j}^k) \right) \frac{Y_{i,j}^k - Y_{i,j-1}^k}{\Delta x}.$$

Simple first order upwinding leads to the approximation of the cross-diffusion flux $F_{i1/2,j}^k$ at the edge shared by the cells (i,j) and $(i+1,j)$,

$$
F_{i+1/2,j}^{k+1} = \begin{cases} u_{i+1/2,j}^k X_{i,j}^{k+1} & \text{if } u_{i+1/2,j}^k \geq 0, \\ u_{i+1/2,j}^k X_{i+1,j}^{k+1} & \text{if } u_{i+1/2,j}^k < 0, \end{cases}
$$

and similarly at the remaining cell edges

$$
F_{i-1/2,j}^{k+1} = \begin{cases} u_{i-1/2,j}^k X_{i-1,j}^{k+1} & \text{if } u_{i-1/2,j}^k \geq 0, \\ u_{i-1/2,j}^k X_{i,j}^{k+1} & \text{if } u_{i-1/2,j}^k < 0, \end{cases}
$$

$$
F_{i,j+1/2}^{k+1} = \begin{cases} v_{i,j+1/2}^k X_{i,j}^{k+1} & \text{if } v_{i,j+1/2}^k \geq 0, \\ v_{i,j+1/2}^k X_{i,j+1}^{k+1} & \text{if } v_{i,j+1/2}^k < 0, \end{cases}
$$

$$
F_{i,j-1/2}^{k+1} = \begin{cases} v_{i,j-1/2}^k X_{i,j-1}^{k+1} & \text{if } v_{i,j-1/2}^k \geq 0, \\ v_{i,j-1/2}^k X_i^{k+1} & \text{if } v_{i,j-1/2}^k < 0. \end{cases}
$$

Putting all of the above together, we arrive for grid cells in the interior of the domain, i.e. for cells with $1 < i < n$, $1 < j < m$ at

$$
\frac{X_{i,j}^{k+1} - X_{i,j}^k}{\tau^k} = \frac{J_{i+1/2,j}^{k+1} - J_{i-1/2,j}^{k+1}}{\Delta x} + \frac{J_{i,j+1/2}^{k+1} - J_{i,j-1/2}^{k+1}}{\Delta x} \tag{13}
$$
$$
- \left(\frac{F_{i+1/2,j}^{k+1} - F_{i-1/2,j}^{k+1}}{\Delta x} + \frac{F_{i,j+1/2}^{k+1} - F_{i,j-1/2}^{k+1}}{\Delta x} \right) + R_1^k X_{i,j}^{k+1}
$$

For grid cells that share at least one of their edges with a boundary of the domain, this formula accesses non existing cells outside the domain. These are eliminated in the usual manner. For example, for the homogeneous Neumann boundary condition $\partial_n X = 0$ at $x = 0$, we eliminate $X_{0,j}^k$ using $\frac{1}{\Delta x}(X_{0,j}^k - X_{1,j}^k) = 0$. Finally, we introduce the lexicographical grid ordering

$$
\pi : \{1, ..., n\} \times \{1, ..., m\} \to \{1, ..., nm\}, (i,j) \mapsto (i-1)m + j
$$

and the vector notation $\mathcal{X} = (\mathcal{X}_1, ..., \mathcal{X}_{nm})^T$ with $\mathcal{X}_{\pi(i,j)}^k = X_{i,j}^k$. This allows us to re-arrange and re-write (13) in the compact matrix vector form

$$
\left(\mathcal{I} - \tau^k \mathcal{D}_X^k + \tau^k \mathcal{F}_X^k - \tau^k \mathcal{R}_X^k \right) \mathcal{X}^{k+1} = \mathcal{X}^k \tag{14}
$$

where $\mathcal{I}$ is the $nm \times nm$ identity matrix, the matrix $\mathcal{D}_X^k$ contains the contributions of self-diffusion, the matrix $\mathcal{F}_X^k$ contains the cross-diffusion contributions, and the diagonal matrix $\mathcal{R}_X^k$ contains the reaction terms.

Remark 1. These matrices depend only on the dependent variables of the previous time-step and have the following properties, which can easily be verified by straightforward calculations:

$\mathcal{D}_X^k$ is sparse with non-positive diagonal entries and non-negative off-diagonal entries if the biomass densities $X_{i,j}^k$ and $Y_{i,j}^k$ are non-negative and $X_{i,j}^k + Y_{i,j}^k < 1$ $\mathcal{F}_X^k$ is sparse with non-negative diagonal entries and non-positive off-diagonal entries if the biomass densities $X_{i,j}^k$ and $Y_{i,j}^k$ are non-negative and $X_{i,j}^k + Y_{i,j}^k < 1$.

We obtain for Y in a similar manner a linear system of the form

$$\left(\mathcal{I} - \tau^k \mathcal{D}_Y^k + \tau^k \mathcal{F}_Y^k\right) \mathcal{Y}^{k+1} = \mathcal{Y}^k + r^k, \tag{15}$$

where vector r^k contains the contributions of R_2^k, and for C and B without cross-diffusion the simpler

$$\left(\mathcal{I} - \tau^k \mathcal{D}_C^k - \tau^k \mathcal{R}_C^k\right) \mathcal{C}^{k+1} = \mathcal{C}^k + b_1 \tag{16}$$

$$\left(\mathcal{I} - \tau^k \mathcal{D}_B^k - \tau^k \mathcal{R}_B^k\right) \mathcal{B}^{k+1} = \mathcal{B}^k + b_2 \tag{17}$$

where vectors b_1, b_2 contain the contributions of the Dirichlet boundary conditions. Note that the matrices $\mathcal{D}$, $\mathcal{F}$, $\mathcal{R}$ are different in (15)–(17) than in (14), but have the same properties as stated above.

In the biological/physical context X and Y as volume fractions need to be non-negative. In the continuous model (1), (2) this is ensured by the specific form of the cross-diffusion terms with $D_{12}(0, \cdot) = D_{21}(\cdot, 0) = 0$. That the numerical method inherits this property shows the following result, which also implies the absence of spurious oscillations at the steep biofilm/water interface, which are often observed in numerical solutions of such interface propagation problems.

Proposition 1. *If τ^k is small enough and $\mathcal{X}^k \geq 0$, $\mathcal{Y}^k \geq 0$, $\mathcal{C}^k \geq 0$, $\mathcal{B}^k \geq 0$, $\mathcal{X}^k + \mathcal{Y}^k < 1$, then $\mathcal{X}^{k+1} \geq 0$, $\mathcal{Y}^{k+1} \geq 0$, $\mathcal{C}^{k+1} \geq 0$, $\mathcal{B}^{k+1} \geq 0$.*

Proof. The matrices $\mathcal{D}^k$ are weakly diagonal dominant with non-positive diagonal entries and non-negative off-diagonal entries and the matrices $\mathcal{F}^k$ are diagonally dominant with non-negative diagonal entries and non-positive off-diagonal entries. The diagonal matrices $\mathcal{R}^k$ in (16), (17) are non-negative. Thus the system matrices of (15)–(17) are M-matrices for every choice $\tau^k > 0$. The signs of the entries of $\mathcal{R}_X^k$ in (14) depend on $\mathcal{C}^k$ and $\mathcal{B}^k$. Let K be the maximum entry of this matrix. Then for the system matrix of (14) to be an M-matrix it suffices if $\tau^k < 1/K$. Per hypothesis $\mathcal{X}^k \geq 0$, $\mathcal{Y}^k \geq 0$, $\mathcal{C}^k \geq 0$, $\mathcal{B}^k \geq 0$, and r^k in (15), $b_{1,2}$ in (16), (17) are non-negative by definition. Since the inverse of M-matrices are non-negative, it follows $\mathcal{X}^{k+1} \geq 0, \mathcal{Y}^{k+1} \geq 0, \mathcal{C}^{k+1} \geq 0$ and $\mathcal{B}^{k+1} \geq 0$. □

Remark 2. K in the above proof is the characteristic time scale for biofilm growth. Hence the constraint $\tau < 1/K$ is not critical for applications.

The numerical method requires in every time-step to solve the linear systems (14)–(17). We use BiCGSTAB [16] from the free Fortran source code library SPARSKIT [15], which we prepared for parallel execution on shared memory computers using OpenMP [12]. This is the computationally most expensive part, and the only part of the code that is parallelized. BiCGSTAB requires in each iteration two matrix-vector products, four inner products with reduction, and a number of vector additions and scalar-vector multiplications. We compiled and tested the code with the Intel, Portland Group and GNU Fortran compilers on a SGI ALTIX 450, a SGI ALTIXUV, a custom workstation and a Toshiba Tecra Laptop. The simulations reported here were conducted on the ALTIXUV, where the number of OpenMP threads was adjusted to grid size. The parallel behaviour of the underlying method, using the same linear algebra routines, for a related model was previously documented in [12].

4   A Typical Simulation and Grid Refinement

We test the numerical method in a case study. The parameters for biofilm growth are taken from the Benchmark Problem 1 of [18], disinfection parameters are adapted from [1]. They are summarized in Table 1.

Initially, three semi-spherical biofilm colonies are placed on the bottom boundary, two of them centred, and a smaller one at the left corner. Substrate is added from the top. After the biofilm reaches a given height, biocide is added from the top. The simulation stops when the biofilm reaches a given size (area or height). The results are shown in Fig. 1. Initially the biofilm grows due to steady substrate supply. The colonies start expanding when the local biomass density approaches the maximum cell density $M \approx 1$. Eventually the neighbouring colonies in the center merge to form a single, larger colony. In the absence of biocides and inactive biomass, the model behaves like the single-species/single-substrate biofilm model of [5]. At $t \approx 6.9$ biocides are added from the top and diffuse toward the biofilm. Inactivation begins. Inert biomass is observed primarily in the outer layers. In the

Table 1. Default model parameters used in the simulations.

Parameter	Symbol	Value	Unit
Domain length	L	10^{-3}	m
Domain height	H	10^{-3}	m
Bulk substrate concentration	C_∞	30	gm^{-3}
Bulk biocide concentration	B_∞	10	gm^{-3}
Maximum specific growth rate	μ	6	d^{-1}
Half saturation concentration for growth	κ_1	4	gm^{-3}
Half saturation concentration for disinfection	κ_2	1	gm^{-3}
Lysis rate	ξ_1	0.4	d^{-1}
Maximum disinfection rate	ξ_2	48	d^{-1}
Maximum biomass density	X_∞	10^4	gm^{-3}
Yield coefficient for growth	Υ_1	0.63	–
Yield coefficient for disinfection	Υ_2	1	–
Nutrient diffusion coefficient in water	d_C^0	$1. \cdot 10^{-4}$	M^2d^{-1}
Nutrient biofilm/water diffusivity ratio	ρ_C	0.9	–
Biocide diffusion coefficient in water	d_B^0	$.8 \cdot 10^{-4}$	M^2d^{-1}
Biocide biofilm/water diffusivity ratio	ρ_B	0.25	–

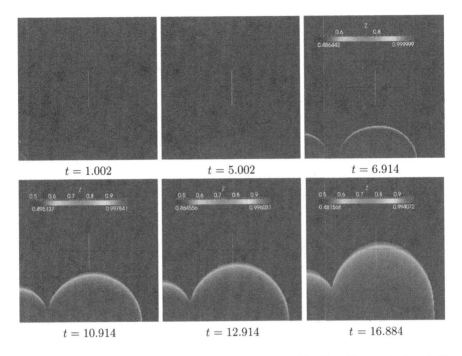

Fig. 1. Simulation of biofilm growth and exposure to biocides. Shown is the biofilm region $X + Y > 0$ for selected t. Biocide is added after the biofilm reaches a certain size. The colour represents active biomass, relative to the overall biomass, $Z = \frac{X}{X+Y}$.

inner layer new biomass continues to be produced. Despite the disinfection rate being significantly larger than the growth rate, this leads to a further expansion of the biofilm in the presence of biocide. Eventually the larger double colony merges with the smaller colony. Throughout the disinfection period, a distinct gradient of inactive biomass relative to the total biomass is observed, from the biofilm/water interface inward. For the parameters used in this simulation, adding the biocide does not prevent the biofilm from growing further or even lead to its decay. This is due to rapid degradation of biocides in the outer layers of the biofilm during inactivation and slow diffusion of biocides into the biofilm.

Due to the nonlinearities, theoretical convergence analysis is difficult. Instead we conduct a computational study on grids of different resolutions, with $n = m = 2^k, k = 5, ..., 11$. The setup is as in Fig. 1, but the colonies are initially bigger, to shorten the phase where no expansion takes place and where cross-diffusion effects do not play a role; see [4,13] for convergence studies of models without cross-diffusion. In Fig. 2 we plot the time course of the following parameters of interest: biofilm size, total active biomass, total inactive biomass, and minimum substrate concentration. We observe grid size effects for the coarsest resolutions, in particular for the measure of biofilm size and the total amount of inactive biomass. Biofilm size is included here also as indicator for the convergence of the

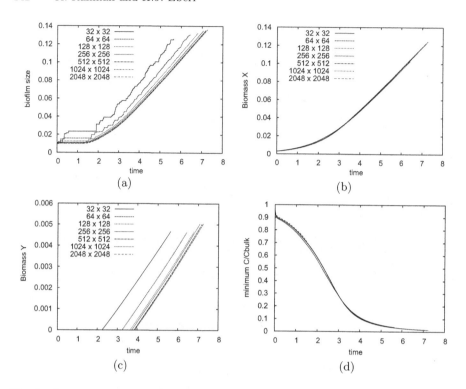

Fig. 2. Output parameters for various grid resolutions: (a) biofilm, (b) total active biomass, (c) total inactive biomass, (c) minimum substrate concentration

biofilm interface. The location of the interface can be accurate at most with Δx which explains the strong differences for the coarsest grids. Disinfection starts when the biofilm reaches a given height. Hence, the onset of disinfection is also closely linked to the accuracy of the interface location. Also here we see big difference for the coarsest grids but note convergence as the grid is refined. Total active biomass and minimum substrate concentration in the system agree well for all grid resolution over the course of the simulation. For practical purposes, a resolution with $n = m = 256$ appears as a good accuracy/cost trade-off. To also account for spatial accuracy explicitly, we compute the difference of the solutions of two subsequent grid resolutions for the last simulation time step, see Table 2. We note a steady decrease in error when refining the grid, indicating convergence of the method. To compute these data the simulation results had to be interpolated between grids. They are subject to interpolation errors and cannot be used to extract the convergence rate. Moreover, these data also are sensitive to the interface location being accurate at most within Δx.

In each time step (14)–(17) are solved. The system matrices change with each instance. Therefore, the number of BiCGSTAB iterations, and hence cost per timestep, is variable over the duration of the simulation (data not shown).

Table 2. Results of the grid refinement study: least square norms for the differences of solutions for grids with $2^k \times 2^k$ and $2^{k-1} \times 2^{k-1}$ cell resolution at the final time step.

k	X	Y	C	B
6	$0.1690 \cdot 10^{-2}$	$0.2473 \cdot 10^{-3}$	$0.3817 \cdot 10^{-3}$	$0.2520 \cdot 10^{-3}$
7	$0.5110 \cdot 10^{-3}$	$0.9279 \cdot 10^{-4}$	$0.7629 \cdot 10^{-4}$	$0.5579 \cdot 10^{-4}$
8	$0.2361 \cdot 10^{-3}$	$0.4853 \cdot 10^{-4}$	$0.2730 \cdot 10^{-4}$	$0.2145 \cdot 10^{-4}$
9	$0.5970 \cdot 10^{-4}$	$0.1427 \cdot 10^{-4}$	$0.4607 \cdot 10^{-5}$	$0.3420 \cdot 10^{-5}$
10	$0.2669 \cdot 10^{-4}$	$0.6714 \cdot 10^{-5}$	$0.1749 \cdot 10^{-5}$	$0.1252 \cdot 10^{-5}$
11	$0.9344 \cdot 10^{-5}$	$0.2428 \cdot 10^{-5}$	$0.4369 \cdot 10^{-6}$	$0.3019 \cdot 10^{-6}$

5 Conclusion

We present a numerical method for a highly nonlinear PDE model of biofilm response to biocides with three nonlinear diffusion effects: (i) porous medium degeneracy, (ii) super-diffusion singularity, (iii) nonlinear cross-diffusion. The scheme extends a previous methods for problems with properties (i) and (ii). The new cross-diffusion terms are treated formally as convective terms with density dependent velocity. The method is based on a Finite Volume discretization in space and semi-implicit, non-local time integration. It preserves positivity. In simulations we showed its convergence with respect to grid refinement.

References

1. Chambless, J.D., Hunt, S.M., Stewart, P.S.: A three-dimensional computer model of four hypothetical mechanisms protecting biofilms from antimicrobials. Appl. Env. Microbiol. **72**, 2005–2013 (2006)
2. Costerton, J.W.: The Biofilm Primer. Springer, Heidelberg (2007)
3. Dimitrov, D.T., Kojouharov, H.V.: Dynamically consistent numerical methods for general productive-destructive systems. J. Diff. Equs. Appls. **17**(12), 1721–1736 (2011)
4. Eberl, H.J., Demaret, L.: A finite difference scheme for a degenerated diffusion equation arising in microbial ecology. El. J. Diff. Equs. CS **15**, 77–95 (2007)
5. Eberl, H.J., Parker, D.F., van Loosdrecht, M.C.M.: A new deterministic spatio-temporal continuum model for biofilm development. J. Theor. Med. **3**, 161–175 (2001)
6. Eberl, H.J., Sudarsan, R.: Exposure of biofilms to slow flow fields: the convective contribution to growth and disinfection. J. Theor. Biol. **253**, 788–807 (2008)
7. Efendiev, M.A., Zelik, S.V., Eberl, H.J.: Existence and longtime behavior of a biofilm model. Comm. Pure Appl. Anal. **8**, 509–531 (2009)
8. Hundsdorfer, W., Verweer, J.G.: Numerical Solution of Time-Dependent Advection-Diffusion-Reaction Equations. Springer, Heidelberg (2003)
9. Jalbert, E., Eberl, H.J.: Numerical computation of sharp travelling wave solutions of a simplified biofilm model. Comm. Nonlin. Sci. Num. Sim. (in press)
10. Mickens, R.E.: Nonstandard finite difference schemes. In: Mickens, R.E. (ed.) Applications of Nonstandard Finite Difference Schemes, pp. 155–180. World Scientific, Singapore (2000)

11. Morton, K.W.: Numerical Solution of Convection-Diffusion Problems. Chapman and Hall, London (1996)
12. Muhammad, N., Eberl, H.J.: OpenMP parallelization of a Mickens time-integration scheme for a mixed-culture biofilm model and its performance on multi-core and multi-processor computers. In: Mewhort, D.J.K., Cann, N.M., Slater, G.W., Naughton, T.J. (eds.) HPCS 2009. LNCS, vol. 5976, pp. 180–195. Springer, Heidelberg (2010)
13. Muhammad, N., Eberl, H.J.: Model parameter uncertainties in a dual-species biofilm competition model affect ecological output parameters much stronger than morphological ones. Math. Biosci. **233**, 1–18 (2011)
14. Rahman, K.A., Eberl, H.J.: Cross-diffusion in biofilms (in preparation)
15. Saad, Y.: SPARSKIT: a basic tool-kit for sparse matrix computations. http://www-users.cs.umn.edu/saad/software/SPARSKIT/index.html (1994)
16. Saad, Y.: Iterative Methods for Sparse Linear Systems, 2nd edn. SIAM, Philadelphia (2003)
17. Samarskii, A.A., Mikhailov, A.P., Galaktionov, V.A., Kurdumov, S.P.: Blow-Up in Quasilinear Parabolic Equations. DeGruyter, Berlin (1995)
18. Wanner, O., Eberl, H., Morgenorth, E., Noguera, D., Picioreanu, D., Rittmann, B., van Loosdrecht, M.: Mathematical Modelling of Biofilms. IWA Publishing, London (2006)

Performance Analysis for Stencil-Based 3D MPDATA Algorithm on GPU Architecture

Krzysztof Rojek[✉], Lukasz Szustak, and Roman Wyrzykowski

Czestochowa University of Technology, Dabrowskiego 73,
42-201 Czestochowa, Poland
{krojek,lszustak,roman}@icis.pcz.pl

Abstract. EULAG (Eulerian/semi-Lagrangian fluid solver) is an established computational model for simulating thermo-fluid flows across a wide range of scales and physical scenarios. The multidimensional positive defined advection transport algorithm (MPDATA) is among the most time-consuming components of EULAG.

The main aim of our work is to design an efficient adaptation of the MPDATA algorithm to the NVIDIA GPU Kepler architecture. We focus on analysis of resources usage in the GPU platform and its influence on performance results. In this paper, a performance model is proposed, which ensures a comprehensive analysis of the resource consumption including registers, shared, global and texture memories. The performance model allows us to identify bottlenecks of the algorithm, and shows directions of optimizations.

The group of the most common bottlenecks is considered in this work. They include data transfers between host memory and GPU global memory, GPU global memory and shared memory, as well as latencies and serialization of instructions, and GPU occupancy. We put the emphasis on providing a fixed memory access pattern, padding, reducing divergent branches and instructions latencies, as well as organizing computation in the MPDATA algorithm in order to provide efficient shared memory and register file reusing.

Keywords: GPGPU · CUDA · EULAG · Stencil · MPDATA · Geophysical flows · Parallel programming

1 Introduction

The multidimensional positive definite advection transport algorithm (MPDATA) is among the most time-consuming calculations of the EULAG model [2,9]. In our previous works [8,10,11] we proposed two decompositions of 2D MPDATA computations, which provide adaptation to CPU and GPU architectures separately. The achieved performance results showed the possibility of achieving high performance both on CPU and GPU platforms.

In the paper [12], we developed a hybrid CPU-GPU version of 2D MPDATA, to fully utilize all the available computing resources by spreading computations across the entire machine. It is the starting point for our current work.

R. Wyrzykowski et al. (Eds.): PPAM 2013, Part I, LNCS 8384, pp. 145–154, 2014.
DOI: 10.1007/978-3-642-55224-3_15, © Springer-Verlag Berlin Heidelberg 2014

In this paper, we focus on parallelization of the 3D MPDATA algorithm, and analysis of resources usage in the GPU platform and its influence on the performance. We detect the bottlenecks and develop the method of efficient distribution of computation across CUDA kernels. Proposed method is based on analysis of memory transactions between GPU global and shared memory.

2   Related Works

Reorganizing stencil calculations to take full advantage of memory hierarchies has been the subject of much investigation over the years.

Modern processor architectures tends to be inherently unbalanced concerning the relation of theoretical peak performance versus memory bandwidth. To reveal performance constraints for MPDATA running on hybrid architectures, we will follow the simple methodology presented in [4], where attainable performance is estimated based on flop and byte ratio.

Memory optimizations for stencil computations have principally focused on different decomposition strategies, like space and blocking techniques [3], that attempt to exploit locality by performing operations on data blocks of a suitable size before moving on to the next block.

The issue of adapting the EULAG model to GPU accelerators was discussed in [5], where the PGI Accelerator compiler was used for the automatic parallelization of selected parts of EULAG on NVIDIA GPUs, including the 2D MPDATA algorithm. However, disadvantage of this approach is relaying entirely on the automatic parallelization, without any efforts to guide the parallelization process taking into account characteristics of target architectures.

In the paper [6], a 3.5D-blocking algorithm that performs 2.5D-spatial blocking of the input grid into on-chip memory for GPUs was discussed. We also employ 2.5D blocking technique to increase data locality, but we propose alternative solution for memory-bounded kernels, which is based on minimizing the number of global memory transactions, rather than applying 3.5D-blocking.

The quite large set of techniques of CUDA optimizations including data parallelism, threads deployment and the GPU memory hierarchy was discussed in [1]. In this work, the authors manually evaluated the best configurations of 2D stencil computations. We offer model-based solution, which automatic configures the code, making our solution more portable.

3   Kepler NVIDIA Architecture

The NVIDIA GTX TITAN GPU [7] is based on the Kepler architecture, and includes 14 streaming multiprocessors (SMX), each consisting of 64 double precision units (DP units) with 48 KB of shared memory and 16 KB of L1 cache. It gives a total number of 896 DP units with the clock rate of 870 MHz. It provides the peak performance of 1.5 TFlop/s in a double precision. This graphics accelerator card includes 6 GB of global memory with the peak bandwidth of 288 GB/s. All the accesses to the global memory go through the L2 cache of size

1.5 MB. This GPU supports two modes of access to data: 32-bit access mode and 64-bit access mode. The number of load/store unit per SMX is 32, so it gives a possibility to load/store 256 bits per clock cycle per SMX.

4 3D MPDATA Overview

Our research includes Multidimensional Positive Definite Advection Transport Algorithm (MPDATA), which is one of the main part of the EULAG geophysical model EULAG (EUlerian/semi-LAGrangian) can be used to simulate: weather prediction; ocean currents; areas of turbulence; urban flows; gravity wave dynamics; micrometeorology; cloud microphysics and dynamics.

The MPDATA algorithm belongs to the group of nonoscillatory forward in time algorithms [9]. The 3D MPDATA is based on the first-order-accurate advection equation:

$$\frac{\partial \Psi}{\partial t} = -\frac{\partial}{\partial x}(u\Psi) - \frac{\partial}{\partial y}(v\Psi) - \frac{\partial}{\partial z}(w\Psi), \tag{1}$$

where x, y and z are space coordinates, t is time, $u, v, w = const$ are flow velocities, and Ψ is a nonnegative scalar field. Equation (1) is approximated according to the donor-cell scheme, which for the $(n+1)$-th time step ($n = 0, 1, 2, \ldots$) gives the following equation:

$$\begin{aligned}
\Psi_{i,j,k}^* = \Psi_{i,j,k}^n &- [F(\Psi_{i,j,k}^n, \Psi_{i+1,j,k}^n, U_{i+1/2,j,k}) - F(\Psi_{i-1,j,k}^n, \Psi_{i,j,k}^n, U_{i-1/2,j,k})] \\
&- [F(\Psi_{i,j,k}^n, \Psi_{i,j+1,k}^n, V_{i,j+1/2,k}) - F(\Psi_{i,j-1,k}^n, \Psi_{i,j,k}^n, V_{i,j-1/2,k})] \\
&- [F(\Psi_{i,j,k}^n, \Psi_{i,j,k+1}^n, W_{i,j,k+1/2}) - F(\Psi_{i,j,k-1}^n, \Psi_{i,j,k}^n, W_{i,j,k-1/2})].
\end{aligned} \tag{2}$$

Here the function F is defined in terms of the local Courant number U:

$$F(\Psi_L, \Psi_R, U) \equiv [U]^+ \Psi_L + [U]^- \Psi_R, \tag{3}$$

$$U \equiv \frac{u\delta t}{\delta x}; \ [U]^+ \equiv 0,5(U + |U|); \ [U]^- \equiv 0,5(U - |U|). \tag{4}$$

The same definition is true for the local Courant numbers V and W.

The first-order-accurate advection equation can be approximated to the second-order in δx, δy and δt, defining the advection-diffusion equation. Such transformation is widely described in literature. For the full description of the main important aspects of the second order equation of MPDATA, the reader is referred to [9].

The 3D MPDATA algorithm consists of 17 stencils that are processed by CUDA kernels on the GPU. Figure 1 shows the mechanism of kernel processing. We employ widely used method of 2.5D blocking [6], where two dimensional CUDA blocks are responsible for computing XY planes of matrices. The loop inside kernel is used to traverse the grid in the Z dimension. Since, the MPDATA algorithm requires to store 3 XY planes at the same time, we use queue of planes placed in registers and shared memory, which firstly copies data from GPU global memory to registers, and then moves data between registers and shared memory. This method allows us to increases data locality significantly.

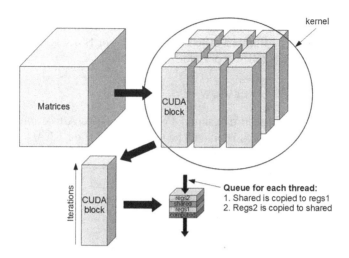

Fig. 1. Kernel processing

5   Analysis of 3D MPDATA with NVIDIA Visual Profiler

The starting point of our considerations is when the 17 stencils are distributed across 6 CUDA kernels. Our analysis begins with detection of bottlenecks of the algorithm. We examine the following potential bottlenecks:

- data transfers between GPU global memory and host memory;
- instructions latency (stall analysis);
- arithmetic, logic, and shared memory operations;
- configuration of the algorithm taking into account size of CUDA block and GPU occupancy.

Our approach is based on the stream processing [12] (Fig. 2) where each stream is responsible for computing a sequence of 3 instructions including: data transfer from host to GPU that occurs only once (before computations); execution the sequence of 6 kernels; data transfer from GPU to host memory (occurs after every time step). Since all streams are processed independently, the computation and data transfers can be overlapped. Table 1 shows the time consumption analysis of MPDATA for the 100 time steps and grid of size 392. Three streams are used in the simulation. The HTOD abbreviation means the data transfer from host to device, while the DTOH means data transfer in the opposite direction.

Based on this analysis, the data transfer takes relatively short time (about 18 % of all execution time). Stream processing decreases execution time by about 0.9 s, which is 2 times more than time of data transfer. We can simply conclude here, that data transfer between host and GPU is not a bottleneck of the MPDATA algorithm.

Now we focus on the analysis of computations. Our research include the most complex part of the MPDATA algorithm. Based on the NVIDIA Visual Profiler,

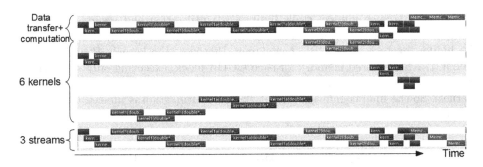

Fig. 2. Utilization of GPU resources by MPDATA

Table 1. Time consumption analysis of the 3D MPDATA algorithm

Operation	Time [s]	Ratio
HTOD	0.023	0.008
DTOH	0.453	0.172
Computation	3.051	1.16
Final time	2.631	1

we estimate two the most time consuming kernels, which are called kernels B and C. These kernels take about 57 % of the execution time. Each of this kernel has 5 input and 3 output matrices and is responsible for computing 3 stencils with 37 flops per each. The flop/B ratio for each kernel is $37 * 3/((5 + 3) * 8) = 1.73$. However, the minimum flop/B ratio required by NVIDIA GTX TITAN to achieve maximum performance is 5.2 [7]. The another conclusion is that the kernels are strongly memory-bounded!

The next analysis is devoted to the stall reasons analysis. Figure 3 shows the main reason of stalls for kernels B and C including execution dependency, data request, texture memory operations, synchronization, and instruction fetching. Based on the analysis, stalls are mostly caused by the execution dependency

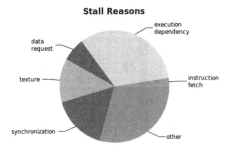

Fig. 3. Analysis of stall reasons for kernels B and C

(about 33 %). Such kind of stalls limits GPU utilization and results from the complex structure of the MPDATA algorithm. The execution dependencies can be hidden by increasing GPU occupancy. However, the kernels B and C use about 47 KB of shared memory for each CUDA block, executing only 768 active threads per SMX. It means, that the GPU occupancy is only 37.5 % for the both kernels. So the final conclusion is that the GPU utilization is limited by shared memory usage!

6   Performance Analysis Based on GPU Global Memory Transactions

We propose a performance analysis based on GPU global memory transactions. Such analysis is particularly helpful, when algorithm is memory-bounded. In our approach, the following scenarios are considered:

– distribution of computation across 2 kernels;
– compression of computation within 1 kernel.

The compression of computation increases hardware requirements for CUDA blocks, and decreases the GPU occupancy. Hence, the second scenario allows us to execute at most 512 active threads per SMX. It means, that the GPU occupancy is even lower than for the first scenario, and it is only 25 %.

At the beginning of our analysis we need to estimate the cost of access to matrix for each scenario. We assume, that CUDA block is of size $g_1 \times g_2$, matrices are processed according with Fig. 1, halo areas are of size 1, and are placed from the four sides of CUDA block (Fig. 4). The number of elements, that need to be transferred from GPU global memory to shared memory or register files is given by the following formula:

$$S_{el} = g_1 * g_2 + 2 * g_2 + 2 * g_1. \tag{5}$$

Taking into account 64-bits access mode, which can be simply enabled on Kepler NVIDIA architecture by calling $cudaDeviceSetSharedMemConfig()$

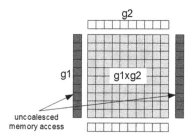

Fig. 4. XY plane of CUDA block with its halo areas

routine with $cudaSharedMemBankSizeEightByte$ parameter, we can estimate the number of required transactions to transfer a single CUDA block:

$$S_{tr} = g_1 * top(g_2/32) + 2 * top(g_2/32) + 2 * g_1, \qquad (6)$$

where $top(x)$ returns rounded up value of x. In this approach, addresses of vertical halo areas are not coalesced.

Table 2 shows the cost of access to matrix for the first scenario. In this analysis, the plane is of size 392×256. Transactions overhead is ratio between required number of transactions to transfer matrix and the naive number of transactions required to transfer matrix assuming unlimited size of shared memory (without halo area). The naive number of transactions for plane of size 392×256 is 3136. Based on our analysis, the minimum number of transactions for the first scenario is $3136 * 185.2\% = 5808$. This analysis also allows us to estimate the most suitable size of CUDA block, which is 6×128.

Table 2. Analysis of GPU global memory transactions: first scenario

g1	g2	Blocks per plane	Transactions per block	Transactions per plane	Transactions overhead [%]
6	128	132	44	5808	185.2
5	128	158	38	6004	191.45
3	256	131	46	6026	192.16
4	128	196	32	6272	200
8	96	147	46	6762	215.63
3	128	262	26	6812	217.22
12	64	132	52	6864	218.88
7	96	168	41	6888	219.64

A similar analysis is made for the second scenario and the results of this analysis are shown in Table 3. The best configuration of CUDA block is 4×64, while the transactions overhead is 250%.

Finally, we estimate the cost of access to all matrices for the both scenarios. Figure 5 shows flow diagram for kernels B and C. There are 5 input matrices for the kernel B and 6 input matrices for the kernel C. Additionally, there are 2 output matrices per each kernel. Transactions overhead for each matrix is 185.2%, so the total cost of access to all matrices is $(5+6+2+2)*1.852 = 27.78$.

The flow diagram for the second scenario is shown in Fig. 6. Here we have 5 input and 3 output matrices. Transactions overhead is 250%. So, the total cost of access to all matrices is $(5+3)*2.5 = 20$.

Table 4 shows the summary of MPDATA analysis, taking into account considered scenarios. Based on our analysis, it is expected to achieve about 1.39 speedup using the second scenario over the first scenario. So the conclusion is that, we should compress kernels B and C into a single kernel.

Table 3. Analysis of GPU global memory transactions: second scenario

g1	g2	Blocks per plane	Transactions per block	Transactions per plane	Transactions overhead [%]
4	64	392	20	7840	250
2	128	392	20	7840	250
3	64	524	16	8384	267.35
2	96	588	16	9408	300
2	64	784	12	9408	300
8	32	392	26	10192	325
1	256	392	26	10192	325
7	32	448	23	10304	328.57

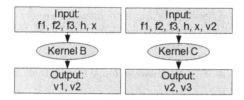

Fig. 5. Flow diagram for kernels B and C

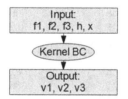

Fig. 6. Flow diagram for BC kernel

Table 4. Summary of MPDATA analysis

	Kernels B and C	Kernel BC	Ratio
Occupancy	37.50 %	25.00 %	1.5
Access overhead	185.20 %	250.00 %	0.74
# of matrices	15	8	1.85
Total cost of access	27.78	20	1.39

7   Performance Results

Table 5 presents performance results for both scenarios. In our tests we used a single NVIDIA GTX TITAN GPU with Intel Core i7–3770 CPU. The MPDATA algorithm was tested for the grid of size $392 \times 256 \times 64$. The achieved results are far from the peak performance due to the complexity of the algorithm, strong instructions and data dependencies, and shared memory size limitations.

Table 5. Performance results for both scenarios

Kernel	Mflops per scenario	Time per scenario [ms]	Performance [Gflop/s]	Speedup
B and C	963.4	15.47	62.29	1
BC	847.8	10.47	80.95	1.48

Our method of stencils distribution across CUDA kernels allows for increasing the MPDATA performance by about 1.48 times. Such a speedup is a little higher than we expected due to the fact that compression of kernels brings some additional advantages, which were not taken into account in our analysis. The main reason of a higher speedup is possibility of applying the common subexpression elimination to reduce the number of MPDATA instructions. It allows us to reduce the number of operation from 963.4 Mflops to 847.8 Mflops.

8   Conclusions and Future Work

The proposed methods allow for estimating "the best" number of kernels, as well as easy selection of CUDA block size for each kernel. The compression of stencils into kernels and improvement of GPU occupancy are mutual excluded. However, the analysis of GPU global memory transactions allows us to find the compromise between these two kinds of optimizations. Moreover, the compression of kernels permits for decreasing the amount of computation. The proposed approach to kernel processing with queues of data placed in registers and shared memory increases the data locality significantly. The performance of kernels in our approach is limited by the number of memory transactions and latency of arithmetic operations. The GPU utilization is mostly limited by the size of shared memory.

Our parallelization of the EULAG model is still under development. The future work will focus on expansion of the implementation across a cluster of CPU-GPU nodes. The particular attention will be paid to implementation of MPDATA using OpenCL in order to ensure the code portability across different devices, as well as development of autotunig mechanisms aiming at providing performance portability.

Acknowledgments. This work was partly supported by the Polish National Science Centre under grant no. UMO-2011/03/B/ST6/03500.

References

1. Cecilia, J.M., García, J.M., Ujaldón, M.: Cuda 2D stencil computations for the Jacobi method. In: Jónasson, K. (ed.) PARA 2010, Part I. LNCS, vol. 7133, pp. 173–183. Springer, Heidelberg (2012)

2. Ciznicki, M., Kopta, P., Kulczewski, M., Kurowski, K., Gepner, P.: Elliptic solver performance evaluation on modern hardware architectures. In: Wyrzykowski, R., Dongarra, J., Karczewski, K., Wasniewski, J. (eds.) PPAM 2013, Part I. LNCS, vol. 8384, pp. 155–165. Springer, Heidelberg (2014)

3. de la Cruz, R., Araya-Polo, M., Cela, J.M.: Introducing the semi-stencil algorithm. In: Wyrzykowski, R., Dongarra, J., Karczewski, K., Wasniewski, J. (eds.) PPAM 2009, Part I. LNCS, vol. 6067, pp. 496–506. Springer, Heidelberg (2010)

4. Hager, A., Wellein, G.: Introduction to High Performance Computing for Science and Engineers. CRC Press, Boca Raton (2011)

5. Kurowski, K., Kulczewski, M., Dobski, M.: Parallel and GPU based strategies for selected CFD and climate modeling models. Environ. Sci. Eng. **3**, 735–747 (2011)

6. Nguyen, A., Satish, N., Chhugani, J., Changkyu, K., Dubey, P.: 3.5-D blocking optimization for stencil computations on modern CPUs and GPUs. In: Proceedings of the 2010 ACM/IEEE International Conference for High Performance Computing, Networking, Storage and Analysis, pp. 1–13 (2010)

7. NVIDIA Kepler Compute Architecture. http://www.nvidia.com/object/nvidia-kepler.html

8. Rojek, K., Szustak, L.: Parallelization of EULAG model on multicore architectures with GPU accelerators. In: Wyrzykowski, R., Dongarra, J., Karczewski, K., Waśniewski, J. (eds.) PPAM 2011, Part II. LNCS, vol. 7204, pp. 391–400. Springer, Heidelberg (2012)

9. Smolarkiewicz, P.: Multidimensional positive definite advection transport algorithm: an overview. Int. J. Numer. Meth. Fluids **50**, 1123–1144 (2006)

10. Szustak, L., Rojek, K., Gepner, P.: Using Intel Xeon Phi coprocessor to accelerate computations in MPDATA algorithm. In: Wyrzykowski, R., Dongarra, J., Karczewski, K., Wasniewski, J. (eds.) PPAM 2013, Part I. LNCS, vol. 8384, pp. 582–592. Springer, Heidelberg (2014)

11. Wyrzykowski, R., Rojek, K., Szustak, L.: Using Blue Gene/P and GPUs to accelerate computations in the EULAG model. In: Lirkov, I., Margenov, S., Waśniewski, J. (eds.) LSSC 2011. LNCS, vol. 7116, pp. 670–677. Springer, Heidelberg (2012)

12. Wyrzykowicz, R., Szustak, L., Rojek, K., Tomas, A.: Towards efficient decomposition and parallelization of MPDATA on hybrid CPU-GPU cluster. In: LSSC 2013. LNCS (in print)

Elliptic Solver Performance Evaluation on Modern Hardware Architectures

Milosz Ciznicki[1][✉], Piotr Kopta[1], Michal Kulczewski[1][✉],
Krzysztof Kurowski[1], and Pawel Gepner[2]

[1] Poznan Supercomputing and Networking Center, Noskowskiego 10 Street,
61-704 Poznan, Poland
{miloszc,michal.kulczewski}@man.poznan.pl
[2] Intel Corporation, Pipers Way, Swindon, Wiltshire SN3 1RJ, UK

Abstract. The recent advent of novel multi- and many-core architectures forces application programmers to deal with hardware-specific implementation details and to be familiar with software optimisation techniques to benefit from new high-performance computing machines. An extra care must be taken for communication-intensive algorithms, which may be a bottleneck for forthcoming era of exascale computing. This paper aims to present performance evaluation of preliminary adaptation techniques to hybrid MPI+OpenMP parallelisation schemes we provided into the EULAG code. Various techniques are discussed, and the results will lead us toward efficient algorithms and methods to scale communication-intensive elliptic solver with preconditioner, including GPU architectures to be provided later in the future.

Keywords: Elliptic solver · Preconditioning · EULAG · High performance computing · Petascale computing · Hybrid parallelisation · Intel Xeon Phi

1 Introduction

The recent advent of novel multi- and many-core architectures, such as GPU and hybrid models, offer notable advantages over traditional supercomputers [1]. However, application programmers have to deal with hardware-specific implementation details and must be familiar with software optimisation techniques to benefit from new high-performance computing machines. It is therefore of great importance to develop expertise in methods and algorithms for porting and adapting the existing and prospective modelling software to these new, yet already established machines [2,3].

Elliptic solvers of an elastic models are usually based on standard iterative algorithms for solving linear systems, e.g. CG, GMRES or GCRK. Numerous reports on porting them to modern architectures are available [4]. However, in an anelastic solver for geophysical flows fast-acting physical processes may enter the elliptic problem implicitly. Furthermore, formulation of the boundary

R. Wyrzykowski et al. (Eds.): PPAM 2013, Part I, LNCS 8384, pp. 155–165, 2014.
DOI: 10.1007/978-3-642-55224-3_16, © Springer-Verlag Berlin Heidelberg 2014

conditions is not trivial and therefore it is not feasible to use standard iterative solvers from linear algebra packages. For simulating physical experiments with a high degree of anisotropy, additional preconditioning is necessary to improve matrix conditioning. Such preconditioned for anisotropic geometries often relies on the direct inversion using the Thomas algorithm. A comprehensive study on implementations of tridiagonal solvers on GPU found that it is possible to implement solvers which perform exceptionally well in the range of grid nodes [5].

EULAG [6], an elastic model for simulating low Mach number flows under gravity, developed in the National Center for Research, is widely used in an international community and has a rich portfolio of applications. It features non-oscillatory forward-in-time (NFT) numerics, which are original and unique. It also employs preconditioned, nonsymmetric, generalised conjugate-residual type "Krylov" scheme [7–10] to solve an elliptic boundary value problem - reported to be among the most effective methods for solving difficult elliptic problems [11]. Based on variational principles, Krylov solvers provide a hierarchical framework, which assures an asymptotic convergence rate in inverse proportion to the square root of the condition number of the linear operator, resulting from the numerical formulation of the model. The hierarchical design of Krylov solvers relies on the operator preconditioning, the goal of which is to accelerate the convergence of the main solver beyond the theoretically optimal limit.

Our research is to provide novel methods to adapt scientific code to novel hardware architectures, taking EULAG as an example. As a result, a prototype parallel formulation of the elliptic solver on modern architectures will become available. In this paper we focus on performance evaluation of preliminary adaption techniques to hybrid MPI+OpenMP parallelisation schemes we provided into the EULAG code. We discuss various techniques, along with results taken from various modern hardware architectures. The results will lead us toward efficient algorithms and methods to scale elliptic solver along with preconditioner, including GPU architectures, to be provided later in the future.

The remainder of the paper is organised as follows. In the next section, parallelisation of the model is discussed, following details on MPI+OpenMP improvements. Design of the parallel experiments and the corresponding results are discussed in Sect. 3. Remarks in Sect. 4 conclude the paper.

2   Model Parallelisation

2.1   Current State

EULAG employs pure MPI programming model for parallelisation between all cores. The computational grid is divided in all three dimensions and each MPI process advances the solution in its subdomain. The computational domain is decomposed evenly so that MPI processes have the same number of grid points and the same computational load. The semi-implicit iterative elliptic solver requires inter-processor communication between all processors in the 3D MPI topology. All-to-all communications are required for calculating global reduction operations, whereas point-to-point communications exchange halo regions between nearest neighbours on x, y and z dimensions.

The latest state-of-the-art scalability results of EULAG model is presented by Piotrowski et al. in [12,13]. The new three-dimensional decomposition has been proposed to increase model performance and scalability. The performance of the new code is demonstrated on the IBM BlueGene/L and Cray XT4/XT5 supercomputers. Form small-scale components of Earth-system models, 3D schemes can provide substantial gains in the model performance over 2D schemes.

The domain decomposition scheme has been recently extended over 2D approach in the vertical direction [13]. The code arrays are explicitly dimensioned to contain a subgrid of a total array corresponding to the entire model grid, plus an extra space for a copy of the neighbouring processors' boundary cells, common referred as "hallo cells". Each subgrid is then assigned to only one processor, though the halo regions may vary through the code being running. To minimise communication, the halo regions information are exchanged partially and only when needed. To exchange such information between processors, the code uses distributed memory message-passing MPI standard.

2.2 Hybrid MPI+OpenMP Improvements

The MPI-all approach provides uniform way of employing the parallelism available in the application. It assumes that the message passing is the correct paradigm for all systems. However, this may not be true especially for the hierarchical systems. Pure MPI ignores the fact that today's computational nodes contain multi- and many-core processors that reside on a shared memory. Furthermore, as the energy cost of moving data across the interconnect is substantially higher than the cost of intra-chip communication [14] we can expect that more emphasis will be put on shared memory programming models. Although some MPI libraries employ shared caches within node to improve communication time, these optimisations are usually hidden from the application programmer. Moreover, our applications of the GCRK solver typically contain small computational grids not larger than 512^3 grid points. Thus, in the pure MPI model adding new cores significantly increases the communication load relatively to the local subdomain size on per rank basis.

OpenMP is a well established shared memory programming model. The application employing it is easy to maintain and debug as the application can still run as a valid serial code. The standard enables to develop performance portable application that would run on multi-core CPUs and many-core coprocessors. Additionally, the recent OpenACC standard allow future development to employ accelerators with small changes to existing OpenMP directives. What is more, any OpenMP compliant compiler can generate shared memory code even if the support for OpenACC is not available. Therefore, the hybrid MPI+OpenMP programming was chosen as a good model that combines the distributed memory inter-node parallelism and the shared memory intra-node parallelism.

The principal logic of the GCRK solver resides within six routines. The *gcrk* main routine advances the solution by iteratively calling other major computational routines. Furthermore, the *gcrk* routine invokes the collective communication to compute grid global values. Second routine *coef0* is called at the beginning of the GCRK solver to calculate required coefficients. The *rhsdiv* routine evaluates right hand side of the governing equations. The pressure forces are calculated by *prforc* routine. The most computational intensive routines are *laplc* and *nablaCnablaxy* that calculate the laplacian and the divergence of $C * gradient$ respectively. Finally, the preconditioner is represented by the *precon* routine. The preconditioner is based on sequential Thomas algorithm to solve tridiagonal system of equations.

The computational loops in the code structure of the GCRK solver represent almost 100 % of the computation. Thus, the OpenMP paradigm should utilise most of the available computational parallelism. The computational loops can be simply divided, with regard to the data access pattern, into two categories: the point wise computations and the stencil computations. The stencil computations consist ordered tasks. At first the message buffers are prepared with necessary data. Next the non-blocking receives and sends are posted and then the inner points are computed. Afterward, when the communication is completed the points that use halo data are updated. This approach clearly separated communication of halo data from computing the inner points thus it allowed to overlap each other. In addition, the boundary conditions that usually involve conditional statements are moved from computations of inner points to evaluation of outer points in order to facilitate the compiler SIMD parallelisation. However, this approach has some constraints. The analysis of the profiling information showed that generally there is no overlap between the communication and the computation. One MPI process is not able to process both tasks in parallel. As a result, the communication is done at the synchronisation point. Moreover, some stencil computations are spreaded between small loops that last less than 1 % of the execution time. Even if the communication is overlapped with the computation there is need to restructure the code to gather more computations within loops. The main advantage of the MPI+OpenMP hybridisation is reduction of the MPI processes and thus the number of exchanged messages.

2.3   Parallelisation of Stencil Computations

To address these issues we restructured the computational routines as shown in the following example:

```
if(th_id.eq.comm_th) then
c     Communication thread starts to transfer halo
      MPI_IRecv(halo data)
      MPI_ISend(halo data)
else
c     Manually distribute iterations
      it_space = (end-start+1+num_ths-1-1) / (num_ths-1)
```

```
        th_start = start+(th_id-1)*it_space
        th_end = MIN(start+th_id*it_space-1,end)
c       Computation
        do i=th_start,th_end
c           Some code
        enddo
endif
```

One OpenMP thread is designated to handle the communication so that the
rest perform the computations. This change allows to fully overlap the communication with the computation. Nevertheless, it complicates the distribution of
iterations as worksharing constructs available in the OpenMP standard operates
only on the whole team of threads. Therefore, the manual work distribution is
employed to remove this obstacle. This workaround has beneficial implications
for the cache reuse as it ensures that each OpenMP thread operates on the same
grid points for the consecutive computational loops. Additionally, the loop iterations are distributed according to the stencil pattern so that the grid points are
located in the local cache. Furthermore, the same work distribution is used for
first touch initialisation of the arrays to ensure the NUMA locality. This technique also significantly reduces overhead related to the parallel DO worksharing
construct. On the other hand, the potential drawback is that directives for loop
worksharing with dynamic or guided schedules cannot be employed and this can
lead to poor load balancing. However, this issue is not relevant for our case as
at some point the addition of new nodes in the strong scaling for small computational domains leads to subdomains that can easily fit in the local cache.
Therefore, the data placement in the local cache is essential for fast computations. The dynamic worksharing constructs typically provide poor cache locality.

2.4   Parallelization of TDMA

The preconditioner is based on the sequential tridiagional Thomas algorithm [15]
and it has two different implementations that depend on the domain decomposition. The naive version of the algorithm is selected for the 2D horizontal decomposition whereas the recurrence doubling version is used with the 3D decomposition. The detailed description of the recurrence doubling version can be found
in [12].

```
c Naive version
c lstep=2 or -2 - parallel recurrence going up or going down
lone=sign(2,lstep)
call get_iterations(ngstart,ngend,mgstart,
& mgend,gstart,gend,ntstart,ntend,mtstart,mtend)
do k=gstart,gend,lone
    do j=mtstart,mtend
        do i=ntstart,ntend
            w(i,j,k) = q(i,j,k) + p(i,j,k)*w(i,j,k-lstep)
```

```
          enddo
       enddo
  enddo

c Pipelined version
c Lowermost MPI ranks in the vertical dimension
c perform naive TDMA until threshold value.
if(lpos.le.threshold) then
c    Wait for the first plane
     call rcvbuff(w)
     call naivetdma(w,p,q)
c    Send the last plane
     call sndbuff(w)
else
c    last=q(lp)+q(lp-2)*p(lp)+q(lp-4)*p(lp-2)*p(lp)
c    Compute products of recurrence coefficients
     call computeqpproduct(q,p,qppro)
c    Wait for the first plane
     call rcvbuff(w)
c    Compute last plane
     call computelastfromfirst(w,qppro)
c    Send the last plane
     call sndbuff(w)
     call naivetdma(w,p,q)
endif
```

The naive version is parallelised with fine granularity in the horizontal dimension nevertheless the dependency on the vertical dimension restrict paralellisation to even and odd planes. In the recurrence doubling version of the algorithm the computation of products of recurrence coefficients and the computation of the last plane is fully parallelisable in the vertical direction.

2.5   Tuning on Xeon Phi

The specific optimisations are employed for the Xeon Phi coprocessor. The profiling data showed that copying the data from MPI buffers to the halo regions contributed to the high number of the cache misses. Particularly, the halo data for the i, j and k dimensions have to be saved in the array of updated variable, when the MPI message is received, with 1, np and $np * mp$ strides respectively. To improve cache hits, the halo region for each dimension was moved to the separate array so that the halo data is linearly ordered within the new array. This change minimised the stride to one for all dimensions and significantly improved cache hits in loops that: update boundary points with the halo data and copy the halo data from MPI buffers. Moreover, it improved cache hits in loops that calculate inner points as the array of updated variable did not contain the halo regions and thus the array was reduced in size. Another important aspect was the memory alignment as it may determine the type of the vector operation used by the compiler. The Intel compiler switch "-vec-report6" generates detailed report about the loop vectorization with aligned or unaligned access. The GCRK solver arrays are included in common blocks and preventing the Intel compiler (v13.0) from aligning arrays on 64 byte boundaries, which is the perfect alignment for Xeon Phi. To partially address this issue the compiler switch "align -zcommons" was used to align block entities on 32 byte boundaries as common blocks could not be removed.

2.6   Mapping Application Topology to Cluster Topology

A considerable amount of effort was spent to improve the application performance by taking advantage of the cluster topology. Current specification of MPI standard (3.0) provides simple primitives such as MPI_Cart_create and MPI_Graph_create to model the application topology. These primitives allow to reorder the MPI ranks based on the information provided by the user. However, the standard makes no assumption about the way of mapping the application topology to the machine topology. The main idea was to implement the method of mapping the application topology to the machine topology to efficiently exploit actual hardware. The first step in our approach consists of discovering the cluster topology. The MPI ping-pong benchmark is used to measure the sustained latency and the sustained bandwidth between all pairs of cluster nodes for different message sizes. After the benchmark the cost matrix is calculated so that each cell represents the cost of sending message between two nodes. The message sent within the same node has the cost equal to zero. The cost is calculated as following:

$$cost = n * latency + msg\_size/bandwidth \tag{1}$$

where n is the number of messages. Afterward, the minimum path is calculated. The procedure starts with selecting the random starting node and then picking the next node with the lowest communication cost until the desired number of nodes is obtained. This procedure is repeated for all starting nodes and the minimum path is selected. Next, the application topology is mapped to the hardware topology. The MPI rank with the position $(0, 0, 0)$ in Cartesian grid is mapped to the first node in the minimum path. Later, the next MPI rank with the closest spatial position is mapped to the same node, if it is not full, or to the next node according to the calculated minimum path. One may want to map more than one rank on each node e.g. one MPI rank per NUMA node. The Hilbert curve [16] is used to calculate the spatial locality. This procedure is repeated until all MPI ranks are mapped.

3   Results

To examine the code scalability the experiments with Held-Suarez idealised climates [17] and the initial-value problem of the Taylor-Green vortex [18], posed

Table 1. Short overview of the tested hardware architectures

Machine name	Inula	Xeon Phi
Processor	AMD Opteron 6234	Xeon Phi 3115A
Cores	24	57
Core frequency (GHz)	2.4	1.1
L1 cache size (data)	16 KB	32 KB
L2 cache size	2 MB	512 KB
L3 cache size	6 MB	–
Memory transfer rate	32 (GB/s)	240 (GB/s)
Interconnect	HyperTransport (6.4 GT/s)	PCI Express 2.0
Number of threads/core	2	4
TDP(W)	115	300

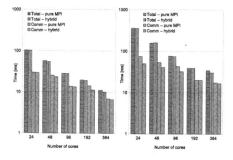

Fig. 1. The strong scaling of two test cases: the Taylor-Green - left and the Held-Suarez - right.

on a triply periodic cube, were preformed. Both tests were conducted on two hardware architectures: Inula cluster with AMD CPUs and Xeon Phi cluster, see Table 1. The number of iterations in the GCRK solver and the number of time steps are fixed in all simulations. The grids employed in tests are composed of 120x120x120 and 240x120x60 points for the HS and the TG tests respectively.

Obviously, to make all tests fair, the mapping of the application topology to the cluster topology was employed for both codes. On runs with the high number of cores this method reduced the execution time by factor of 2 relatively to the random mapping. The MPI processes and the OpenMP threads were pinned to the cores. The hybrid code used four MPI ranks per Inula node so that each MPI process was pinned to one NUMA node to reduce the off-chip data movement. The scaling performance was very sensitive to the particular grid decomposition used for a given number of cores so the best decomposition was selected for each run. The execution times are averaged over all cores.

Figure 1 shows the strong scaling for the HS and the TG test cases with runs from 1 to 16 nodes on the Inula cluster. Both the pure MPI and the hybrid MPI+OpenMP codes used form 24 to 384 cores.

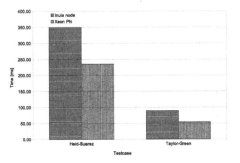

Fig. 2. Test cases execution times using one Inula node and one Xeon Phi card.

As one can see the hybrid MPI+OpenMP code scales better than the pure MPI code. The communication overhead is reduced up to 25 % as the number of the MPI ranks in the hybrid version is decreased by a factor of 6. Furthermore, there is improved overlap of the communication with the computation. For instance, the run with 384 cores reduced the MPI communication overhead by only 5 %, still the total execution time decreased by 10 %. On the other hand, the hybrid code introduces the OpenMP overhead and the percentage of computation that is parallelised is not 100 % whereas the computation in the pure MPI code is parallelised by 100 %. Figure 2 presents the execution times for both test cases with the hybrid MPI+OpenMP code. The hybrid code is executed on one Inula node with 24 OpenMP threads and on one Xeon Phi coprocessor with 112 OpenMP threads. The speedup for the Xeon Phi card is about x1.6 relatively to the dual socket AMD CPU for both test cases. The MPI+OpenMP hybridisation allows to efficiently utilise multi- and many-core architectures.

4 Summary

We have demonstrated various parallel software improvements to deal with multi-level parallelism and hybrid programming models as well. In order to obtain high-level of parallel scalability various modifications and tuning procedures are required as hardware configurations, including processors characteristics, interconnects and topologies, have a great influence on large-scale simulations. We have shown an example method to efficiently map the application topology to the cluster topology that can be used to improve the application scalability. Furthermore we have shown that a significant amount of work is required to efficiently port the pure MPI code to the hybrid MPI+OpenMP code to utilise many- and multi-core architectures. As next step, the restructured code enables us to adapt more easily the graphic accelerators by adding OpenACC directives.

Acknowledgements. We gratefully acknowledge the help and support provided by Jamie Wilcox from Intel EMEA Technical Marketing HPC Lab. This work is supported by the Polish National Center of Science under Grant No. UMO-2011/03/B/ST6/03500.

References

1. Kurzak, J., Bader, D., Dongarra, J.: Scientific Computing with Multicore and Accelerators. Computer and Information Science Series. Chapmann and Hall/CRC, Boca Raton (2010)
2. Rojek, K., Szustak, L., Wyrzykowski, R.: Using Intel Xeon Phi coprocessor to accelerate computations in MPDATA algorithm. In: Wyrzykowski, R., Dongarra, J., Karczewski, K., Waśniewski, J. (eds.) PPAM 2013, Part I. LNCS, vol. 8384, pp. 582–592. Springer, Heidelberg (2014)
3. Rojek, K., Szustak, L., Wyrzykowski, R.: Performance analysis for stencil-based 3D MPDATA algorithm on GPU architecture. In: Wyrzykowski, R., Dongarra, J., Karczewski, K., Waśniewski, J. (eds.) PPAM 2013, Part I. LNCS, vol. 8384, pp. 145–154. Springer, Heidelberg (2014)
4. Georgescu, S., Okuda, H.: Conjugate gradients on multiple GPUs. Int J. Numer. Meth. Fluids **64**, 1254–1273 (2010)
5. Zhang, Y., Cohen, J.M., Owens, J.D.: Fast tridiagonal solvers on GPU. In: Newsletter ACM SIGPLAN Notices - PPoPP'10, vol. 45, No. 5 (2010)
6. Prusa, J.M., Smolarkiewicz, P.K., Wyszogrodzki, A.: EULAG, a computational model for multiscale flows. Comput. Fluids **37**(9), 1193–1207 (2008)
7. Smolarkiewicz, P.K., Margolin, L.G.: Variational elliptic solver for atmospheric applications. Appl. Math. Comp. Sci. **4**, 527–551 (1994)
8. Smolarkiewicz, P.K., Grubisic, V., Margolin, L.G.: On forward-in-time differencing for fluids: stopping criteria for iterative solutions of anelastic pressure equations. Mon. Wea. Rev. **125**, 647–654 (1997)
9. Skamarock, W.C., Smolarkiewicz, P.K., Klemp, J.B.: Preconditioned conjugate-residual solvers for helmholtz equations in nonhydrostatic models. Mon. Wea. Rev. **125**, 587–599 (1997)
10. Smolarkiewicz, P. K., Margolin, L.G.: Variational methods for elliptic problems in fluid models. In: Proceeding of ECMWF Workshop on Developments in numerical methods for very high resolution global models, ECMWF, pp. 137–159. Reading, UK, 5–7, June 2000
11. Thomas, S.J., Hacker, J.P., Smolarkiewicz, P.K., Stull, R.B.: Spectral pre conditioners for non hydrostatic atmospheric models. Mon. Wea. Rev. **131**, 2464–2478 (2003)
12. Wyszogrodzki, A.A., Piotrowski, Z.P., Grabowski, W.W.: Parallel implementation and scalability of cloud resolving EULAG model. In: Wyrzykowski, R., Dongarra, J., Karczewski, K., Waśniewski, J. (eds.) PPAM 2011, Part II. LNCS, vol. 7204, pp. 252–261. Springer, Heidelberg (2012)
13. Piotrowski, Z.P., Wyszogrodzki, A., Smolarkiewicz, P.K.: Towards petascale simulation of atmospheric circulations with soundproof equations. Acta Geophys. **59**(6), 1294–1311 (2011)
14. Shalf, J., Dosanjh, S., Morrison, J.: Exascale computing technology challenges. In: Palma, J., Daydé, M., Marques, O., Lopes, J. (eds.) VECPAR 2010. LNCS, vol. 6449, pp. 1–25. Springer, Heidelberg (2011)
15. Strikwerda, J.: Finite difference schemes and partial differential equations. 2nd Edn. SIAM: Society for Industrial and Applied Mathematics (2004). ISBN: 978-0-89871-567-5. doi:10.1137/1.9780898717938
16. Kamata, S.I., Eason, R.O., Bandou, Y.: A new algorithm for n-dimensional hilbert scanning. IEEE Trans. Image Process. **8**(7), 964–973 (1999)

17. Smolarkiewicz, P.K., Margolin, L.G., Wyszogrodzki, A.A.: A class of nonhydrosta-tic global models. J. Atmos. Sci. **58**(4), 349–364 (2001)
18. Drikakis, D., Fureby, C., Grinstein, F.F., Youngs, D.: Simulation of transition and turbulence decay in the Taylor Green vortex. J. Turbul. **8**(20), 1–12 (2007)

Parallel Geometric Multigrid Preconditioner for 3D FEM in NuscaS Software Package

Tomasz Olas[✉]

Czestochowa University of Technology,
Dabrowskiego 69, 42-201 Czestochowa, Poland
olas@icis.pcz.pl

Abstract. Multigrid methods are among the fastest numerical algorithms for solving large sparse linear systems. The Conjugate Gradient method with Multigrid as a preconditioner (MGCG) features a good convergence even when the Multigrid solver itself is not efficient.

The parallel FEM package *NuscaS* allows us to solve adaptive FEM problems with 3D unstructured meshes on parallel computers such as PC-clusters. The parallel version of the library is based on the geometric decomposition applied for computing nodes of a parallel system; the distributed-memory architecture and message-passing model of parallel programming are assumed. In our previous works, we extend the *NuscaS* functionality by introducing parallel adaptation of tetrahedral FEM meshes and dynamic load balancing capabilities.

In this work we focus on efficient implementation of Geometric Multigrid as a parallel preconditioner for the Conjugate Gradient iterative solver used in the *NuscaS* package. Based on the geometric decomposition, for each level of Multigrid, meshes are partitioned and assigned to processors of a parallel architecture. Fine-grid levels are constructed by subdivision of mesh elements using the parallel 8-tetrahedra longest-edge refinement mesh algorithm, where every process keeps the assigned part of mesh on each level of Multigrid. The efficiency of the proposed implementation is investigated experimentally.

Keywords: Geometric Multigrid · Conjugate Gradient method · Preconditioner · FEM · Parallel adaptation · Parallel software package · MPI

1 Introduction

The finite element method (FEM) is a powerful tool for studying different phenomena in various areas. Parallel computing allows FEM users to overcome computational and/or memory bottlenecks of sequential applications. In particular, an object-oriented environment for the parallel FEM modeling, called *NuscaS*, was developed at the Czestochowa University of Technology [15]. This package allows for solving adaptive FEM problems with 3D unstructured meshes on distributed-memory parallel computers such as PC-clusters [7].

R. Wyrzykowski et al. (Eds.): PPAM 2013, Part I, LNCS 8384, pp. 166–177, 2014.
DOI: 10.1007/978-3-642-55224-3_17, © Springer-Verlag Berlin Heidelberg 2014

Multigrid (MG) methods are among the fastest numerical algorithms for solving large sparse systems of linear equations [2,3,13]. They are motivated by the observation that simple iterative methods are efficient in reducing high-frequency error components, but they can not efficiently reduce the slowly oscillating content of error. These methods were initially designed for the solution of elliptic partial differential equations (PDEs). MG algorithms were later adopted to solve other PDEs, as well as problems not described by PDEs.

In our previous works, we extend the *NuscaS* functionality by introducing parallel adaptation of tetrahedral FEM meshes and dynamic load balancing capabilities [7–9]. This paper is devoted to efficient implementation of Geometric Multigrid as a parallel preconditioner for the Conjugate Gradient (CG) iterative solver used in the *NuscaS* package. Based on the geometric decomposition, for each level of Multigrid, meshes are partitioned and assigned to processors of a parallel architecture. Fine-grid levels are constructed by subdivision of mesh elements using the previously proposed [8], parallel 8-tetrahedra longest-edge refinement mesh algorithm.

The materials of this paper is organized as follows. In Sect. 2, the basic concepts and architecture of the *NuscaS* package are introduced shortly, including parallel computing capabilities of the package. Section 3 gives an overview of the Geometric Multigrid, while Sect. 4 is devoted to parallel implementation of Multigrid in *NuscaS*. Performance of the Conjugate Gradient method with the parallel Geometric Multigrid preconditioner is studied in Sect. 5. Conclusions and future work are presented in Sect. 6.

2 *NuscaS* Package

The basic part of the *NuscaS* package is the kernel (NSC library) - a class library which provides an object-oriented framework for developing finite element applications. It consists of basic classes necessary to implement FEM modeling, irrespective of the type of a problem being solved. The inheritance is used to develop new applications classes for a particular modeling problem.

We assume a distributed memory architecture and message-passing model of parallel programming (MPI). The parallel version of the library is based on the geometric decomposition applied for nodes of a parallel system. In this case, a FEM mesh is divided into p submeshes (domains), which are assigned to separate processors (cores) of a parallel architecture. Every processor (or process) keeps the assigned part of the mesh. In consequence, the j-th domain will own a set of N_j nodes selected from all the N nodes of the mesh. For an arbitrary domain with index j, we have three types of nodes [8]: (i) N_j^i of internal nodes; (ii) N_j^b of boundary nodes; (iii) N_j^e of external nodes. Internal and boundary nodes are called *local* ones, so the number of local nodes is $N_j^l = N_j^i + N_j^b$;

When solving a system of equations in parallel using iterative methods, values of unknowns computed in boundary nodes are exchanged between neighbor domains. For this aim, in each domain (process) j for every neighbor process k the following sets are stored:

– S_j^k - set of those indexes of boundary nodes in process j that are external nodes for process k;
– R_j^k - set of those indexes of external nodes in process j that are assigned to process k.

In the NSC library, object oriented techniques are used to hide details of parallel processing. To solve system of equations with sparse matrices, which are the results of FEM discretization, we use iterative algorithms based on Krylov subspace methods [14], such as the CG method. The computational kernel of this methods is the matrix-vector multiplication with sparse matrices. When implementing this operation in parallel, the overlapping of computation and communication is exploited to reduce execution time of the algorithm [6]. Also, METIS and ParMETIS packages [4] are used for mesh partitioning.

3   Multigrid Method

Multigrid is motivated [3] by the observation that simple iterative methods are effective at reducing the high frequency error, but are ineffectual in reducing the low frequency content of the error. On the coarse grid relaxation is performed to reduce high frequency errors, followed by the projection of a correction equation on yet a coarser grid and so on. In this work, we use the Geometric Multigrid, which involves a hierarchy of computational grids of different mesh resolution.

The simplest original formulation of Geometric Multigrid is based on the following two-grid scheme:

1. perform some iterations of the basic iterative method (smoother) with fast convergence for high frequencies in order to smooth out the error;
2. project the (low-frequency) error onto a coarse grid (restriction);
3. solve the resulting projected problem on the coarsest grid;
4. interpolate the coarse grid solution back to the finest grid (interpolation);
5. update the solution on the finer grid and reapply the smoother to get rid of the new high-frequencies introduced.

In the above scheme, the linear system on the coarsest grid is too large to be solved efficiently. Therefore, steps 1, 2 and 3, 4 are applied recursively until the degrees of freedom have been small enough to solve the system efficiently. This multigrid cycling scheme is called V-cycle (Fig. 1). To inject the error from the fine grid Ω_h to the coarse grid Ω_H, it is necessary to apply a restriction operator, while its inverse, called the interpolation (prolongation) operator, is used to map the residual from the coarse grid to the fine grid. There are also other cycling schemes such as W-cycle, and Full Multigrid Algorithm [3].

4   Parallel Implementation of Multigrid in *NuscaS*

In the *NuscaS* package, the Geometric Multigrid is used for unstructured meshes. Fine-grid levels are automatically constructed by subdivision of mesh elements

$$MGV(A_h, b_h, x_h, h)$$

- $x_h \leftarrow S_h^{v_1}(A_h, b_h, x_h)$ - perform v_1 iterations of the smoother (pre-smoothing)
- $r_h \leftarrow b_h - A_h x_h$ - compute the residual on Ω_h
- $f_H \leftarrow R_h r_h$ - restriction of residual to coarse grid
- $x_H \leftarrow 0$ - initialise the coarse grid approximation
- $A_H \leftarrow R_h A_h P_h$ - variational coarse operator
- if Ω_H is the coarsest grid of the hierarchy then
    - $A_H x_H = f_H$ - direct solve of coarsest grid
- else
    - $MGV(A_H, f_H, x_H, H)$ - next level
- $e_h \leftarrow P_h x_H$ - interpolate the coarse grid approximation
- $x_h \leftarrow x_h + e_h$ - correction
- $x_h \leftarrow S_h^{v_2}(A_h, b_h, x_h)$ - post-smoothing

Fig. 1. Recursive definition of multigrid V-cycle

using our refinement algorithm [8,9], where the refinement is performed for all the elements of FEM meshes. At this moment, the V-cycle and Full Multigrid Algorithm have been implemented. Furthermore, we use the weighted Jacobi algorithm as a smoother.

When solving a system of equations in parallel, every processor (process) keeps the assigned part of a mesh on each level of Geometric Multigrid. For each level, every process stores information concerning the assigned nodes, including the splitting on internal, boundary and external nodes. In addition, for each level the process stores information about its coupling with neighbor processes, in order to organize the data exchange between processes.

4.1   Generation of Mesh Hierarchy in Parallel

The mesh distribution is perform only once, for the input mesh, which is assigned to p processors of a parallel architecture. In our refinement algorithm, finer mesh levels are constructed in parallel for every processor. The finer levels inherit the communication pattern from the coarser mesh decomposition (Fig. 2).

To minimize the parallel overhead, parts of submeshes are duplicated, and calculations are performed locally as much as possible, instead of exchanging data between processors. We use the same communication patterns and mechanisms as for the standard parallel CG solver [5]. In consequence, the coarser mesh for level h_{l-1} is generated from the finer mesh level h_l by subdivision of its elements. Moreover, meshes for subsequent levels of Geometric Multigrid are assigned to processes (processors) based on the decomposition performed for the coarsest mesh, in such a way that divisions on domains for subsequent levels were as close as possible.

Using such a decomposition scheme provides a simple way of constructing the prolongation and restriction operators. In particular, the data exchange between

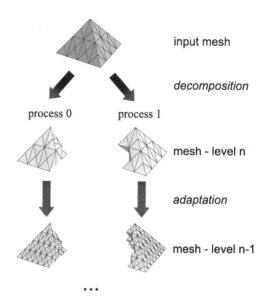

Fig. 2. Generation of mesh hierarchy in parallel

processes is avoided. At the same time, this solution has an important disadvantage; it is difficult to implement an efficient method of load balancing.

In this approach, the communication scheme for the Geometric Multigrid is the same as in the standard case of implementing the CG method. The communication take place when solving linear systems using an iterative method, e.g., during the sparse matrix-vector multiplication and computing the global dot product.

4.2   Data Structures

In order to implement the Geometric Multigrid, the way of storing FEM meshes in the NSC library has to be changed. For subsequent levels of Multigrid, nodes and elements of FEM meshes are placed in the same structures which are used in the standard case: vectors **nodes**, **elements**, **connectivity** in the class **Mesh**. Additionally, the array Emg of size $s(L, n_l)$ is introduced to store identifiers of elements corresponding to a certain level l of Multigrid, where L denotes the number of levels, and n_l is the number of elements for the level l. Also, for each level of Multigrid an inverse array Egm is required. For an element stored in the structure **elements**, the array Egm gives the identifier of this element on the level l, or -1 if this element does not belong to the given level. The analogous structures are created for nodes (Nmg and Ngm).

Figure 3 shows structures used to store dependencies between elements, edges, faces and nodes. Each element has an attribute **level** pointing out to the level of Multigrid which this element belongs to, as well as an attribute **parent** identifying the parent element from which this element has been created by subdivision.

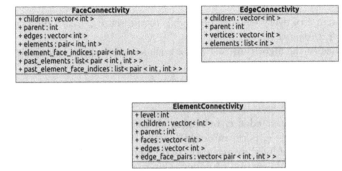

Fig. 3. Data structures for mesh connectivity

In turn, the parent element has information about all the elements which has been created by its subdivision. The analogous data structures are stored for faces (`FaceConnectivity`) and edges (`EdgeConnectivity`). Also, each element has attributes identifying its faces and edges; these attributes are used for setting boundary conditions. In the structure `EdgeConnectivity`, each edge is described by corresponding vertices (`verticles`). Additionally, this structure stores relations between elements and edges (`elements`). In the case of faces, besides information about their hierarchy, the structure `FaceConnectivity` identifies edges which describe each face. Also, this structure stores information about elements corresponding to each face, and identifiers of faces inside these elements.

4.3   Mesh Refinement

To generate FEM meshes for the subsequent levels of Geometric Multigrid, we use the parallel refinement algorithm previously developed by us for the FEM adaptation [8]. For the multigrid method, the refinement is performed for all the elements of FEM meshes. To implement this algorithm in parallel, we utilize the Longest-Edge Propagation Path (LEPP) method [11]. The partitioning of tetrahedral elements is performed based on the iterative algorithm of 8-tetrahedra longest edge partition (8T-LT in short). Our solution is based on a decentralized approach, so it is more scalable in comparison with previous implementations [1,12], where a centralized synchronizing node is required.

The proposed parallel algorithm of mesh refinement is presented in (Fig. 4). Communication between neighbor processes takes place in steps 3 and 4, based on information about edges which are exchanged between processes. This information is stored in sets e^i_{send} ($i = 0, \ldots, p - 1$), after performing the procedure *SelectEdge* (Fig. 5). As a result, in step 3 which is responsible for providing coherency of local propagation paths between neighbor processes, pairs of global indexes (n^1_g, n^2_g) describing edges are sent, using the non-blocking MPI routine `MPI_Isend`. After receiving this information, the mapping from the global enumeration to local one is performed, to allow for placing edges in local data structures.

1. for each edge e of every element belonging to the set $E_{selected}$, perform the procedure *SelectEdge(e)*
2. perform locally the algorithm *LEPP*: for each selected edge e belonging to the set $e_{selected}$
   - for each element E which uses the edge e
     - add the element E to the set $E_{selected}$ of already selected elements (unless this element was added before)
     - for the longest edge e_l belonging to faces of the element which uses the selected edge e, perform *SelectEdge(e_l)*
3. provide coherency of local propagation paths between neighbor processes
4. derive the global enumeration of nodes taking into account newly created nodes, together with assigning newly created nodes to separate processes
5. perform partitioning of elements which belong to $E_{selected}$
6. modify enumeration of nodes in such a way that internal nodes are located in the first place, then boundary, and finally external; then upgrade data structures used for communication between neighbor processes

Fig. 4. Parallel algorithm for mesh refinement

- add the edge e to the list $e_{selected}$ of already selected edges
- check if the edge e is located on the boundary with a neighbor domain (process); if so, add this edge to the list e^i_{send} of edges which are sent to this neighbor
- divide the edge e

Fig. 5. Procedure *SelectEdge(e)*

When performing FEM computation in parallel, elements which are located on boundaries of domains are duplicated in neighbor processes. This solution allows for avoiding communications at the cost of extra computation. We follow this concept when performing the mesh refinement. In this case, the partitioning of elements must be realized in the same way in neighbor processes. The only difficulty emerges when during the element partitioning the algorithm selects two or more edges with the same length. To avoid this difficulty, we propose a solution based on the global indexes of nodes. So, when determining the longest edge in the case of edges with the same length, we additionally compare the global indexes of edges to choose an edge with the highest value of global indexes. This solution allows us to avoid communication between processes. It is sufficient to derive the global enumeration of nodes.

The implementation of step 4 starts with the parallel procedure of deriving a new global enumeration of nodes, taking into account newly created nodes. This procedure includes the following three stages:

1. **Determine interval of global indexes for each process, as well as assign global indexes to the local nodes of processes:**

For this aim, the information about the number n_l^i of local nodes assigned to each process is distributed among all the processes, using the MPI_Allgather routine. The global index n_g^i of a node i in process j is determined by adding the local index n^i of this node to the sum of numbers of nodes in all the processes from 0 do $j - 1$:

$$n_g^i = n^i + \sum_{k=0}^{j-1} N_k^l. \tag{1}$$

2. **Exchange global indexes of nodes located on boundaries of domains**
3. **Exchange global indexes of newly created nodes.**

4.4   Construction of Subsequent Levels of Multigrid

After performing the mesh refinement, a new level of Multigrid has to be constructed, starting with defining internal, boundary and external nodes for each process. At the same time, some nodes could not belong to any of these categories since they are assigned to other processes, and are not connected with the local nodes of a given process; these nodes are not taken into consideration in the rest of the algorithm.

Based on the information about assignment of nodes to processes, the structure nodes is completed by new nodes created as a result of mesh refinement. All the nodes have to be ordered in such a way that at the beginning there are internal nodes followed by boundary and external nodes. The structure Nmg is constructed in the same way. The next stage is creating structures which describe which boundary nodes will be sent to various processes, as well as structures specifying processes from which external nodes will be received. Since this stage is implemented in parallel by all the processes, it is essential to provide the same order for nodes sent from process s_j, and nodes received in another process. For this aim the global enumeration n_g is used, so all the nodes in the sending and receiving processes are sorted according to global identifiers of nodes.

For the given domain, the results of mesh refinement are also used to create mesh elements on the succeeding level of Multigrid; these elements are connected with local nodes of the process. Also, the structures Emg and Egm are created and filled in. Objects of classes ElementConnectivtiy are built as well, for newly created elements. These objects are completed by information about hierarchy of elements (attributes parent and children).

The available information about assignment of nodes to processors allows also for specifying edges belonging to the given domain. For these edges, objects of class FaceConnectivity are created, and hierarchies of edges and faces are set. Finally, the connectivity between elements, faces and edges are completed.

4.5   Details of Implementation

To solve sparse linear systems using the Multigrid, the class MultigridSolver is created; it inherits from the template class IterativeLinearSystemSolver.

As a parameter of constructor of this class, the reference to an object of the class FiniteElementMesh is used. This object stores the whole information about the FEM mesh, including objects representing elements and boundary conditions.

In order to use the multigrid solver, it is necessary to have FEM meshes for all the levels of Multigrid. This is provided by the mesh refinement, which is implemented using the method generate_multilevel_mesh_bottom_up of the class MultigridSolver. The number of levels of Multigrid is one of parameter of this method, which is currently implemented for tetrahedral FEM elements. Furthermore, the class MultigridSolver is responsible for storing sparse matrices for all the levels (attribute level_matrices).

The interpolation and restriction operators are stored as sparse matrices P_- and R_-, which are created by the method prepare_projection_matrices. Then the method project_vector is used to project a vector from level l to level $l - 1$, while the method restrict_vector is applied to restrict a vector from level $l - 1$ to level l. As mentioned before, both the V-cycle scheme and Full Multigrid algorithm have been implemented by this time.

5   Performance of Conjugate Gradient Method with Parallel Multigrid Preconditioner

5.1   Parallel Multigrid Preconditioner

A multigrid method with an intentionally reduced tolerance can be used as an efficient preconditioner for Krylov iterative solvers [3,14]. The CG method with the Geometric Multigrid as a preconditioner (MGCG) features a good convergence even when the multigrid solver itself is not efficient. In comparison with other preconditioners (ICC, ILU, ...), the main benefit of using the Geometric Multigrid as a preconditioner is independence of the convergence rate from the problem size N. The solution may still be obtained in O(N) time as well as in the case where the multigrid method is used as a solver.

For a given residual r, the preconditioner step $z = M^{-1}r$ can be implemented as a call $z = MG(A, r, z, level)$. Depending on the selected multigrid cycling scheme, a suitable method is invoked, for example, the selection of MGV invokes V-cycle.

In the NSC library, the template of the class Preconditioner is responsible for implementation of preconditioning when solving sparse linear systems by iterative methods. One of attributes of this class is object of the class Multigrid, which is responsible for the implementation of Geometric Multigrid.

5.2   Performance Results

At this stage of research, our main goal is to study performance characteristics of the proposed parallel solution, and determining possible bottlenecks. For this aim, a rather simple problem of simulating the heat transfer is used; this simulation is executed for different sizes of FEM matrices and various numbers

Table 1. The number of nodes in FEM meshes used in experiments, for different levels of multigrid

Mesh name	Level 2	Level 1	Level 0
1000	1147	8245	62569
8000	8245	62569	487633
20000	19253	141307	1096157

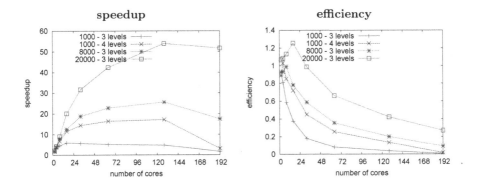

Fig. 6. Speedup and efficiency for CG method with multigrid preconditioner for different FEM meshes versus number of processors

of processors (Table 1). The multigrid method with V-cycle scheme is used as a preconditioner for the CG iterative method. We utilize the weighted Jacobi algorithm as a smoother with two pre-smoothing and two post-smoothing iterations.

These experiments are performed on a cluster with 16 nodes connected by the Infiniband network. Each node contains two Ivy Bridge E5 V2 2.90 GHz processors (12 cores each) with 128 GB RAM. The achieved performance results (Fig. 6) show a clear increase in speedup with increasing the problem size. In consequence, for the largest problem the maximum speedup of more than 50 is achieved for 128 nodes. At the same time, increasing the number of processors to 192 results in decreasing the speedup in this case.

To identify reasons for the observed scalability problems, the load imbalance across processors is studied first of all. The load imbalance factor η_p is defined as:

$$\eta_p = \frac{max_{0 \leq j < p} N_j^l}{\frac{N}{p}}, \tag{2}$$

where N_j^i is the number of internal nodes in subdomain j (they are coupled only with nodes belonging to this subdomain), N denotes the number of nodes in the FEM mesh, and p is the number of processors

Table 2 presents values of η_p for different levels of Multigrid and various numbers of processors. It can be seen that the imbalance factor is increased

Table 2. The load imbalance factors for different numbers of processes, for different levels of multigrid

Level	Number of nodes	η_2	η_4	η_8	η_{16}	η_{32}	η_{64}
Level 3	1147	1.03	1.03	1.03	1.02	1.06	1.17
Level 2	8245	1.07	1.05	1.19	1.26	1.42	1.66
Level 1	62569	1.10	1.09	1.31	1.39	1.67	2.04
Level 0	487633	1.11	1.11	1.38	1.47	1.81	2.24

after each refinement step, and with increasing the number of processors. For the largest number of processors in this table ($p = 64$), this factor reaches even 2.24 for the coarsest mesh. These results confirm the importance of load imbalance as a reason for observed scalability problems. At the same time, they show the necessity of further work on this issue. In particular, this works will aim at adapting the load balancing algorithm, which was previously proposed [9] for the dynamic load balancing when solving 3D FEM problems on clusters, to take into account properties of Geometric Multigrid.

6 Conclusions and Further Work

In this work, the implementation and investigation of using the Multigrid as a parallel preconditioner for the CG iterative solver were presented. The Geometric Multigrid is utilized for 3D unstructured FEM meshes. Fine-grain levels of Multigrid are automatically constructed in parallel by subdivision of mesh elements using the previously proposed refinement algorithm. In consequence, the distributed memory architecture and message-passing model of parallel programming can be utilized. The preliminary performance results are rather promising, but they indicate scalability problems for larger numbers of processors.

One of reasons for these problems is load imbalance arising during refinement for finer levels of Geometric Multigird. A possible solution could be adapting the load balancing algorithm, which was previously proposed [9] for the dynamic load balancing when solving 3D FEM problems, to take into account properties of Geometric Multigrid.

Acknowledgments. We gratefully acknowledge the help and support provided by Jamie Wilcox from Intel EMEA Technical Marketing HPC Lab.

References

1. Balman, M.: Tetrahedral mesh refinement in distributed environments. In: 2006 International Conference on Parallel Processing Workshops (ICPPW'06), pp. 497–504. IEEE Computer Soc. (2006)
2. Banaś, K.: Scalability analysis for a multigrid linear equations solver. In: Wyrzykowski, R., Dongarra, J., Karczewski, K., Wasniewski, J. (eds.) PPAM 2007. LNCS, vol. 4967, pp. 1265–1274. Springer, Heidelberg (2008)

3. Hulsemann, F., Kowarschik, M., Mohr, M., Rude, U.: Parallel geometric multigrid. In: Bruaset, A.M., Tveito, A. (eds.) Numerical Solution of Partial Differential Equations on Parallel Computers. Lecture Notes in Computational Science and Engineering, vol. 51, pp. 165–208. Springer, Heidelberg (2006)
4. Karypis, G., Schloegel, K., Kumar, V.: PARMETIS Parallel Graph Partitioning and Sparse Matrix Ordering Library Version 3.1. Univ. Minnesota, Army HPC Research Center. http://glaros.dtc.umn.edu/gkhome/fetch/sw/parmetis/manual. pdf (2003)
5. Olas, T., Karczewski, K., Tomas, A., Wyrzykowski, R.: Fem computations on clusters using different models of parallel programming. In: Wyrzykowski, R., Dongarra, J., Paprzycki, M., Waśniewski, J. (eds.) PPAM 2001. LNCS, vol. 2328, pp. 170–182. Springer, Heidelberg (2002)
6. Olas, T., Wyrzykowski, R., Tomas, A., Karczewski, K.: Performance modeling of parallel FEM computations on clusters. In: Wyrzykowski, R., Dongarra, J., Paprzycki, M., Waśniewski, J. (eds.) PPAM 2004. LNCS, vol. 3019, pp. 189–200. Springer, Heidelberg (2004)
7. Olas, T., Leśniak, R., Wyrzykowski, R., Gepner, P.: Parallel adaptive finite element package with dynamic load balancing for 3d thermo-mechanical problems. In: Wyrzykowski, R., Dongarra, J., Karczewski, K., Wasniewski, J. (eds.) PPAM 2009, Part I. LNCS, vol. 6067, pp. 299–311. Springer, Heidelberg (2010)
8. Olas, T., Wyrzykowski, R.: Adaptive fem package with decentralized parallel adaptation of tetrahedral meshes. In: Lirkov, I., Margenov, S., Waśniewski, J. (eds.) LSSC 2011. LNCS, vol. 7116, pp. 622–629. Springer, Heidelberg (2012)
9. Olas, T., Wyrzykowski, R., Gepner, P.: Parallel FEM adaptation on hierarchical architectures. In: Wyrzykowski, R., Dongarra, J., Karczewski, K., Waśniewski, J. (eds.) PPAM 2011, Part I. LNCS, vol. 7203, pp. 194–205. Springer, Heidelberg (2012)
10. Patzak, B., Rypl, D.: A framework for parallel adaptive finite element computations with dynamic load balancing. In: Proceedings of the First International Conference on Parallel, Distributed and Grid Computing for Engineering, Paper 31. Civil-Comp Press (2009)
11. Plaza, A., Rivara M.: Mesh refinement based on the 8-tetrahedra longest-edge partition. In: Proceedings of the 12th International Meshing Roundtable, pp. 67–78. Sandia National Laboratories (2003)
12. Rivara, M., Pizarro, D., Chrisochoides, N.: Parallel refinement of tetrahedral meshes using terminal-edge bisection algorithm. In: Proceedings of the 13th International Meshing Roundtable, pp. 427–436. Sandia National Labs (2004)
13. Romanazzi, G., Jimack, P.K.: Performance prediction for multigrid codes implemented with different parallel strategies. In: Proceedings of the First International Conference on Parallel, Distributed and Grid Computing for Engineering, Paper 43. Civil-Comp Press (2009)
14. Saad, Y.: Iterative Methods for Sparse Linear Systems. SIAM, Philadelphia (2003)
15. Wyrzykowski, R., Olas, T., Sczygiol, N.: Object-oriented approach to finite element modeling on clusters. In: Sørevik, T., Manne, F., Moe, R., Gebremedhin, A.H. (eds.) PARA 2000. LNCS, vol. 1947, pp. 250–257. Springer, Heidelberg (2001)

Scalable Parallel Generation
of Very Large Sparse Benchmark Matrices

Daniel Langr[1]([⊠]), Ivan Šimeček[1], Pavel Tvrdík[1], and Tomáš Dytrych[2]

[1] Department of Computer Systems, Faculty of Information Technology,
Czech Technical University in Prague, Thákurova 9, 160 00 Praha, Czech Republic
[2] Department of Physics and Astronomy, Louisiana State University,
Baton Rouge, LA 70803, USA
langrd@fit.cvut.cz

Abstract. We present a method and an accompanying algorithm for scalable parallel generation of sparse matrices intended primarily for benchmarking purposes, namely for evaluation of performance and scalability of generic massively parallel algorithms that involve sparse matrices. The proposed method is based on enlargement of small input matrices, which are supposed to be obtained from public sparse matrix collections containing numerous matrices arising in different application domains and thus having different structural and numerical properties. The resulting matrices are distributed among processors of a parallel computer system. The enlargement process is designed so its users may easily control structural and numerical properties of resulting matrices as well as the distribution of their nonzero elements to particular processors.

Keywords: Sparse matrix · Benchmark matrix · Enlargement · Parallel algorithm · Scalability

1 Introduction

Public collections of sparse matrices, such as the Matrix Market [3] or the University of Florida Sparse Matrix Collection (UFSMC) [4], represent useful sources of benchmark matrices that can be utilized to evaluate generic algorithms which involve sparse-matrix computations. These collections contain numerous matrices that originate from various application domains and therefore have generally different structural and numerical properties. However, the matrices in these collections are available in a form of downloadable files, which, in effect, limits their sizes. This hinders their use for evaluation of performance, robustness, and especially scalability of *massively parallel algorithms* that involve computations with *very large sparse matrices*. By very large sparse matrices we denote sparse matrices that due to their size need to be processed by massively parallel computer systems with (generally) distributed memory architectures, which we further call *clusters*.

R. Wyrzykowski et al. (Eds.): PPAM 2013, Part I, LNCS 8384, pp. 178–187, 2014.
DOI: 10.1007/978-3-642-55224-3_18, © Springer-Verlag Berlin Heidelberg 2014

To evaluate such massively parallel algorithms, we would like to employ a scalable parallel generator capable of producing, in a uniform way, sparse matrices of arbitrary sizes *representative* of various real-world scientific and engineering problems. To our best knowledge, no such generator exists.

There are generators of sparse matrices, such that Matgen, Matrix Market Deli, SPARSKIT, Test Matrix Toolbox for MATLAB, XLATMR, MATRAN, some of which are capable to produce scalable representative matrices. However, to our best knowledge, none of these generators works in parallel neither is suitable for parallel processing.

Since each sparse matrix corresponds to a graph and vice versa, we can also use existing generators of large graphs as generators of benchmark sparse matrices (Erdös–Rényi models, Stochastic Kronecker graphs, etc.). However, graph generators typically produce random graphs, hence, the obtained matrices have random properties and are not representative of real-world problems (possibly with the exception of network-simulation field).

Another option is to utilize various open-source *high performance computing* (HPC) programs that involve sparse-matrix computations and extract matrices from them. However, such a non-uniform approach might be extremely complicated and highly impractical for many reasons (for instance, such programs usually do not have APIs for extraction of sparse matrices, thus, such a task would require to get familiar with the source code of each utilized program).

We propose here a different solution. It is based on a controlled enlargement of existing sparse matrices, e.g., matrices obtained from public collections. This solution does not provide authentic representative matrices (such as those extracted from real-world HPC programs), however, it allows us to easily obtain matrices of arbitrary sizes with known structural and numerical properties.

2   Methodology

Let $A = (a_{i,j})$ be an $m_A \times n_A$ matrix with z_A nonzero elements. We call A the *seed matrix*. Let further $Q = (q_{i,j})$ be an $m_Q \times n_Q$ matrix with z_Q nonzero elements. We call Q the *enlargement matrix*. Let $\otimes$ denote the *Kronecker product* operation. Then,

$$A' = (a'_{i,j}) = A \otimes Q = \begin{bmatrix} a_{1,1}Q & \cdots & a_{1,n_A}Q \\ \vdots & \ddots & \vdots \\ a_{m_A,1}Q & \cdots & a_{m_A,n_A}Q \end{bmatrix} \tag{1}$$

is an $m_A m_Q \times n_A n_Q$ matrix with $z_{A'} = z_A z_Q$ nonzero elements. We call A' the *enlarged matrix*.

If both A and Q are square matrices, let $\mu_1, \ldots, \mu_{n_A}$ and $\nu_1, \ldots, \nu_{n_Q}$ denote their eigenvalues associated with eigenvectors $\boldsymbol{u}_1, \ldots, \boldsymbol{u}_{n_A}$ and $\boldsymbol{v}_1, \ldots, \boldsymbol{v}_{n_Q}$, respectively. Then the eigenvalues of A' have the following form:

$$\omega_{i,j} = \mu_i \nu_j, \quad 1 \le i \le n_A, \quad 1 \le j \le n_Q,$$

where the corresponding eigenvectors are given by $\boldsymbol{w}_{i,j} = \boldsymbol{u}_i \otimes \boldsymbol{v}_j$ (see for instance [9, Chap. 13] for details).

Let $p_1, \ldots, p_P$ denote a particular subset of processors of some cluster. By a *mapping function* we denote an arbitrary function $\mathcal{M}$ of the following form:

$$\mathcal{M} : (i, j) \rightarrow k, \quad 1 \leq i \leq m_A m_Q, \quad 1 \leq j \leq n_A n_Q, \quad 1 \leq k \leq P.$$

Problem 1. We are looking for a parallel algorithm that generates an enlarged matrix A' distributed among P processors $p_1, \ldots, p_P$ according to a mapping function $\mathcal{M}$ such that $a'_{i,j}$ is generated by processor p_k if $\mathcal{M}(i, j) = k$. The additional requirements are following:

Req. 1: The algorithm should not depend on a particular computer representation of A, Q, and A'.

Req. 2: The algorithm should be able to use an arbitrary mapping function $\mathcal{M}$.

Req. 3: The algorithm should have a minimal memory footprint.

Req. 4: We primarily assume that A is read from a file.

The consequences of these requirements are as follows:

Con. 1: Req. 3 precludes storing A and Q in computer memory. Rather, we regard A as a sequentially accessible set of its nonzero elements, in any order, denoted by $(i, j, a_{i,j})_1, \ldots, (i, j, a_{i,j})_{z_A}$,

$$A := \{ m_A, n_A, z_A, (i, j, a_{i,j})_1, \ldots, (i, j, a_{i,j})_{z_A} \}, \tag{2}$$

and we consider Q to be a function $\mathcal{Q}(i, j) = q_{i,j}$,

$$Q := \{ m_Q, n_Q, z_Q, \mathcal{Q} \}.$$

We call $\mathcal{Q}$ the *enlargement function.* Ideally, $\mathcal{Q}$ would be defined analytically (e.g., $\mathcal{Q}_1(i, j) = 1$), which would ensure zero data-segment requirements for its implementation. Some considerations regarding the choice of a suitable enlargement function are discussed in Sect. 2.2.

Con. 2: To satisfy Req. 1, the file reading operation cannot constitute a part of the algorithm. Instead, it is up to a user to provide A in the form of (2). Moreover, matrices available in public collections are usually stored in text-based file formats [2,5,8], which are not suitable for parallel processing. We hence assume that A is available to the algorithm only on a single processor, namely p_1.

Con. 3: The output of the algorithm is the enlarged matrix A' whose nonzero elements are distributed among P processors according to the mapping function $\mathcal{M}$. The computer representation of A' as well as $\mathcal{M}$ are application-dependent. To make the algorithm application-independent (Req. 1 and Req. 2), we assume that a user provides a subroutine, denoted as $\mathcal{O}$, that stores the generated nonzero elements of A' in memory. We call this subroutine the *output function.*

Algorithm 1. Enlargement of a sparse matrix

Input: $A = \{m_A, n_A, z_A, (i, j, a_{i,j})_1, \ldots, (i, j, a_{i,j})_{z_A}\}$ on p_1 // seed matrix
Input: $Q = \{m_Q, n_Q, z_Q, \mathcal{Q}\}$ // enlargement matrix
Input: $\mathcal{M}, \mathcal{O}$ // mapping and output functions
Output: A' such that $\mathcal{O}(i, j, a'_{i,j})$ is called on processor p_k for all $a'_{i,j} \neq 0$ and
$\qquad \mathcal{M}(i, j) = k$ // enlarged matrix
Data: $\alpha, i', j', i'', j'', r$;   // auxiliary variables (local to a processor)

> **for** all processors $p_1, \ldots, p_P$ **do in parallel**
>> $r \leftarrow$ actual processor number ;
>> broadcast m_A, n_A, and z_A from p_1 to all other processors ;
>> **for** $k \leftarrow 1$ **to** z_A **do**
>>> broadcast $(i, j, a_{i,j})_k$ from p_1 to all other processors ;
>>> **for** $i' \leftarrow 1$ **to** m_Q **do**
>>>> **for** $j' \leftarrow 1$ **to** n_Q **do**
>>>>> $i'' \leftarrow (i - 1) \cdot m_Q + i'$;
>>>>> $j'' \leftarrow (j - 1) \cdot n_Q + j'$;
>>>>> **if** $\mathcal{M}(i'', j'') = r$ **then**
>>>>>> $\alpha \leftarrow a_{i,j} \cdot \mathcal{Q}(i', j')$;
>>>>>> **if** $\alpha \neq 0$ **then** call $\mathcal{O}(i'', j'', \alpha)$

2.1 Algorithm

Let us present Algorithm 1 that solves Problem 1 and that conforms to the previous analysis. Its *computational complexity* is determined by:

1. $(z_A \cdot m_Q \cdot n_Q)$ calls of the mapping function $\mathcal{M}$ by each processor,
2. $(z_A \cdot m_Q \cdot n_Q)$ calls of the enlargement function $\mathcal{Q}$ by all processors,
3. $(z_A \cdot z_Q)$ calls of the output function $\mathcal{O}$ by all processors.

As to the *space complexity*, the algorithm needs only few auxiliary variables and a small buffer for broadcasting nonzero elements of A allocated by each processor in its local memory.

The *communication complexity* involves $z_A + 1$ broadcast operations. However, the broadcasting of nonzero elements of A in an element-by-element fashion would be inefficient. Within an implementation of Algorithm 1, one should transfer data in chunks, which would considerably reduce the number of broadcast operations, albeit at the expense of an additional memory overhead associated with larger communication buffers.

Note that Algorithm 1 needs to access each nonzero element of A only once and it does not depend on the order in which are the elements provided.

2.2 Enlargement Functions

The choice of a suitable enlargement function has a significant impact on the properties of the enlarged matrix A'. Let us further restrict our analysis to square enlargement matrices Q. Recall that, ideally, we would like to define an enlargement function analytically. In the most simple case, we can use the constant function $Q_1(i, j) = 1$, which however gives rise to a singular enlarged matrix A'.

In cases where singularity is undesirable, an alternative enlargement function might be defined as follows:

$$Q_2(i, j) = \begin{cases} 1, & \text{if } i \leq j, \\ 0, & \text{otherwise,} \end{cases} \tag{3}$$

which corresponds to Q being an upper triangular matrix of ones and thus $z_Q = n_Q(n_Q + 1)/2$. If A is a square matrix, then the usage of Q_2 results in A' with degenerated eigenvalues $\omega_1, \ldots, \omega_{n_A}$, each with multiplicity n_Q. To prevent such a degeneration of eigenvalues, we may adjust definition (3), e.g., as follows:

$$Q_3(i, j) = \begin{cases} 1/i, & \text{if } i \leq j, \\ 0, & \text{otherwise,} \end{cases} \tag{4}$$

which corresponds to Q being an upper triangular matrix with eigenvalues $1, 1/2, \ldots, 1/n_Q$. Thus, if A is a square matrix, then, generally, the usage of Q_3 generates A' with distinct eigenvalues.

The suitable enlargement function is application-dependent and a more general discussion of its choice is beyond the scope of this paper. However, let us analyze the choice of the size of the enlargement matrix Q.

Question 1. How to choose the size of Q in order to generate the largest possible A' that would fit into the available amount of memory?

We further show how to answer this question for an example case defined as follows.

Example 1.

- Let A and Q be real matrices.
- Suppose that nonzero elements of A' will be stored on each processor in the *coordinate storage format* (see for instance [10, Sect. 3.4] for details).
- Assume that row/column indexes of nonzero elements of A' are stored using the d_i-bit unsigned integer data type and their values using the d_f-bit floating-point data type. The storage of a single nonzero element thus requires $d = (2d_i + d_f)/8$ bytes.
- Consider the Q_3 enlargement function (4).
- Suppose that $\mathcal{M}$ produces a balanced distribution of A' among P processors, i.e., that each processor handles approximately its $z_{A'}/P$ nonzero elements.
- Let L denote the amount of memory in bytes that is available to each processor for storing its local nonzero elements of A'.

Solution 1. The local nonzero elements must fit into the local memory of each processor, therefore $d \cdot z_{A'}/P \leq L$. By substituting $z_{A'} = z_A z_Q$, we get

$$z_Q \leq \frac{LP}{z_A d}.$$

To generate the largest possible A' using $\mathcal{Q}_3$, which implies $z_Q = n_Q(n_Q + 1)/2$, this inequality turns into the equation

$$\frac{n_Q(n_Q + 1)}{2} = \frac{LP}{z_A d},$$

which has the following positive solution:

$$n_Q = -\frac{1}{2} + \left(\frac{1}{4} + \frac{2LP}{z_A d}\right)^{1/2}. \tag{5}$$

This result gives us the maximum size of the enlargement matrix Q, and thus provides an answer to Question 1 for Example 1. Note that the number of nonzero elements of A' is proportional to P in this case.

2.3 Mapping Functions

Recall that the mapping function is application-dependent, i.e., it depends on the intended usage of the enlarged matrix A'. The justifiable requirement for a good mapping function would be that:

1. its usage results in a balanced distribution of the nonzero elements of A' among processors,
2. its implementation requires no or only small amount of memory,
3. it is computationally inexpensive.

It is a common approach for many HPC codes to map matrices onto processors by continuous chunks of rows or columns. (Such row-wise mapping is, for instance, utilized by widely-used PETSc/SLEPc [1,6] and Trilinos [7] sparse matrix libraries.) The simplest mapping of this type is to assign each processor the same number of rows/columns, which is provided by the following mapping functions:

$$\mathcal{M}_R(i, j) = \left\lfloor \frac{i-1}{m_A m_Q} P \right\rfloor + 1, \quad \mathcal{M}_C(i, j) = \left\lfloor \frac{j-1}{n_A n_Q} P \right\rfloor + 1. \tag{6}$$

Such functions are memory-optimal, since they are defined purely analytically. However, due to (1) the block nonzero pattern of A' is the same as the nonzero pattern of A. Thus, if the nonzero elements of A are located in rows/columns unevenly, the usage of these functions might result in an unbalanced distribution of the nonzero elements of A' among processors.

Let $\mathcal{M}_{IR}$ and $\mathcal{M}_{IC}$ denote *ideal* mapping functions that provide balanced row/column-wise distributions of A' to processors, respectively. Generally, these functions cannot be defined analytically, rather, they involve algorithmic processes that compute a number of nonzero elements of A for each of its rows/columns.

A more detailed discussion of the choice of a suitable mapping function is beyond the scope of this paper. However, note that to construct a mapping function that would provide a balanced distribution of nonzero elements of A' to processors (such as $\mathcal{M}_{\mathrm{IR}}$ or $\mathcal{M}_{\mathrm{IC}}$), the nonzero pattern of A might need to be analyzed before the execution of Algorithm 1, which would require more than one iteration over the sequence of the nonzero elements $(i, j, a_{i,j})_1, \ldots, (i, j, a_{i,j})_{z_A}$. If A is read from a file, this would imply either multiple reading of such a file or storing of its elements in memory.

3   Experiments

We have implemented Algorithm 1 in C++ using the MPI parallel programming model for evaluation of its performance and scalability. Within the performed experiments, seed matrices were read from input files in the Matrix Market file format [2]. Particularly, we used the seed matrices obtained from UFSMC that are listed in Table 1.

Table 1. Seed matrices used for the experiments.

Matrix	m_A	n_A	z_A	File size [MB]
ex25	$8.48 \cdot 10^2$	$8.48 \cdot 10^2$	$2.46 \cdot 10^4$	0.7
cage12	$1.30 \cdot 10^5$	$1.30 \cdot 10^5$	$2.03 \cdot 10^6$	61.5
Freescale1	$3.43 \cdot 10^6$	$3.43 \cdot 10^6$	$1.89 \cdot 10^7$	648.0

We utilized the $\mathcal{Q}_3$ enlargement function (4) and the $\mathcal{M}_{\mathrm{C}}$ mapping function (6). We evaluated the weak scalability of Algorithm 1 (fixed per-processor problem size), hence we set the enlargement matrix Q to be square with size n_Q computed via (5), where we set $L = 600\,\mathrm{MB}$ and used 32-bit row/column indexes and 32-bit floating point values, thus $d = 12$. Since we wanted to measure primarily the performance of Algorithm 1 itself, we used by default a void (do-nothing) output function.

Unless stated otherwise, we buffered the broadcasting operations and we used the buffer size of 1024 elements by default.

We measured the running times of Algorithm 1 using the QueenBee/LONI cluster, which has the listed peak performance 50 TFLOPS. The obtained results are presented in Fig. 1, where the running times are labeled with the names of the corresponding seed matrices. The running times labeled with the (w/o buffering) suffix correspond to the measurements where the buffering of broadcast operations was turned off. The running times labeled with the (storage) suffix correspond to the measurements where generated nonzero matrix elements were additionally stored in memory in the coordinate storage format using the C++/STL vector container and its push_back() member function. Note that in this case, the measured results include running times of the in-memory storage operations, which does not constitute a part of Algorithm 1.

Considering the default settings, i.e., the void output function and the broadcast buffer allocated for 1024 elements, **the running times of Algorithm 1**

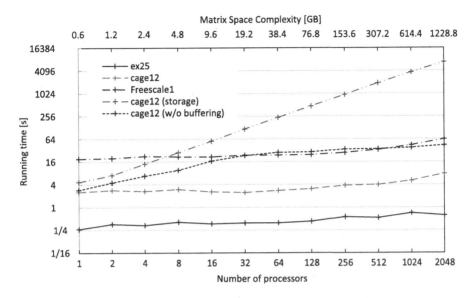

Fig. 1. Running times of performed experiments.

were either approximately constant or slightly raising with the increasing number of processors (and thus with the corresponding increase of the size of the generated matrix). The resulting times include reading of seed matrices from input files, and thus the I/O overhead imposed by this operation depends on their file sizes. This is the reason for considerably different running times for all 3 matrices (see the file sizes in Table 1).

Comparing the results for the cage12 matrix with turned on and turned off buffering of the broadcast operations, we can see that the buffering considerably reduced the algorithm running time. The corresponding memory requirements (12 kB per processor within our experiments) are negligible on modern clusters.

As for the (storage) case, the operation of storing the resulting nonzero elements in memory took most of the running time of the corresponding experiment (recall that this operation does not constitute a part of Algorithm 1). The linear

Table 2. Large-scale results for an experimental variant of Algorithm 1 adapted for balanced row-wise partitioning of matrices among processors. The running times include the in-memory storage of generated nonzero elements in the coordinate storage format.

Number of processors	Space complexity of A' [TB]	Number of nonzeros of A'	Running time [s]
128	0.38	$1.7 \cdot 10^{10}$	17.6
1024	3	$1.3 \cdot 10^{11}$	17.3
16384	48	$2.1 \cdot 10^{12}$	16.6
131072	384	$1.7 \cdot 10^{13}$	17.7

growth of the running time with the increasing number of processors P was here caused by the fact that the nonzero elements of the cage12 seed matrix were stored in the input file in the order given by the growing column index. Thus, due to the utilization of the $\mathcal{M}_C$ mapping function, the generated nonzero elements of A' were first stored only in the local memory of processor p_1, then of p_2, etc., which resulted in zero scalability of this operation.

4   Discussion

The drawback of the presented method/algorithm is that, under some circumstances, the storage of the generated nonzero elements in memory can break its scalability. This is the price for the versatility of the algorithm, which is application-independent, i.e., it works with an arbitrary mapping function, an arbitrary enlargement function, an arbitrary output routine, and an arbitrary ordering of nonzero elements of seed matrices.

However, in practice, matrices are often partitioned among processors according to contiguous chunks of rows/columns/blocks. For such restricted situations, we might be able to develop more efficient variants of Algorithm 1 (at a price of loosing versatility). To verify this assumption, we have experimentally developed, implemented, and tuned such an algorithm variant for balanced row-wise partitioning of matrices among processors. The results of the experiments performed with this adapted algorithm on the 13-PFLOPS BlueWaters/NCSA cluster are shown in Table 2. Using this algorithm, we were able to generate enormously large matrices in less than 20 s independently of the number of utilized processors. Note that the running times include the storage of generated nonzero elements of the enlarged matrices in memory.

5   Conclusions

This paper presents a versatile method along with an accompanying algorithm for scalable parallel generation of very large sparse benchmark matrices with known numerical and structural properties. In the experiments performed with the implementation of this algorithm, we were able to generate such matrices within tens of seconds using a 50-TFLOPS cluster. However, we have shown that, under some circumstances, storing of generated nonzero elements in memory can break the scalability of the whole enlargement process.

Considering todays most powerful (PFLOPS) clusters, is crucial to preserve the scalability of the enlargement process even with in-memory storage of the generated nonzero elements. This can be achieved by adapting the presented algorithm for some particular mappings of matrices to processors. Hence, in our future work, we want to focus on the development of such adapted algorithm variants.

Acknowledgements. This work was supported by the Czech Science Foundation under Grant No. P202/12/2011, by the U.S. National Science Foundation under Grant No. OCI-0904874, and by the U.S. Department of Energy under Grant No. DOE-0904874. D.L. acknowledges support from Jerry P. Draayer and the Louisiana State University (LSU). We acknowledge the Louisiana Optical Network Initiative (LONI) for providing HPC resources. This research is part of the Blue Waters sustained-petascale computing project, which is supported by the National Science Foundation (award number OCI 07-25070) and the state of Illinois. Blue Waters is a joint effort of the University of Illinois at Urbana-Champaign and its National Center for Supercomputing Applications.

References

1. Balay, S., Brown, J., Buschelman, K., Eijkhout, V., Gropp, W.D., Kaushik, D., Knepley, M.G., McInnes, L.C., Smith, B.F., Zhang, H.: PETSc users manual. Technical report ANL-95/11 - Revision 3.2, Argonne National Laboratory (2010)
2. Boisvert, R.F., Pozo, R., Remington, K.: The matrix market exchange formats: initial design. Technical report NISTIR 5935, National Institute of Standards and Technology (1996)
3. Boisvert, R.F., Pozo, R., Remington, K., Barrett, R.F., Dongarra, J.J.: Matrix market: a web resource for test matrix collections. In: Boisvert, R.F. (ed.) The Quality of Numerical Software: Assessment and Enhancement, pp. 125–137. Chapman & Hall, London (1997)
4. Davis, T.A., Hu, Y.F.: The University of Florida sparse matrix collection. ACM Trans. Math. Softw. **38**(1), 1:1–1:25 (2011)
5. Duff, I., Grimes, R., Lewi, J.: User's guide for the Harwell-Boeing sparse matrix collection (Release I). Technical report TR/PA/92/86, CERFACS. http://people.sc.fsu.edu/~jburkardt/pdf/hbsmc.pdf (1992). Accessed 27 March 2011
6. Hernandez, V., Roman, J.E., Vidal, V.: SLEPc: a scalable and flexible toolkit for the solution of eigenvalue problems. ACM Trans. Math. Software **31**(3), 351–362 (2005)
7. Heroux, M., Bartlett, R., Hoekstra, V.H.R., Hu, J., Kolda, T., Lehoucq, R., Long, K., Pawlowski, R., Phipps, E., Salinger, A., Thornquist, H., Tuminaro, R., Willenbring, J., Williams, A.: An overview of trilinos. Technical report SAND2003-2927, Sandia National Laboratories (2003)
8. Hoemmen, M.: Matlab (ASCII) sparse matrix format, berkeley Benchmarking and Optimization Group. http://bebop.cs.berkeley.edu/smc/formats/matlab.html (2008). Accessed 27 April 2011
9. Laub, A.J.: Matrix Analysis for Scientists and Engineers. SIAM, Philadelphia (2005)
10. Saad, Y.: Iterative Methods for Sparse Linear Systems, 2nd edn. Society for Industrial and Applied Mathematics, Philadelphia (2003)

Parallel Non-Numerical Algorithms

Co-operation Schemes
for the Parallel Memetic Algorithm

Jakub Nalepa[1]([✉]), Miroslaw Blocho[1,2], and Zbigniew J. Czech[1,3]

[1] Silesian University of Technology, Gliwice, Poland
{jakub.nalepa,zbigniew.czech}@polsl.pl
[2] ABB ISDC, Krakow, Poland
miroslaw.blocho@pl.abb.com
[3] University of Silesia, Sosnowiec, Poland

Abstract. This paper presents a study of co-operation schemes for the parallel memetic algorithm to solve the vehicle routing problem with time windows. In the parallel co-operative search algorithms the processes communicate to exchange the up-to-date solutions, which may guide the search and improve the results. The interactions between processes are defined by the content of the exchanged data, timing, connectivity and mode. We show how co-operation schemes influence the search convergence and solutions quality. The quality of a solution is defined as its proximity to the best, currently-known one. We present the experimental study for the well-known Gehring and Homberger's benchmark. The new world's best solutions obtained in the study confirm that the co-operation scheme has a strong impact on the quality of final solutions.

Keywords: Parallel memetic algorithm · Co-operation scheme · Hybrid genetic algorithm · Vehicle routing problem with time windows

1 Introduction

The vehicle routing problem with time windows (VRPTW) is a two-objective NP-hard discrete optimization problem. Its main objective is to determine the minimal number of homogeneous vehicles to serve customers dispersed on the map. In addition, the total travel distance is to be minimized. A solution of the VRPTW is feasible if (i) all customers are visited within their time windows and (ii) the capacities of the vehicles are not exceeded.

The applications of the VRPTW are of wide range, including food, cash and parcels delivering, school bus and airline fleet routing, rail distributions and more. Thus, a number of exact and approximate algorithms to solve the VRPTW have been proposed over the years. The former approaches incorporate dynamic programming, branch-and-bound, greedy algorithms and more [5]. Due to the NP-hardness of the VRPTW, numerous approximate algorithms have emerged to solve it in acceptable time. Heuristic algorithms improving an initial solution (improvement heuristics) and constructing a feasible solution from scratch

R. Wyrzykowski et al. (Eds.): PPAM 2013, Part I, LNCS 8384, pp. 191–201, 2014.
DOI: 10.1007/978-3-642-55224-3_19, © Springer-Verlag Berlin Heidelberg 2014

(construction heuristics) have been explored [4,6,11,12]. Metaheuristics, incorporating search space exploration and exploitation mechanisms and allowing for temporary deterioration of a solution, evolutionary, genetic and memetic algorithms (GAs and MAs), both sequential and parallel, have emerged recently [8, 13,14]. MAs are the population-based methods combining the GAs to explore the solution space with local search algorithms for the exploitation of already found solutions.

Here we study the co-operation schemes for our two-stage parallel memetic algorithm to solve the VRPTW [1,2,9,10]. We present the experimental results obtained for selected tests from the Gehring and Homberger's benchmark. The paper is organized as follows. Section 2 formulates the VRPTW. The parallel memetic algorithm and co-operation schemes are described in Sect. 3. The experimental study is reported in Sect. 4. Section 5 concludes the paper.

2   Problem Formulation

Let $G = (V, E)$ be a directed graph with a set V of $C + 1$ vertices and a set of edges $E = \{(v_i, v_{i+1}) | v_i, v_{i+1} \in V, v_i \neq v_{i+1}\}$. The VRPTW is defined on the graph G, where each customer v_i, $i \in \{1, 2, ..., C\}$ (and the depot v_0) is given as a vertex, and each edge represents a travel connection with the cost $c_{i,j}$, $i, j \in \{0, 1, ..., C\}$, $i \neq j$. Customers define their demands d_i, $i \in \{0, 1, ..., C\}$, $d_i \geq 0$, $d_0 = 0$, time windows $[e_i, l_i]$, $i \in \{0, 1, ..., C\}$, and service times s_i, $i \in \{1, 2, ..., C\}$. Let Q be a constant vehicle capacity and K denote a number of vehicles. Each route r, $r = (v_0, v_1, ..., v_{n+1})$, starts and finishes at the depot, thus $v_0 = v_{n+1}$. A solution σ (with K routes) is feasible if (i) the vehicle capacity Q is not exceeded for any vehicle, (ii) the service of every customer starts within its time window, (iii) every customer is served in exactly one route, and (iv) every vehicle leaves and returns to the depot within the time window $[e_0, l_0]$.

The primary objective of the VRPTW is to minimize the number of routes K ($K \leq \lceil D/Q \rceil$, where $D = \sum_{i=1}^{C} d_i$). In addition, the total travel distance $T = \sum_{i=1}^{K} T_i$ is to be minimized, where T_i is the travel distance of the i-th route. Let σ_A and σ_B be the feasible solutions. Then, σ_A is of higher quality if $(K(\sigma_A) < K(\sigma_B))$ or $(K(\sigma_A) = K(\sigma_B)$ and $T(\sigma_A) < T(\sigma_B))$.

3   Parallel Memetic Algorithm and Co-operation Schemes

We had proposed and later improved [1,3,9] a two-stage parallel memetic algorithm (PMA) in which the number of routes is minimized by a parallel heuristic algorithm based on the approach suggested by Nagata and Bräysy [7]. The total travel distance is minimized using a parallel memetic algorithm. The sequential algorithm was proposed by Nagata et al. [8] and was subjected to our improvements [9,10]. In this section we give an overview of the PMA and describe proposed co-operation schemes which are experimentally evaluated.

Algorithm 1. Parallel memetic algorithm (PMA)

1: $K \leftarrow$ RouteNumber(RM(0));	$\triangleright$ First stage
2: **for** $i \leftarrow 1$ **to** N **do**	
3: Generate a population of N_p solutions with K routes each for each process P_i;	
4: **end for**	
5: **parfor** $P_i \leftarrow P_1$ **to** P_N **do**	$\triangleright$ Second stage
6: *finished* $\leftarrow$ false;	
7: **while not** *finished* **do**	$\triangleright$ Creating the next generation of solutions
8: Determine N_p pairs (σ_A, σ_B);	$\triangleright$ Selecting the parent solutions
9: **for all** (σ_A, σ_B) **do**	
10: $\sigma_c^b \leftarrow \sigma_A$;	
11: **for** $i \leftarrow 1$ **to** N_c **do**	
12: $\sigma_c \leftarrow$ EAX(σ_A, σ_B);	$\triangleright$ Creating the child solution
13: $\sigma_c \leftarrow$ Repair(σ_c); $\sigma_c \leftarrow$ LocalSearch(σ_c);	
14: **if** $T(\sigma_c) < T(\sigma_c^b)$ **then**	
15: $\sigma_c^b \leftarrow \sigma_c$;	
16: **end if**	
17: **end for**	
18: **end for**	
19: Form the next population;	
20: Co-operate according to the co-operation scheme;	
21: *finished* $\leftarrow$ CheckStoppingCondition();	
22: **end while**	
23: **end parfor**	
24: **return** best solution;	

3.1 Parallel Memetic Algorithm Outline

The PMA consists of two stages in which the number of vehicles and the total travel distance are minimized independently (Algorithm 1). First, the number of routes is minimized by the parallel heuristic algorithm executed by N processes (line 1). Then, an initial population of N_p feasible solutions, each consisting of K routes, is generated for each parallel process (lines 2–4). A solution with the minimal total travel distance is found using a parallel memetic algorithm (lines 5–24).

In the first stage (Algorithm 2), an initial solution in which every customer is served in a separate route is subject to the route minimization (RM). At each step, a random route r is selected in σ, its customers are inserted into the ejection pool (EP) (line 5) and the penalty counters p indicating the re-insertion difficulty are initialized (line 6). Then, a customer v is popped from the EP (line 8). If there exist several feasible positions for v, then a random one is chosen (lines 9 and 10) – $S(v, \sigma)$ is a set of feasible insertions of v. Otherwise v is inserted infeasibly, and σ is squeezed to restore its feasibility (line 12). If the squeezing fails, then the additional customer ejections minimizing the sum of their penalty counters are tested (lines 15–17) and σ is perturbed by the local search moves (LSM) for search diversification (line 18). Let $\mathcal{N}(v, \sigma)$ be the neighborhood of v in σ, which is obtained by applying the following operators (moves): 2-opt*, OR-opt,

---

Algorithm 2. Parallel heuristic algorithm for minimizing the number of routes

1: **function** RM(K)
2:    **parfor** $P_i \leftarrow P_1$ **to** P_N **do**
3:       $\sigma \leftarrow \sigma_I$, where σ_I is an initial solution; $finished \leftarrow$ false;
4:       **while not** $finished$ **do**
5:          Initialize the EP with M customers from a random route r from σ;
6:          $p_i \leftarrow 0, i = 1, 2, \ldots, M$;   ▷ Initializing penalty counters
7:          **while** (**not** $finished$) **and** (EP$\neq \emptyset$) **do**   ▷ Inserting ejected customers
8:             $v \leftarrow$ EP.pop();
9:             **if** $S(v, \sigma) \neq \emptyset$ **then**   ▷ v is being inserted
10:                $\sigma \leftarrow \sigma'$ selected randomly from $S(v, \sigma)$;
11:             **else**
12:                $\sigma \leftarrow$ Squeeze(v, σ);
13:             **end if**
14:             **if** $v \notin \sigma$ **then**
15:                $p_i \leftarrow p_i + 1$;   ▷ Increasing the penalty counter
16:                $\sigma \leftarrow \sigma'$ from $S_e(v, \sigma)$ such that $\sum_{j=1}^{M_e} p_j$ is minimal;
17:                Insert M_e ejected customers into the EP;
18:                $\sigma \leftarrow$ Perturb(σ);
19:             **end if**
20:             Co-operate according to the co-operation scheme;
21:             $finished \leftarrow$ CheckStoppingCondition();
22:          **end while**
23:       **end while**
24:    **end parfor**
25:    **return** best solution among all processes after additional local search;
26: **end function**

---

out-relocate, in-relocate and exchange [1]. In order to decrease the search space, only $N_N = 50$ customers nearest to v are considered. Then, the best feasible solution σ', $\sigma' \in \mathcal{N}(v, \sigma)$, replaces σ, and the processes co-operate (line 20).

Once a minimal number of routes K is found, a population of N_p feasible solutions is created for each process P_i (Algorithm 1, lines 2–4). Then, according to a pre-selection scheme [10], N_p pairs of chromosomes (σ_A, σ_B) are determined (line 8) and for each pair N_c children σ_c are generated using the edge assembly crossover (EAX) operator (line 12). If σ_c is infeasible then it is repaired (if possible) and enhanced by the LSM (line 13). The best feasible child σ_c^b, i.e. with the shortest total travel distance, is found for each (σ_A, σ_B) (lines 14–16). The next population is formed (line 19) according to the post-selection scheme [10], and the processes co-operate (line 20). Finally, the best individual is returned (line 24). The detailed descriptions of the algorithms can be found in [1,10].

3.2   Co-operation Schemes

The N processes in the PMA co-operate periodically in order to exchange solutions found up-to-date and to guide the search towards solutions of better quality. Let δ_R and δ_D denote the periods of process co-operations during the number of routes and the total travel distance minimization stages. Here we present the proposed co-operation schemes that were explored and experimentally evaluated in this work. The characteristics of the schemes are as follows:

1. **Rigid.** Each process P_i runs both optimization stages independently. After each stage all solutions are compared and the best one is chosen. This approach can be considered as the independent multi-search method since the processes do not co-operate during the PMA execution.

2. **Knowledge synchronization (KS).** In the first stage, the processes send their best solutions to process P_1[1] which determines the best one σ_b and broadcasts it back to each process. In the second stage, each process P_i sends the best solution σ_i^b to the master. It sorts N received solutions and selects randomly N_b, $1 \leq N_b < N$, ones for each process P_i. Finally, every process P_i replaces its N_b worst solutions with these received from the master.

3. **Ring.** Let $(P_1, P_2, \ldots, P_N)$ be the order in which processes co-operate. Then, the ring co-operation scheme is given as: $P_1 \to P_2 \to \cdots \to P_N \to P_1$. The process P_1 sends its best solution (in both stages) to P_2. It replaces its best solution if the received one is of higher quality and sends it to P_3. As the scheme constitutes a ring (P_N co-operates with P_1 at last) it is ensured that the best solution found so far is kept by the master process P_1.

4. **Randomized EAX (R-EAX).** The scheme is similar to the ring, however, if the solution received by process P_i is of lower quality than the current best one of P_i, then the EAX operator is performed on these solutions to generate a child σ_c. If it is not feasible, then the feasibility is restored by the repairing moves. Finally, if σ_c is feasible then it replaces the best solution kept by P_i in both stages. The order of co-operation is randomly determined by the master process P_1 at the beginning of each co-operation phase.

5. **Pool.** In the first stage, process P_1 handles the pool of size s, $1 \leq s \leq N$, of solutions with the currently minimal number of routes K_P. If σ is the best solution kept by process P_i, and $K(\sigma) > K_P$, then it is replaced by a random pool solution σ_P. If $K(\sigma) = K_P$, then σ_P replaces σ with a probability p, $0 < p < 1$. If process P_i finds a feasible solution σ, and $K(\sigma) < K_P$, then the pool is emptied, σ is sent to P_1 to form the pool of size 1. Each process P_i sends to P_1 its best σ with K_P routes to replace a random pool solution, only if σ does not exist in the pool. In the second stage, each process P_i sends to P_1 the best σ in its population (of size N_p), and it is inserted into the pool as in the first stage. The process P_1 determines ηN, $0 < \eta < 1$, $\eta N \leq N_p$, best pool solutions. They replace ηN random ones in the population of P_i.

6. **Pool with EAX (P-EAX).** The regular pool is enhanced by applying the EAX operator. In the first stage, if the solution σ_P received by process P_i contains the same number of routes as the best solution σ kept by P_i, and $\sigma_P \neq \sigma$, then the EAX operator is applied to generate a child σ_c and its feasibility is restored if necessary. At most N_c children are generated, until a feasible one σ_c is obtained. Then, σ_c replaces σ. In the second stage, the EAX operator is employed while replacing ηN received solutions by process

---

[1] We distinguish the process P_1 as the master process.

P_i – a solution σ_a that is being replaced is crossed-over with the replacing solution σ_b, $\sigma_a \neq \sigma_b$, to generate σ_c. If σ_c is feasible, then it replaces σ_b.

The time complexities of the considered co-operation schemes (i.e. of a single co-operation phase) are presented in Table 1 (it can be shown [1] that the EAX operator takes $O(C^2)$ time, where C is the number of customers in each parent solution). It is worth noting that data are transferred asychronously between the processes. This means that the processes proceed with the algorithm execution while the communication progresses, i.e. computation and communication overlap. Furthermore, a two-step data passing is applied: the complete solution is transferred only if its quality has been improved since the previous co-operation.

Table 1. Time complexities of a single co-operation phase for the first ($\mathcal{T}_1$) and the second ($\mathcal{T}_2$) optimization stages; N – number of processes, C – number of customers.

Scheme	$\mathcal{T}_1(N, C)$	$\mathcal{T}_2(N, C)$
Rigid	$C \cdot N$	$C \cdot N$
KS	$2C \cdot N$	$2C \cdot N + N \log N$
Ring	$C \cdot N$	$C \cdot N$
R-EAX	$C^2 \cdot N$	$C^2 \cdot N$
Pool	$2C \cdot N$	$2C \cdot N + N \log N$
P-EAX	$C^2 \cdot N$	$C^2 \cdot N + N \log N$

4   Experimental Results

In this section we present the settings of the PMA, describe the Gehring and Homberger's (GH) tests solved by the PMA, and discuss the experimental results obtained for each proposed co-operation scheme. The analyses of the quality of solutions, search convergence and execution time of the PMA are also provided.

4.1   Settings

The proposed co-operation schemes were tested in the PMA for solving the 400-customers GH tests. Also, we ran the PMA for some larger tests with 600 and 800 customers. However, to draw the meaningful conclusions about them, it is necessary to investigate all instances in a subclass, thus we focus on analysis of the results obtained for the 400-customers problems. The GH tests are divided into six groups: C1, C2, R1, R2, RC1 and RC2. Customers are: (i) clustered (C class), (ii) randomly dispersed on the map (R class), (iii) mixed – both clustered and random (RC class). The subclasses C1, R1 and RC1 have smaller vehicle capacities and shorter time windows than C2, R2 and RC2. There are 10 problem instances in a subclass, resulting in 300 GH instances in total. Tests can be distinguished by their unique names, α_β_γ, where α denotes the subclass (C1, C2, R1, R2, RC1, RC2), β relates to the number of customers (2 for 200 customers, 4 for 400, and so forth) and γ is the instance number ($\gamma = 1, 2, \ldots, 10$).

The PMA was implemented in C++ using the Message Passing Interface (MPI) library and the experiments were performed on the Galera[2] supercomputer. Let $t_T = t_R + t_P + t_D$, $t_T \leq t_{max}$, be the total execution time of the PMA, where t_R is the time necessary to find the minimum number of routes (Algorithm 2), t_P is the time of generating a population of solutions of size N_p (Algorithm 1, lines 2–4), and t_D is the time of the total distance minimization (Algorithm 1, lines 5–24). The PMA was run on 32 cores (4 nodes equipped with 8 cores each), each running a single process, and $t_{max} = 240$ min. The PMA parameters were tuned experimentally to the following values: $N_p = 60$, $N_c = 10$, $\eta N_p = N_b = 4$ ($\eta \approx 0.07$), $p = 0.5$, $\delta_R = 200000$, $\delta_D = 40$. The values of other PMA parameters along with the discussion on their influence on final results can be found in [1,9].

4.2 Analysis and Discussion

In order to verify the speed of search convergence and quality of solutions obtained using the PMA, each 400-customers GH test (60 tests in total) was run n times, $7 \leq n \leq 10$, using each co-operation scheme. As mentioned earlier, the first objective of the VRPTW is to minimize the number of routes. Thus, the number of routes K obtained from the first stage of the PMA was compared with the currently best-known one K_b[3]. The minimal number of routes K_b was achieved for all 400-customers tests using the PMA. However, in case of the independent multi-search method, i.e. the rigid co-operation, the PMA did not converge to the best K_b within time t_{max}, and K was larger by one route for several tests[4] from C1, C2 and RC2 subclasses. It indicates that the co-operation of processes is crucial for converging to highest-quality solutions.

The average best total travel distance obtained using the proposed co-operation schemes and allowing for the evaluation of the quality of solutions containing the same number of routes K, was calculated for each GH test. Then, these averages were used to calculate the average total distance of each GH subclass. The averaged best distances achieved for all 400-customers GH tests (tests for which $K > K_b$ are omitted for the rigid co-operation) are presented in Table 2. The experiments showed that the ring and KS co-operations outperformed other schemes and gave the best average total travel distances. The percentages of the average total travel distances (%Best) shown in Table 2 are calculated with respect to the best currently-known ones. If within a certain subclass a better, i.e. with a shorter total travel distance, solution than this obtained using the PMA exists, then the percentage is larger than 100.00 %. The R1 and RC1 subclasses turned out to be the most difficult with %Best equal to 101.07 and 100.69 respectively. The detailed results for R1 subclass given in Table 3 shows that the ring, KS and R-EAX schemes outperformed other co-operation schemes.

---

[2] More details can be found at: http://www.task.gda.pl/english/hpc

[3] The best-known solutions of the GH tests are published at: http://www.sintef.no/ Projectweb/TOP/VRPTW/Homberger-benchmark/; reference date: April 27, 2013.

[4] These were: C1_4_8, C2_4_3, C2_4_4, C2_4_10, RC2_4_1, RC2_4_2, RC2_4_5, RC2_4_6.

Table 2. Average total travel distances of all 400-customers GH tests (the best average total travel distance within a subclass is marked in boldface).

Scheme	C1	C2	R1	R2	RC1	RC2
Rigid	—	—	8569.73	6188.83	8074.94	—
KS	7182.05	3907.47	8480.41	6174.17	7997.52	**5301.41**
Ring	**7178.94**	**3905.91**	**8478.69**	**6157.35**	**7996.45**	5306.01
R-EAX	7186.47	3914.56	8484.82	6163.99	8004.98	5314.27
Pool	7245.33	3919.83	8653.49	6212.85	8181.13	5378.28
P-EAX	7235.77	3931.50	8638.59	6224.70	8194.22	5403.72
Best GH	7169.31	3897.91	8388.64	6146.17	7941.85	5288.9
%Best	100.13	100.21	101.07	100.18	100.69	100.24

Table 3. Average total travel distances for the R1 tests (the best average total travel distance for each test is marked in boldface).

Test	GH best	Rigid	KS	Ring	R-EAX	Pool	P-EAX
R1_4_1	10372.31	10397.18	10391.91	10391.91	**10391.62**	10405.21	10405.21
R1_4_2	8955.50	9156.59	9027.83	**9026.92**	9055.03	9252.09	9273.93
R1_4_3	7826.71	8064.23	7993.51	**7935.95**	7963.86	8170.70	8077.30
R1_4_4	7318.62	7512.17	7412.08	**7397.63**	7416.84	7577.34	7518.02
R1_4_5	9242.43	9356.58	9320.95	**9313.12**	9317.08	9448.14	9430.42
R1_4_6	8373.71	8626.37	8513.00	**8503.31**	8527.90	8739.17	8788.30
R1_4_7	7645.88	7871.13	7739.90	7762.46	**7698.84**	7912.49	7849.95
R1_4_8	7288.95	7418.83	**7353.29**	7400.92	7391.00	7583.91	7546.35
R1_4_9	8742.41	8915.21	8812.72	**8809.56**	8835.96	9082.86	9065.89
R1_4_10	8119.88	8378.99	**8238.94**	8245.08	8250.08	8362.95	8430.55

A number of factors, e.g. the population size or the number of ejected customers M_e (Algorithm 2, line 17), can make the execution time of the PMA arbitrarily large. Here we compare the average execution time t after which the best total travel distance (among N processes) was not further improved and the search converged, for each GH subclass (Fig. 1(a–c)). The total average execution time t of the PMA is given in Fig. 1(d). We also present the average number of generations g. Intuitively, the PMA with the KS scheme requires the shortest execution time to converge, since the search is guided towards the most promising parts of the solution space which are intensively exploited. This, however, leads to getting stuck in local minima (KS was outperformed by the ring scheme for most GH tests, see Table 2). The best solutions propagate in case of the ring and R-EAX schemes, but the search is diversified due to updating only the best individuals. Applying the EAX operator in the former scheme results in increasing the execution time of the PMA. However, it is easy to see that t is smaller for P-EAX scheme than for the regular pool. Thus, the quality of initial populations and guiding the search, strongly influence the convergence of the total distance minimization stage. In the PMA with the rigid co-operation, each

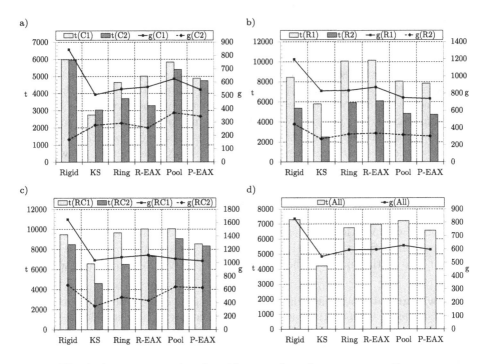

Fig. 1. Average execution time t in seconds and average generation g.

process minimizes the travel distance independently. Clearly, the average number of generations is significantly larger than in case of the co-operative methods (Fig. 1).

The study indicated that the GH tests with wider time windows and larger vehicle capacities (C2, R2 and RC2) can be solved faster than these with shorter time windows and smaller capacities (C1, R1 and RC1). The tests with random customers (R1, R2) are more difficult to solve to good quality in comparison with the clustered-customers tests (C1, C2). It is worth noting, that applying the R-EAX scheme resulted in improving the world's best total travel distances T_b for two GH tests and decreasing the best-known number of routes for two other ones, whereas the ring scheme decreased T_b in one GH test (Table 4).

Table 4. New world's best results for Gehring and Homberger's tests; in expression x/y, x denotes the number of routes and y the total travel distance.

Test	New GH best	Old GH best	Scheme
C1_4_2	36/7686.97	36/7687.38	R-EAX
C1_6_7	57/15997.59	58/14816.55	R-EAX
C2_4_8	11/4352.95	12/3787.08	R-EAX
R1_8_2	72/32817.67	72/32942.77	R-EAX
R2_4_5	8/7129.03	8/7136.90	Ring

5   Conclusions and Future Work

In this work we proposed new co-operation schemes used in the parallel memetic algorithm for the NP-hard vehicle routing problem with time windows. We showed how the solutions quality, execution time and number of generations are influenced by the choice of the scheme. The experiments performed on Gehring and Homberger's benchmark proved the ring and knowledge synchronization schemes to be the best in terms of the search convergence speed and quality of final solutions. However, it is the randomized EAX scheme that allowed for finding four new world's best solutions (one was found using the ring scheme). We showed that the independent multi-search method gave the worst results, what indicates that the search should be guided during the algorithm execution.

Our ongoing research includes performing full GH tests, investigating the influence of co-operation frequency on the quality of solutions and execution time and conducting the sensitivity analysis for its automatic tuning. Finally, our aim is to further improve the rigid, KS and R-EAX co-operation schemes.

Acknowledgments. We thank the following computing centers where the computations of our project were carried out: Academic Computer Centre in Gdańsk TASK, Academic Computer Centre CYFRONET AGH, Kraków, Interdisciplinary Centre for Mathematical and Computational Modeling, Warsaw University, Wrocław Centre for Networking and Supercomputing.

References

1. Blocho, M.: A parallel memetic algorithm for the vehicle routing problem with time windows. Ph.D. thesis, Silesian University of Technology (2013) (in Polish)
2. Blocho, M., Czech, Z.J.: An improved route minimization algorithm for the vehicle routing problem with time windows. Stud. Informatica **32**(99), 5–19 (2010)
3. Blocho, M., Czech, Z.J.: A parallel EAX-based algorithm for minimizing the number of routes in the vehicle routing problem with time windows. In: Proceedings of the IEEE HPCC-ICESS, pp. 1239–1246 (2012)
4. Bräysy, O., Gendreau, M.: Vehicle routing problem with time windows, part I: route construction and local search algorithms. Trans. Sc. **39**(1), 104–118 (2005)
5. Kallehauge, B.: Formulations and exact algorithms for the vehicle routing problem with time windows. Comput. Oper. Res. **35**(7), 2307–2330 (2008)
6. Mester, D., Bräysy, O.: Active guided evolution strategies for large-scale vehicle routing problems with time windows. Comput. Oper. Res. **32**(6), 1593–1614 (2005)
7. Nagata, Y., Bräysy, O.: A powerful route minimization heuristic for the vehicle routing problem with time windows. Oper. Res. Lett. **37**(5), 333–338 (2009)
8. Nagata, Y., Bräysy, O., Dullaert, W.: A penalty-based edge assembly memetic algorithm for the vehicle routing problem with time windows. Comput. Oper. Res. **37**(4), 724–737 (2010)
9. Nalepa, J., Czech, Z.J.: A parallel heuristic algorithm to solve the vehicle routing problem with time windows. Stud. Informatica **33**(1), 91–106 (2012)

10. Nalepa, J., Czech, Z.J.: New selection schemes in a memetic algorithm for the vehicle routing problem with time windows. In: Tomassini, M., Antonioni, A., Daolio, F., Buesser, P. (eds.) ICANNGA 2013. LNCS, vol. 7824, pp. 396–405. Springer, Heidelberg (2013)

11. Russell, R.: Hybrid heuristics for the vehicle routing problem with time windows. Trans. Sc. **29**(2), 156 (1995)

12. Thompson, P.M., Psaraftis, H.N.: Cyclic transfer algorithms for multivehicle routing and scheduling problems. Oper. Res. **41**(5), 935–946 (1993)

13. Ursani, Z., Essam, D., Cornforth, D., Stocker, R.: Localized genetic algorithm for VRPTW. Appl. Soft Comput. **11**(8), 5375–5390 (2011)

14. Zhong, Y., Pan, X.: A hybrid optimization solution to VRPTW based on simulated annealing. In: Proceedings of the IEEE ICAL, pp. 3113–3117 (2007)

Scalable and Efficient Parallel Selection

Christian Siebert$^{(\boxtimes)}$

Laboratory for Parallel Programming, Department of Computer Science,
RWTH Aachen University, Aachen, Germany
christian.siebert@rwth-aachen.de

Abstract. Selection algorithms find the kth smallest element from a set of elements. Although there are optimal parallel selection algorithms available for theoretical machines, these algorithms are not only difficult to implement but also inefficient in practice. Consequently, scalable applications can only use few special cases such as minimum and maximum, where efficient implementations exist. To overcome such limitations, we propose a general parallel selection algorithm that scales even on today's largest supercomputers. Our approach is based on an efficient, unbiased median approximation method, recently introduced as *median-of-3 reduction*, and *Hoare*'s sequential *QuickSelect* idea from 1961. The resulting algorithm scales with a time complexity of $\mathcal{O}(\log^2 n)$ for n distributed elements while needing only $\mathcal{O}(1)$ space. Furthermore, we prove it to be a practical solution by explaining implementation details and showing performance results for up to $458,752$ processor cores.

Keywords: Selection · QuickSelect · Median · Parallel algorithms · MPI

1 Introduction

Complex problems such as the search for the nearest neighbor or shortest path require a solution for a subproblem, namely *selection*. We consider this problem of selecting the kth smallest element from a set of n given elements. All elements have a key, and an ordering relation denoted by $\leq$ is defined on the keys. Therefore, we can compare two keys and determine which key is smaller than the other. If all elements would be sorted according to $\leq$ then the kth smallest element would be at position[1] k. As massively parallel supercomputers are becoming widespread, there is a growing need for efficient parallel solutions to the selection problem, where the n elements are distributed over p processors. A common form of the parallel selection problem in practice is the special case with a single element per processor. This forms also the base case where the other two cases with $n < p$ and $n > p$ can be reduced to. Consequently, this paper focuses

---

[1] In statistics, the kth order statistic of a sample is equal to its kth smallest value, and the position of this value is called *rank*. Unfortunately, *rank* is also used in MPI to identify a process. To disambiguate, we use the terms *position* and *MPI rank*.

R. Wyrzykowski et al. (Eds.): PPAM 2013, Part I, LNCS 8384, pp. 202–213, 2014.
DOI: 10.1007/978-3-642-55224-3_20, © Springer-Verlag Berlin Heidelberg 2014

on an efficient, comparison-based solution for the parallel selection problem with $n = p$. A few special cases of selection such as finding the minimum (i.e., $k = 1$) and the maximum (i.e., $k = n$) are properly solved and available in most parallel programming environments (e.g., `MPI_Reduce(...MPI_MIN...)`). Unfortunately, this is not the case for the general selection problem with an arbitrary k. Even the frequent task of finding the *median* element $\tilde{x}$ (i.e., $k = \lfloor n/2 \rfloor$) efficiently remains unsolved in practice. Instead, a common solution is to sort all elements and then simply extract the desired element. However, this workaround is inefficient as it requires more work than necessary. For example: optimal sequential sorting requires $\mathcal{O}(n \log n)$ time (e.g., with *MergeSort*), but there exist sequential selection algorithms such as *BFPRT* [1] that work in $\mathcal{O}(n)$ time. In parallel, the situation is even worse: many scalable sorting algorithms cannot handle few elements per process (e.g., *SampleSort* [3]) let alone scale to more than a couple of thousand processes. The usual workaround, gathering all distributed elements for sequential processing, is not feasible when we approach *Exascale* [8]. Interestingly, there exist parallel selection algorithms for theoretical machines such as PRAM (Parallel Random Access Machine, see [4]), but from their complexity it can be expected that they are highly inefficient in practice. In fact, we have not encountered any implementation of such parallel selection algorithms.

The rest of this paper is organized as follows: Sect. 2 presents a well-known sequential selection algorithm and an idea for a minor improvement. By transferring both into the parallel world, Sect. 3 proposes a parallel selection algorithm. Its two main ingredients are a median approximation and a parallel partitioning scheme, respectively presented in Sects. 3.1 and 3.2. Both parts achieve a running time of $\mathcal{O}(\log p)$, turning the complete parallel algorithm into a solution that scales well with $\mathcal{O}(\log^2 p)$ while requiring only a constant amount of space. Section 4 provides practical and theoretical evidence that the approximated median indeed leads to suitable partitions. A practical performance evaluation considering different inputs and up to $458, 752$ processor cores demonstrates the efficiency of our implementation in Sect. 5. Finally, Sect. 6 concludes our findings related to parallel selection.

2   Sequential Selection

In 1961, *C.A.R. Hoare* published a partition-based general selection algorithm [5], also known as *QuickSelect*. Similar to *QuickSort*, *QuickSelect* is a randomized algorithm as it chooses a "pivot" element uniformly at random. Once this pivot is chosen, the algorithm partitions the input according to smaller and larger elements. Based on the size of those partitions, one can infer where the target element must reside. Contrary to *QuickSort*, which proceeds recursively in both partitions, the *QuickSelect* algorithm proceeds recursively only within this target partition. Eventually, recursion stops when the kth smallest element is found. Pseudocode of a sequential *QuickSelect* implementation is shown in Algorithm 1. Although, *QuickSelect* has an expected $\mathcal{O}(n)$ running time, unlucky choices of the pivot make it as slow as $\mathcal{O}(n^2)$ in the worst case. In 1995, Kirschenhofer et al.

Algorithm 1. Seq_QuickSelect($A[1 \ldots n], k$) finds the kth smallest element.

1: $r \leftarrow \text{random}(1 \ldots n)$
2: $pivot \leftarrow A[r]$
3: {partition A into smaller and larger elements}
4: **for** $i \leftarrow 1$ **to** n **do**
5:    **if** $A[i] < pivot$ **then**
6:        append $A[i]$ to A_1
7:    **else if** $A[i] > pivot$ **then**
8:        append $A[i]$ to A_2
9:    **end if**
10: **end for**
11: **if** $k \leq \text{length}(A_1)$ **then**
12:    {target element is among the smaller elements}
13:    **return** Seq_QuickSelect(A_1, k)
14: **else if** $k > \text{length}(A) - \text{length}(A_2)$ **then**
15:    {target element is among the larger elements}
16:    **return** Seq_QuickSelect($A_2, k - (\text{length}(A) - \text{length}(A_2))$)
17: **else**
18:    {target element is the pivot}
19:    **return** $pivot$
20: **end if**

---

improved the odds for a good running time by using a median-of-3 pivot [6]. This approach selects not only one but three sample elements at random and picks the median of these three samples as pivot. Although this improves the probability of selecting a more suitable pivot, the worst case running time still remains $\mathcal{O}(n^2)$. In fact, the authors of a more recent analysis [2] clearly state "[...] that *median-of-three does not yield a significant improvement over the classic rule: the lower bounds for the classic rule carry over to median-of-three*". The problem is that only a constant number of elements are considered in the pivot selection.

Although Kirschenhofer's median-of-3 pivot selection improves the sequential *QuickSelect* performance only insignificantly, we will turn his main idea into an effective parallel pivot selection algorithm. The advantage there is that not a constant number of samples but instead all elements are taken into consideration.

3   Parallel Selection

In summary, the *QuickSelect* algorithm consists of three major steps:

(1) chose a pivot element (preferably close to the median)
(2) partition all elements according to the pivot
(3) recursively proceed in the corresponding partition (until target is found)

The main challenge for a practical parallel selection algorithm is to provide efficient parallel implementation options for all three steps.

3.1   Median Approximation

The actual median of all elements would be optimal for a perfect partitioning into subsets of equal size. Han [4] showed that a parallel median algorithm exists with an optimal running time of $\mathcal{O}(\log p)$. However, this asymptotic complexity hides a constant that is too large for such an algorithm to be applicable in practice. In fact, he also showed that finding the median is just as complex as solving the selection problem in general. Fortunately, any element close to the median is also a suitable pivot. Therefore, we suggest the use of our median approximation approach, which was only briefly introduced in [10]. This approach also achieves the optimal $\mathcal{O}(\log p)$ running time, but in contrast to Han's algorithm is highly efficient in practice. We now explain the details of our approach and subsequently evaluate the quality of the selected median approximation in Sect. 4.

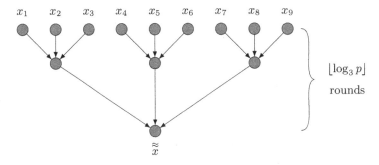

Fig. 1. Median-of-3 reduction scheme to approximate the median of elements x_i as $\tilde{\tilde{x}}$.

Our parallel median approximation approach groups elements into subgroups of three elements each, finds their median and recursively proceeds with the results of all other subgroups until a single result is obtained. Thus, this method is essentially a median-of-3 reduction scheme within a complete ternary tree topology as shown in Fig. 1. All processes start as leaf nodes and send their single input value x_i to a specific process, which corresponds to an inner node. Except for the last (i.e., rightmost) node which can be ignored when it has less than three children, all these inner-node processes receive the values from two other processes, select the median of these values (i.e., drop the smallest and the largest value) and forward the result to the next level. We implemented the necessary data exchange via MPI point-to-point communication, more precisely via MPI_Isend, MPI_Irecv and MPI_Waitall. Although such a communication tree is a recursive data structure, the actual implementation can be done in an iterative fashion, thus requiring only $\mathcal{O}(1)$ space. After $\lfloor \log_3 p \rfloor$ rounds of communication, the process at the root node obtains the final result $\tilde{\tilde{x}}$. This is regarded as approximation to the median of all input values x_i and broadcast as pivot to all participating processes. These processes can then compare their own element with this pivot and proceed with the partitioning step.

Note: As a median of smaller subgroups is not suitable for such a reduction scheme, the *Median-of-3* case represents the base case of a more general *Median-of-k* reduction. At first glance, a general scheme seems beneficial because increasing k (e.g., $k \in \{5, 7, \dots\}$) improves the accuracy of the approximated median until $k \geq n$ where it selects always the exact median. However, though larger values of k yield slightly fewer number of communication rounds according to $\lfloor \log_k p \rfloor$, the optimal time needed to select the median for each subgroup grows linearly with k and therefore easily dominates the overall running time of $\mathcal{O}(k \cdot \log_k p)$. Although occasionally our implementation with $k = 5$ was slightly faster than the base case, we usually observed a performance degradation, especially with higher k. Therefore, we decided to use only the fast *Median-of-3* reduction. Section 4 shows that even this base case suits our approach perfectly.

3.2   Partitioning

Once a pivot is chosen, all elements need to be partitioned into three sets: the elements that are smaller than, equal to, and larger than this pivot. While sequential partitioning is straightforward, doing so in parallel is more difficult. We utilize an auxiliary vector $\vec{v}_i$ consisting of three integers, each representing either $<$, $=$ or $>$. These integers are initialized with zeros. After comparing the process-local element x_i with the globally chosen pivot element, the corresponding vector element is set to one. Using vector notation, this can be written as

$$\vec{v}_i = \begin{cases} (1\ 0\ 0) & \text{if } x_i < \text{pivot,} \\ (0\ 1\ 0) & \text{if } x_i = \text{pivot,} \\ (0\ 0\ 1) & \text{if } x_i > \text{pivot.} \end{cases}$$

Adding these auxiliary vectors element-wise in parallel yields a vector $\vec{s}$ with the total number of elements that are smaller than, equal to or larger than the pivot, respectively. Similarly, using a parallel prefix sum we can determine the actual offsets $\vec{o}$ within each partition, that is, the number of lower-ranked processes owning elements that are smaller than, equal to, or larger than the pivot. With a linear combination of both results, we can compute the destination process d_i for each element. Using MPI terminology[2], this can be expressed as

$$\vec{s} = \text{Allreduce}(\vec{v}_i, +)$$
$$\vec{o}_i = \text{Exscan}(\vec{v}_i, +)$$
$$d_i = \vec{v}_i \cdot \left(\begin{pmatrix} 0\ 0\ 0 \\ 1\ 0\ 0 \\ 1\ 1\ 0 \end{pmatrix} \cdot \vec{s}^T + \vec{o}_i^T \right).$$

Both, `Allreduce` and `Exscan` exist as collective operations in the MPI standard and can be implemented efficiently using only $\mathcal{O}(\log p)$ individual communications [7,9]. After sending every element to its destination process, the

---

[2] We use MPI terminology: assuming x_i is the input at MPI rank i then `Allreduce` computes the sum $\sum_{j=0}^{p-1} x_j$ and `Exscan` computes the prefix sum $\sum_{j=0}^{i-1} x_j$ in parallel.

global partitioning is completed. This can be accomplished in $\mathcal{O}(1)$ time with a single MPI_Sendrecv() operation using MPI_ANY_SOURCE for the receive part. A pseudocode implementation of the partitioning step is shown in Algorithm 2. Selecting the pivot via our median approximation scheme from Sect. 3.1 ensures that the partitions for $<$ and $>$ have roughly equal size. Therefore, at most close to half of the elements remain to be searched after a partitioning. Consequently, this will lead to a total of $\mathcal{O}(\log p)$ divide-and-conquer rounds.

Algorithm 2. Par_Partition$(x, pivot)$ partitions all elements according to *pivot*.

1: {initialize auxiliary array}
2: int $v[3] \leftarrow \{0, 0, 0\}$
3: {compare a process' own element with the pivot element}
4: **if** $x < pivot$ **then**
5:    $v[0] \leftarrow 1$
6: **else if** $x == pivot$ **then**
7:    $v[1] \leftarrow 1$
8: **else if** $x > pivot$ **then**
9:    $v[2] \leftarrow 1$
10: **end if**
11: {determine the sizes of the resulting partitions}
12: $s \leftarrow$ Allreduce$(v, 3, +)$
13: {determine the offsets within each partition}
14: $o \leftarrow$ Exscan$(v, 3, +)$
15: {combine the results to compute the destination of x}
16: **if** $x < pivot$ **then**
17:    $d \leftarrow o[0]$
18: **else if** $x == pivot$ **then**
19:    $d \leftarrow s[0] + o[1]$
20: **else if** $x > pivot$ **then**
21:    $d \leftarrow s[0] + s[1] + o[2]$
22: **end if**
23: {send the element to its destination process and receive a new element}
24: Sendrecv$(x, d, x_{new}, \text{ANY\_SOURCE})$

3.3   Proceed in the Target Partition

Once all elements are partitioned according to the pivot element, our selection algorithm needs to proceed with the partition where the target element must reside. Similar to the sequential *QuickSelect* algorithm, this target partition is determined by comparing k with the partition sizes in $\vec{s}$. In parallel however, continuing only in the target partition comes with the challenge that only those processes which are responsible for that particular partition can participate in subsequent steps. This is especially problematic for the required collectives Bcast, Allreduce and Exscan, because MPI mandates that all processes in a communicator have to participate in a collective operation. A possible solution could create one new communicator per target partition (e.g., with MPI_Comm_split) including only those processes that are responsible for such a partition. Unfortunately, the existing communicator creation's complexity of $\Omega(p)$ is too expensive for an efficient implementation. Instead, we propose special "range collectives" originally mentioned and briefly introduced in [10]. These range collectives are conceptually identical to their *MPI* counterparts but work only on a continuous sub-range of all processes. As such their interface provides two additional integer arguments firstproc and numprocs to specify the desired range of participating processes. Only those processes within this range must call these collectives and are actually involved in the operation; all other processes outside the specified range do not need to call the collective, and are ignored by our implementation even if they do. We have implemented all necessary range collectives for our algorithm: they work with constant space and a time complexity of $\mathcal{O}(\log p)$. Therefore, they enable an efficient parallel selection in $\mathcal{O}(\log^2 p)$ time. Moreover, using these range collectives proceeding in the target partition becomes as simple as adjusting the processor range accordingly.

4   Quality of Our Median Approximation

Although our parallel median approximation approach does not necessarily find the exact median, it always finds elements close to it. To support this claim, we quantify now the accuracy of our median approximation approach.

4.1   Simulation

First, we evaluate the accuracy of our median approximation approach in simulation experiments. For this purpose, it is applied numerous times to pseudo-random input. Since we are mainly interested in the position of an element, we use the *Scalable Parallel Random Number Generators Library* to generate input permutations of $\{1, \ldots, n\}$ uniformly at random. Figure 2 depicts the outcome of 244 billion such simulations for $n = 3^7 = 2187$ and plots for each $\tilde{\tilde{x}}$ the number of times it occurred. From these resulting frequencies, we derive individual probability estimations for each possible outcome x_i of a median approximation as $p_i = \frac{\text{frequency}(x_i)}{\text{totalruns}}$. For our $2.44 \cdot 10^{11}$ independent simulations, the resulting expected value for the outcome $E[X] = \sum x_i p_i$ is 1092.9989, which is consistent with the actual position of the true median position$(\tilde{\tilde{x}}) = 1093$. In statistical terms, our median approximation method is therefore called an unbiased estimator for the true median. The resulting variance $\sum p_i \cdot (x_i - E(X))^2$ of this simulation is 2101.27, and thus the standard deviation is only 45.83. In other words, in 50 % of all simulations the position of the approximated median value is at most ± 30 elements or 1.37 % off the actual median. Even when using the most extreme outliers (cf. marked regions of unencountered results in Fig. 2) as pivots, the ratio of partition sizes would never be worse than 34 : 66. We conducted simulations also for 6561 and 19683 elements, confirming the general trend that our median approximation leads to suitable partitions in practice.

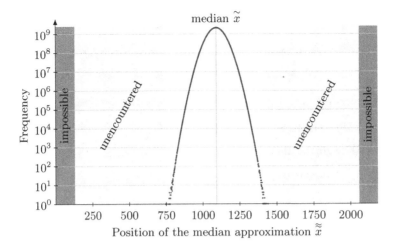

Fig. 2. Simulation of 244 billion median approximations with 2187 random elements.

4.2 Worst Case

While the previous simulation results provide a practical insight into the quality of our median approximation method, we are also interested in the theoretically possible worst case. Our construction of such a worst case is based on two properties. First, observe that the algorithm is insensitive to the order in which the elements enter a node in the ternary reduction tree (cf. Fig. 1); This follows from the commutative property of the median operation. Therefore, we are free to arbitrarily exchange the children (of course with their connected subtrees) of any node without changing the result $\tilde{\tilde{x}}$. For a systematic analysis, we choose a sorted order of the children: the leftmost child is therefore smaller than the middle child, which itself is smaller than the rightmost child. Second, we retrace the median-of-3 reduction from the root node at the bottom towards the leaf nodes at the top. While doing so, we keep track of the relationship between a parent and its children: either a child is identical to the parent or it is smaller or larger than the parent. As comparisons are transitive, this method establishes for many nodes a relationship to the result element $\tilde{\tilde{x}}$. Both, the reordering of children and the relationships with respect to $\tilde{\tilde{x}}$ are illustrated in Fig. 3.

Although the related types of nodes (i.e., ◁, □ and ▷) are settled, we can change the unrelated nodes (i.e., '?'). This enables the construction of a worst case input for our median-of-3 reduction scheme: by inserting elements that are larger than $\tilde{\tilde{x}}$ into the unrelated nodes, we can maximize the bias of the reduction towards selecting a small element. Solving recurrence relations starting at the root node reveals the number of elements for each node type: for $n = 3^k$ ($\forall\, k \in \mathbb{N}$ and $k > 0$) induction shows that there are $2^k - 1$ elements larger than $\tilde{\tilde{x}}$ and $3^k - 2^{k+1} + 1$ elements unrelated to it. Thus in the worst case, only $2^k - 1$ elements

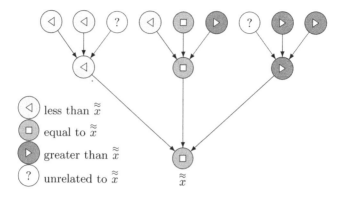

Fig. 3. "Sorted" median-of-3 reduction tree showing the node relationship with $\overset{\approx}{x}$.

are guaranteed to be smaller but up to $3^k - 2^k$ elements can be larger than $\overset{\approx}{x}$. We conclude that the $2^{\lfloor \log_3 n \rfloor} - 1$ largest keys and the $2^{\lfloor \log_3 n \rfloor} - 1$ smallest keys can never be chosen as an approximate median. These two border-zone areas of impossible outcomes of the median approximation are also sketched in Fig. 2. For practical values of p, even such a worst case input results only in a minor degradation of the algorithm's overall running time.

5   Performance Evaluation

This section shows the evaluation results for our parallel selection implementation. All experiments were conducted on *Juqueen*, an IBM BlueGene/Q system consisting of 28,672 compute nodes, each providing 16 IBM PowerPC A2 cores running at 1.6 GHz. Individual nodes communicate within a 5D torus network. Our implementation was built using the IBM XL C/C++ compiler V12.1 and the vendor-supplied MPI library V1.5. All presented measurements utilized 16 MPI processes per node, which corresponds to one process per core.

5.1   Different Inputs

The parallel selection algorithm recursively searches until the target element is found, which can potentially be already after a single round but also after $\mathcal{O}(\log n)$ rounds. As the number of rounds directly influences the performance of our implementation, a time variation can be expected depending on the actual input. Therefore, we measured the individual running time of 10,000 executions with 64 MPI processes for different input arguments k. Figure 4 shows that the minimum, average and maximum (except for few outliers) times over the many measurements are similar for each particular input, indicating reproducible performance for fixed inputs. In contrast, the running time with different input values of k can differ a lot: in our measurements it varies from $208\,\mu s$ for 1 round up to $736\,\mu s$ for 7 rounds. Similar performance differences can be observed by changing not only k but also the input elements themselves or both.

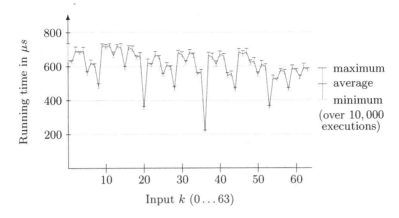

Fig. 4. Selection performance with 64 MPI processes and different values of k.

5.2 Scalability

For our scalability evaluation, we measured the running time of our parallel selection implementation for all possible values of k (i.e., $0 \leq k < p$), and recorded five statistical properties: the minimum, lower quartile, median, upper quartile, and maximum. The most extreme outliers and the necessary percentiles (25th, 50th and 75th) are—of course—determined efficiently and with constant space requirements using our scalable selection solution itself. We observed that the quartiles are too close to the median values to be clearly visible. Therefore, Fig. 5 shows only the minimum, the median and the maximum timings. While the minimum values are very sensitive to the input, the maximum and especially the median values show a smooth scalability curve, representing the $\mathcal{O}(\log^2 p)$ complexity of our parallel selection algorithm. Even utilizing $458{,}752$ MPI processes,

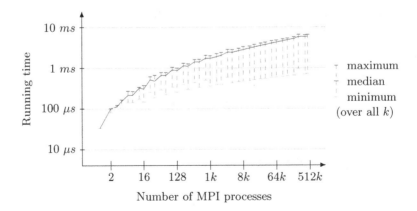

Fig. 5. Selection performance with different number of MPI processes.

representing the full *Juqueen* supercomputer, our implementation needs at most only 6.54 ms in all 458, 752 measurements to select an arbitrary element.

In addition, the closely spaced quartiles show that there are only few outliers (e.g., the quick ones that find the target element in early rounds) while most executions are similarly fast. We also noticed that an average timing can be misleading as the few but distant outliers carry a heavy weight to the average and can distort it even beyond the quartile boundaries.

6   Conclusion

This paper presented a scalable and efficient solution for the parallel selection problem. The proposed algorithm is based on the original *QuickSelect* idea with a high-quality median approximation scheme for the pivot selection, an efficient partitioning using parallel prefixes and sums, and an iterative scheme employing range collectives. We provided both practical and theoretical evidence to prove that our pivot selection leads to suitable partitions. The resulting overall time complexity of $\mathcal{O}(\log^2 p)$ and the minimal space requirement makes this parallel selection solution viable for the largest supercomputers. Although it is not asymptotically optimal, it is very efficient in practice: Performance evaluation with up to 458, 752 processor cores shows that it can select an arbitrary element in less than 6.6 ms, regardless of the input. As such, it is a scalable solution to the selection problem and therefore applicable to more complex parallel problems. In the future, we want to extend our base algorithm to include the case with multiple elements per process, apply it to some of the many use cases and consider different implementations as well as further optimizations.

References

1. Blum, M., Floyd, R.W., Pratt, V., Rivest, R.L., Tarjan, R.E.: Time bounds for selection. J. Comput. Syst. Sci. **7**(4), 448–461 (1973)
2. Fouz, M., Kufleitner, M., Manthey, B., Jahromi, N.Z.: On smoothed analysis of quicksort and Hoare's find. Comput. Comb. **5609**, 158–167 (2009)
3. Frazer, W.D., McKellar, A.C.: Samplesort: a sampling approach to minimal storage tree sorting. J. ACM **17**(3), 496–507 (1970)
4. Han, Y.: Optimal parallel selection. ACM Trans. Algorithms **3**(4) (2007)
5. Hoare, C.A.R.: Algorithm 63 (Partition) and Algorithm 65 (Find). Commun. ACM **4**(7), 321–322 (1961)
6. Kirschenhofer, P., Prodinger, H., Martínez, C.: Analysis of Hoare's FIND algorithm with Median-of-Three partition. Random Struct. Alg. **10**, 143–156 (1997)
7. Rabenseifner, R.: Optimization of collective reduction operations. In: Bubak, M., van Albada, G.D., Sloot, P.M.A., Dongarra, J. (eds.) ICCS 2004. LNCS, vol. 3036, pp. 1–9. Springer, Heidelberg (2004)
8. Sack, P., Gropp, W.: A scalable MPI_Comm_split algorithm for exascale computing. In: Keller, R., Gabriel, E., Resch, M., Dongarra, J. (eds.) EuroMPI 2010. LNCS, vol. 6305, pp. 1–10. Springer, Heidelberg (2010)

9. Sanders, P., Träff, J.L.: Parallel Prefix (Scan) algorithms for MPI. In: Mohr, B., Träff, J.L., Worringen, J., Dongarra, J. (eds.) PVM/MPI 2006. LNCS, vol. 4192, pp. 49–57. Springer, Heidelberg (2006)
10. Siebert, C., Wolf, F.: Parallel sorting with minimal data. In: Cotronis, Y., Danalis, A., Nikolopoulos, D.S., Dongarra, J. (eds.) EuroMPI 2011. LNCS, vol. 6960, pp. 170–177. Springer, Heidelberg (2011)

Optimal Diffusion for Load Balancing in Heterogeneous Networks

Katerina A. Dimitrakopoulou[(⊠)] and Nikolaos M. Missirlis

Department of Informatics and Telecommunications, National and Kapodistrian University of Athens, Panepistimiopolis, 157 84 Athens, Greece
{kdim,nmis}@di.uoa.gr

Abstract. In [7] we studied the local Extrapolated Diffusion (EDF) method for the load balancing problem in case of homogeneous torus networks. The present paper develops the convergence theory of the local EDF for heterogeneous torus networks. In particular, we determine its quasi-optimal iteration parameters and the corresponding quasi-optimal convergence factor using local mode analysis. As a result dynamic load balancing becomes an efficient procedure since the parameters of local EDF are computed via a closed form formulae resulting in the maximization of its rate of convergence. Moreover, it is shown how the convergence factor depends upon the communication edge weights and the processor speeds of the network.

Keywords: Laplacian matrix · Load balancing · Weighted torus · Fourier analysis · Diffusion method · Heterogeneous networks

1 Introduction

In the present work we apply the approach of [7] to determine the optimum local Extrapolated Diffusion (EDF) method for heterogeneous networks and in particular for 2d-torus. Heterogeneous networks consist of processors with different computing speeds and different inter-processor communication links. In [5,6] a hydrodynamic approach is proposed for a heterogeneous environment characterized by different computing speeds and uniform communication. Further, in [11] implicit diffusion schemes are considered, whereas in [3] diffusion schemes for a computational environment characterized by uniform computing speeds and different communication characteristics are studied. In [4,9,10] these schemes are extended for heterogeneous computational environments both with respect to the processing performances and the communication speeds. In all the above studies the optimum values of the involved parameters were either computed numerically via the eigenvalues of the Laplacian matrix of the communication graph or they used empirical formulae [9,10]. However, computing numerically

This research was partially funded by the University of Athens Special Account of Research Grants no 10812.

R. Wyrzykowski et al. (Eds.): PPAM 2013, Part I, LNCS 8384, pp. 214–223, 2014.
DOI: 10.1007/978-3-642-55224-3_21, © Springer-Verlag Berlin Heidelberg 2014

eigenvalues is a time consuming process and the use of empirical formulas does not produce the best performance of the Diffusion method. It is therefore of vital importance to determine optimum values for the parameters of the local EDF using closed form formulae in order to (i) maximize its rate of convergence and (ii) make efficient the process of redistributing the load due to changes in the communication graph.

Local mode analysis is used extensively for studying multigrid methods [8]. We use this analysis to determine good approximations to the eigenvalues of the weighted Laplacian as now this matrix is not circulant and the theory developed in [13] does not apply. In this way we are able to find a closed form formula for the set of the parameters of local EDF in the sense that its rate of convergence is maximized for heterogeneous torus networks. These quasi-optimum values depend only upon the speed and the communication edge weights of neighbor processors and so their computation requires only local communication.

The rest of the paper is organized as follows. In Sect. 2, we introduce the heterogeneous EDF method. In Sect. 3, we define the local heterogeneous EDF method and determine the eigenvalues of its iteration matrix for torus graphs applying local mode analysis [1,8]. In Sect. 4, we develop the convergence analysis of the local heterogeneous EDF method. Section 5 presents our numerical experiments and conclusions.

2   The Heterogeneous Extrapolated Diffusion Method

Let $G = (V, E)$ be a connected, undirected graph with $|V|$ nodes and $|E|$ edges, which maps the processor network. Let $w_{ij} > 0 \in \mathbb{R}$ be the weight of edge $e_{ij} \in E$ associated to the communication link capacity and $s_i > 0 \in \mathbb{R}$ be the weight of node $v_i \in V$ associated to the processor speed. Let us consider the following iterative scheme that requires communication with adjacent nodes only

$$u_i^{(n+1)} = u_i^{(n)} - \tau \sum_{j \in N(i)} w_{ij}\Big(\frac{u_i^{(n)}}{s_i} - \frac{u_j^{(n)}}{s_j}\Big), \tag{1}$$

where $\tau \in \mathbb{R} \setminus \{0\}$ is the extrapolation parameter, $w_{ij} > 0$ are the edge weights to be determined and $N(i)$ is the set of nearest neighbors of node $v_i \in V$. Then, the overall workload distribution at step n, denoted by $u^{(n)}$, is the transpose of the vector $(u_1^{(n)}, u_2^{(n)}, \ldots, u_{|V|}^{(n)})$ and $u^{(0)}$ is the initial workload distribution. The iterative scheme (1) will be referred to as the Heterogeneous Extrapolated Diffusion (HEDF) method. Note that when $w_{ij} = 1$ and $s_i = 1$ then (1) is the homogeneous EDF method [2,7]. Our goal is to determine the amount of workload to transfer between neighbor processors in the communication graph such that to balance the load proportionally to the speed of each processor. Therefore, let $\bar{u}_i = (\sum_{j=1}^{|V|} u_j^{(0)})/(\sum_{j=1}^{|V|} s_j)s_i$, $i = 1, 2, \ldots, |V|$ denotes the proportionally balanced load. In matrix form (1) becomes

$$u^{(n+1)} = Mu^{(n)}, \tag{2}$$

where M is called the *diffusion matrix*. The elements of M, m_{ij}, are equal to

$$m_{ij} = \begin{cases} \tau \frac{w_{ij}}{s_j}, & \text{if } j \in N(i), \\ 1 - \tau \sum_{j \in N(i)} \frac{w_{ij}}{s_i}, & \text{if } i = j, \\ 0, & \text{otherwise.} \end{cases} \tag{3}$$

With this formulation, the features of diffusive load balancing are fully captured by the iterative process (1) governed by the diffusion matrix M. The diffusion matrix of HEDF can be written as

$$M = I - \tau \hat{L}, \quad \hat{L} = LS^{-1}, \tag{4}$$

where $S = diag(s_i)$ and L is the generalized Laplacian matrix.

3   The 2D-Torus

We consider the following version of HEDF, which involves a set of parameters $\tau_i, i = 1, 2, \ldots, |V|$

$$u_i^{(n+1)} = (1 - \tau_i \sum_{j \in N(i)} \frac{w_{ij}}{s_i}) u_i^{(n)} + \tau_i \sum_{j \in N(i)} \frac{w_{ij}}{s_j} u_j^{(n)}, i = 1, 2, \ldots, |V|. \tag{5}$$

Note that if $\tau_i = \tau$, $i = 1, 2, \ldots, |V|$, then (5) yields the HEDF method. The iterative scheme (5) will be referred to as the local HEDF method. We define M_{ij} as the local HEDF operator for the $N_1 \times N_2$ torus. Then, the local HEDF scheme at a node (i, j) can be written as

$$u_{ij}^{(n+1)} = (1 - \tau_{ij} d_{ij}) u_{ij}^{(n)} + \tau_{ij} (l_{ij} u_{i-1,j}^{(n)} + r_{ij} u_{i+1,j}^{(n)} + t_{ij} u_{i,j+1}^{(n)} + b_{ij} u_{i,j-1}^{(n)}) \tag{6}$$

with

$$d_{ij} = l_{ij} + r_{ij} + t_{ij} + b_{ij}, \tag{7}$$

where

$$t_{ij} = \frac{w_{i,j+1}^{(2)}}{s_{i,j+1}}, \quad b_{ij} = \frac{w_{i,j-1}^{(2)}}{s_{i,j-1}},$$

$$l_{ij} = \frac{w_{i-1,j}^{(1)}}{s_{i-1,j}}, \quad r_{ij} = \frac{w_{i+1,j}^{(1)}}{s_{i+1,j}}, \tag{8}$$

where the $w_{ij}^{(1)}$ and $w_{ij}^{(2)}$ denote the row and column edge weights, respectively. The matrix $\hat{L}$ is symmetric, when

$$l_{i+1,j} = r_{i,j} \text{ and } b_{i,j+1} = t_{ij}. \tag{9}$$

In addition $\hat{L}$ is nonnegative definite since it is irreducible diagonal dominant [14]. From (9), because of (8), it follows that

$$\frac{w_{ij}^{(1)}}{s_{ij}} = \frac{w_{i+1,j}^{(1)}}{s_{i+1,j}} \quad \text{and} \quad \frac{w_{ij}^{(2)}}{s_{ij}} = \frac{w_{i,j+1}^{(2)}}{s_{i,j+1}} \tag{10}$$

From (10) it follows that for any j

$$w_{ij}^{(1)} = c_i^{(1)} s_{ij}, i = 1, 2, ..., N_1, \tag{11}$$

where $c_i^{(1)} = constant$. Moreover for any i

$$w_{ij}^{(2)} = c_i^{(2)} s_{ij}, j = 1, 2, ..., N_2, \tag{12}$$

where $c_i^{(2)} = $ constant. Therefore, $\hat{L}$ is symmetric if (11) and (12) hold. Otherwise $\hat{L}$ and hence M is not symmetric. In the sequel we will refer to the symmetric and not symmetric $\hat{L}$ as the symmetric and not symmetric case, respectively. If $\hat{L}$ is symmetric and because it also has nonnegative diagonal elements and weak diagonal dominance it follows that the matrix $\hat{L}$ is nonnegative definite (Theorem 5.5, p. 41 of [14]).

Next, we define M_{ij} as the local HEDF operator for the $N_1 \times N_2$ torus. The local HEDF scheme (6) at a node (i, j) can be written as

$$u_{ij}^{(n+1)} = M_{ij} u_{ij}^{(n)}, \tag{13}$$

where

$$M_{ij} = 1 - \tau_{ij} \hat{L}_{ij} \tag{14}$$

with

$$\hat{L}_{ij} = d_{ij} - (l_{ij} E_1^{-1} + r_{ij} E_1 + t_{ij} E_2 + b_{ij} E_2^{-1}) \tag{15}$$

the local operator of the generalized Laplacian matrix. The operators E_1, E_1^{-1}, E_2, E_2^{-1} are defined as $E_1 u_{ij} = u_{i+1,j}$, $E_1^{-1} u_{ij} = u_{i-1,j}$, $E_2 u_{ij} = u_{i,j+1}$, $E_2^{-1} u_{ij} = u_{i,j-1}$, which are the *forward-shift* and *backward-shift* operators in the x_1-direction, x_2-direction, respectively with $u_{ij} = u(ih_1, jh_2) = u(x_1, x_2)$, where $x_1 = ih_1$, $x_2 = jh_2$, $h_1 = \frac{1}{N_1}$ and $h_2 = \frac{1}{N_2}$.

Lemma 1. *The spectrum of the local generalized Laplacian operator $\hat{L}_{ij}$ is given by*

$$\lambda_{ij}(k_1, k_2) = (l_{ij} + r_{ij})(1 - \cos k_1 h_1) + (b_{ij} + t_{ij})(1 - \cos k_2 h_2)$$
$$+ i(l_{ij} - r_{ij}) \sin k_1 h_1 + i(b_{ij} - t_{ij}) \sin k_2 h_2, \tag{16}$$

where

$$i = 1, 2, \ldots, N_1, \; j = 1, 2, \ldots, N_2, \; k_1 = 2\pi\ell_1, \; \ell_1 = 0, 1, 2, \ldots, N_1 - 1,$$
$$k_2 = 2\pi\ell_2 \quad \text{and} \quad \ell_2 = 0, 1, 2, \ldots, N_2 - 1.$$

Proof. By assuming that an eigenfuction of the local operator $\hat{L}_{ij}$ is the complex sinusoid $e^{i(k_1x_1+k_2x_2)}$ we have $L_{ij}e^{i(k_1x_1+k_2x_2)} = \lambda_{ij}(k_1,k_2)e^{i(k_1x_1+k_2x_2)}$, where

$$\lambda_{ij}(k_1,k_2) = d_{ij} - \left(l_{ij}e^{-ik_1h_1} + r_{ij}e^{ik_1h_1} + t_{ij}e^{-ik_2h_2} + b_{ij}e^{ik_2h_2}\right). \quad (17)$$

So, we may view $e^{i(k_1x_1+k_2x_2)}$ as an eigenfunction of $\hat{L}_{ij}$ with eigenvalues λ_{ij} (k_1,k_2) given by (17). It is easily verified that (17) yields (16).   □

In the sequel we will assume that the coefficients r_{ij}, l_{ij}, t_{ij} and b_{ij} are smooth in the sense that $r_{ij} - l_{ij} = l_{i+1,j} - l_{ij} = O(h)$ and $t_{ij} - b_{ij} = b_{i,j+1} - b_{ij} = O(h)$. Then, the two cosine terms in (16) are the dominant terms and we can approximate (16) by keeping only these terms, thus

$$\lambda_{ij}(k_1,k_2) = 2[r_{ij}(1 - \cos k_1h_1) + t_{ij}(1 - \cos k_2h_2)] \quad (18)$$

where

$$i = 1,2,\ldots,N_1,\ j = 1,2,\ldots,N_2,\ k_1 = 2\pi\ell_1,\ \ell_1 = 0,1,2,\ldots,N_1 - 1,$$
$$k_2 = 2\pi\ell_2\ \text{and}\ \ell_2 = 0,1,2,\ldots,N_2 - 1.$$

In (18) we could have used l_{ij} and b_{ij} instead of r_{ij} and t_{ij}, respectively.

The convergence rate of the HEDF method depends on the convergence factor $\gamma(M)$ which is the second largest eigenvalue in absolute value of the matrix M. Following the same streamline we define the convergence factor $\gamma_{ij}(M_{ij})$ of the operator M_{ij} as the second largest eigenvalue in absolute value of M_{ij}. The above approach known as local Fourier analysis [1] has two implicit assumptions. First, M_{ij} should be space-invariant. Secondly, the problem domain should be either extended to infinity or be rectangular with Dirichlet or periodic boundary conditions. In our case γ_{ij} is a spatially varying function (see (16)) and generally is not equal to the convergence factor $\gamma(M)$ of the HEDF method. Nevertheless, if the edge weights and the speeds are all equal to a constant value, then M_{ij} and hence γ_{ij} are space invariant in which case γ_{ij} is equal to $\gamma(M)$. This is verified by the fact that in the homogeneous case $l_{ij} = r_{ij} = \frac{1}{4}$, $i = 1,2,\ldots,N_1, j = 1,2,\ldots,N_2$ and (18) becomes

$$\lambda_{ij}(k_1,k_2) = 2 - \cos k_1h_1 - \cos k_2h_2 \quad (19)$$

which coincides with the eigenvalues of the Laplacian matrix L determined in [12,13] using matrix analysis.

4   Optimum τ_{ij}

Let μ_{ij} denote an eigenvalue of the operator M_{ij}. Since

$$\gamma_{ij}(M_{ij}) = \max_{k_1,k_2} |\mu_{ij}(k_1,k_2)|, \quad (20)$$

where not both k_1, k_2 can take the value zero and

$$\mu_{ij} = 1 - \tau_{ij}\lambda_{ij}. \quad (21)$$

Next, we will study the symmetric case. This means that we will assume that (18) holds. Under this assumption the minimum value of γ_{ij} with respect to τ_{ij} is attained at

$$\tau_{ij}^{opt} = \frac{2}{\lambda_{ij,2} + \lambda_{ij,N}}, \tag{22}$$

where $\lambda_{ij,2}, \lambda_{ij,N}$ are the smallest positive and largest eigenvalues of the operator $\hat{L}_{ij}$, respectively. Moreover, the corresponding minimum value of $\gamma_{ij}(M_{ij})$ is given by

$$\gamma_{ij}^{opt} = \frac{P_{ij} - 1}{P_{ij} + 1}, \tag{23}$$

where

$$P_{ij} = \frac{\lambda_{ij,N}}{\lambda_{ij,2}} \tag{24}$$

is the P-condition number of $\hat{L}_{ij}$. The last quantity plays an important role in the behavior of γ_{ij}^{opt}. Indeed, from (23) it follows that γ_{ij}^{opt} is a decreasing function of P_{ij}. Therefore, minimization of P_{ij} has the effect of maximizing R(LHEDF), the rate of convergence of the local HEDF method, defined by

$$R(LHEDF) = -\log \gamma_{ij}^{opt}. \tag{25}$$

Theorem 1. *If N_1, N_2 are even, then the convergence factor $\gamma_{ij}(M_{ij})$ is minimized at*

$$\tau_{ij}^{opt} = \begin{cases} \dfrac{2}{(r_{ij} + l_{ij})\left(3 + 2\sigma_{ij} - \cos\dfrac{2\pi}{N_1}\right)}, & \sigma_{ij} \geq \sigma_2 \\[3em] \dfrac{2\sigma_{ij}}{(t_{ij} + b_{ij})\left[2 + \sigma_{ij}\left(3 - \cos\dfrac{2\pi}{N_2}\right)\right]}, & \sigma_{ij} \leq \sigma_2 \end{cases} \tag{26}$$

and its corresponding minimum is

$$\gamma_{ij}^{opt} = \begin{cases} \dfrac{1 + 2\sigma_{ij} + \cos\dfrac{2\pi}{N_1}}{3 + 2\sigma_{ij} - \cos\dfrac{2\pi}{N_1}}, & \sigma_{ij} \geq \sigma_2 \\[3em] \dfrac{2 + 2\sigma_{ij} + \cos\dfrac{2\pi}{N_2}}{2 + 3\sigma_{ij} - \cos\dfrac{2\pi}{N_2}}, & \sigma_{ij} \leq \sigma_2. \end{cases} \tag{27}$$

If N_1, N_2 are odd, then the convergence factor $\gamma_{ij}(M_{ij})$ is minimized at

$$
\tau_{ij}^{opt} =
\begin{cases}
\dfrac{2}{(r_{ij} + l_{ij})\left[2 + \cos\dfrac{\pi}{N_1} - \cos\dfrac{2\pi}{N_1} + \sigma_{ij}\left(1 + \cos\dfrac{\pi}{N_2}\right)\right]}, & \sigma_{ij} \geq \sigma_2 \\[4mm]
\dfrac{2\sigma_{ij}}{(t_{ij} + b_{ij})\left[1 + \cos\dfrac{\pi}{N_1} + \sigma_{ij}\left(2 + \cos\dfrac{\pi}{N_2} - \cos\dfrac{2\pi}{N_2}\right)\right]}, & \sigma_{ij} \leq \sigma_2
\end{cases}
\tag{28}
$$

and its corresponding minimum is

$$
\gamma_{ij}^{opt} =
\begin{cases}
\dfrac{\cos\dfrac{\pi}{N_1} + \cos\dfrac{2\pi}{N_1} + \sigma_{ij}\left(1 + \cos\dfrac{\pi}{N_2}\right)}{2 + \cos\dfrac{\pi}{N_1} - \cos\dfrac{2\pi}{N_1} + \sigma_{ij}\left(1 + \cos\dfrac{\pi}{N_2}\right)}, & \sigma_{ij} \geq \sigma_2 \\[6mm]
\dfrac{1 + \cos\dfrac{\pi}{N_1} + \sigma_{ij}\left(\cos\dfrac{\pi}{N_2} + \cos\dfrac{2\pi}{N_2}\right)}{1 + \cos\dfrac{\pi}{N_1} + \sigma_{ij}\left(2 + \cos\dfrac{\pi}{N_2} - \cos\dfrac{2\pi}{N_2}\right)}, & \sigma_{ij} \leq \sigma_2.
\end{cases}
\tag{29}
$$

where σ_{ij} and σ_2 are given by (31).

Proof. The optimum value for τ_{ij} will be determined by (22), while the minimum value of γ_{ij}^{opt} by (23) and (24). It is therefore necessary to determine $\lambda_{i,j,2}$ and $\lambda_{ij,N}$. For the determination of $\lambda_{ij,2}$ we let $\ell_1 = 0$ and $\ell_2 = 1$, or $\ell_1 = 1$ and $\ell_2 = 0$ in (18) which lead to the following

$$
\lambda_{ij,2} =
\begin{cases}
(r_{ij} + l_{ij})\left(1 - \cos\dfrac{2\pi}{N_1}\right), & \sigma_{ij} \geq \sigma_2 \\[4mm]
(t_{ij} + b_{ij})\left(1 - \cos\dfrac{2\pi}{N_2}\right), & \sigma_{ij} \leq \sigma_2,
\end{cases}
\tag{30}
$$

where

$$
\sigma_{ij} = \frac{t_{ij} + b_{ij}}{r_{ij} + l_{ij}} \quad \text{and} \quad \sigma_2 = \frac{1 - \cos\frac{2\pi}{N_1}}{1 - \cos\frac{2\pi}{N_2}}.
\tag{31}
$$

The maximum eigenvalue $\lambda_{ij,N}$ is determined by letting $\ell_1 = \lceil\frac{N_1}{2}\rceil$ and $\ell_2 = \lceil\frac{N_2}{2}\rceil$ in (18). Next, we distinguish two cases according to whether N_1, N_2 are even or odd.

Case I: Both N_1 and N_2 are even numbers
In this case $\ell_1 = \frac{N_1}{2}$ and $\ell_2 = \frac{N_2}{2}$, hence (18) yields

$$
\lambda_{ij,N} = 2d_{ij}.
\tag{32}
$$

Case II: Both N_1 and N_2 are odd numbers

In this case $\ell_1 = \frac{N_1+1}{2}$ and $\ell_2 = \frac{N_2+1}{2}$, hence (18) yields

$$\lambda_{ij,N} = (r_{ij} + l_{ij})\left(1 + \cos\frac{\pi}{N_1}\right) + (t_{ij} + b_{ij})\left(1 + \cos\frac{\pi}{N_2}\right). \tag{33}$$

Using the expressions of $\lambda_{ij,2}$ and $\lambda_{ij,N}$ given by (30) and (32), respectively in (22), (23) and (24), we verify the results given in (26), (27), (28) and (29).   □

Analogous results can be derived for the mixed cases, where one of N_1, N_2 is even and the other is odd.

5   Optimum Edge Weights and Speeds

Up to this point we were concerned with the determination of optimum values for the set of parameters τ_{ij}, in terms of the coefficients r_{ij}, l_{ij}, t_{ij} and b_{ij}, $i = 1, 2, \ldots, N_1$, $j = 1, 2, \ldots, N_2$, such that γ_{ij} is minimized. Nevertheless, these values are quasi-optimum with respect to minimizing $\gamma(M)$ unless the coefficients r_{ij}, l_{ij}, t_{ij} and b_{ij} are spatially invariant in which case become optimum. Next, we will attempt to determine r_{ij}, l_{ij}, t_{ij} and b_{ij} such that P_{ij} (and hence γ_{ij}) is minimized.

Theorem 2. *The convergence factor γ_{ij} is minimized at*

$$\sigma_{ij} = \sigma_2. \tag{34}$$

If N_1, N_2 are even, then γ_{ij} is minimized at

$$\tau_{ij}^{opt} = \frac{2}{(r_{ij} + l_{ij})\left(3 + 2\sigma_2 - \cos\dfrac{2\pi}{N_1}\right)} \tag{35}$$

and its corresponding minimum is

$$\gamma_{ij}^{opt} = \frac{1 + 2\sigma_2 + \cos\dfrac{2\pi}{N_1}}{3 + 2\sigma_2 - \cos\dfrac{2\pi}{N_1}}. \tag{36}$$

If N_1, N_2 are odd, then γ_{ij} is minimized at

$$\tau_{ij}^{opt} = \frac{2}{(r_{ij} + l_{ij})\left[2 + \cos\dfrac{\pi}{N_1} - \cos\dfrac{2\pi}{N_1} + \sigma_2\left(1 + \cos\dfrac{\pi}{N_2}\right)\right]} \tag{37}$$

and its corresponding minimum is

$$\gamma_{ij}^{opt} = \frac{\cos\dfrac{\pi}{N_1} + \cos\dfrac{2\pi}{N_1} + \sigma_2\left(1 + \cos\dfrac{\pi}{N_2}\right)}{2 + \cos\dfrac{\pi}{N_1} - \cos\dfrac{2\pi}{N_1} + \sigma_2\left(1 + \cos\dfrac{\pi}{N_2}\right)}. \tag{38}$$

Proof. When N_1, N_2 are even, it follows from (30) and (32), the P-condition number of $\hat{L}_{ij}$ is given by

$$
P_{ij}(\hat{L}_{ij}) =
\begin{cases}
\dfrac{2\,(1+\sigma_{ij})}{1 - \cos \dfrac{2\pi}{N_1}}, & \sigma_{ij} \geq \sigma_2 \\[4ex]
\dfrac{2\,(1+\sigma_{ij})}{\sigma_{ij}\left(1 - \cos \dfrac{2\pi}{N_2}\right)}, & \sigma_{ij} \leq \sigma_2.
\end{cases}
\tag{39}
$$

When N_1, N_2 are odd, it follows from (30) and (32), that the P-condition number of $\hat{L}_{ij}$ is given by

$$
P_{ij}(\hat{L}_{ij}) =
\begin{cases}
\dfrac{1 + \cos \dfrac{\pi}{N_1} + \sigma_{ij}\left(1 + \cos \dfrac{\pi}{N_2}\right)}{1 - \cos \dfrac{2\pi}{N_1}}, & \sigma_{ij} \geq \sigma_2 \\[5ex]
\dfrac{1 + \cos \dfrac{\pi}{N_1} + \sigma_{ij}\left(1 + \cos \dfrac{\pi}{N_2}\right)}{\sigma_{ij}\left(1 - \cos \dfrac{2\pi}{N_2}\right)}, & \sigma_{ij} \leq \sigma_2.
\end{cases}
\tag{40}
$$

□

Studying the behaviour of (39) and (40) with respect to σ_{ij} we can easily verify that $P_{ij}(\hat{L}_{ij})$ is minimized at σ_2. Therefore, (22) because of (30) and (33), yields (35), whereas (23) yields (36). In a similar way we obtain (37) and (38) when N_1, N_2 are odd.

6 Numerical Experiments

In order to test our theoretical results obtained so far we applied local HEDF for different sizes of 2d-tori. The initial load of the network was placed on a single node of the graph, while we normalized the balanced load $\bar{u} = 1$. Hence, the total number of amount of load was equal to the total number of nodes in the graph. We carried out three experiments. We select the coefficients l_{ij}, r_{ij}, t_{ij} and b_{ij} according to the following criteria:

1. $b_{ij} = l_{ij} = e^{10h(i+j)}$ and $t_{ij} = r_{ij} = e^{20h(2N)} - b_{ij}$
2. $b_{ij} = l_{ij} = 100h^2(i^2 + j^2)$ and $t_{ij} = r_{ij} = 200h^2 2N^2 - b_{ij}$
3. $b_{ij} = e^{10h(i+j)}$, $t_{ij} = e^{20h(2N)} - b_{ij}$, $l_{ij} = 100h(i + j)$ and $r_{ij} = e^{20h(2N)} - b_{ij}$

We kept iterating until an almost evenly distributing flow was calculated. The iteration were terminated when the criterion $\|u^{(n)} - \bar{u}\|_2 / \|u^{(0)} - \bar{u}\|_2 < \epsilon$ for some small ϵ was satisfied. For the 1 and 3 cases we have the same results as presented

Table 1. Number of iterations of LHEDF method for $N \times N$ tori.

$N \times N$	Exp. 1, 3	$l_{ij} = 0.5$	$l_{ij} = 0.6$	$l_{ij} = 0.7$	$l_{ij} = 0.8$	$l_{ij} = 0.9$
10×10	1691	153	180	180	184	214
20×20	23033	569	670	671	687	806
40×40	339216	2149	2531	2534	2599	3049

in the Table 1, whereas for the 2 case LHEDF did not converge. These results clearly show that as long as the coefficients do not differ drastically (cases 1, 3) the LHEDF method converges otherwise it diverges. This verifies our theory which holds under the assumption that the coefficients l_{ij}, r_{ij}, t_{ij} and b_{ij} should be smooth functions. Optimality condition guarantees best performance. Indeed, the results in columns 3–7 of Table 1 show that the minimum number of iterations is achieved when the optimality condition holds. In the optimum case we have $b_{ij} = 0.2, t_{ij} = 1 - b_{ij}$ and $r_{ij} = 1 - l_{ij}$.

References

1. Brandt, A.: Multi-level adaptive solutions to boundary-value problems. Math. Comput. **31**, 333–390 (1977)
2. Cybenko, G.: Dynamic load balancing for distributed memory multi-processors. J. Parallel Distrib. Comput. **7**, 279–301 (1989)
3. Diekmann, R., Frommer, A., Monien, B.: Efficient schemes for nearest neighbour load balancing. Parallel Comput. **25**, 789–812 (1999)
4. Elsässer, R., Monien, B., Preis, R.: Diffusion schemes for load balancing on heterogeneous networks. Theory Comput. Syst. **35**, 305–320 (2002)
5. Hui, C.C., Chanson, S.T.: Theoretical analysis of the heterogeneous dynamic load balancing problem using a hydro-dynamic approach. J. Parallel Distrib. Comput. **43**, 139–146 (1997)
6. Hui, C.C., Chanson, S.T.: Hydrodynamic load balancing. IEEE Trans. Parallel Distrib. Syst. **10**(11), 1118–1137 (1999)
7. Karagiorgos, G., Missirlis, N.M.: Convergence of the diffusion method for weighted torus graphs using Fourier analysis. Theor. Comp. Sci. **401**(1–3), 1–16 (2008)
8. Trottenberg, U., Oosterlee, C.W., Schüller, A.: Multigrid. Academic Press, New York (1971)
9. Rotaru, T., Nägeli, H.H.: Dynamic load balancing by diffusion in heterogeneous systems. J. Parallel Distrib. Comput. **64**, 481–197 (2004)
10. Rotaru, T., Nägeli, H.H.: Fast algorithms for fair dynamic load distribution in heterogeneous environments. Appl. Numer. Math. **49**, 81–95 (2004)
11. Watts, J., Taylor, S.: A practical approach to dynamic load balancing. IEEE Trans. Parallel Distrib. Systems **9**, 235–248 (1998)
12. Xu, C.Z., Lau, F.C.M.: Load Balancing in Parallel Computers: Theory and Practice. Kluwer Academic Publishers, Dordrecht (1997)
13. Xu, C.Z., Lau, F.C.M.: Optimal parameters for load balancing the diffusion method in k-ary n-cube networks. Inf. Process. Lett. **47**, 181–187 (1993)
14. Young, D.M.: Iterative Solution of Large Linear Systems. Academic Press, New York (1971)

Parallel Bounded Model Checking
of Security Protocols

Mirosław Kurkowski[1,2]([⊠]), Olga Siedlecka-Lamch[2],
Sabina Szymoniak[2], and Henryk Piech[2]

[1] Computer Science and Communication, University of Luxembourg,
6, rue Richard Coudenhove-Kalergi, 1359 Luxembourg, Luxembourg
[2] Institute of Computer and Information Sciences, Czestochowa University of
Technology, Dabrowskiego 73, Czestochowa, Poland
miroslaw.kurkowski@uni.lu, olga.siedlecka@icis.pcz.pl

Abstract. The verification of security protocols is a difficult process
taking into consideration a concept and computations. The difficulties
start just during the appropriate adequate protocol specification, and
during studying its properties. In case of the computation connected
with constructing and searching of the modeling structures of protocol
execution and scattered knowledge of the users, the problems are the sizes
of those structures. For small values of parameters such as numbers of
sessions, users, or encryption keys the proper models are usually not very
big, and searching them is not a problem, however in case of increasing
the values of the above mentioned parameters, the models are sometimes
too big, and there is no possibility to construct them nor search properly. In order to increase the values of studying protocol parameters, and
necessary increase the computation effectiveness, the appropriate solutions must be introduced. In the article, the solutions which enable full
and effective parallelization of the computations during automatic verification of security protocols are introduced. The suitable experimental
results are also presented.

Keywords: Security protocols · Model checking · Parallel computations

1 Introduction

The security of systems and computer networks is nowadays one of the most
important problems which must be solved by the designers of such systems.
Many examples confirm the fact that improperly secured systems can expose

The first author acknowledges the support of the FNR (National Research Fund)
Luxembourg under project GALOT – INTER/DFG/12/06. The second and the
third author acknowledge that this research was co-financed by the European
Union under the European Social Fund. Project PO KL "Information technologies:
Research and their interdisciplinary applications", Agreement UDA-POKL.04.01.01-
00-05110-00.

R. Wyrzykowski et al. (Eds.): PPAM 2013, Part I, LNCS 8384, pp. 224–234, 2014.
DOI: 10.1007/978-3-642-55224-3_22, © Springer-Verlag Berlin Heidelberg 2014

the users to different types of losses. Security protocols are the key elements of security systems, including communication protocols. Some of them are responsible for user's authentication that is equivalent to people identification in real biometric, security systems [4]. As concurrent algorithms, they should ensure the confidentiality of the transmitted data, confirm the identity of the communication participants, or execute the distribution of the new session keys. At the planning stage of the protocol, some faults which can lead to serious errors, as well as seizing information by outsiders can occur. There is a necessity to model and verify the protocol executions, including the initial knowledge, the changes of the knowledge of the communicating parts, and also the Intruder in order to find the errors of the protocol, and to fix them. The study of the correctness of such protocols is an essential problem of today's IT infrastructure.

The most effective methods in this domain are model checking methods, however, automatic or semi-automatic verification of protocol correctness is not easy. The biggest problems are caused by the exponential explosion of states of the examined models representing the protocol's executions. For low values of protocol parameters (the number of the users, encryption keys, sessions, etc.) the proper models are usually fast and automatically constructed and searched by specialized tools. Among them are: TAPS [5], NRL [14], Isabelle [2,3,16], VerICS [9], recently developed Scyther [7] or AVISPA [2]. The last one, AVISPA, can be described as the leader among model checking tools such as, for example, the number of the studied sessions of the protocol, constructed models are complicated, and the problems connected with them are in such a way very hard to solve even by the best tools. In all the above cases, various types of a formal model of the tested protocol are constructed. The corresponding spaces of states in these models are then respectively automatically searched for the states corresponding to the tested properties, and especially the security properties of the protocol. Descriptions of the mentioned methods can be found, for example, in [12].

One of the most natural methods of dealing with the necessity of executions of a large number of computations (too big for a single processor) is the use of parallel solutions. However, in the case of model checking solutions for complicated systems, it is sometimes not possible. Searching the model cannot be divided into subtasks, because the next paths of the investigated model depend on other fragments of the whole model.

In this article, the authors' method of security protocols of modeling verification is presented. It is worth noting that in the suggested approach all phases of verification can be processed parallelly. This approach is based on previous works of the authors of [10,11,17]. The proposed ideas connected with the protocol verification focus on suitable protocol specification (the execution of the pattern) in the language proposed in [10], and next in the automatic execution consisted in the following steps:

- generating the hypothetical executions of the protocol by computing the simple substitution from the considered space to the pattern (as in the works [10,11]),

- generating the chains of states coding particular actions, making up the protocol (see [17]),
- constructing and searching on the fly a tree of real protocol executions, in searching the states corresponding to verified properties (as in [17]).

The solutions proposed in previous works connected with constructing the models of protocol executions allow the use of the parallel computation in each of the three above mentioned phases of the automatic verification. It allows to increase the sizes of the parameters of the protocol during the verification of its properties. This is important according to verification with many sessions of the protocol (see [6]).

2   The Needham-Schroeder Public Key Protocol

The designing of the communication protocols is a very hard task, which is connected with the possibility of many problems occurring with the later use of a made protocol. As an essential example of a scheme which has such faults, the NSPK protocol can be presented [14]. In the notation of this protocol, there are designations $i(A)$ and $i(B)$, which identify the participants of the communication, respectively the A and B participants. The expression $\langle X \rangle_{K_A}$ stands for the X message encrypted by the public key of A user. Similarly, the message encrypted by the public key of B participant - $\langle X \rangle_{K_B}$.

The A user has a very important role- it starts the run of the protocol. The aim of the execution of this protocol is to achieve the mutual confirmation of the identity (the authentication) between the communication participants. The designation $A \rightarrow B : X$ refers to sending the X message from A participant to B participant. We assume that sending the message causes the operation of receiving it by the suitable person. The concatenation of the elements in the message was determined by the operator $\cdot$.

The scheme of NSPK protocol proposed in [15] is as follows:

Example 1

$$\alpha_1 \quad A \rightarrow B \quad : \quad \langle N_A \cdot i(A) \rangle_{K_B},$$
$$\alpha_2 \quad B \rightarrow A \quad : \quad \langle N_A \cdot N_B \rangle_{K_A},$$
$$\alpha_3 \quad A \rightarrow B \quad : \quad \langle N_B \rangle_{K_B}.$$

In first stage of the protocol execution, the A participant creates the random number (nonce) N_A, then sends the message which is encrypted by the public key B to B participant. It consists the drawn number N_A and the identifier of A participant. The next stage of NSPK protocol execution starts from the drawing of its own random number N_B by B participant. Next, it creates the message consisting of nonces of both communication participants, encrypts this message by the public key K_A, and sends it to A participant.

In the further part of the protocol execution, A runs the decrypting operation of the received message, and the operation of comparing the N_A number, which was received from B with the number N_A prepared by himself. If both numbers are correct, A considers B as authenticated. In the last step of the protocol execution, A sends its nonce N_B encrypted by the K_B key to B. After decrypting and comparing the numbers, B participant can consider A participant as authenticated. Executing this protocol should guarantee both participants the identity of the communicating person.

The NSPK protocol was used throughout 17 years. In 1995, Gavin Lowe discovered a version of the protocol execution which consisted a possible attack, showing that the NSPK protocol is susceptible to a break-in [13]. The Intruder attacks, and the additional participant has his own identity card and keys determined by ι symbol. The Intruder does not execute the protocol according to the scheme but, using own ways, pretends to be different subscribers, and deceives them. The attack presented by Lowe is as follows:

Example 2

$$\alpha_1^1 \quad A \rightarrow \iota \; : \; \langle N_A \cdot i(A) \rangle_{K_\iota},$$
$$\alpha_1^2 \quad \iota(A) \rightarrow B \; : \; \langle N_A \cdot i(A) \rangle_{K_B},$$
$$\alpha_2^2 \quad B \rightarrow \iota(A) \; : \; \langle N_A \cdot N_B \rangle_{K_A},$$
$$\alpha_2^1 \quad \iota \rightarrow A \; : \; \langle N_A \cdot N_B \rangle_{K_A},$$
$$\alpha_3^1 \quad A \rightarrow \iota \; : \; \langle N_B \rangle_{K_\iota},$$
$$\alpha_3^2 \quad \iota(A) \rightarrow B \; : \; \langle N_B \rangle_{K_B}.$$

The Lowe scheme consists of two different simultaneous NSPK protocol executions. The α^1 execution refers to the communication between A participant and the Intruder ι. The α^2 execution refers to the situation where the Intruder pretends to be the A participant ($\iota(A)$) and in his name communicates with B.

To have a correct protocol, which cannot be broken, it is enough to add the identifier of the sender to sent message in the second step of original NSPK protocol execution: $\alpha_2 \; B \rightarrow A \; : \; \langle N_A \cdot N_B \cdot i(B) \rangle_{K_A}$. All known verification methods and tools confirm the safety of this version of NSPK.

3   Idea of the Chains of States

A new method of modeling executions, as well as a new method of verification of security protocols' properties through encoding to the chains of states, and a way of their effective parallelization during modeling and verifying is now going to be presented.

In the work [10], a new idea of specifying the security protocols has been proposed. To the description used in the Common Language, containing a scheme of the process of sending the messages while performing the protocol, an addition of a description of internal actions executed by the performers of the protocol has been made . A step of the protocol α is defined as 5-tuple: $\alpha = (\mathcal{P}, \mathcal{X}, \mathcal{G}, \mathcal{L}, \mathcal{Q})$, where $\mathcal{P}$ is a party sending a message in step α, $\mathcal{X}$ is a set of information which $\mathcal{P}$ needs for composing a message $\mathcal{L}$ sent in the step, $\mathcal{G}$ is a set of confidential information generated by $\mathcal{P}$ needed for composing messages $\mathcal{L}$ ($\mathcal{G} \subseteq \mathcal{X}$), and $\mathcal{Q}$ is a party receiving a message $\mathcal{L}$. In the following parts of the article, the elements $\mathcal{P}, \mathcal{X}, \mathcal{G}, \mathcal{L}, \mathcal{Q}$ are determined as $Send, Comp, Gen, Lett$ and Rec. As an example we consider the following specification of NSPK protocol.

Example 3

$\Sigma_{NSPK} = (\alpha_1, \alpha_2, \alpha_3)$, where:

$\alpha_1 = (\mathcal{A}, \{\mathcal{N_A}, \mathcal{I_A}, \mathcal{K_B}\}, \{\mathcal{N_A}\}, \mathcal{B}, \langle \mathcal{N_A} \cdot \mathcal{I_A} \rangle_{\mathcal{K_B}})$,

$\alpha_2 = (\mathcal{B}, \{\mathcal{N_A}, \mathcal{N_B}, \mathcal{K_A}\}, \{\mathcal{N_B}\}, \mathcal{A}, \langle \mathcal{N_A} \cdot \mathcal{N_B} \rangle_{\mathcal{K_A}})$,

$\alpha_3 = (\mathcal{A}, \{\mathcal{N_B}, \mathcal{K_B}\}, \emptyset, \mathcal{B}, \langle \mathcal{N_B} \rangle_{\mathcal{K_B}})$.

As can be seen, particular steps of the protocol are thoroughly specified. Precise information is defined, information which is needed by the party in order to compose a sending message, and generate new confidential data.

As proposed in [10] and the following works, this method requires the generation of a set of hypothetical executions of a tested protocol in order to construct a tree of its real executions. In this tree, the states corresponding to the states of threat to the protocol during its executions will be searched. During a generation, a defined space of considered parameters Π is used (parameters such as the number of users, their type, number of secured information, keys etc.).

The generation of the executions here depends on a simple substitution of a protocol of its respective parameters from the space Π to the specification mentioned above. Therefore, if we consider a space containing three honest users A, B, C, then, for example, we can generate the two following executions substituting $\mathcal{A} \leftarrow a, \mathcal{B} \leftarrow c$ and so on, according to the natural conditions precisely defined in [10].

Two different executions of NSPK are as follows.

Example 4

$\Sigma^1_{nSPK} = (\alpha^1_1, \alpha^1_2, \alpha^1_3)$, where :

$\alpha^1_1 = (a, \{n_a, i_a, k_c\}, \{n_a\}, c, \langle n_a \cdot I_a \rangle_{k_c})$,

$\alpha^1_2 = (c, \{n_a, n_c, k_a\}, \{n_c\}, a, \langle n_a \cdot n_c \rangle_{k_a})$,

$\alpha^1_3 = (a, \{n_c, k_c\}, \emptyset, c, \langle n_c \rangle_{k_c})$.

$\Sigma^2_{nSPK} = (\alpha^2_1, \alpha^2_2, \alpha^2_3)$, where :

$\alpha^2_1 = (c, \{n_c, i_c, k_b\}, \{n_c\}, b, \langle n_c \cdot I_c \rangle_{k_b})$,

$\alpha^2_2 = (b, \{n_c, n_b, k_c\}, \{n_b\}, c, \langle n_c \cdot n_b \rangle_{k_c})$,

$\alpha^2_3 = (c, \{n_b, k_b\}, \emptyset, b, \langle n_b \rangle_{k_b})$.

As shown, the first execution is performed between users a and c, where a is an initiator of a protocol. The second execution is performed between c and b, where c is an initiator.

Generating such executions (as basic model) in the case of testing one or a few sessions as it is usually done, a generating a model is not time consuming, therefore, usually no attention is paid to the effectiveness of this part of verification. However, if we want to verify a protocol for more than a few values of the parameters, a parallel approach may seem to be necessary in order to increase the efficiency of the computations. Of course, generating the above substitutions-executions may be carried out in parallel. Separate threads may choose initiators in a combinatorial way and generate the executions by choosing the receivers of an execution for them. Such an approach turns out to be very efficient. Respective results will be presented and explained in the last section of this work.

Another phase of the verification method proposed in [17] is generating respective chains of states coding the executions. According to the concepts proposed in [17], four types of the states encoding steps of executions are defined:

1. steps representing particular executions of the steps (as in Example 3). States of this type will be described by: S_j^i, where parameter i defines the number of step in the appropriate execution of the protocol, and parameter j defines the number of execution.
2. states representing generation of confidential information by the users (nonces, encryption keys). These states will be denoted: G_U^X, where this state defines generating a secret X by user U (for example $G_a^{n_a}$ denotes generating the random number n_a by the user a).
3. states representing the gaining of knowledge about its particular elements (there might be cryptograms) by the receiver of a particular message. These states are expressed as K_U^X - gaining knowledge about X by user U (for example $K_a^{n_b}$ - gaining by a the number n_b.
4. states representing the necessity of having the proper knowledge necessary for composing and sending a letter in a modeled step, these states will be tagged P_U^X - U must have knowledge about X. For example, $P_a^{k_b}$ means that a must know the key k_b.

A method of the automatic generation of chains (sequences) for the execution steps of a tested protocol will now be described.

If we consider ith-execution of a protocol, and its kth-step α_k^i, then we place the state S_k^i in the constructed sequence for this step. Before this state, we place the states of the $P_{Send_k^i}^X$-type ($Send_k^i$ means a sender in a kth-step of a ith-execution of a protocol) beside the elements, which are in the set Gen_k^i. We encode these last ones as the states of $G_{Send_k^i}^X$-type, and also place them before the S_k^i state. In the constructed chain, after the state S_k^i we place the state K_U^X-type for all information which the user Rec_k^i can obtain from the message $Lett_k^i$ by the use of his own keys. In case of an Intruder there are as many generated chains as the number of sets of generators for the message $Lett_k^i$ (see [17]).

Let us consider the space of n-executions of a protocol consisting of k-steps. Algorithm 1, which generates the chains is as follows:

Algorithm 1. Chains of states generation

```
 1 : for i = 1 to n do
 2 :     for j = 1 to k do
 3 :         G ← G_j^i
 4 :         l ← 1
 5 :         s(l)_j^i ← S_j^i
 6 :         Add_needed()
 7 :         Add_generated()
 8 :         Add_knowledge()
 9 :         return s_j^i = s(1)_j^i, ..., s(l)_j^i
10 :     end for
11 : end for
```

As can be seen, the loops 1–11 and 2–10 choose the steps of particular, considered executions of protocols. Procedures from lines 6, 7 and 8 add to the sequence the states connected to the needs, generating or gaining the knowledge by the users.

Below, as an example, we present the procedure *Add_needed*() which assigns the states (connected with the knowledge needed to create a sent letter) to the chain.

Algorithm 2. Chains of states generation

```
 1 :  while G_j^i ≠ ∅ do
 2 :      if x ∈ G_j^i then
 3 :          for t = l to 1 do
 4 :              s(t + 1)_j^i ← s(t)_j^i
 5 :          end for
 6 :          l ← l + 1
 7 :          s(1)_j^i ← P_{Send_j^i}^x
 8 :          G_j^i ← G_j^i \ {x}
 9 :      end if
10 :  end while
```

It must be stressed that all procedures of the assignment of the states can be performed in parallel by the accessible plots. A suitable synchronicity is necessary but the obtained experimental results indicate a rather big acceleration of this verification phase.

3.1   Intruder and Attacks

The definitions introduced before do not include the presence of the Intruder in the network. In our investigations, we can consider many different models of Intruder. In this paper, we explore and give experimental results for one of the versions of a well-known Intruder's model called Dolev-Yao model [8]. In our approach, we assume that there is only one Intruder, who actively tries to deceive the others by executing the protocols against their assumptions. The Intruder may use any information obtained from a network, and impersonate other users. Note that the Intruder can compose sent messages contrary to the idea and assumptions of the protocol. This can be done in many ways described precisely in [10].

Using this we can generate protocol executions with the Intruder. A good example is one of the executions of Lowe's attack upon NSPK Protocol mentioned before:

Example 5
$$\alpha_1^2 \quad \iota(a) \to b \; : \; \langle n_a \cdot i(a) \rangle_{k_b},$$
$$\alpha_2^2 \quad b \to \iota(a) \; : \; \langle n_a \cdot n_b \rangle_{k_a},$$
$$\alpha_3^2 \quad \iota(a) \to b \; : \; \langle n_b \rangle_{K_b}.$$

In some of such executions an Intruder can appear in an honest role, and execute the protocol according to the algorithm. However, according to the model an Intruder can also impersonate the other users, and provided that he knows them, can use their parameters (keys, nonces) as in the execution from example 5. We call these last executions the attacking executions. In the final phase of the verification in the constructed tree of real executions of a protocol, we will search for the states corresponding to the attacking executions.

3.2   Correct Chains of States

Various hypothetical executions of a protocol (see [10]) can be generated in the above described way. However, some of them cannot be executed at all, and some cannot be executed by themselves because of lack of suitable knowledge of users, and especially of an Intruder. The last phase of verification executed according to the methodology in [17] is constructing chains that encode the real executions of protocols in the computer networks. These chains will consist of the states described above. However, of course, not all possible sequences of states represent the real executions of protocols.

We are now going to define the sequences of states representing the real executions of protocols in the networks. These chains correspond to computations defined in the work [10].

Let Π be a space consisting of a defined number of users and their attributes (identifiers, nonces, cryptographic keys, etc.). Furthermore, all executions of a protocol in this space, as well as all the chains of states representing all these executions have to be considered. Subsequently, in the set of all chains of states we can define the states which correspond to the runs in the computational structure from the work [10].

Definition 1. *The chain of states* $\mathfrak{s} = s_1, s_2, \ldots, s_p$ *will be called* **correct chain of states** *if and only if the following conditions hold:*

1. *if* $s_i = S_j^k$ *for some* $j, k \leq p$, *then* $j = 1 \vee \exists_{t<i}(s_t = S_{j-1}^k)$ *and* $PreCond(S_j^k) \subseteq \{s_1, \ldots, s_{i-1}\} \wedge PostCond(S_j^k) \subseteq \{s_{i+1}, \ldots, s_p\}$,
2. *if* $s_i = G_U^X$, *then* $\forall_{t\neq i}(s_t \neq G_U^X)$,
3. *if* $s_i = P_U^X$, *then* $\exists_{t<i}(s_t = G_U^X \vee s_t = K_U^X)$.

The first point provides a proper relation in order to the realization of particular steps in a given execution. Points 2 and 3 guarantee a proper relation of knowledge of the users necessary for the realization of particular actions in a given step of the protocol execution.

In the work [17], a theorem has been given justifying the adequacy of constructed chains with runs in the structure from paper [10], which represent the real executions of protocols in the computer networks.

3.3   Verification Algorithm

Below, Algorithm 3 constructs and searches on the fly a tree of real protocol executions. It is based on the chains of states encoding particular steps of the executions of the protocols and is generated in the previous phases. On an initially empty stack, at first the chains not containing the initial conditions (states P type) are placed. Respectively the chains representing the subsequent states according to the conditions defined in the Definition 1 are added. A constructed tree is controlled during its creation in terms of containing the chains representing the attacking executions. Creating and searching the tree at the same time finishes up in two cases. Firstly, if one of the created paths contains an attacking execution, and secondly, if all possibilities are depleted, and none of the paths represent an attack.

Algorithm 3. Construction and searching a tree

```
 1 :  for state in sch do
 2 :      if (check_cond(state, knowledge)) then
 3 :          stack.push(state)
 4 :          update_knowledge(state)
 5 :          state.visited = true
 6 :          if (state.isEnd()) then
 7 :              print_path(stack)
 8 :              return
 9 :          else if ¬(state.isEnd()) then
10 :              find_path(stack, knowledge)
11 :          end if
12 :      end if
13 :  end for
```

The main part of the algorithm is executed in the function *find_path* which constructs a path of a potential attack (represented by the sequence of particular states). In this function the following objects are used: *stack* - local stack, on which the following states fulfilling the initial conditions are placed, *knowledge* - this function gathers the information about the gained knowledge of all the parties executed in the protocol, *state* - state in which there is a considered protocol execution (the information about events which take place in every considered step of the protocol execution is recorded).

The most important methods and functions used in implementation are as follows: *check_cond* - checks if the state represented by the appropriate parameter fulfills the initial conditions (comparing the events with the actual state of the knowledge), *update_knowledge* depending on the next states added to the stack, it increases the knowledge of the protocol users.

The executions of a full function *find_path* depend on current checking all the states. If anyone fulfills the initial conditions, it can be added to the stack. At the same time, the knowledge of the users is updated, and the state added to the stack is designed as visited. Finally, the function *find_path* is executed again taking as the arguments the copy of actual knowledge, and current state of the stack.

Although the pessimistic time complexity of algorithm is exponential because of all parameters of the tested protocol: the number of steps, executions, users and objects used by the users, the obtained and presented experimental results are very good. The below mentioned results show that presented approach is nowadays one of the most effective.

It is important to note that all phases of the proposed method of verification: generation of executions, chains, and construction and searching the tree can be implemented in parallel.

4   Experimental Results and Summary

For protocols NSPK and $NSPK_{Lowe}$, we examined their vulnerability to attack on authentication and confidentiality. A tree of real executions of the protocol was constructed starting from states representing the initial steps of the executions up to the states representing the attack (the last states from executions corresponding to the attacks). Taking into account the existence of the attack, we obtained the expected results: an attack exists for the NSPK, and an attack does not exist for $NSPK_{Lowe}$. However, when considering the effectiveness of the calculation it can be seen that only parallelization of the above-mentioned procedures gives us interesting experimental results.

Table 1 shows the experimental results for sequential and parallel implementation, carried out on a computer equipped with the quad core processor Intel Pentium D (3000 MHz), 2 GB main memory, and the operating system Linux. We observe 30 to 40 % speed up while increasing the number of sessions for protocols NSPK and $NPSK_{Lowe}$, for both executions and chains of states generation, as well as for the construction and searching the tree. All results are expressed in seconds.

Summarizing, we developed a method which uses chains of states for representing executions and automatic verification of security protocols. The proposed approach enables the testing of the protocol in several steps: the generation of hypothetical executions of the protocol, their coding, and constructing a tree of real executions. It is important to note that it provides a feature difficult to obtain for other methods: the ability to easily parallelize all the phases of our approach. Therefore, we obtained very good experimental results, especially worth noting are the excellent results for the examination of executions for multiple sessions of the protocol. In our opinion, the proposed method is very simple, intuitive and flexible. The results are promising and lead us to further development of the method: survey other protocols, adding of time into consideration, or considering other types of attacks and other security properties. We also would like to thank for the practical implementation of our ideas to Paweł Dudek.

Table 1. Sequential and parallel experimental results for NSPK and $NSPK_{Lowe}$.

Generating of executions and chains					
Implementation	Protocol	10 Sessions	20 Sessions	50 Sessions	100 Sessions
Sequential	NSPK	0,072	0,344	1,969	8,185
Parallel	NSPK	0,066	0,188	1,424	5,659
Sequential	$NSPK_{Lowe}$	0,074	0,261	1,793	7,946
Parallel	$NSPK_{Lowe}$	0,060	0,210	1,434	5,714
Constructing and searching the tree					
Implementation	Protocol	10 Sessions	15 Sessions	20 Sessions	25 Sessions
Sequential	NSPK	5,300	62,729	351,620	1231,003
Parallel	NSPK	3,543	39,420	212,701	778,650
Sequential	$NSPK_{Lowe}$	5,339	60,503	350,430	1233,304
Parallel	$NSPK_{Lowe}$	3,518	40,288	246,255	765,290

References

1. Armando, A., et al.: The avispa tool for the automated validation of internet security protocols and applications. In: Etessami, K., Rajamani, S.K. (eds.) CAV 2005. LNCS, vol. 3576, pp. 281–285. Springer, Heidelberg (2005)
2. Bella, G., Massacci, F., Paulson, L.C.: Verifying the set registration protocols. IEEE J. Sel. Areas Commun. **20**(1), 77–87 (2003)
3. Bella, G., Paulson, L.C.: Using Isabelle to prove properties of the kerberos authentication system. In: Orman, H., Meadows, C. (eds.) Proceedings of the DIMACS Workshop (1997)
4. Bobulski, J., Kubanek, M.: Person identification system using an identikit picture of the suspect. Opt. Appl. **42**(4), 865–873 (2012)
5. Cohen, E.: Taps: A first-order verifier for cryptographic protocols. In: CSFW'00: Proceedings of the 13th IEEE Computer Security Foundations Workshop (CSFW'00), p. 144, Washington, DC, USA, IEEE Computer Society (2000)
6. Comon-Lundh, H., Cortier, V.: Security properties: two agents are sufficient. In: Degano, P. (ed.) ESOP 2003. LNCS, vol. 2618, pp. 99–113. Springer, Heidelberg (2003)
7. Cremers, C., Mauw, S.: Operational Semantics and Verification of Security Protocols. Information Security and Cryptography, pp. 1–155. Springer, Berlin (2012)
8. Dolev, D., Yao, A.: On the security of public key protocols. IEEE Trans. Inf. Theory **29**(2), 198–207 (1983)
9. Kacprzak, M., et al.: Verics 2007 - a model checker for knowledge and real-time. Fundam. Inform. **85**(1–4), 313–328 (2008)
10. Kurkowski, M., Penczek, W.: Verifying security protocols modeled by networks of automata. Fund. Inform. **79**(3–4), 453–471 (2007)
11. Kurkowski, M., Penczek, W.: Verifying timed security protocols via translation to timed automata. Fund. Inform. **93**(1–3), 245–259 (2009)
12. Kurkowski, M., Penczek, W.: Applying timed automata to model checking of security protocols. In: Wang, J. (ed.) Handbook of Finite State Based Models and Applications, pp. 223–254. CRC Press, Boca Raton (2012)
13. Lowe, G.: Breaking and fixing the needham-schroeder public-key protocol using fdr. In: Margaria, T., Steffen, B. (eds.) TACAS 1996. LNCS, vol. 1055, pp. 147–166. Springer, Heidelberg (1996)
14. Meadows, C.: The nrl protocol analyzer: an overview. J. Logic Program. **26**(2), 13–131 (1996)
15. Needham, R.M., Schroeder, M.D.: Using encryption for authentication in large networks of computers. Commun. ACM **21**(12), 993–999 (1978)
16. Paulson, L.C.: Inductive analysis of the internet protocol tls. ACM Trans. Inf. Syst. Secur. **2**(3), 332–351 (1999)
17. Siedlecka-Lamch, O., et al.: A New Effective Approach for Modelling and Verification of Security Protocols. In: Proceedings of CS&P'2012 Humboldt University Press, Berlin, Germany, pp. 191–202 (2012)

Tools and Environments
for Parallel/Distributed/Cloud
Computing

Development of Domain-Specific Solutions Within the Polish Infrastructure for Advanced Scientific Research

J. Kitowski[1,2](✉), K. Wiatr[2], P. Bała[5,6], M. Borcz[5], A. Czyżewski[4],
Ł. Dutka[2], R. Kluszczyński[5], J. Kotus[4], P. Kustra[3], N. Meyer[7],
A. Milenin[3], Z. Mosurska[2], R. Pająk[2], L. Rauch[3],
M. Sterzel[2], D. Stokłosa[7], and T. Szepieniec[2]

[1] Department of Computer Science, AGH University, Krakow, Poland
[2] ACC Cyfronet AGH, AGH University, Krakow, Poland
kito@agh.edu.pl
[3] Department of Applied Computer Science and Modelling, AGH University,
Krakow, Poland
[4] ETI Faculty, Multimedia Systems Department, Gdansk University of Technology,
Gdansk, Poland
[5] Interdisciplinary Centre for Mathematical and Computational Modelling,
University of Warsaw, Warsaw, Poland
[6] Department of Mathematics and Computer Science,
Nicolaus Copernicus University, Torun, Poland
[7] Poznan Supercomputing and Networking Center, Poznan, Poland

Abstract. The Polish Grid computing infrastructure was established during the PL-Grid project (2009–2012). The main purpose of this Project was to provide the Polish scientists with an IT basic platform, allowing them to conduct interdisciplinary research on a national scale, and giving them transparent access to international grid resources via international grid infrastructures. Currently, the infrastructure is maintained and extended within a follow-up PLGrid Plus project (2011–2014). Its main objective is to increase the potential of the Polish Science by providing necessary IT services for research teams in Poland, in line with European solutions. The paper presents several examples of the domain-specific computational environments, developed within the Project. For particular environments, specialized IT solutions are prepared, i.e. dedicated software implementation and infrastructure adaptation, suited for particular researchers groups' demands.

Keywords: PL-Grid · PLGrid Plus · Domain-specific solutions · Computing infrastructure · Computational environment

1 Introduction

In the last two decades a number of big infrastructure (grid) projects and initiatives have been developed in US, Europe and Japan based on EU, national and

R. Wyrzykowski et al. (Eds.): PPAM 2013, Part I, LNCS 8384, pp. 237–250, 2014.
DOI: 10.1007/978-3-642-55224-3_23, © Springer-Verlag Berlin Heidelberg 2014

other resources. Currently, in Europe one of the important initiatives for developing European computing infrastructure is performed by the EGI.eu organization [1] with one of its main outcome – the EGI InSPIRE project [2]. EGI.eu is a not-for-profit foundation established to coordinate and manage the European Grid Infrastructure (EGI) federation on behalf of its participants: National Grid Initiatives (NGIs) and European International Research Organizations (EIROs). The NGIs are organizations set up by individual countries to manage the computing resources they provide to the European Grid Infrastructure (EGI). Each of the European countries (and some outside) has been involved in developing their own NGI in order to become a part of the European research space.

In the paper, we shortly describe the process of building the national computing infrastructure for the Polish scientists, allowing them to conduct interdisciplinary research. Next, we move to the domain specific solutions created within this infrastructure and we present several examples of scientific results. In the conclusions, we make a summary of the development together with future trends of extension.

2   Polish Grid Infrastructure

The Polish Grid (PL-Grid) Infrastructure [3] has been created during the PL-Grid project (2009–2012) [4–6], when the basic infrastructure has been developed. Soon, in March 2010, the PL-Grid infrastructure turned out to be the first operational NGI in Europe. Since then, the users not only were able to conduct interdisciplinary research on a national scale, but also they have been given transparent access to international grid resources. Next, the PLGrid Plus project (2011–2014) [7] was started, aiming at the infrastructure extension with specific environments, solutions and services, according to identified needs of different groups of scientists (cf. Fig. 1). Both projects have been maintained by the Polish Grid Consortium and co-funded by the European Regional Development Fund as part of the Innovative Economy program.

It is worth to mention that access to computing resources of the PL-Grid Infrastructure is free to Polish researchers and all others engaged in scientific activities associated with a university or research institute in Poland.

3   Domain-Specific Solutions

Contemporary science has great need for e-Infrastructures: networking, data storage, computing. However, from the point of view of a domain scientist, using modern computing systems, services and tools often becomes relatively difficult. Therefore, the scientist requires assistance and close collaboration with service providers. Due to diversity of scientists' requirements, preparation of domain-specific computing environments, i.e., solutions, services and extended infrastructure (including software), tailored to the needs of different groups of scientists (see Fig. 1) has been undertaken in order to fit the PL-Grid Computational Infrastructure to the problems being a subject of research.

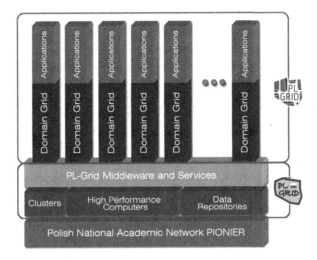

Fig. 1. The layered view of the PL-Grid infrastructure.

To help researchers from different areas of science understand and use the potential of the Polish Grid Infrastructure and to define their requirements and expectations, the following 13 pilot communities have been included in the PLGrid Plus project and – at the same time – involved in the infrastructure development: AstroGrid-PL, HEPGrid, Nanotechnologies, Acoustics, Life Science, Quantum Chemistry and Molecular Physics, Ecology, SynchroGrid, Energetics, Bioinformatics, Health, Materials, and Metallurgy.

Introduction of domain-oriented solutions for these 13 communities opens the scope of use of development results to various research groups. However, the scope is not limited to the selected domains. Subsequently, within the Project it is foreseen to launch more IT services, also for teams of researchers representing other scientific disciplines, who plan experiments supported by large-scale simulations or work with large databases (collections) of data.

4   Use Cases

Considering the fact that there are as many as 13 pilot communities included in the Project, it is impossible to provide the detailed view of all domain services already developed in the infrastructure or being planned for implementation. Therefore, only four sample use cases are presented within this section, introducing the research conducted by the scientists in Bioinformatics, Metallurgy, Acoustics and Ecology domains.

4.1   Bioinformatics – Processing Genetic Data

The modern life sciences research widely uses large data sets, either stored in the existing databases or obtained in high throughput experiments. The data is of different type and origin but its amount is significant and is rapidly growing as experimental equipment becomes more affordable. In result, we observe increasing demand for disk space and computer power to store and process data. Emerging technologies, such as Next Generation Sequencing (NGS), put additional stress on this demand. The amount of data, which biologists have to handle, is usually too big to be stored and processed using desktop or even more powerful single computers in the laboratory. Up to now, the research teams were building their own infrastructures, sometimes quite extensive ones, to store and process data. In addition, they had to provide dedicated IT staff to operate and maintain it. Building and maintaining dedicated computer infrastructure, capable to process data in required time, is very costly and simply not feasible. In addition, in the last few years we have observed that high throughput systems are widely spread and are used by smaller groups or even become diagnostics tools at hospitals. Thus, new solutions such as Grids are needed.

From its beginning, the Grid has been considered as possibility to provide disk space and computational resources for the life sciences community. There has been number of different grid infrastructures used. The solution built using UNICORE middleware [8] is an example of the most successful one [9].

UNICORE infrastructure in PL-Grid is built based on the European Middleware Initiative (EMI) [10] releases, which are distributed from EMI repository as RPM packages [11]. This allows for easy installation and configuration with only few additional dependencies.

The UNICORE has typical 3-layer architecture, covering target system infrastructure, middleware and user interfaces. There are three types of interfaces available for the end user: UNICORE Rich Client (URC), UNICORE Commandline Client (UCC) and High Level API (HiLA).

For the data transfer between user workstation and grid infrastructure, and between grid nodes, a new protocol called UNICORE File Transfer Protocol (UFTP) [12] is used.

Answering the needs of the groups running genetic research and thus facing the problem of storing and processing large data sets coming from the high throughput sequencing, we have created dedicated solution allowing to store and process genetic data on the Grid. This way we have integrated GS FLX Instrument available at the Collegium Medicum, N. Copernicus University with the PL-Grid distributed infrastructure as the data storage and processing system [13]. Similar work for the Illumina sequencer located at the Warsaw Medical University is in progress and should be finalized soon.

During the sequencing reaction, digital images are captured by the CCD camera. The raw image data from a single experiment (approximately 30 GB in size) is than converted with the use of GS Run Processor (Roche Diagnostics) into base-called results. Reports of the base-calling analysis are generated with GS Reporter (Roche Diagnostics). The obtained reads are aligned to a revised Cambridge Reference Sequence (rCRS) with the use of a GS Reference Mapper (Roche Diagnostics), which enables mtDNA mutations to be detected. For this process, we have developed a workflow, which consists of the abovementioned programs from the FLX program set.

First, the GS Run Processor processes raw images generated by the FLX Instrument (Roche Diagnostics). This data, which can exceed tens of gigabytes, is put into the UNICORE storage in an automatic way. Once a new file or directory with data is created as a result of experiment, the UNICORE Commandline Client program is used to put automatically data to the PL-Grid UNICORE target system storage. Because of the size of data, a UFTP protocol is used for transfers. Appropriate access rights to the group storage ensure privacy. For the details of the system, refer to Fig. 2.

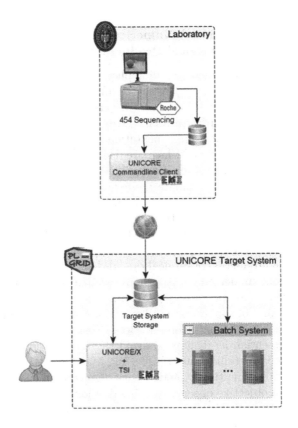

Fig. 2. Schematic view of the data processing system.

The GS Run Processor has several components. We run the `runAnalysisPipe` script for full processing of the acquired data and use the GS_LAUNCH_MODE environmental variable to set MPI mode enabling us to use multiple worker nodes. The next part of the workflow can be run simultaneously on the available resources. The GS Reporter generates all reports files from files created in the previous job. The GS Reference Mapper consists of two steps, which create the mapping project and align the reads against a reference sequence.

The usage of the storage and processing power of the distributed infrastructures allowed for significant reduction (from several days to hours) of the analysis time, which was the bottleneck in the process. Available distributed storage allowed for tracking multiple data, which opened up field for detailed statistical analysis. The data processing has been automated and simplified. Practically all stages of the data processing are run automatically, reducing manual work and unnecessary delays. Created solution allows biological and medical users to focus on their research instead of mastering computer science details necessary to process data.

4.2   Metallurgy – Grid-Based Numerical Modeling Dedicated to Simulation of Metallurgical Production Processes

Numerical simulations, implemented and applied in computational science dedicated to metallurgy, became very sophisticated and computationally demanding. In many cases they require huge computing resources as well as creation of new algorithms for innovative hardware architectures. Therefore, the main objective of this work is design and implementation of new grid-based software in form of Client-Server applications, which are dedicated to numerical simulation of sophisticated metallurgical production processes. Currently developed grid-based applications are used to simulate the following industrial processes based on metal forming: (a) Continuous Steel Casting (CSC), which is fundamental production process of steel manufacturing [14]; proposed computer application is based on ProCast software, allowing reliable modeling of casting and prediction of material properties, (b) Extrusion, being applied in production of metallic profiles as well as in production of surgical threads [15]; the computational service offers 3D simulations implemented as in-house source codes, while there exists no professional dedicated software, (c) Welding and other processes based on liquid steel – this kind of sophisticated phenomena cannot be applied in industrial practice because of its high computational cost [16], hence usage of grid infrastructure is essential, as it aims to avoid these restrictions, (d) Stamping concerning innovative metallic materials like DP, Mart or TRIP steels, for which concurrent or upscaling multiscale models with high computational demand are required. The main objective of services being developed, based on Statistically Similar Representatives Volume Element [17] and Monte Carlo simulations [18], is to reduce high computational cost of multiscale simulations.

For the purposes of this paper, numerical simulations of extrusion process in 3D are selected and described in some details.

The proposed grid-based software is devoted to simulation of extrusion of thin profiles and rods of special alloys of magnesium, containing calcium supplements (MgCa08, Ax30). These alloys are characterized by extremely low technological plasticity during metal forming. For this reason, the range of parameters of extrusion, not leading to the fraction of the wire during deformation, is not wide. Therefore, the dedicated mathematical model of the processes of extrusion, taking into account the processes of fracture, has been proposed. This model is based on the Finite Element Method (FEM).

In the present work the material is considered as incompressible rigid-viscoplastic continua and elastic deformation are neglected.

The reliable numerical simulation gives the possibility of production technology design through optimization. Because of need of a number of 3D FEM simulations, parallelization of the model is developed [19]. This offers opportunity for parametric study, including a range of possible changes in the pressing speed, extrusion temperature and different geometry of the channel profile. All generated variants are performed in parallel by using grid infrastructure and on the basis of obtained results the optimal solution is selected.

Example of geometry optimization for profile extrusion is shown in Fig. 3. Optimal variant for extrusion was selected on the basis of the smallest torsion and bending of the profile.

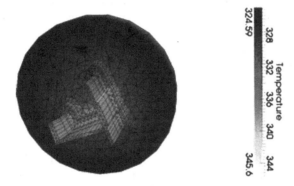

Fig. 3. Example of optimization in Extrusion-Grid software for PL-Grid computation system.

The programs were developed as Extrusion3D computing service. Implementation of proposed parallel version is based on the Intel Fortran compiler and OpenMP library, which allowed using full capabilities of grid infrastructure computing nodes. The calculations were carried out in parallel way without any barrier, thus optimization time is equal to the time of simulation of one particular case, offering significant reduction of computational time.

Performed optimization was verified experimentally in the laboratory. The extrusion of rods was performed in accordance with the calculated parameters (temperature, velocity of extrusion, shape of die). The resulting rods contain no fracture and have high mechanical properties.

4.3 Acoustics – New Services for Urban Planning, Research and Education

This interdisciplinary domain is extremely important in many areas of science, technology and engineering. The main purpose of the presented design is twofold, namely: providing detailed information about the noise threats that occur every day in city areas and preventing the noise induced hearing loss, especially among young people. The assessment of environmental threats is performed based on online data, acquired through a grid of engineered monitoring stations, employing some selected psychoacoustic properties of the human hearing system. Another aim is to make available efficient computational tools for the community of acousticians engaged in the noise threat combating.

In the "Acoustics" domain grid two kinds of services were prepared. The first one can be used to calculate the noise map of large city areas and it is called the "Noise Map" [20]. The second one, called the "Hearing" service, enables simulations of noise impact on the human hearing system.

Two scenarios of use can be distinguished. In the first scenario, a user has to provide the input data by uploading it into the storage. The input data has to be prepared by the user locally with dedicated software. The user specifies the location of input data in configuration options of the program. The user manages the computations with PL-Grid tools: QCG text client or Unicore. When calculation process is completed, the user downloads the output data and performs post-processing and visualization on the local terminal. In the second scenario, the dynamic noise map and estimate of influence of noise on hearing are produced periodically.

The engineered "Noise Map" service is intended for creating maps of noise threats for roads, railways and industrial sources. Integration of the software service with the network of distributed sensors brings a possibility of making automatic updates of noise maps for a specified time period. Illustration of an application of the developed solution is the urban area noise mapping. The fragment of the calculated noise map of the city of Gdansk is presented in Fig. 4. The map can be updated completely within relatively short period of time, employing the PL-Grid Infrastructure. Operations are performed employing a dedicated noise prediction model, optimized for a computer cluster. In addition, predicted maps may be adjusted, using real noise level measurements.

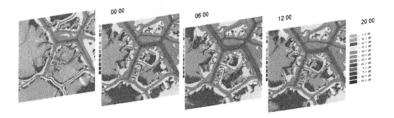

Fig. 4. The dynamic noise map.

The unique feature of the developed "Hearing" service is estimation of auditory effects, which are caused by the exposure to excessive noise. The main part of the "Hearing" service is the Psychoacoustical Noise Dosimeter, which is based on utilizing the modified psychoacoustic model of hearing. The primary function of the Dosimeter is to estimate, in a real time, auditory effects, which are caused by exposure to noise. Owing to that, it is possible to recognize a character of the auditory threshold shift for a given type of noise. The user can define detailed conditions of exposure to noise such as: noise level, exposure time and energy distribution in the frequency domain. The calculations by means of real sound data are also possible. The outcomes are presented in a form of the cumulated noise dose and the characteristic of temporary shift of the hearing threshold.

An example of noise-induced temporary threshold shift (TTS) simulation during outdoor concert was presented in Fig. 5 [21]. The considered auditory area was of 100×100 m. The stage width was 20 m. Two loudspeakers were located at both sides of the stage. The assumed duration of the concert was 3 hours. The spectrum distribution of the acoustic energy and TTS effect evoked by the exposure to music were expressed in critical bands of hearing as a function of the distance from the stage. The observed temporary threshold shift exceeding 20 dB extends in radius of about 20 m from the center of the stage. The hearing recovery time required for the people being present in this area is about 450 min.

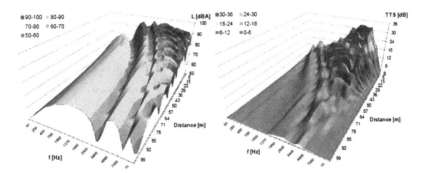

Fig. 5. Spectrum distribution of acoustic energy of noise source (left) and TTS effect evoked by the exposure to that noise expressed in critical bands as a function of distance from the noise source (right).

The infrastructure and the software developed can be utilized for urban or sound enforcement planning, and for research and education purposes.

4.4 Ecology – Phenology Observations Automated by IT Platforms

Phenology is the study of the timing of life cycle events of plants and animals [22]. The presence of advanced weather forecasting systems pushed aside this area of research, however phenological records can be a useful proxy for temperature analysis in historical climatology [23]. Phenological observations have a long tradition in Poland and are one of the oldest in Europe [24].

The main objective of the Ecology domain grid within PLGrid Plus project is to develop the Automatic Phenology Observations Service (APheS) that gives an opportunity to observe the flora, together with important processes occurring in it. The service will be based on the KIWI Remote Instrumentation Platform [25]. The KIWI Platform (as a successor of the Virtual Laboratory [26]) is a framework for building remote instrumentation systems. The platform provides a set of components to control and manage scientific equipment or sensors like cameras, weather, air pollution and water flows sensors and others. All the equipment connected to the KIWI Platform can be controlled remotely with one unique user interface. It allows users to design and run so-called observation workflows. It is possible to set up a sequence of operations starting from data acquisition, data processing and finally visualization. This workflow can be launched periodically by the KIWI Workflow Manager component with a desired frequency and time.

Devices and sensors produce different output data. Most of the sensors produce numerical data, which is to be visualized by the PLGrid Plus Ecogrid Web Portal. To make the system flexible, we have designed one common and generic interface for retrieving data from the instruments using OGC [27] standards with the same interface used when dealing with different equipment.

The ApheS service will provide a set of meteorological sensors and KIWI Eye monitoring system, cf. Fig. 6.

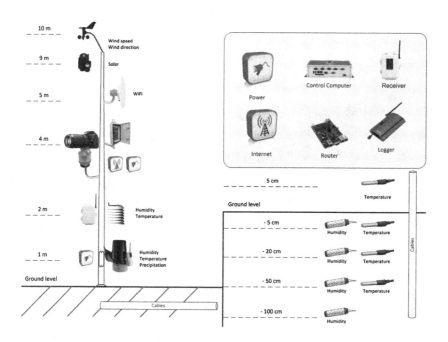

Fig. 6. Automatic phenology observations – infrastructure.

All the sensors and devices composing APheS are installed on site chosen by a scientist from National Park of Wielkopolska and National Research Institute of Meteorology and Water Management, consulted by experts from the natural sciences domain. Client side tools can be used to browse all data freely and compare different data sets with each other in elastic manner. Different views are to be provided for meteorological data with high responsiveness (using AJAX technology [28]) – allowing a user for manipulation of views. Another view will be available for phenology data. This is due to the nature of work with great number of graphical data that needs to be compared not only with other pictures, but also confronted with meteorological data.

Experts from the Institute of Meteorology and Water Management – National Research Institute and Botanical Gardens in Poznan, a phenology Research Institute of Adam Mickiewicz University keep the data gathered under constant surveillance – examining their accuracy, analyzing trends for any anomaly indicating problems with equipment.

There are several scenarios designed to fully utilize installation's potential and provide scientists with data, otherwise requiring great human effort to acquire. Among them are: defining: phenological year parts, climate local to the data-gathering site, phenophases for plants cultivated on the nearby arable area, radiation conditions' influence on plants growth, relation between sum of heath and plants growth, examining effects of cultivation conditions on duration time of vegetation, determining relation between air humidity and wild plants growth, and much more potentially.

For each scenario, to deliver results, it takes analysis of raw and post processed meteorological data and/or photos. As a result of studies, pictures accessible through KIWI system may be tagged with information determined by the scientist. This brings information sharing to another level – results produced by other studies are accessible not only e.g. through articles, but are spread through whole system's data with tags, approved by the scientists. This approach makes new experiments easier, as they are built on top of results from other scenarios, directly linked with raw data.

Automatic Phenology Observations Service described above composes unique and innovative tool, introducing phenological observations to automated, remote measurements. This solution, polished and installed within PLGrid Plus project, is of great value, especially as it was developed in close collaboration with the scientists from natural sciences domain. Data acquired by those tools is both of high quality and meaningful from the scientific point of view. Generic design of KIWI platform, on which APheS was built, allows for adapting solution presented for other sensors and devices, also from other domains.

5   Summary

The described process of developing the PL-Grid Infrastructure by the PL-Grid Consortium in the framework of PL-Grid and PLGrid Plus projects fits well with the need of the development of the advanced IT infrastructure designed for the implementation of modern scientific research, and providing Polish researchers with capability for collaboration with international research organizations.

The PL-Grid Infrastructure does not mean just computers and storage resources – installed in the five partner centers that are part of the PL-Grid

Consortium. The infrastructure also includes specialized software as well as the services and tools – developed by the PL-Grid Consortium – which support both users and administrators of computers. Access to the PL-Grid Infrastructure enables scientists to increase the scale of the calculations carried out within the scientific research, which would be impossible to achieve using separate computers, and to place demands for resources and services available within the infrastructure.

The presented approach proves that the most important goal of the PLGrid Plus project – expansion of the existing computational infrastructure towards domain-specific solutions for research teams – allowed for conducting more effective and valuable research. The results of most of these scientific calculations can be applied in various branches of science and technology.

Pilot introduction of 13 domain grids opened the scope of use of the Project results by various research communities. Moreover, by using the developed general base services and experience in building the domain ones, the integration of new groups should proceed smoothly and at much lower cost.

In the future, we plan to focus on some additional specific domains and wider offerings on cloud [29–31] and big data [32,33] services. Modernization of both, software environments (i.e. toward EMI components) and hardware solutions (with more GPGPU and Intel Phi) are foreseen. We also plan to reuse results of our previous achievements like, for example, [34] and [35].

Acknowledgements. This work was made possible thanks to the PL-Grid and PLGrid Plus projects POIG.02.03.00-00-007/08-00 and POIG.02.03.00-00-096/10. This research was supported in part by the PL-Grid Infrastructure. Acknowledgments are due to all members of the PL-Grid Consortium.

References

1. European Grid Infrastructure web site, http://www.egi.eu
2. EGI InSPIRE Project web site, http://www.egi.eu/about/egi-inspire/
3. The Polish Grid Infrastructure web site, http://www.plgrid.pl
4. The PL-Grid project web site, http://projekt.plgrid.pl
5. Kitowski, J., et al.: Polish computational research space for international scientific collaborations. In: Wyrzykowski, R., Dongarra, J., Karczewski, K., Waśniewski, J. (eds.) PPAM 2011, Part I. LNCS, vol. 7203, pp. 317–326. Springer, Heidelberg (2012)
6. Bubak, M., Szepieniec, T., Wiatr, K. (eds.): PL-Grid 2011. LNCS, vol. 7136. Springer, Heidelberg (2012)
7. The PLGrid Plus project web site, http://www.plgrid.pl/plus
8. UNICORE middleware website, http://unicore.eu
9. Bala, P., Baldridge, K., Benfenati, E., Casalegno, M., Maran, U., Rasch, K., Schuller, B.: UNICORE - a middleware for life sciences grids. In: Cannataro, M. (ed.) Handbook of Research on Computational Grid Technologies for Life Sciences, Biomedicine and HealthCare, IGI 2009, pp. 615–643 (2009)
10. European Middleware Initiative (EMI) website, http://www.eu-emi.eu
11. EMI Software Repository, http://emisoft.web.cern.ch/emisoft

12. Schuller, B., Pohlmann, T.: UFTP: high-performance data transfer for UNICORE. In: Romberg, M., Bala, P., Müller-Pfefferkorn, R., Mallmann, D. (eds.) Proceedings of UNICORE Summit 2011, Forschungszentrums Jülich, IAS Series, vol. 9, pp. 135–142 (2011). ISBN 978-3-89336-750-4

13. Borcz, M., Kluszczyński, R., Skonieczna, K., Grzybowski, T., Bała, P.: Processing the biomedical data on the grid using the UNICORE workflow system. In: Caragiannis, I., et al. (eds.) Euro-Par Workshops 2012. LNCS, vol. 7640, pp. 263–272. Springer, Heidelberg (2013)

14. Buczek, A., Burbelko, A., Drożdż, P., Dziarmagowski, M., Falkus, J., Karbowniczek, M., Kargul, T., Mołkowska-Piszczek, K., Rywotycki, M., Sołek, K., Ślęzak, W., Telejko, T., Trębacz, L., Wielgosz, E.: Modelling of continuous casting process of steel: monograph. Wydawnictwo Naukowe Instytutu Technologii Eksploatacji (2012)

15. Milenin, A., Kustra, P., Seitz, J.-M., Bach, F.-W., Bormann, D.: Production of thin wires of magnesium alloys for surgical applications. In: Conference Proceedings of the Wire Association International, pp. 61–70 (2010)

16. Siwek, A., Rońda, J., Banaś, K.: Model of convective heat transfer in keyhole mode laser welding. Comput. Meth. Mater. Sci. **11**(1), 179–184 (2011)

17. Rauch, L., Pernach, M., Bzowski, K., Pietrzyk, M.: On application of shape coefficients to creation of the statistically similar representative element of DP steels. Comput. Meth. Mater. Sci. **11**(4), 531–541 (2011)

18. Szyndler, J., Madej, L.: Monte Carlo method in application to generation of the digital material representation. Hutnik - Wiadomości Hutnicze **80**(2), 172–176 (2013)

19. Milenin, A., Kustra, P.: Optimization of extrusion and wire drawing of magnesium alloys using the finite element method and distributed computing. In: Proceedings of the InterWire 2013 International Conference (in press)

20. Czyzewski, A., Szczodrak, M., Kotus, J.: Creating acoustic maps employing supercomputing cluster. Arch. Acoust. **36**(2), 124 (2011)

21. Kotus, J., Szczodrak, M., Czyzewski, A., Kostek, B.: Distributed system for noise threat evaluation based on psychoacoustic measurements. Metrol. Meas. Syst. **XIX**, 219–230 (2012)

22. About Phenology. http://www.usanpn.org/about/phenology. Accessed 28 April 2010

23. Molga, M.: Meteorologia rolnicza. PWRiL (1983)

24. Piotrowicz, K.: Historia obserwacji fenologicznych w Galicji. IMiGW (2007)

25. KIWI Remote Instrumentation Platform, http://kiwi.psnc.pl/

26. Virtual Laboratory PSNC. http://vlab.psnc.pl/

27. OGC - Open Geospatial Consortium, http://www.opengeospatial.org/

28. Ajax Technology, https://en.wikipedia.org/wiki/Ajax_(programming)

29. Kryza, B., Król, D., Wrzeszcz, M., Dutka, L., Kitowski, J.: Interactive cloud data farming environment for military mission planning support. Comput. Sci. **13**(3), 89–100 (2012)

30. Malawski, M., Juve, G., Deelman, E., Nabrzyski, J.: Cost- and deadline-constrained provisioning for scientific workflow ensembles in IaaS clouds. In: SC'12 Proceedings of the International Conference on High Performance Computing, Networking, Storage and Analysis, Article No. 22. IEEE Computer Society Press, Los Alamitos (2012). ISBN: 978-1-4673-0804-5

31. Bubak, M., Kasztelnik, M., Malawski, M., Meizner, J., Nowakowski, P., Varma, S.: Evaluation of Cloud Providers for VPH Applications. Accepted for CCGrid 2013 (2013)

32. Słota, R., Król, D., Skałkowski, K., Orzechowski, M., Nikolow, D., Kryza, B., Wrzeszcz, M., Kitowski, J.: A toolkit for storage QoS provisioning for data-intensive applications. Comput. Sci. **13**(1), 63–73 (2012)
33. Słota, R., Nikolow, D., Kitowski, J., Krol, D., Kryza, B.: FiVO/QStorMan semantic toolkit for supporting data-intensive applications in distributed environments. Comput. Inf. **31**, 1003–1024 (2012)
34. Dutka, L., Kitowski, J.: Application of component-expert technology for selection of data-handlers in CrossGrid. In: Kranzlmüller, D., Kacsuk, P., Dongarra, J., Volkert, J. (eds.) Euro PVM/MPI 2002. LNCS, vol. 2474, pp. 25–32. Springer, Heidelberg (2002)
35. Marco, J., Campos, I., Coterillo, I., et al.: The interactive european grid: project objectives and achievements. Comput. Inform. **27**(2), 161–171 (2008)

Cost Optimization of Execution
of Multi-level Deadline-Constrained
Scientific Workflows on Clouds

Maciej Malawski[1]([✉]), Kamil Figiela[1], Marian Bubak[1,2],
Ewa Deelman[3], and Jarek Nabrzyski[4]

[1] Department of Computer Science, AGH, al. Mickiewicza 30, 30-059 Kraków, Poland
{malawski,kfigiela,bubak}@agh.edu.pl
[2] ACC CYFRONET AGH, ul. Nawojki 11, 30-950 Kraków, Poland
[3] USC Information Sciences Institute,
Admiralty Way, Marina Del Rey, CA 4676, USA
deelman@isi.edu
[4] Center for Research Computing, University of Notre Dame, Notre Dame, IN, USA
naber@nd.edu

Abstract. This paper introduces a cost optimization model for scientific workflows on IaaS clouds such as Amazon EC2 or RackSpace. We assume multiple IaaS clouds with heterogeneous VM instances, with limited number of instances per cloud and hourly billing. Input and output data are stored on a Cloud Object Store such as Amazon S3. Applications are scientific workflows modeled as DAGs as in the Pegasus Workflow Management System. We assume that tasks in the workflows are grouped into levels of identical tasks. Our model is specified in AMPL modeling language and allows us to minimize the cost of workflow execution under deadline constraints. We present results obtained using our model and the benchmark workflows representing real scientific applications such as Montage, Epigenomics, LIGO. We indicate how this model can be used for scenarios that require resource planning for scientific workflows and their ensembles.

Keywords: AMPL optimization · Cloud computing · Scientific workflows

1 Introduction

Nowadays, science requires processing of large amounts of data and use of hosted services for compute-intensive tasks [10]. Cloud services are used not only to provide resources, but also for hosting scientific datasets, as in the case of AWS public datasets [2]. Scientific applications that run on these clouds have often the structure of workflows or workflow ensembles that are groups of inter-related workflows [16]. Infrastructure as a Service (IaaS) cloud providers offer services, where virtual machine instances differ by performance and price [7]. Planning

R. Wyrzykowski et al. (Eds.): PPAM 2013, Part I, LNCS 8384, pp. 251–260, 2014.
DOI: 10.1007/978-3-642-55224-3_24, © Springer-Verlag Berlin Heidelberg 2014

scientific experiments requires optimization decisions that take into account both execution time and cost.

Research presented in this paper can be seen as a step towards developing a cloud resource calculator for scientific applications in the hosted science model [10]. Specifically, we address the cost optimization problem of large-scale scientific workflows running on multiple heterogeneous clouds, using mathematical modeling with AMPL [12] and mixed integer programming. This approach allows to describe the model mathematically and use a set of available optimization solvers. On the other hand, an attempt to apply this method to the general problem of scheduling large-scale workflows on heterogeneous cloud resources would be impractical due to the problem complexity, therefore simplified models need to be developed. In our previous work [15], we used a similar technique to solve the problem where the application consists of tasks that are either identical or vary in size within a small range. As observed in [4,11], large-scale scientific workflows often consist of multiple parallel stages or levels, each of which has a structure of bag of tasks, i.e. the tasks in each level are similar. In the case of large workflows, when the number of tasks in the level is high, it becomes more practical to optimize the execution of the whole level instead of looking at each task individually, as many scheduling algorithms do [17]. Therefore, in this paper, we extend our model to deal with applications that are workflows represented as DAGs consisting of levels or layers of uniform tasks.

After outlining the related work in Sect. 2, we introduce the application and infrastructure model in Sect. 3. In Sect. 4 we provide the problem formulation in AMPL. Section 5 describes the evaluation of our model on a set of benchmark workflows, while Sect. 6 gives conclusions and future work.

2   Related Work

Our work is related to heuristic algorithms for workflow scheduling on IaaS clouds, such as the ones described in [1,3,5,17]. Our infrastructure model differs in that we assume multiple heterogeneous clouds with object storage attached to them, instead of individual machines with peer-to-peer data transfers between them. Instead of scheduling each task individually, our approach proposes a global optimization of placement of workflow tasks and data.

The deadline-constrained cost optimization of scientific workloads on heterogeneous IaaS described in [6] addresses multiple providers and data transfers between them, where the application is a bag of tasks. The global cost minimization problem on clouds, addressed in [18] focuses on data transfer costs and does not address workflows. Other approaches presented in [8,14] consider unpredictable dynamic workloads on IaaS clouds and optimize the objectives such as cost, runtime or utility function by autoscaling the resource pool at runtime.

Pipelined workflows consisting of stages are addressed in [19], where the processing model is a data flow and multiple instances of the same workflow are executed on the same set of cloud resources. Our work is different in that our goal is cost optimization instead of meeting the QoS constraints.

3   Application and Infrastructure Model

We assume that a workflow is divided into several levels (layers) that can be executed sequentially and tasks within one level do not depend on each other (see Fig. 1). Each layer represents a bag of tasks that can be partitioned in several groups (e.g. application A, B, etc.) that share computational cost and input/output size. We assume that only one task group is executed on a specific cloud instance (VM). This forbids instance sharing between multiple layers, which means that each application needs its own specific VM template.

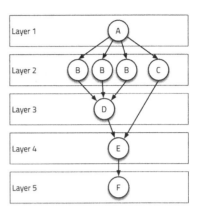

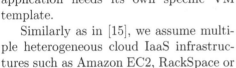

Fig. 1. Example application structure

Similarly as in [15], we assume multiple heterogeneous cloud IaaS infrastructures such as Amazon EC2, RackSpace or ElasticHosts. Clouds have heterogeneous VM instance types, with limits on the number of instances per cloud, e.g. 20 for EC2, 15 for RackSpace, etc. Input and output data are stored on Cloud Object Store such as Amazon S3 or RackSpace CloudFiles. In our model, all VM instances are billed per hour of usage, and there are fees for data transfers. In the model we can also have a private cloud where costs are set to 0.

4   Problem Formulation Using AMPL

To perform optimization of the total cost, Mixed Integer Problem (MIP) is formulated and implemented in A Mathematical Programming Language (AMPL) [12]. AMPL requires us to specify input data sets and variables to define the search space, as well as constraints and objective function to be optimized.

Input data. The formulation requires the following input sets, which represent the infrastructure model, in a similar way as we approached the problem in [15]:

– $S = \{s3, cloudfiles\}$ – defines available cloud storage sites,
– $P = \{amazon, rackspace, \ldots\}$ – defines possible computing cloud providers,
– $I = \{m1.small, \ldots, gg.1gb, \ldots\}$ – defines instance types,
– $PI_p \subset I$ – instances that belong to provider P_p,
– $LS_s \subset P$ – compute cloud providers that are local to storage platform S_s.

Each instance type I_i is described by the following parameters:

– p_i^I – fee in \$ for running instance I_i for one hour,
– ccu_i – performance of instance in CloudHarmony Compute Units (CCU) [9],

– p_i^{Iout} and p_i^{Iin} – price for non-local data transfer to and from the instance, in
   $ per MiB (1 MiB = $1024 * 1024$ B).

Storage sites are characterized by:

– p_s^{Sout} and p_s^{Sin} characterize price in $ per MiB for non local data transfer.

Additionally we need to provide data transfer rates in MiB per second between
storage and instances by defining function $r_{i,s} > 0$.

   Our application model is different from the one in [15] because it groups tasks
into layers:

– L – set of layers,
– G – set of tasks groups,
– G_l – set of tasks groups belonging to layer l,
– A_t^{tot} – number of tasks in group t,
– t_t^x – execution time in hours of a single task of group t on 1 CCU machine,
– d_t^{in} and d_t^{out} – data size for input and output of one task t in MiB,
– p^R – price per request for queuing service, such as Amazon SQS, required to
   execute a single task,
– t^D – total time for completing workflow (deadline).

Auxiliary parameters. A set of precomputed parameters, which are derived from
the main input parameters of the model includes:

– $t_{i,s}^{net} = \frac{d^{in}+d^{out}}{r_{i,s} \cdot 3600}$ – *transfer time*: time for data transfer between I_i and S_s,
– $t_{i,s}^u = \frac{t^x}{ccu_i} + t_{i,s}^{net}$ – *unit time*: time for processing a task on instance I_i using
   storage S_s that includes computing and data transfer time (in hours),
– $c_{i,s}^T = (d^{out} \cdot (p_i^{Iout}+p_s^{Sin})+d^{in} \cdot (p_s^{Sout}+p_i^{Iin}))$ – cost of data transfer between
   instance I_i and storage S_s,
– I_i^{idx} – set of possible instance I_i indexes (from 0 to $n_i^{Imax} - 1$).

Variables. Variables that will be optimized and define the solution space are:

– $A_{t,i,x}$ – binary, 1 iff (if and only if) instance I_i with index x is launched to
   process task group G_t, otherwise 0;
– $H_{t,i,x}$ – int, for how many hours is instance launched;
– $T_{t,i,x}$ – int, how many tasks of G_t are processed on that instance,
– D_l^t – actual computation time for L_l,
– D_l – int, maximal number of hours that instances are allowed to run in L_l.

Objectives. Cost of running one task including instance and transfer cost is:

$$(t^{net} + t^u) \cdot p^I + d^{in} \cdot (p^{Sout} + p^{Iin}) + d^{out} \cdot (p^{Iout} + p^{Sin}) + p^R, \qquad (1)$$

while the objective function represents the total cost of running multiple tasks
of the application on the cloud infrastructure is defined as:

$$\underset{\substack{\text{minimize} \\ \text{total cost}}}{} \sum_{t \in G, i \in I, x \in I_i^{idx}} ((p_i^I * H_{t,i,x} + p^R + c_{i,s}^T) * T_{t,i,x}), \qquad (2)$$

subject to the constraints:

1. $\sum_{l \in L} D_l \leq t^D$ ensures that workflow finishes in the given deadline,
2. to fix that $D = \lceil D^t \rceil$ we require that: $\forall_{l \in L} D_l^t \leq D_l \leq D_l^t + 1$,
3. $\forall_{t \in G, i \in I, x \in I_i^{idx}} A_{t,i,x} \leq H_{t,i,x} \leq A_{t,i,x} \cdot t^D$ ensures that H may be allocated only iff A is 1,
4. $\forall_{t \in G, i \in I, x \in I_i^{idx}} A_{t,i,x} \leq T_{t,i,x} A_{t,i,x} \cdot A_t^{tot}$ ensures that T may be allocated only iff A is 1,
5. $\forall_{t \in G, i \in I, x \in I_i^{idx}} H_{t,i,x} \leq D_l$ enforces layer deadline on instances runtime,
6. $\forall_{l \in L, t \in G_l, i \in I, x \in I_i^{idx}} T_{t,i,x} \cdot t_{t,i,s}^u \leq D_l^t$ enforces that a layer finishes work in D^t,
7. to make sure that all the instances run for enough time to process all tasks allocated to them we require: $\forall_{t \in G, i \in I, x \in I_i^{idx}} T_{t,i,x} \cdot t_{t,i,s}^u \leq H_{t,i,x} T_{t,i,x} \cdot t_{t,i,s}^u + 1$, which adjusts H respectively to T,
8. $\forall_{t \in G} \sum_{i \in I, x \in I_i^{idx}} T_{t,i,x} = A_t^{tot}$ ensures that all tasks are processed,
9. To reject symmetric solutions, we add three constraints:
   (a) $\forall_{t \in G, i \in I, x \in \{1..(n_i^{I max}-1)\}} H_{t,i,x} \leq H_{t,i,x-1}$,
   (b) $\forall_{t \in G, i \in I, x \in \{1..(n_i^{I max}-1)\}} A_{t,i,x} \leq A_{t,i,x-1}$, and:
   (c) $\forall_{t \in G, i \in I, x \in \{1..(n_i^{I max}-1)\}} T_{t,i,x} \leq T_{t,i,x-1}$.
10. $\forall_{l \in L, p \in P} \sum_{i \in PI_p, t \in G_l, x \in I_i^{idx}} A_{t,i,x} \leq n_p^{Pmax}$ enforces instance limits per cloud.

To keep this model in MIP class we had to take a different approach than in previous model, and schedule each virtual machine instance separately. The drawback of this approach is that we need to increase the number of decision variables. We also divided the search space by storage provider. Additionally, the deadline becomes a variable with upper bound as it may happen that shorter deadline may actually give a cheaper solution (see Fig. 3 and its discussion).

5 Evaluation

To evaluate our model on realistic data, we use CloudHarmony [9] benchmarks to parameterize the infrastructure model, and we use the Workflow Generator Gallery workflows [4] as test applications. In the infrastructure model we assumed that we have 4 public cloud providers (Amazon EC2, RackSpace, GoGrid and ElasticHosts) and a private cloud with 0 cost. The infrastructure has two storage services, S3 that is local to EC2 and CloudFiles that is local to RackSpace, so data transfers between local compute and storage are free. We tested our model with all applications from the gallery: Montage, CyberShake, Epigenomics, LIGO and SIPHT for all available workflow sizes (from 50 to 1000 tasks per workflows, up to 5000 tasks in the case of SIPHT workflow). We varied the deadline from 1 to 30 h with 1-h increment. We solve the problem for two cases, depending on whether the data is stored on S3 or on CloudFiles.

Figure 2 shows the example results obtained for the Epigenomics application and workflows of two sizes (400 and 500 tasks). For longer deadlines the private cloud instances and the cheapest RackSpace instances are used so the cost is low when using CloudFiles. For shorter deadlines the cost grows rapidly, since we reach the limit of 15 instances per cloud and additional instances must be

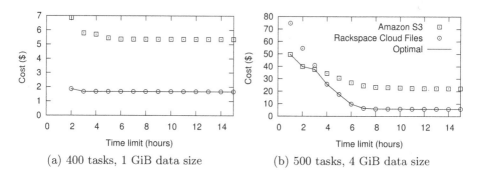

Fig. 2. Result of the optimization procedure for the Epigenomics application.

spawned on a different provider, making the transfer costs higher. This effect is amplified in Fig. 2b, which differs from Fig. 2a not only by the number of tasks but also by the data size of one layer. This means that the transfer costs are growing more rapidly, so it becomes more economical to store the data on Amazon EC2 that provides more powerful instances required for short deadlines.

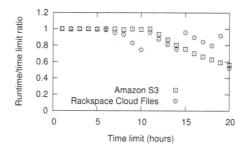

Fig. 3. Ratio of actual completion time to deadline for Epigenomics workflow with 500 tasks.

One interesting feature of our model is that for longer deadlines it can find the cost-optimal solutions that have a shorter workflow completion time than the requested deadline. This effect can be observed in Fig. 3 and is caused by the fact that for long deadlines the simple solution is to run the application on a set of the least expensive machines.

Figure 4a–d show results obtained for other workflows. These workflows are relatively small and even for short deadlines the model is able to schedule tasks on cheapest instances on a single cloud, thus resulting in flat characteristics.

To investigate how the model behaves for workflows with the same structure, but with much longer run times of tasks, we run the optimization for Montage workflow with tasks 1000× longer. This corresponds to the scenario where tasks are in the order of hours instead of seconds. The sample results in Fig. 5a show

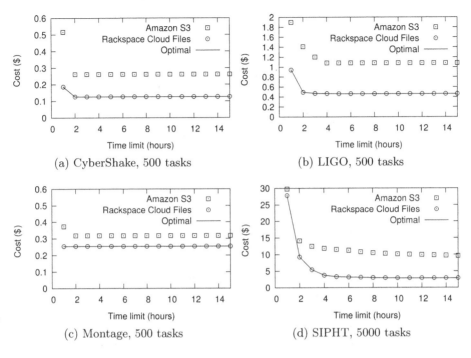

Fig. 4. Optimal cost found by the model for different applications.

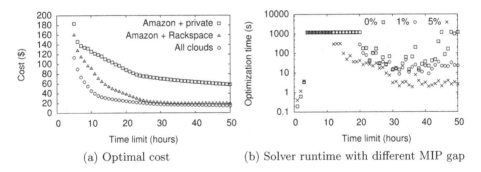

Fig. 5. Results obtained by the model for Montage 500 workflow with tasks runtimes artificially multiplied by 1000 for different cloud infrastructures.

how the cost increases much steeply with shorter deadlines, illustrating the trade-off between time and cost. The difference between Figs. 4c and 5a illustrates that the model is more useful for workflows when tasks are of granularity that is similar to the granularity of the (hourly) billing cycle of cloud providers. Additionally, Fig. 5a shows how the optimal cost depends on the cloud available.

The run time of the optimization algorithm for workflows with up to 1000 tasks ranges from few seconds up to 4 min using the CPLEX [13] solver running on a server with 4 16-core 2.3 GHz AMD Opteron processors (model 6276), with

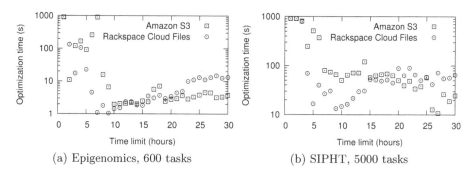

(a) Epigenomics, 600 tasks   (b) SIPHT, 5000 tasks

Fig. 6. Solver execution wall time.

a limit set to 32 cores. Figure 6a shows that the time becomes much higher for shorter deadlines and increases for very long deadlines. This is correlated with size of search space: the longer the deadline, the search space is larger, while for shorter deadlines the problem has a very small set of acceptable solutions. The problem becomes more severe for bigger and more complex workflows like SIPHT as optimization time becomes very high (Fig. 6b).

Figure 5b illustrates how the optimization time depends on MIP gap solver setting [13]. Applying a relative MIP gap of 1 % or 5 % instead of default 0.01 % shortens optimization time in orders of magnitude. Increasing the MIP gap to 5 % did not decrease the quality of the result noticeably: the minimum cost obtained for the gap of 5 % was higher only by 3.63 % in the worst case.

6   Conclusions and Future Work

In this paper, we presented a cost optimization model for scientific workflows executing on multiple heterogeneous clouds. The model, formulated in AMPL, allows us to find the optimal assignment of workflow tasks, grouped into layers, to cloud instances. We tested our model on a set of benchmark workflows and we observed that it gives useful solutions in a reasonable amount of computing time. By solving the model for multiple deadlines, we can produce trade-off plots, showing how the cost depends on the deadline. We believe that such plots are a step towards a scientific cloud workflow calculator, supporting resource management decisions for both end-users and workflow-as-a-service providers.

In future work we plan to apply this model to the problem of provisioning cloud resources for workflow ensembles [16], where the optimization of cost can drive the workflow admission decisions. We also plan to refine the model to better support smaller workflows by reusing instances between layers, to fine-tune the model, and to test different solver configurations to reduce the computing time.

Acknowledgement. This research was partially supported by the EC ICT VPH-Share Project (contract 269978), the KI AGH grant, and by the National Science Foundation under grant OCI-1148515.

References

1. Abrishami, S., Naghibzadeh, M., Epema, D.H.: Deadline-constrained workflow scheduling algorithms for infrastructure as a service clouds. Future Gener. Comput. Syst. **29**(1), 158–169 (2013). http://www.sciencedirect.com/science/article/pii/S0167739X12001008
2. AWS: AWS public datasets. http://aws.amazon.com/publicdatasets/ (2013)
3. Barrionuevo, J.J.D., Fard, H.M., Prodan, R.: Moheft: a multi-objective list-based method for workflow scheduling. In: 4th IEEE International Conference on Cloud Computing Technology and Science Proceedings, CloudCom 2012, Taipei, Taiwan, 3–6 December 2012, pp. 185–192 (2012)
4. Bharathi, S., Chervenak, A., Deelman, E., Mehta, G., Su, M.H., Vahi, K.: Characterization of scientific workflows. In: Third Workshop on Workflows in Support of Large-Scale Science, WORKS 2008, pp. 1–10. IEEE (2008). http://dx.doi.org/10.1109/WORKS.2008.4723958
5. Bittencourt, L.F., Madeira, E.R.M.: Hcoc: a cost optimization algorithm for workflow scheduling in hybrid clouds. J. Internet Serv. Appl. **2**(3), 207–227 (2011)
6. den Bossche, R.V., Vanmechelen, K., Broeckhove, J.: Online cost-efficient scheduling of deadline-constrained workloads on hybrid clouds. Future Gener. Comput. Syst. **29**(4), 973–985 (2013). http://www.sciencedirect.com/science/article/pii/S0167739X12002324
7. Bubak, M., Kasztelnik, M., Malawski, M., Meizner, J., Nowakowski, P., Varma, S.: Evaluation of cloud providers for VPH applications. In: CCGrid2013 - 13th IEEE/ACM International Symposium on Cluster, Cloud and Grid, Computing, May 2013. http://ieeexplore.ieee.org/xpl/articleDetails.jsp?arnumber=6546092
8. Chen, J., Wang, C., Zhou, B.B., Sun, L., Lee, Y.C., Zomaya, A.Y.: Tradeoffs between profit and customer satisfaction for service provisioning in the cloud. In: Proceedings of the 20th International Symposium on High Performance Distributed Computing, HPDC '11, pp. 229–238. ACM, New York (2011)
9. CloudHarmony: Benchmarks. http://cloudharmony.com/benchmarks (2011)
10. Deelman, E., Juve, G., Malawski, M., Nabrzyski, J.: Hosted science: managing computational workflows in the cloud. Parallel Process. Lett. **23**(2), June 2013. http://www.worldscientific.com/doi/abs/10.1142/S0129626413400045
11. Duan, R., Prodan, R., Li, X.: A sequential cooperative game theoretic approach to storage-aware scheduling of multiple large-scale workflow applications in grids. In: 2012 ACM/IEEE 13th International Conference on Grid Computing (GRID), pp. 31–39. IEEE (2012). http://dx.doi.org/10.1109/Grid.2012.14
12. Fourer, R., Gay, D.M., Kernighan, B.W.: AMPL: A Modeling Language for Mathematical Programming. Duxbury Press, Belmont (2002)
13. IBM: IBM ILOG CPLEX Optimization Studio - CPLEX User's Manual. http://pic.dhe.ibm.com/infocenter/cosinfoc/v12r5/index.jsp (2013)
14. Kim, H., El-Khamra, Y., Rodero, I., Jha, S., Parashar, M.: Autonomic management of application workflows on hybrid computing infrastructure. Sci. Program. **19**, 75–89 (2011)
15. Malawski, M., Figiela, K., Nabrzyski, J.: Cost minimization for computational applications on hybrid cloud infrastructures. Future Gener. Comput. Syst. **29**(7), 1786–1794 (2013). http://www.sciencedirect.com/science/article/pii/S0167739X13000186
16. Malawski, M., Juve, G., Deelman, E., Nabrzyski, J.: Cost- and deadline-constrained provisioning for scientific workflow ensembles in IaaS clouds. In: Proceedings of the

International Conference on High Performance Computing, Networking, Storage and Analysis, SC '12. IEEE Computer Society Press (2012). http://portal.acm.org/citation.cfm?id=2389026

17. Mao, M., Humphrey, M.: Auto-scaling to minimize cost and meet application deadlines in cloud workflows. In: Proceedings of 2011 International Conference for High Performance Computing, Networking, Storage and Analysis, SC '11. ACM, New York (2011). http://dx.doi.org/10.1145/2063384.2063449

18. Pandey, S., Barker, A., Gupta, K.K., Buyya, R.: Minimizing execution costs when using globally distributed cloud services. In: 24th IEEE International Conference on Advanced Information Networking and Applications, pp. 222–229. IEEE (2010)

19. Tolosana-Calasanz, R., Banares, J.A., Pham, C., Rana, O.F.: Enforcing QoS in scientific workflow systems enacted over cloud infrastructures. J. Comput. Syst. Sci. **78**(5), 1300–1315 (2012). http://www.sciencedirect.com/science/article/pii/S0022000011001607

Parallel Computations in the Volunteer–Based Comcute System

Paweł Czarnul$^{(\boxtimes)}$, Jarosław Kuchta, and Mariusz Matuszek

Faculty of Electronics, Telecommunications and Informatics,
Gdansk University of Technology, Gdańsk, Poland
{pczarnul,qhta,mrm}@eti.pg.gda.pl

Abstract. The paper presents Comcute which is a novel multi-level implementation of the volunteer based computing paradigm. Comcute was designed to let users donate the computing power of their PCs in a simplified manner, requiring only pointing their web browser at a specific web address and clicking a mouse. The server side appoints several servers to be in charge of execution of particular tasks. Thanks to that the system can survive failures of individual computers and allow definition of redundancy of desired order. On the client side, computations are executed within web browsers using technologies such as Java, JavaScript, Adobe Flash etc. without the need for installation of additional software. This paper presents results of scalability experiments carried on the Comcute system.

Keywords: Volunteer computing · Parallel computations · Scalability · Reliability

1 Introduction

Many areas of modern science rely heavily on supercomputing power availability. In fact, computing demand from just Materials Science, Biology, Astronomy and Medicine [1] outpaces the supply and it is an ongoing effort to keep up with the demand.

Part of this effort is designing systems which combine the power of many distributed personal computers and make it available for science. Several such systems exist e.g. BOINC[1] [2], but their common property is need for a dedicated computing module to be installed and configured by the user, who wishes to make their computer available. This task is somewhat technical in nature and often intimidates potential volunteers. Volunteer computing plays an important role in supplying the computational power demanded by science. Harnessing the power of personal computers connected to the Internet requires dedicated systems, which distribute computations and collect results. Usually such systems require the user of a personal computer to install and run a dedicated client software, which often presents a difficulty to less technical-savvy users.

---

[1] http://boinc.berkeley.edu/

R. Wyrzykowski et al. (Eds.): PPAM 2013, Part I, LNCS 8384, pp. 261–271, 2014.
DOI: 10.1007/978-3-642-55224-3_25, © Springer-Verlag Berlin Heidelberg 2014

Recognizing this difficulty led to design and implementation of a Comcute system, which allows volunteers to make their computing resources available by just pointing their web browser at a web address and clicking a mouse without installation of additional software. Development of the Comcute project[2] [3] took place in years 2010–2012 and was supported under Grant OR00010811 by the Polish Ministry of Science and Higher Education.

2 Related Work

There is a variety of paradigms and tools for parallel computations proposed and implemented at various levels:

- shared memory systems: GPGPU with NVIDIA CUDA and OpenCL [4,5], OpenMP, Pthreads, Java Threads for multithreaded programming on SMP systems [6],
- distributed memory systems:
  - dedicated HPC systems: MPI [7], PVM [8],
  - collection of HPC systems: MPICH-G2 [9], PACX-MPI [10],
  - distributed systems, including HPC in various Virtual Organizations: grid systems implemented on top of grid middlewares such as Globus Toolkit, Unicore, Gridbus with scheduling and management of resources [11–13],
  - frameworks such as Hadoop[3],
  - workflow systems such as the one in BeesyCluster [14],
  - volunteer-based systems such as BOINC [2] in which distributed volunteers donate computing power of their own computers to shared projects. Paper [15] demonstrates how a new WeevilScout prototype framework can be used to engage thousands of Internet browsers with JavaScript support for computations in the master-slave fashion for a bio-informatics task.

Reference [16] presents a framework using the master-slave model for computations built on top of Google App Engine that allows free of charge execution using the TaskQueue scheme. The master and slaves are implemented behind a Web interface and then use the TaskQueue for execution. Compared to [16], Comcute was designed in order not to require access to Google or any other infrastructure and rely on computing power of Internet users' computers instead.

Compared to volunteer computing such as BOINC [2], Comcute was created to offer several new unique features:

- ability to run the client within a web browser supporting many technologies such as Java, JavaScript, Adobe Flash etc. not just JavaScript as addressed by WeevilScout [15],
- advanced management of computations at the server side supporting:

---

[2] http://comcute.eti.pg.gda.pl
[3] http://hadoop.apache.org/

- redundancy of desired order, i.e. requesting redundant computations of data chunks by volunteers,
- ability to partition input data and integration on-the-fly as data chunks come,
- distributed management of computations on the server side that is able to survive failures of individual servers (not addressed by WeevilScout [15] nor BOINC [2]).

3   Proposed Solution

The Comcute system uses the same volunteer-based paradigm of calculation as BOINC. A computational task (code and partitioned data) is distributed to a great number of volunteers (Internet users). On the other hand, Comcute differs from other volunteer-based computing systems in a level of calculation flexibility, a level of reliability and ease of use. First, the volunteers may be recruited from the users of common public services, such as e-government or e-administration services, video-sharing services (e.g. YouTube) or social networking services (e.g. Facebook). Calculation tasks (code and data) are loaded to their computers in a simple one-click fashion. Calculations are performed at their computers in a web browser context which should be safe for the client. The code is matched to the capabilities supported by the volunteer's browser (e.g. Java Script, Java, Flash). A single user may process many data packs for various tasks in a single session. In this way, the Comcute system can process various computing tasks ordered by customers at the same time.

3.1   Architecture

A quad-layered architecture of the Comcute system ensures the efficiency and reliability of calculations. Usually a multi-layered architecture is presented by vertically divided regions, but the quad-layered architecture of the Comcute system is presented in a form of concentric regions (Fig. 1) with Z-layer representing user interface in the center, internal W-layer with system core nodes, external S-layer with distributing servers and surrounding "layer" containing grouped sets of Internet users' computers.

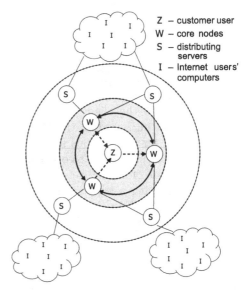

Fig. 1. Multi-layered Comcute architecture

There is no central S nor W node in the system. The Z-layer gives access to a number of W-nodes organized in a load-balanced grid. The first W-node contacted by a customer forms a set of other W-nodes (W'-set) in the number needed to complete a task, as requested by the customer. The task (code and data) is distributed to all the W'-nodes, where task data is divided into a set of data packs. The W'-nodes divide task data using the same algorithm and its parameters. The task code along with data packs is sent as independent packets from W'-nodes through S-servers to computers of Internet users (Is).

The S-servers are placed in the public domain of the World Wide Web. They may be set up as public administration (government) servers, video (movie) servers, social network servers and so on. Beside their normal activity, they offer participation in the Comcute project to their users. By joining the Comcute project, users agree to download and execute computing tasks. Each data packet is processed for a short time, but a great number of processors gives the effect of a huge computational scale. The S-servers also separate the Internet users from the W-nodes, which are located in a protected network. Locations of W and S servers depend on the system administrator so that W-S links may offer large bandwidths.

The code is executed at the computers of Internet users in the context of a web browser. The calculation results are sent back through the S-servers to the W'-nodes, where they are assembled into aggregated result. This aggregation is carried out in cooperation between the W'-nodes.

The functionality of the W-layer grid is designed to withstand attacks on the system and the computations. Each task is processed by a set of nodes which exchange and compare the results. An optional verification of the results can be provided in this way. If the S-servers are not responding for a long time or communication between the W-nodes is broken, the W-node grid may reconfigure itself. This way the system can perform its task as long as a single W-node is in an operational state.

3.2 Distributed Volunteer Task Execution

The W-layer is composed of independent but collaborating W-nodes driven by the same algorithm. At the beginning of the computational cycle the first W-node responding to the client request takes the task orders from the clients and authorizes them. Based on the task parameters it estimates the number of W-nodes needed for calculations and invites them to form the W' set. It then distributes the accepted tasks to other nodes of the W' set.

Then each node of the W' set divides task data into packs according to the task parameters using a partition algorithm specified by the client. If the customer requested a high level of reliability, the W'-set of nodes is further divided into smaller groups G (e.g. three nodes in each group) for tight collaboration. If not, the nodes work in loose collaboration only. Subsequently each W'-node offers the code of the calculation tasks and packs of data as independent packets on demand to S-servers.

During calculations W'-nodes offer data packs in a random order from the whole data set. Collaboration among the W'-nodes means that they share partial results of calculations so each node has a full set of results, obtained not only from its cooperating S-servers, but from the other W'-nodes as well. Tight collaboration means that the W'-nodes within a G group offer the data packs in the same order thus forming a separate subset of redundantly calculated data. When gathering the results in the tight collaboration mode the W'-nodes within each group exchange and compare results among each other to avoid calculation errors (incidental or intentional).

The users of public Internet services located at the S-servers may volunteer to the Comcute project and agree to load task code and packs of data to their computers. The S-servers may buffer task packets taken from the W'-node and distribute the task data along with ordinal service data. The task code is contained within original service web pages.

As the service web page is loaded to the Internet user computer, the user's web browser executes the task code loader (among the other web code) which reports the browser capabilities to the S-server. The server chooses which form of the task code to load (e.g. JavaScript, Java, Silverlight). As the loader receives the task and data, it launches the code execution and after it stops the loader sends the result back to the S-server.

The W'-node gathers the results of each data pack calculations reported by S-servers from the Internet computers. It then exchanges the results with the other W'-nodes. Thus each W'-node independently merges the partial results into a final result in accordance with the algorithm specified by the customer.

If a high level of reliability was requested, each node first compares partial results obtained from other nodes of the same G group in accordance with arbitration logic specified by the customer. If partial results differ, the nodes may repeat the calculation cycle. If there is a sufficient number of consistent partial results, W'-nodes aggregate them into a final result. If some nodes within the G group do not report partial results to the other members of the group, the operating W'-nodes try to invite and join new nodes from the whole pool of W-nodes (beyond W'-set). If not possible, they continue to operate at a lower level of reliability.

Once each G group completes calculations of their subset of data they return to the pool, ready to join other groups to help them to complete calculations. However, when there are no requests for help from other groups, they try to form a new group taking control over another subset of remaining data.

This way the calculations will complete as long as one W'-node is able to operate. Each operating W'-node completes the whole set of partial results and forms the final result. In the end it stores the result in its own repository and makes it available to the other nodes and subsequently to the customer.

3.3 Performance Factors

In the process described above the number of W'-nodes ($N_{W'}$) is the first factor of concurrency. It depends on the total number of W-nodes (N_W) and the mean W-node load factor L_W ($0 \leq L_W \leq 1$). This number is divided by the cardinality of the G group ($|G|$), which is 1 if there is no need to form the groups, and 2 or more if a higher level of reliability was requested. Concurrency may be degraded by a factor $0 \leq \kappa \leq 1$ dependent, among other things, on computation overlap between W' nodes or G groups.

A mean number of S-servers cooperating with each W'-node is the second main factor (N_S). The third factor is a mean number of people using each public S-service at the same time (N_I). Here we consider the level of readiness (R_L) i.e. how much the users are willing to participate in the Comcute system and share the calculation power of their computers. This level may be leveraged by a set of marketing means (e.g. free movies). Finally, the probability of calculation completing at each Internet user computer (P_C) depends on the mean time of single data pack calculation (T_P) and the mean time when a user remains connected to the service (T_S). The practical concurrency factor (C_F), defined as how much Comcute can speed up computations taking into account redundancy and willingness of Internet clients, can be estimated as:

$$C_F = \frac{N_{W'}}{|G|}(1 - \kappa)N_S N_I R_L P_C \tag{1}$$

$$|G| = \begin{cases} 1 & \text{- if there is no need for higher reliability} \\ \geq 2 & \text{- if a higher level of reliability was requested} \end{cases} \tag{2}$$

$$N_{W'} = N_W(1 - L_W) \quad P_C = \begin{cases} \frac{T_S}{T_P} & \text{- if } T_S < T_P \\ 1 & \text{- if } T_S \geq T_P \end{cases} \tag{3}$$

3.4   A Versatile Client Template

In order to test the Comcute system a versatile Internet client template, nicknamed *iRobot* was implemented. Internal structure of iRobot is illustrated in Fig. 2. Many instances of iRobots can be deployed simultaneously on hosting computers and controlled remotely. The remote control capability allows the operator to:

– switch each iRobot from a standby state to an active state,
– switch back from an active state to a standby mode,
– command every iRobot to complete its running task and exit.

In its active state, each iRobot loops a series of transitions: query S node for a task → execute the returned task → sleep. Each query of an S node is directed at the generic DNS address, which in turn gets resolved to a specific S node by a round robin load balancing algorithm, located in a DNS server. If tasks are available for execution, the S node queried will respond with a task implemented in a technology supported by iRobot instance, as determined by an availability of a iTask executor module. For this mechanism to work,

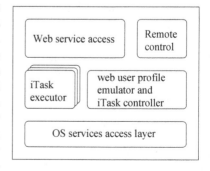

Fig. 2. *iRobot* web user emulator structure

every task query form an iRobot contains a JSON-encoded list of technologies supported by the iRobot. This guarantees that only tasks which can be executed by an iRobot will be sent to it.

In addition to a remote control capability, also parameters for tasks executed by iRobots (*iTasks*) can be supplied remotely from a central control location. Once supplied, these parameter sets are matched against task names being run, thus allowing for very flexible adaptation of the testing environment to different test patterns.

Once an iTask is started, its execution is supervised by the iTask controller. The controller is governed by a set of timing parameters, which determine the maximum execution time of a task T_e and a delay time T_d after execution is finished, before iRobot will query S node for another task. This allows iRobot to mimic a behavior of an average web user browsing the Web [17]. Both times are calculated using a general normal (Gaussian) distribution:

$$T_{\{e,d\}} = |f(x)| \quad \text{where} \quad f(x) = \frac{1}{\sigma}\phi\left(\frac{x-\mu}{\sigma}\right) \tag{4}$$

Both μ (mean) and σ (variance) parameters can be controlled by the test operator.

4   Experiments

We designed experiments in order to test the scalability of Comcute and obtain timelines of particular Internet clients in order to observe characteristics of processing and interaction with S servers.

4.1   Testbed Application and Configurations

For the following experiments, we used the client template described in Sect. 3.4 with the following parameters:

- probability of returning correct results by a volunteer equal to 1 – this allows comparison of adequate execution times of various configurations for scalability tests,
- processing of a data chunk by a volunteer equal to 10 s – this corresponds to values discussed in [17]. In summary, [17] states that clients often leave pages after 10–20 s and present probabilities of them doing so. From this perspective, 10 s seems adequate for our tests. It is also clear that the first 10 s are critical for the client to decide whether to stay on the page or leave.
- the size of the data chunk sent from Comcute to the client equal to 5000 bytes; this corresponds to input data such as a text fragment to search, definition of a subspace to search by the client, coefficients of a set of equations to solve etc.

– the size of results equal to 1000 bytes. In many aforementioned applications, results are smaller than the input data packets e.g. the following ones would correspond to the applications above: returning location in a text fragment or search subspace where matches have been found, solutions to a set of equations etc.

We assumed 10000 data chunks which gives sufficient granularity to balance the load among the numbers of Internet clients tested (up to 256). In fact, the system used $|G| = 2$ which means that Comcute created a copy of each data packet for a total number of packets equal to 20000.

The testbed code used on the client side contacts the S server access URL and is automatically redirected to a particular S server by a DNS system. The DNS system was modified so that the client can contact any of the S servers available using the round robin scheme. The client downloads a Java client code as a `jar` file. It is executed on the client side and fetches data packets from an S server as long as the data is available. Upon termination of processing, the client contacts the S server access URL again and repeats the procedure.

4.2 Testbed Environment

We used the following environment for tests: two W servers, each with 48 GB RAM 2 x Intel(R) Xeon(R) E5640 2,66 GHz CPUs (4 cores, 8 threads each) CentOS 6.2, four S servers, each with 24 GB RAM, 2 x Intel(R) Xeon(R) E5640 2,66 GHz CPUs (4 cores, 8 threads each) CentOS 6.2. Internet clients ran on a cluster of 8 nodes, each with 4 GB RAM, 2 x Intel(R) Xeon(TM) CPU 2.80 GHz CPUs (2 cores, 4 threads each).

We used Ethernet network connection between the components. For the sizes of the data packets, the startup time played a crucial role in the communication time which is much shorter than processing in this case anyway.

4.3 Simulation Results

Firstly, we aimed at assessment of the system scalability i.e. the ability of the system to decrease the execution time of a task of a given size with an increasing number of volunteers. Figure 3 presents the execution times we obtained for the aforementioned system parameters. Figure 4 shows the obtained speed up compared to theoretically ideal values. The latter is computed assuming all the data packets are processed sequentially on a single machine without communication. Consequently, these ideal values should be regarded as a theoretical upper bound that cannot be obtained in a distributed system. The system scales well for the tested numbers of volunteers up to 256. It should be noted that the ideal theoretical speed-up refers to the total number of data packets used i.e. 20000 in this case. On the other hand $C_F = 128$ for 256 clients because $|G| = 2$.

The following conclusions could be drawn from this experiment:

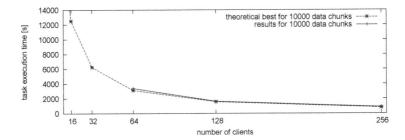

Fig. 3. Execution times of the testbed task

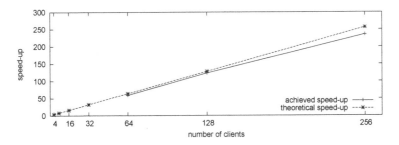

Fig. 4. Speed-up of the testbed task vs the number of Internet clients

1. The system scales well with volunteers not limited by the resources.
2. The practical limit on the number of volunteers tested per our cluster node is around 32. At this point with the number of volunteers higher than 32 per cluster node, we started to observe shortage of system resources for running the volunteers, mainly memory limitations.

We also obtained individual timelines of particular Internet clients to observe:

1. delay in taking up the task compared to other clients,
2. potential idle times during the task execution. This would indicate a temporary lack of data chunks on S servers; on the other hand constant supply of data chunks would indicate correct prefetching of data chunks by S servers from W servers.

Figure 5 presents a timeline for 64 Internet clients. Clients compute the task in parallel from about time step 100 s up to around 3500 s. It can be seen that usually the delays in starting processing the data are within 10–30 s. Before processing of data packets starts i.e. before time step 80 s, clients query S servers for computational codes. Around time steps 2300 and 3400 s, some S servers ran out of data packets (fetched from W servers) that caused delay in processing on the client side. Prefetching data from W servers is good to a certain degree because in case server S fails, W will need to wait before sending the lost packets to other S servers.

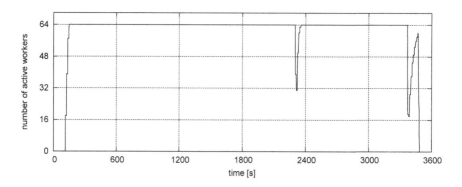

Fig. 5. Timeline for 64 Internet clients

5   Conclusions and Future Work

The main contribution of the Comcute system shown in this paper is its capability to balance between efficiency of concurrent calculations and reliability of volunteer computing. It is important as each web-open system is exposed to attacks. Comcute is resistant both to the attacks on the system itself (DDoS attacks) and results falsification.

Another contribution and the novelty of the system for Internet users is very simple usage (no need for installation) and ability to use many technologies within web browsers so this idea may be used by almost any web public service.

The experiments have proven that Comcute scales well. A slight difference between real and ideal concurrency factor results from Amdahl's law.

In the future, we want to extend the work to a higher number of client computers. Additionally, experiments with tasks with various data packet priorities and integration with workflow management in BeesyCluster [18] will be performed.

Acknowledgments. The work was performed within grant "Modeling efficiency, reliability and power consumption of multilevel parallel HPC systems using CPUs and GPUs" sponsored by and covered by funds from the National Science Center in Poland based on decision no DEC-2012/07/B/ST6/01516.

We would like to thank W. Korlub for his help in the environment configuration.

References

1. Czarnul, P., Grzeda, K.: Parallel simulations of electrophysiological phenomena in myocardium on large 32 and 64-bit linux clusters. In: Kranzlmüller, D., Kacsuk, P., Dongarra, J. (eds.) EuroPVM/MPI 2004. LNCS, vol. 3241, pp. 234–241. Springer, Heidelberg (2004)
2. Anderson, D.P.: Boinc: a system for public-resource computing and storage. In: Proceedings of 5th IEEE/ACM International Workshop on Grid Computing, Pittsburgh, USA (2004)

3. Balicki, J., Krawczyk, H., Nawarecki, E. (eds.): Grid and Volunteer Computing. Gdansk University of Technology, Faculty of Electronics, Telecommunication and Informatics Press, Gdansk (2012). ISBN: 978-83-60779-17-0

4. Kirk, D.B., Hwu, W.W.: Programming Massively Parallel Processors. A Hands-on Approach, 2nd edn. Morgan Kaufmann, San Francisco (2012). ISBN-13: 978–0124159921.

5. Sanders, J., Kandrot, E.: CUDA by Example: An Introduction to General-Purpose GPU Programming. Addison-Wesley Professional, Reading (2010). ISBN-13: 978–0131387683

6. Buyya, R. (ed.): High Performance Cluster Computing, Programming and Applications. Prentice Hall, Upper Saddle River (1999)

7. Wilkinson, B., Allen, M.: Parallel Programming: Techniques and Applications Using Networked Workstations and Parallel Computers. Prentice Hall, Upper Saddle River (1999)

8. Geist, A., Beguelin, A., Dongarra, J., Jiang, W., Mancheck, R., Sunderam, V.: PVM Parallel Virtual Machine. A Users Guide and Tutorial for Networked Parallel Computing. MIT Press, Cambridge (1994)

9. Karonis, N.T., Toonen, B., Foster, I.: Mpich-g2: a grid-enabled implementation of the message passing interface. J. Parallel Distrib. Comput. 63, 551–563 (2003). (Special Issue on Computational Grids)

10. Keller, R., Müller, M.: The Grid-Computing library PACX-MPI: Extending MPI for Computational Grids. www.hlrs.de/organization/amt/projects/pacx-mpi/

11. Garg, S.K., Buyya, R., Siegel, H.J.: Time and cost trade-off management for scheduling parallel applications on utility grids. Future Gener. Comput. Syst. 26, 1344–1355 (2010)

12. Chin, S.H., Suh, T., Yu, H.C.: Adaptive service scheduling for workflow applications in service-oriented grid. J. Supercomput. 52, 253–283 (2010)

13. Yu, J., Buyya, R., Ramamohanarao, K.: Workflow scheduling algorithms for grid computing. In: Xhafa, F., Abraham, A. (eds.) Meta. for Sched. in Distri. Comp. Envi. SCI, vol. 146, pp. 173–214. Springer, Heidelberg (2008)

14. Czarnul, P.: Integration of compute-intensive tasks into scientific workflows in BeesyCluster, In: Alexandrov, V.N., van Albada, G.D., Sloot, P.M.A., Dongarra, J. (eds.) ICCS 2006. LNCS, vol. 3993, pp. 944–947. Springer, Heidelberg (2006)

15. Cushing, R., Putra, G., Koulouzis, S., Belloum, A., Bubak, M., de Laat, C.: Distributed computing on an ensemble of browsers. IEEE Internet Comput. 17, 54–61 (2013)

16. Malawski, M., Kuzniar, M., Wojcik, P., Bubak, M.: How to use google app engine for free computing. IEEE Internet Comput. 17, 50–59 (2013)

17. Nielsen, J.: How long do users stay on web pages? Nielsen Norman Group (2011). http://www.nngroup.com/articles/how-long-do-users-stay-on-web-pages/

18. Czarnul, P.: Modeling, run-time optimization and execution of distributed workflow applications in the jee-based beesycluster environment. J. Supercomput. 63, 46–71 (2013)

Secure Storage and Processing of Confidential Data on Public Clouds

Jan Meizner[1]([✉]), Marian Bubak[2,3], Maciej Malawski[2], and Piotr Nowakowski[1]

[1] ACC Cyfronet AGH, AGH University of Science and Technology, Krakow, Poland
jan.meizner@cyfronet.pl
[2] Department of Computer Science, AGH University of Science and Technology,
Krakow, Poland
[3] Informatics Institute, University of Amsterdam, Amsterdam, The Netherlands

Abstract. The goal of this paper is to describe problems associated with storage and processing of confidential data in public clouds, and to propose relevant mitigation strategies. In our opinion many types of data in the commercial and scientific worlds require special attention to protect them against possible mishandling. This issue affects highly valuable data, such as trade secrets and financial information, as well as personal data including medical records used in scientific research. We analyse situations which require special care and next we propose a set of solutions for ensuring data security and describe feasibility studies based on tests performed using popular cryptographic software (OpenSSL). Those solutions would allow to fullfil objectives of the paper which are to analyze the requirements of scientific software regarding protection of confidential data, the nature of the data itself and threats against those assets.

Keywords: Security · Data · Clouds · Hybrid · AES · 3DES · OpenSSL

1 Introduction

The most commonly mentioned drawbacks of cloud computing (especially in public clouds) are related to security and trust. Due to the nature of such infrastructures users have limited control over data being processed and stored by the system. As a result, multiple potential risks arise. These risks may be caused by:

1. use of vulnerable software,
2. use of insecure infrastructures, vulnerable to eavesdropping during transmission, (D)DOS attacks etc.,
3. mishandling of data storage devices (e.g. turning over HDDs with potentially recoverable client data to an unverified third party), which becomes especially critical in large, complex storage systems where the user might not be allowed to physically overwrite data,

R. Wyrzykowski et al. (Eds.): PPAM 2013, Part I, LNCS 8384, pp. 272–282, 2014.
DOI: 10.1007/978-3-642-55224-3_26, © Springer-Verlag Berlin Heidelberg 2014

4. human error leading to infrastructures being unintentionally compromised (including social engineering),
5. malicious actions taken by the provider against the user (economic espionage).

Those risks are critical if the data being processed is confidential. There are multiple use cases that could benefit from cloud computing, but are (or might be) restricted due to security issues. This applies both to the commercial world (e.g. financial institutions) as well as to research (e.g. clinical studies involving unanonymized medical records). A good example of a research project which requires special provisions for handling sensitive medical data is the VPH-Share initiative [1]. Such data must be strictly protected at every step - this includes processing, storage and ensuring that the data can be provably erased once no longer needed. Of course, any custom infrastructure may suffer issues similar to those described above; however some security aspects remain unique to public services, including clouds. A notable example is the inability to separate a single storage device (disk drive) from an array so that it is not shared with other users and can be physically destroyed (e.g. demagnetized) when the owner has no further need for its contents. Such an action would be certainly possible in the case of e.g. a standalone server (even one collocated at a third-party data center), but not in the shared environment of virtualized cloud hardware.

On the most basic level we can categorize clouds based on ownership:

1. public - owned by an external entity (such as a commercial provider) and made available to any interested party in exchange for payment,
2. community - owned by any organization and made available to a specific group of users, such as a collaboration of scientists (physicists, clinicians, etc.),
3. private - owned and used by a single entity (in-house infrastructure), yet providing greater manageability and flexibility than in the case of a legacy infrastructure.

Clearly, data security aspects depend on the category in question. Public clouds are open to all clients - including potential intruders who, by gaining (legal) access to the infrastructure, might try to exploit its vulnerabilities [2] to maliciously escalate privileges e.g. by:

1. eavesdropping on traffic due to insufficient network separation,
2. attempting to defeat hypervisor isolation by exploiting some vulnerability (e.g. CVE-2012-0217 [3]),
3. disrupting other cloud users either by generating local load affecting other VMs (such as high I/O load) or by exploiting DoS vulnerabilities against the host (for example allowing a guest VM to crash the host/hypervisor - e.g. CVE-2012-6030 [4]).

Additionally (with some exceptions described later on), communication between the user and the public cloud is carried over the public Internet, which might also affect security. Similar issues may arise in relation to community clouds, although "invited" users are less likely to be hostile. It is also easier to maintain traffic separation between partners in a community cloud by combining VPNs (e.g. IPSec, OpenVPN, MPLS) - which may also be offered by public cloud providers - with the benefits of a dedicated network infrastructure (whether physical or virtual). Finally, the private cloud infrastructure is, by its nature, exclusive to the entity which owns it. This limits the number of potential intruders and enables traffic to be restricted to the enterprise LAN without extra effort, assuming all branches which host parts of the cloud are already securely interconnected.

The final key aspect of security relates to actions performed on the data. If the data needs to be stored in a public/community cloud but can be processed in-house, it is easier to ensure its security e.g. through encryption. However, if the data is processed in a potentially hostile environment, additional problems arise, since, at a minimum, some portion of data must be temporarily decrypted. We will show how such problems can be mitigated.

The main scientific objective of this paper is to analyze the requirements of scientific software regarding protection of confidential data as well as the nature of the data itself, determine threats against those assets stemming from the use of clouds (especially public ones) and, finally, propose and validate a suitable mitigation strategy.

The structure of the paper is as follows. Section 2 presents policies and technical solutions applicable to public and private clouds in the context of data security. Section 3 shows a possible solution that might be constructed on the basis of this research. Section 4 presents the validation of the presented solution including efficiency testing. Section 5 contains conclusions and closing remarks.

2   Related Work

As already described, public clouds are, by their nature, highly prone to attacks or other actions leading to misuse of confidential information. As such, it becomes very important to analyze all third-party audits confirming quality of the infrastructure as well as any solutions offered by the provider to enhance cloud platform security. In this section we will discuss these certificates and technologies on the example of the Amazon cloud which is one of the leading public cloud platforms. In addition to that we will also mention solutions offered by Rackspace which is another world-wide big player in this field.

Amazon provides a dedicated page [5] describing security standards and certificates the cloud is in compliance with. These have been listed in Table 1.

In addition, Amazon defines a set of procedures aimed at enhancing platform security. This includes a set of user rights preventing AWS users from initiating any potentially dangerous actions, which, in combination with monitoring, should ensure fair use of the infrastructure and prevent attacks, whether internal or external, e.g. by blocking the malicious user or firewalling suspicious external parties. Those procedures are so strict that even legitimate penetration testing needs to be reported to Amazon [6] in order to receive permission (and not

Table 1. Security standards and certificates with which Amazon claims compliance.

Standard	Description
SOC 1/SSAE 16/ISAE 3402	SOC 1 report published by Amazon, documenting internal audits based on SSAE 16 and ISAE 3402 standards
SOC 2	Amazon report defining practices relevant to basic security aspects (such as confidentiality, integrity and privacy) of customer data
FISMA, DIACAP, FedRAMP	Amazon declares that its infrastructure enables US agencies to remain compliant with the Federal Information Security Management Act (FISMA) and that numerous organizations have positively passed the Assurance Certification and Accreditation Process (DIACAP) as defined by NIST and US DoD using the Amazon cloud. Finally, Amazon claims to be working with a third-party auditing company to ensure compliance with the Federal Risk and Authorization Management Program (FedRAMP)
PCI DSS Level 1	Amazon declares that it has been validated as compliant with the Payment Card Industry (PCI) Data Security Standard (DSS) which certifies that the infrastructure is ready for storage and processing of credit card information
ISO 27001	Global, periodic audit-based security standard granted to Amazon
International Traffic In Arms Compliance	Related to Amazon GovCloud dedicated to US Government usage (not commercial or open) - confirms (through a third-party audit) that Amazon is capable of restricting access to specific personnel (in this case, US residents) as well as a specific physical location (US territory).
FIPS 140-2	(US GovCloud only) Confirms high-quality encryption mechanism in use (required for confidential/secret data processing)

be blocked). There is also a dedicated contact point for software vulnerability reporting [7] with the promised reaction time (SLA) of 24 h, which seems quite acceptable.

Reckspace also offers multiple informations related to security both in general and specifically related to data on their website [8]. This information describes both formal procedures such as certifications and auditing as well as technical solutions. On the technical side Rackspace offers interesting hybrid solutions, in contrast to Amazon which at present focuses strictly on pure cloud services (with possibility to use VPC with in-house or third party dedicated servers). Those includes abilities to interconnect public IaaS services with dedicated servers hosted in Rackspace on dedicated and separated hardware both in legacy non-cloud mode as well as private cloud (based on OpenStack). The solution is acompanied with additional glue services such as firewalls (traditional and web app), dedicated secured links (such as site-to-site VPN - branded as Rackconnect service), Intrusion Detection Systems etc.

Table 2. Cryptographic libraries that can be potentially used to provide server-side encryption and tools for end users. Certificate numbers are provided for FIPS validated implementations, where available.

Name	Description	FIPS Cert Nr
Libraries		
OpenSSL	Well known advanced free cryptographic library including symmetric/asymmetric enryption	#1747, #1758
Network Security Services (NSS)	Another cross-platform, open source (MPL/GPL/LGPL) library	#1837
GnuTLS	Most commonly used LGPL alternative to OpenSSL (which carries a free license but is incompatible with GPL)	N/A
libgcrypt	Simple C library developed as part of GnuPG	#1757
Bouncy Castle	Cryptographic library for Java and C#	N/A
Tools		
OpenSSL Client Tools	In addition to the core library, OpenSSL includes tools that can be used to generate cryptographic material (keys) as well as perform data encryption/decryption/signing/verification	N/A
GnuPG	Offers solution for signing and encrypting data, comply used to protect e-mail messages and applicable to any arbitrary data	N/A
LUKS [13] and dm-crypt [14]	Manages encrypted disks (including real and loopback devices) preventing access to data without a valid key when the disk is unmounted	#1933 (for dm-crypt)

Most of the security arrangements in an IaaS cloud (going beyond simple credential/key injection) need to be taken care of by the user. However, some providers offer solutions that could assist the users in this respect. One such solution is the AWS Virtual Private Cloud [9]. It enables users to aggregate instances and connect them using a virtual network which can be separated from the Internet, with the ability to control aspects such as firewalls. It is also possible to interconnect a corporate network with the VPC using IPSec VPNs and BGP dynamic routing protocol. Another solution offered by Amazon is built-in encryption for S3. Users can set a flag ensuring that data moved to S3 is encrypted on the provider's end [10]. Encryption may also be handled on the client side - Amazon provides a suitable helper library as part of its Java SDK [11].

Similarly to public clouds, private infrastructures may also implement solutions assisting users in protecting their data on the server/provider side, or on the client side. Examples from both groups are described below.

There are multiple cryptographic libraries that could potentially be used to provide server-side encryption, as well as tools suitable for end users. Some notable examples are listed in Table 2. Where FIPS 140-2 validated implementations [12] are available, their numbers are also listed.

This section shows that service providers offer a wide range of security-related assurances, and that specific solutions such as server-side encryption are also offered. Even so, some scientific use cases (e.g. in the previously mentioned medical domain) may require an independent cryptographic solution. To support this requirement we propose a mechanism that could be deployed in a secure (custom/third-party) infrastructure or directly applied by the end user. Users who are not comfortable with trusting external providers should benefit from a centralized service based on our solution and deployed by an organization they're more likely to trust (such as a partner involved in a collaborative project, or even their own organization). Similar solutions used to verify computation done on untrusted infrastructure with singe trusted service [15] have been described in literature. If, however, the end user decides not to trust anyone else, they can still benefit from the results presented below when choosing tools suitable for custom end-to-end encryption/decryption mechanisms.

3   New Approaches to Secure Data Handling in Clouds

Even moderately confidential data can be stored in both private and public infrastructures, given the use of strong cryptographic mechanisms such as those based on the AES encryption schema which is considered by NIST to be strong enough for protection of classified information (AES-128 up to SECRET level, AES-192 and AES-256 up to TOP SECRET). The key used for the decryption process needs to be stored securely (outside of the public cloud). In this scenario even if provider trust is abused, data should not be compromised due to the protection afforded by the encryption mechanism.

As mentioned above, encryption is the most basic mechanism for securing data over which the user does not have full control, and must therefore assume that stored data might be leaked. Below we suggest three possible solutions, balancing the required level of trust with the implied user/organizational effort:

1. provider-level encryption - requires very little or no effort on the part of the user, however, the user also needs to trust the provider in that encryption will be performed properly and no unencrypted data will be stored. This situation is, of course, preferable to complete lack of server-side encryption which might leave data unencrypted for an indefinite amount of time (possibly even after its removal is requested, due to the previously mentioned issues with erasing data from cloud storage),

2. organization-level encryption, as shown in Fig. 1 - this assumes that users do not access the cloud directly; rather they're expected to use middleware provided by their organization (e.g. scientific institution or business) built on top of cloud resources leased from a single provider or multiple providers (possibly both public and private, such as in a hybrid cloud). In this model data is encrypted prior to being dispatched to the public cloud provider (who

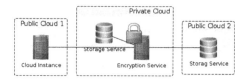

Fig. 1. Encryption service usable by end users to secure data before they are stored in a public cloud.

therefore does not need to be trusted), yet there is a new entity (organization) that has to be trusted instead. In this scenario, just like in the preceding one, the user does not have to personally handle data encryption / decryption,
3. client-level encryption - in this case data encryption and decryption is done using tools controlled exclusively by the end user (such as his/her PC). This precludes the need to trust external organizations or companies but, of course, requires the user to manually handle the encryption process, which might be too complex (especially for non-IT users) and, in any case, might consume a considerable amount of local resources, especially if the user is processing a large volume of data.

It is also possible to store chunks of data (preferably fragmented in such a way as to be unrecoverable - e.g. every n-th pixel of every image - providing n is sufficiently large) on different public clouds. This would permit storage of unencrypted data in a form that remains moderately secure if the attacker cannot compromise multiple locations. Although this solution alone would most likely not be sufficient to protect confidential data, it might be applied in addition to encryption to further reduce the odds of data being compromised.

The data processing problem cannot be completely solved assuming a lack of trust for the provider as the data needs to be decrypted prior to processing. As a result, a malicious entity, either internal or external to the provider, given access to the machine, may hijack any decrypted data and/or obtain the security keys and use them to decrypt data. The problem can, however, be mitigated in some situations by:

1. passing the key along with the request (through a secure channel), store it only in memory and purge it when no longer needed. It could then be used to decrypt chunks of input data on the fly; Intermediate results should also be kept in memory and sanitized. Finally, results should be encrypted prior to being stored on any potentially unsecured storage device,
2. streaming it through a secure channel to a service in a private infrastructure, which would be considered trusted and might also be responsible for immediate encryption of data with a symmetric key (not stored in the public cloud). It might also be possible for the service to provide a protocol that could be mounted e.g. using FUSE [16] and, by emulating a regular filesystem, allow seamless access to data - such as in the case of a regular hard drive,
3. performing calculations on a selection of data (and recombining results as a final step), then encrypting different chunks of data with different keys

and possibly store them in separate locations (note, however, that this is not mandatory - indeed, it is sufficient to ensure that a single instance has access only to part of the data). In this case each instance performs computations on a subset of data (whose theft would not be as critical as in case of a complete data set). Partial results can be treated as described above (point 1 or 2), while the final part of the computation can be performed on trusted resources (such as a private cloud) with access to all intermediate results.

There are many issues that could prevent the data from being permanently erased. This includes possible (usually undisclosed) optimization in the cloud middleware layer or hardware components (such as storage arrays) that would preclude the ability to overwrite data. In addition, certain magnetic storage devices might allow retrieval of data even after it has been overwritten.

For the reasons stated above we should conclude that the only way to prevent exposure of data to unauthorized entities is to use strong encryption mechanisms before data is written to any non-volatile media. Then, by destroying the key, we would be able to ensure, with a high degree of certainty, that no data can be recovered in a realistic timeframe. The key itself needs to be stored in the private cloud, preferably on a non-magnetic device (such as flash or RAM) which would allow secure erasing.

Another possible solution is to avoid writing data to any permanent storage altogether and simply keep it in memory. It is generally inadvisable to store critical data in RAM as it could easily be destroyed by accident; however, this solution might become viable given sufficient redundancy. Multiple copies of data could be maintained at different physical locations so that even a large-scale disaster at a single site would not bring down the entire infrastructure. Instead of plain data copies more advanced structures, such as hashes, might be constructed, similarly to methods used by more advanced RAID levels (like 5 or 6); although such a mechanism would have to take into account the increased probability of failure of a given instance of volatile memory compared to a single drive in a RAID array. The obvious merits of this solutions include (in addition to efficiency) ease of sanitization and near-zero risk of data being retrieved from decommissioned media. It is worth noting that while the theoretical refresh time for modern DRAM is in the millisecond range, practical studies mentioned in [17] have shown that this type of memory might potentially maintain coherence for several seconds in normal temperature, and up to several minutes in cryogenic conditions. Fortunately, such conditions should not be expected to occur under any normal circumstances (i.e. by accident) and would involve severe breaches of cloud provider security (as the attacker would need to be able to freeze the DRAM modules shortly after power is cut, i.e. at the data center itself).

4   Validation

The goal of this section is to validate the feasibility of encryption as a data protection measure for public cloud computations using confidential data, as well as determining the best cipher to use for this purpose. From the list of possible tools and libraries described above we have chosen OpenSSL (version 1.0.1) and GnuPG (version 1.4.11) as the most popular, reliable and well optimized pieces of software available both as a command-line tool that can be applied directly by end users. During our test we first tested the I/O performance of the used storage device (ramdisk) first, and then the encryption time for different ciphers and block sizes. The following ciphers where used:

1. 3DES in CBC mode (des-ede3-cbc) - as an example of a legacy symmetric cipher which still enjoys significant popularity,
2. AES-128 in CBC mode (aes-128-cbc) - as the most basic version of the currently suggested algorithm,
3. AES-192 in CBC mode (aes-192-cbc) - as a more advanced variant of AES,
4. AES-256 in CBC mode (aes-256-cbc) - as the strongest AES variant available.

All calculations where performed on chunks containing 10 MB, 100 MB and 1000 MB of pseudo-random data (generated using /dev/urandom under Linux). Each measurement was performed 10 times, with the average values and standard deviations calculated for each sample. All tests were ran on a standard PC equipped with Intel Core i7 CPU model 860 (4 cores, 8 threads, 2.80 GHz) and 16 GB of memory (DDR3). The test system used Linux Ubuntu 12.04.1 LTS. A ramdisk was used for storage of input and output (encrypted) data to minimize read/write overhead. Additionally, to obtain more accurate measurements of the

Table 3. The time needed to perform I/O operations (for a 1 GB sample file) on the ramdisk used for storage of input and output data and the one needed to encrypt chunks of data (10 MB, 100 MB and 1000 MB) using 3DES and different variations of AES ciphers for OpenSSL and GnuPG.

	I/O operations measurement			
	Avg. Time (s)	Std. dev.	Transfer (MB/s)	
Read (1 GB chunk)	0.1432	0.0028	6981	
Write (1 GB chunk)	0.3696	0.0061	2706	
	OpenSSL		**GnuPG**	
	Avg. Time (s)	Std. dev.	Avg. Time (s)	Std. dev.
3DES - 10 MB	0.412	0.004	0.724	0.003
3DES - 100 MB	3.977	0.017	6.905	0.026
3DES - 1000 MB	39.810	0.275	68.760	0.271
AES 128 - 10 MB	0.049	0.004	0.370	0.025
AES 128 - 100 MB	0.385	0.005	3.425	0.005
AES 128 - 1000 MB	3.722	0.030	34.095	0.099
AES 192 - 10 MB	0.058	0.003	0.370	0.004
AES 192 - 100 MB	0.439	0.004	3.526	0.013
AES 192 - 1000 MB	4.296	0.050	35.198	0.202
AES 256 - 10 MB	0.063	0.003	0.380	0.006
AES 256 - 100 MB	0.505	0.004	3.631	0.014
AES 256 - 1000 MB	4.906	0.038	36.088	0.099

encryption time, I/O efficiency tests were performed for the above mentioned drive, as shown in Table 3. This table also presents average values and standard deviations for different data chunk sizes and ciphers.

As can be expected, encryption introduced considerable overhead - on the order of 10–100 times the baseline value. However, if the proper cipher and tool is applied the solution appears sufficient for many scientific use cases, even when large volumes of data (like medical images) need to be processed. Even with a very strong cipher (AES-256) it is still possible to achieve throughputs of ca. 200 MB/s, which is greater than typical network bandwidth available to most cloud instances (up to 1 GBit/s = 128 MB/s), as well as typical consumer HDDs. While insufficient for data-intensive HPC, these solutions would still solve many scientific problems up to the terabyte range (e.g. processing multiple medical images). It's important to mention that more advanced ciphers might be handled much more efficiently than legacy ones: for instance, AES (using OpenSSL implementation) is approximately 10 times faster than 3DES. The relatively small differences between AES variants suggest that even for large blocks of data AES-256 might be preferable to weaker ciphers. Finally, the results above clearly have shown that choice of the right implementation is critical as GnuPG performed not so good as OpenSSL. The AES implementation in the selected version of GnuPG seams to be poorly optimised.

5   Conclusions and Future Work

In this paper we described problems related to the storage and processing of confidential data in public clouds. We have shown possible ways to mitigate those problems and concluded that the use of encryption appears to be the most promising strategy. Consequently, we have decided to analyze and suggest possible software modules that might be used for data encryption and decryption, and propose potential architectures which strike a balance between the implied trust in third-party providers and the effort needed on the part of the user. Finally, we have measured and analyzed various encryption mechanisms using sample data and different ciphers offered by the OpenSSL library and GnuPG tool as a representative examples of the proposed solutions. We conclude that the proposed solution is technically feasible even in conjunction with a strong cryptographic cipher (AES-256) providing that the right choice of used tools has been made. Additionally, we have shown that selecting right tool is critical so the described benchmark should be performed on a target system to choose the best configuration. We plan to perform more in-depth testing of the available solutions including tests involving AES-NI hardware accelerator mechanisms available in modern CPUs.

Acknowledgments. This research was partially funded by EC ICT VPH-Share Project (contract no. 269978) and the corresponding KI AGH grant. The authors wish to thank Dario Ruiz Lopez and Dmitry A. Vasunin for their helpful advice.

References

1. VPH-Share. http://vph-share.org/
2. Ristenpart, T., Tromer, E., Shacham, H., Savage, S.: Hey, you, get off of my cloud: exploring information leakage in third-party compute clouds. In: Proceedings of the 16th ACM conference on Computer and Communications Security (CCS '09), pp. 199–212. ACM, New York (2009)
3. Vulnerability Summary for CVE-2012-0217, National Vulnerability Database, NIST. http://web.nvd.nist.gov/view/vuln/detail?vulnId=CVE-2012-0217 (2012)
4. Vulnerability Summary for CVE-2012-6030, National Vulnerability Database, NIST. http://web.nvd.nist.gov/view/vuln/detail?vulnId=CVE-2012-6030 (2012)
5. AWS Security and Compliance Center. https://aws.amazon.com/security/
6. Penetration Testing. https://aws.amazon.com/security/penetration-testing/
7. Vulnerability Reporting.
   https://aws.amazon.com/security/vulnerability-reporting/
8. Security in Rackspace. http://www.rackspace.com/security/
9. Amazon VPC. http://aws.amazon.com/vpc/
10. Using Data Encryption. http://docs.aws.amazon.com/AmazonS3/latest/dev/UsingEncryption.html
11. Client-Side Data Encryption with the AWS SDK for Java and Amazon S3. http://aws.amazon.com/articles/2850096021478074
12. Validated FIPS 140-1 and FIPS 140-2 Cryptographic Modules. http://csrc.nist.gov/groups/STM/cmvp/documents/140-1/140val-all.htm
13. LUKS: Linux Unified Key Setup. http://code.google.com/p/cryptsetup/
14. dm-crypt: Linux kernel device-mapper crypto target. http://code.google.com/p/cryptsetup/wiki/DMCrypt
15. Canetti, R., Riva, B., Rothblum, G.N.: Refereed delegation of computation. Inf. Comput. **226**, 16–36 (2013). ISSN: 0890-5401
16. FUSE: Filesystem in Userspace. http://fuse.sourceforge.net/
17. Alex Halderman, J., Schoen, S.D., Heninger, N., Clarkson, W., Paul, W., Calandrino, J.A., Feldman, A.J., Appelbaum, J., Felten, E.W.: Lest we remember: cold boot attacks on encryption keys. In: Proceedings of the 2008 USENIX Security Symposium, 21 February 2008

Efficient Service Delivery in Complex Heterogeneous and Distributed Environment

Mariusz Fras and Jan Kwiatkowski[(✉)]

Institute of Informatics, Wroclaw University of Technology,
Wybrzeze Wyspianskiego 27, 50-370 Wroclaw, Poland
{mariusz.fras,jan.kwiatkowski}@pwr.wroc.pl

Abstract. The problem of providing quality of service (QoS) guarantees is studied in many areas of information technologies. For network services three attributes are directly related to everyday perception of the QoS by the end user: availability, usability, and performance. The paper focuses on performance issues of service delivery with use of virtualization of services and processing resources. There are presented general issues of efficient service delivery, and proposed solutions with different formulation of guaranties for service processing. The best effort and SLA-based approaches are considered. The selected aspects of utilizing processing resources virtualization are also discussed.

Keywords: Quality of services · Service request distribution · Virtualization

1 Introduction

The quality of service delivery is one of the very current challenges for most of the service providers. There are three attributes that directly relate to everyday perception of the quality of service for the end user: availability, usability, and performance [6]. Particularly for the last one it is very difficult to fulfil sufficient values of non-functional parameters. There are a number of metrics that are used during service delivery evaluation process. However, there is a common agreement that an essential non-functional parameter for the assessment of quality of service delivery is the response time. When services are offered using global heterogeneous distributed system the response time consists of two components, request processing time on the server where the service is offered and the data transfer time.

To improve the quality of service delivery the common used solution is redundancy of offered services (e.g. CDN solutions) and service request distribution. In the context of Web services useful standards for quality of service mechanisms, such as WS-* and others have been proposed. For example in the work [8] a comprehensive overall infrastructure to guarantee SLA (Service Level Agreement) for services has been proposed, including general scheme of the runtime environment, specifications and procedures for handling requests, and measurement of services.

R. Wyrzykowski et al. (Eds.): PPAM 2013, Part I, LNCS 8384, pp. 283–292, 2014.
DOI: 10.1007/978-3-642-55224-3_27, © Springer-Verlag Berlin Heidelberg 2014

The paper presents the architecture of the Virtual Service Delivery System (VSDS) that uses virtualization of services and virtualization of processing resources for efficient processing of service requests. There are presented general issues of service delivery in the considered environment, and proposed solutions with different formulation of guaranties - best effort as well as SLA-based.

The paper is organized as follows. Section 2 briefly describes the main ideas used during designing and developing of the VSDS. The service request control is presented in Sect. 3. In the Sect. 4 procedures related to allocation of execution resources to service request are discussed. Finally, Sect. 5 outlines the work and discusses ongoing work.

2   The General Concept of Virtual Service Delivery

The concept of quality-aware service delivery is based on the idea of virtualization of services and resources and distribution of client's requests according to actual service instance non-functional parameters. The real services are hidden from client point of view. The client deals with virtual service (so called atomic service) that can be executed (by so called service instances) at different locations. The main components of the system called Virtual Service Delivery System (VSDS) are network service broker (further called Broker) and Virtual Resource Manager (VRM). The Broker advertises virtual services and handles client request for services. The main assumptions for operation of both modules are as follows:

- the Broker delivers to clients the set of J atomic services as_j, $j \in [1, J]$,
- the Broker knows execution systems es_m, $m \in [1, M]$, where real services (service instances $is_{j,m}$) are available,
- the Broker acts as a service proxy - distribute client requests for services to proper instances according to some distribution policy,
- the Virtual Resource Manager is responsible for service execution,
- the Virtual Resource Manager is responsible for management of processing resources (virtualization) and service instances (creation, maintenance and removing).

The Broker implements the Virtual Service Layer (Fig. 1). The VRM manages virtualized computational resources. Both layers are defined as the tuple $<ES, CL, AS, IS>$, where: $ES = \{es_1, ..., es_m, ..., es_M\}$ is the set of execution systems es_m, $CL = \{cl_1, ..., cl_m, ..., cl_M\}$ is the set of communication links cl_m from the Broker to execution systems, $AS = \{as_1, ..., as_j, ..., as_J\}$ is the set of J atomic services, $IS = \{IS_1, ..., IS_j, ..., IS_J\}$ is the set of all instances of services, where: IS_j - the subset of all instances of service as_j, $is_{j,m}$ - the m-th instance of j-th service as_j localized in given execution system es_m.

Both virtual services and virtual resources can be flexible managed to improve system efficiency as well as to support client service tasks. This is achieved with use of virtualization techniques and building the system in accordance with SOA paradigm. The real resources are available on virtual servers (execution systems) that run service instances, and can be instantiated and tuned by VRM (Fig. 2). The execution systems and instances are registered in the Broker. The Broker maps atomic service to instances of that service. From now registered services can by requested by clients.

The information about services, execution systems, values of its parameter, and values of parameters of request processing are collected. Using this data, and functionalities of system components, the quality-aware request distribution according to different algorithms is possible.

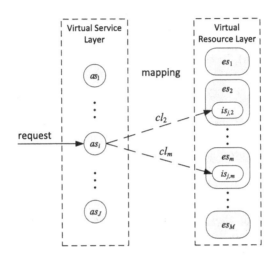

Fig. 1. The layers of Virtual Service Delivery System

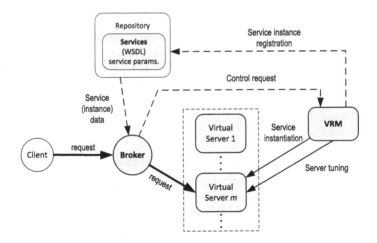

Fig. 2. Virtual service and resource management

3   Service Request Control

The quality of network services can be variously defined and different approaches to this problem can be proposed according to problem formulation. One of the often used approaches is to perform such request handling that quality of user perception of service processing on the basis of non-functional service parameters is achieved.

The instances of given atomic service are functionally the same and differ only in values of non-functional parameters $\{\psi^1(is_{j,m}), ..., \psi^f(is_{j,m}), ..., \psi^F(is_{j,m})\}$, where $\psi^f(is_{j,m})$ is f-th non-functional parameter of m-th instance of j-th atomic service. The service parameters can be static - constant in long period of time (e.g. service price), or dynamic - variable in short period of time (e.g. the completion time of execution). In context of static parameters different methods based on classic approaches (e.g. integer programming) are proposed, e.g. [9]. From the client point of view, the very sensitive parameter is service response time, which is often very dynamic parameter.

To satisfy proper values of service parameters distribution of the request to proper service instance must be performed. The problem of service request distribution can be stated using criterion function Q:

$$is_{j,m^*} \leftarrow \arg_m Q(\psi^1(is_{j,m}), ..., \psi^f(is_{j,m}), ..., \psi^F(is_{j,m})) \tag{1}$$

Later in this work the problem of dynamic parameters is mainly considered, and proposed algorithms are formulated for guaranties for response time, however the general description is sometimes used.

3.1   Local and Global Distribution

Taking into account localization of system components two basic cases can be distinguished. The Broker can be localized in the service processing center, where execution systems are run (so called the local Broker), or near the client, and execution systems are accessible via internet links (the global Broker). From the service request control point of view the basic difference is to take into account specific impact of data transfer via internet links on values of service parameters. The main factors are:

– the data transfer time can be significant in overall cost of service processing,
– the estimation of values of certain service parameters is a challenge,
– usually the Internet communication links are out of hand i.e. there is lack of guarantees of constant link performance, especially available bandwidth.

Another issue is that local Broker assure the full knowledge about all processed requests. All above determine how to perform service request control.

The best effort approach can be adopted in both cases. Considering only dynamic parameters the incoming request to the Broker at the moment n is distributed to selected service instance $is_{j,m}$ using criterion function of forecasted or calculated values of service instance parameters:

$$is_{j,m^*} \leftarrow \arg_m Q(\hat{\psi}_n^1(is_{j,m}), ..., \hat{\psi}_n^f(is_{j,m}), ..., \hat{\psi}_n^F(is_{j,m})) \qquad (2)$$

where $\hat{\psi}_n^f(is_{j,m})$ is forecasted or calculated value of f-th parameter of instance $is_{j,m}$ at the moment n. An example of such dynamic parameter is request processing time which depends on server load.

Calculation of values of service parameters is possible under some consideration. E.g. system resources must be dedicated only for requests known for system control unit (here the Broker), characteristic of resource consumption must be known, etc. Forecasting the values of parameters can be performed in various ways. The common techniques used are time series analysis and methods based on artificial intelligence approaches.

For service parameters that should be minimized, such as usually most important response time, fuzzy neural forecasting is solid solution [1,3]. Each service instance is modelled as two stage fuzzy-neural network described in detail in [3]. An output of the model is forecasted value of parameter of the instance $is_{j,m}$ at the moment n (here request processing time $\hat{t}_{Pj,m}$ in execution system), on the basis of two input values (Fig. 3a). One input value is the number of serviced requests l_n. The second input is execution system state indicator s_n. The indicator may be constructed variously to best reflect the system state. E.g. for computational tasks the processors load can be used, for intensive file data processing tasks the number of I/O operations can be used, or any combination of these and other suitable parameters. In case of global Broker the model of communication link for each instance is used. The transfer time $\hat{t}_{Tj,m}$ is derived with use of link load ll_n and TCP Connect time $t_{TCPC,n}$ inputs (Fig. 3b).

Both, calculation and forecasting need monitoring of values of service parameters and execution system state. For processing and data transfer times forecasting, the real values of previous requests are used. The local Broker can use data received from monitoring on execution system level. The global Broker can derive estimation of this values with use of the monitoring of TCP session which handles service request to the processing system [2].

The SLA-based distribution with use of integrated services (IntServ) and differential services (DiffServ) approaches depends on several assumptions about the processing environment and processing schema. As mentioned before, the data transfer control over the Internet links is difficult. Some task of other than best effort approaches can be used only with dedicated links, what is rather rare

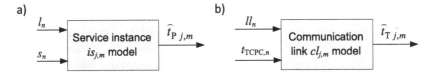

Fig. 3. Model for forecasting of values of service parameters

in the Internet environment. In terms of execution systems' resources the specific control is possible in two cases:

- the local Broker is used,
- the system serves client requests only via registration in global Broker.

The presented VSDS architecture supports this two cases and the request distribution can be performed according to utilization of dedicated execution resources. Hereafter, it is assumed that communication between the Broker and execution systems is not the bottleneck and the transfer time is not considered separately. To simplify further reading the index of the moment n is omitted later on.

3.2   SLA-Based Distribution

Considering request service manner two general cases of request processing can be distinguished. In the case of one-by-one request processing by execution system the well-known queuing methods can be used. In the case of parallel processing requests by one execution system (typical for web services) and running many service instances, the handling of service request can be performed with aid of specific parameters characterizing processing.

For both IntServ and DiffServ approaches resource reservation is required. For the designed method of response time evaluation one or two parameters characterizing system usage can be taken under consideration: execution system state indicator and especially number of requests being processed at given moment. Many studies show that response time of typical web service server increases slightly to a certain (characteristic for that server) request limit, and above this value the increase is much more rapid. This limit value l_m^* for every m-th execution system determine DiffServ and IntServ algorithms thresholds.

For IntServ approach it is assumed that each incoming request clearly indicates the requirements for service non-functional parameters $req(\psi(as_j))$. Considering response time of service as_j requirement $req(t_j)$ the request distribution that saves system resources for future requests, selects the worse instance guaranteeing proper processing, i.e.:

$$is_{j,m^*} \leftarrow \arg \max_m \hat{t}_{j,m} \tag{3}$$

with respect to $\hat{t}_{j,m} < req(t_j)$.

For DiffServ approach the distribution algorithms are formulated depending on the assumptions about guaranties for request processing. The simplest solution is for assumption that the request requirements are met when the delivery system is not overloaded in the sense that the number of serviced requests l_m is less or equal the limit value l_m^* for every execution system m. Let $C = \{c_1, ..., c_k, ..., c_K\}$ be the set of K classes of served requests. Assume that request priority of class c_k is higher than class c_{k+1} (class c_k is higher than c_{k+1}) and for each higher class the system must guarantee correct processing of minimal amount of $l^*_{c_k}$ requests. Denote by l_{c_k} the number of actually processed requests of class c_k. The basic request distribution algorithm is to use best effort approach with constraints on number of handled requests of given class. For the request of class c_k for the service as_j, the instance is selected according to:

$$is_{j,m^*} \leftarrow \arg \min_m \hat{t}_{j,m} \qquad (4)$$

with respect to:

$$
\begin{cases}
l_m \leq l^*_m \\
\sum_{m=1}^{M} l^*_m > \left(\sum_{j=1}^{K} l_{c_j} + \sum_{j=1}^{k-1} \max \left[\left(l^*_{c_j} - l_{c_j} \right), 0 \right] \right)
\end{cases} \qquad (5)
$$

The constraint rules (5) are: (a) the limits of processed requests l^*_m must not be exceeded, (b) the number of actually served requests plus the number of possible to appear requests of higher classes then class c_k that must be processed is less than the total number of requests that can be processed.

Another case is the demand to guarantee better values of service parameters (here response times) for established number of requests $l^*_{c_k}$ of higher classes. Denote as t^*_m the maximal time of processing any request within limit l^*_m on the system m. The time of processing l^*_m-th request for most demanding service is assumed at this value. The distribution algorithm that performs this scenario is as (4), but with other constraints:

$$\sum_{j=1}^{k-1} \max \left[\left(l^*_{c_j} - l_{c_j} \right), 0 \right] < \sum_{m:t^*_m \leq t^*_{m^*}} (l^*_m - l_m) \qquad (6)$$

The constraint (6) means that the number of additional requests that can be processed on servers with value t^*_m lower or equal then value $t^*_{m^*}$ for selected execution server m^* is sufficient to serve possible to appear requests of higher classes then class c_k that must be processed.

The last, most demanding, scenario is to assure processing times for individual classes. Let $TH = \{th_1, ...th_k, ...th_K\}$ be the required times of processing for each class $\{c_1, ...c_k, ...c_K\}$, respectively. There are defined the following requirements for each class c_k the system must fulfil:

– the minimal amount of requests $l^*_{c_k}$ for every class c_k processed in parallel,
– the maximal processing time th_k of request of class c_k,
– the class c_k always has priority over class c_{k+1} within the limits $l^*_{c_k}$ of specified requirements.

Let denote $M^*_k = \{m : t^*_m \leq th_k\}$ as the set of execution systems that satisfy processing times for class c_k. The distribution algorithm that assures above requirements is instance selection according to (4) with the following constraints:

$$
\begin{cases}
\forall_{i \leq k} \sum_{m \in M_i^*} (l_m^* - l_m) \geq \sum_{j=1}^{i} \max \left[\left(l_{c_j}^* - l_{c_j} \right), 0 \right] \\
\exists_{i \leq k} \sum_{m \in M_i^*} (l_m^* - l_m) > \sum_{j=1}^{k} \max \left[\left(l_{c_j}^* - l_{c_j} \right), 0 \right]
\end{cases}
\tag{7}
$$

The first rule tells that for each class higher then class c_k of given request, the sum of additional requests that can be processed on servers within the limits l_m^* and that satisfy time requirements (i.e. $t_m^* \leq th_i$) must be not less than possible to appear requests of all higher classes that must be processed. The second rule tells that for at least one of above classes the number of additional request that can be processed is bigger than required.

The presented algorithms are based on some simplifying assumptions, e.g. that server load caused by given number of processed requests is close to the medium value. In fact, this can fluctuate. To avoid the impact of such fluctuation the second parameter s_n characterizing execution system state should be used. Moreover, the parameter s_n and the number of processed requests l_m can be used on virtual resource layer for managing execution system resources to improve request processing, especially when services are near its performance boundaries.

4   Virtual Resource Management

In general, the Virtual Resource Manager (VRM) is responsible for service execution and efficient resources utilization. To increase resource utilization exploits the capabilities offered by virtualization techniques. The VRM uses the information collected by its monitoring system and stored in the service repository. The VRM is composed of a number of independent modules that provide separate functionalities and interacting with each other using defined interfaces (XML-RPC and SOAP). The following modules constitute the resource management part of the VRM: VRM-Controller, VRM-Matchmaker and VRM-Virtualizer. Basic duties of these modules are as follows: VRM-Controller is responsible for assignment the proper amount of resources (RAM memory, percentage of CPU time, etc.) to virtual servers. The VRM-Matchmaker module is used to match existing set of available hardware resources with required ones when the new service instance has to be launched and finally VRM-Virtualizer is responsible for management of virtual machines. The VRM-Virtualizer uses *libvirt* library to execute commands which make it independent on a particular *hypervisor*. More information about the general architecture of the VRM can be found in [4]. In conclusion the VRM is responsible for: minimization of resource usage during service delivery and maintenance of the fulfilment request requirements.

To improve the work of proposed algorithms the VRM can directly affect two important parameters limit value l_m^* of virtual servers and resource usage s_m. The ways the VRM supports request processing and takes decision can be divided into a number of scenarios depending on current set of processed requests and current state of execution systems:

- the service request is received and all resources are satisfied,
- the service request is received but the VRM detects that resource requirements are violated,
- the Broker request for new service instance is received.

The VRM-Controller is able to react when will be notified by the monitoring system that certain service is under-performing, violating the requested non-functional requirements. The possible actions are realized by the VRM-Virtualizer that is able to create the new service instance, change resources allocation to existing service instances, migrate service instances to different location, hibernate and de-hibernate them and finally close any existing service instances. The VRM always checks the current state of the virtual server on which the service instance is called. The current number of served requests l_m and available other resources (e.g. RAM) are compared with the limits. When the results are positive the request is passed to the selected by the Broker instance - first scenario happens.

In the second scenario, when no correct amount of resources is allocated, mainly less resources is allocated to the server, the VRM can perform two actions. In the first step VRM-Virtualizer tries to allocate additional required amount of resources to the server. When it is not possible VRM-Controller takes decision that the new service instance should be instantiated and request redirected to it. However the request redirection must be specially handled by the Broker due to service modelling needs. In that case the Broker is informed about the address of the service instance that is used as response for the request.

The service instance can be instantiated in two ways. The instance can be created using an service image stored in the service repository or the hibernated instance can be de-hibernated. The implemented by the VRM-Controller procedure prefers the second approach because of the time that is needed for both specified actions. The new service instance is then registered in the Broker.

VRM-Controller should receive from VRM-Matchmaker information about the possible location (available servers) that can fulfil the specified in the request requirements. For implementation of VRM-Matchmaker the idea used by the HTCondor and the ClassAd Language was used [7]. However unlike in Condor the matching is done regardless of used parameters, what enables usage of parameters that have not been defined during creation of ontology. Design and implemented for the VRM language PL (Property language) consists of three instructions: substitution ($<property\_id> = <expression>$) is used to define the property of Information Units (IU) - a set of properties specific to the particular element of our system, for example server, service instance, etc., assertion ($assert <logical\_expression>$) used for defining requirements and conditional statement ($if <logical\_expression> then <property\_id> = <expression>$). An example how it works is presented in [5].

In the third scenario the hibernated service instance is de-hibernated or, when it is not possible, the new service instance is created in the way presented above. When specified in the request requirements cannot be fulfilled the request is rejected.

5   Conclusions

Quality of service delivery takes important role during the process of service delivery system designing. It is especially important in the complex and distributed environment when a large number of processing resources are used. Using the virtualization of delivered services and virtualization of execution resources (namely using virtual servers) the effective service request processing can be flexible supported. The presented architecture permits applying different quality-aware distribution algorithms, as well as improving operation of algorithms by low level resource management at the virtual resource layer.

The presented solutions assume some simplifications for the processing rules. In future works, the proposed methods should take into consideration request distribution control with separate impact of transfer times on algorithm decisions. Also using the execution system state indicator should be exploited.

References

1. Borzemski, L., Zatwarnicka, A., Zatwarnicki, K.: Global distribution of HTTP requests using the fuzzy-neural decision-making mechanism. In: Nguyen, N.T., Kowalczyk, R., Chen, S.-M. (eds.) ICCCI 2009. LNCS, vol. 5796, pp. 752–763. Springer, Heidelberg (2009)
2. Fras, M., Kwiatkowski, J.: Quality aware virtual service delivery system. J. Telecommun. Inf. Technol. **3**, 29–37 (2013)
3. Fras, M., Zatwarnicka, A., Zatwarnicki, K.: Fuzzy-neural controller in service requests distribution broker for SOA-based systems. In: Kwiecień, A., Gaj, P., Stera, P. (eds.) CN 2010. CCIS, vol. 79, pp. 121–130. Springer, Heidelberg (2010)
4. Kwiatkowski, J., Fras, M.: Request distribution toolkit for virtual resources allocation. In: Wyrzykowski, R., Dongarra, J., Karczewski, K., Waśniewski, J. (eds.) PPAM 2011, Part I. LNCS, vol. 7203, pp. 327–336. Springer, Heidelberg (2012)
5. Kwiatkowski, J., Pawlik, M., Fras, M., Konieczny, D., Wasilewski, A.: Design of SOA-based distribution system. In: Ambroszkiewicz, S., Brzeziński, J., Cellary, W., Grzech, A., Zieliński, K. (eds.) SOA Infrastructure Tools: Concepts and Methods, pp. 263–288. Poznan University of Economics Publishing House, Poznan (2010)
6. O'Brien, L., Merson, P., Bass, L.: Quality attributes for service-oriented architectures. In: Proceedings of the International Workshop on Systems Development in SOA Environments. IEEE Computer Society, Washington DC (2007)
7. Solomon, M.: The ClassAd Language Reference Manual Version 2.4, May 2004. http://www.cs.wisc.edu/condor/classad/refman/
8. Schmietendorf, A., Dumke, R., Reitz, D.: SLA management - challenges in the context of web-service-based infrastructures. In: Proceedings of the IEEE International Conference on Web Services, San Diego, California (2004)
9. Zeng, L., Benatallah, B., Ngu, A.H.H., Dumas, M., Kalagnanam, J., Chang, H.: QoS-aware middleware for Web services composition. IEEE Trans. Softw. Eng. **30**(5), 311–327 (2004)

Domain-Driven Visual Query Formulation over RDF Data Sets

Bartosz Balis[1](✉), Tomasz Grabiec[2], and Marian Bubak[1,2]

[1] Department of Computer Science, AGH University of Science and Technology,
Krakow, Poland
[2] ACC Cyfronet AGH, AGH University of Science and Technology, Krakow, Poland
{balis,bubak}@agh.edu.pl

Abstract. Semantic Web technologies, such as RDF, SPARQL and OWL, are increasingly used for data representation and information retrieval in real-world applications including those from the e-Science domain. Visual query formulation has been recognized as a useful approach facilitating information retrieval for domain-experts not familiar with RDF query languages. We propose a visual query approach over RDF data sets based on an abstract domain-driven (conceptual) query language. The basis for the query model are ontologies describing the RDF data sets. We have built the QUaTRO2 tool which implements this query approach and provides a high usability graphical user interface to assist domain experts in constructing complex queries and browsing their results. The concepts and their implementation are validated by applying the QUaTRO2 tool to query the UniProt protein database.

Keywords: Semantic Web · RDF · SPARQL · Visual query formulation · OWL · Ontologies

1 Introduction and Motivation

The Semantic Web is a vision of machine-understandable organization of information in the Web. Thanks to the maturation of associated technologies over the past decade – notably the W3C's semantic web technology stack (RDF, SPARQL, OWL and others) – Semantic Web is increasingly adopted for real-world complex applications such as social data analysis [7]. In e-Science semantic web has been investigated as a tool for scientific data integration [5]. Also, advantages of representing scientific data using RDF / OWL in comparison to the relational model have been pointed out [9]. Large scientific data sets are being published in the RDF/OWL representation, notably the protein database Uniprot which currently consists of nearly 7.5B triples [1].

A potential barrier to effective information retrieval from RDF data sets is the lack of query formulation methods and tools which on the one hand enable one to design advanced queries, on the other hand do not require expert IT skills such as the knowledge of SPARQL. We propose a visual approach to formulating

R. Wyrzykowski et al. (Eds.): PPAM 2013, Part I, LNCS 8384, pp. 293–301, 2014.
DOI: 10.1007/978-3-642-55224-3_28, © Springer-Verlag Berlin Heidelberg 2014

semantic queries over data sets represented in RDF and described by ontologies in OWL. The approach consists in using ontologies as a basis for the visual query language and the underlying query model, effectively providing a visual domain-driven (conceptual) query language. The principal design goals for the proposed query methods comprise *domain-expert orientation*, so that query formulation does not require a significant IT expertise; *domain independence* to avoid any assumptions as to the contents or structure of the data and thus potentially support any RDF data set; and *high query expressiveness* so that formulation of advanced queries is possible despite a high-level visual query language. Finally, not the least of our goals was to provide a *high-usability graphical user interface*.

These objectives have led us to implementing the QUaTRO2 tool (preliminarily described in an abstract [2]). As a validation of the proposed concepts, we have applied QUaTRO2 to query the UniProt RDF database. Our previous experience with visual querying concerned querying provenance data [3,4]. The development of QUaTRO2 was driven by significantly different goals and does not reuse any algorithms, design or code from the previous tool.

This paper is organized as follows. Section 2 presents related work. Section 3 explains the concept for a domain-driven query language, its representation, translation and execution. In Sect. 4, the visual query formulation approach is presented along with example queries. Section 5 concludes the paper.

2   Related Work

The need for visual tools oriented towards domain experts, as opposed to computer science engineers, has been recognized in various aspects related to utilization of semantic web technologies [7,13]. A number of visual approaches have been developed to facilitate domain experts in querying RDF data sets.

RDF-GL [8] is a SPARQL-based graphical query language which covers most of the SPARQL syntax and is very powerful in terms of query expressiveness. The authors argue that the tool allows for an intuitive way of building complex queries. However, because of a tight coupling to the SPARQL language, a non-expert may find it difficult to intuitively express some relations. For example, it is not obvious how to express a negation of a pattern. Using the tool effectively requires a deep technical knowledge of the query language.

TAMBIS [11] is a tool for biologists to query multiple data sources using a uniform interface. The tool utilizes an ontology to describe the domain and formulate queries. The query is a tree where nodes represent concepts while branches are their relationships. The query construction starts with a single root concept (*Protein* by default). Next, restrictions on data or object properties can be added. In the latter case, the target concept of an object property can further be restricted. Some limitations of the tool include the lack of logical expressions and operators in data property restrictions, which significantly limits the expressiveness of the query language. Also, the tool is designed to handle a specific domain and it is unclear if it is capable of supporting different ontologies.

Eros [12] is an ontology browsing and query construction tool. The goal of the tool is to provide a user interface to the ontology so that the user can browse it and get familiar with it before building queries. Another goal is to leverage ontology to assist the user in query construction. Eros presents the ontology using two trees laid out next to each other. Each tree presents a class hierarchy. The left tree is a so-called 'domain tree' whereas the right tree is called the 'range tree'. An arrow connecting a concept from the left tree and a concept from the right tree represents a domain – property – range relationship between these concepts. Eros provides also another mode of exploring the ontology, the property view. In this mode the left tree contains a hierarchy of properties. When given property is selected, arrows are drawn to the right tree pointing domain and range of the property. Overall, the query construction requires a significant knowledge of RDF and its query language, RQL in this case.

VQS (Visual Query System) [6] is a set of tools assisting in constructing queries for RDF data sets for which no ontologies exist. The methods featured in VQS include RDF and SPARQL visual editors, a browser-like query creator, and a condensed data viewer which is capable of visualizing a complex RDF graph in a compact manner by grouping similar resources. The advantage of this approach is the capability to handle data not described by ontologies. At the same time, this positions the approach within specific area of applications, such as social web data analysis, for which ontologies are difficult to design.

NITELIGENT [10] is a Web-based tool for semantic query construction based on SPARQL. The proposed visual query language – vSPARQL – is a set of graphical notations which correspond to SPARQL syntactic elements. The tool is complemented with an ontology browser which facilitates query formulation by allowing the users to drag and drop ontology classes and properties into the actual query design canvas. As in the case of some other tools, the tight coupling to SPARQL is an advantage in terms of query expressiveness, but it may affect the usability for domain experts who are not familiar with this query language.

Overall, main novelties of our approach with respect to existing solutions lie in the following areas. Firstly, the existing approaches provide rather a low-level query formulation interface which directly maps the underlying RDF query language. In contrast, our approach relies on ontologies which leads to a domain-driven query language and, consequently, a domain-expert-oriented tool. Secondly, tools which do offer a more abstract query interface, tend to be tied to a particular domain. Our approach is different in that it offers a generic framework which can be adapted for any RDF data set for which ontologies exist; moreover, it achieves that while providing advanced query capabilities (e.g. the ability to combine multiple query trees with logical operators into complex query graphs), and without sacrificing domain-expert-orientation.

3   Semantics-Driven Query Design and Execution

We propose an abstract, conceptual query model which is best explained using a simple query example shown in Fig. 1. This query uses concepts from the

UniProt ontologies and its purpose is to select genes encoding proteins which also cause human diseases, as well as the descriptions of these diseases. The query is basically a graph of domain concepts (ontology classes) connected by relationships: *Protein* which is encoded by a *Gene* which in turn exists in an organism that belongs to *Taxon* commonly named 'Human'. Also, the Protein should have an annotation about a related disease (*Disease_Annotation*). The result of the query can be any subset of attributes from all classes in the query graph. In this particular case three attributes are selected: the URI of the *Protein* RDF resource, attribute `core#prefLabel` from the associated *Gene* resource (preferred name of the Gene), and attribute `comment` from the associated *Disease_Annotation* resource (description of the disease).

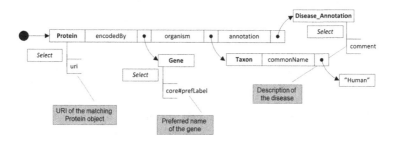

Fig. 1. Semantic query as a tree of domain ontology concepts and properties.

A textual representation of this query is shown below.

Listing 1.1. "Textual representation of a sample semantic query."

```
ConceptSelector{<http://purl.uniprot.org/core/Protein>, (URI)}(
  PropertySelector{<http://purl.uniprot.org/core/encodedBy>}(
    ConceptSelector{<http://purl.uniprot.org/core/Gene>, (core#
        prefLabel)}
  ),
  PropertySelector{<http://purl.uniprot.org/core/organism>}(
    ConceptSelector{<http://purl.uniprot.org/core/Taxon>}(
      PropertySelector{<http://purl.uniprot.org/core/commonName>}(
        PropertyValueConstraint{EQUALS, 'Human'}
      )
    )
  ),
  PropertySelector{<http://purl.uniprot.org/core/annotation>}(
    ConceptSelector{<http://purl.uniprot.org/core/Disease_Annotation>,
        (comment)}
  )
)
```

The query model supports multiple logical operators (alternative, conjunction, negation, etc.) which can be applied at both resource level and property level. At the resource level, logical operators join multiple property selectors each of which can be bound to a different resource. In this way, a query can be branched into multiple subgraphs joined with complex logical relationships. At the property level, logical operators join multiple value constraints of a given property. Examples of queries with logical expressions are provided in Sect. 4.

We have designed and implemented a translation engine which converts the abstract query graph into the target query language. Given the complexity of query graphs introduced by the logical operators, this turned out to be a challenging task. One approach would be to generate and execute multiple subqueries for all branches of the query graph, and subsequently merge their result subsets (union for the alternative operator, intersection for the conjunction operator, etc.). However, this approach is very inefficient: intermediate results may be of significant volume which would increases the query execution time, as well as both client- and server-side CPU and memory consumption. Multiple request round trips to the RDF store would also increase the query response time. Finally, this approach would prevent the engine of the RDF store from applying global query optimizations.

Consequently, we have devised an algorithm (whose detailed description is impossible due to space limitations) which applies a series of transformations in order to create a single query in the target language of the RDF store. Currently only support for SPARQL is implemented. However, the code is designed for extensibility: new query languages can be supported by providing implementations of additional translation engines.

4   Visual Query Formulation

4.1   The QUaTRO2 Tool

QUaTRO2 is a visual query construction tool and a RDF data browser. Its main components, shown in Fig. 2, include: (1) a Web-based Graphical User Interface with visual query construction and data browsing capabilities; (2) server side components: query executor & web server with QUaTRO2 web application; and (3) a user, session and query persistence database.

The main configuration of a particular QUaTRO2 setup includes the connection with an external RDF store, and ontologies describing the structure of data in this store. The ontologies are used both at the visual query level and the query execution level, as a basis for abstract queries. By providing different ontologies and underlying RDF stores QUaTRO2 can be easily set up as a query and browser service for a particular RDF data set. We have deployed such service for part of the UniProt database. It is publicly available at the following URL: http://149.156.9.71:8080/quatro.

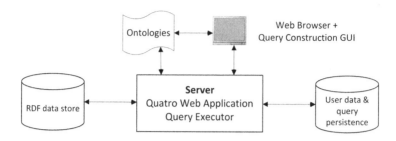

Fig. 2. Main components of QUaTRO2.

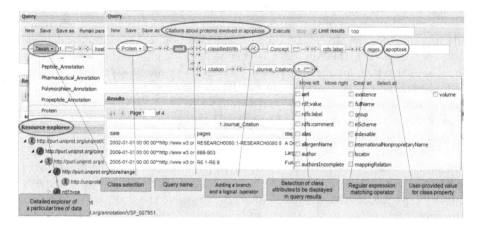

Fig. 3. Query design GUI and its features in the QUaTRO2 tool. Example query selects citations related to proteins involved in apoptosis.

4.2   GUI and Query Examples

High usability of the graphical user interface for query construction is arguably as much important as query expressiveness or other functional requirements. Figure 3 presents an example query constructed in QUaTRO GUI. The query selects citations related to proteins involved in apoptosis.

Query construction starts by selecting the initial concept. The GUI automatically displays a selection list with all properties relevant for this concept. Selecting a property further expands the query graph. When a data property is selected, constraints on its value can be defined. In the case of selecting an object property, another class (target of the property) is added to the query graph along with its property list.

A small fork icon allows to create a new query branch at any time and at any of the designated points. When creating a new branch, a logical operator connecting the subgraphs should be selected. For existing logical operators it is possible to add new operands (subgraphs), remove existing ones, change the operator itself, or even change the sequence of its operands (subgraphs).

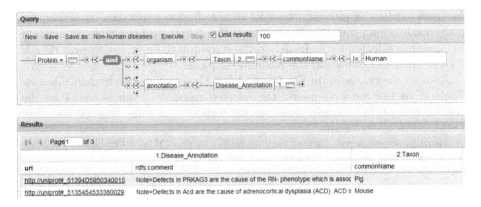

Fig. 4. Example query: non-human diseases.

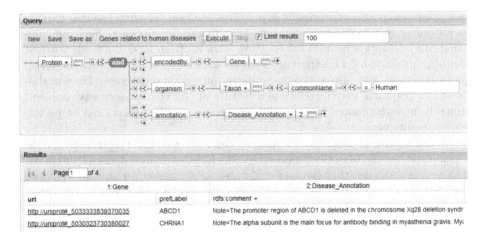

Fig. 5. Example query: genes related to human diseases.

For any class in the query graph, a subset of their attributes can be selected to appear in a table with query results; this resembles the SELECT/JOIN operation from the relational model. Among attributes that can be selected is the URI of a matching resource; it may be clicked to display the resource in a Resource Explorer and further explore the underlying resource tree. QUaTRO automatically retrieves resource properties from the database as the tree in Resource Explorer is expanded. This model allows to construct queries which return only a few key properties while exploring all properties for those results which turn out to be of particular interest.

A simple query which features a logical operator is shown in Fig. 4. This query searches for non-human diseases. To this end, the query is actually started from the Protein concept and selects a resource graph including a Taxon with name different than 'Human', and a Disease_Annotation. Becasue diseases are the main subject of interest, the URI of the disease annotation is chosen for query results along with the description of the disease and the common name of the organism affected by it. Based on these results, the user can inspect details of each particular disease in the Resource Explorer.

Another query, shown in Fig. 5, selects all Genes related to human diseases. Similarly to the previous query, the selected resource graph is based on the Protein concept. In this case, the information about Genes and associated disease is selected as query results.

More pre-configured queries can be found in the QUaTRO2-based UniProt service deployed at http://149.156.9.71:8080/quatro.

5   Conclusion

Visual construction of queries over RDF scientific data sets, using ontologies as a query language, can provide a powerful solution enabling domain experts to formulate advanced queries using domain-specific concepts. Such an approach can facilitate advanced information retrieval from RDF data sets without requiring professional IT expertise, such as knowledge of RDF data model and query language, but also without sacrificing query expressiveness. High usability of the graphical user interface is a feature of the visual query environment, arguably underestimated, yet equally important as other functional or performance requirements.

Initial results of usability evaluation confirm that the target user groups which could benefit from the proposed tool include, as expected, domain experts. However, it turns out that the tool could also be of interest for anyone who does not know the target data set in which case the guidance features of the query formulation interface prove to be very useful in exploring the structure of the data. This user feedback collected so far suggests an interesting direction for future development of the tool: a SPARQL query construction assistant. In such an application, the graphical interface would facilitate the generation of an initial version of a query which would be subsequently refined manually by an expert.

Future work involves further usability evaluation within target users, as well as extensions to visual query capabilities and the abstract query model in order to enhance the query expressiveness. Also, quantitative study of QUaTRO2 regarding query performance will be conducted.

QUaTRO2 is available as open source. Installation and user manuals, including information about customization for particular RDF ontologies and data sets, are available at http://dice.cyfronet.pl/products/quatro.

Acknowledgments. This work is partially supported by the European Union Regional Development Fund, POIG.02.03.00-00-096/10 as part of the PLGrid Plus Project. AGH Grant 11.11.230.015 is also acknowledged.

References

1. Apweiler, R., Martin, M.J., Donovan, C., Magrane, M., Alam-Faruque, Y., Antunes, R., Barrell, D., Bely, B., Bingley, M., Binns, D., et al.: The universal protein resource (UniProt) in 2010. Nucleic Acids Res. **38**, D142–D148 (2010)
2. Balis, B., Bubak, M., Grabiec, T.: Graphical Query Construction over Scientific Data Sets using Semantic Technologies. http://www.ci.uchicago.edu/escience2012/pdf/escience2012_submission_198.pdf
3. Balis, B., Bubak, M., Pelczar, M., Wach, J.: Provenance Tracking and End-User Oriented Query Construction. In: Cannataro, M. (ed.) Handbook of Research on Computational Grid Technologies for Life Sciences, Biomedicine, and Healthcare, chap. 4, pp. 60–75. Medical Information Science Reference (2009)
4. Bubak, M., Gubala, T., Malawski, M., Balis, B., Funika, W., Bartynski, T., Ciepiela, E., Harezlak, D., Kasztelnik, M., Kocot, J., Król, D., Nowakowski, P., Pelczar, M., Wach, J., Assel, M., Tirado-Ramos, A.: Virtual laboratory for development and execution of biomedical collaborative applications. In: Proceedings of the Twenty-First IEEE International Symposium on Computer-Based Medical Systems, 17–19 June 2008, Jyväskylä, Finland, pp. 373–378. IEEE Computer Society (2008)
5. Goble, C., Stevens, R., et al.: State of the nation in data integration for bioinformatics. J. Biomed. Inform. **41**(5), 687–693 (2008)
6. Groppe, J., Groppe, S., Schleifer, A.: Visual query system for analyzing social semantic Web. In: Proceedings of the 20th International Conference Companion on World Wide Web, pp. 217–220. ACM (2011)
7. Groppe, S.: Data management and query processing in semantic web databases. Springer, Berlin (2011)
8. Milea, F.H.V., Frasincar, F., Kaymak, U.: RDF-GL: a SPARQL-based graphical query language for RDF. Emergent Web Intelligence: Advanced Information Retrieval, pp. 87–116. Springer, London (2010)
9. Roure, D.D., Frey, J.: Three perspectives on collaborative knowledge acquisition in e-science. In: Workshop on Semantic Web for Collaborative Knowledge Acquisition (SWeCKa 2007) (2007)
10. Smart, P.R., Russell, A., Braines, D., Kalfoglou, Y., Bao, J., Shadbolt, N.R.: A visual approach to semantic query design using a web-based graphical query designer. In: Gangemi, A., Euzenat, J. (eds.) EKAW 2008. LNCS (LNAI), vol. 5268, pp. 275–291. Springer, Heidelberg (2008)
11. Stevens, R., Goble, C., Paton, N.W., Bechhofer, S., Ng, G., Baker, P., Brass, A.: Complex query formulation over diverse information sources in TAMBIS, chap. 7. In: Lacroix, Z., Critchlow, T. (eds.) Bioinformatics: Managing Scientific Data, pp. 189–223. Morgan Kaufmann, San Francisco (2003)
12. Vdovjak, R., Barna, P., Houben, G.J.: EROS: Explorer for RDFS-based ontologies. In: Proceedings of the 8th International Conference on Intelligent User Interfaces, pp. 330–330. ACM (2003)
13. Wibisono, A., Koning, R., Grosso, P., Belloum, A., Bubak, M., de Laat, C.: OIntEd: online ontology instance editor enabling a new approach to ontology development. Softw. Pract. Exp. **43**, 1319–1335 (2012)

Distributed Program Execution Control Based on Application Global States Monitoring in PEGASUS DA Framework

Damian Kopański[1], Łukasz Maśko[2], Eryk Laskowski[2], Adam Smyk[1],
Janusz Borkowski[1], and Marek Tudruj[1,2(✉)]

[1] Polish-Japanese Institute of Information Technology, ul. Koszykowa 86,
02-008 Warsaw, Poland
[2] Institute of Computer Science, Polish Academy of Sciences, ul. Jana Kazimierza 5,
01-248 Warsaw, Poland
{damian,asmyk,janb,tudruj}@pjwstk.edu.pl

Abstract. This paper presents control implementation methods for an original distributed program design framework PEGASUS DA (Program Execution Governed by Asynchronous SUpervision of States in Distributed Applications) which provides automated design of distributed program execution control based on program global states monitoring. The framework includes a built in support for handling local and global application states as well as automatic construction and use of strongly consistent application global states for program execution control. In particular, the paper presents methods used to implement distributed program control inside the PEGASUS DA framework run on clusters of contemporary multicore processors based on multithreading. The program design method is illustrated on a distributed multithreaded application executed with load balancing in a multicore system.

Keywords: Distributed program execution control · Distributed program design · Global application states · Strongly consistent global states · Program design tools

1 Introduction

Parallel and distributed programs execution control quite frequently has to be organized based on the monitoring of global computation states in sets of application components. Such needs occur in distributed program run-time optimization, parallel event-driven simulation, industrial process control, scientific distributed computing based on divide and conquer or branch and bound methods. Unfortunately, existing commercial distributed system have no support for automated control of global program states. So, this type of control has to be organized at the expense of tricky programming done by programmers. Some initial attempts were reported in 1990'ies [3,4] to include some formalisms and infrastructure for program control based on global states, however not supported

R. Wyrzykowski et al. (Eds.): PPAM 2013, Part I, LNCS 8384, pp. 302–314, 2014.
DOI: 10.1007/978-3-642-55224-3_29, © Springer-Verlag Berlin Heidelberg 2014

by an efficient usable framework. First contemporary runnable programming framework, which included an infrastructure for automated detection of global program states and the design of the respective program execution control was PS-GRADE [5]. In this framework, asynchronous control of the behaviour of distributed processes based on global states was added to the run-time environment for parallel program graphical design in C language.

This paper describes implementation of the internal control in the PEGA-SUS DA framework in which ideas of distributed program execution control based on global application states have been extended to cover not only internal process/thread behaviour but also the control flow in distributed programs [7]. The framework provides GUI for the design of the program control flow graph dependent on the analysis of global program states. It also provides an API of an extended programming in C /C++ languages to be used by the programmer to design distributed program execution control based on application global states. This API is supported by a run-time system which provides automated global states construction and respective control design including communication.

The PEGASUS DA framework strongly extends system hardware support for global states-dependent program execution control to reduce time overheads of global control implementation. We assume that the hosting system is equipped with a triple communication network, each with features adjusted to performed functions such as exchange of short messages for local states collection and transfers of control signals, processor clock synchronization for global states detection and user data transmission.

The basic text of the paper is organized as follows. First, general features of the program global control in PEGASUS DA framework are described. Next, the general structure of the framework is presented. Internal PEGASUS communication library is outlined in turn. Then, the design of main elements of the internal control in PEGASUS DA is presented. At the end, the structure of an exemplary application – Travelling Salesman Problem (TSP) by the Branch & Bound (B&B) method is described.

2   Features of the Applied Global States Monitoring

The PEGASUS DA implements automated global states monitoring by special runtime infrastructure. To enable monitoring of global states processes/threads communicate with a number of synchronizers. Synchronizers learn local state information (events), construct global application states [1], evaluate predicates on global states and send back control signals based on predicate values, Fig. 1.

Two types of global states are considered: Strongly Consistent Global States (SCGSs) and Observed Global States (OGS). Strongly Consistent Global States are identified by the synchronizers as composed of local states, which have occurred precisely in the same time in relevant distributed processes/ threads of an application. Observed Global States are local states of relevant processes/threads identified by synchronizers without checking any concurrency with other states. The states are identified by some starting and ending events that have to be known or estimated by synchronizers.

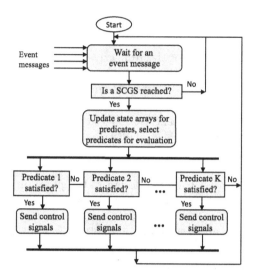

Fig. 1. Synchronizer activities for SCGs-based control.

If processor local clocks are synchronized with a known accuracy, then Strongly Consistent Global States can be detected based on the clock timestamps attached to event messages sent to synchronizers to identify event occurrence time [2]. Events are sorted by the synchronizer according to timestamps and are projected on the common time axis. An SCGS is constructed by local states occurring simultaneously and not covered by any clock uncertainty interval. The causality relation has to be preserved when constructing global states. If clock synchronization accuracy (timestamp uncertainty interval) is smaller than half of the message transfer time, then causality is preserved in the SCGS monitoring.

In the design of application programs execution control based on global states monitoring we apply two kinds of control: asynchronous and synchronous ones.

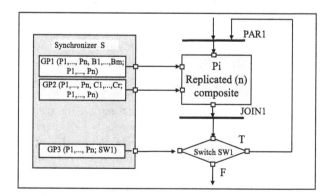

Fig. 2. Parallel do-until loop for replicated, composite program block.

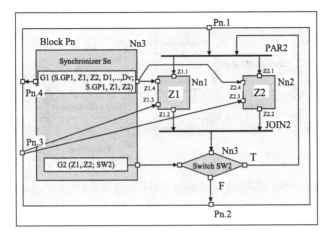

Fig. 3. Nested parallel do-until loop inside the composite block Pi from Fig. 2.

Application asynchronous control based on monitoring global states is applied to asynchronously modify the behaviour of processes/threads as a result of a predicate fulfilled on a global state. For this Unix signals are sent which are handled by additional computation activation (a kind of distributed interrupts) or by computation cancellation if received inside the signal sensitivity regions marked in the process/thread code. Otherwise signals are neglected. To apply synchronous program execution control we provide global distributed synchronous control flow structures in the distributed application code. They are combinations of global control constructs on program blocks and process/thread activation on specified computational resources. The structures are graphically designed using the GUI provided in the framework. The global control constructs can be replicated and nested (composite).

Exemplary PEGASUS DA control flow graphs are shown in Figs. 2 and 3. Rectangles inside the synchronizer blocks represent predicates with outgoing signal arrows. Other arrows represent control flow. Control flow in paths in program graphs can be additionally synchronized using four synchronization primitives: AND, OR, BARRIER and EUREKA. Process blocks which do not contain nested control constructs called terminal blocks contain a thread synchronizer and a number of thread blocks, Fig. 9(b). Thread blocks contain parallel threads with the same body, created in Pthreads library. For more details of the programming model, GUI and API in PEGASUS DA see [6–8].

3   General Structure of the PEGASUS DA Framework

The general flow diagram of the PEGASUS DA framework is shown in Fig. 4. Program Graph Editor is a GUI-based tool for editing the global control flow graph and specifying global control elements of an application. The graph is composed of application components (synchronizers, processes, nested composite structures etc.). The tool also enables editing details of an application control

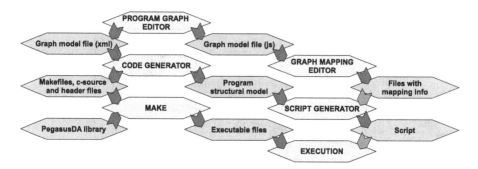

Fig. 4. General flow diagram of the PEGASUS DA framework.

governed by the global state analysis in synchronizers (local events, global states, predicates and detectors inside synchronizers, signals and their relationship). Parallel applications designed using the graph editor are stored in XML-based and JavaScript data files. They are the input for next components of the framework, such as the code generator module and the graph mapping editor.

The code generator transforms the program description given as an XML code into a set of C files, which are later compiled and linked with the Pegasus DA library. It also creates the unrolled program graph structure, which is a basis for the script generator. The included code parser is composed of two classes, which implement analysis of main graph structure (PGSControlflowParser class) and global states control structure (PGSWiringParser).

The basic structure of the graph is returned as an instance of PGSControlflow class. To represent internal structures of the graph, a set of classes was created. They implement fork-join constructs, conditional constructs, loops, synchronizers, processes, thread blocks in the process. Graph description structure includes also the control communication. It is returned by the XML parser as an instance of PGSWiring class. This information represents a graph of states and signals communication dependencies and is implemented as another set of classes.

Information from both parts of the input XML file are merged to obtain the final program graph. The objects are created during compile-time, while the XML program description file is read. The created objects are linked in such way, which depicts the unrolled program graph structure. They are then used for automatic generation of source files for each process in the application.

The initial program graph structure is used to create another graph representation, Program Structural Model. In this graph, replicated structures from the initial graph are recursively substituted with their copies to create the unrolled program graph structure. The unrolled graph is used to determine the unique numbering of all the processes (ranks) in the distributed application, required by the MPI library. The unrolled graph is also used to create the name space used by the internal communication library to deliver event messages/signals to correct destinations.

The unrolled program graph is also used by the Script Generator to create the application execution script, which starts execution of the whole application. This generator also uses the mapping information delivered by the Graph Mapping Editor. During compilation, executable files which implement the global control primitives (PAR, JOIN, SWITCH, OR, AND) and executable process files containing synchronizer and multithreaded process code are generated.

Before program execution, application components are mapped onto executive processors/cores. The only input data for this stage is the application skeleton graph mentioned above. First, the program graph is enriched with elements that are required to run and control the application execution. Next, a tree containing the definition of the logical executive system is created. The system can be considered at different levels of detail: computers, processors, cores and hardware threads. The mapping assigns elements of the program graph to elements of a logical system. The next step is to define a tree containing a description of the physical system. It is supported by additional software called *hwloc* which collects information on system architecture. In the last step, a mapping of the logical system into the physical system is done.

4 Internal PEGASUS DA Communication Library

The internal PEGASUS DA communication library (peg-lib) is a set of functions used to perform control data transfers and process/thread activations necessary to implement program execution control based on the global application states monitoring. These functions are used in the PEGASUS DA run-time code and also in the user application code. The peg-lib library contains the following function layers:

- inter-process-communication layer: based on message passing, it implements messages between application processes,
- intra-process communication layer: based on message queueing, it implements communication between threads and local (thread) synchronizers,
- system parameters configuration layer,
- application configuration layer,
- process configuration layer,
- internal process control functions.

  System parameters configuration layer is composed of two layers: application configuration layer and process configuration layer. Application configuration layer contains functions, which enable definitions of application process and control blocks, synchronous control, signal and partial states and creating a map for partial state transfers. Process configuration layer contains functions which enable defining threads blocks, synchronizers, state detectors, predicates, local events and local signals. It contains also functions which enable binding local events with global state detectors, partial states with state detectors, state detectors and remote signals with predicates. Internal process control functions enable process and synchronizers handling, including implementation of control messages and user threads activities.

The following types of API global control functions are available for the user: event messaging at the thread level, event messaging at process level, asynchronous signal handling at thread level, predicate defining support, state messaging at the predicate level (partial SCGSs and process SCGSs), remote signal transfers, state vector accessing.

5   Internal Control Implementation in PEGASUS DA

Application threads or processes send messages to synchronizers on current events selected to be relevant for the program execution global control. These events are defined by a programmer by instrumenting application thread code for sending a thread event message or instrumenting a predicate for sending a process event message. Event messages have single timestamps defined by readouts of the processor timestamp clock TS or by readouts of processor timestamp counter TSC. Synchronizers can detect SCGSs of sets of threads of a process, SCGSs of sets of threads from a set distributed processes, SCGSs of sets of processes, partial SCGSs for hierarchical detection of SCGSs for global sets of processes [6] or they can monitor observed states. For different types of monitored states, different kinds of event and state messages have to be sent. After reception and identification, the messages are processed by different types of state detectors included in synchronizers. The correspondence of states, message types and detectors used in the PEGASUS DA framework is shown in Table 1.

Depending on a predicate evaluation and checking result a control signal can be sent by the predicate. We distinguish: local control signals – sent to the threads of the same process in which the synchronizer containing the predicate is embedded and remote control signals when the signal is sent to a synchronizer in another process. A local signal is sent by a predicate using a local signal sending function of the peg-lib library in which a thread id identifies the signal receiver. A remote control signal can be sent from any predicate to a predicate in another process. The signal is addressed by the receiver process id and the signal id.

To deliver remote signals to proper threads and process level states to synchronizers, communication dispatchers are provided in application processes. A communication dispatcher is a default thread in each application process which communicates with other processes at the synchronizer level. A dispatcher receives messages representing partial global states and global states coming from other process synchronizers as a result of their predicates and messages representing remote signals from other synchronizers (via InfiniBand network). A communication dispatcher sends the received states/signals to message queues of synchronizers selected based on the dispatcher configuration data (memory write). Based on their configuration data, the synchronizers activate the relevant predicates, which invoke the remote signal handling by sending Unix-type signals to appropriate threads. Internal communication inside PEGASUS DA is illustrated in Fig. 5.

Table 1. Control messages in PEGASUS DA

What is monitored	Source of the message	Message semantics	Message type	Timestamps used in the event	Detector type
SCGS on threads	Thread	Thread event inside a process	SCGS-TSC	TSC	SCGS-TSC
Partial SCGS or global state of processes	Predicate	Partial state in SCGS hierarchic detection or process state	HSCGS	TS, TS	HSCGS
SCGS on threads or processes	Thread or predicate	Thread event inside application or process event	SCGS-TS	TS	SCGS-TS
Observed states of threads or processes	Thread or predicate	Observed event	Observed	None	Observed

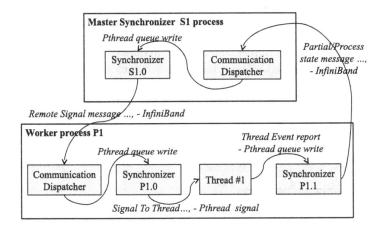

Fig. 5. Internal communication inside PEGASUS DA.

6   Pegasus DA Application Execution Control Flow

Each PEGASUS DA application is executed by the use of the mpirun function of the MPI library. Each application process is accompanied by its rank and a name of a processor host. The mapping of the application components and additional parameters are generated from the information provided by a programmer during the program graph edition, code component edition and application graph mapping by means of the GUI and API provided in the framework.

The application execution general control flow is shown in Fig. 6. All constituent application processes are created by the mpirun function. In each created process all application-specific and process-specific data structures are first initialized. Next, the communication dispatchers and all synchronizers are created as threads in each process by means of the pthread_create functions of the pthread library. After threads creation gets synchronized for all processes (barrier), the application processes start their runtime cycles shown in Fig. 8. In the process runtime cycle, processes wait for activation (ACT)/termination (END) control messages coming from other processes. After ACT message, a process starts all computational thread blocks defined inside the process and then, all threads in the blocks. After execution of its computational threads, the process can send ACT/END messages to other processes. The process cycle can iterate on received ACT messages until a termination message (END) arrives. After such message, all active process threads including communication dispatcher and synchronizers are cancelled and the process is terminated.

After compilation, a process activation structure exists in the application code based on the inserted activation and cancellation orders to be sent by means of control messages. Figure 7 shows the process activation graph for the exemplary program control flow graph shown in Figs. 2 and 3.

In many thread-oriented control actions it is necessary to kill threads based on some global predicates. Each application process contains at least one thread synchronizer (#0) which besides user definable predicates has a number of default mechanisms necessary to implement the proper execution control inside applications. The mechanisms are implemented using predicates. In the synchronizer #0 there is a hidden predicate DefCancel, see Fig. 8, which sends the SIG_CANCEL local signals to threads. Each thread has a default handler for the SIG_CANCEL signal with the function pgs_def_signal_cancel (to be activated inside the sensitivity region to SIG_CANCEL signal), which makes the receiver thread terminate.

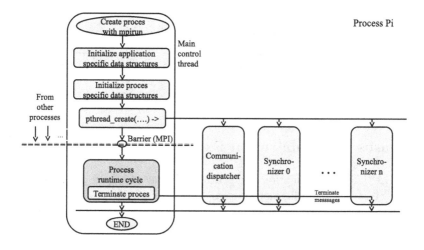

Fig. 6. Application process control flow.

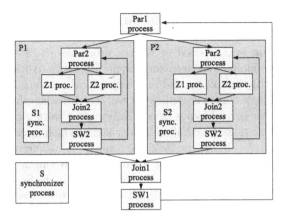

Fig. 7. Exemplary process activation graph.

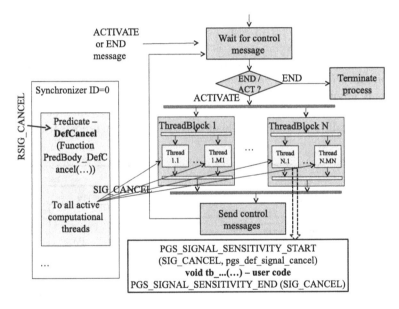

Fig. 8. Application process runtime cycle and thread cancelling method.

The DefCancel predicate is activated by the remote signals RSIG_CANCEL sent from other synchronizers. The RSIG_CANCEL signals can also be generated in the result of the global synchronizing construct OR and from any user-defined predicate. The SIG_CANCEL signals can also be sent from predicates defined by the user.

7   Example: Travelling Salesman Problem

The TSP application is composed of worker processes P1, P2, P3, controlled
by predicates of the global synchronizer GSync, Fig. 9(a). Worker processes are
composed of worker thread blocks and the thread synchronizers TSync, Fig. 9(b).
A worker thread block contains replicated worker threads, which are assigned
to the same processor core. TSyncs periodically report worker processes current
load to GSync. GSync takes load balancing decisions, which direct pools of search
subtasks to TSyncs, or, it orders task migrations to balance loads in worker
processes. A subtask is a trajectory of towns of known length to be developed
by adding towns. TSyncs report best solutions to GSync to be validated.

GSync synchronizer contains a number of predicates. NewSubtask – computes the global mean load in the system and compares it to the loads reported by
TSyncs. Based on the load deviation of the given worker process, load balancing
decisions are taken. New subtasks can be sent to the underloaded processes, i.e.
their TSyncs, to be distributed between threads. NewBest – maintains the best
known solution found so far in response to min_dist messages sent by TSyncs.
The valid new best solution min-dist is broadcasted to all TSyncs. StopIteration
– takes care of the quality of the solutions found so far. If a sufficiently good solution is found, further search can be stopped by a broadcast of the stop_iteration
signal to all TSyncs.

Predicates of TSync synchronizers are as follows. NS (from NewSubtask) –
reports periodically current process load to GSync based on reports from threads,

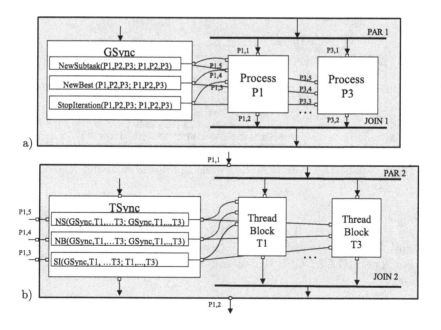

Fig. 9. Application (a) and terminal process (b) control graphs.

manages subtask reception i.e. distributes subtasks to threads based on their reported load. NB (from NewBest) – maintains the best known solutions found in the process by threads or sent by GSync. It broadcasts valid new best solutions min-dist to all threads in the process. SI (from StopIteration) – stops search in process threads by broadcast of the stop-iteration signal to all threads.

Threads add towns to the current trajectory one by one and compute the trajectory length after each added town. They compare each new trajectory length with the current known min-dist. A partial trajectory not shorter than min-dist is not perspective and is rejected from further development (B&B bounding). A new full trajectory not shorter than min-dist can be also rejected. A full trajectory shorter than min-dist is sent to TSync as the new best.

When a thread receives a subtask of t towns out of total N, the process load is computed as the number of possible search steps i.e. the factorial $S = (N - t)!$. The current load S is decreased by 1 for each perspective and by $p!$ for each not perspective trajectory, p is N minus the number of towns in the not perspective trajectory. The sum of concurrent values of S for all subtasks in a thread (an SCGS) is sent to TSync. The sum of concurrent values of S in all threads in a TSync is sent to GSync as the process load. GSync computes the current average load in the system based on values of S corresponding to a SCGS of all processes.

8 Conclusions

Essential control solutions in the PEGASUS DA distributed program design framework based on global state monitoring have been presented. The framework provides a runtime infrastructure for global state monitoring in the control design for application processes/threads. It supports global control constructs for the global control flow design based on graph representation. The PEGASUS DA framework implementation is based on NetBeans graphical environment.

Acknowledgments. This paper has been partially sponsored by the MNiSW grant No. NN 516 367 536.

References

1. Babaoglu, O., Marzullo, K.: Consistent global states of distributed systems: fundamental concepts and mechanisms. In: Mullender, S.J. (ed.) Distributed Systems. Addison-Wesley, Reading (1995)
2. Stoller, S.D.: Detecting global predicates in distributed systems with clocks. Distrib. Comput. **13**(2), 85–98 (2000)
3. Marzullo, K., Wood, D.: Tools for constructing distributed reactive systems. Technical report 14853, Cornell University, Department of Computer Science, Feb. 1991
4. Tudruj, M.: Fine-grained global control constructs for parallel programming environments. In: Bakkers, A. (ed.) Parallel Programming and Java: WoTUG-20, pp. 229–243. IOS, Amsterdam (1997)

5. Tudruj, M., Borkowski, J., Kopański, D.: Graphical design of parallel programs with control based on global application states using an extended P-GRADE system. In: Juhász, Z., Kacsuk, P., Kranzlmüller, D. (eds.) Distributed and Parallel Systems: Cluster and GRID Computing. Kluwer International Series in Engineering and Computer Science, vol. 777, pp. 113–120. Springer, New York (2004)
6. Borkowski, J.: Hierarchical detection of strongly consistent global states. In: Proceedings ISPDC 2004, pp. 256–261. IEEE CS (2004)
7. Tudruj, M., Borkowski, J., Maśko, Ł., Smyk, A., Kopański, D., Laskowski, E.: Program design environment for multicore processor systems with program execution controlled by global states monitoring. In: ISPDC 2011, pp. 102–109. IEEE CS (2011)
8. Borkowski, J., Kopański, D., Laskowski, E., Olejnik, R., Tudruj, M.: A distributed program global execution control environment applied to load balancing. Sacl. Comput. Pract. Exp. **13**(3), 269–280 (2012)

Application of Parallel Computing

New Scalable SIMD-Based Ray Caster Implementation for Virtual Machining

Alexander Leutgeb, Torsten Welsch[✉], and Michael Hava

RISC Software GmbH, Softwarepark 35, 4232 Hagenberg, Austria
{alexander.leutgeb,torsten.welsch,michael.hava}@risc-software.at

Abstract. We present a highly efficient ray casting system for the visualization of subtractive manufacturing, combining state-of-the-art results of various active research fields. Besides popular techniques like acceleration structures, coherent traversal and frustum culling, we integrated the novel surface cell evaluation (SCE) algorithm, allowing the elimination of surfaces that have no effect on the final workpiece's shape. Thus, our ray caster allows an interactive, non-approximate visualization of thousands of Boolean subtraction operations between a stock and arbitrary triangular swept volumes. Compared to image-space based approaches for virtual machining, such as *z-maps* [3], *dexels* [4] or *layered depth images* [10], our scalable SIMD-based implementation offers a higher rendering performance as well as a view-independent workpiece modeling. Hence, it is perfectly suited for of both simulation and verification in computer-aided manufacturing (CAM) applications.

Keywords: Ray casting · SIMD · CAM · Boolean subtraction operations · Subtractive manufacturing · Space partitioning strategy · Multi-axis milling

1 Introduction and Related Work

With CAM being extensively applied in industry nowadays, simulations of subtractive manufacturing - material removal processes, such as milling, turning or drilling - have become complex tasks, requiring efficient systems for their computation and visualization. In our development scenario, for example, complex multi-axis milling processes, that result in workpieces with a high-quality surface finish, consist of thousands of material removal steps. Hence, the manufacturing industry has a strong need for both real-time and offline visualization as well as verification during the whole machining process. Especially in terms of verification, not only a continuous but an interactive simulation is essential. To develop a system that is capable of meeting the outlined requirements, one has to ensure that state-of-the-art research results of various fields are combined in an efficient manner, lending itself to both the available infrastructure and the application scenario.

R. Wyrzykowski et al. (Eds.): PPAM 2013, Part I, LNCS 8384, pp. 317–326, 2014.
DOI: 10.1007/978-3-642-55224-3_30, © Springer-Verlag Berlin Heidelberg 2014

With that respect, our implementation is related to: the STAR about ray tracing by Wald et al. [9], the coherent grid traversal by Wald et al. [8], the SIMD frustum culling by Dmitriev et al. [2], the fast ray/triangle intersection test by Moeller and Trumbore [5] and the very efficient SCE algorithm presented in this paper. As the generation of swept volumes is not the concern of this paper, we assume that all swept volumes mentioned in the following have triangulated, closed ("watertight") and well oriented surfaces.

2   Basic Idea

In the context of material removal simulations, the workpiece is visualized while a machine tool virtually cuts layer-by-layer into it. Along its path, the tool's movement is discretized and combined to swept volumes that are further applied to the original stock geometry as Boolean subtraction operations.

Roth [7] firstly presented ray casting as an algorithm for rendering Boolean operations between volumes. To apply this algorithm to our subtractive manufacturing scenario, an intersection counter i_r has to be added per ray r, recording whether r enters or leaves a swept volume along its way through the virtual three-dimensional (3D) space. After all intersections between the rays and the volumes have been determined, we sort the intersections per r in ascending order, depending on their distance to r's origin. Given an arbitrary r and the $k + 1$-th intersection along its way, we update i_r according to the following rule:

$$
\begin{aligned}
i_r^0 &:= -1 \\
i_r^{k+1} &:= \begin{cases} i_r^k - 1 \, , \, r \text{ enters a swept volume} \\ i_r^k + 1 \, , \, r \text{ leaves a swept volume.} \end{cases}
\end{aligned}
\tag{1}
$$

Note that we model the stock geometry as a subtraction from the completely solid 3D space as well, enabling us to treat all inserted volumes equally. Thus, each i_r has to be initialized with -1 so that the final surface hit of r is found iff $i_r = 0$ holds. In this case, r does not lie inside a swept volume any more, as all entered swept volumes have been left.

3   High-Level Optimizations

3.1   Acceleration Structure

Considering only volumes with triangulated surfaces, we accelerate the original boolean ray casting algorithm by reducing the number of triangles per ray that have to be tested for intersections. This is done by subdividing the 3D scene into multiple cells that refer to triangles lying inside them. Hence, a ray has to test only those nearby triangles for an intersection that lie inside cells the ray traverses. With respect to the requirements of virtual machining, consisting of thousands of successive Boolean subtraction operations, we need a space partitioning strategy that allows us to apply new swept volumes to the workpiece

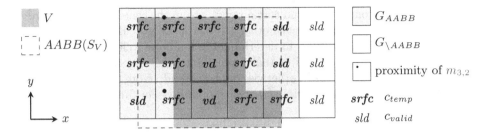

Fig. 1. Illustration of the temporary classification algorithm.

rapidly at low computational cost. As regular grids are fastest to build [9], we implement a coarse regular grid whose cubical cells we call *macrocells*.

Additionally, we add a second regular grid level per macrocell that consists of so-called *microcells*, supporting the elimination of triangles in macrocells that are enclosed by the union of two or more swept volumes, see Sect. 3.3. The rays' grid traversal, however, is not affected by this extension as its computation takes place on the macrocell level only, see Sect. 3.5.

3.2 Cell Classification

As the correct evaluation of the rays' intersection counters (1) depends on "watertight" swept volumes, we have to keep even those triangles available for intersection tests that have no effect on the final workpiece's shape. To deal with this performance issue, we first have to introduce a grid cell classification: A grid cell can lie completely inside (*void*) or outside (*solid*) of a swept volume or it can contain triangles of this volume (*surface*). For simplicity's sake, we describe the following algorithms in the 2D space only.

Given the grid's microcells, $G = \{$microcells $m_{x,y} \mid 1 \leq x \leq k \ \wedge \ 1 \leq y \leq l\}$, with $k = $ the number of microcells in x- and $l = $ the number of microcells in y-direction. Given further an arbitrary, triangulated swept volume V that has to be inserted into that grid. The swept volume's triangles, $S_V = \{$triangles $t \mid t$ belongs to $V\}$, are mapped into the grid by initially computing the axis-aligned bounding box (AABB) of S_V. Then, we determine all microcells that are affected by the insertion of V, $G_{AABB} = \{$microcells $m_{x,y} \mid m_{x,y} \in G \ \wedge \ m_{x,y} \cap AABB(S_V) \neq \emptyset\}$. Finally, the G_{AABB} store references to all $t \in S_V$ that (partially) lie inside them, if any.

Depending on the 8-connected proximity relationship between the $m_{x,y} \in G$ and on the microcells that are not affected by the insertion of V, $G_{\backslash AABB} = G \setminus G_{AABB}$, the grid cells are classified as described in the following: Starting with the $m_{x,y} \in G_{AABB}$, we determine their individual temporary classification, $c_{temp} \in \{void, solid, surface\}$ (or vd, sld and $srfc$) by labeling all $m_{x,y}$ that store $t \in S_V$ as *surface*. Afterwards, if a $m_{x,y}$ has a neighbor $n \in G$ of classification type either *void* or *solid*, it gets the same label. Otherwise, if either $m_{x,y}$ has no neighbors or the label *surface* is the the only classification available in $m_{x,y}$'s proximity, we take one $t \in S_V$ and cast a ray r_{mt} from the center of $m_{x,y}$ toward that t. Given the intersections with other $t \in S_V$ on its way, if any, we use the one nearest to the origin of r_{mt} to determine the temporary classification c_{temp} of $m_{x,y}$: If r_{mt} and the normal of that t point into the same half-space of t's plane, $m_{x,y}$ lies inside V (*void*), otherwise outside (*solid*). Note that we define all triangle normals of swept volumes as outward-pointing. For further illustration, see Fig. 1.

Given a microcell's c_{temp} and its original valid classification c_{valid}, the updated valid classification for that microcell, c_{valid}^{+}, is given by (2). Note that all microcells of the grid are initially labeled with $c_{valid} = solid$ as the grid is considered to be completely solid before applying the first swept volume onto it.

$$c_{valid}^{+} := \begin{cases} sld & , c_{temp} = sld \land c_{valid} = sld \\ vd & , c_{temp} = vd \lor c_{valid} = vd \\ srfc & , else \end{cases} \qquad (2)$$

After c_{valid}^{+} has been determined for all $m_{x,y} \in G_{AABB}$, the classification of their corresponding macrocells is assigned in the following way: Every macrocell containing at least one $m_{x,y}$ of the type *surface* is labeled as *surface* as well and stores references to all $t \in S_V$ inside its $m_{x,y}$. Otherwise, as one can easily see, a macrocell can contain only *void* or *solid* microcells exclusively and therefore gets the same attribute as its uniform microcells.

3.3 Triangle Elimination

Given both the new valid macro- and microcell classification for all cells that are affected by the insertion of a swept volume, all triangles inside *void* macrocells can be dismissed as they reside completely inside swept volumes and therefore have no effect on the current workpiece's shape. With this elimination, we get a first grid level that holds triangles in *surface* macrocells only, significantly reducing the number of triangles per ray that have to be tested for intersections during the ray casting, see Fig. 6/2–3 and Sect. 5. As we handle *surface* macrocells isolatedly during the ray's grid traversal, there is one fundamental condition for the ray casting of Boolean subtraction operations that has to hold if we want to eliminate triangles that way: The ray caster must be able to determine if the intersection point between an entering ray and a *surface* macrocell lies inside or outside of swept volumes with triangles stored in that cell. Otherwise, as this triangle removal results in non-watertight swept volumes, there is no way to evaluate the ray's intersection counter (1) correctly.

3.4 Surface Cell Evaluation

The basic idea is to determine if the point p_{rM} where a ray r enters a *surface* macrocell $M_{x,y}$ lies inside or outside of swept volumes V with triangles $t \in S_V$ stored in $M_{x,y}$. Given such an r that enters an $M_{x,y}$, we start with the computation of all intersection points p_{rt} between r and all triangles stored in

$M_{x,y}$. If r intersects with at least one triangle per V, we can determine whether r enters or leaves these V by comparing r with the normals of those intersected $t \in S_V$ with the shortest distance to r's origin, see Sect. 3.2.

A problem occurs when there is a V with $t \in S_V$ stored in $M_{x,y}$ not intersecting with the ray r: To solve this, we chose an arbitrary triangle $t \in S_V$ stored in $M_{x,y}$ and determine a reference point p_V that is guaranteed to lie on t and inside $M_{x,y}$. Given p_V, we create a so-called positioning ray r_{pos} from p_{rM} toward p_V. As r_{pos} is guaranteed to intersect with t at least, we can use it to determine if p_{rM} lies inside or outside of V, again by comparing r_{pos} with the normal of the intersected $t \in S_V$ nearest to p_{rM}, see Sect. 3.2.

After thus proceeding with all V with $t \in S_V$ stored in $M_{x,y}$, we know the relative location of r in p_{rM} in relation to all these V, allowing us to preset the initial intersection counter i_r^0 as follows: This time starting from $i_r^0 := 0$, i_r^0 is decreased by one for each V with p_{rM} inside, as r must have entered these V anywhere before the current $M_{x,y}$ during its grid traversal. For further illustration, see Fig. 2.

Given the initial i_r^0, we sort the determined intersections between r and the V with $t \in S_V$ stored in $M_{x,y}$ in ascending order, depending on their distance to r's origin. Starting with the first intersection, we update i_r according to (1) while checking these intersections one by one. Finally, at the point k where $i_r^k = 0$ holds, we have found the final surface hit of r. Otherwise, if no such k exists in the current $M_{x,y}$, we continue with the

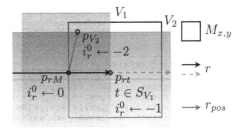

Fig. 2. Illustration of the intersection counter presetting algorithm.

SCE algorithm in the next *surface* macrocell along the ray's way through the grid.

3.5   Coherent Grid Traversal

Given the standard 3D-DDA traversal algorithm [1] for a single ray, we have two seemingly opposing problems to solve: Firstly, we want the grid's macrocell level to be as fine as possible to decrease the number of triangles that have to be tested per ray. Secondly, we want the grid to be as coarse as possible to reduce the time spent on grid traversal. To solve both problems at once, we join nearby rays into disjunct ray packets similarly to Wald et al. [8]: As the four corner rays of a ray packet define a frustum, we can incrementally compute the overlap of that frustum with the grid's macrocells, resulting in a slice-wise traversal of the grid along the packet's major traversal axis. In combination with our SCE algorithm in Sect. 3.4, we test a complete ray packet against all swept volumes' triangles that are stored in *surface* macrocells of that slice. Hence, we are able to utilize the SIMD paradigm, see Sect. 4.4: We load the *surface* macrocell's triangles only once and compute the intersection tests for all rays inside the packet in parallel afterwards. If all the packet's rays either leave the grid or hit the final workpiece's surface after all *surface* macrocells of the current slice have been tested, we stop the ray casting for that packet. Otherwise, the next slice along the major traversal axis has to be determined and investigated in the same way. Furthermore, we can apply the well-known frustum culling on these triangles, see Sect. 4.2. Thus, we can reduce the number of triangles to be tested against a ray packet without refining the macrocell level of the grid, preserving the fast packets' macrocell traversal.

4   Low-Level Optimizations

4.1   Single Instruction, Multiple Data (SIMD)

Intel introduced MMX in 1997, enabling CPUs to simultaneously perform the same operations on multiple data. Nowadays, the advanced vector extensions (AVX) extend both the original MMX instructions and register width (AVX: 256 bit). For performance reasons, we utilize SIMD explicitly with C/C++ intrinsics instead of letting the compiler decide where to use SIMD, for example:

```
#include <immintrin.h>

#define VECTOR3F_SUB(operand0, operand1, result) do { \
    result.x = _mm256_sub_ps(operand0.x, operand1.x); \
    result.y = _mm256_sub_ps(operand0.y, operand1.y); \
    result.z = _mm256_sub_ps(operand0.z, operand1.z); \
} while (0)

struct Vector3f {__m256 x; __m256 y; __m256 z;};
```

We arrange our data so that it scales with the registers' width, meaning that eight 3D vectors are stored vertically in three registers. In our implementation, we use this layout triangle-parallel during the frustum culling, see Sect. 4.2, and ray-parallel while computing the ray/triangle intersection tests, see Sect. 4.4.

4.2   SIMD Frustum Culling

Given a *surface* macrocell that is traversed by a ray packet and the packet's four corner rays defining its frustum, we compute the frustum culling for triangles stored in that macrocell similarly to the SIMD-based approach by Dmitriev et al. [2]: We use a triangle's barycentric coordinates to determine whether it lies inside the packet's frustum or not, reducing the number of ray/triangle intersection tests in *surface* macrocells that are not completely overlapped by the frustum. Note that we apply frustum culling to all swept volumes' triangles S_V stored in a traversed *surface* macrocell $M_{x,y}$ before testing the triangles $t \in S_V$ against any ray r of a ray packet P. As our SCE algorithm may need a valid reference point p_V per swept volume V to determine whether p_{rM}, the point where r enters $M_{x,y}$, lies inside or outside of these V, see Sect. 3.4, we have to distinguish between three cases here:

if p_V lies outside of P's frustum (a $t \in S_V$ between p_{rM} and p_V may have been culled)
   if $\exists t \in S_V : t$ intersects with P's frustum **then**
      use t to determine a new p_V inside both $M_{x,y}$ and P's frustum
   else use p_V and all $t \in S_V$ inside $M_{x,y}$ to determine whether P lies inside V or
      not, and if so, skip all remaining intersections tests with other V inside $M_{x,y}$
else p_V is valid as it lies inside the P's frustum

4.3 Ray Packet Repacking

As mentioned by Wald et al. [8], the slice-wise grid traversal yields the disadvantage of testing rays for intersections with triangles stored in *surface* macrocells they would never have traversed in the non-packet case. Hence, we repack the rays of a packet after the frustum culling so that only rays that really intersect with the current *surface* macrocell have to be checked for triangle intersections, resulting in a reduced number of tests while preserving SIMD compatibility.

Additionally, we record the intersecting rays per swept volume inside the current *surface* macrocell. Before using positioning rays to determine the relative location for rays that have not intersected with one or more swept volumes, see Sect. 3.4, we repack the packet's rays - with respect to the swept volumes they have missed - a second time: Given a swept volume and a list of rays that have no intersection with this volume, we have to build positioning rays to the valid reference point of that swept volume only for rays on the list. Again, we get rid of some dispensable work while supporting the simultaneously computation of the same operations on all repacked rays.

4.4 SIMD Intersection Test

Considering its good parallelization capability, we decided to adapt the algorithm presented by Moeller and Trumbore [5] to the SIMD paradigm. In their algorithm, the triangle is translated to the origin and transformed to a unity triangle in y and z, where the rays are aligned with x. In this state, the test whether the rays intersect with the triangle or not is easy to compute, see Fig. 3.

5 Results

The experimental results presented in this section are based on the following hardware configuration: Intel Core i5–2400 CPU (four cores @ 3.1 GHz, four threads, Turbo disabled, 198 GFlops) and 16 GB DDR-1333 RAM.

Using a ray packet size of eight times eight rays, 100 macrocells in the longest grid dimension and zero microcells in one macrocell dimension, (100,0), our implementation is tested against a simple scene first, see Fig. 5/1. Although being no reproduction, this scene is used to demonstrate our performance speedup compared to the recently published approximative image-space based modeling approach by Zhao et al. [11] that uses layered depth images. By adding the times for mapping, classification and rendering in Table 1, one can see that our implementation outperforms [11] by factor 10.4 with respect to [#triangles/(second · GFlops)] ([11]: 525 k triangles, 585 ms, Nvidia GeForce GTX 480, 1344 GFlops).

```
__forceinline void determineRaysTriangleIntersection(__m256& mask, const
    Vector3f& origin, const Vector3f& directions, const Vector3f& vertex0,
    const Vector3f& edge1, const Vector3f& edge2, const Vector3f& normal)
{
  Vector3f  tVec, qVec, pVec, uvw;
  __m256 det;
  mask = c_false;

  VECTOR3F_SUB(origins, vertex0, tVec);
  VECTOR3F_CP(directions, edge2, pVec);
  VECTOR3F_DP(edge1, pVec, det);
  VECTOR3F_DP(tVec, pVec, uvw.x);
  if ((MOVEMASK(uvw.x) ^ MOVEMASK(det)) == 255) return;

  VECTOR3F_CP(tVec, edge1, qVec);
  VECTOR3F_DP(directions, qVec, uvw.y);
  if ((MOVEMASK(uvw.y) ^ MOVEMASK(det)) == 255) return;

  uvw.z = SUB(SUB(det, uvw.x), uvw.y);
  if ((MOVEMASK(uvw.z) ^ MOVEMASK(det)) == 255) return;

  mask = XOR(c_true, XOR(CMPLT(uvw.x, c_zero), CMPLT(det, c_zero)));
  mask = ANDNOT(XOR(CMPLT(uvw.y, c_zero), CMPLT(det, c_zero)), mask);
  mask = ANDNOT(XOR(CMPLT(uvw.z, c_zero), CMPLT(det, c_zero)), mask);
}
```

Fig. 3. SIMD intersection test between eight rays (`origin`, `directions`) and one triangle (`vertex0`, `edge1`, `edge2`, `normal`). A valid intersection for the ray at position `index` is found iff `int (mask.m256_f32[index]) != 0` holds.

To compare our implementation with an object-space based modeling approach as well, we created Fig. 5/2 similar to the scene Romeiro et al. [6] used in their paper. With a macro-/microcell configuration of (100,3), our overall performance - the sum of the times for mapping, classification and rendering in Table 1 - is faster by a factor of 9.2 compared to their kd-tree-based ray caster that is limited to convex primitives only; with respect to [#rays/(second · GFlops)] ([6]: 640 × 480 rays, 2314 ms, Nvidia GeForce 6800 GT, 67 GFlops).

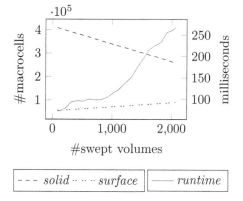

Fig. 4. Sequence of the subtractive manufacturing process of Fig. 6/4.

The last scene - as an example of complex subtractive manufacturing - is an impeller that is machined by swept volumes generated from 5-axis milling with a macro-/microcell configuration of (125,3), see Fig. 6/4. Compared to an implementation with an accelerating first grid level but without classification and elimination (68.8m grid triangles), our SCE-based approach that adds both macrocell classification and elimination reduces the number of triangles inside the grid's macrocells by 90 % (6.4 m grid triangles), resulting in a performance boost of factor 10.2. With the second

Table 1. Overview over the benchmark scenes' results. LTR: figure reference, number of used CPU cores, total number of *surface* macrocells, total number of *solid* macrocells, total number of *void* macrocells, total number of triangle references inside *surface* macrocells, used memory in [MB], average time for mapping per swept volume in [ms], average time for classification per swept volume in [ms], ray casting rendering time for 0.25 M pixels in [ms], ray casting rendering time for 1.00 M pixels in [ms], average speedup with respect to the number of used CPU cores.

fgr	#crs	#srfc	#sld	#vd	#ctrngls	mmry	mppng	clssfctn	0.25 M	1.00 M	spdp
5/1	1	4,046	7,464	374,990	7,529,401	2,359	8,019	392	1,341	2,493	-
	2						6,474	235	687	1,271	1.3
	4						5,515	157	356	657	1.7
5/2	1	63,875	22,984	413,141	1,620,027	1,283	22.6	2.9	406	978	-
	2						16.0	1.9	207	501	2.0
	4						13.8	3.0	107	259	3.6
6/4	1	90,812	259,157	449,906	4,222,313	2,794	78.9	29.6	1,043	2,554	-
	2						47.4	17.7	531	1,297	1.9
	4						29.0	10.8	270	663	3.8

grid level's microcells, the total number of triangles is reduced by another 34 % (4.2 m grid triangles), additionally increasing the rendering performance by one third. Figure 4 illustrates the generation sequence of Fig. 6/4, showing the total number of *surface* and *solid* macrocells and the rendering runtime (0.25 M pixels) of our implementation on four CPU cores, see also Table 1. The last column in Table 1 shows our implementation's scalability for all scenes: In our target scenario (Fig. 6), it is nearly linear. Furthermore, note that for a memory-bound algorithm, a higher resolution yields a higher spatial coherence of the rays which leads to an increase of cache coherent operations. Hence, the runtime of our SCE-based ray caster implementation scales sublinear with higher resolutions.

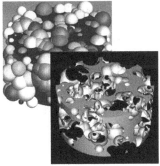

Fig. 5. LTR: *Asian Dragon* minus *Happy Buddha* (8,306,513 triangles) and cylinder minus 1,000 random spheres (5,177,870 triangles), both original volumes and ray casted Boolean subtraction result.

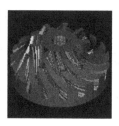

Fig. 6. Scene details: 2,062 volumes, 12,553,756 triangles. LTR: original stock and swept volumes, grid's macrocells containing all triangles (no classification) and remaining *surface* macrocells (with classification), ray casted completely machined impeller.

Acknowledgments. The models *Asian Dragon* and *Happy Buddha* were taken from *The Stanford 3D Scanning Repository*. Furthermore, this research was funded within the scope of the program *Regionale Wettbewerbsfähigkeit OÖ 2007–2013* by the *European Regional Development Fund* and the state *Upper Austria*.

References

1. Amanatides, J., Woo, A.: A fast voxel traversal algorithm for ray tracing. In: EUROGRAPHICS Proceedings, pp. 3–10 (1987)
2. Dmitriev, K., Havran, V., Seidel, H.P.: Faster ray tracing with SIMD shaft culling. Research Report MPI-I-2004-4-006, Max-Planck-Institut für Informatik, Stuhlsatzenhausweg 85, 66123 Saarbrücken, Germany (2004)
3. Goldfeather, J., Hultquist, J.P.M., Fuchs, H.: Fast constructive solid geometry display in the pixel-powers graphics system. In: ACM SIGGRAPH Proceedings of the 13th Annual Conference on Computer Graphics and Interactive Techniques, pp. 107–116 (1986)
4. Hook, T.V.: Real-time shaded NC milling display. In: ACM SIGGRAPH Proceedings of the 13th Annual Conference on Computer Graphics and Interactive Techniques, pp. 15–20 (1986)
5. Möller, T., Trumbore, B.: Fast, minimum storage ray/triangle intersection. J. Graph. Tools **2**(1), 21–28 (1997)
6. Romeiro, F., Velho, L., de Figueiredo, L.H.: Hardware-assisted rendering of CSG models. In: SIBGRAPI Proceedings of the 19th Brazilian Symposium on Computer Graphics and Image Processing, pp. 139–146 (2006)
7. Roth, S.D.: Ray casting for modeling solids. J. Comput. Graph. Image Proc. **18**(2), 109–144 (1982)
8. Wald, I., Ize, T., Kensler, A., Knoll, A., Parker, S.G.: Ray tracing animated scenes using coherent grid traversal. ACM J. Trans. Graph. **25**(3), 485–493 (2006)
9. Wald, I., Mark, W.R., Günther, J., Boulos, S., Ize, T., Hunt, W., Parker, S.G., Shirley, P.: State of the art in ray tracing animated scenes. Comput. Graph. Forum J. **28**(6), 1691–1722 (2009)
10. Wang, C.C.L., Leung, Y.S., Chen, Y.: Solid modeling of polyhedral objects by layered depth-normal images on the GPU. Int. J. Comput.-Aided Des. **42**(6), 535–544 (2010)
11. Zhao, H., Wang, C.C.L., Chen, Y., Jin, X.: Parallel and efficient boolean on polygonal solids. Int. J. Comput. Graph. **27**(6–8), 507–517 (2011)

Parallelization of Permuting XML Compressors

Tyler Corbin[1], Tomasz Müldner[1]([⊠]), and Jan Krzysztof Miziołek[2]

[1] Jodrey School of Computer Science, Acadia University,
Wolfville, NS B4P 2A9, Canada
{094568c,tomasz.muldner}@acadiau.ca
[2] IBI AL, University of Warsaw, Warsaw, Poland
jkm@ibi.uw.edu.pl

Abstract. The verbose nature of XML results in overheads in storage and network transfers, which may be overcome by using parallel computing. This paper presents four permuting parallel XML compressors, based on an existing XML compressor, called XSAQCT. Tests were performed on multi-core machines using a test suite incorporating XML documents with various characteristics, and results were analyzed to find upper bounds given by Amdahl's law, the actual speedup, and compression ratios.

Keywords: XML · XML compression · Parallelization · Java

1 Introduction

The eXtensible Markup Language, XML [9], is a World Wide Web Consortium (W3C) endorsed standard, which has become a popular representation standard for 'big data' because of its application-agnostic and human-understandable flexibility. Therefore, there has been considerable research on *XML-conscious* compressors, i.e., compression that is aware of specific syntactical or redundant features of XML, e.g., XMill [4]. XML-conscious compressors are said to be *permutation-based* when the document is re-arranged before performing the actual compression. These compressors separate structure from content, and then apply a partitioning strategy to group content nodes into a series of data containers that are compressed using general-purpose compressor (a *back-end compressor*). The grouping philosophy of permutation compressors is forward adaptive, i.e., it builds an entire model, based on the syntax of the XML document, before any actual compression of the semantic data.

With the recent availability of multi-core processors, parallel programming environments designed to take advantage of multiprocessors and shared memory have allowed applications to take advantage of the more computationally expensive compression algorithms (e.g., context mixing and PPM), with a marginal cost to the performance of the original application. While there has been plenty of research of parallel general-purpose compression algorithms, to date, there has been no generalized research on applying parallelization techniques to XML-conscious compressors.

R. Wyrzykowski et al. (Eds.): PPAM 2013, Part I, LNCS 8384, pp. 327–337, 2014.
DOI: 10.1007/978-3-642-55224-3_31, © Springer-Verlag Berlin Heidelberg 2014

Contributions. (1) Description of four parallel approaches; (2) Analysis that examines the cause and effect of (a) CPU Utilization with respect to thread synchronization, permutation uniformity, and back-end compressor complexity; (b) Disk Utilization and its relation to compression complexity; and (c) The benefits of each specific algorithm for other domains of XML; (3) Implementation and tests of all algorithms to determine their speedups using a specially designed XML corpus; and (4) Scalability analysis in terms of file size, the type of the backend compressor, and the number of cores.

Organization. Section 2 briefly describes a permuting XML compressor, XSAQCT, and Sect. 3 provides parallel algorithms based on XSAQCT. Section 4 describes the implementation, results of testing and their analysis, and finally Sect. 5 provides conclusions and describes future work.

2   Overview of Sequential XSAQCT

The algorithms described in this paper can be used for almost any permuting compressor, but here they have been applied to the XSAQCT compression process briefly described in this section. An XML document can be logically represented as a tree composed of many highly redundant subtrees that can be transformed to a more compressible form. An *annotated tree* of an XML document is a tree in which all *similar paths* (i.e., paths that are identical) are merged into a single path labeled by its tag name. Each node is labeled with a sequence of integers, called an *annotation list*, representing the number of occurrences of this node's children. In addition, the mapping from the document D to its annotated tree is one-to-one if D satisfies the so-called full mixed-content property; see Fig. 1. For the treatment of cycles, e.g., consecutive siblings of the form x→y→x and for more details see [6]. The complete XML compression process involves creating *text containers* for each unique path of an XML document, storing a delimited and possibly indexed list of *character data* for each similar path. Character data is a general term for all the characters not defined in the syntax of XML, mainly text elements and structure (whitespace) data. For example, in Fig. 1 the text container of $c[1,1,1]$ has three text elements t_{14}, t_{15}, and t_{19}.

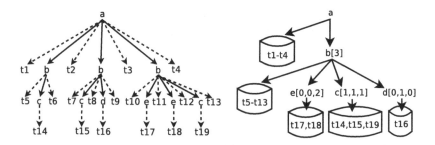

Fig. 1. A document tree (left side) and its annotated tree (right side).

In summary, XSAQCT is a schema-less, single-pass process that produces a semantically lossless representation of an XML document in a more compressed form, allowing query, and update functionalities. In the original version of XSAQCT, denoted by XSAQCT-F, when the annotated tree is being constructed, annotation lists are kept in-memory, while the contents of text containers are forced to be temporarily written to a backing-store because it is often impossible to buffer each partition in memory. Since frequent interactions with a backing store may drastically reduce CPU utilization, *especially* with a single-threaded process, XSAQCT will often pre-compress the data, i.e., the character data of the text containers will be encoded to bytes and compressed on the fly when written to the backing store. When the "compressed output" is to be created, only a copy operation is needed. This implementation proved to be quite efficient in terms of compression and decompression times, compression ratios, and computational overheads, while using only a single-pass. However, for querying, there is a non-uniformity to data access times, as it takes a longer time to access elements at the end of a text container in comparison to elements at the beginning.

3   Parallel XSAQCT

The work described in [5] showed that frequent interfacing with the backing store is often a bottleneck in sequential compression, and pre-compression should be adopted. Since pre-compression is the most computationally expensive process, multiple text containers need to be compressed concurrently on the fly. Only a single input (file or socket) stream is needed; one thread will be parsing and annotating the data and the other threads will work in tandem to create the compressed text containers. This study of different relations among threads and paths of an annotated tree showed two key points: (1) The ordering of text values is important, e.g., in a text container the bytes of a/b/c[313] must always directly precede the bytes of a/b/c[314]. This feature defines the "critical section", the portion of code that requires mutual exclusion; and (2) Indexes and centralized storage (i.e., a database) can drastically decrease synchronization overheads. The reason for this speedup is that the ordering of text values is no longer a requirement as long as the index can relate a/b/c[314] to a primary key in a database, and the synchronization overheads of non-uniform data sources (discussed in the next section) can be drastically reduced. Using this analysis, the following four new versions of parallel XML compressors (based on XSAQCT) have been designed:

XSAQCT-F: Complete XML compressor based on the use of writing and locking strategies *limited to the parsing stage*, as described in [5]. A text container buffers bytes, and when the buffer is full, it notifies a thread pool that it has data to compress. A thread eventually services this request, by locking the text container from other threads; it compresses and writes all data in this container (stealing work from other threads).

Table 1. Overview of XML corpus.

XML file	Size	Depth	Paths*	Elements	Attributes	Breakdown	Semantic	Syntax
e.w.latest [3]	37.73	6	37	1.84(8)	0.19(8)	89.3/10.7	33.69	4.04
posts [2]	11.98	4	22	0.10(8)	1.17(8)	84.6/15.4	10.14	1.85
uniprot_sprot [1]	4.90	7	216	0.99(8)	1.44(8)	36.8/63.2	1.80	3.09
1gig [10]	1.09	12	548	0.17(8)	0.04(8)	72.4/27.6	0.79	0.30
e.w.books	0.15	5	29	0.53(6)	0.05(6)	91.4/8.6	0.13	0.01
dblp	0.12	6	145	3.33(6)	0.40(6)	53.5/46.5	0.07	0.06
SwissProt	0.11	5	264	2.98(6)	2.19(6)	36.8/63.2	0.04	0.07
e.w.news	0.04	5	29	2.79(5)	0.25(5)	84.7/15.3	0.04	0.01
lineitem	0.03	3	19	10.23(5)	1	19.3/80.7	0.01	0.02
shakespeare	0.01	7	58	1.80(5)	0	61.5/38.5	0.004	0.003
uwm	0.002	5	22	0.67(5)	6	21/79	0.0005	0.0017
BaseBall	0.0006	6	46	0.28(5)	0	10.6/89.4	0.0001	0.0006
macbeth	0.0002	6	22	0.04(5)	0	59.8/40.2	0.0001	0.0001

* = Number of Unique Paths

XSAQCT-FB: A variant of XSAQCT-F designed for parallelization, which is similar to generic parallel compressors in that each text container consists of fixed sized blocks (compressed individually, so the *order* of blocks matters). Synchronization is needed when multiple threads are processing in the same container.

XSAQCT-D: XSAQCT-D is designed to act as a database front end rather than a generic XML file compressor because of the potential overheads required by a database, and the storage mechanisms are completely anonymous to XSAQCT. Instead of using *file I/O*, the annotated tree is transformed into a series of SQL statements, with the database schema representing the tree in a right-sibling-left-child structure. All of the data, including the text data, is stored in a single database instance. Moreover, XSAQCT-D does not produce a single compressed output file, which is not necessarily a fast process. Annotation lists, and more specifically, text containers consist of a series of compressed blocks. Each block is given a unique primary key referring to data stored in a database and a text container is only required to store a list of those keys, *defining the block order*. Unlike XSAQCT-FB, the *order* of blocks does not matter, which are compressed and stored in the schema-table of the underlying database.

XSAQCT-FI: A lightweight database implementation. which provides all of the necessary functionalities of a database using simple file I/O with a compression mentality. It adopts the block-based philosophy of XSAQCT-FB and the unordered nature of XSAQCT-D, where each text container is indexed and blocks are compressed. *Each* text container is an indexed list using a B-tree (or Hash) indexer. Blocks have a fixed size and data can span multiple blocks, which can be compressed individually. Annotations can be held entirely in memory, or be written to another indexed file. The default size of blocks is defined by the text compressor, e.g., if the blocks are compressed with BWT/BZIP2, the block size and BWT transform size are identical.

4   Implementation, Results of Testing and Their Analysis

4.1   Test Suite and Testing Environment

To analyze the characteristics of XML, we designed an *extension* of the Wratislavia XML Corpus [8] consisting of 13 files, provided in Table 1 (files not referred to in the table can be found in that corpus). The suite describes many *general* characteristics of XML, e.g., the "Breakdown" column shows the amount of semantic data versus the amount of syntax data in each XML file. The file size and the semantic and syntax data are in GB; $v(n)$ for the number of elements and attributes denotes $v * 10^n$. The *speedup, theoretical upper bound*, and *definite lower bound* terminology are from [5].

Path Uniformity. Since XSAQCT derives its permutations using 'similar paths', the amount of data associated with each path can play a significant role in the compression performance. For instance, for XSAQCT-F and XSAQCT-FB (the ordering of specific datum is necessary) performance issues arise when the majority of the data is local to only a few paths since many threads will require synchronization mechanisms. However, in the case of XSAQCT-D and XSAQCT-FI, the use of an index, *at the expense of compression ratio*, allows unordered writing at the cost of simple synchronization. We considered, but did not analyze, *Temporal Uniformity*, with globally uniform XML source and all the similar subtrees are locally clustered. Figure 2a depicts the data distribution of enwiki-latest.xml [3], with the path /mediawiki/page/revision/text storing 31 GB of character data, while for the next largest path there is only 0.71 GB of data. unitprot_sprot.xml has a more evenly distribution of data, see Fig. 2b.

Testing. Tests were implemented with Java (Version 1.7) on an Intel Core i7–3770 Processor, with four physical and eight logical cores, and an 8 MB cache. Each test was ran with three backend compressors: GZIP, BZIP, XZ, and no compression at all. The Berkeley Sleepycat Key-Value Database [7] was used by XSAQCT-D.

4.2   Compression, Speedups, and Analysis

A comparison of the theoretical upper bound, using Amdahl's law, to the actual speedup (obtained using eight threads) is provided in Fig. 2c–f (for files from the corpus in Sect. 4.1, with one of the three back-end compressors or no compression).

Factors in Performance. Although XSAQCT-F will typically produce the best compression ratio, Fig. 2c depicts why this technique is not inherently scalable to XML data of any type. Looking at the results of XZ, we can correctly hypothesize the "uniformed-ness" of each XML file by calculating the amount of speedup achieved. Note that XSAQCT-FB and XSAQCT-FI (see Fig. 2e, f respectively), both share the general trend that as the amount of semantic data increases, the greater speedup achieved. XSAQCT-D, see Fig. 2d, also shares many of the trends with XSAQCT-FB and XSAQCT-FI. Subtle differences arise with

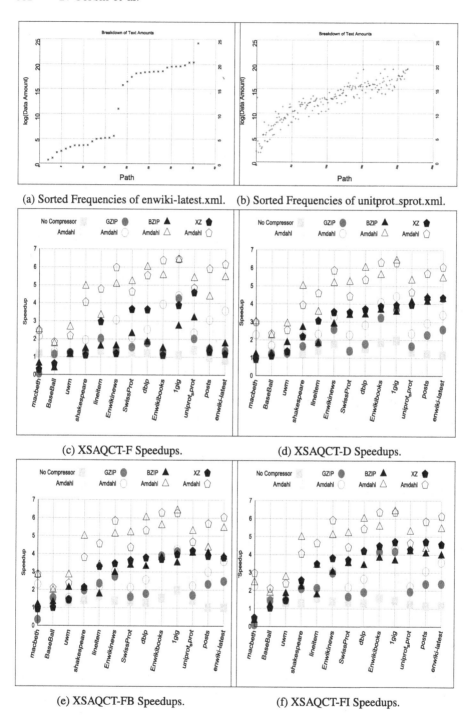

(a) Sorted Frequencies of enwiki-latest.xml.   (b) Sorted Frequencies of unitprot_sprot.xml.

(c) XSAQCT-F Speedups.   (d) XSAQCT-D Speedups.

(e) XSAQCT-FB Speedups.   (f) XSAQCT-FI Speedups.

Fig. 2. Data distribution and speedups of XSAQCT using various back-end compressors.

additional overheads offered by the database, but one major difference arises in the "time of execution" of XSAQCT-D versus the other three algorithms. Since XSAQCT-D is a centralized writing mechanism, it requires no "build phase". Comparing XSAQCT-FI to XSAQCT-D, our indexing and file I/O scheme is as scalable as a regular database system. The four smallest files (all less than 10 MB), do not scale accordingly (with eight threads) because the parser can essentially buffer the entire file (in memory). Uniformity is not a concern because the amount of data is unsubstantial. Based on the values derived by Amdahl's law, we hypothesize that building the annotated tree is as computationally expensive as compressing the data.

CPU Utilization and Compressor Complexity. To explain path uniformity and its effect on CPU utilization, consider the areas of mutual exclusion for each variant of XSAQCT: (1) for XSAQCT-F, each *unique path* has its own atomic test-and-set, which can cause starvation in heavily non-uniform XML sources. However, in highly uniform data distributions, it will excel because each thread can steal compression requests thereby reducing the overhead in thread-pool scheduling; (2) for XSAQCT-FB, since the blocks of a text-container have to be written in order, the scalability will often depend on how many threads are forced to work together in one text container. Multiple threads working together in a single container may result in certain threads waiting for some other thread to write its data. The amount of time to wait is solely dependent on the back-end compressor complexity; (3) XSAQCT-FI, while being similar to XSAQCT-FB, uses a FCFS method allowed by the underlying index; the only computational overhead is the updating of the index; and (4) XSAQCT-D follows the same methodology as XSAQCT-FI, but the synchronization is completely dependent on the database implementation and not our own implementation. Figure 3a plots the XSAQCT-F CPU usage of a uniform unitprot_sprot.xml (top) and enwiki-latest.xml (bottom). The graphs show that GZIP, or no compression, do not fully exploit computational resources, thereby hurting compression efficiency, and ultimately making it less capable of achieving a better speedup (regardless of XSAQCT-version). This implies that the threads were often *starved* because the parser was not feeding the thread pool fast enough, and not because of synchronization overheads (in this case, CPU-usage is independent of uniformity). However, for BZIP2 (or XZ, which is generally more complex), the CPU usage was often near 100 % using the uniform source, with the parser thread often waiting for the compressors to catch up, but for a non-uniform source, the CPU usage oscillations are directly related to lack-of-uniformity.

Parser and Disk Utilization. The percentage of CPU time during which I/O requests were issued to the device is another important factor in the detriment of permuting XML compressors because creating these permutations may not be an entirely in-memory process. Note that the underlying data store must be able to efficiently handle input (reading and parsing the XML data), and multiple threads writing (temporarily buffering the temporary data). Figure 3b provides the read and write transfer rates of the backing store during an XSAQCT-F compression of a uniform data source uniprot_sprot, using GZIP (top) and

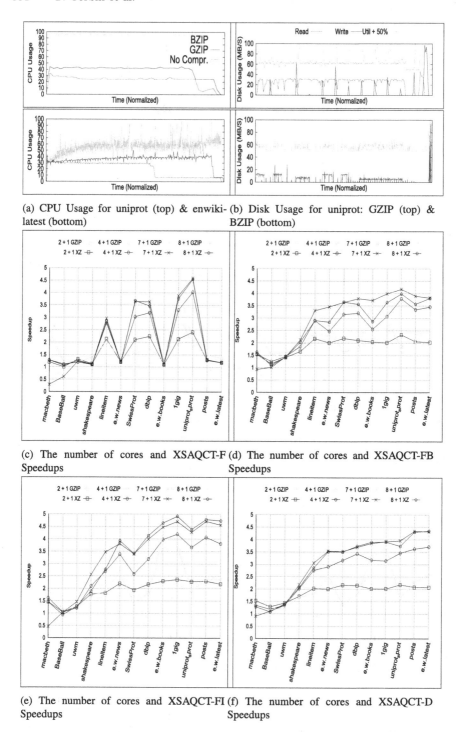

(a) CPU Usage for uniprot (top) & enwiki- (b) Disk Usage for uniprot: GZIP (top) &
latest (bottom)   BZIP (bottom)

(c) The number of cores and XSAQCT-F (d) The number of cores and XSAQCT-FB
Speedups   Speedups

(e) The number of cores and XSAQCT-FI (f) The number of cores and XSAQCT-D
Speedups   Speedups

Fig. 3. Examples of CPU and Disk usages, and impact of the number of cores on speedup.

BZIP (bottom). These graphs confirm the analysis showing that a fast compressor GZIP is not causing the parsing to slow down substantially. However, when large bursts of write requests are handled by the operating system, the parsing performance diminishes by more than 50 %. For an inherently slow compressor BZIP (also shown in Fig. 3b), the parser is always playing catch-up – waiting for some data to be fully processed – and the writing bursts are less substantial. Therefore, in some situations it is indeed possible to over-allocate threads and receive performance gains since the parser will be blocked and waiting for the compressors.

Effects of Scaling the Number of Cores. Figures 3c–f show the speedup achieved for all XML files, using one core operating the parsing thread, respectively two, four, seven, and eight cores performing compression, with GZIP and XZ as the back-end compressors. With XSAQCT-F (specifically XZ), we clearly see the prominence of path uniformity and its effect on performance. However, even with the most uniform of data sources, there is a clear point of diminishing returns after we double the number of compressor threads from four to eight, as opposed to increasing from one to two, and to four threads (compression is still complex). Comparing the smaller XML files leads to more interesting results, i.e., there is no inherent scaling, except for the smallest file and GZIP, which seems to be a consistent anomaly across all flavours of XSAQCT. Analyzing the reduction of compressor complexity shows that only 1gig.xml scales (which also happens to be a temporally uniform data file, i.e., each thread is dedicated to different nodes of the same subtree). In conclusion, adding four cores to eight cores will provide no benefit to compression, only more starvation. Finally, when we over-allocate the number of compressing threads by one (to make up for the lack of CPU-complexity in the parser), there are some files in which some slight performance is achieved and other files in which there is a performance detriment. Next, consider XSAQCT-FB. The first noticeable feature is that scaling from one to two threads doubles performance for all files of sufficient size, and in addition, it scales much more consistently than XSAQCT-F, using XZ. Reducing compressor complexity, yields better results than XSAQCT-F, but once again lacks consistent scalability. Compressing uniprot_sprot.xml with XSAQCT-FB and XSAQCT-F shows that highly uniform sources (and lack of temporal uniformity) can reduce the amount of synchronization overhead to a point, where container locking (XSAQCT-F) proves to be more efficient. Finally, over-allocating the number of compressor threads by one often decreases the performance of compression (more threads waiting for other threads). With XSAQCT-FI, the first noticeable trend is how well it scales with two and four compressing threads using XZ. Moving from one to two threads doubles the relative speedup, and doubling two threads to four once again almost doubles the relative speedup for XML files of substantial size. Finally, XSAQCT-FB, XSAQCT-FI, and XSAQCT-D all have the same linear trend with the smaller files. For macbeth.xml, baseball.xml, uwm.xml, shakespeare.xml, and lineitem.xml (ordered by size), a linear trend is noticed between file-size and core scalability. Once the files become of sufficiently large sizes, a more plateauing trend occurs (all of these trends seem to be independent of the characteristics of the file), and this defines a point in which parallelization (w.r.t file size) using a specific XSAQCT technique would be most efficient.

Additional Results. While the analysis above was described in the context of compression speedup, a plethora of benefits and detriments arise in other domains. The first obvious issue is *compression ratio*. XSAQCT-F will often produce the best compression ratios, especially with GZIP and XZ, because it builds a model backward adaptively on the entire input (on average XSAQCT-F compresses its data 8–12 % better for large data files—recall that blocks are compressed individually for other versions). XSAQCT-F and XSAQCT-FB produce the same compression ratios using BZIP because BZIP is a blocking compressor. Comparing XSAQCT-FI, which incorporates an additional index for each path, to XSAQCT-FB, our testing has shown a 2 % increase in size for files of size greater than 25 MB. However, this approach alleviates the problem with non-uniform query access as hinted in Sect. 2. For example, if an entire text container is compressed with a single GZIP instance, as with XSAQCT-F, to access the ith text element, all previous $i-1$ texts must be decompressed. On the other hand using XSAQCT-FI to access the same text element, the index is first traversed to find the block that contains the specific element. The block is then decompressed, and then traversed, searching for the string. Thus, for certain applications this overhead is certainly worth it. Finally, the size of XSAQCT-D files tend to be much more bloated (in the case of SleepyCat, there is only a slight overhead) because of the underlying architectural components of a database.

5   Conclusions and Future Work

This paper presented four new parallel compression algorithms for permuting XML compressors, as well as results of tests to determine their speedups (using a specially designed XML corpus). Regardless of XSAQCT version, the use of GZIP (or no compression) as a backend compressor resulted in frequent thread starvation and the disk/parser utilization was maximized. Using XZ or BZIP2, the CPU usage was maximized, with significant fluctuation of disk/parser utilization. We found that for XML files with more centralized data, the compression speeds are lower using XSAQCT-F and XSAQCT-FB than using XSAQCT-F and XSAQCT-D; however, in the latter case the speedup was achieved at the expense of the compression ratio. For large files, pre-compression, while saving space, did indeed increase performance. Compression-Indexing increased parallel compression speed because it minimized the amount of mutual exclusion and synchronization. In conclusion, the analysis provided an excellent example of "compression time versus compression ratio".

Preliminary work on decompression has shown that in comparison to compression there are more subtle issues with scalability, memory overheads, and using the threads efficiently (mostly to deal with pre-decompressing specific portions of each text container). Therefore, we will work on application of a divide-and-conquer approach to compression approach to XSAQCT.

Acknowledgments. The work of the first author is partially supported by NSERC CSG-M and the work of the second author by the NSERC RGPIN grant.

References

1. Consortium, T.U.: Update on activities at the Universal Protein Resource (UniProt) in 2013. http://dx.doi.org/10.1093/nar/gks1068 (2013). Accessed 20 June 2013
2. CreativeCommons: Stack Overflow Creative Commons data dump. http://blog.stackoverflow.com/?s=Data+Dump (2011). Accessed 20 June 2013
3. enwiki dumps: enwiki-latest.xml. http://dumps.wikimedia.org/enwiki/latest/ (2012). Accessed 20 June 2013
4. Liefke, H., Suciu, D.: XMill: an efficient compressor for XML data. In: Proceedings of the 2000 ACM SIGMOD International Conference on Management of Data, SIGMOD '00, pp. 153–164. ACM, New York. http://doi.acm.org/10.1145/342009.335405 (2000)
5. Müldner, T., Fry, C., Corbin, T., Miziołek, J.K.: Parallelization of an xml data compressor on multi-cores. In: Wyrzykowski, R., Dongarra, J., Karczewski, K., Waśniewski, J. (eds.) PPAM 2011, Part II. LNCS, vol. 7204, pp. 101–110. Springer, Heidelberg (2012). http://dx.doi.org/10.1007/978-3-642-31500-8_11
6. Müldner, T., Fry, C., Miziołek, J., Durno, S.: SXSAQCT and XSAQCT: XML queryable compressors. In: S. Böttcher, M. Lohrey, S.M., Rytter, W. (eds.) Structure-Based Compression of Complex Massive Data. No. 08261 in Dagstuhl Seminar Proceedings. http://drops.dagstuhl.de/opus/volltexte/2008/1673 (2008)
7. Oracle: Berkeley DB Java edition architecture. http://www.oracle.com/technetwork/database/berkeleydb/overview/index-093405.html (2013) Accessed 20 June 2013
8. Wratislavia: Wratislavia XML corpus. http://www.ii.uni.wroc.pl/~inikep/research/Wratislavia/ (2012). Accessed 20 June 2013
9. XML: Extensible markup language (XML) 1.0, 5th edn. http://www.w3.org/TR/REC-xml/ (2013). Accessed 20 June 2013
10. xmlgen: The Benchmark Data Generator. http://www.xml-benchmark.org/generator.html (2012). Accessed 20 June 2013

Parallel Processing Model for Syntactic Pattern Recognition-Based Electrical Load Forecast

Mariusz Flasiński[✉], Janusz Jurek, and Tomasz Peszek

Information Technology Systems Department, Jagiellonian University,
Ul. Prof. St. Lojasiewicza 4, 30–348 Cracow, Poland
mariusz.flasinski@uj.edu.pl

Abstract. A model of a recognition of distorted/fuzzy patterns for a electrical load forecast is presented in the paper. The model is based on a syntactic pattern recognition approach. Since a system implemented on the basis of the model is to perform in a real-time mode, it is parallelized. An architecture for parallel processing and a method of tasks distribution is proposed. First experimental results are also provided and discussed.

Keywords: Syntactic pattern recognition · Distorted/fuzzy patterns · Grammar · GDPLL(k) · Parallel parser · Electrical load forecast

1 Introduction

There are two main approaches to pattern recognition: decision-theoretic and syntactic. A syntactic approach is applied, if patterns considered are of a structural nature. Therefore, applications of this approach include e.g. analysis and recognition of chromosome shapes, bubble chamber tracks, contours in 2D pictures, ECG signals, EEG signals, a speech treated as a signal [2,8–10,22,23,28, 29]. A syntax analysis is made on the basis of theory of formal languages. It is performed in two phases. In the first phase a structural representation of a pattern is generated. Firstly, a pattern is segmented in order to identify elementary patterns, called *primitives*. Secondly, the symbolic representation of the pattern in the form of a string (a structure), which consists of symbols representing primitives is defined. This representation is treated as a word belonging to a formal language. During the next phase such a word is analyzed by a formal automaton, strictly speaking by its implementation i.e. a parsers, which is constructed on the basis of a formal grammar generating the corresponding formal language. In result a derivation of the analyzed word is obtained. The derivation is used by the syntactic pattern recognition system for describing structural features of the pattern and for recognizing (classifying) it.

Since 1980s parallel techniques and environments have been successfully applied for increasing an efficiency of pattern recognition systems [3,7,12,20,21, 25]. They have been used mainly in the first phase. However, the second phase can be also time-consuming, especially in case of a recognition of distorted/fuzzy

R. Wyrzykowski et al. (Eds.): PPAM 2013, Part I, LNCS 8384, pp. 338–347, 2014.
DOI: 10.1007/978-3-642-55224-3_32, © Springer-Verlag Berlin Heidelberg 2014

patterns [10]. Then, the use of parallel techniques in this phase could increase time efficiency.

A research into applying syntactic pattern recognition for short-term electrical load forecasting (STLF) has resulted in constructing the forecasting system [18,24]. The use of the system has revealed that distortions and fuzziness of signals analyzed influence a precision of forecasting performed in the second phase of an analysis remarkably. Therefore, enhanced syntactic pattern recognition models have been developed [6,10] in this research area. Unfortunately, in case of a problem of short-term electrical load forecasting none of the known models fulfils *all* the following conditions, which influence a forecasting precision:

– a grammar has strong generating power, i.e. it is stronger than context-free grammars,
– both algorithms of: grammar inference and syntax analysis are computationally efficient,
– fuzziness and distortions of patterns is handled not only in the primitive recognition phase, but also during a syntax analysis.

Therefore, in order to fulfil these conditions a hybrid model based on probabilistic neural networks and GDPLL(k) grammar-based parsers [5,15] has been developed recently [19]. In the model a syntax analysis is performed repeatedly in order to increase a precision of a recognition. On the other hand, it results in increasing a processing time. A parallelization of performing the analysis tasks is the best way to cope with this dilemma. For this purpose single tasks are to be identified and a suitable architecture for a task distribution is to be defined. In this paper we present a modification of the basic model of short-term electrical load forecasting, which allows us to run parsing processes in a parallel manner.

The model of the parallel analysis of distorted/fuzzy string patterns, which are used to represent electrical load signals is presented in Sect. 2. Section 3 contains the experimental results of the model. Concluding remarks are included in Sect. 4.

2 Syntactic Pattern Recognition Model for Distorted String Pattern Analysis

The model of the parallel recognition of distorted/fuzzy patterns is presented in this section. The model is an enhanced version of a model introduced in [19]. It is a hybrid model, i.e. it is based on probabilistic neural networks and GDPLL(k) grammar-based parsers. It covers both phases of a syntactic pattern recognition process, namely a primitive extraction and a syntax analysis.

2.1 Primitive Extraction

The first phase of a syntactic pattern recognition consists in a primitive extraction. In our model, the system module responsible for this task delivers an additional information related to an uncertainty factor. As a result we obtain the *extended* symbolic representation. Each element of the representation contains many possible symbols together with the recognition probability (instead of a single symbol). In this way a distortion/fuzziness nature of a primitive is characterized.

The primitive extraction is performed in our system with the help of probabilistic neural networks (PNN) [26], which have been successfully applied in a pattern recognition area [4,11,27]. A detailed description of this phase can be found in [19]. (We do not present it here, since it is out of the scope of our considerations in the paper.) It is worth to point that the primitive extraction functionality can be implemented with the use of different techniques that would fit into the designed framework and deliver a representation of the required form.

2.2 Syntax Analysis

As we have mentioned it in Introduction a grammar used for description for electrical load signals should be stronger, if a descriptive power is concerned, than context-free grammars. It results from the fact that a formal language consisting of structural representations of such signals is more complex than context-free languages. On the other hand, one cannot use a context-sensitive grammar, since the corresponding automaton, i.e. a linear bounded automaton is of the non-polynomial complexity. Therefore, a special class of quasi-context sensitive grammars, called GDPLL(k) grammars has been defined [5,15]. The GDPLL(k) grammars are strong enough to generate structural representations of electrical load signals. Moreover, a GDPLL(k) automaton is of the linear time complexity [15]. Now, we introduce GDPLL(k) grammars in a formal way.

Definition 1. A *generalized dynamically programmed context-free grammar* is a six-tuple $G = (V, \Sigma, O, P, S, M)$, where: V is a finite, nonempty alphabet; $\Sigma \subset V$ is a finite, nonempty set of terminal symbols (let $N = V \setminus \Sigma$); O is a set of basic operations on the values stored in the memory; $S \in N$ is the starting symbol; M is the memory; P is a finite set of productions of the form: $p_i = (\mu_i, L_i, R_i, A_i)$ in which $\mu_i : M \longrightarrow \{TRUE, FALSE\}$ is the predicate of applicability of the production p_i defined with the use of operations ($\in O$) performed over M; $L_i \in N$ and $R_i \in V^*$ are left- and right-hand sides of p_i respectively; A_i is the sequence of operations ($\in O$) over M, which should be performed if the production is to be applied. □

Definition 2. Let $G = (V, \Sigma, O, P, S, M)$ be a dynamically programmed context-free grammar. The grammar G is called a *GDPLL(k) grammar*, if the following two conditions are fulfilled.

1. Stearns's condition of LL(k) grammars. (The top-down left-hand side derivation is deterministic if it is allowed to look at k input symbols to the right of the current position of the input head in the string).
2. There exists a certain number ξ such that after the application of ξ productions in a left-hand side derivation we get at the "left-hand side" of a sentence at least one new terminal symbol. □

GDPLL(k) grammars have the following important features:

- a strong descriptive power (stronger than context-free grammars) [5,13],
- an efficient parsing algorithm [15],
- an efficient grammar inference algorithm [14,16,17].

2.3   Parallel Model for Distorted String Pattern Recognition

A parallel version of the model has been constructed in order to fulfil real-time processing requirements. An algorithm proposed works in an iterative way. It receives and processes vectors of the form $((v_1, p_1), \ldots, (v_n, p_n))$, where v_i is a terminal symbol denoting a primitive received, and p_i a probability that the primitive is represented by v_i. As the parser performs many parsing trials in parallel, there is a lot of tasks to process in every iteration, all of which have the following form:

$Task(PP, a) = \{Try\ to\ derive\ terminal\ symbol\ a\ with\ help\ of\ parser\ no.\ PP\}$.

The domain of (PP, a) runs over the Cartesian product of the set of all active parsing trials (PP) and all components of a currently processed vector (a). Since the domain size can be large, the number of tasks to process can increase an analysis time remarkably. On the other hand, skipping some tasks in an analysis process can lead to a less precise recognition. A parallelization of processing tasks is a good way to handle the problem. A task dispatcher assigns a derivation task to separate processors/cores/threads, which are processed in a parallel way.

The architecture of a parallel GDPLL(k) parser is shown in Fig. 1. Let us describe its basic elements.

- The **input tape** contains a string of vectors.
- The **collection of GDPLL(k) parsing processes.** For each process the following information is stored: the current stack, the current production list, the current automaton state, the string already analyzed etc. On the basis of this information resuming the analysis process is possible.
- The **control module** governs the whole analysis process. It involves managing the following two elements.
  - The **task queue.** The task queue is created by the control module before each iteration on the basis of the collection of parsing processes and components of the input vector.
  - The **processor/thread/core pool.** The control module assigns subsequent tasks from the queue to execute to available parsers (processors/cores/threads) and collects analysis results. As a result of each analysis, existing parsing processes can change their state or can be terminated (acceptance, rejection).
- The **pool of GDPLL(k) parsers.** For each parser the control function is defined. This function is used by the control module for deciding whether the parser should continue its analysis. The control module sends tasks from the task queue to the parsers. The parser, which has finished processing,

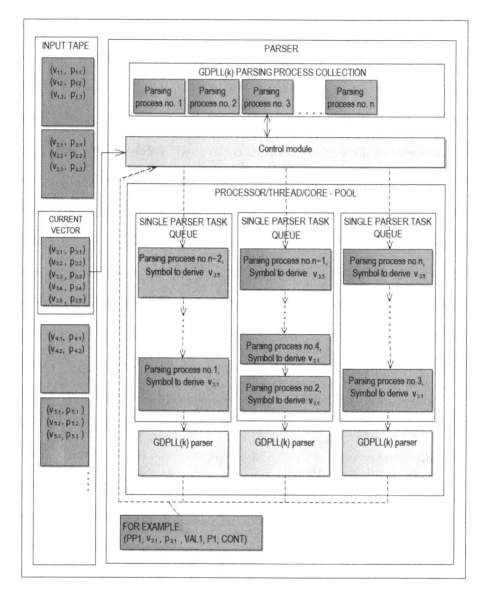

Fig. 1. The architecture of the parser.

requests the control module for assigning the next task from the queue. During an analysis performed by the parser, a value of its control function is computed. The parser continuously delivers messages of the following form: $(PP, v, p, val, prod, action)$, where: PP is an index of the parsing process, (v, p) is a component of the input vector, val is the current value of the control function, *prod* is the production to be used in the derivation process, and *action* is one of the following possible control actions

- *ACC*: A symbol v is derived and the word related to the analysis path is accepted.
- *CONT*: A symbol v is derived. A derivation starts with the production *prod*. This path should be a subject of further analysis (the value of the control function is big enough).
- *TERM*: A symbol v is derived. However, further analysis along this path should not be continued (the value of the control function is too small).
- *REJ*: A symbol v cannot be derived and the word related to the analysis path is rejected.

- The **output of the control module** delivering information on symbols read and productions applied to derive these symbols.

The parser performs the following steps in an iterative way:

- read next vector (or k vectors) from the input tape,
- create a task queue,
- execute tasks from the queue with the help of the pool of available parsers,
- collect the results of the analysis and update the collection of parsing processes,
- send messages on termination/acceptance/rejection of parsing processes to the output of the control module.

A recognition of an input symbolic vector, which represents an electric load signal is a result of the syntax analysis performed by the parser. A value of a probability of such a recognition is also delivered by the parser.

3   Experimental Results

In this section we present experimental results obtained with the system implemented on the basis of the model presented above. Let us consider a language $L = \{a^n b^n c^n : n > 0\}$. Since the language is context-sensitive, it is difficult to construct an efficient syntax analyzer for it. A GDPLL(k) parser recognizes such a language in a polynomial time thanks to its very good computational properties, mentioned in a previous section.

For our experiments we have used words represented by the input vectors, which consist of two components. Consequently, the number of the possible resolutions is exponential with the base equal to 2. Let us assume that the input vector to be parsed is defined in the following way:

$$\begin{bmatrix} a, 0.7 \\ b, 0.3 \end{bmatrix} \begin{bmatrix} a, 0.75 \\ b, 0.25 \end{bmatrix} \begin{bmatrix} b, 0.6 \\ c, 0.4 \end{bmatrix} \begin{bmatrix} b, 0.5 \\ c, 0.5 \end{bmatrix} \begin{bmatrix} b, 0.2 \\ c, 0.8 \end{bmatrix} \begin{bmatrix} b, 0.2 \\ c, 0.8 \end{bmatrix}$$

Thus, the number of possible resolutions is equal to $2^6 = 64$. The size of the input data depends on both: the number of resolutions and the length of the input vector word. These parameters influence the number of operations that have to be made for complete parsing. Fortunately, many of this resolutions are rejected at the initial phase of the syntax analysis.

A parser efficiency has been tested with a four core processor machine (Intel(R) Core(TM) 2Quad CPU Q6600 2,4 GHz) with 3 GB of an available RAM.

A computation time as a function of the input word length is shown in Fig. 2. (The word length means here the number of vectors, which belong to a particular word.) Let us notice that in spite of a big number of possible resolutions, the function is still linear-like. This is a consequence of eliminating whole "branches" of the resolutions during a derivation by the control module.

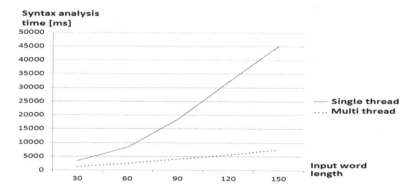

Fig. 2. A syntax analysis time as a function of the input vector word length (an analysis with a one- and four-thread environment).

Table 1 contains times of computations performed with the help of a one- and four-threaded processes.

Table 1. Time required for a complete analysis of a vector word as a function of its length

Word length	Single threaded (ms)	Multi(4) threaded (ms)
30	3479	1310
60	8286	2434
90	18413	4099
120	32041	5612
150	45143	7497

The real-world experiments concerning the language of structural shapes, which represent an electrical load demand have been performed as well. Structural patterns of load demand shapes are described with a set of primitives shown in Fig. 3. For example, a language word $f^4c^2b^4e^2d^5fdb^2cf$ is a string coding of a structural pattern of the electrical load demand, which is shown in Fig. 3.

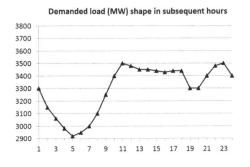
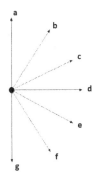

Fig. 3. A structural pattern of an electrical load demand and a set of primitives used for its string coding

Words of various lengths containing structural distortions have been analyzed during experiments. The results are included in Table 2.

Table 2. Time required for a complete analysis of a vector word representing structural shapes of an electrical load demand as a function of its length

Word length	Single threaded (ms)	Multi(4) threaded (ms)
18	1988	529
30	3751	946
42	5471	1376
54	7200	1810
66	8932	2243

One can easily see that the use of a parallel environment results in a big reduction of the syntax analysis time. A difference between a single-threaded process and a multi-threaded process increases as the length of the input vector is longer and longer. A reduction of a time of the syntax analysis time relates approximately to the number of available threads.

4   Concluding Remarks

A parallel model of syntactic recognition of distorted/fuzzy string patterns has been presented in the paper. The model has been developed to match require-ments of the application for the short-term electrical load forecasting (STLF) [1,18,24,30]. STLF is of a great significance as far as the optimal management of resources required for energy production is concerned.

Good points of the model can be sum up as follows. The model is based on GDPLL(k) grammars, which are of a big generating/descriptive power. Secondly, a GDPLL(k) parser is computationally efficient, i.e. it is of the linear time complexity. Nevertheless, such a complexity has been obtained for model patterns, i.e. for patterns, which are neither distorted nor fuzzy. In case of an analysis of electrical load signals such an assumption is too strong. Of course, when distorted/fuzzy patterns are to be analyzed, a parsing time increases remarkably. Therefore the model had to be parallelized to meet real-time requirements. The experiments have shown that the use of the parallel model of GDPLL(k) parsing reduces processing time considerably.

References

1. Alfares, H.K., Nazeeruddin, M.: Electric load forecasting: literature survey and classifcation of methods. Int. J. Syst. Sci. **33**, 23–34 (2002)
2. Bunke, H.O., Sanfeliu, A. (eds.): Syntactic and Structural Pattern Recognition Theory and Applications. World Scientific, Singapore (1990)
3. Chiang Y., Fu K.S.: Parallel parsing algorithms and VLSI implementation for syntactic pattern recognition. IEEE Trans. Pattern Anal. Machine Intell. PAMI-6, 302–313 (1984)
4. Emary, I.M., Ramakrishnan, S.: On the application of various probabilistic neural networks in solving different pattern classification problems. World Appl. Sci. J. **4**, 772–780 (2008)
5. Flasiński, M., Jurek, J.: Dynamically programmed automata for quasi context sensitive languages as a tool for inference support in pattern recognition-based real-time control expert systems. Pattern Recogn. **32**, 671–690 (1999)
6. Flasiński, M., Jurek, J.: On the analysis of fuzzy string patterns with the help of extended and stochastic GDPLL(k) grammars. Fundamenta Informaticae **71**, 1–14 (2006). IOS Press, Amsterdam
7. Flasiński, M., Jurek, J., Myśliński, S.: Multi-agent system for recognition of hand postures. In: Allen, G., Nabrzyski, J., Seidel, E., van Albada, G.D., Dongarra, J., Sloot, P.M.A. (eds.) ICCS 2009, Part II. LNCS, vol. 5545, pp. 815–824. Springer, Heidelberg (2009)
8. Flasiński, M., Jurek, J.: Syntactic pattern recognition: survey of frontiers and crucial methodological issues. Adv. Intell. Soft Comput. **95**, 187–196 (2011). Springer, Berlin
9. Flasiński M., Jurek J., Fundamental methodological issues of syntactic pattern recognition. Pattern Anal. Appl. (2013) (in print). Springer
10. Fu, K.S.: Syntactic Pattern Recognition and Applications. Prentice Hall, Englewood Cliffs (1982)
11. Goodman, R.M., Higgins, C.M., Miller, J.W.: Rule-based neural networks for classification and probability estimation. Neural Comput. **4**, 781–804 (1992)
12. Guerra, C.: 2d object recognition on a reconfigurable mesh. Pattern Recogn. **31**, 83–88 (1998)
13. Jurek, J.: Syntactic pattern recognition-based agents for real-time expert systems. In: Dunin-Keplicz, B., Nawarecki, E. (eds.) CEEMAS 2001. LNCS (LNAI), vol. 2296, pp. 161–168. Springer, Heidelberg (2002)
14. Jurek, J.: Towards grammatical inferencing of GDPLL(k) grammars for applications in syntactic pattern recognition-based expert systems. In: Rutkowski, L., Siekmann, J., Tadeusiewicz, R., Zadeh, L.A. (eds.) ICAISC 2004. LNCS (LNAI), vol. 3070, pp. 604–609. Springer, Heidelberg (2004)

15. Jurek, J.: Recent developments of the syntactic pattern recognition model based on quasi-context sensitive languages. Pattern Recogn. Lett. **26**, 1011–1018 (2005). Elsevier, Amsterdam

16. Jurek, J.: Generalisation of a language sample for grammatical inference of GDPLL(k) grammars. Adv. Soft Comput. **45**, 282–288 (2007). Springer, Berlin

17. Jurek, J.: Grammatical inference as a tool for constructing self-learning syntactic pattern recognition-based agents. In: Bubak, M., van Albada, G.D., Dongarra, J., Sloot, P.M.A. (eds.) ICCS 2008, Part III. LNCS, vol. 5103, pp. 712–721. Springer, Heidelberg (2008)

18. Jurek, J., Peszek, T.: On the use of syntactic pattern recognition methods, neural networks, and fuzzy systems for short-term electrical load forecasting. Adv. Soft Comput. **30**, 851–858 (2005). Springer, Berlin

19. Jurek, J., Peszek, T.: Model of syntactic recognition of distorted string patterns with the help of GDPLL(k)-based automata. Adv. Intell. Soft Comput. **226**, 101–110 (2013)

20. Lee S.S., Tanaka H.T.: Parallel image segmentation with adaptive mesh. In: Proceedings of the 15th International Conference Pattern Recognition, Barcelona, Spain, vol. 1, pp. 635–639 (2000)

21. Miguet, S., Montanvert, A., Wang, P.S.P.: Parallel Image Analysis. World Scientific, Singapore (1998)

22. Ogiela, M.R., Ogiela, U.: Dna-like linguistic secret sharing for strategic information systems. Int. J. Inf. Manag. **32**, 175–181 (2012)

23. Pavlidis, T.: Structural Pattern Recognition. Springer, New York (1977)

24. Peszek, T.: Neuro–fuzzy prediction systems in energetics. Schedae Informaticae **15**, 73–94 (2006)

25. Ranganathan, N.: VLSI and Parallel Computing for Pattern Recognition and Artificial Intelligence. World Scientific, Singapore (1995)

26. Specht, D.F.: Probabilistic neural networks. Neural Netw. **3**, 109–118 (1990)

27. Tadeusiewicz, R.: Sieci neuronowe. Akademicka Oficyna Wydawnicza, Warszawa (1993)

28. Tadeusiewicz, R., Flasiński, M.: Rozpoznawanie Obrazów. Państwowe Wydawnictwo Naukowe PWN, Warszawa (1991)

29. Tadeusiewicz, R., Ogiela, M.R.: Medical Image Understanding Technology. Springer, Berlin (2004)

30. Taylor, J., McSharry, P.: Short-term load forecasting methods: an evaluation based on European data. IEEE Trans. Power Syst. **22**, 2213–2219 (2008)

Parallel Event–Driven Simulation
Based on Application Global State Monitoring

Łukasz Maśko[1]([✉]) and Marek Tudruj[1,2]

[1] Institute of Computer Science of the Polish Academy of Sciences,
ul. Jana Kazimierza 5, 01-248 Warsaw, Poland
[2] Polish–Japanese Institute of Information Technology,
ul. Koszykowa 86, 02-008 Warsaw, Poland
{masko,tudruj}@ipipan.waw.pl

Abstract. Discrete event simulation is a well known technique used
for modeling and simulating complex parallel systems. Parallel simulation introduces multiple simulated event queues processed in parallel.
A proper synchronization between parallel queues must be introduced.
Program global state monitoring is a natural way to organize global
simulation state monitoring and control. Every queue process reports
its progress state, being the timestamp of the most recently processed
event, to a global synchronizer. Reporting is done asynchronously and
has no influence on the simulation process. A global simulation state
can be defined as the vector containing timestamps of the most recently
processed event in every queue. The paper presents the principles of parallel simulation designed by the use of a system infrastructure for global
states monitoring. Comparison to existing parallel simulation methods
is provided.

Keywords: Parallel simulation · Strongly consistent global states

1 Introduction

Discrete–Event Simulation (DES) is a well known technique used for modeling
and simulating complex parallel systems [1,2]. In this approach, each simulated
entity reacts to events, which are addressed to it, and as a result produces new
events, which may be sent as messages to any system component. Each such
message is marked with a timestamp, which corresponds to the event occurence
time. To preserve correctness of simulation, all event messages are examined in
respect to their timestamps. Such approach may be easily adjusted to desired
system granularity. It can be used for cycle–accurate simulations and for simulation with lower precision requirements.

DES can be implemented using a simple serial algorithm. This algorithm uses
a single event priority queue. All events are sorted according to their timestamps.
In each step, the event with the lowest timestamp is selected and processed. The
resulting events are then stored in the same event queue, according to their

R. Wyrzykowski et al. (Eds.): PPAM 2013, Part I, LNCS 8384, pp. 348–357, 2014.
DOI: 10.1007/978-3-642-55224-3_33, © Springer-Verlag Berlin Heidelberg 2014

timestamps. Because the processed event has always timestamp which is not higher than the timestamps of the resulting events, simulation is always correct. The main drawback of this method is its strictly sequential nature, which makes it hard to parallelize.

Parallel DES (PDES) introduces multiple event queues [1,2], which are distributed between Logical Processes (LPs), which process events in parallel. The events, which are results of processing, are exchanged between queues using messages. The distribution of the events between queues depends on the assumed algorithm and the simulated system architecture. It can depend on the connection topology between simulated components, so that messages between queues are migrated only if they address communication between disjoint parts of the simulated system [3]. Simulation methods determine possible synchronization mechanisms between parallel queues. If this problem is not properly solved, a queue may receive a remote message from another queue with a timestamp lower than the events it had already processed, which breaks the correctness rule.

Instead of strict synchronization of LPs in the PDES, which leads to decrease of simultarion performance, simulation progress may be controlled with an asynchronous method, which uses strongly consistent global states (SGCS). SCGS is a set of fully concurrent local states detected by a synchronizer. The paper proposes the simulation algorithm, which encorporates the idea of global–state driven program control into the PDES algorithm in order to reduce the probability of rollbacks.

The paper is composed of 4 parts. First, the current state of the art in the field of PDES methods is briefly shown. Then, the idea of global–state driven program control model is presented. The third part introduces the algorithm using global states to control the parallel discrete event simulation progress. Finally, an eamplary simulation is depicted.

2 Related Work

The conservative approach to PDES [4] assumes that it is not allowed that a message with a timestamp lower then events already processed arrives in a queue. Such approach requires exact synchronization between event queues. In general, any queue cannot process its events, if the timestamp of the earliest one is higher than timestamps of first events in other queues. Such conservative rule leads to sequential execution and parallelism is available only if there are events with the same timestamp in front of different queues. This limit can be relaxed using information about the scheduled system. For instance, if queues correspond to separate parts of the system and sending a message between these parts takes a known amount of time, the difference between clocks in these queues may reach this value. Such relaxation mechanism was assumed in [3].

Unlike in conservative approach, the optimistic approach assumes, that the progress of distributed queues doesn't have to be strictly synchronized [1,2,5–7]. The basic optimistic parallel simulation method is called Time Warp [5]. Time Warp allows different queues to proceed with their computations as far as possible. At the same time, to preserve consistency of simulation, in the case when events are processed in the wrong order (one of the queues receives a remote event generated in a different queue with timestamp lower that its own clock value), the queue, in which such error occurred, performs a rollback, i.e. restores its state from the point of time, which is equal to timestamp of the remote event, which caused rollback.

Each rollback imposes reduction of simulation performance. Restoring a state of a queue to a recorded state means not only removal of some events from the queue, but also requires cancellation of messages beeing results of processing of these events, sent to other queues. If such events had been already processed, those queues must also perform rollbacks, which may cause more cancellations and more simulation performance degradationas a result of rollback avalanche. The probability of rollback depends on the simulated system architecture and on the distribution of simulated components between event queues. It also depends on the time difference of simulation between queues.

In order to lower the probability of rollback, the distance between the least and the most advanced queue in the whole simulated system can be limited. Some solutions of this problem have been proposed. In Moving Time Windows [6], each process is allowed to proceed with its simulation only up to a fixed time limit, after which all the processes synchronize and proceed with the next step of concurrent computations. The Time Bucket [7] approach also proposes work in phases, but the length of each phase depends on new messages generated. After each such phase, a global synchronization takes place. In the Breathing Time Warp algorithm [7], in each phase first a standard Time Warp algorithm is used for a fixed number of messages, then simulation is switched to Time Bucket algorithm and finally, synchronization of all the processes is performed before the next simulation step starts. All of these algorithms require central control or additional synchronization phases, which constitute a possible bottleneck and also may decrease simulation speed.

3   Global–State Driven Program Control Model

The idea of asynchronous program execution control based on global states monitoring is illustrated in Fig. 1 [8,9]. Application consists of computational processes and threads that can be executed in parallel in a system with shared or distributed memory. All these processing elements can exchange data by using standard communication mechanisms like sockets, MPI or shared memory, implemented manually by a programmer.

In order to perform global synchronization operations, a special control infrastructure has been delivered. Its main element is called a synchronizer. It gathers local state messages from all computational elements. It detects, if the application has reached a strongly consistent global state SCGS. SCGS is a set of fully concurrent local states detected by a synchronizer. The construction of strongly consistent global states is based on projecting local states of all relevant processes or threads on a common time axis and finding time intervals undoubtedly covered by local states of all these processes/threads (Fig. 2).

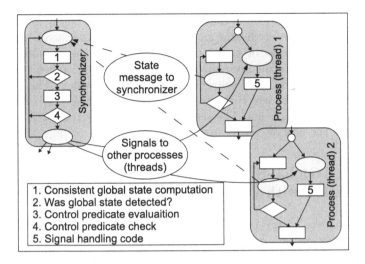

Fig. 1. Synchronizer co-operation with processes or threads.

The system run-time framework, working on-line in parallel with the application, can provide the infrastructure for sending the process local states, reception of the state messages by synchronizers, reconstruction of different kinds of global application states including Strongly Consistent Global States (SCGS) and using them to define the execution control of the application. An SCGS is a state, which has occurred for sure in all involved processes.

For the assumed SCGS generation algorithm [8,9], the application processes send local state messages accompanied by real-time event timestamps obtained with the use of partially synchronized local processor clocks, globally synchronized with a known accuracy. It requires a synchronization facility installed in the distributed system.

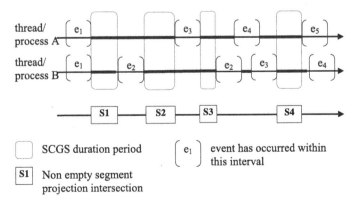

Fig. 2. SGCSs detection.

When an SGCS is detected, a synchronizer evaluates, based on them, control predicates. If some predicates are true, the synchronizer sends control signals to selected processes or threads. The signals must be handled by these processes/threads and some desired reaction should be undertaken.

Two types of control by synchronizer signals are provided. The first is the asynchronous control in which the reaction consists in triggering or cancelling some computation fragments. The second is synchronous control in which the signals change the flow of control in the program graph.

4 Parallel Global-State Controlled Event-Driven Simulation with Optimistic Approach

Global-state driven program control paradigm may be used to improve the efficiency for optimistic approach to parallel DES. The queues used for event processing are distributed between LPs. To control the whole simulation, a single global synchronizer or a hierarchy of synchronizers may be used. Every Logical Process reports a progress state of its queue to a global synchronizer. The state of LP is the timestamp of the most recently processed event. Reporting is done asynchronously at the beginning and end of processing of each event and has no influence on the simulation process. To improve efficiency of asynchronous control, we also assume that whenever the state caused by the start of such procedure is sent, LP also attaches to it the predicted logical end time of this procedure, if known. It can be obtained if each signal handling block has a known length. Knowledge of this value may allow the synchronizer to react even before an LP advances its logical clock beyond the threshold value.

The global simulation state can be defined as the vector containing timestamps of the most recently processed events in every queue. The global synchronizer gathers progress states of all the queues in the simulated system. If a global state is detected, the synchronizer can compute the differences in progress of simulation of all the queues, as shown in Algorithm 1. Whenever this difference is higher than the assumed threshold ΔT, the synchronizer sends the SUSPEND signal to those queues, in which progress has evaluated more then ΔT time units, comparing to the slowest queues.The suspended LP is supposed to stop all its activities after it finishes execution of a current event handling procedure (if any), including sending of the resulting messages. It should accept all the incoming events, but it must wait with processing its elements, until it receives the RESUME signal or if it must perform a rollback due to a delayed message, which arrives. In the same time, the synchronizer sends the RESUME signal to all the already suspended queues, for which the difference in their progress, comparing to the slowest queues, is smaller then ΔT.

The proposed algorithm requires a threshold ΔT, which influences the simulation performance. A small value may limit the parallelism in a simulation, because many queues will be suspended even if they proceed with their events only a little, when compared to the most delayed one. On the other hand, increase of ΔT improves parallelism in the simulation, but also increases the probability of rollbacks, which reduce simulation performance. Therefore, the proper value for each simulated system must be determined using profiling.

Progress control may benefit from the knowledge of the simulated system structure. If parts of the system assigned to the two queues are closely coupled and the messages sent between them introduce a small delay, the difference between progresses of these queues should be smaller. In the same time it can be bigger if the two parts of the system are distant in terms of time needed for a message to travel between them.

The presented algorithm does not fully prevent from rollbacks. Therefore, whenever they happen, the global synchronizer must be informed about this fact and it must update its local LPclock and LPstate arrays to match the current state of the simulated system.

Algorithm 1. Predicate code for strongly consistent state driven simulation progress control

---

Input:
– LPclock[]: array containing local clocks determined from messages received from all LPs. This array is assembled by the predicate.
– LPclockP[]: array containing the predicted logical times of completion of the currently executed event handling procedure in LPs (-1 if not reported).
– LPstate[]: array containing information about the state of each LP (running or suspended). At the beginning, all the LPs are in the "running" state which means, that they process messages. If a message receives a SUSPEND signal, its state is changed to "suspended" until the RESUME signal is sent by the synchronizer. Then the state of such LP is set back to "running".

```
 1: {Compute minimal local clock value over values from LP_clocks}
 2: minClk = LPclock[1];
 3: for lp=2 to N do
 4:     minClk = min(minClk, LPclock[lp])
 5: end for
 6: for lp=1 to N do
 7:     if LPclock[lp] > minClk + threshold or
        (LPclockP[lp] <> -1 and LPclockP[lp] > minClk + threshold) then
 8:         {LP number lp has advanced too much...}
 9:         if LPstate[lp] = "running" then
10:             {...and it's still running – suspend it}
11:             send SUSPEND signal to LP number lp
12:             LPstate[lp] = "suspended"
13:         end if
14:     else
15:         if LPState[lp] = "suspended" then
16:             {LP has been suspended, now it should proceed with simulation}
17:             send RESUME signal to LP number lp
18:             LPstate[lp] = "running"
19:         end if
20:     end if
21: end for
```

4.1   Simulation Example

As an example, we propose a simulation of a simple parallel system, divided into 3 logical processes LP1, LP2 and LP3. Processes LP2 and LP3 periodically generate events, which are sent to LP1. LP1 processes these messages and sends back a reply message to the corresponding LP, which begins generation of the next message and operation repeats.

In our example, we assume that generation of a message consumes 4 logical time units in LP2 and 10 logical time units in LP3. Message processing on LP1 takes 1 logical time unit. In the same time, we assume that simulation of message generation takes 4 physical time units in LP2 and 3 physical time units in LP3. Simulation of response generation in LP1 takes 1 physical time unit. Message delivery in the simulating system also takes 1 physical time unit. We assume that sending/receiving of a physical message is included in the message processing block and data transfer operation is asynchronous. We assume that rolling back of a block takes twice as much time as simulation of this block. The threshold ΔT required for the presented algorithm was set to 5.

Figure 3 depicts simulation of the proposed case. Due to disproportions in physical and logical processing times between logical processes LP2 and LP3, LP1 will detect a misplaced message at time 5. This message has the timestamp equal to 4 (while the local logical clock value is 10) and will force LP1 to rollback previously processed event (no anti-messages messages are sent for no responses

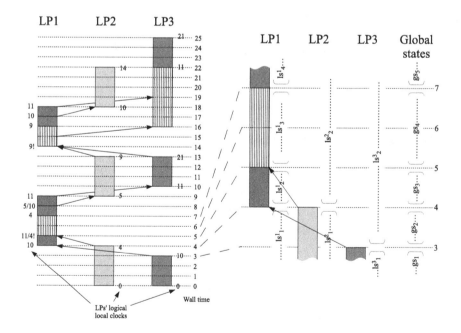

Fig. 3. Diagram of the simulation with the Time-Warp method with rollbacks (left) and local and global states in the part of the simulation (right).

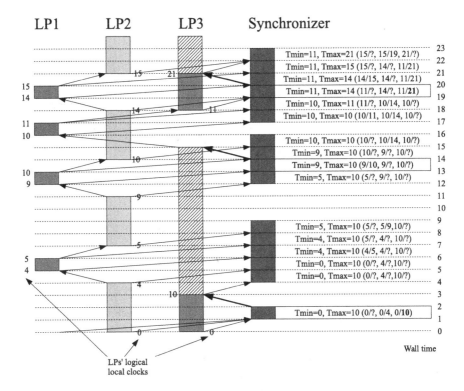

Fig. 4. Diagram of the simulation with asynchronous control.

were sent yet). The same situation may be spotted at timestamp 14, but here an anti-message must be sent to LP3 to force it to also perform a rollback and cancel computations which were performed after LP3 received previous response. In the presented diagram, each rollback is represented as vertically lined rectangle.

Figure 4 depicts the same simulation, but with a synchronizer added. We assume that the synchronizer is capable of determining the simulation state within 1 physical time unit. State report sending is done asynchronously by LPs and its delivery consumes 1 physical time unit. Also control signals sent from the synchronizer to LPs are delivered after 1 physical time unit. We assume that many state messages and control signals may be sent and delivered in parallel.

The simulation states are represented by means of records shown for the physical time points at the right hand side in Fig. 4. $Tmin$ and $Tmax$ correspond to the minimal and the maximal local logical clock value over all LPs. $Tmin$ is equal to GVT (Global Virtual Time) of the simulated system. The values $(T_1/T_1^P, T_2/T_2^P, T_3/T_3^P)$ in brackets represent the global system state and correspond to local logical clock values of LPs (T_i corresponds to the logical clock value of LPi) and predicted logical block simulation completion time (T_i^P, '?' means that this value is unknown).

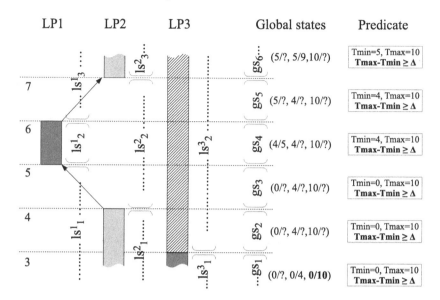

Fig. 5. Global states and predicates in simulation with asynchronous control.

The synchronizer implements the proposed simulation progress control algorithm. The synchronizer receives from LPs state report messages, which contain local logical clock values of LPs. The messages are sent at each beginning and end of message processing block (arrows leading from LPs to the synchronizer). The synchronizer collects local state messages and, based on them, it generates strongly consistent global states (SCGSs). Whenever it detects a SCGS for which the difference between $Tmax$ and $Tmin$ exceeds the assumed threshold ΔT, it sends the "SUSPEND" signal to such LPs for which $T_i \geq Tmin + \Delta T$. It also sends such signal if predicted logical time of completion of the currently simulated block exceeds such limit (for instance, the message to LP3 at time 2), which prevents an LP from sending the resulting messages too early. On the other hand, the synchronizer sends the "RESUME" signal to all such suspended processes LPi for which $T_i < Tmin + \Delta T$.

Figure 5 depicts strongly consistent global states as they are created in a part of the simulation from Fig. 4. When the global state gs_1 is detected, the synchronizer evaluates its predicate, which reports that the condition $Tmax < Tmin + \Delta T$ does not hold. This triggers a SUSPEND signal, which is sent to LP3 (Fig. 4). As a result, simulation in LP3 is suspended. All the activities are resumed in LP3 after two other LPs increase their LVTs above 5.

In the presented case, the proposed algorithm leads to elimination of rollbacks. Analysis of the diagram reveals that the overall algorithm efficiency strictly depends on the speed of the control subsystem. Efficient synchronization, although asynchronous, will lead to performance improvement – if the synchronizer is capable for activity closer to synchronous control, its efficiency improves.

5   Conclusions

The paper presents a new approach to parallel simulation, which introduces asynchronous control based on global consistent states of the simulation. Simulation execution control is performed by the global synchronizer (or hierarchy of synchronizers), in which a global synchronizer collects information about simulation progress from LPs in the simulation system and controls their progress by sending SUSPEND and CONTINUE signals. This control is done asynchronously and aims in rollback reduction, not elimination, therefore, due to network latencies, the LPs must be ready to perform rollback operations, like in a standard Time Warp algorithm. The asynchronous control efficiency strictly depends on efficiency of network communication. Therefore, the state reports and control signals should be sent via a fast, dedicated network, which is not used for standard data transfers. Depending on the simulated system properties, the presented approach may lead to significant reduce in the number of performed rollbacks.

References

1. Fujimoto, R.M.: Parallel discrete event simulation. Commun. ACM - Special Issue on Simulation **33**(10), 30–53 (1990). (ACM, New York)
2. Ferscha, A., Tripathi, S.K.: Parallel and distributed simulation of discrete event systems, technical report, UM Computer Science Department; CS-TR-3336, UMIACS; UMIACS-TR-94-100 (1998)
3. Lv, H., Cheng, Y., Bai, L., Chen, M., Fan, D., Sun, N.: P-GAS: parallelizing a cycle-accurate event-driven many-core processor simulator using parallel discrete event simulation. In: PADS '10 Proceedings of the 2010 IEEE Workshop on Principles of Advanced and Distributed Simulation, pp. 89–96. IEEE Computer Society, Washington, DC (2010)
4. Chandry, K.M., Misra, J.: Distributed simulation: a case study in design and verification of distributed programs. IEEE Trans. Softw. Eng. **5**(5), 440–452 (1979)
5. Jefferson, D.R.: Virtual time. ACM Trans. Prog. Lang. Syst. **7**, 404–425 (1985)
6. Sokol, L., Briscoe, D., Wieland, A.: MTW: a strategy for scheduling discrete simulation events for concurrent execution. In: Proceedings Distributed Simulation Conference (1988)
7. Steinman, J.S.: Breathing time warp. In: PADS '93 Proceedings of the Seventh Workshop on Parallel and Distributed Simulation. ACM, New York (1993)
8. Tudruj, M., Borkowski, J., Maśko Ł., Smyk, A., Kopański, D., Laskowski, E.: Program design environment for multicore processor systems with program execution controlled by global states monitoring. In: ISPDC 2011, Cluj-Napoca, pp. 102–109, IEEE CS, July 2011
9. Kopański, D., Maśko, Ł., Laskowski, E., Smyk, A., Borkowski, J., Tudruj, M.: Distributed program execution control based on application global states monitoring in PEGASUS DA framework. In: Wyrzykowski, R., Dongarra, J., Karczewski, K., Waśniewski, J. (eds.) PPAM 2013, Part I. LNCS, vol. 8384, pp. 302–314. Springer, Heidelberg (2014)

Applied Mathematics, Evolutionary Computing and Metaheuristics

It's Not a Bug, It's a Feature: Wait-Free Asynchronous Cellular Genetic Algorithm

Frédéric Pinel[1]([✉]), Bernabé Dorronsoro[2], Pascal Bouvry[1],
and Samee U. Khan[3]

[1] FSTC/CSC/ILIAS, University of Luxembourg, Kirchberg, Luxembourg
{frederic.pinel,pascal.bouvry}@uni.lu
[2] University of Lille, Lille, France
bernabe.dorronsoro_diaz@inria.fr
[3] Department of Electrical and Computer Engineering,
North Dakota State University, Fargo, USA
samee.khan@ndsu.edu

Abstract. In this paper, we simplify a Parallel Asynchronous Cellular Genetic Algorithm, by removing thread locks for shared memory access. This deliberate error aims to accelerate the algorithm, while preserving its search capability. Experiments with three benchmark problems show an acceleration, and even a slight improvement in search capability, with statistical significance.

Keywords: Cellular Genetic Algorithm · Parallelism

1 Introduction

Evolutionary Algorithms (EAs) have been used for many years to solve hard combinatorial and continous optimization problems. These nature-inspired algorithms iteratively apply transformations to solutions, and converge to an optimal or near-optimal solution. However, EAs require many iterations to conduct their search. This motivates the design of concurrent versions of these algorithms, in order to exploit the parallelism available in current computers. Moreover, parallelism can also improve the search capability of the algorithms [3,5].

In this paper, we propose a new Parallel Asynchronous Cellular Genetic Algorithm (PA-CGA). Our PA-CGA deliberately includes an error, that simplifies the design of the algorithm but also improves the speed of the algorithm. We compare this new PA-CGA with two known PA-CGAs, in terms of execution speed but also search capability.

Section 2 defines our parallelism objective. Section 3 presents the different models and how they are compared. Sections 4 describes the experiments.

2 Problem Description

The problem addressed in this paper is the parallelization of PA-CGA to improve its scalability: how does a PA-CGA *behave* as the number of threads increases.

R. Wyrzykowski et al. (Eds.): PPAM 2013, Part I, LNCS 8384, pp. 361–370, 2014.
DOI: 10.1007/978-3-642-55224-3_34, © Springer-Verlag Berlin Heidelberg 2014

The behavior of a PA-CGA should not be limited to runtime, but also include how well the algorithm searches solutions.

The next sections provide background information on parallel EAs and presents the PA-CGA.

2.1   Background

A survey of parallel genetic algorithm can be found in [9]. Concurrency in genetic algorithms is often introduced at the population level, because the evolutionary steps can be applied independently across a population of solutions. An evolutionary step is a sequence of operations, which generate new solutions (called children). The sequence is: parent selection (choosing individuals), crossover (generating a child from the parents), mutation (applying a small random change to the child) and replacement (criteria for the child to join the population). One evolution of all individuals in a population is called a generation.

The concurrency in the population generally occurs in three ways: master-slave, island and cellular. The master-slave model dispatches the operators' work to a number of slaves. In the island model, the population is partitioned into isolated evolutionary processes, which periodically exchange individuals. The cellular model, implemented in Cellular Genetic Algorithm (CGA) [1], is a fine-grain island model, where the periodical exchange of individuals is replaced with a more frequent update to a shared population. The CGA imposes a structure on the population of candidate solutions, usually a two-dimensional grid, and the parents for crossover are selected from the neighborhood of an individual. This increases the diversity in the population. The population structure in a CGA provides a fine-grain control over the evolution, which facilitates the exploration of different concurrency models [4].

Individuals evolving in parallel across the population usually evolve together, which requires synchronization. Asynchronous evolution relaxes this global time constraint [10,11]: individuals evolve independently and the population is not of the same age (underwent the same number of evolutions). Asynchronous models are also known to improve search capability [2].

2.2   Parallel Asynchronous Cellular Genetic Algorithm

The PA-CGA we study was presented in [13,14]. Parallelism in the PA-CGA is introduced at the population level. The population partition model of our PA-CGAs is inspired from [6,7,12]. As shown in Fig. 1, the population is partitioned into a number of contiguous sub-populations, with a similar number of individuals. Each partition contains $pop\_size/\#threads$ individuals, where $\#threads$ represents the number of threads launched. The neighborhood of an individual may cross partition boundaries. The threads in a PA-CGA evolve their partition independently: they do not wait on the other threads in order to pursue their evolution. The combination of a concurrent execution model with overlapping neighborhoods leads to concurrent access to shared memory, and requires synchronization.

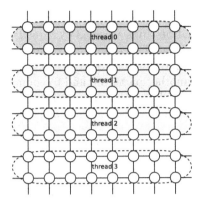

Fig. 1. Partition of an 8×8 population over 4 threads.

3 Approach

We present in Sect. 3.1, three parallel models for a PA-CGA: the *Island* [9], *Lock* [13] and *Free*, based on the principles of Sect. 2. The Free model is our contribution. It is an incorrect implementation of a PA-CGA: the thread locks protecting the shared population are removed, thus data consistency is not ensured. This is meant to improve the runtime and scalability of the PA-CGA. However, this change should impact the search capability of the PA-CGA. To investigate the behavior of the Free model, we compare the models across a selection of benchmark problems, presented in Sect. 3.2. The behavior of the models is observed across several metrics, presented in Sect. 3.3.

3.1 PA-CGA Models

In the PA-CGA *Island* model, presented in Algorithm 1, each thread operates on two populations: (a) its local partition, and (b) the global population. Once 100 generations are completed, the thread-local partition (a) is copied to the

---

Algorithm 1. Island model

---
while $< max\_gens$ **do**
    while $< max\_gens$ & not every 100 gens **do**   ▷ evolve local partition 100×
        for all *individual* in local partition **do**
            individual ← *evolved(individual)*   ▷ "←" follows replacement policy
        end for
    end while
    for all *individual* in global partition **do**   ▷ update the global partition
        rw_lock(global individual)
        global individual ← *local individual*
        rw_unlock(global individual)
    end for
end while

---

global population (b). This copy is performed asynchronously (one individual at a time). The global population (b) is accessed by threads when they require individuals from another partition. This occurs at crossover, when a parent selected belongs to another partition. POSIX read-write locks [8] are used by the threads to read individuals from another partition, and to commit their partition to the global population. The Island model aims to reduce contention on the shared population by operating on a thread-local data as much as possible.

Algorithm 2. Lock model

---
for all gens **do**
    for all *individual* in global partition **do**
        child ← *evolved*(*individual*)
        rw_lock(*global individual*)
        global individual ← *child*   ▷ "←" follows replacement policy
        rw_unlock(*global individual*)
    end for
end for

The PA-CGA *Lock* model, presented in Algorithm 2, is the closest to the classic asynchronous CGA. The only difference is that each thread evolves the individuals of its partition only. Each individual is protected with a POSIX read-write lock. This allows for concurrent read access. When an individual can be replaced with a better child, the change occurs immediately (provided a thread lock is acquired), and is then visible to all other threads. The Lock model requires more communication across threads than the Island model, however, changes are reflected immediately.

Algorithm 3. Free model

---
for all gens **do**
    for all *individual* in global partition **do**
        child ← *evolved*(*individual*)
        global individual ← *child*   ▷ "←" follows replacement policy
    end for
end for

The PA-CGA *Free* model is the simplest of all models, as per Algorithm 3. A thread evolves its partition, and updates the global population immediately. However changes in the global population are made without thread locking. This is apparently an error, because a thread may read an individual that is currently being updated (dirty read). This is possible because of the representation of an individual; usually a large array of word size elements. This model is considered wait-free, because a thread's progress is bounded by a number of steps it has to wait before progress resumes. Increasing the number of threads makes dirty reads more frequent, because it reduces the size of each partition, each thread evolves its partition faster, and more individuals lie on the border of a partition.

Table 1. Benchmark of combinatorial optimization problems

Problem	Fitness function	n	Optimum
MTTP	$f_{MTTP}(\boldsymbol{x}) = \sum_{i=1}^{n} x_i \cdot w_i$	200	-400.0
PPEAKS	$f_{PPEAKS}(\boldsymbol{x}) =$ $\frac{1}{N} \max_{1 \leq i \leq p}(N - HammingD(\boldsymbol{x}, Peak_i))$	100	100.0
MMDP	$f_{MMDP}(\boldsymbol{s}) = \sum_{i=1}^{k} fitness_{s_i}$ $fitness_{s_i} = 1.0$ if s_i has 0 or 6 ones $fitness_{s_i} = 0.0$ if s_i has 1 or 5 ones $fitness_{s_i} = 0360384$ if s_i has 2 or 4 ones $fitness_{s_i} = 0.640576$ if s_i has 3 ones	240	40.0

3.2 Benchmarks

The benchmark problems selected for our comparison are well-known combinatorial optimization problems, displaying different features like multi-modality, epistasis, large search space, etc. Due to the lack of space it is not possible to give details on these problems, but the reader is referred to [1]. They are summarized in Table 1 (name, fitness value, number of variables $-n-$, and optimum). They are the Massively Multi-modal Deceptive Problem (MMDP) –instance of 40 subproblems of 6 variables each–, the Minimum Tardiness Task Problem (MTTP) –instances of 200 tasks–, and the PPEAKS problem, with 100 peaks.

3.3 Metrics

The metrics for the comparison aim to capture the behavior of the three algorithms as the number of threads increases.

Our first metric is execution speed. It is the wall-clock runtime of the algorithms for the maximum number of generations. However, increased speed is useless if the search capability is degraded such that it requires more generations, therefore we add the following metrics:

- Success rate: the number of experiments when the optimum was found.
- Evaluation-efficiency: the number of evaluations required to find the optimum, when found. This is measured in evaluations (calculation of the fitness of an individual) instead of generations, because of the concurrent evolution in each partition.
- Time-efficiency: speed and evaluation-efficiency are combined by measuring the wall-clock time required to find the optimum (when found). This is useful from the perspective of a potential user of the algorithm.

4 Experiments

This section defines the parameters, the environment and results for the experiments.

Table 2. PA-CGA parameters

Parameter	Value
Population size	40 × 40
Asynchronous mode	Fixed line sweep
Selection operator	L5, binary tournament
Crossover operator	Two-point crossover
Crossover probability	1.0
Mutation operator	×2 flips
Mutation probability	1.0
Maximum generations	2500
Island synchronization period	100 generations
Runs	100

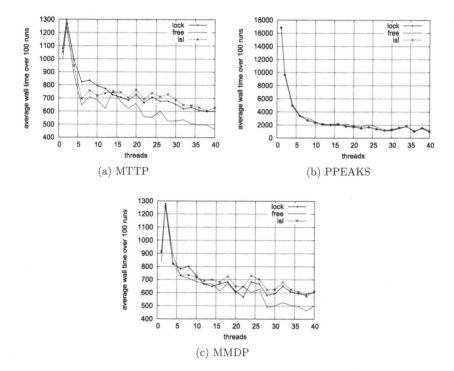

(a) MTTP (b) PPEAKS

(c) MMDP

Fig. 2. Runtime

4.1   Experimental Setup

Table 2 summarizes the various parameters for the PA-CGA. The asynchronous mode sets the order in which the threads evolve the individuals in their partition. This is consistent with [2]. Mutation consists in randomly flipping two bits in the individual. The maximum number of generations is the stop condition per thread. The Island synchronization period specifies when the thread-local

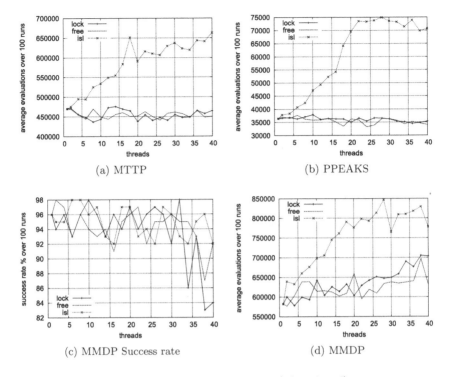

(a) MTTP
(b) PPEAKS
(c) MMDP Success rate
(d) MMDP

Fig. 3. Evaluations to optimum (when found)

partition is committed to the global population (for other threads to access). For each benchmark, 100 searches or runs are performed. The individuals are randomly generated for each run.

The computer used for the experiments is a Bullx S6030, where one board holds four Intel Xeon E7-4850@2GHz processors of 10 cores each. We use one board for the experiments (up to 40 cores). The operating system is GNU/Linux 2.6.32-5-amd64 (Debian), GCC is version 4.4.5.

4.2 Experimental Results

In this section, we present the results from the benchmark problems, grouped by metric.

Runtime. Figure 2 plots the average runtime (wall-clock) over the 100 runs in msec, as defined in Sect. 3.3. We can observe that all models reduce their runtime as the number of threads increases. The Free model is the fastest and scales the best, which is expected given the wait-free design, although not significantly for PPEAKS, Fig. 2b. The small difference between models for PPEAKS is due to the fitness function of PPEAKS, which is more time consuming than MTTP and MMDP and therefore minors the synchronization delays. The speedup observed

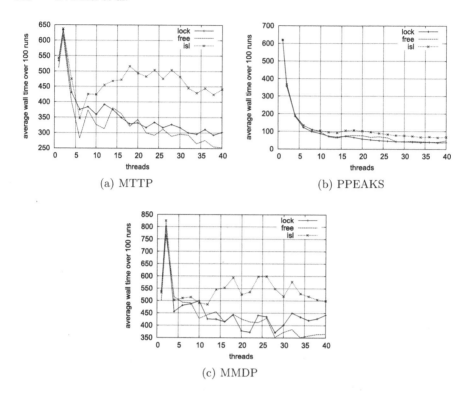

Fig. 4. Time to optimum (when found)

may seem low (especially for MTTP and MMDP), but the load is essentially due to synchronization.

Evaluation Efficiency. Figure 3 show the average number of evaluations needed to find the optimum, as defined in Sect. 3.3. Because this metric measures runs when the optimum is found, we first discuss the success rate.

The success rate for the different PA-CGA models for MTTP and PPEAKS is 100 % across the runs (and is not plotted). For MMDP, Fig. 3c, the rate is below 100 %. All models display about the same success rate, which also decreases from 35 threads and up. At this point, the partitions become too small, the generations too fast, thus reducing diversity in the partitions, which hurts the search.

Regarding evaluation-efficiency, the Free model obtains similar or better results than Lock (Wilcoxon Signed-Rank test). On MTTP, PPEAKS and MMDP, Free is better in respectively 5, 10 and 20 % of the cases. Also, the Lock and Free obtain constant results with the number of threads. The dirty reads in the Free model slightly help its evaluation-efficiency. The other observation is that the Island model does not scale well.

Time Efficiency. Figure. 4 show the time elapsed to reach the optimum, when found, as defined in Sect. 3.3.

For the Island model, Fig. 4a,c show that the gain in runtime is offset by the loss in evaluation-efficiency. For PPEAKS, the gain in runtime is so high, that time-efficiency manages to improve. The Lock and Free models do improve their time-efficiency with a greater number of threads, mainly because of the gain in speed. The Free model obtains the best results. This is due to the surprisingly good evaluation-efficiency, which means that the dirty reads do not harm the search, and may actually help.

5 Conclusions

We proposed a new PA-CGA parallel model, called Free. The Free model is based on a deliberate design error in the PA-CGA: all thread locks are removed, and access to the shared population leads to dirty reads. The absence of thread locks makes it wait-free. It is also the simplest PA-CGA design. This new model was compared to existing models: Island and Lock. The evaluation consisted in solving three benchmark problems (MTTP, PPEAKS, MMDP) using 1 to 40 threads, on a 40-core machine. These benchmarks are not computationally intensive, therefore the differences between models is more apparent. Experiments show that the Free model scales the best, and provides better or equal search capability, compared to the previously published Island and Lock models.

Future work includes exploring other sources of randomness such as operating system thread scheduling, and removing partition borders.

Acknowledgment. This work is supported by the Fonds National de la Recherche Luxembourg: CORE Project Green-IT, INTER Project Green@cloud (i2r-dir-tfn-12grcl) and AFR contract no 4017742.

References

1. Alba, E., Dorronsoro, B.: Cellular Genetic Algorithms. Operations Research/Compuer Science Interfaces. Springer, Heidelberg (2008)
2. Alba, E., Giacobini, M., Tomassini, M., Romero, S.: Comparing synchronous and asynchronous cellular genetic algorithms. In: Guervós, J.J.M., Adamidis, P.A., Beyer, H.-G., Fernández-Villacañas, J.-L., Schwefel, H.-P. (eds.) PPSN 2002. LNCS, vol. 2439, pp. 601–610. Springer, Heidelberg (2002)
3. Alba, E., Tomassini, M.: Parallelism and evolutionary algorithms. IEEE Trans. Evol. Comput. **6**(5), 443–462 (2002)
4. Alba, E., Blum, C., Asasi, P., Leon, C., Gomez, J.A.: Optimization Techniques for Solving Complex Problems, vol. 76. Wiley, New York (2009)
5. Cantú-Paz, E.: Efficient and Accurate Parallel Genetic Algorithms. Book Series on Genetic Algorithms and Evolutionary Computation, vol. 1, 2nd edn. Kluwer Academic, Dordrecht (2000)
6. Folino, G., Pizzuti, C., Spezzano, G.: Parallel hybrid method for SAT that couples genetic algorithms and local search. IEEE Trans. Evol. Comput. **5**(4), 323–334 (2001)

7. Folino, G., Pizzuti, C., Spezzano, G.: A scalable cellular implementation of parallel genetic programming. IEEE Trans. Evol. Comput. **7**(1), 37–53 (2003)
8. IEEE and The Open Group: POSIX (ieee std 1003.1-2008, open group base specifications issue 7). http://www.unix.org (2008)
9. Luque, G., Alba, E., Dorronsoro, B.: Parallel genetic algorithms (Chap. 5). In: Alba, E. (ed.) Parallel Metaheuristics: A New Class of Algorithms, pp. 107–125. Wiley, New York (2005)
10. Maruyama, T., Konagaya, A., Konishi, K.: An asynchronous fine-grained parallel genetic algorithm. In: Proceedings of the International Conference on Parallel Problem Solving from Nature II (PPSN-II). pp. 563–572. Lecture Notes in Computer Science (LNCS), North-Holland (1992)
11. Muhlenbein, H.: Evolution in time and space - the parallel genetic algorithm. In: Rawlins, G. (ed.) Foundations of Genetic Algorithms, pp. 316–337. Morgan Kaufmann, San Mateo (1991)
12. Nakashima, T., Ariyama, T., Ishibuchi, H.: Combining multiple cellular genetic algorithms for efficient search. In: Proceedings of the Asia-Pacific Conference on Simulated Evolution and Learning (SEAL), pp. 712–716 (2002)
13. Pinel, F., Dorronsoro, B., Bouvry, P.: A new parallel asynchronous cellular genetic algorithm for de novo genomic sequencing. In: Proceedings of the 2009 IEEE International Conference of Soft Computing and Pattern Recognition, pp. 178–183. IEEE Press (2009)
14. Pinel, F., Dorronsoro, B., Bouvry, P.: A new parallel asynchronous cellular genetic algorithm for scheduling in grids. In: Nature Inspired Distributed Computing (NIDISC) Sessions of the International Parallel and Distributed Processing Symposium (IPDPS) 2010 Workshop, p. 206b. IEEE Press (2010)

Genetic Programming in Automatic Discovery of Relationships in Computer System Monitoring Data

Wlodzimierz Funika[1,2(✉)] and Pawel Koperek[1]

[1] Faculty of Computer Science, Electronics and Telecommunications,
Department of Computer Science, AGH University of Science and Technology,
al. Mickiewicza 30, 30-059 Kraków, Poland
funika@agh.edu.pl, pkoperek@gmail.com
[2] ACC CYFRONET AGH, AGH University of Science and Technology,
ul. Nawojki 11, 30-950 Kraków, Poland

Abstract. Modern computer systems have become very complex. Analyzing and modifying them requires substantial experience and knowledge. To make administrative tasks easier, automated methods for discovering relationships between system components are required. In this paper we discuss the use of genetic programming as a method for identification of meaningful relationships between computer system components. We present our implementation of evolutionary computations environment and compare it with an already existing solution. Next we analyze results of a sample experiment and share our conclusions. The final section provides directions for future work.

Keywords: Distributed systems · Automatic modeling · Monitoring · Genetic programming · Automatic system management

1 Introduction

Nowadays scientific experiments provide enormous amounts of data. Due to sophisticated measurement methods scientists are able to monitor various aspects of observed environment. Creating theoretical models and looking for meaningful relationships between variables in such a vast amount of data is a real challenge. Until recently such a kind of work was mostly a domain for humans. Today as computer techniques become more and more advanced, automated analysis gains more and more attention. According to [1] computers are now used at different stages of research process: from gathering knowledge about related work and similar experiments, through automatic data analysis [2], up to complete automatic systems capable of creating and verifying new hypotheses on their own [3].

Similar situations can be observed in the management of computer systems. They become more and more complex. Usually, they are built from many components grouped in layers and distributed over a number of physical machines.

R. Wyrzykowski et al. (Eds.): PPAM 2013, Part I, LNCS 8384, pp. 371–380, 2014.
DOI: 10.1007/978-3-642-55224-3_35, © Springer-Verlag Berlin Heidelberg 2014

There are also many tools which gather and present very detailed information about various aspects of observed systems. It is possible to analyze both the real-time behavior and a state at a certain moment in the past. However, at some point, the amount of gathered data becomes so big that it is very hard to determine the actual status and relationships between different system elements. Therefore we cannot create a useful abstract model of monitored resources.

Discussed environments have similar problems, however solutions created in other scientific domains have not been widely used in the domain of computer systems monitoring. One of them is *symbolic regression* based on evolutionary programming [4]. It can be used to automatically search for relationships between variables in a data set - in other words to create a model that could explain the observations conducted. The results of this method are mathematical expressions which can be easily interpreted and verified by humans. Such a method can be a great convenience and accelerate work with measurement data. In this paper we propose to adapt this idea to automate the analysis of a computer system and create its model. In our approach, we attempt to automatically find relations between system's resources. The created model is being constantly updated and provides both knowledge about system's dynamics and predictions on how the system would behave under given conditions.

In the next section we provide more data about genetic programming and symbolic regression. Then we explain details of an algorithm implementation, compare it with an already existing one and present some results of a sample monitoring data analysis. The last sections are conclusions and directions of future work.

2   Background and Related Work

Genetic programming [4] is an example of algorithms inspired by natural evolution. It attempts to solve problems defined by the user through generating and improving a population of possible solutions. Initially they are random, but in subsequent iterations the best of them get *mutated* and *recombined (crossed-over)* to make them closer to an ideal solution. Quality of each such *individual* is measured by a *fitness function*. For the purpose of mutation and crossing-over processes, individuals are built from smaller building blocks - which are analogs of genes observed in nature.

Symbolic regression is a method used to find a mathematical expression meeting particular criteria defined as an error metric. Contrary to traditional regression methods, it searches for both the form of equations and their parameters. Genetic programming is often used to conduct symbolic regression. Each individual is then a mathematical expression and the fitness function can be, e.g., the mean squared error over a provided data set. Each expression is built from specified primitive elements such algebraic operations $(+, -, *, /)$, variables $(x, y \dots)$, constants $(3.1415, 2.71 \dots)$, etc.

Thanks to such a general definition, this method can be applied to solve a magnitude of different problems by changing only a part of the above-mentioned parameters. Particularly, it even can be used to find relationships between variables in measurement data sets. In [5] authors discuss various issues related to such a problem statement and provide a fitness function definition which forces the algorithm to discover **implicit relationships**. An implicit relationship is a function of form $f(x, y) = 0$ whereas the explicit function is represented as $y = f(x)$).

We investigated the functionality of some genetic programming tools and libraries regarding the applicability to the domain of computer systems monitoring. We focused on tools which implemented fitness functions based on the idea presented in [5] and on the libraries which could be used to create such a tool in Java. **Eureqa** [6] is the first one we examined. It is a ready to use software package which can analyze any provided data sets. It is possible to generate solutions by using many fitness metrics (implicit derivatives, absolute error etc.). Input can be pre-processed for better results (e.g. divide series by standard deviation, subtract a constant). Computations are probably accelerated with use of fitness-predictors [7], however due to the closed-source nature of the tool, the exact algorithm is unknown. Bigger and more complicated data sets can be processed with use of cloud services or a private grid of servers with use of the dedicated server feature. Finally, the solution is presented in form of a report containing information about the best individuals creating a size-fitness Pareto front of population. However, it is still impossible to analyze a constantly changing stream of data, e.g. one generated by a monitoring facility.

Java Genetic Algorithms Package (JGAP) [8] is a Java-based open source library for genetic computations. It focuses on providing a simple yet powerful interface which can be easily used to create a particular problem solution by applying evolution principles. The design is highly modular and allows for easy customization of various computation engine elements. Its authors provide numerous examples of usage and a tool to graphically visualize results of evolution. ECJ [9] is another example of evolutionary computation system implemented in Java. It provides a rich set of features including support for the asynchronous island model, checkpointing, coevolution and parsimony pressure. Similarly to JGAP it was created with extensibility and modularity in mind as well as the efficiency. A feature which distinguishes ECJ from JGAP is the genotype held in memory. The former system uses a pure tree representation while the latter one stores the tree as an array. Both frameworks provide some extensions for genetic programming and could be used to create systems supporting dynamically changing data sets. Unfortunately, they lack fitness functions one could use and are not able to evolve the population according to the principles defined in [5].

3 Concept and Implementation Details

We decided to create a proof-of-concept implementation of evolution engine which could be used with regard to monitoring data streams. To ensure simplicity, our implementation is based on Java platform and Java Genetic Algorithms Package [8].

Each individual is represented as an expression tree built from primitive blocks $(+, -, \times, /, sin, cos$, variables defined in input data set, constants). The number of nodes of the tree can be interpreted as a measure of how complicated the particular solution is. We call this parameter *complexity* and use as an indicator of individual's generality.

As the fitness function we used the formula proposed in [5]. To evaluate a particular candidate expression f, we compute numerically partial derivatives of a pair of variables x, y - $\frac{dx}{dt}$ and $\frac{dy}{dt}$. Then we find symbolically partial derivatives of candidate expression $\frac{\delta f}{\delta x}$ and $\frac{\delta f}{\delta y}$. To compute the actual fitness value we combine those elements in formula 1. Its value can be interpreted as an error rate of a solution. Therefore, the algorithm attempts to minimize it (i.e. individuals with a lower values are considered better).

$$\frac{1}{N} \sum_{i=1}^{N} \log \left(1 + abs \left(\frac{\Delta x_i}{\Delta y_i} - \frac{\delta x_i}{\delta y_i} \right) \right) \tag{1}$$

where: $\frac{\delta x}{\delta y} = \frac{\delta f}{\delta y} \Big/ \frac{\delta f}{\delta x}$.

To counter the problem of stagnation of computations, we used the evolution method proposed in [10]. The population is evolved using the following list of steps:

1. Randomly initialize a population of a given size
2. Randomly group individuals in pairs of *parent* individuals
3. Create *children* individuals by crossing-over of created pairs
4. Conduct mutations on children individuals
5. Add mutated individuals to the population
6. Repeat until the population size return to the initial size
   (a) Select randomly two individuals
   (b) Form an age-fitness Pareto front from these individuals
   (c) Discard dominated individual
   (d) If there are no more dominated individuals - break the loop

It is theoretically possible that the Pareto front is larger than the initial population. In this case all non-dominated individuals should be stored and used in a next algorithm iteration. This case is handled by a additional test from point 6d.

To simulate the conditions of computer system monitoring, we implemented a mechanism of *moving window* over a given data set. The user needs to specify the following parameters:

1. **window's size** - the number of data samples used to evaluate equations
2. **time of progression** - an interval of time after which we increment the position of the first row (relative to the complete data set)

Such a facility simulates a constantly updated stream of data, e.g., measurements coming from the observed software or hardware resources.

In order to be able to compare different aspects of the algorithm we established three criteria of stop:

- *time* - the computations are stopped after a fixed amount of time
- *target fitness fun* - the computations are stopped once the best solution's fitness value is lower than the target value
- *number of iterations* - the computations are stopped after a fixed number of iterations

Once the computations are finished, the formulas describing the relationships found are provided to the user. They are presented in form of a complexity-fitness Pareto front. In our experiments we validated the output manually and chose which formulas provide the best description of the observed system. A lower complexity indicated that the individual provided a more general description of the monitored system. On the other hand it would not be possible to create a meaningful expression with a too small number of elements. The system is more precisely described by the individuals with lower fitness values, but if the fitness is too low, it means that part of the population got over-fitted. We picked solutions with the smallest complexity and with the fitness values relatively low compared with other elements of the Pareto front.

4 Experiments

To verify the usefulness of evolutionary algorithms for the monitoring/performance analysis of computer systems, we tried to automatically discover relationships in a sample test case. In the following sections we describe how algorithm's input data was generated and compare results from our implementation with those obtained from Eureqa. Finally we discuss the algorithm's behavior when simulating a constantly changing data set.

4.1 Test Case and Input Data

The test case is a very simple WWW server serving some simple HTML pages. The workload is a simple application which creates a specific number of HTTP requests to the mentioned server per minute. The exact amount of requests is described by formula 2.

$$requests(t) = sin\left(\frac{3.1415 \times t}{20}\right) * 200 + 200 \qquad (2)$$

where t is the number of minutes elapsed since the beginning of the test. The number of requests sent each minute is presented in Fig. 1.

During the request generation we measured the CPU time used by the server process. The algorithm input data set contains two variables (*timestamp*,

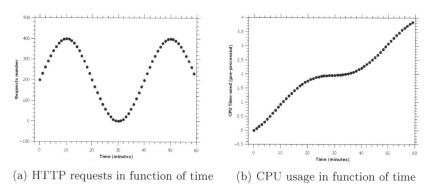

(a) HTTP requests in function of time   (b) CPU usage in function of time

Fig. 1. Input data

cputime), each comprising 60 measurements. The data for the input data set were gathered with nanosecond precision and had values over 10^9. Using them directly in computations led to overflow errors. Therefore we preprocessed input series with use of the following formulas:

$$input\_timestamp_i = \frac{raw\_timestamp_i - raw\_timestamp_0}{std\_dev(raw\_timestamp)} \tag{3}$$

$$input\_cputime_i = \frac{raw\_cputime_i - raw\_cputime_0}{std\_dev(raw\_cputime)} \tag{4}$$

The pre-processed CPU time data series is presented in Fig. 1.

4.2 Comparison with Eureqa

To the best of our knowledge, the only publicly available tool which provides implementation of fitness function proposed in [5] is Eureqa. We used it as a reference for our results to verify if the analyzed data set contains meaningful relationships and what their possible form is. Both algorithms were run for around 25 minutes with the complete data set. Table 1 presents the left sides of equations treated as functions. Figure 2 presents their values obtained by substituting the variables with the corresponding values coming from subsequent samples from the input data set.

Comparing the output formulas from Eureqa and our implementation is not straightforward. Evolutionary computations are based on subsequent random transformations of output formulas. Based on the acquired results (Fig. 2), we can only conclude that both implementations are capable of finding solutions of a similar quality in the analyzed test case. Since we are looking for a solution for $f(\boldsymbol{x}) = 0$ equation, the ideal individual $f(\boldsymbol{x})$ would evaluate to 0 for all input values (depicted by $y = 0$ in Fig. 2). Formulas E and A seem to fluctuate around a constant negative error value over the whole data set. Solution B has a low average error rate and therefore could also be considered as one

Table 1. Output equations. Equation IDs correspond to the legend from Fig. 2.

Equation ID	Source	Equation
E	Eureqa	$3.115 \times t + cos(3.08138 + 2.88161 \times t) - 3.37825 * c$
A	Prototype	$cos\left(\dfrac{(t+2 \times c)}{cos\left(\dfrac{cos(4.718734)}{c \times t}\right) \times c}\right)$
B	Prototype	$\dfrac{c}{6.182605} - \left(\dfrac{sin\left(\dfrac{sin(6.182605)/c}{t*(8.556137+c)}\right)}{5.7461066}\right)$

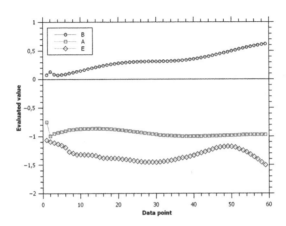

Fig. 2. Sample equations evaluated for raw data values

correctly describing the system under observation. On the other hand it is clear that its error values constantly grow as its verification progresses through the data set.

4.3   Simulation of Changing Data Set

To verify the usefulness of the algorithm in the monitoring domain we compared the results of the algorithm in two cases. First, the sample data set was treated as a whole - all the measurements were available at the same time. In the second case only a fragment of a predefined size was visible - each minute the oldest data sample was discarded and a new sample added.

All the experiments were executed on the same machine (AMD Phenom 9550 with 4 cores, 8 GB RAM). Each time the prototype was able to utilize 2 cores. The genetic programming parameters are given in Table 2.

Since the new data rows were inserted at a specific interval, computations were limited by a fixed amount of time (1860 s). In case of the constantly changing stream of data (as when monitoring a real computer system) we want to constantly adapt the solution to the incoming data instead of stopping the evolution and restarting it with each new data sample.

Table 2. Genetic programming parameters used in the experiments

Parameter	Value
Population size	64
Max individual length	64
Cross-over probability	0.75
Mutation probability	0.05
New data row insertion interval	60 s
Moving window size	30 data samples

4.4   Results

To set a baseline we first analyzed the approach with using the whole data set. For each solution complexity we computed the average fitness over all the runs. The result is illustrated in Fig. 3(a). It is clearly visible that there exists a number of formulas that have low error values, for which the number of nodes in tree representations (their *complexity*) was varying from 15 to 25. This agrees with the result obtained with the Eureqa tool. Solutions with higher complexity have even better fitness. Often they could be simplified with algebraic transformations to a form which has fewer elements in the tree representation.

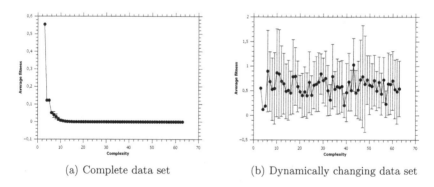

(a) Complete data set   (b) Dynamically changing data set

Fig. 3. Average fitness per formula complexity

The average fitness of the second test case (the dynamically changing data set) is depicted in Fig. 3(b). It is evident that there is no straightforward pattern. However, the low minimum values of fitness indicate that the solutions properly modelling the complete data set are found (Fig. 4(a)).

The analysis of individual runs revealed that in most of them (70 %) at least one *good enough* solution was found. The poor average statistics is caused therefore by the fact that in each execution, the solutions with good fitness had different complexity values. To determine which of the found solutions can be considered *good enough*, we compared their fitness with the average fitness

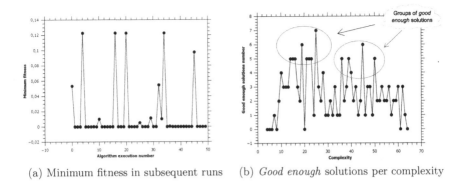

(a) Minimum fitness in subsequent runs   (b) *Good enough* solutions per complexity

Fig. 4. Results obtained for the case of *changing data set*

of the solutions obtained for the first test case (the complete data set). If the solution's fitness was less or equal to the average value for a given complexity - it was considered correct. Figure 4(b) presents how many solutions were found per complexity. There are two areas where the number of "good enough" solutions is significantly bigger than the average (2.6) - complexity varying from 15 to 25 and from 35 to 50.

5   Conclusions and Future Work

In this paper we explained details of the concept and implementation of an evolutionary engine capable of finding the relationships in the data coming from computer system monitoring. We analyzed its usefulness by comparing a new implementation with the already existing commercial-grade product and discussing the experiments conducted.

The presented results show that both Eureqa and our prototype implementation were able to find interesting results in a relatively simple sample data set. The relationships discovered are not trivial. We collected only the information about the CPU usage and real timestamps of measurements. The number of HTTP requests sent by a workload generating application had direct impact on the amount of computations; on the other hand the number of requests was a function of time. The found relationships bind linear time progression with CPU time changes - so indirectly they describe the server's input load.

The test case with a dynamically changed data set proves that the algorithm exploited can be used to analyze a constantly changing data stream. Genetic programming was able to find globally correct solutions, however it required an additional effort to determine which of the presented formulas is the proper one.

Our research opens new capabilities for automatic computer system modeling domain. The presented prototype, based only on a limited amount of data, provides deep insight into the observed system and enables predicting its future state. Thanks to the iterative nature of the process used to discover the formulas, output is always updated with the most recent observations. Finally the amount of time used to create or update the results can be easily adjusted with use of one of the presented stop criteria. The work on the prototype is ongoing: we are focusing on extending the set of basic building blocks and introducing a mechanism to simplify the evolved equations to speedup the evaluation. We also plan to investigate possibility of integration with semantic-oriented and autonomous monitoring systems [11,12].

Acknowledgements. This research is partly supported by EU VPH-Share project and the AGH grant 11.11.230.015.

References

1. Evans, J., Rzhetsky, A.: Machine science. Science **329**, 399–400 (2010)
2. Schmidt, M.D., Lipson, H.: Data-mining dynamical systems: automated symbolic system identification for exploratory analysis. In: ASME Conference Proceedings, vol. 2008(48364), pp. 643–649 (2008)
3. King, R.D., et al.: The automation of science. Science **324**, 85–89 (2009)
4. Koza, J.R.: Genetic Programming - On the Programming of Computers by Means of Natural Selection. Complex adaptive systems. MIT Press, Cambridge (1993)
5. Schmidt, M., Lipson, H.: Distilling free-form natural laws from experimental data. Science **324**, 81–85 (2009)
6. Nutonian Inc.: Eureqa(tm). http://www.nutonian.com/
7. Schmidt, M.D., Lipson, H.: Co-evolving fitness predictors for accelerating and reducing evaluations. In: Riolo, R.L., Soule, T., Worzel, B. (eds.): Genetic Programming Theory and Practice IV. Genetic and Evolutionary Computation, vol. 5. Springer, Ann Arbor, 11–13 May 2006
8. Meffert, K., et al.: Java genetic algorithms package. http://jgap.sourceforge.net/
9. Luke, S., et al.: Ecj 20. http://cs.gmu.edu/eclab/projects/ecj/
10. Schmidt, M.D., Lipson, H.: Age-fitness pareto optimization. In: Pelikan, M., Branke, J. (eds.) GECCO, pp. 543–544. ACM (2010)
11. Funika, W., Kupisz, M., Koperek, P.: Towards autonomic semantic-based management of distributed applications. Comput. Sci. J. - AGH-UST **11**, 51–63 (2013)
12. Funika, W., Godowski, P., Pegiel, P., Krol, D.: Semantic-oriented performance monitoring of distributed applications. Comput. Inform. **31**(2), 427–446 (2012)

Genetic Algorithms Execution Control Under a Global Application State Monitoring Infrastructure

Adam Smyk[1]([✉]) and Marek Tudruj[1,2]

[1] Polish-Japanese Institute of Information Technology, 86 Koszykowa Str.,
02-008 Warsaw, Poland
[2] Institute of Computer Science, Polish Academy of Sciences, 21 Ordona Str.,
01-237 Warsaw, Poland
{asmyk,tudruj}@pjwstk.edu.pl

Abstract. In the paper a new approach to the design of parallel genetic algorithms for execution in distributed systems with multicore processors is presented. The use of a distributed genetic algorithm based on new control implementation principles is proposed for an optimized irregular computational mesh partitioning for the *FDTD* (Finite-Difference Time-Domain) problem. The algorithm defines computational mesh partitions based on two objectives: load balancing and "min-cut" – the minimal number of edges between partition elements. The control in the parallel genetic algorithm assumes the use of program execution global control functions based on global application states monitoring. A control design infrastructure is provided to a programmer based on generalized synchronization/control processes called synchronizers. They collect local states of program computational elements, compute global control predicates and send back control signals. The paper describes how the assumed infrastructure can be used for convenient global program execution control at thread and process levels applied in the proposed genetic algorithm.

Keywords: Distributed program design paradigms · Global application states monitoring · Distributed program design tools · Mesh partitioning · Tiling · *FDTD*

1   Introduction

The efficiency of numerical applications in a parallel and distributed environment strongly depends on a computational data distribution among available processing elements. To do this, there are two obvious problems to solve. We should ensure a proper load distribution of all computational data and simultaneously minimal data transfers time between parts of the algorithm placed on different machines in a distributed environment. In shared memory systems, data transfer cost can be expressed as a synchronization overhead, which also

R. Wyrzykowski et al. (Eds.): PPAM 2013, Part I, LNCS 8384, pp. 381–391, 2014.
DOI: 10.1007/978-3-642-55224-3_36, © Springer-Verlag Berlin Heidelberg 2014

should be minimized. A widely discussed problem is a mesh partitioning, which appears in many numerical mesh-based algorithms. Optimal mesh partitioning is a NP-complete problem and in general is not possible to be obtained. There are many heuristic partitioning methods based on direct techniques (focusing on cut-min optimization) [9] or iterative algorithms [5,6,8,9]. In the paper, we present a genetic algorithm which is used for optimization of partitioning obtained by standard mechanisms, which are not sufficient for irregular computational areas. Convergence of genetic algorithms strongly depends on definitions of chromosome, genetic operators and also on the evaluation of the fitness function.

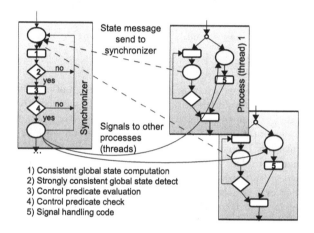

1) Consistent global state computation
2) Strongly consistent global state detect
3) Control predicate evaluation
4) Control predicate check
5) Signal handling code

Fig. 1. Synchronizer co-operation with processes or threads.

We consider a parallel implementation of the genetic algorithm (GA), so the whole population of individuals (solutions) will be distributed among computational processors. Fitness functions for individuals can be computed independently by threads on separate cores. But, since genetic operations in many cases must be done globally, so some data should be exchanged between remote instances of GA. To avoid this, we have decreased optimized computational area into a bundle of computations, performed on one physical processor. All individuals from one population are processed locally by threads. To speed-up all the process and to increase the convergence, evaluation of genetic metrics is done in an approximate way. Only every given number of iterations, each local instance of GA broadcast its state to remote instances of GA located in the distributed system. Global convergence depends on the global state of the application which should be dynamically monitored at runtime [14]. To perform such monitoring, a special control design infrastructure is created. Such an asynchronous program execution control mechanism is explained in Fig. 1, [2]. It has been embedded in the PEGASUS DA parallel program design framework (from Program Execution Governed by Asynchronous SUpervision of States in Distributed Applications) [11,12]. We have assumed that an application consists of computational multithreaded processes that can be executed in parallel in a system based on shared-distributed memory. Local communication and synchronization is done by shared memory primitives, while remotely all these processing elements can send and receive data by using standard communication mechanisms like sockets or MPI. Communication depends on the computational process and should be implemented manually by the programmer. In order to perform local or global synchronization operations, a special control infrastructure is delivered.

The main element of the control infrastructure is called a synchronizer. It collects local thread state messages from all computational elements and it determines, if the application has reached a strongly consistent global state SCGS [1]. SCGS is defined as a set of fully concurrent local states detected without doubt by a synchronizer. The construction of strongly consistent global states is based on projecting the local states of all processes or threads on a common time axis and finding time intervals which are covered by local states in all participating processes or threads [3,13]. The synchronizer, computes control predicates on global states and undertakes predefined control actions. If some predicate is true, then the synchronizer sends control signals (by network or by inter-process communication subsystem) to selected processes or threads, see Fig. 2. The signals are handled by these processes and some desired actions can be performed. Synchronizers can work locally (inside a process) or globally (inside an application), so their actions can be visible by whole computational process, or only by its part.

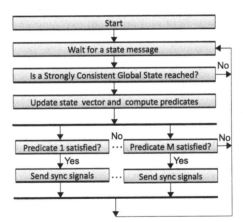

Fig. 2. Control flow diagram of a synchronizer.

In the code of a process or thread, which is connected to a synchronizer, we can distinguish regions sensitive to incoming signals. They are indicated by special tags, similarly to try and catch tags from e.g. C++ programming language. If the signal arrives when the process is in the signal sensitive region, reactions to this signal defined in the process code will take place. Otherwise, the signal is ignored and the reaction will not happen. All control messages are physically separated from data used by computational elements. Such dual communication network is described in [4] and it increases the global control performance.

The rest of the paper is ordered as follows. Section 2 contains an overview of the *FDTD* computational method. Section 3 explains computational mesh partitioning using the proposed genetic algorithm. Section 4 describes the genetic algorithm parallelization based on global program state monitoring.

2 FDTD Method Overview

As an example of numerical application, we present the *FDTD* (Finite Difference Time Domain method) computational method. It is used for simulation of high frequency electromagnetic wave propagation by solving Maxwell equations (1). The global asynchronous control system described in previous section will be used to find an optimal partitioning of the *FDTD* data flow graph among executive processors.

In the paper, we have assumed that a wave propagation simulation area is represented by two dimensional irregular shape, see Fig. 3. Some parts of this shape is characterized by strong irregularity. Before the simulation, a computational mesh is created. Its internal structure depends on the *FDTD* theory and for a two dimensional problem it is defined by Maxwell equations (1) transformed into their vector forms (2). The number of simulation points in a mesh depends on the frequency of simulation and each point describes alternately electric component Ez of electromagnetic field and one from two magnetic field components Hx or Hy (depending on its coordinates).

$$\nabla \times H = \gamma E + \varepsilon \frac{\partial E}{\partial t}, \qquad \nabla \times E = -\mu \frac{\partial H}{\partial t}, \tag{1}$$

$$\begin{cases} \overline{H}_y^n(i,j) = \overline{H}_y^{n-1}(i,j) + RC \cdot [\overline{E}_z^{n-0.5}(i,j-1) - \overline{E}_z^{n-0.5}(i,j+1)], \\ \overline{H}_x^n(i,j) = \overline{H}_x^{n-1}(i,j) + RC \cdot [\overline{E}_z^{n-0.5}(i-1,j) - \overline{E}_z^{n-0.5}(i+1,j)], \\ \overline{E}_z^n(i,j) = CA_z(i,j) \cdot \overline{E}_z^{n-1}(i,j) + CB_z(i,j) \cdot [\overline{H}_y^{n-0.5}(i+1,j) - \overline{H}_y^{n-0.5}(i-1,j) + \overline{H}_x^{n-0.5}(i,j-1) - \overline{H}_x^{n-0.5}(i,j+1)] \end{cases} \tag{2}$$

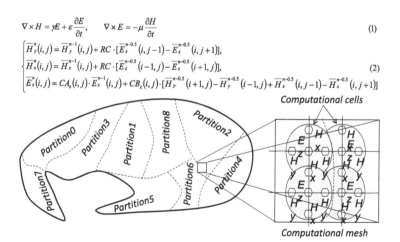

Fig. 3. Irregular computational area with FDTD computational mesh.

Simulation is divided into a given number of iterations. Each iteration consists of the following steps:

1. Preparing for boundary conditions computation;
2. Computation of the values for electric components;
3. Computation of the values for magnetic components;
4. Computation of boundary conditions;
5. Go to step 1 or end the computation.

For regular computational areas, *FDTD* computations can be easily parallelized (e.g. by stripe, block partitioning) but when an irregular shape of computational area is considered, the decomposition is much more complicated. In this case, we can use some advanced tool for efficient partitioning, like Metis, Jostle or Scotch. Big disadvantage of such solution is that mentioned partitioners, can produce very irregular structures. In such structures, despite a better data distribution, the cost of computations is higher. Instead of the standard loop, it requires the implementation of a complex iterator, what is impossible in numerical problems, due to their scale. Alternative techniques is loop tiling widely discussed in numerical area [7]. A tile can be defined, as a part of computation, which can be well defined in terms of typical programming structures, like a loop for.

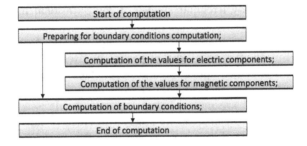

Fig. 4. Macro data flow graph for single tile for FDTD computations.

Shape of the tiles depends on the relations between computational data and usually is defined as rectangular, hexagonal, rhomboid, or equilateral. Because of data dependency (2), in our case the rectangular tile shape has been chosen. For the *FDTD* computation we have used a thread pool concurrency pattern. Each macro data flow node can be proceed by a single thread according to the dependency presented in Fig. 4. Such a processing pattern can be useful only in case when the number of threads is greater than number of tiles. A macro node can be fired for execution only if all input data have been delivered to the physical processor on which this macro node has been mapped. The data dependencies are described by edges between macro nodes. Usually edges can be attributed with weights which give the amount of data, sent from one macro node to another. The weight depends on the length of the boundary line between two adjacent cell sub-areas. Additional cost should be also considered, if the boundary line separates two tiles proceed on two different machines.

3   Computation Partitioning Using a Genetic Algorithm

In this section, we describe main assumptions on repartitioning a genetic algorithm data flow graph used to improve of the quality of a computational mesh partitioning, which assures time-balanced simulation execution in processors. The main steps of the genetic algorithm are as follows:

1. Creation of population P and initial maro data flow graph for computational area with initial partitioning;
2. Set iteration i = 1;
3. Population of selected individuals for iteration i;
4. Operations according to given chromosomes and Fitness function evaluation
5. If is the last iteration, go to step 8;
6. Selection, crossover and mutation operations
7. i = i + 1; go to step 3;
8. Finish

We don't want to use GA for the whole mesh partitioning, because of the time limitation. Only chosen parts of the mesh will be taken into consideration. Before the genetic algorithm will start, we need to create an initial population. It is done either by Metis partitioner, or by loop tiling splitter. Both of these methods have problems with highly irregular shapes of computational mesh. In case of Metis, it is very fast and efficient partitioner but it tends to create islands of nodes belonging to the same partition located inside another partition. Another, already discussed problem is regularity of partitions. Regularity is so high, that it is quite difficult to find a computational pattern which would be easily adopted for each node. In this case the GA should work on a very fine grain level – a single computational cell are moved between partitions. It is a very expensive and time consuming approach. Even we receive satisfactory load balancing, the implementation needs to be done according the data flow paradigm, which affects big machine code with separate instruction for separate data.

Another, much more flexible solution is loop tiling. It is very fast method producing partitions with regular, well defined patterns of computations. The irregularity appears only on the boundary of the computational mesh. Each tile contains up to 4 macro nodes (Fig. 4). Computations in each macro nodes are done according to Eq. (2) and can be easily implemented by standard loops, and optimize by typical compiler mechanisms. A genetic algorithm can change the size of a tile, but it does not change its shape (the regularity of computations depends on the tile shape). Tile shape can be changed only on the boundary area, where for irregular shape of computational area, cannot be matched to the shape of tile. The genetic algorithm, expands (or collapse) chosen tiles by moving boundary line cells between two adjacent tiles.

The definition of a chromosome is presented in Fig. 5. It contains a set of considered tiles and move operations performed for all tiles. We distinguish 5 operations, move: *UP*, *DOWN*, *LEFT* or *RIGHT* and no-move. To speed-up the GA, we have extended the definition of the chromosome, and in general it can specify more than one move operation executed in one iteration of GA. It makes more difficult to execute classical genetic operators, so to introduce additional control for moving operations (to avoid completely random movements) all involved tiles are sorted from the most irregular tile to the most regular. At several first iterations the GA works only on the most irregular tiles, so any move operations are defined only for the smallest numbers of tiles. As the iteration number increases, the number of considered tiles in this iteration increases as well. At the end of each iteration, we have to evaluate the fitness function for all individuals. The value of the fitness function indicates the best candidates for reproduction process. In our experiments we have used several fitness functions:

- the estimated execution time of n partitioned tiles computed by n threads – it requires implementing an simplified architectural model that can be used to estimate the execution time;
- the estimated execution time of n partitioned tiles computed by m threads where m < n – it requires implementing a thread pool model;
- the min-cut value;
- the difference between the maximally and minimally loaded tiles;

When the fitness function is evaluated we can select the best candidates for reproduction and we can perform crossover and mutation operations.

Tile number	0	1	2	3	4	5	6
Operation 1	Move to 1	Move to 2	Move to 3	Move to 4	Move to 1	Move to 4	Move to 5
Operation 2	Move to 2	Move to 2	Move to 3	Move to 4	Move to 1	Move to 6	Move to 4
...	...	...	...	...	...	...	...
Operation M	Move to 2	Move to 3	Move to 1	Move to 2	Move to 2	Move to 6	Move to 2

Fig. 5. Single chromosome meaning.

4 Parallelization of a Genetic Algorithm Based on Global Program State Monitoring

Genetic algorithms, can be easily implemented in a parallel way especially when particular instances of GA work on loosely dependent data. An instance of the GA is working on one processor and it processes a separate group of tiles. It can create an isolated population without any interference of individuals from populations created on other processors.

Each instance contains a given number of individuals that represent operations of moves for selected parts of tiles belonging to one group. All genetic operators are implemented by threads (one thread per individual). All information used by threads during the execution of GA, is stored in a shared memory. So one thread can use the results produced by another threads. To decrease the synchronization cost (granularity) they don't use this information directly. To access shared data stored in a shared memory, specially designed infrastructure

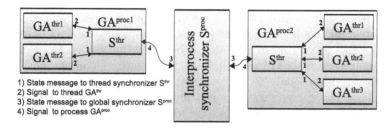

1) State message to thread synchronizer S^{thr}
2) Signal to thread GA^{thr}
3) State message to global synchronizer S^{proc}
4) Signal to process GA^{proc}

Fig. 6. Co-operation between instances of a genetic algorithm and inter-process and inter-thread synchronizers.

of synchronizers is used. We have assumed that each instance of GA can be executed as a process (e.g. GA^{proc2}), see Fig. 6. During execution, all information concerning states of individuals and of the whole local population is stored in local structures. State of a single individual is described by its chromosome, a total execution time for current partitioning and by local (between tiles from one GA instances) and remote (between tiles from different GA instances) min-cut values. The local min-cut value is used to control the regularity of tiles, while the remote min-cut for estimation of the communication cost. Both of these values are used by synchronizers to control the whole *FDTD* computational process. We have assumed that partitioning phases will be executed with computational phases alternately. It will be done by the same threads that are switched from one phase to another by a synchronizer. We distinguish synchronizers for threads (S^{th})and for processes (S^{p}). Each thread synchronizer is a special thread that co-operates with group of computational threads created inside one process. Each process synchronizer is a special process that co-operates with a group of computational processes.

A process synchronizer doesn't directly exchange information with a group of computational threads. In Fig. 7 we present a control flow diagram of a

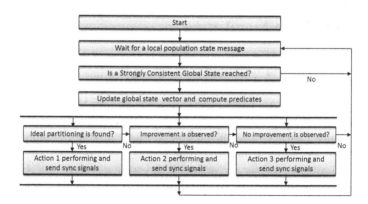

Fig. 7. Control flow diagram of a synchronizer for the parallel genetic algorithm.

synchronizer for the parallel genetic algorithm implementation. The thread synchronizer waits for state messages from all computation elements connected to it, containing reports on their local states. In our case all states are kept in shared memory, so partitioning threads sends only the location of their state in a memory. When all messages have been received, a local thread synchronizer starts their analysis. It updates the quality report concerning each individual based on the knowledge from last iteration. It sets the global rank for all individuals and next is starts to compute the predefined predicates. In the case of our parallel GA, we have specified the following predicates:

1. the synchronizer has detected an "ideal" partitioning ($IDEAPART$) with accepted load imbalance or min-cut values;
2. an improvement has been observed – no strategy changing is requested ($IMOBS$);
3. there is no improvement observed – a change strategy is requested ($NIMOBS$).

If the first predicate ($IDEAPART$) is met, the local partitioning paused and the thread synchronizer sends a message to the inter process synchronizer. It coordinates partitioning phases performed in different parts of the analyzed computational area. If the second predicate ($IMOBS$) is met, then the thread synchronizer sends a signal to cell partitioning threads that information about the quality of the individuals form last iteration has been update and are stored in shared memory. Each thread receives a signal and starts next iteration of GA. Information stored by the synchronizer is mainly used during the selection operation. If the last predicate ($NIMOBS$) is met, there is no significant improvement observed, no better individuals are formed, so the signal sent to cell partitioning threads by the synchronizer suggests to change the strategy of cell movement operations. It is quite easy to implement, because the last N iterations are recorded in memory, and they can be easily modified and introduced by the partitioning threads. After given number of $NIMOBS$ actions, the partitioning process can be canceled and retiling can be done. Retiling is a very expensive process because it is done globally and it causes canceling of all partitioning processes for all instances of GA. Thread synchronizers send their states to the process synchronizer every given number of iterations, or when the "ideal" partitioning is observed, or when the improvement cannot be achieved. The process synchronizer collects their states and evaluates the following predicates: "in most of GA instances improvement has been observed" and "global no improvement has been observed".

In the first case, synchronizer is switched from the partitioning phase to the computational phase, to estimate the efficiency of communication and load imbalance for current partitioning. According to the estimated state information, inter-process synchronizer definitely finishes the partitioning phase, and switches to the $FDTD$ computation phase. Another action taken by the synchronizer is to enter into a repartitioning phase, with or without a retiling phase.

5   Conclusions

In this paper, we have presented an outline of the implementation of a genetic algorithm in distributed systems with multicore processors. A genetic algorithm is used to improve partitioning of partially irregular computational meshes. It is done by a modification of initial partitions. In our case, an initial partitioning has been obtained by a tiling method. For each irregular part of a computational shape, a separate instance of genetic algorithm has been created. Execution of each genetic algorithm is globally supervised at the level of threads and also processes. This supervision is done by synchronizers, which undertake pre-defined actions according to the information received from processing elements (threads or processes). Such a control infrastructure, embedded in the PEGASUS DA framework, is very convenient for the implementation of genetic algorithms, which in our case are used in both phases, the partitioning one and the *FDTD* one. We have shown that such infrastructure can be very easily used for implementation of different numerical problems based on the genetic algorithm approach.

References

1. Babaoglu, O., Marzullo, K.: Consistent global states of distributed systems: fundamental concepts and mechanisms. In: Mullender, S.J. (ed.) Distributed Systems. Addison-Wesley, Reading (1993)
2. Borkowski, J.: Interrupt and cancellation as synchronization methods. In: Wyrzykowski, R., Dongarra, J., Paprzycki, M., Waśniewski, J. (eds.) PPAM 2001. LNCS, vol. 2328, pp. 3–9. Springer, Heidelberg (2002)
3. Borkowski, J.: Strongly consistent global state detection for on-line control of distributed applications. In: PDP 2004, pp. 126–133. IEEE CS, February 2004
4. Borkowski, J., Tudruj, M.: Dual communication network in program control based on global application state monitoring. In: ISPDC 2007, Hagenberg, Austria, pp. 37–44. IEEE CS, July 2007
5. Fiduccia, C.M., Mattheyses, R.M.: A linear time heuristic for improving network partitions. In: Proceedings of the Nineteenth Design Automation Conference, pp. 175–181 (1982)
6. Garey, M., Johnson, D., Stockmeyer, L.: Some simplified NP-complete graph problems. Theor. Comput. Sci. **1**, 237–267 (1976)
7. Jingling, X.: Loop Tiling for Parallelism. Kluwer Academic Publishers, Dordrecht (2000)
8. Karypis, G., Kumar, V.: Unstructured graph partitioning and sparse matrix ordering, Technical Report, Department of Computer Science, University of Minesota (1995). http://www.cs.umn.edu/~kumar
9. Khan, M.S., Li, K.F.: Fast graph partitioning algorithms. In: Proceedings of IEEE Pacific Rim Conference on Communications, Computers, and Signal Processing, Victoria, B.C., Canada, pp. 337–342, May 1995
10. Smyk, A., Tudruj, M.: Parallel implementation of FDTD computations based on macro data flow paradigm. In: PARELEC 2004, Dresden, Germany, 7–10 September 2004

11. Tudruj, M., Borkowski, J., Masko, L., Smyk, A., Kopanski, D., Laskowski, E.: Program design environment for multicore processor systems with program execution controlled by global states monitoring. In: ISPDC 2011, Cluj-Napoca, pp. 102–109. IEEE CS, July 2011
12. Kopanski, D., Maśko, Ł., Laskowski, E., Smyk, A., Borkowski, J., Tudruj, M.: Distributed program execution control based on application global states monitoring in PEGASUS DA framework. In: Wyrzykowski, R., Dongarra, J., Karczewski, K., Waśniewski, J. (eds.) PPAM 2013, Part I. LNCS, vol. 8384, pp. 302–314. Springer, Heidelberg (2014)
13. Stoller, S.D.: Detecting global predicates in distributed systems with clocks. Distrib. Comput. **13**(2), 85–98 (2000)
14. Tudruj, M., Kacsuk, P.: Extending grade towards explicit process synchronization in parallel programs. Comput. Artif. Intell. **17**, 507–516 (1998)

Evolutionary Algorithms for Abstract Planning

Jaroslaw Skaruz[1]([✉]), Artur Niewiadomski[1], and Wojciech Penczek[1,2]

[1] ICS, Siedlce University of Natural Sciences and Humanities,
3-Maja 54, 08-110 Siedlce, Poland
jaroslaw.skaruz@uph.edu.pl, artur@ii.uph.edu.pl
[2] Institute of Computer Science, Polish Academy of Sciences,
Jana Kazimierza 5, 01-248 Warsaw, Poland
penczek@ipipan.waw.pl

Abstract. The paper presents a new approach based on evolutionary algorithms to an abstract planning problem, which is the first stage of the web service composition problem. An abstract plan is defined as an equivalence class of sequences of service types that satisfy a user query. Two sequences are equivalent if they are composed of the same service types, but not necessarily occurring in the same order. The objective of our genetic algorithm (GA) is to return representatives of abstract plans without generating all the equivalent sequences. Experimental results are presented and compared with these obtained using an SMT-solver, showing that GA finds solutions for very large sets of service types in a reasonable and shorter time.

Keywords: Genetic algorithm · Web service composition · Abstract planning

1 Introduction

The number of web services available in the Internet has recently increased tremendously. The users may want to achieve some goals taking advantage of these services, but they also demand more sophisticated functionality from computer systems. Frequently, a simple web service does not realize the user objective, so a composition of services need to be executed to this aim. The problem of finding such a composition is NP-hard [8] and well known as the Web Service Composition Problem (WSCP) [10].

There is a number of various approaches to solve WSCP [2]. Here, we follow the approach of the system PlanICS [3,4], which has been inspired by [1]. The main assumption is that all the web services in the domain of interest as well as the objects processed by them, can be strictly classified in a hierarchy of *classes*, organised in an *ontology*. Another key idea consists in having several stages of planning. The first phase deals with types (classes), while the second one - with

The research described in this paper has been supported by the National Science Centre under the grant No. 2011/01/B/ST6/01477.

R. Wyrzykowski et al. (Eds.): PPAM 2013, Part I, LNCS 8384, pp. 392–401, 2014.
DOI: 10.1007/978-3-642-55224-3_37, © Springer-Verlag Berlin Heidelberg 2014

the *concrete services* (instances of classes). The first stage produces an *abstract plan*, which becomes a *concrete plan* in the second phase. Such an approach enables to reduce the number of concrete services, which are taken into account.

This paper focuses on the abstract planning problem only. We propose a new approach based on an application of genetic algorithms. An individual of GA represents a multiset of service types and all the operations of GA are performed on this multiset. This feature of GA constitutes a great improvement in comparison to a linear representation of an individual as the algorithm does not need to care about the correct order of the service types represented. A linearization of an individual is generated only in order to compute its fitness value. An abstract plan is defined by a multiset of service types such that it has a linearization satisfying a user query. The algorithm stores each newly found abstract plan. In the subsequent iterations all similar individuals are 'punished' by decreasing their fitness value, proportionally to the similarity to all the abstract plans stored. To the best of our knowledge, the above approach is novel, and as our experiments show is also very promising.

As far as the related work is concerned, some approaches to WSCP are listed below. However, none of the existing algorithms for abstract planning uses genetic algorithms. The existing solutions to WSCP can be divided into several groups. Following [11] our approach belongs to AI planning methods, including also approaches based on: automata theory [12], Petri nets [14], theorem proving [13], and model checking [15]. The approach closest to ours is given in [7], where a genetic algorithm is used to one phase planning, which combines an abstract and a concrete one. Our idea of a multiset representation of a GA individual is not entirely new as it was already suggested in [5]. However, contrary to our approach, no linearization of a multiset is generated in order to compute the fitness value. In [9] the authors model a problem of non-coding DNA in biological systems in a form of genes, where their positions in an individual are not fixed, which leads to better experimental results. The constrained optimization problem, considered by us, was also studied in [6], where the penalty function does not need any parameter too, but ours is defined as a *similarity measure*.

The rest of the paper is organized as follows. Section 2 defines the abstract planning problem. Section 3 presents an application of GA to finding abstract plans. Section 4 discusses experimental results and provides a comparison with an SMT-based algorithm. The last section summarizes the results.

2   Abstract Planning Problem

This section introduces the Abstract Planning Phase (APP). APP makes intensive use of the *service types* and the *object types* defined in a given *ontology*. In what follows, let $\mathbb{S}$ denote a set of all the service types defined in the ontology. A service type represents a set of web services with similar capabilities, while object types are used to represent data processed by the service types. The *attributes* are components of the object types. The ontology defines the inheritance relation, such that a subtype of some base object type retains all the attributes of the base type, and optionally introduces some new attributes. The *objects* are instances of the object types. The values of the attributes of an object determine its state. A set of the objects in a certain state is called a *world*.

User queries and service types. The main aim of PlanICS is to find a composition of web services, which allows to achieve a user goal. The user requirements are specified in a form of a *user query*. Its specification, as well as a service type specification, consists of three sets of objects: *in*, *inout*, and *out*, and two Boolean formulas, namely *preCondition* and *postCondition* (*pre* and *post*, for short). *Pre* is defined over attributes of the objects from *in* and *inout*, while *post* can involve also attributes of the objects from *out*. Since for APP there is no need to know the exact states of the objects, the values of the attributes are mapped to the two abstract values: **set** or **null**, denoting whether an attribute does have some value or it does not. An *abstract world* is a set of objects, which attributes have abstract values. In what follows we use the notions *worlds* and *values* instead of *abstract worlds* and *abstract values*, respectively.

A service type s is a pair of world sets (W^s_{pre}, W^s_{post}), called the *input* and the *output worlds*, respectively. That is, a service type is an interpretation of its specification, such that the input worlds are defined by *in*, *inout*, and *pre*, while the output worlds are determined by *in*, *inout*, *out*, and *post*. A user query q is a pair of world sets (W^q_{init}, W^q_{exp}), called the *initial* and the *expected worlds*, respectively, defined similarly to the service types.

Example 1 (Service type). Consider an object type $Ware$ containing the attributes *name*, *weight*, *owner*, and *location*. Let $Transport$ (T) be a service type able to deliver any instance of $Ware$ to the requested destination, specified as: $in_T = out_T = \emptyset$, $inout_T = \{w : Ware\}$, $pre_T = isNull(w.location)$, $post_T = isSet(w.location)$. Thus, W^T_{pre} is the set of worlds containing one instance of $Ware$ with **null** *location* and any valuation of the remaining attributes, while in the worlds of W^T_{post} the *location* attribute is **set**.

Example 2 (User query). Assume that $Doghouse$ is an object type extending $Ware$. Consider that the user wants to obtain a doghouse. An example query q could be: $in_q = inout_q = \emptyset$, $out_q = \{d : Doghouse\}$, $pre_q = true$, $post_q = isSet(d.location)$. The interpretation of q is an empty initial world, and a set of expected worlds containing an instance of $Doghouse$ with the attribute *location* **set** and any valuation of the remaining attributes.

World transformations. Assume we have two objects o_1 and o_2 of some worlds. The state of the object o_1 *is compatible* with the state of the object o_2, if o_1 contains all the attributes of o_2 (thus both objects are of the same type, or o_1 is a subtype of o_2 type), and they agree on valuations of all common attributes. A world w_1 *is compatible* with a world w_2, if both of them contain the same number of objects and every object from w_2 corresponds to a compatible object from w_1. Finally, by a *sub-world* of a world w we mean a restriction of w to some subset of objects from w, and by the *size* of w we mean the number of the objects in w, denoted by $|w|$. We say that a service of type s transforms a world w into w', denoted by $w \xrightarrow{s} w'$, if all of the following conditions hold:

- w contains a sub-world IN compatible with a sub-world of some input world of s, restricted to the objects from in,
- the objects from IN, as well as the objects not involved in the transformation, do not change their states,
- w contains a sub-world IO compatible with a sub-world of some input world of s, restricted to the objects from $inout$,
- w' contains a sub-world IO compatible with a sub-world of some output world of s, restricted to the objects from $inout$,
- w' contains a sub-world OU compatible with a sub-world of some output world of s, restricted to the objects from out,
- the sets of objects from IN, IO, OU are mutually disjoint, w does not contain any of the objects from OU, and $|w'| = |w| + |OU|$.

We refer to a world transformation by a service type s also as an *execution* of s.

Transformation sequences. Let $seq = (s_1, \ldots, s_k)$ be a sequence of service types of length k, and let w_0 and w_k be worlds, for some $k \in \mathbb{N}$. We say that the sequence seq transforms the world w_0 into w_k, denoted by $w_0 \overset{seq}{\leadsto} w_k$, if there exist worlds $w_1, \ldots, w_{k-1}$, such that $w_{i-1} \overset{s_i}{\to} w_i$, for every $i = 1, \ldots, k$.

A sequence seq of service types is called a *transformation sequence*, if there are worlds w, w' such that $w \overset{seq}{\leadsto} w'$. The world w' is called the *final world* of seq while M_{seq} denotes the multiset of the service types $[s_1 + \cdots + s_k]$ of the transformation sequence seq. A transformation sequence seq that transforms a given world w is called a *transformation sequence for w*, and the process of transforming w by seq is called the *execution of seq in w*. wo transformation sequences are called *equivalent* if they are built over the same multiset of service types.

User query solutions and abstract plans. Let $seq = (s_1, \ldots, s_k)$ be a transformation sequence of length k, and $q = (W^q_{init}, W^q_{exp})$ be a user query. We say that seq is a *solution of the user query q*, if there are worlds w, w', such that $w \overset{seq}{\leadsto} w'$, $w \in W^q_{init}$, and $w' \in W^q_{exp}$. By QS_q we denote the set of all the solutions of q. An abstract plan is defined as a set of all the solutions equivalent to some $seq \in QS_q$ and is represented by the multiset of the service types M_{seq} for q.

Example 3. Assume that *Selling* (S), *Transport* (T), *Assembly* (A) are service types, while *Boards*, *Nails*, and *Doghouse* are object types extending the object type *Ware*. The service type *Selling* is able to provide any *Ware*, *Transport* can deliver any *Ware* to the requested destination, while *Assembly* is able to build a doghouse using nails and boards. If the user wants to obtain a doghouse, then there are several possibilities to achieve this goal. The shortest solution is the sequence (S, T), which is the only solution of the abstract plan represented by the multiset $[S + T]$. Another possibility is (S, T, S, T, A), where the first pair (S, T) provides and transports boards and the second pair provides and delivers nails, which are finally assembled by A providing a doghouse. This solution constitutes another abstract plan represented by $[A + 2S + 2T]$. Note that there exists another equivalent solution, namely the sequence (S, S, T, T, A).

3   Application of GA to Abstract Planning

The objective of GA is to find abstract plans (as many as possible) for a user query q. While GA maintains a population of individuals, each representing a multiset M of service types, it is essential to check whether M represents an abstract plan. To this aim, for M a sequence of service types seq_M is constructed. If seq_M is a solution of the user query q, then M represents a new abstract plan.

The initial population of GA is generated randomly. As each gene of an individual models a service type, the number of all the genes is equal to the length of an abstract plan searched for. While we take advantage of the multiset representation, the order of the genes in an individual is irrelevant. This non-standard form of a GA individual allows for performing genetic operations in such a way that we do not have to receive offspring containing service types in the correct order. A linearization of an actual multiset is generated only in order to compute the fitness value of the corresponding individual.

Generating a sequence of service types seq_M from a multiset M. The iterative procedure starts with an empty sequence and some initial world w_0, randomly selected at the start of GA from W_{init}^q. In the successive iterations a resulting sequence is built by removing from M a service type s, which is able to transform[1] a current world w. Then, the current world becomes the one obtained from the transformation of w by s, and s is appended to seq_M. If none of the service types remaining in the multiset can be executed in the current world, then they are copied in a random order at the end of the sequence. Besides seq_M, the procedure returns also the length l_M of the maximal executable prefix of seq_M, as well as the world w_M, obtained after transformation of w by this prefix.

Example 4. Consider the multiset $M = [2A+S+T]$, where A, S, T are the same service types as in Example 3, and let w be an empty world. Only S is able to transform the empty world, so it is appended to seq_M. Then, the current world contains a single instance of $Ware$ (or its subclass). In this case T has to be chosen as the next service type, because A needs at least two objects (*Nails* and *Boards*), in order to be executed. Finally, two occurrences of A are appended to the resulting sequence, and the procedure returns $seq_M = (S, T, A, A)$, $l_M = 2$, and the world w_M containing a $Ware$.

Fitness function. In order to evaluate an individual, its fitness value is calculated. To this aim the notion of a *good service type* is used. A service type is *good*, if it produces objects of types from the expected worlds, or of types from input worlds of other *good* service types. After an individual M has been transformed to a sequence of service types seq_M, the fitness function takes the triple (seq_M, l_M, w_M) and an expected world[2] w_q as the arguments, and is calculated according to (1):

---

[1] If there are more than one such a service type, then one of them is chosen randomly.
[2] Selected randomly from W_{exp}^q at the start of GA.

$$fitness(M) = \frac{f_{w_M} * \delta + c_{w_M} * \alpha + l_M * \beta + g_{seq_M} * \gamma}{|w_q| * \delta + |w_q| * \alpha + k * \beta + k * \gamma} \tag{1}$$

where: $f_{w_M} = |w_{sub}|$ with w_{sub} being a maximal sub-world of w_M compatible with a sub-world of w_q; c_{w_M} is the number of the objects from w_M, which types are consistent with the types of objects from w_q; g_{seq_M} is the number of the good service types occurring in seq_M; k is the length of seq_M; and α, β, γ are parameters of the fitness function. In all the experiments presented in Sect. 4 we use the following values: $\alpha = 0.7$, $\beta = 0.1$, $\gamma = 0.2$, and $\delta = 0.1$. These values of the parameters ensure that building a proper sequence of service types starts with a service type which produces an object required by a user. In the next steps, GA finds service types producing objects required by a user or needed as an input for the other service types in a given sequence.

After the first solution has been found, the aim of GA is to find other solutions remaining in the search space. On the other hand, each individual that represents a solution equivalent to some of the already found, should be eliminated. This is the task of the *measure of similarity* (the coefficient in (2)), between the currently rated individual and the plans found so far, used for modifying the fitness value of the individual. Let *Sol* denote a non-empty set of plans (in a form of multisets) found at some point of GA. Then, the modified fitness value of the individual M is calculated according to (2):

$$fitness_{Sol}(M) = \left(1.0 - max\left(\left\{\frac{|M \cap S|}{|M|} \;\middle|\; S \in Sol\right\}\right)\right) * fitness(M) \tag{2}$$

Mutation operator. Another original contributions of this paper consists in defining a mutation operator specialized for the discussed problem, which takes advantage of the *good service type* concept. So, a gene is mutated only if it does not represent a good service type, and if there exists a *good* service type for the considered sequence. To this aim, one has to compute a set of all service types good for this sequence. If this set is not empty, then a randomly selected element of the set replaces the mutated gene. Notice that the mutation operator is not deterministic and it does not work in a greedy way.

4   Experimental Results

We have evaluated our algorithm using the ontologies, the user queries, and the abstract plans generated by our software - Ontology Generator (OG, for short). Each ontology contains an information about the services and the object types. OG generates the ontologies in a random manner such that semantic rules are met. Moreover, OG provides us with a user query which corresponds to services and object types contained in the ontology. Each query is also generated randomly in such a way that the number of various abstract plans equals to the value of a special parameter of OG. This guarantees that we know a priori whether GA finds all solutions. The remaining parameters of the generator are: the number of various object types, the minimal and maximal number of the object attributes, the number of service types, the minimal and maximal number of objects in the sets *in*, *inout*, and *out* of the service types, the number of the objects required by a user, and the number of the services in an abstract plan. Thanks to many different settings of OG, one can receive such data, which are helpful for checking how well GA scales for finding optimal solutions. The scalability can be examined by fixing different sizes of services in the ontology and the number of services in the abstract plans.

The experimental study was divided into two stages. In the first one, we have tuned the values of all GA parameters.

The tuning procedure is as follows. We select a parameter, the other parameters are set to the typical values, and several experiments are conducted in order to find the best value of the selected parameter, which is then fixed and set to this value. This procedure is repeated in the same way for all the remaining parameters, where the values of the fixed parameters are not changed anymore. The number of the individuals is equal to 1000, probability of mutation and two-point crossover are 5 % and 95 %, respectively. The roulette selection operator was used in all experiments. Each benchmark has been stopped after 50 iterations. The experiments were run on a standard PC computer with two cores 2.8 GHz CPU and 8 GB RAM.

Table 1 presents the summary of all the 18 experiments comparing the efficiency of our GA with an SMT-based planner [8]. The columns from left to right display: the experiment labels, the number of service types in the plans, the number of the existing plans, the total number of the service types, and the search space size. The next six columns contain the following GA results: the probability of finding a solution, the maximum and average number of the solutions found, the number of iterations needed to find the first and the second abstract plan, and the total GA runtime. The last four columns contain the times consumed by the SMT-based planner, in order to: find the first and the second solution, search the whole state space to ensure that there is no more plans, and the total SMT-planner runtime.

The experimental results can be summarized in the following way. As far as the time needed to find the first plan is concerned, the approach based on GA outperforms that based on SMT, because GA finds it dozen of times faster. However, the probability of finding a solution by GA decreases along with the increase of the length of abstract plans, similarly as for the average and the maximum number of the solutions found. Obviously, the more service types in an abstract plan the longer runtime of both the planners. In all the experiments the time required to find a solution by GA is below 21 s, while SMT needs even over 500 s. On the other hand, the SMT-based planner finds all the solutions in each run. Moreover, it is able to check that all possible abstract plans have been found.

Table 1. Experimental results

Exp	Plan len.	Sol	Serv. types	Sp. Serv.	Prob [%]	Max sol.	Avg sol.	GA first	GA next	GA [s]	SMT first [s]	SMT next [s]	SMT unsat [s]	SMT total [s]
1	6	1	64	2^{36}	100	1	1	6	—	6	4.28	—	2.9	8.09
2			128	2^{42}	100	1	1	6	—	7.5	7.76	—	5.87	14.83
3			256	2^{48}	100	1	1	8	—	10	11.19	—	7.57	20.57
4		10	64	2^{36}	100	6	4.1	9	11	6	4.89	4.99	4.95	12.86
5			128	2^{42}	100	6	3.7	7	8	7.5	5.95	6.0	9.63	20.5
6			256	2^{48}	100	4	2.6	11	13	11	13.85	14.68	17.56	38.47
7	9	1	64	2^{54}	100	1	1	10	—	11	21.05	—	25.57	47.63
8			128	2^{63}	100	1	1	10	—	15	39.48	—	41.75	83.04
9			256	2^{72}	100	1	1	12	—	22	94.55	—	77.65	174.5
10		10	64	2^{54}	80	2	1	15	17	12	17.84	19.44	161.1	239.2
11			128	2^{63}	60	2	0.8	18	23	15	34.19	44.76	276.1	341.8
12			256	2^{72}	50	1	0.5	26	—	22	65.85	67.87	542.5	669.7
13	12	1	64	2^{72}	100	1	1	15	—	16	91.9	—	429.8	523.2
14			128	2^{84}	100	1	1	21	—	22	126.4	—	1141	1270
15			256	2^{96}	90	1	0.9	29	—	34	213.1	—	>1800	
16	15	1	64	2^{90}	80	1	0.8	21	—	28	191.7	—		
17			128	2^{105}	20	1	0.2	22	—	35	425.7	—		
18			256	2^{120}	20	1	0.2	21	—	49	552.2	—		

Figure 1 presents the fitness value of the best individuals obtained in the experiments *exp1-exp12*. The most important observation resulting from the interpretation of the charts is a significant reduction of the best individual fitness value just after finding the solution, by the similarity measure. The fitness of the best individual chart shape can be viewed as a proof that our algorithm works as we have expected.

In the case of the first three experiments (Fig. 1a) the plans were found quite quickly and in all the runs of GA we have obtained solutions. Since the similarity measure works nicely, in the experiments *exp4 − exp6* (Fig. 1b) we obtained a number of solutions within the first 20 iterations. In the experiments *exp7 − exp9* (Fig. 1c) the fitness values of the initially generated individuals are in the range between 0.5 and 0.62. In the subsequent iterations GA finds better potential solutions. Finally, optimum is found in the 10th and the 12th iteration. In the experiments *exp10 − exp12* (Fig. 1d) GA obtains solutions before the 26th iteration. After one solution has been found the algorithm tries to find the next one, as the fitness value of the best individual increases in subsequent iterations. However, due to a much larger search space than in the experiments *exp4 − exp6*, only in the experiment *exp10* the next solution has been found.

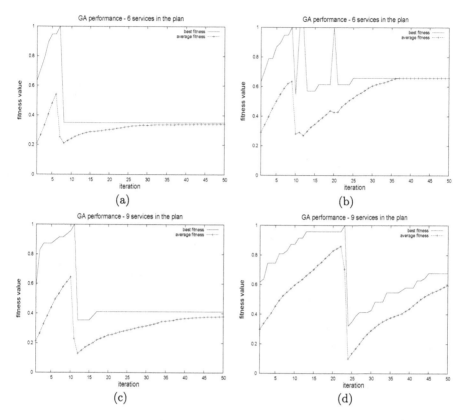

Fig. 1. GA performance for 64, 128, and 256 service types. $exp1-exp3$ (a), $exp4-exp6$ (b), $exp7-exp9$ (c), $exp10-exp12$ (d)

5   Conclusions

In the paper we presented a novel approach to the abstract planning problem with use of a genetic algorithm. Optimal solutions representing abstract plans have been found in each instance of the problem. This was achieved thanks to the special forms of the fitness function and the mutation operator. To overcome the problem of generating similar abstract plans, we have used multisets of service types for representing abstract plans as well as individuals of GA. Such a representation allows to generate only one solution from the set of all the equivalent ones. The experimental results give a clear evidence that our approach is quite efficient and allows to find abstract plans containing as many as 15 service types. In comparison to the results obtained using an SMT-solver, GA finds solutions in a much shorter time, which makes it a suitable tool for deployment in information systems.

References

1. Ambroszkiewicz, S.: Entish: a language for describing data processing in open distributed systems. Fundam. Inform. **60**, 41–66 (2004)
2. Ching-Seh, W., Khoury, I.: Tree-based search algorithm for web service composition in SaaS. In: 9th International Conference on Information Technology: New Generations (ITNG), pp. 132–138 (2012)
3. Doliwa, D., Horzelski, W., Jarocki, M., Niewiadomski, A., Penczek, W., Półrola, A., Skaruz, J.: HarmonICS - a tool for composing medical services. In: 4th Central-European Workshop on Services and Their Composition (ZEUS-2012), pp. 25–33 (2012)
4. Doliwa, D., Horzelski, W., Jarocki, M., Niewiadomski, A., Penczek, W., Półrola, A., Szreter, M., Zbrzezny, A.: PlanICS - a web service composition toolset. Fundam. Inform. **112**(1), 47–71 (2011)
5. Garibay, I., Wu, A.S., Garibay, O.: Emergence of genomic self-similarity in location independent representations. Genet. Program. Evolvable Mach. **7**(1), 55–80 (2006)
6. Kalyanmoy, D.: An efficient constraint handling method for genetic algorithms. Comput. Methods Appl. Mech. Eng. **186**, 311–338 (2000)
7. Lécué, F.: Optimizing QoS-aware semantic web service composition. In: Bernstein, A., Karger, D.R., Heath, T., Feigenbaum, L., Maynard, D., Motta, E., Thirunarayan, K. (eds.) ISWC 2009. LNCS, vol. 5823, pp. 375–391. Springer, Heidelberg (2009)
8. Niewiadomski, A., Penczek, W., Półrola, A.: SMT-based abstract planning in PlanICS ontology. ICS PAS Rep. **127**, 1–62 (2012)
9. Wu, A.S., Lindsay, R.K.: A comparison of the fixed and floating building block representation in the genetic algorithm. Evol. Comput. **4**(2), 169–193 (1996)
10. Rao, J., Su, X.: A survey of automated web service composition methods. In: Cardoso, J., Sheth, A.P. (eds.) SWSWPC 2004. LNCS, vol. 3387, pp. 43–54. Springer, Heidelberg (2005)
11. Li, Z., O'Brien, L., Keung, J., Xu, X.: Effort-oriented classification matrix of web service composition. In: 5th International Conference on Internet and Web Applications and Services, pp. 357–362 (2010)
12. Mitra, S., Kumar, R., Basu, S.: Automated choreographer synthesis for web services composition using I/O automata. In: 2007 IEEE International Conference on Web Services (ICWS 2007), pp. 364–371 (2007)
13. Rao, X., Jinghai, W., Küngas, P., Peep, A., Matskin, W., Mihhail, M..: Composition of semantic web services using linear logic theorem proving. Inf. Syst. **31**(4–5), 340–360 (2006)
14. Gehlot, V., Edupuganti, K.: Use of colored petri nets to model, analyze, and evaluate service composition and orchestration. In: 42nd Hawaii International Conference on System Sciences (HICSS'09), pp. 1–8 (2009)
15. Traverso, P., Pistore, M.: Automated composition of semantic web services into executable processes. In: McIlraith, S.A., Plexousakis, D., van Harmelen, F. (eds.) ISWC 2004. LNCS, vol. 3298, pp. 380–394. Springer, Heidelberg (2004)

Solution of the Inverse Continuous Casting Problem with the Aid of Modified Harmony Search Algorithm

Edyta Hetmaniok, Damian Słota$^{(\boxtimes)}$, and Adam Zielonka

Institute of Mathematics, Silesian University of Technology, Kaszubska 23,
44-100 Gliwice, Poland
{edyta.hetmaniok,damian.slota,adam.zielonka}@polsl.pl

Abstract. In the paper a description of procedure for solving the inverse problem of continuous casting is given. The problem consists in reconstruction of the cooling conditions of solidified ingot and is based on minimization of the appropriate functional by using the modified Harmony Search algorithm – the algorithm of artificial intelligence inspired by process of composing the jazz music.

Keywords: Artificial intelligence · Harmony Search algorithm · Inverse continuous casting problem

1 Introduction

Mathematical optimization consists in finding the best solution of a problem from among the set of admissible solutions with regard to some criteria. This branch of mathematics has found many applications in mechanics, engineering, economics, operations research and in any other cases requiring a selection of the best solution satisfying some conditions. There is a number of available classical optimization algorithms, however most of them need to fulfil complex assumptions about optimized function or its domain, that is why they are often complicated and can be applied for solving only specific groups of problems. Elaboration of algorithms of almost universal character and, simultaneously, very easy to implement became possible after taking inspirations from the natural behaviors existing in the surrounding world.

The artificial intelligence algorithms, about which is the talk, are the advanced mathematical methods based on the ability of learning and taking actions in order to improve the chances for success. This ability is realized in various ways, which divides the artificial intelligence optimization algorithms in many subgroups, including, for example, the evolutionary algorithms based on the best adaptation to the given conditions by applying the evolutionary mechanisms, like natural selection, mutation or recombination [1], the immune algorithms inspired by the mechanisms functioning in the immunological systems of living organisms [2] or the swarm intelligence algorithms based on the collective

R. Wyrzykowski et al. (Eds.): PPAM 2013, Part I, LNCS 8384, pp. 402–411, 2014.
DOI: 10.1007/978-3-642-55224-3_38, © Springer-Verlag Berlin Heidelberg 2014

behavior of swarms of individuals exploring the considered space of solutions and communicating among themselves [3,4].

Interesting example of the artificial intelligence optimization algorithm inspired by the human behavior is the Harmony Search algorithm, proposed by Zong Woo Geem [5,6], based on the process of searching for the harmony of sounds by the musicians taking part in the act of jazz improvisation. Jazz improvisation consists in finding the best state of music harmony, similarly as the optimization algorithm consists in finding the argument realizing minimum of the function. This analogy led to elaborating the useful procedure for optimizing a function which has found an application in many fields of computer science and engineering. Group of problems solved with the aid of this algorithm includes, for instance, tour planning [7], vehicle routing [8] and water network design [9].

Authors of this paper have already used the swarm intelligence algorithms [3, 4], as well as the HS algorithm for solving selected problems of heat conduction, in classical version of the algorithm in paper [10] and in modified version in paper [11]. Proposed modification consists in dynamic change of the value of an important parameter in case when several successive executions of the algorithm do not improve the result. It appears that such small modification improves significantly the convergence of algorithm.

In this paper the modified Harmony Search algorithm will be used for solving the inverse problem of continuous casting. Continuous casting is the process whereby the molten metal is poured in the controlled way into the crystallizer where it solidifies by taking the appropriate form and then is consecutively moved out from there [12–17]. Word "inverse" means that in mathematical model describing this process some input information is unknown and must be reconstructed, which is possible thanks to some additional information about the effects caused by the input data. In presented approach we need to determine the cooling conditions of continuous casting process (the heat flux in crystallizer and the heat transfer coefficient in the secondary cooling zone described by means of boundary conditions defining the process) in case when the values of temperature in selected points of the ingot are known. Problem formulated in this way belongs to the group of ill-posed problems because its solution may not exist or it exists but is unstable. In previous works we have used the deterministic methods for solving that kind of problem, however they gave worse reconstruction of boundary conditions in comparison with the currently used heuristic stochastic approach.

2   Problem Description

Let us consider the continuous casting of pure metal on the vertical device (see Fig. 1) working in the undisturbed cycle. We assume that the cooling conditions, changing with reference to direction of the ingot forming, are identical in the entire perimeter of the ingot and the dimensions of the ingot cross section satisfy condition $a \ll b$, where a denotes the ingot thickness and $2b$ describes its width.

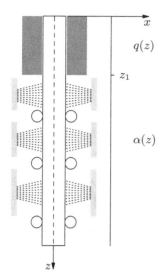 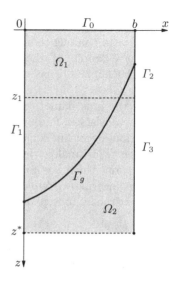

Fig. 1. Scheme of the problem (left figure) and domain of the two-dimensional problem (right figure)

Moreover, we assume that the heat flows only in direction perpendicular to the ingot axis. Such assumption results from the fact that the amount of heat conducted in direction of the ingot move, in comparison with the amount of heat conducted in the direction perpendicular to the ingot axis, is slight [12]. We consider the apparently steady field of temperature generated in the course of undisturbed cycle of working of the continuous casting device.

Under the above assumptions and because of the heat symmetry the region Ω of the ingot can be considered as the two-dimensional region consisted of two subregions: Ω_1 taken by the liquid phase and Ω_2 taken by the solid phase, separated by the freezing front Γ_g (described by means of function $x = \xi(t)$). In these subregions, with the space orientation taken as in Fig. 1, the heat transfer process, including the apparently steady field of temperature and location of the freezing front, can be described by the two-phase Stefan problem [12].

Boundary of region $\Omega = [0, b] \times [0, z^*] \subset \mathbb{R}^2$ is divided into four subsets (Fig. 1):

$$\Gamma_0 = \{(x, 0); \ x \in [0, b]\}, \tag{1}$$
$$\Gamma_1 = \{(0, z); \ z \in [0, z^*]\}, \tag{2}$$
$$\Gamma_2 = \{(b, z); \ z \in [0, z_1]\}, \tag{3}$$
$$\Gamma_3 = \{(b, z); \ z \in (z_1, z^*]\}, \tag{4}$$

where the boundary conditions are defined.

Discussed problem consists in determination of the cooling conditions for the ingot in such way that the temperature in selected points of the solid phase takes the given values $((x_i, z_j) \in \Omega_2)$:

$$T_2(x_i, z_j) = U_{ij}, \qquad i = 1, 2, \ldots, N_1, \quad j = 1, 2, \ldots, N_2, \tag{5}$$

where N_1 denotes the number of sensors and N_2 describes the number of measurements taken from each sensor. Another elements which should be determined are the functions of temperature T_k in regions Ω_k ($k = 1, 2$). Functions of temperature within regions Ω_k (for $k = 1, 2$) satisfy the heat conduction equation

$$c_k \, \varrho_k \, w \, \frac{\partial T_k}{\partial z}(x, z) = \frac{\partial}{\partial x}\left(\lambda_k \frac{\partial T_k}{\partial x}(x, z) \right), \tag{6}$$

where c_k, ϱ_k and λ_k denote, respectively, the specific heat, mass density and thermal conductivity in liquid phase ($k = 1$) and solid phase ($k = 2$), w is the velocity of continuous casting, and, lastly, x and z denote the spatial variables.

On the respective parts of boundary the appropriate boundary conditions must be satisfied – on boundary Γ_0 the boundary condition of the first kind with the given pouring temperature ($T_z > T^*$):

$$T_1(x, 0) = T_z, \tag{7}$$

on boundary Γ_1 the homogeneous boundary condition of the second kind

$$\frac{\partial T_k}{\partial x}(0, z) = 0, \tag{8}$$

on boundary Γ_2 (crystallizer) the boundary condition of the second kind

$$-\lambda_k \frac{\partial T_k}{\partial x}(b, z) = q(z), \tag{9}$$

on boundary Γ_3 (secondary cooling zone) the boundary condition of the third kind

$$-\lambda_k \frac{\partial T_k}{\partial x}(b, z) = \alpha(z) \left(T_k(b, z) - T_\infty \right) \tag{10}$$

and, finally, on the interface Γ_g the continuity condition and the Stefan condition

$$T_1\big(\xi(z), z\big) = T_2\big(\xi(z), z\big) = T^*, \tag{11}$$

$$L \, \varrho_2 \, w \, \frac{d\xi(z)}{dz} = -\lambda_1 \left.\frac{\partial T_1(x, z)}{\partial x}\right|_{x=\xi(z)} + \lambda_2 \left.\frac{\partial T_2(x, z)}{\partial x}\right|_{x=\xi(z)}, \tag{12}$$

where function ξ describes the freezing front location, α is the heat transfer coefficient, q denotes the heat flux, T_z is the pouring temperature, T_∞ is the ambient temperature, T^* is the solidification temperature and L describes the latent heat of fusion.

In considered approach the sought elements are the heat flux in crystallizer and the heat transfer coefficient in the secondary cooling zone, it means the following function f should be determined

$$f(z) = \begin{cases} q(z) & \text{for } z \leq z_1, \\ \alpha(z) & \text{for } z > z_1. \end{cases} \tag{13}$$

For the fixed form of function f problem (6)–(12) turns into the direct Stefan problem, solving of which enables to find the courses of temperature $T_{ij} = T_2(x_i, z_j)$ corresponding to function f. By using the calculated temperatures T_{ij} and the given temperatures U_{ij} the following functional is constructed

$$J(q, \alpha) = \Big(\sum_{i=1}^{N_1} \sum_{j=1}^{N_2} \big(T_{ij} - U_{ij} \big)^2 \Big)^{1/2}, \tag{14}$$

representing the error of approximate solution. Since our goal is to find such form of function f that the reconstructed temperatures will be as close as possible to the measurement values, solving of considered problem reduces to minimization of functional (14). For minimizing this functional we intend to use the modified Harmony Search algorithm, paying attention to the fact that each running of the procedure requires to solve for many times the direct Stefan problem, appropriate for taken conditions. To solve the Stefan problem we use the finite difference method with application of the alternating phase truncation method [12,15,17].

3 Modified Harmony Search Algorithm

Optimization of jazz composition runs in the following way: one of the musicians plays a note, the others remember its sound and select the next notes such that a harmonic music arises. Successively, the musicians remember the notes played before, add the next notes and improve them such that the most beautiful music is composed from the chaos. Similar idea is applied for optimizing a function – arguments of this function play the role of the notes and the values are considered as the tones of instruments caused by these notes. Similarly like the musicians are searching for the combination of notes giving the best harmony of music, the procedure seeks the argument in which minimum of the function is taken.

The algorithm starts by selecting the random set of notes (arguments) and ordering them with regard to the values of minimized function in the harmony memory vector (HM). Next, the harmony given by the combination of selected notes is randomly improved. One can choose the note already included in the harmony memory vector and test it one more time or change it slightly in hope of improving the general harmony. One can also try to find the completely new notes. Each note is put in the right order in the HM vector. After the assumed number of iterations the first element of HM vector is taken as the solution.

In details, the algorithm is formed from the following steps.

1. Initial data:
   - minimized function $J(x_1, \ldots, x_n)$;
   - range of the variables $a_i \le x_i \le b_i, \quad i = 1, \ldots, n$;
   - size of the harmony memory vector HMS (1–100);
   - harmony memory considering rate coefficient $HMCR$ (0.7–0.99);
   - pitch adjusting rate coefficient PAR (0.1–0.5);
   - number of iterations IT.

2. Preparation of the harmony memory vector HM – random selecting of HMS vectors $(x_1, \ldots, x_n)$ and ordering them in vector HM according to the increasing values $J(x_1, \ldots, x_n)$:

$$
HM = \begin{bmatrix} x_1^1, \ldots, x_n^1 & \Big| J(\mathbf{x}^1) \\ \vdots & \Big| \vdots \\ x_1^{HMS}, \ldots, x_n^{HMS} & \Big| J(\mathbf{x}^{HMS}) \end{bmatrix}.
$$

3. Selection of the new harmony $\mathbf{x}' = (x_1', \ldots, x_n')$.
   – For each $i = 1, \ldots, n$ the element x_i' is selected:
     • with probability equal to $HMCR$ from among numbers x_i collected in the harmony memory vector HM;
     • with probability equal to $1 - HMCR$ randomly, with the uniform probability distribution, from the assumed range $a_i \leq x_i \leq b_i$.
   – If in the previous step the element x_i' is selected from the harmony memory vector HM then:
     • with probability equal to PAR the sound of note is regulated, it means the element x_i' is modified in the following way: $x_i' \rightarrow x_i' + \Delta$ for $\Delta = bw \cdot u$, where bw denotes the bandwidth – part of range of the variables and u is the randomly selected, with the uniform probability distribution, number from interval $[-1, 1]$;
     • with probability equal to $1 - PAR$ nothing is done.
4. If $J(\mathbf{x}') < J(\mathbf{x}^{HMS})$ then element $\mathbf{x}'$ is put into harmony memory vector HM in place of element $\mathbf{x}^{HMS}$ and vector HM is ordered according to the increasing values of minimized function.
5. If the successive 5 iterations do not bring any improvement of the result, the bandwidth is updated: $bw \rightarrow 0.5 \cdot bw$. The bandwidth parameter participates in regulation of the sound of note, so this action imitates the frets on the neck of a guitar representing the semitones.
6. Steps 3–5 are repeated IT number of times. The first element of vector HM gives the solution.

4 Verifying Example

To verify the proposed approach let us solve the inverse problem of continuous casting of aluminium. Considered process is described by means of Eqs. (6)–(12) with the following values of parameters: $\lambda_1 = 104 \, [\mathrm{W/(m\,K)}]$, $\lambda_2 = 204 \, [\mathrm{W/(m\,K)}]$, $c_1 = 1290 \, [\mathrm{J/(kg\,K)}]$, $c_2 = 1000 \, [\mathrm{J/(kg\,K)}]$, $\varrho_1 = 2380 \, [\mathrm{kg/m^3}]$, $\varrho_2 = 2679 \, [\mathrm{kg/m^3}]$, $L = 390000 \, [\mathrm{J/kg}]$, velocity of casting $w = 0.002 \, [\mathrm{m/s}]$, solidification temperature $T^* = 930 \, [\mathrm{K}]$, ambient temperature $T_\infty = 298 \, [\mathrm{K}]$, pouring temperature $T_z = 1013 \, [\mathrm{K}]$ and $b = 0.1 \, [\mathrm{m}]$.

To solve the investigated problem we need to determine the temperature distribution in considered domain and to reconstruct the cooling conditions – heat flux $q(z)$ in crystallizer and heat transfer coefficient $\alpha(z)$ in the secondary cooling zone. Exact values of the sought elements are known:

$$q(z) = 400000 \,[\mathrm{W/m^2}],$$
$$\alpha(z) = 4000 \,[\mathrm{W/(m^2\,K)}].$$

Measured values of temperature, required for constructing functional (14), are taken from two thermocouples ($N_1 = 2$) located 0.001 and 0.002 m away from boundary of the region. From each thermocouple we took 100 measurements ($N_1 = 100$) and distance along the Oz axis between the successive measurements was equal to 0.002 m. To execute the calculations we used the exact values of temperature and values burdened by the random error of normal distribution and values 1, 2 and 5 %.

Modified Harmony Search algorithm was executed for size of the harmony memory vector $HMS = 25$, harmony memory considering rate $HMCR = 0.85$, pitch adjusting rate $PAR = 0.3$ and number of iterations $IT = 250$. Elements of the initial harmony memory vector HM were randomly selected from different range for each of identified parameters because of the big difference between expected values of reconstructed parameters (for reconstructing q from interval $[250000, 500000]$ and for reconstructing α from interval $[1000, 5000]$). Initial value of bandwidth parameter bw corresponded with 10 % of the range of variables. We have decided about such values of parameters in result of many testing calculations and it is important to notice that slight change of these values influences the convergence of procedure. Moreover, to take into account the heuristic nature of Harmony Search algorithm, meaning that each execution of the procedure can give slightly different results, we evaluated the calculations in each considered case for 20 times and as the approximate values of reconstructed elements we accepted the best of obtained results.

Selected results of executed calculations are presented in figures and table given below. Figure 2 display the relative errors of the heat flux q and the heat transfer coefficient α identification in dependence on the number of iterations obtained for input data burdened by 2 % and 5 % error, respectively. We can see that in both cases about 25 iterations is needed to obtain very good reconstructions, after about 50 iterations the results stabilize on some level and any further iterations do not improve them significantly. To improve the results probably some other techniques should be used, which is planned for the future, however in this moment we find obtained reconstructions as satisfying. Stability of the procedure is confirmed by the reconstruction errors for unburdened input data converging very quickly to zero.

Except the cooling conditions, another sought element is the distribution of temperature. Distributions of temperature in measurement points located, respectively, 0.001 m and 0.002 m away from boundary of the region are compared with the known exact distributions in Figs. 3 and 4. The results are obtained for input data burdened by 5 % error and one can observe that in both cases the reconstructed and known courses of temperature almost cover and the absolute errors of these reconstructions are at the level of 0.6 [K].

Statistical elaboration of results obtained in 20 executions of the procedure for various noises of input data is compiled in Table 1. Relative errors of the heat flux and the heat transfer coefficient reconstructions for unburdened input

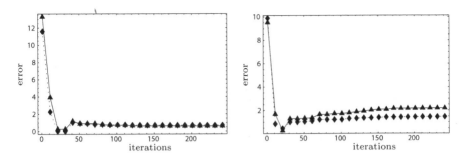

Fig. 2. Relative errors of reconstructing the boundary conditions for the successive iterations (▲ – for q, ◆ – for α) obtained for 2 % (left figure) and 5 % (right figure) noise of input data

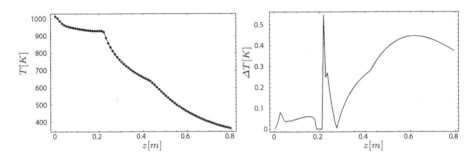

Fig. 3. Exact (solid line) and reconstructed (dots) distributions of temperature (left figure) in control point located 0.001 m away from the boundary obtained for 5 % noise of input data and absolute error of this reconstruction (right figure)

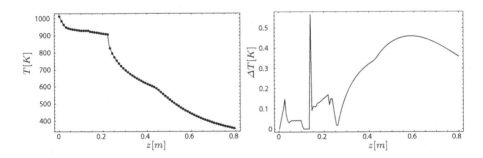

Fig. 4. Exact (solid line) and reconstructed (dots) distributions of temperature (left figure) in control point located 0.002 m away from the boundary obtained for 5 % noise of input data and absolute error of this reconstruction (right figure)

Table 1. Reconstructed values of boundary conditions (f), relative errors (δ_f), standard deviations (σ) and standard deviations (σ^p) expressed as a percent of mean values of these reconstructions, together with the maximal relative (δ_T^{max}) and absolute (Δ_T^{max}) errors of temperature reconstruction, obtained for various noises of input data

Noise (%)	f	δ_f [%]	σ	σ^p [%]	δ_T^{max}[%]	Δ_T^{max} [K]
0	400201.06	0.05016	388.500	0.09713	0.03082	0.172
	4002.39	0.05973	1.613	0.04032		
1	397542.20	0.61445	3228.581	0.80714	0.51576	4.796
	4028.21	0.61032	24.445	0.61112		
2	397543.76	0.61445	2339.822	0.58496	0.51457	4.761
	4028.20	0.70514	27.807	0.69517		
5	408822.12	2.20553	8863.732	2.21593	0.59914	5.572
	3941.77	1.45581	58.474	1.46184		

data are at the level of 0.05 %. The errors increase obviously with the increasing value of input data perturbation but in each case is much smaller than the input data error, as well as the maximal relative error of temperature reconstruction which is in each case insignificant. Standard deviations of the sought parameters reconstruction, expressed as a percent of mean values of these reconstructions, in the worst case is equal to 2.2 % which confirms stability of the procedure. At the end let us notice that execution of the procedure took approximately 40 min.

5   Conclusions

Aim of this paper was the examination of possibilities of applying the modified Harmony Search algorithm for solving the technical problem consisted in identification of the cooling conditions in continuous casting process such that the reconstructed values of temperature are as close as possible to the measurement values. Experimental verification indicates that the elaborated method gives satisfying results with regard to their exactness as well as to their stability and advantage of using the proposed approach, in comparison with classical methods, is lack of particular assumptions needed to be satisfied by minimized functional. Additional advantage is the respective reliability of investigated approach. For appropriately selected parameters and adequate number of iterations the method based on HS algorithm gives in any case the best possible solution, in other words the method is the most resistant to falling into the local minima from among other investigated heuristic algorithms. Whereas the disadvantage of discussed approach is longer, however acceptable, time of computations and the necessity of appropriate selection of the parameters (on the way of testing calculations). More detailed comparison of the described method with other approaches based on the heuristic algorithms is planned for the future.

Acknowledgements. This project has been financed from the funds of the National Science Centre granted on the basis of decision DEC-2011/03/B/ST8/06004.

References

1. Bäck, T.: Evolutionary Algorithms in Theory and Practice: Evolution Strategies, Evolutionary Programming, Genetic Algorithms. Oxford University Press, Oxford (1996)
2. Cutello, V., Nicosia, G.: An immunological approach to combinatorial optimization problems. In: Garijo, F.J., Riquelme, J.-C., Toro, M. (eds.) IBERAMIA 2002. LNCS (LNAI), vol. 2527, pp. 361–370. Springer, Heidelberg (2002)
3. Grzymkowski, R., Hetmaniok, E., Słota, D., Zielonka, A.: Application of the ant colony optimization algorithm in solving the inverse stefan problem. Steel Res. Int. Special Edition: Metal Forming, 1287–1290 (2012)
4. Hetmaniok, E., Słota, D., Zielonka, A.: Solution of the inverse heat conduction problem by using the abc algorithm. In: Szczuka, M., Kryszkiewicz, M., Ramanna, S., Jensen, R., Hu, Q. (eds.) RSCTC 2010. LNCS, vol. 6086, pp. 659–668. Springer, Heidelberg (2010)
5. Geem, Z.W., Kim, J.H., Loganathan, G.V.: A new heuristic optimization algorithm: Harmony Search. Simulation **76**, 60–68 (2001)
6. Geem, Z.W.: Improved harmony search from ensemble of music players. In: Gabrys, B., Howlett, R.J., Jain, L.C. (eds.) KES 2006. LNCS (LNAI), vol. 4251, pp. 86–93. Springer, Heidelberg (2006)
7. Geem, Z.W., Tseng, C.-L., Park, Y.-J.: Harmony search for generalized orienteering problem: best touring in China. In: Wang, L., Chen, K., Ong, Y.S. (eds.) ICNC 2005. LNCS, vol. 3612, pp. 741–750. Springer, Heidelberg (2005)
8. Geem, Z.W., Lee, K.S., Park, Y.: Application of harmony search to vehicle routing. Amer. J. Appl. Sci. **2**, 1552–1557 (2005)
9. Geem, Z.W.: Optimal cost design of water distribution networks using harmony search. Eng. Optim. **38**, 259–280 (2006)
10. Hetmaniok, E., Jama, D., Słota, D., Zielonka, A.: Application of the Harmony Search algorithm in solving the inverse heat conduction. Scientific notes of Silesian University of Technology (Zesz. Nauk. Pol. Śl.). Appl. Math. **1**, 99–108 (2011)
11. Hetmaniok, E., Słota, D., Zielonka, A.: Identification of the heat transfer coefficient by using the modified Harmony Search algorithm. Steel Res. Int. Special Edition: Metal Forming, 1039–1042 (2012)
12. Mochnacki, B., Suchy, J.: Numerical Methods in Computations of Foundry Processes. PFTA, Cracow (1995)
13. Nowak, I., Nowak, A.J., Wrobel, L.C.: Inverse analysis of continuous casting processes. Int. J. Numer. Meth. Heat Fluid Flow **13**, 547–564 (2003)
14. Santos, C.A., Garcia, A., Frick, C.R., Spim, J.A.: Evaluation of heat transfer coefficient along the secondary cooling zoones in the continuous casting of steel billets. Inverse Probl. Sci. Eng. **14**, 687–700 (2006)
15. Słota, D.: Identification of the cooling condition in 2-d and 3-d continuous casting processes. Numer. Heat Transfer B **55**, 155–176 (2009)
16. Nowak, I., Smolka, J., Nowak, A.J.: Application of Bezier surfaces to the 3-D inverse geometry problem in continuous casting. Inverse Probl. Sci. Eng. **19**, 75–86 (2011)
17. Słota, D.: Restoring boundary conditions in the solidification of pure metals. Comput. Struct. **89**, 48–54 (2011)

Influence of a Topology of a Spring Network on its Ability to Learn Mechanical Behaviour

Maja Czoków[1]([✉]) and Jacek Miękisz[2]

[1] Faculty of Mathematics and Computer Science,
Nicolaus Copernicus University, Toruń, Poland
maja@mat.umk.pl
[2] Institute of Applied Mathematics and Mechanics,
University of Warsaw, Warsaw, Poland
miekisz@mimuw.edu.pl

Abstract. We discuss how the topology of the spring system/network affects its ability to learn a desired mechanical behaviour. To ensure such a behaviour, physical parameters of springs of the system are adjusted by an appropriate gradient descent learning algorithm. We find the betweenness centrality measure particularly convenient to describe topology of the spring system structure with the best mechanical properties. We apply our results to refine an algorithm generating the structure of a spring network. We also present numerical results confirming our statements.

Keywords: Spring system · Mechanical behaviour problem · Betweenness centrality measure

1 Introduction

In this paper we propose heuristics which can be applied during the construction of a spring system, in order to enhance its efficiency in learning mechanical behaviour. It appears that the ability of adaptation of a spring system depends on the topology of the graph representing its structure. We discuss the betweenness centrality measure and conduct numerical analyses to see how it affects the efficiency. The obtained results are applied to determine a proper structure of connections between nodes in the spring system in order to construct relatively small networks which are capable of reproducing an expected physical behaviour.

In this context, the aspect of learning of the spring system can be twofold. The first one is the parametric learning, where we are given both training examples and a structure of the spring system and we are only required to find unknown spring parameters. The algorithm implementing such an approach was discussed in [4]. This algorithm exhibits many common features with classical methods of machine learning [12]. The second one is the structural learning, where both a structure of the spring system and its parameters are required to

R. Wyrzykowski et al. (Eds.): PPAM 2013, Part I, LNCS 8384, pp. 412–422, 2014.
DOI: 10.1007/978-3-642-55224-3_39, © Springer-Verlag Berlin Heidelberg 2014

be found, with the possible constraints on limited resources. A simple 'brute-force' method was put forward in [5].

It turns out that it is not only the amount of available resources, which have an influence on the quality of learning, but also the topology of the system. We note that graph centrality measures, e.g. the betweenness, are surprisingly accurate heuristics describing the capability of learning the structure.

Spring systems are widely used for modelling properties of microscopic and macroscopic objects. They are for example applied to model disordered media in material sciences [11], elastic properties of physical systems [8], self-organisation [9], and system design in material and architectural sciences [10]. We are going to employ our model to detect intermediate conformations between two known protein conformations [2] and to identify key residues (amino acids, groups of atoms) during transitions from one conformation to another one [6]. To the best of our knowledge our model is the first one, which uses spring systems as devices capable of learning predefined mechanical behaviour.

The work is organised as follows. In Sect. 2, we briefly describe the formal model of the spring system and its dynamics. The stochastic algorithm for generating an optimal spring network architecture for mechanical behaviour problems is presented in Sect. 3. Results of our analysis are provided in Sect. 4. Finally, we give concluding remarks in Sect. 5.

2   Formal Definition of the Spring System

Formally, we represent a spring system by an undirected rigid graph $\mathcal{G} = (\mathcal{V}, \mathcal{E})$, where $\mathcal{V}$ is a finite set of nodes/vertices and $\mathcal{E}$ is a set of edges. We say that graph is rigid if it is impossible to change distances between two nodes without the modification of the lengths of the edges, see [3] for more details. The node set $\mathcal{V} \subset \mathbb{R}^3$ is partitioned into

- Set $\mathcal{V}_{\text{in}}$ of control nodes. These are nodes, whose positions are determined by external intervention, such as user interaction and thus it is regarded as system's *input*.
- Set $\mathcal{V}_{\text{out}}$ of observed nodes, whose positions are regarded as system's *output*.
- Set $\mathcal{V}_{\text{fixed}}$ of immobilised nodes, whose positions are kept fixed in the course of system's evolution.
- The remaining set $\mathcal{V}_*$ of auxiliary movable nodes.

The coordinates of a vertex $v \in \mathcal{V}$ are denoted by $(x_v^{(1)}, x_v^{(2)}, x_v^{(3)})$. With each edge $e = \{u, v\} \in \mathcal{E}$, $u, v \in \mathcal{V}$ we associate its equilibrium (rest) length $\ell_0[e]$ and we write $\ell[e]$ for its actual length (the Euclidean distance between vertices u and v). Moreover, the spring constant $k[e] \geq 0$ is ascribed to each edge $e \in \mathcal{E}$, determining elastic properties of the spring represented by the edge e. The energy (Hamiltonian) of a spring system configuration $\bar{x}_\mathcal{V} = ((x_v^{(1)}, x_v^{(2)}, x_v^{(3)})_{v \in \mathcal{V}})$ is given by the usual formula [3]

$$\mathcal{H}(\bar{x}_\mathcal{V}) := \frac{1}{2} \sum_{e \in \mathcal{E}} k[e] (\ell[e] - \ell_0[e])^2 . \tag{1}$$

We are interested in configurations, which are local minima of $\mathcal{H}$ and are called *local equilibrium points*. Let us assume that we are given a non-equilibrium configuration $\bar{x}_{\mathcal{V}}^0$. To determine the local minimum of the spring system, denoted by $G[\bar{x}_{\mathcal{V}}^0]$, we let it evolve in time according to standard gradient descent dynamics

$$\frac{d}{dt}\bar{x}_{\mathcal{V}} := -\nabla\mathcal{H}(\bar{x}_{\mathcal{V}}) . \tag{2}$$

Since we put constraints on the positions of immobilised nodes $\mathcal{V}_{\text{fixed}} \cup \mathcal{V}_{\text{in}}$ it implicates taking the derivative with respect to only movable nodes $\mathcal{V}_* \cup \mathcal{V}_{\text{out}}$. Because of that and the fact that fixed nodes $\mathcal{V}_{\text{fixed}}$ are always in the same positions we are allowed to rewrite the notation $G[\bar{x}_{\mathcal{V}}^0]$ to the form $G[\bar{x}_{\mathcal{V}_* \cup \mathcal{V}_{\text{out}}}^0 ; \bar{x}_{\mathcal{V}_{\text{in}}}^0]$.

Now we are given a set $(E^{(i)})_{i=1}^N$ of *training examples*, each example $E^{(i)} = (\bar{y}_{\mathcal{V}_{\text{in}}}^{(i)}, \bar{y}_{\mathcal{V}_{\text{out}}}^{(i)})$ consisting of

1. input part $\bar{y}_{\mathcal{V}_{\text{in}}}^{(i)}$ specifying the locations of input nodes,
2. output part $\bar{y}_{\mathcal{V}_{\text{out}}}^{(i)}$ specifying the desired locations of output nodes.

The learning mechanical behaviour problem with a fixed/predefined structure $\mathcal{G} = (\mathcal{V}, \mathcal{E})$ is to find parameters $k[e]$ and $\ell_0[e]$, $e \in \mathcal{E}$, so that the positions $G[\bar{x}_{\mathcal{V}_* \cup \mathcal{V}_{\text{out}}}^0 ; \bar{y}_{\mathcal{V}_{\text{in}}}^{(i)}]_{\mathcal{V}_{\text{out}}}$ of output vertices in equilibrium $G[\bar{x}_{\mathcal{V}_* \cup \mathcal{V}_{\text{out}}}^0 ; \bar{y}_{\mathcal{V}_{\text{in}}}^{(i)}]$ are as close as possible to the desired output locations part $\bar{y}_{\mathcal{V}_{\text{out}}}^{(i)}$. To this end, we define the mean squared error function

$$\Phi := \frac{1}{N}\sum_{i=1}^N \Phi^{(i)} \tag{3}$$

where

$$\Phi^{(i)} := \sum_{v \in \mathcal{V}_{\text{out}}} \text{dist}\left(y_v^{(i)}, G[\bar{x}_{\mathcal{V}_* \cup \mathcal{V}_{\text{out}}}^0 ; \bar{y}_{\mathcal{V}_{\text{in}}}^{(i)}]_v\right)^2 . \tag{4}$$

The adaptation of parameters $k[e]$ and $\ell_0[e]$ is performed according to the following gradient descent scheme:
1. Start from a predefined configuration $\bar{x}_{\mathcal{V}}$
2. Reiterate T times, where T is a predefined positive integer number
3.   Choose cyclically a subsequent example $E^{(i)}$ and set $\bar{x}_{\mathcal{V}_{\text{in}}} := \bar{y}_{\mathcal{V}_{\text{in}}}^{(i)}$
4.   Set $\bar{x}_{\mathcal{V}}^0 := \bar{x}_{\mathcal{V}}$ and find $\bar{x}_{\mathcal{V}} := G[\bar{x}_{\mathcal{V}_* \cup \mathcal{V}_{\text{out}}}^0 ; \bar{y}_{\mathcal{V}_{\text{in}}}^{(i)}]$
5.   Calculate $\Phi^{(i)}$ and denote it as $\Phi_{before}^{(i)}$
6.   Iterate through parameters $p \in k[e] \cup \ell_0[e]$ for all $e \in \mathcal{E}$
7.   Increase p by a small constant $\delta > 0$ and find the equilibrium state
8.   Recalculate $\Phi^{(i)}$ and denote it as $\Phi_{after;p}^{(i)}$
9.   Decrease p by δ
10.   Set $p := p - \rho \cdot \frac{\Phi_{after;p}^{(i)} - \Phi_{before}^{(i)}}{\delta}$, where ρ is a small learning constant.

The main focus of this work is to extend the learning problem to a more general case when a structure of the spring network is not known as it was already suggested in [4]. Formally, we are given only the training examples $E^{(i)} = (\bar{y}_{\mathcal{V}_{\text{in}}}^{(i)}, \bar{y}_{\mathcal{V}_{\text{out}}}^{(i)})$, but the structure of the network $\mathcal{G} = (\mathcal{V}, \mathcal{E})$ is no longer provided. In fact, finding the graph $\mathcal{G}$ along with spring elastic constants $k[e]$ and spring rest lengths $\ell_0[e]$ is our explicit target. The set of input nodes $\mathcal{V}_{\text{in}}$ and observed nodes $\mathcal{V}_{\text{out}}$ are predefined by the set of the training examples $(E^{(i)})_i^N$. The goal is to find a set of auxiliary nodes $\mathcal{V}_*$, fixed nodes $\mathcal{V}_{\text{fixed}}$ and a set of edges $\mathcal{E}$, for which adjusting parameters will lead to a low error Φ.

3   Structure Searching for a Set of Training Examples

In this section we briefly reiterate the algorithm for generating a spring network architecture for learning mechanical behaviour problems formulated earlier. The thorough description of the algorithm can be found in [5]. The algorithm is based on the Henneberg construction method [1], which assures us that the graph $\mathcal{G}$ returned by the algorithm is rigid in d (in our case 3) dimensional space. The generation of the graph is based on the set of the training examples $(E^{(i)})_{i=1}^N$ according to the following scheme:

1. Find the mean value $m \in \mathbb{R}^3$ of all input and output locations specified in $(E^{(i)})_{i=1}^N$.
2. Randomly pick 4 different nodes uniformly distributed in the ball centred at the point m, and with the diameter equals to maximal distance between point m and any location described in $(E^{(i)})_{i=1}^N$.
3. Add the random nodes to the set $\mathcal{V}_*$, link each pair of these nodes with an edge.
4. Add aux_const subsequent graph nodes to the set $\mathcal{V}_*$ always requiring that the new node is connected to $edge\_const \geq 3$ already existing ones.
5. Fix $fixed\_const$ random nodes; as a result of this operation $fixed\_const$ nodes are moved from the set $\mathcal{V}_*$ to the set $\mathcal{V}_{\text{fixed}}$.
6. Sequentially add output nodes to the set $\mathcal{V}_{out}$. The number of these nodes is determined by the set of training example. Each new node is connected to $edge\_const \geq 3$ already existing ones and the respective location is specified by $\bar{y}_{\mathcal{V}_{\text{out}}}^{(1)}$.
7. Sequentially add input nodes to the set $\mathcal{V}_{in}$. The number of these nodes is determined by the set of training example. Each new node is connected to $edge\_const \geq 3$ already existing ones belonging to the set $\mathcal{V}_* \cup \mathcal{V}_{out}$ and the respective location are specified by $\bar{y}_{\mathcal{V}_{\text{in}}}^{(1)}$.
8. Perform the adaptation of parameters $k[e], \ell_0[e]$, $e \in \mathcal{E}$ according to the gradient descent algorithm in order to minimise the value of the error function Φ.
9. Return the obtained structure.

The satisfaction of the condition $edge\_const \geq 3$ is required rigidity of graph, for more details see again [1]. One must keep in mind that 3 noncollinear *immobilised* nodes are required to prevent the system from rotating, translating or a combination of these two operations (moves), see Sect. 4 in [7].

There is one difference between this algorithm and the one presented in the previous paper [5]. The order between the steps 6 and 7 is changed. In the next section we present numerical analysis which shown that adding output nodes before input nodes is much more effective then reversely.

4   Results

We start our numerical analysis from employing the structure generating algorithm to create a large set of samples of spring systems, for a given set of training examples and for various combinations of pairs $(|\mathcal{V}_*|, edge\_const)$. We create two collections of such architectures for both versions of structure searching algorithm: the old one with input nodes added to the graph before output nodes and the new one, with output nodes added before input nodes (with points 6 and 7 inversed). Next, the learning algorithm is applied in order to adapt spring systems to the desired mechanical behaviour. The algorithm, adjusting parameters of the springs, applies the same number of iterations T, for all learning simulations. The value of T is sufficiently large to let us tell, that the systems obtain adaptation near to the best possible. Plots shown in the Fig. 1 present the mean error Φ obtained in 50 simulations, by the both versions of the algorithm, versus $|\mathcal{V}_*| \in \{3, .., 19\}$ for $edge\_const$ equal to 3 (in Fig. 1a) and 5 (in Fig. 1b). The variant of the algorithm, which adds output nodes before input ones, for a given $|\mathcal{V}_*|$ creates a graph with more edges, than the version which add input nodes before output ones, only if $|\mathcal{V}_*| < edge\_const$, so in the tests it is true for $|\mathcal{V}_*| \in 3, 4$ and $edge\_const = 5$. After the modification, the value of error function decreases on average 4.181 ± 2.35 times for $edge\_const = 3$ and 2.120 ± 0.685 times for $edge\_const = 5$, for tests, for which results are presented in Fig. 1. As we can see, the correction of the algorithm particularly improves results for spring system with the set $\mathcal{V}_*$ with small size. Throughout the rest of the paper we assume that tests are carried with the more efficient version (step 6 before step 7).

Since we know that the graph topology has the relevant influence on the adaptation ability of the spring system, we decided to exploit this fact during the construction of the structure of a spring system. In this context we will discuss *betweenness centrality measure*. The betweenness centrality measure of a vertex is defined by

$$B(v) := \sum_{s \neq v} \sum_{t \neq v, t \neq s} \frac{\sigma_{st}(v)}{\sigma_{st}}, \tag{5}$$

where σ_{st} is total number of shortest paths from node s to node t and $\sigma_{st}(v)$ is the number of those paths pass through v.

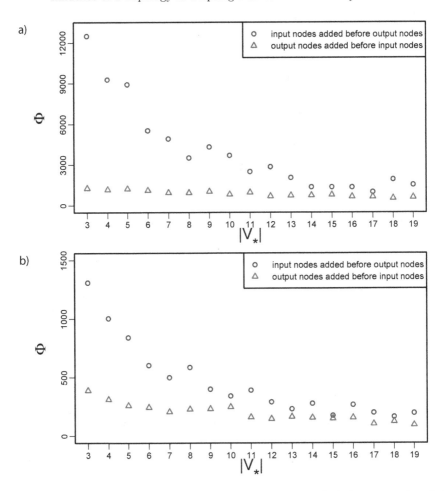

Fig. 1. Plots depict error Φ vs the number of auxiliary vertices $|\mathcal{V}_*|$, for mean value of all 50 independent simulations. The network is adapted to three training examples, $|\mathcal{V}_{in}| = 6$, $|\mathcal{V}_{out}| = 4$, $|\mathcal{V}_{fixed}| = 2$, the initial error is of the order 10^4; (a) for $edge\_const = 3$; (b) for $edge\_const = 5$

In Fig. 2 we present a graph with values of betweenness of particular nodes. In Fig. 3, for one exemplary graph $\mathcal{G}$ built by the enhanced structure searching algorithm, we present sequence of attachment of nodes versus values of betweenness for the respective nodes. It is clear that nodes added later have on average lower values of betweenness centrality measure than the ones added earlier. Having introduced the notion of betweenness and conducted comparison of numerical results returned by two versions of the structure searching algorithm, we conclude that a spring system has better learning capability when its input nodes have lower values of betweenness than output nodes.

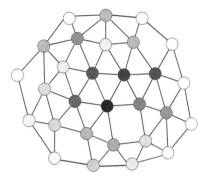

Fig. 2. Brightness shows the node betweenness, the darkest node has the highest value of betweenness, the lightest node has the lowest value of betweenness

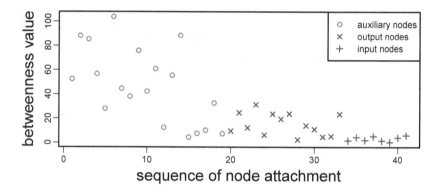

Fig. 3. Sequence of attachment of nodes versus values of betweenness for the respective nodes, for one exemplary graph $\mathcal{G}$ built by the structure searching algorithm

We start next analysis from creating a random sample of spring systems. For a given set of training examples and for each pair of different values ($|\mathcal{V}_*|$, $edge\_const$) we construct and adapt 50 spring systems. For each of these networks we calculate $\overline{B_{in}}$, $\overline{B_{out}}$, $\overline{B_*}$, which are mean values of betweenness of the nodes belonging to the sets $\mathcal{V}_{in}$, $\mathcal{V}_{out}$ and $\mathcal{V}_*$ respectively. Next, for each sample of 50 systems constructed for pair ($|\mathcal{V}_*|$ and $edge\_const$) we calculate Pearson correlations $(\rho(X,Y) := \sum_i (x_i - EX)(y_i - EY)/\sqrt{D^2 X D^2 Y})$ $\rho_{in} := \rho(\overline{B_{in}}, \Phi)$, $\rho_{out} := \rho(\overline{B_{out}}, \Phi)$, $\rho_* := \rho(\overline{B_*}, \Phi)$. On average statistics ρ_{in}, ρ_{out} are positive and ρ_* is negative. The absolute value of ρ_{out} is on average greater than absolute value of ρ_{in}. In Table 1 are presented values of ρ_{out}, ρ_{in} and ρ_* for network constructed for exemplary set of training examples, for $|\mathcal{V}_*| \in \{4, ..., 13\}$ and $edge\_const \in \{4, 6\}$.

In order to obtain systems with lower values of error function Φ, we decided to adapt to training examples only the chosen networks. Criteria of selection

Table 1. Values of ρ_{in}, ρ_{out} and ρ_* for $|\mathcal{V}_*| \in \{4, ..., 13\}$ and $edge\_const \in \{4, 6\}$. The network applied in analysis were adapted to three training examples, $|\mathcal{V}_{in}| = 8$, $|\mathcal{V}_{out}| = 14$, $|\mathcal{V}_{fixed}| = 2$

	4	5	6	7	8	9	10	11	12	13	$E\rho$
	For $edge\_const = 4$										
ρ_{out}	0.103	0.019	0.013	0.154	0.186	0.170	0.099	0.153	0.211	0.352	0.146
ρ_{in}	0.063	-0.207	0.010	0.007	0.043	-0.089	0.336	0.075	-0.004	0.088	0.032
ρ_*	-0.207	-0.112	-0.097	-0.210	-0.107	-0.216	-0.188	-0.147	-0.243	-0.257	-0.178
	For $edge\_const = 6$										
	4	5	6	7	8	9	10	11	12	13	$E\rho$
ρ_{out}	0.374	0.329	0.043	0.296	0.204	0.048	0.145	0.117	0.063	0.239	0.185
ρ_{in}	0.165	0.028	0.095	0.075	0.115	0.068	0.003	0.162	-0.120	-0.027	0.056
ρ_*	-0.335	-0.342	-0.035	-0.199	-0.220	-0.141	-0.112	-0.152	-0.127	-0.181	-0.184

are established on the base of $\overline{B_{out}}$. Statistics ρ_{out}, ρ_{in} and ρ_* illustrate, that systems, which are better adapted to required mechanical behaviour, have on average lower values of $\overline{B_{out}}, \overline{B_{in}}$ and higher of $\overline{B_*}$ than systems, which are worse adapted. Correlation between Φ and $\overline{B_{out}}$ is stronger than between Φ and $\overline{B_{in}}$. So, eventually, as a criterion of network selection we decided to pick the one with minimal value of $\overline{B_{out}}$. It is unequivocal with selection graphs with high $\overline{B_*}$, because Pearson correlation between $\overline{B_{out}}$ and $\overline{B_*}$ turns out to be near to -1.

Now we are ready to employ networks with low output betweenness to perform numerical analysis. To this end, we create new collections of 50 samples of adapted spring systems constructed for each pair ($|\mathcal{V}_*|$, $edge\_const$) and for the set of the training examples. But this time selection of structures of the graphs is conducted. Namely, each sample of 50 structures is picked up out of 500 random architectures with the lowest mean values of betweenness of the output nodes. In Fig. 4 we can see how the selection of the graphs enhances obtained results. The plots present the results for two different sets of training examples. For the first set (Fig. 4a), after the modification, the value of the error function decreases on average 1.256 ± 0.208 times for $edge\_const = 3$ and 1.594 ± 0.410 times for $edge\_const = 5$. In turn, for the second set of training examples (Fig. 4b), after the modification, the value of the error function decreases on average 1.153 ± 0.130 times for $edge\_const = 4$ and 1.239 ± 0.127 times for $edge\_const = 6$. This modification slightly decreases the value of an error generated by networks, but since the adaptation of spring systems to learning examples is extremely complex and time-consuming, even small improvements are significant.

The applied method of selection of structures with a low mean value of betweenness of the output nodes is convenient for many reasons. First of all, this keeps graph $\mathcal{G}$ rigid. The auxiliary and the input nodes preserve their large and low values of betweenness respectively. The generation of one structure has the time complexity $O(|\mathcal{V}| \cdot |\mathcal{E}|)$. In addition we note that for most applications the spring network is a sparse graph, so this estimation can be improve

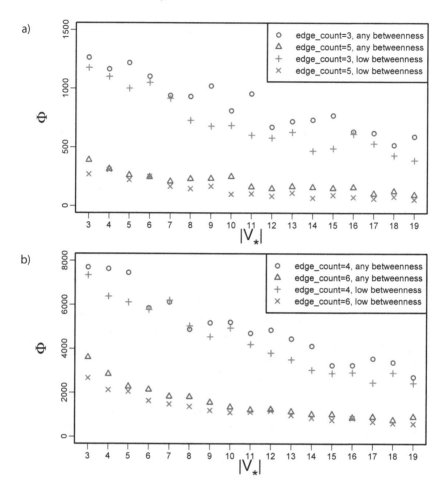

Fig. 4. A plot of error Φ vs the number of auxiliary vertices $|\mathcal{V}_*|$, for mean value of all 50 independent simulations. (a) The network is adapted to three training examples, $|\mathcal{V}_{in}| = 6$, $|\mathcal{V}_{out}| = 4$, $|\mathcal{V}_{fixed}| = 2$,   results for *edge_const* = $\{3,5\}$; the initial error is of the order 10^4. (b) The network is adapted to three training examples, $|\mathcal{V}_{in}| = 8$, $|\mathcal{V}_{out}| = 14$, $|\mathcal{V}_{fixed}| = 2$,   results for *edge_const* = $\{4,6\}$; the initial error is of the order 10^5

to $O(|\mathcal{V}|^2)$ The algorithm is sufficiently fast even if before learning we want to randomly pick several spring structures and choose one of them for the learning process. The presented analyses were replicated for 50 structures picked up out of 5000. Such increase of the number of random samples has not introduced further improvements of the ability of learning of mechanical behaviour by spring systems.

5 Conclusions

We discussed how the topology of the spring system/network influences its ability of learning mechanical behaviour. Betweenness centrality measure was employed in order to describe structures with the required topology. It turned out that it is especially advantageous to keep low values of betweenness of input nodes for networks with a small set of auxiliary nodes. This property can be vital for real-world problems whose character demands a reduction of the set $|\mathcal{V}_*|$. Low values of betweenness of input vertices are achieved by adding these vertices at the end of the process of network construction.

Additional enhancement of the learning ability of spring system is obtained by building a group of graphs and selecting the one with the lowest mean value of betweenness of output nodes. This modification slightly decreases the value of an error generated by networks, but since the adaptation of spring systems to learning examples is so complex, even small improvements are significant. Perhaps it could be interesting to apply an algorithm which builds a rigid graph with minimal values of betweenness of some group of vertices (input and output ones) and maximal values for remaining vertices (auxiliary vertices). To the best of our knowledge such a method is not known.

Setting different values of betweenness for different types of nodes is vital in spring systems which are applied to mimic dynamics of real-word objects. It is known for example that atoms which play a key role in conformational movements of proteins have on average higher values of betweenness than remaining ones [6]. This is important in the context of our ongoing work where we are going to apply the model of spring systems and the results presented in this paper to explore conformational transitions of proteins.

References

1. Anderson, B.D.O., Belhumeur, P.N., Morse, A.S., Eren, T.: A framework for maintaining formations based on rigidity. In: Proceedings of the IFAC World Congress, Barcelona (2002)
2. Chirikjian, G.S., Jernigan, R.L., Moon, K.K.: Elastic models of conformational transitions in macromolecules. J. Mol. Graph. Model. **21**, 151–160 (2002)
3. Connelly, R.: Rigidity and energy. Invent. Math. **66**, 11–33 (1982)
4. Czoków, M., Schreiber, T.: Adaptive spring systems for shape programming. In: Rutkowski, L., Scherer, R., Tadeusiewicz, R., Zadeh, L.A., Zurada, J.M. (eds.) ICAISC 2010, Part II. LNCS, vol. 6114, pp. 420–427. Springer, Heidelberg (2010)
5. Czoków, M., Schreiber, T.: Structure searching for adaptive spring networks for shape programming in $3D$. In: Rutkowski, L., Korytkowski, M., Scherer, R., Tadeusiewicz, R., Zadeh, L.A., Zurada, J.M. (eds.) ICAISC 2012, Part II. LNCS, vol. 7268, pp. 207–215. Springer, Heidelberg (2012)
6. Dokholyan, N.V., Karplus, M., Paci, E., Vendruscolo, M.: A small-world view of the amino acids that play a key role in protein folding. Phys. Rev. E **65**, 061910 (2002)
7. Greiner, W.: Classical Mechanics: Systems of Particles and Hamiltonian Dynamics. Classical Theoretical Physics. Springer, New York (2010)

8. Gusev, A.A.: Finite element mapping for spring network representations of the mechanics of solids. Phys. Rev. Lett. **93**, 034302 (2004)

9. Kanellos, A.: Topological self-organisation: using a particle-spring system simulation to generate structural space-filling lattices. Masters thesis, UCL (2007)

10. Kilian, A., Ochsendorf, J.: Particle-spring systems for structural form finding. J. IASS **46**, 77–84 (2005)

11. Ostoja-Starzewski, M.: Lattice models in micromechanics. Appl. Mech. Rev. **55**, 35–60 (2002)

12. Trawinski, B., Smetek, M., Telec, Z., Lasota, T.: Nonparametric statistical analysis for multiple comparison of machine learning regression algorithms. Int. J. Appl. Math. Comput. Sci. **22**(4), 867–882 (2012)

Comparing Images Based on Histograms of Local Interest Points

Tomasz Nowak, Marcin Gabryel, Marcin Korytkowski, and Rafał Scherer[(⊠)]

Institute of Computational Intelligence, Czestochowa University of Technology,
al. Armii Krajowej 36, 42–200 Czestochowa, Poland
{tomasz.nowak,marcin.gabryel,marcin.korytkowski,
rafal.scherer}@iisi.pcz.pl
http://iisi.pcz.pl

Abstract. One of the key unresolved issues of image processing is the lack of methods for searching images similar to the reference image. This paper focuses on objects that there are in images and presents a method to compare the objects and search for images that contain objects belonging to the same classes. Taking advantage of the fact that local keypoints of images constitute a very good basis for further processing images, we use them for objects comparison. More precisely, the comparison of images is based on histograms, that are generated on the basis of the keypoints of objects contained in images. We present results of experiments which have been conducted for various classes of objects and histograms generated using the proposed method.

Keywords: Content-based image processing · Histogram comparison

1 Introduction

Digital image processing is a very complex issue. Humans looking at a picture at the very first moment are able to recognize objects on the image, and see the dependencies between them so they can conclude what the image shows. In addition, they see differences between images, even if they represent the same object, for example car racing or car dealer - such an interpretation is called a semantic image analysis. The image is a collection of pixels, which are arranged in a corresponding manner to show given situation. A collection of these pixels is really a huge amount of data to be processed and their combinations is infinite, therefore it is not possible to make a dictionary, which contains an index of all the words. Therefore, image analysis is so difficult and complicated for computers and a very serious challenge for researchers. For the time being there is no method which would always be effective. There are many algorithms used at the different stages of image analysis [1,4,5,7,10], however, there is not one coherent algorithm allowing to perform fully automated image search in a large variety of graphics data sets. The correctness of many methods depends on the input data, which can be: distribution of colors, edges, shapes, groups or keypoints. The method presented in this paper is based mainly on keypoints.

R. Wyrzykowski et al. (Eds.): PPAM 2013, Part I, LNCS 8384, pp. 423–432, 2014.
DOI: 10.1007/978-3-642-55224-3_40, © Springer-Verlag Berlin Heidelberg 2014

There exists a large number of methods for image processing used in specific cases, e.g. face detection [8], signature recognition [15,16], or smile detection for photo cameras [14]. There are also many specialized methods used e.g. in medical imaging. They are able today to locate a specific cell [12], to detect bones [11] or other changes in the cells [2]. At the moment there is no solution that would allow for automatic search of similar objects in images which are semantically similar. The paper presents a method developed on the basis of histograms generated from image keypoints. The authors note that each class of objects generates similar histograms. Using this relationship it is possible to extract from a database the images consisting of similar objects.

The paper is structured as follows. Section 2 shows an overview of popular existing solutions image comparison. Section 3 shows the new method of comparing objects in images. The process of comparison is based on an analysis of histograms generated for individual objects. Section 4 shows the results of experiments carried out using the proposed method. Section 5 presents conclusions.

2   Previous Works

The base of content-based image retrieval in many cases are image keypoints. One of the most popular algorithms used to generate the keypoints of a picture is SIFT (Scale Invariant Feature Transform) [9] which was presented in 1999 by D. Lowe. The newer version of SIFT is the SURF (Speeded Up Robust Features) [3] algorithm which was presented in 2006 by H. Bay. SIFT is much more accurate, but somewhat slower than SURF. The method proposed in this paper is based on the keypoints that were generated with the help of the SURF algorithm.

The main element that makes SURF runs faster is the structure named integral images, what allows to significantly reduce the number of operations. This structure is represented by the sum of pixels in any rectangular area in the input image I

$$I_{\sum}(x,y) = \sum_{i=0}^{i \leq x} \sum_{j=0}^{j \leq y} I(x,y) \, , \tag{1}$$

where I is a processed image, $I_{\sum}(x,y)$ - the sum of all pixels in the image. Calculation of the sum of the pixels in the selected area of the image (integral image) is presented in Fig. 1, and is described by (2).

$$\sum = A - B - C + D \, , \tag{2}$$

where A, B, C and D are the coordinates of the vertices of the selected rectangular area in the image. Thanks to integral images, it takes only three additions and four memory accesses to calculate the sum of intensities inside a rectangular region of any size. Calculation time is independent of its size, which translates to the performance of the algorithm. SURF uses different filters (simpler ones) than SIFT thus is faster and also somewhat less accurate (Fig. 2a) thanks to using less complex filters. However, in many applications high accuracy is not

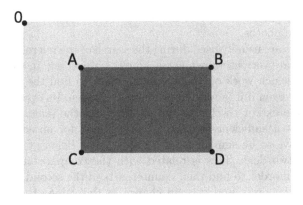

Fig. 1. Thanks to the integral images concept, it takes only three additions and four memory accesses to calculate the sum of intensities inside a rectangular region of any size.

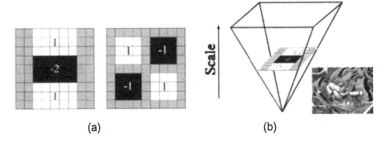

Fig. 2. The SURF algorithm uses filters of simple construction, which are scaled during searching keypoints. This approach allows to speed up the algorithm [3].

needed, and we can sacrifice it for the speed of the algorithm. SURF searches interesting points at different scales. In the SIFT, scale-spaces are created from pyramids consisting resized images which requires more resources, whereas in SURF only filters are scaled while the image remains unchanged. This approach allowed to substantially speed up calculations (Fig. 2b). Object recognition based on keypoints and their descriptors is done by matching keypoints belonging to the reference image to the keypoints contained in the test images. Many of these comparisons will be invalid due to various noise present in the test images. Matching is done by finding the nearest neighbour in a database of keypoints. The nearest neighbour is defined as the keypoint with minimum Euclidean distance for the invariant descriptor vector as was described in [9]. The RANSAC algorithm (Random Sample Consensus) [6] is also frequently used to compare images based on keypoints. The algorithm randomly selects samples from a small set of matching candidates and estimates homography between these points by minimizing the least squared error. The algorithm is used usually during search operations of the pattern object in another image, or combining images e.g. to create panoramic photos.

The keypoints are usually used during the search of the reference object in the test image, they are very useful while tracking the object e.g. in video sequences. The algorithms which work on the keypoints allow to find the searched object in another image even if it is rotated or partially hidden. Its descriptors are also resistant to some extent to change the lighting and the scale. The keypoints, however, are not a sufficient solution when we look for objects that are not identical, but have to be similar to each other - they belong to the same class of objects, e.g. two dogs. This is related with the comparison method of the group of points in order to find their counterparts in the second image, they are compared point by point and a small change in the value of their descriptors contributes to the fact that they are not taken into account. Two object that are similar to each other but not identical, contain different sets of keypoints, therefore it is not possible to use them to find similar objects.

3   Proposed Method

Starting point of the proposed method is a set of keypoints generated using the SURF algorithm. The main task of the algorithm is to sort keypoints that belong to an object in the image, depending on the area of the coordinate system where they are located. An important part is also the angle which is created between a vector passing through the origin and the keypoint and the X-axis of coordinate system. All the resulting vectors are grouped on the basis of the formed angles with increments of $5°$, then the average vector is calculated for all vectors included in one group. This operation is aimed at reducing the amount of processed data. The next step is to generate a histogram, which is the basis for the comparison of images. Below is a detailed description of the individual steps of the proposed method.

3.1   Construction of Coordinate System

In the first step of the algorithm we determine an additional point that specifies the center of the test object. Then, we move the origin of the coordinate system to this point location. The new point shall be calculated on the basis of all the keypoints found by the SURF algorithm. Determining position of the new point to the object takes place according to the following expression

$$\bar{x} = \frac{\sum_{i=1}^{n} w_i x_i}{\sum_{i=1}^{n} w_i}, \quad \bar{y} = \frac{\sum_{i=1}^{n} w_i y_i}{\sum_{i=1}^{n} w_i} \;, \tag{3}$$

where $[x_1, x_2, ..., x_n]$, $[y_1, y_2, ..., y_n]$ are coordinates of the points and $[w_1, w_2, ..., w_n]$ are weights of the points. If we have the center point and other keypoints of the object, we can proceed to classify the keypoints by angles (Fig. 3).

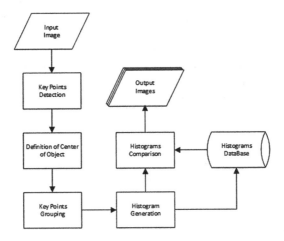

Fig. 3. General block diagram of the proposed method.

3.2   Classification of Keypoints on the Basis of Angles

In order to compare the keypoints using histograms they need to be correctly grouped. In order to do this, we must move the entire coordinate system relative to the test object. After moving the coordinate system, we create vectors between the beginning of the coordinate system and subsequent points designated by the SURF algorithm. Then, we normalize the length of each vector and calculate the angles between them and the X-axis. For this purpose we use the dependency of tangent tg of the angle α of the inclination relative to the X-axis on the linear function directional coefficient

$$y = ax + b, \ a = tg\alpha .$$
(4)

Having calculated all the angles which are formed by the vectors between the X-axis and the keypoints, we can proceed to their grouping (Fig. 4). The keypoints

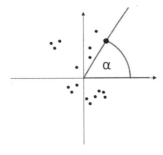

Fig. 4. The keypoints are divided into groups depending on the angle formed between the vector that passes through the keypoint and the X-axis.

are divided into groups depending on the angle formed between the vector that passes through the keypoint as well as the origin of coordinate system and the X-axis. As the interval we assumed circle sector of the angle $5°$. The circle was divided into the 72 equal sectors, the first sector begins with the X-axis (from $0°$ to $5°$), each next sector has $5°$, until value $360°$ is reached. All the length of the vectors and the keypoints that are in the specified sector are counted, and on the basis of these values a histogram is built (Fig. 5). Storing keypoints in this way causes, that objects of similar class generate similar histograms. The similarity of histograms allows coarse finding objects similar to each other. Transforming keypoints into histogram allows also to reduce the amount of data that must be stored in data base in order to perform future comparison and retrieval of objects.

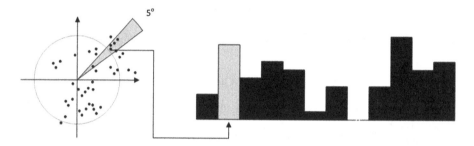

Fig. 5. In the presented method the keypoints are grouped depending on the circle sector in which they are. On the basis of such aggregated points histogram is created, by means of which it is possible to find similar objects to the reference object. In addition, storing keypoints transformed into a histogram greatly reduces the amount of required calculations needed to compare two objects as well as memory requirements.

3.3   Comparison of Histograms

For the comparison purposes the histogram is divided into four parts, in accordance with the relevant quarters of the circle. The first quarter includes angles from $0°$ to $90°$ degrees, the second from $90.1°$ to $180°$, etc. In each separate part we calculate the absolute value of the differences of individual sectors, and the average difference in each sector (5). Then the results from each of the quarters W_l are compared with a threshold value P which was determined during the experiments empirically, and on the basis of this comparison image is classified as similar to a reference image or not (6)

$$W_l = \frac{1}{k} \sum_{n=1}^{k} |z_n|, \ |z_n| = |x_n - y_n| \,, \tag{5}$$

$$\sum_{l=1}^{j} W_l \geq P \,, \tag{6}$$

where x_n and y_n are number of points found in the data range appropriate for the reference image and test image, i is the number of sectors, k the number of sectors in the selected quarter (j).

4   Experimental Results

The database on which we have tested our method consists of various objects belonging to different classes: different kind of animals and everyday objects. Namely, for our research we used the VOC2012 [13] database, from which we selected 10 classes of objects (dog, horse, butterfly, car, computer, etc.). We need to indicate that the focus is on developing methods to help search similar objects, hence the database contains images of objects, mostly without the background, we used more than 500 images (Fig. 6). The main task was to find similar objects in the database, acting only on the previously generated histograms for all images. The proposed method is based on the assumption that similar histograms are generated for the objects belonging to the same class. Figure 7 shows a set of generated histograms for several examples of objects contained in the tested database. As we can see, most of the objects of the same class generate similar data distribution in the histogram. The main task was to find similar images based on the reference image. The reference image was chosen at random, and all found images were different from the reference image. Table 1 shows the results of the experiment for each of the 10 tested classes, and they prove that the search based on the histogram created on the basis of keypoints gives satisfactory results. An average of 77 % of objects belonging to the same class was found in the database on the basis of the proposed method. Worse results were obtained comparing the ratio of the quantity of good images to all found images. On average, every third image from the obtained was correct. But it must be taken into account that the method of analysis of histograms presented in detail in Sect. 3 is based on a very simple solution, which translates to so many additional images in the obtained results. Using other features of images and using additional methods of filtration can significantly reduce the amount of incorrect images in the obtained results.

Fig. 6. Examples of images used in the experiments.

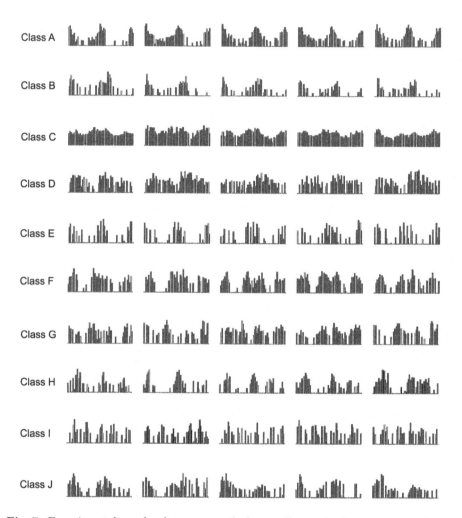

Fig. 7. Experimental results for ten tested classes. For each class, we selected five examples of objects to illustrate the results. As we can see, each of the classes generates different histogram. However, they are relatively similar for every class.

Table 1. Test results. The first column shows the class label for which the experiment was conducted. The second column shows the number of found images in a database from of all the images of the class in the database. The third column shows the number of correct matches for tested class against the background of all the obtained results.

Class label	Found images from the selected class [%]	The correct images to the entire result of the search [%]
A	91	24
B	100	45
C	60	50
D	67	32
E	100	32
F	80	29
G	64	43
H	90	27
I	60	13
J	56	17
Avg.	77	31

5   Conclusions and Future Work

In this paper we presented a new method for content-based image retrieval. Namely, we proposed a quick search of objects belonging to the same class (e.g. dogs, cars, computers) from a set of pictures using histograms of image interest point descriptors. On the basis of the generated histograms and the appropriate method of comparing them, we are able to find in the database objects similar to the presented pattern. Using methods based only on the comparison of keypoints it is not possible to find similar objects, because keypoints are very useful to search for the same objects, their fragments, or to track selected objects e.g. in video sequences. This is caused by necessity to exact match pairs of keypoints of the respective images. However, in the situation when the test objects are not identical, the use of only keypoints is not sufficient. Our method extends the applicability of keypoints towards the possibility of finding objects similar to each other. In the future, the authors plan to streamline the process of comparing histograms and reduce the amount of redundant images in the obtained results by including other important features of images, such as color saturation or the shapes of objects.

Acknowledgments. The project was funded by the National Center for Science under decision number DEC-2011/01/D/ST6/06957.

References

1. Achanta, R., Süsstrunk, S.: Saliency detection for content-aware image resizing. In: IEEE International Conference on Image Processing, pp. 1005–1008 (2009)

2. Akgül, C.B., Rubin, D.L., Napel, S., Beaulieu, C.F., Greenspan, H., Acar, B.: Content-based image retrieval in radiology: current status and future directions. J. Digit. Imaging **2**, 208–222 (2011)

3. Bay, H., Ess, A., Tuytelaars, T., Van Gool, L.: Speeded-Up Robust Features (SURF). Int. J. Comput. Vis. Image Underst. (CVIU) **110**(3), 346–359 (2008)

4. Bazarganigilani, M.: Optimized image feature selection using pairwise classifiers. J. Artif. Intell. Soft Comput. Res. **1**(2), 147–153 (2011)

5. Chang, Y., Wang, Y., Chen, C., Ricanek, K.: Improved image-based automatic gender classification by feature selection. J. Artif. Intell. Soft Comput. Res. **1**(3), 241–253 (2011)

6. Fischler, M.A., Bolles, R.C.: Random sample consensus: a paradigm for model fitting with applications to image analysis and automated cartography. Comm. ACM **24**, 381–395 (1981)

7. Górecki P., Artiemjew P., Drozda P., Sopyła K.: Categorization of similar objects using bag of visual words and support vector machines. In: Proceedings of 4th International Conference on Agents and Artificial Intelligence, ICAART'12, pp. 231–236 (2012)

8. Kisku, D.R., Rattani, A., Grosso, E., Tistarelli, M.: Face identification by SIFT-based complete graph topology. In: IEEE Workshop on Automatic Identification Advanced Technologies 2007, pp. 63–68 (2007)

9. Lowe, D.: Distinctive image features from scale-invariant keypoints. Int. J. Comput. Vis. **2**(60), 91–110 (2004)

10. Rygał, J., Najgebauer, P., Romanowski, J., Scherer, R.: Extraction of objects from images using density of edges as basis for GrabCut algorithm. In: Rutkowski, L., Korytkowski, M., Scherer, R., Tadeusiewicz, R., Zadeh, L.A., Zurada, J.M. (eds.) ICAISC 2013, Part I. LNCS, vol. 7894, pp. 613–623. Springer, Heidelberg (2013)

11. Shubhangi, D.C., Raghavendra, S., Chinchansoor, P., Hiremath, S.: Edge detection of femur bones in X-ray images - a comparative study of edge detectors. Int. J. Comput. Appl. **42**(2), 13–16 (2012)

12. Tek, F.B., Dempster, A.G., Kale, I.: Malaria parasite detection in peripheral blood images. In: British Machine Vision Conference (2006)

13. Visual Object Classes Challenge 2012. http://pascallin.ecs.soton.ac.uk/challenges/VOC/voc2012/

14. Whitehill, J., Littlewort, G., Fasel, I., Bartlett, M., Movellan, J.: Toward practical smile detection. IEEE Trans. Pattern Anal. Mach. Intell. **31**(11), 2106–2111 (2009)

15. Zalasiński, M., Łapa, K., Cpałka, K.: New algorithm for evolutionary selection of the dynamic signature global features. In: Rutkowski, L., Korytkowski, M., Scherer, R., Tadeusiewicz, R., Zadeh, L.A., Zurada, J.M. (eds.) ICAISC 2013, Part II. LNCS, vol. 7895, pp. 113–121. Springer, Heidelberg (2013)

16. Zalasiński, M., Cpałka, K.: Novel algorithm for the on-line signature verification using selected discretization points groups. In: Rutkowski, L., Korytkowski, M., Scherer, R., Tadeusiewicz, R., Zadeh, L.A., Zurada, J.M. (eds.) ICAISC 2013, Part I. LNCS, vol. 7894, pp. 493–502. Springer, Heidelberg (2013)

Improved Digital Image Segmentation Based on Stereo Vision and Mean Shift Algorithm

Rafał Grycuk, Marcin Gabryel, Marcin Korytkowski, Jakub Romanowski,
and Rafał Scherer$^{(\boxtimes)}$

Institute of Computational Intelligence, Częstochowa University of Technology,
Al. Armii Krajowej 36, 42-200 Częstochowa, Poland
{rafal.grycuk,marcin.gabryel,marcin.korytkowski,jakub.romanowski,
rafal.scherer}@iisi.pcz.pl
http://iisi.pcz.pl

Abstract. Segmentation of digital images is an important issue of object recognition. This method of image processing allows to determine single object areas in images. This paper presents an improved segmentation method which gives a possibility to detect single objects in images by using the disparity map algorithm in connection with the mean shift pixel grouping algorithm. Images are processed in grayscale where range of colors is in from 0 to 255. Grayscale allows to detect objects on the basis of pixels brightness. To achieve this purpose we used one of grouping algorithms known as mean shift. Images obtained from mean shift are in the form of separated images which could be subject of further processing. Important feature of mean shift processing is that we obtain the results in the form of backgroundless images containing important objects from the input image.

Keywords: Content-based image processing · Stereo vision · Image segmentation

1 Introduction

The main goal of this research is to perform segmentation of image based on stereo vision images. Segmentation is the process of dividing the input image on the homogeneous objects having similar properties. Segmentation is one of the most complicated image processing fields, as it is extremely difficult to obtain a uniform, homogeneous objects from images containing background or overlapping areas. There are many algorithms for object extraction, but none of them is universal for all types of images [1]. This paper aim is to present an algorithm that uses stereo-vision images to extract objects. This section presents some of the methods of image segmentation. The proposed algorithm uses several methods which are presented in Sect. 2. Sections 3 and 4 describe the proposed method and obtained results. In the literature, there are many methods of image segmentation, which can be split into the following groups:

R. Wyrzykowski et al. (Eds.): PPAM 2013, Part I, LNCS 8384, pp. 433–443, 2014.
DOI: 10.1007/978-3-642-55224-3_41, © Springer-Verlag Berlin Heidelberg 2014

– Methods that operate on image areas - homogeneity of pixel neighborhoods,
– Methods based on thresholding (edge detection), consisting in determining the boundaries and contour of objects,
– Segmentation based on pattern matching to a specified object [18].

The first one consists in determining the coherent areas. i.e. direct (B-neighbors) and indirect (N-neighbors) neighbors of the selected point.

Fig. 1. Determining of neighborhood

Figure 1 illustrates the method of determining the pixels neighbourhood. For the selected pixel (x, y) in its direct neighborhood there are pixels with common sides of the pixel. In this case all pixels are marked in gray (0,2,4,6). As can be seen, B-neighborhood pixels are even numbers. In contrast, N-neighborhood are pixels having common corner with the selected pixels (1,3,5,7). There are two basic types of neighborhood: 4-neighborhood and 8-neighborhood.

Fig. 2. A: 4-neighborhood, B: 8-neighborhood [16]

Figure 2A,B illustrate the neighborhood. Two points p and q are neighbors, if p is included the 4-neighborhood of point $qN_4(q)$ and if q is included in 4-neighborhood of the point $pN_4(p)$. Similarly, in the case of 8-neighborhood. Coherence in the sense 4- or 8-neighborhood can refer as well to the contours and areas [11,16].

Another type of segmentation is based on merging areas. It divides the image into individual areas on the basis of thresholding techniques. The main advantage of this solution is the simplicity of implementation. There is however a problem with the selection of thresholds and it usually requires additional logic filtration to remove isolated pixels and operations related to the anti-aliasing or standardization inside them [12,20,21].

1.1   Selected Segmentation Methods

Use of stereo-vision methods in image segmentation has been already proposed by Katto et al. [15]. However, their method uses image fusion and a multi-camera system. Their algorithms require images from four cameras and they also used k-means clustering. In this case a certain number of groups is required. In our approach we use the mean shift algorithm which does not require setting k parameter [13,15]. Another interesting algorithm referring to this topic is the work of Toru in which segmentation uses split and merge algorithm [5,19]. However in the case of multiple objects, this method does not perform perfectly. They also did not show the results of tests for objects in a heterogeneous background.

2   Algorithms Used in the Proposed Approach

During the experiments, a number of algorithms and methods were used to achieve the results. This section describes the most important of them.

2.1   Stereo Vision and Disparity Map

An exceptionally interesting field of image processing is stereo vision and related algorithms which can be generally divided into global and local algorithms. The first group is characterized by a high cost of computing and is usually outside of scientific research. The second group of algorithm is less precise, but much faster and is used in many applications, for example to create 3D scenes. Local algorithms are also known as algorithms based on areas. They calculate the difference for each pixel based on the photometric properties of neighboring pixels. In stereo vision we could create a map of distances (disparity map). It is created on the basis of two video cameras positioned in one axis but offset relative to each other, as well as arrangement in a human eye. For most of the pixels in the left camera image there are equivalent pixels in the right camera image (with suitable offset). Differences of coordinates of the pixels is called disparity. It is inversely proportional to the distance from the camera. Disparity can be expressed by the formula $d = \frac{bf}{z}$ [3], where z is the distance between camera and the observed object, b is the distance between cameras, while f is the focal length of the camera. Figure 3 presents a diagram illustrating the above values [3].

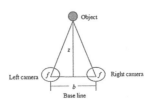

Fig. 3. The method of calculating the disparity map [8]

2.2 Mean Shift Clustering Algorithm

Mean shift is a clustering algorithm which does not require setting the number of output clusters or their shapes. The number of parameters of the algorithm is, in fact, limited to the radius [3]. The basic version of the algorithm was presented in the two-dimensional Euclidean space. The task of the algorithm is to compute the Euclidean distance between each point and all other points. Then the groups measures are calculated and assignment to the selected group. Mean shift determines the points in d-dimensional space as a probability density function, where the denser regions correspond to local maxima [2]. For each point in this space there is performed a procedure of gradient increase until coverage. Points assigned to one agent of group (stationary point) are considered to be a part of the group and form a single cluster (group) [9]. Given n points $x_i \in R^d$, multivariate kernel density function $K(x)$ is expressed using the following equation [4,9]

$$\widehat{f_k} = \frac{1}{nh^d} \sum_{i=1}^{n} \left(\frac{x - x_i}{h} \right), \tag{1}$$

where h is a radius of the kernel function. The kernel function is defined as follows [4,9] $K(x) = c_k k(\| x \|^2)$, where c_k is a normalization constant. If estimator density gradient is given it is possible to make the following calculations [6]

$$\nabla \widehat{f}(x) = \frac{2c_{k,d}}{nh^{d+2}} \underbrace{\left[\sum_{i=1}^{n} g \left(\| \frac{x - x_i}{h} \|^2 \right) \right]}_{term1} \underbrace{\left[\frac{\sum_{i=1}^{n} x_i g \left(\| \frac{x-x_i}{h} \|^2 \right)}{\sum_{i=1}^{n} g \left(\| \frac{x-x_i}{h} \|^2 \right)} - x \right]}_{term2}, \tag{2}$$

where $g(x) = -k'(x)$ is derivative of selected kernel of function. First term ($term1$) of formula 2 allows to define the density, instead second term ($term2$) is named as mean shift vector $m(x)$. Points in the direction of maximum gain and proportional to the density gradient can be determined at the point x obtained with the kernel function K.

2.3 Blob Extraction

Blob extraction is one of the basic methods of image processing. It allows to detect and extract a list of blobs (objects) in the image. Unfortunately, obtaining homogeneous objects from an image as a list of pixels is extremely complicated. Especially when we are dealing with a heterogeneous background. In other words, the objects containing multicolored background. There are many methods for extracting objects (blobs) from image [5]. In this paper we use methods implemented in the AForge.NET library. These algorithms are described by Andrew Kirillov [14]. There are four types of algorithms: Convex full, Left/Right Edges, Top/Bottom Edges, Quadrilateral. Figure 4 describes the blob detection methods. Figure 6A illustrates Quadrilateral method. As can be seen, round the edges of the objects are not detected correctly. Much better results are obtained by

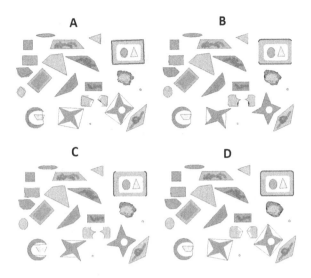

Fig. 4. Comparison of methods for blob detection used in the AForge.NET library [14]

the algorithm Top/Bottom Edges (Fig. 6C). Edges of objects are detected mostly correctly, with individual exceptions. The Left/Right Edges method behaves similarly (Fig. 6B). The last method has a problem with the detection of vertices inside figures, e.g. star type objects (Fig. 6D).

3 Proposed Method for Stereo Images Segmentation

The proposed segmentation algorithm can be divided into several stages. In this subsection we present them in a simplified block diagram.

We can distinguish seven stages of the algorithm. On the input of the algorithm we specify stereo images (left, right). Based on these images we create a disparity map as a grayscale image. The next step is to create a dictionary structure that stores the following items (key - value). The key is a pixel intensity, defined in grayscale while the value is a list of all pixels with an intensity equal

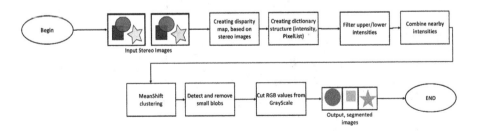

Fig. 5. Block diagram of the proposed algorithm

to the value of the key. So formed structure can be suitable for further image processing. The next step of the algorithm is to filter the upper or lower values of the dictionary (depending on the image type). This step allows to remove unnecessary pixels from the image. Objects colors closer to the observer (cameras), are shifted toward white, while distant objects have color shifted toward black, it is possible to cut out insignificant pixels. In obtained image we combine neighbor intensities. This step is based on combining the nearest pixels. This is possible due to the equal distance of objects in the image. Only the pixels of the same object can be combined. Another element of the proposed method is the clustering algorithm (mean shift). This element of the algorithm is extremely important because it significantly increase the division of the image into objects. At the output we obtain subjected pixels grouping, which allows to uniquely identify objects. The penultimate step is the detection and removal of unnecessary objects. As a result of the creation of the disparity map we obtain small groups of pixels that do not fit into any objects. They must be removed. The final step is to cut out objects from the input image and to impose them to those obtained in the previous steps of the algorithm. As a result of the proposed algorithm we obtained segmented images. Steps of proposed algorithmic solution:

1. Creating a disparity map based on the stereo-images.
2. Create of the dictionary structure (intensity, pixel count),
3. Filter intensities,
4. Combine neighboring intensities,
5. Run cluster algorithm on obtained intensities,
6. Find and remove insignificant (small) objects.
7. Extract grayscale objects from original image in the RGB scale.

Now we will analyse the proposed method in a more detailed way. For two input stereo-images it is possible to create a disparity map. This is a grayscale image, reflecting the position of objects on stereo images. In the proposed framerowrk, we use the differential mapping implementation written by E. Georgiou [7].

The next step is to create a dictionary structure described earlier. The next stage of the algorithm is to filter out unnecessary (insignificant) intensities. For

Fig. 6. Generation disparity map. The left (A) and middle (B) are stereo images. Right (C) is the disparity map

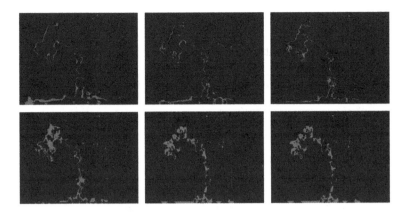

Fig. 7. Combining similar pixel clusters

this purpose we use high/low pass filter. The algorithm parameter g is a threshold value – intensities of value below g are filtered out. Then, pixels are combined by similar intensity. Figure 7 shows the result of this process.

This step is very important, because we need distinctly different data at the input of the mean shift algorithm (the next step). Clusters must have clear boundaries. The next step is extremely interesting as we use mean shift to cluster pixel intensities. After this process we obtain pixel clusters representing specific patterns of objects (blobs). The only parameter of this step is value h (see Sect. 2.2) and it has a significant impact on the quantity and size of the groups. The result of the algorithm is shown in Fig. 8A.

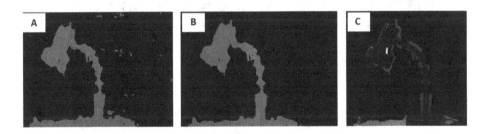

Fig. 8. The results of the mean shift algorithm

The structure of objects is clearly extracted. However, it can also be observed that there are pixels not associated with the object. In other words, sometimes blobs not connected with extracted object (e.g. lamp) are sometimes detected. In order to remove these groups of pixels, we use blob detection methods. Blobs with the area less then the specified value of the image are removed to extract only the object. This step was implemented with the support of the AForge.Net library (Computer Vision module). Figure 8B illustrates the effect of this part of the algorithm.

The final step of the algorithm is to extract objects in grayscale from original RGB image. In other words, RGB pixel values must be replaced by pixels in grayscale. This step is not particularly interesting, but it allows to obtain the RGB scale results. The result of the last stage of the proposed algorithm is shown in Fig. 8C.

4   Experimental Results

The proposed algorithm has been tested on a number of stereo images. Some of the simulations are presented in this section. The images are arranged in the following order: Fig. 9(A) and (B) - stereo images, Fig. 9(C) - disparity map, Fig. 9(D) - the segmented objects.

4.1   Segmentation of Multiple Objects

First image used in experiments contains many objects. They are in fairly close proximity relative to the cameras. Thus the algorithm has been tested for this type of images. Figure 9 illustrates important steps of the algorithm on the selected image.

As can be seen, the algorithm correctly segmented input image into objects. The resulting files containing images are presented in Fig. 9D.

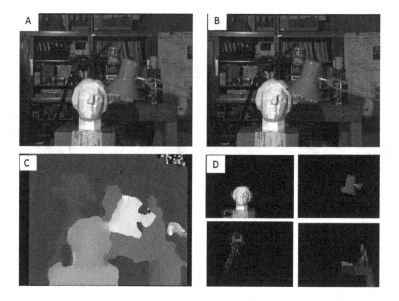

Fig. 9. The output images, objects: head, lamp, camera, table

4.2   Extraction of Distant Objects

During the simulations we have tested the case where the distance to the object is considerable. In this experiment we use different pair of images from that presented and analyzed in Sect. 3. Similarly, as in the previous case we presented algorithm steps in Fig. 10. The result of the experiment is also correct and consistent with the objectives of the proposed method. Since only distinctive object is a lamp, it has been extracted from the image. It can be considered, that the proposed method is valid for both types of objects, i.e. that are near and far away from the observer (cameras).

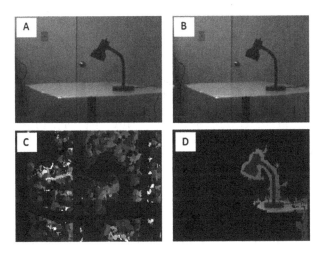

Fig. 10. The output image, object: distant lamp

5   Final Remarks

In this paper we proposed a novel method for object extraction from stereo images. The proposed method of stereo image segmentation shows high flexibility. The experiments demonstrated the presented method is suitable for both types of images, close and far away from the observer (cameras). The algorithm works well when there is a heterogeneous background of the analyzed image. During the simulations we have encountered the problem of determining the value of the parameter h (mean shift algorithm). Although the mean shift does not require determining the cluster count, it require the parameter h. However, it is possible to develop a method for estimation of this parameter. The algorithm allows to extract multiple objects from image, so it is possible to further use of the newly created objects. The next step in our research will be export of segmented images to the database and develop method for fast content-based image search. We will also study the mean shift algorithm working in a time-varying environment using techniques proposed in [10,17].

Acknowledgments. The project was funded by the National Center for Science under decision number DEC-2011/01/D/ST6/06957.

References

1. Chang, Y., Wang, Y., Chen, C., Ricanek, K.: Improved image-based automatic gender classification by feature selection. J. Artif. Intell. Soft. Comput. Res. **1**(3), 241–253 (2011)
2. Cheng, Y.: Mean shift, mode seeking, and clustering. IEEE Trans. Pattern Anal. Mach. Intell. **17**(8), 790–799 (1995)
3. Chowdhury, M.M.H., Bhuiyan, M.A.A.: A new approach for disparity map determination. Daffodil Int. Univ. J. Sci. Technol. **4**(1), 9–13 (2009)
4. Comanciu, D., Meer, P.: Mean shift analysis and applications, computer vision. In: The Proceedings of the 7th IEEE International Conference, pp. 1197–1203 (1999)
5. Damiand, G., Resch, P.: Split and merge algorithms defined on topological maps for 3D image segmentation. Graph. Models **65**(1–3), 149–167 (2003)
6. Derpanis, K.G.: Mean Shift Clustering. http://www.cse.yorku.ca/kosta/CompVis_Notes/mean_shift.pdf (2005)
7. Evangelos, G.: Stereo Correspondence Disparity Map with Emgu CV, http://mymobilerobots.com/myblog/ (2012)
8. Fukunaga, K., Hostetler, L.: The estimation of the gradient of a density function, with applications in pattern recognition. IEEE Trans. Inf. Theor. **21**(1), 32–40 (1975)
9. Georgescu, B., Shimshoni, I., Meer, P.: Mean shift based clustering in high dimensions: a texture classification example. In: Ninth IEEE International Conference on Computer Vision, vol. 1, pp. 456–463 (2003)
10. Greblicki, W., Rutkowska, D., Rutkowski, L.: An orthogonal series estimate of time-varying regression. Ann. Inst. Stat. Math. **35**(2), 215–228 (1983)
11. Haralick, R.H., Shapiro, L.G.: Image segmentation techniques. Comput. Vis. Graph. Image Process. **29**(1), 100–132 (1985)
12. Jiang, X., Bunke, H.: Edge detection in range images based on scan line approximation. Comput. Vis. Image Underst. **73**(2), 183–199 (1999)
13. Katto, J.; Ohta, M.: Novel algorithms for object extraction using multiple camera inputs. In: Proceedings of International Conference on Image Processing, pp. 863–866 (1996)
14. Kirillov, A.: Detecting some simple shapes in images. AForge.NET. http://www.aforgenet.com/articles/shape_checker/ (2010)
15. Marugame, A., Yamada, A., Ohta, M.: Focused object extraction with multiple cameras. IEEE Trans. Circ. Syst. Video technol. **10**(4), 530–540 (2000)
16. Nakib, A., Najman, L., Talbot, H., Siarry, P.: Application of graph partitioning to image segmentation. In: Bichot, C.-E., Siarry, P. (eds.) Graph Partitioning, pp. 251–274. ISTE Wiley, London (2011)
17. Rutkowski, L.: On Bayes risk consistent pattern recognition procedures in a quasi-stationary environment. IEEE Trans. Pattern Anal. Mach. Intell. **4**(1), 84–87 (1982)
18. Schreiber, J., Schubert, R., Kuhn, V.: Femur Detection in Radiographs Using Template-Based Registration, Bildverarbeitung fur die Medizin. Springer, Heidelberg (2006)

19. Tamaki, T., Yamamura, T., Ohnishi, N.: Image segmentation and object extraction based on geometric features of regions. In: Proceedings of SPIE - The International Society for Optical Engineering. vol. 3653, pp. 937–945 (1999)
20. Wani, M.A., Batchelor, B.G.: Edge-region-based segmentation of range images. IEEE Trans. Pattern Anal. Mach. Intell. **16**, 314–319 (1994)
21. Wu, Q., Yu, Y.: Two-level lmage segmentation based on region and edge integration. In: Sun C., Talbot H., Ourselin S., Adriaansen, T. (eds.) Proceedings of the VIIth Digital Image Computing: Techniques and Applications, pp. 957–966 (2003)

Minisymposium on GPU Computing

Evaluation of Autoparallelization Toolkits for Commodity GPUs

David Williams[1]([✉]), Valeriu Codreanu[1], Po Yang[2], Baoquan Liu[2], Feng Dong[2], Burhan Yasar[3], Babak Mahdian[4], Alessandro Chiarini[5], Xia Zhao[6], and Jos B.T.M. Roerdink[1]

[1] University of Groningen, Groningen, The Netherlands
d.p.williams@rug.nl
[2] University of Bedfordshire, Luton, UK
[3] RotaSoft Ltd, Ankara, Turkey
[4] ImageMetry, Prague, Czech Republic
[5] Super Computing Solutions, Bologna, Italy
[6] AnSmart, Wembley, UK

Abstract. In this paper we evaluate the performance of the OpenACC and Mint toolkits against C and CUDA implementations of the standard PolyBench test suite. Our analysis reveals that performance is similar in many cases, but that a certain set of code constructs impede the ability of Mint to generate optimal code. We then present some small improvements which we integrate into our own GPSME toolkit (which is derived from Mint) and show that our toolkit now out-performs OpenACC in the majority of tests.

Keywords: GPU computing · Autoparallelization · Evaluation

1 Introduction

The last ten years have seen the widespread adoption of parallel computing hardware in the form of Graphics Processing Units (GPUs). These GPUs have steadily increased in programmability and have found widespread application in a number of specialist fields [1], but typically require a developer to be experienced with OpenCL or CUDA and to have a strong understanding of the GPU's parallel architecture. This acts as a barrier to adoption in environments where such specialised knowledge is not readily available.

The development of automatic parallelization tools [2–5] has the potential to shift this balance and encourage more wide-spread adoption of GPU acceleration. The OpenMP standard [8] has already had a significant impact on the use of multiple cores in CPU applications, and we believe that such semi-automatic approaches are also the most promising approach to easily bringing code to the GPU. These tools typically work by augmenting the input C/C++ code with compiler directives which mark regions to be parallelized, and then automatically generating the required OpenCL/CUDA code for device initialization, memory

R. Wyrzykowski et al. (Eds.): PPAM 2013, Part I, LNCS 8384, pp. 447–457, 2014.
DOI: 10.1007/978-3-642-55224-3_42, © Springer-Verlag Berlin Heidelberg 2014

transfers and kernel implementation. Based on this we have adopted the open source Mint tool [6] as a foundation of our research efforts.

In this paper we provide an evaluation of Mint against OpenACC [7] (which is emerging as the standard for directive-based autoparallelization) with the particular aim of identifying and implementing improvements to Mint. We perform this evaluation against the PolyBench [11] test suite after adding OpenACC, Mint, and OpenMP directives. We then identify the areas in which Mint is not competitive, and in the second half of the paper we present some changes which we include in our enhanced version of Mint (known as the GPSME toolkit [9]).

2   Related Work

OpenACC is a relatively new technology, with the first version of the standard being finalised in 2011. Coupled with the lack of freely available and mature implementations, this means it has not been widely evaluated by the academic community. The small number of available evaluations have focused primarily on small test cases [10,18] though some application to real-world code has also been performed [19,20]. In all cases significant speedup was observed on sections of parallelizable code, and it should be noted that the OpenACC compilers are still undergoing rapid development due to the standard being so new.

To our knowledge, work evaluating Mint has been limited to that undertaken by its authors. The original Mint paper [6] claimed its performance was twice that of the PGI Accelerator model when using 3D heat simulation as a test case. The PGI Accelerator model was the predecessor to PGI's OpenACC compiler, so it is interesting to see how the performance has changed. A later paper presented an application of Mint to earthquake simulation and demonstrated an order of magnitude performance increase over the CPU reference implementation [13].

We have chosen to use an existing benchmark suite rather than to design our own, in order to minimise the bias which would be implicit in such a process. The PolyBench polyhedral benchmark suite [11] was designed to test the performance of a number of kernels from various application domains, and was recently extended with GPU implementations of most of the tests [12]. This provides an ideal basis for evaluating the performance of autoparallelization tools.

3   Methodology

The PolyBench test suite contains 15 test programs in the domains of convolution, linear algebra, data mining and stencil operations. The original implementations [11] were in C but GPU implementations in OpenCL and CUDA were added later [12] by Grauer-Gray *et al.* We do not consider these GPU implementations to be optimal as we were able to obtain equal or better results than most of them using our autoparallelization tools (see Sect. 4), but they still provide a useful reference point in the performance analysis. We consider only the CUDA implementation as there are known differences between OpenCL and CUDA performance [14,15], and our tools use CUDA as a backend anyway.

Table 1. Optimal thread block sizes were selected by testing a range of parameters and selecting the best via an auto-tuning process. The sizes are similar between OpenACC and GPSME because they are running on the same hardware. Thread blocks are two-dimensional in most cases, except for those ending with '×1' which are one-dimensional.

Benchmar	OpenACC	GPSME	Benchmark	OpenACC	GPSME
2DCONV	32×16	32×16	FDTD-2D	64×2	128×2
2MM	32×32	32×32	GEMM	32×32	32×32
3DCONV	8×128	8×128	GESUMMV	8×1	32×1
3MM	32×32	32×32	GRAMSCHM	16×16	16×16
ATAX	32×1	16×1	MVT	128×1	128×1
BICG	16×1	32×1	SYR2K	8×8	8×8
CORR	64×8	64×16	SYRK	8×128	8×128
COVAR	64×2	64×16			

Within this work we will sometimes need to refer to individual tests within PolyBench. We do this by using the name which PolyBench assigns to each test, written in upper case letters. For example, the 'ATAX' test uses matrix transpose and vector multiplications while the 'SYRK' test contains symmetric rank-k operations. A full list of test names and descriptions is provided with PolyBench [11].

We are primarily concerned with measuring the performance of Mint and OpenACC with respect to each other, but a comparison to CPU performance is also useful as a baseline. In the interests of fairness this CPU implementation should take advantage of all available cores and threading opportunities. We have therefore added OpenMP directives to each of the PolyBench tests.

The performance of CUDA programs (and, by extension, the output of our tools) is often highly dependent on the way in which the problem space is partitioned into thread blocks within the application. The optimal size for these thread blocks depends on a number of factors including the nature of the work done in the kernel, the need for synchronization between threads, the data access pattern, and the target hardware. Prior to performing the evaluation we wrote an automated system to compile each test in a large number of thread block configurations, and then chose to perform our tests using the configuration which showed the greatest performance (see Table 1). A similar benchmarking approach is described in [17].

3.1   Modifications to PolyBench

A few of the benchmarks required minor modifications in order for their computational pattern to be successfully captured by the Mint and OpenACC programming models. These modifications affected the ATAX, BICG, and GRAMSCHMIDT tests as they contained two or three-level nested loops with inter-loop dependencies. These were resolved by splitting the loop into two

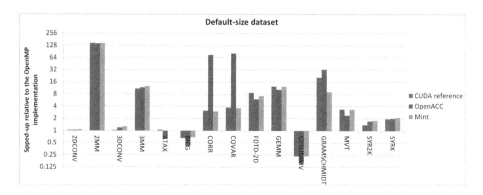

Fig. 1. Speed-up of GPU implementations compared to OpenMP on the default dataset. The values are in logarithmic scale.

successive nested loops. The modifications affected all the parallelization tools equally as each test is only implemented once but with multiple sets of pragmas applied.

Additionally, we adjusted the timing code for the CUDA manual implementations to be inclusive of the data transfer time as this was previously omitted. The time taken to transfer data to and from the GPU is often significant and can dominate algorithm runtime in some cases [16]. By accounting for this we help ensure a fair comparison.

3.2   Test Configuration

The test machine is comprised of a quad-core Intel Core i7-2600K 3.4 GHz CPU and an NVidia GTX680 GPU. All tests were performed under Ubuntu 12.04 LTS using GCC 4.6 as the C/OpenMP compiler, NVCC (from the CUDA 5.0 SDK) as the CUDA compiler, and PGI 13.1 as the OpenACC compiler. All tests were compiled on maximum optimization settings.

4   Initial Results

Initial performance measurements for the CUDA, Mint and OpenACC versions of each of the 15 tests are shown in Fig. 1. These measurements are all shown relative to the baseline set by the OpenMP implementation running on the CPU and using all cores. Note that the tests are performed using the default dataset size provided by the PolyBench suite.

It can be seen that for many tests the performance of the three implementations is similar. Among these cases there is usually a small amount of variation in the exact performance distribution. When the CUDA version is *faster* it can be attributed to missed optimization opportunities in the autogenerated code, and when the CUDA version is *slower* we found that autogenerated code was using pitched memory allocations to reduce the memory coalescing penalties. This

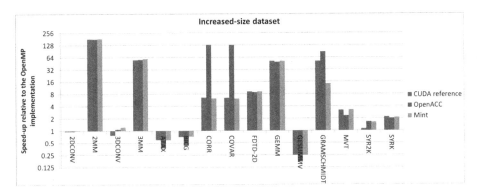

Fig. 2. Speed-up of GPU implementations compared to OpenMP on the enlarged dataset. The values are in logarithmic scale.

penalty is heavily incurred in the case of SYR2K, the manual version being more than 30 % slower than the automatic ones, with the kernel code being similar.

Furthermore we can observe that the performance of Mint and OpenACC is generally very similar, but that Mint has a slight edge in the majority of cases. An analysis of the output code has suggested that this is due to better register usage in the Mint code, as well as some redundant instructions in the OpenACC version (probably reflecting its more general-purpose nature).

More interestingly, there are a few tests which are notably different from the generalizations described above. Most striking is the large performance difference between OpenACC and Mint in the CORR, COVAR and GRAMSCHMIDT tests. This can be explained by the *triangular* nature of the nested loops in these tests, and is further discussed in Sect. 5.1 where we address this problem.

Three of the tests (ATAX, BICG, GESUMMV) actually showed reduced performance when running on the GPU. These tests made use of a one-dimensional thread block which did not contain enough work to benefit from offloading to the GPU, and the additional communication overhead caused an overall slowdown. However, it can be observed that Mint performed significantly better than OpenACC in these cases, and this is due to its support of the *tile* and *chunksize* parameters. OpenACC could not have the same degree of control with the *vector*, *worker* and *gang* parameters, and will include a tile parameter in OpenACC version 2.0.

There are other differences arising from the organization of the parallel loops. The 3DCONV benchmark is composed of a three-level nested for-loop. In the manual CUDA version the outer for-loop is iterated on the CPU, and only the inner two levels of the loop nest are offloaded as a 2D thread block. The automatic approaches use a 3D thread block, reducing thus the CPU−GPU communication. The same happens with the CORR, COVAR and GRAMSCHMIDT examples.

Before moving on to make improvements to Mint in the next section, we first ran the tests again with all the dataset sizes doubled in each dimension. All other parameters were left the same and the results can be seen in Fig. 2.

```
#pragma mint copy(data,toDevice, M, N)
#pragma mint copy(mean,toDevice, M)
#pragma mint copy(symmat,toDevice, M, N)
#pragma mint parallel
{
  ... //Some code omitted for brevity

  /* Calculate the m * m covariance matrix. */
  #pragma mint for nest(2) tile(16, 16)
  for (j1 = 0; j1 < M; j1++)
  {
    for (j2 = j1; j2 < M; j2++)
    {
      ... //Some code omitted for brevity
    }
  }
}
#pragma mint copy(symmat,fromDevice, M, N)
```

Algorithm 1. The covariance code from PolyBench with Mint directives added.

In these results for the enlarged dataset we can see that the speed difference between the GPU approaches and the CPU implementation is increasing when compared to the default dataset. This happens because most of the benchmarked problems have a complexity which is quadratic with respect to the dataset size and this increases the amount of work performed by each thread. Even at this dataset size, the ATAX, BICG and GESUMMV tests perform better on the CPU, but the Mint model continues to provide faster code than OpenACC.

5   Mint Enhancements

As part of our GPSME project [9] we have developed a number of extensions to Mint to create a new tool known as the 'GPSME toolkit'. We have added significant functionality (C++ and multi-file support, preliminary OpenCL output, etc.) but this is not used for the PolyBench tests and is not the focus of this paper. Instead, we wish to use the insight we have gained in Sect. 4 to improve the performance characteristics of our toolkit relative to OpenACC.

5.1   Supporting Triangular Loops

In Sect. 4 it was stated that three of the tests (CORR, COVAR and GRAM-SCHMIDT) suffered from poor performance in Mint due to the usage of *triangular loops*. A problematic section of code from the COVAR test is shown in Algorithm 1, complete with the Mint directives which were added.

The key issue is the dependency of the initial value of $j2$ in the inner loop on the current value of $j1$ in the outer loop. Although Mint does process this loop

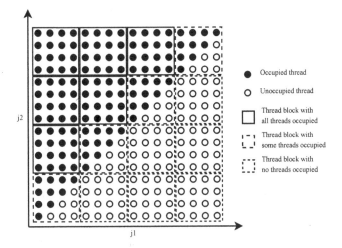

Fig. 3. Iteration space of the two-level covariance loop

it fails to understand the dependencies, and so generates non-compilable output code in which variables are initialised by other variables which have not yet been declared. Replacing 'nest(2)' with 'nest(1)' allowed parallelization of only the outermost loop to proceed as expected, but the performance was significantly less than that obtained from OpenACC (see Sect. 4).

An extension in our GPSME toolkit has allowed this situation to be handled naturally. A rectangular iteration space is defined by the full range of values which $j1$ and $j2$ can assume, and just over half of these points fall within the triangular region processed by the test (see Fig. 3). A grid of CUDA thread blocks is overlaid on the rectangular iteration, and the CUDA kernel contains a test to determine whether a given set of thread indices actually form part of the triangular region. With this in mind, a thread block can be categorized as being in one of three states with respect to the number of threads which need to execute:

- **Full:** All threads are part of the triangular iteration space and must be executed. No processing capability is wasted in this scenario.
- **Empty:** None of the threads are part of the triangular iteration space. All threads will fail the membership test implemented in the kernel and return immediately.
- **Half-full:** In this case the running time of the thread block is determined by the threads which do need to run. Threads which do not need to run must still wait upon those that do, and this represents some wasted processing capability.

With this new addition, the performance of the output code is greater than the one generated with OpenACC, and is more than 30 times faster than the one generated by the base Mint (full results presented later in Sect. 6).

Table 2. Timing improvement for the 2MM benchmark

	2MM-1D [s]	2MM-2D [s]	SYR2K-1D [s]	SYR2K-2D [s]
OpenACC	3.921	8.927	16.671	32.272
GPSME	3.814	2.812	17.01	12.08

The GPU's streaming architecture is not well-suited to processing conditional logic, but in our case only the half-full blocks exhibit the problematic divergent behavior. The proportion of blocks which are half-full decreases as the problem size grows and this results in a net gain overall.

The number of idle processing elements in the 'half-full' category is dependant upon the size of the thread block, and so a smaller thread block size results in better utilization. However, CUDA applications in general benefit from making thread blocks rather large (in the absence of synchronization concerns) and this outweighs the benefits of better utilization for the CORR, COVAR and GRAMSCHMIDT examples.

5.2 Single-Dimensional vs. Multi-dimensional Arrays

Another advantage to the GPSME model is that it finds more optimization opportunities when applied to code that uses multi-dimensional arrays. The optimizations are in terms of better register reuse, as well as better shared memory usage. We've tested this assumption on some of the tests. For the 2MM and SYR2K tests, a further 25 % performance increase is obtained when using two-dimensional addressing instead of the default flattened array addressing.

The changes from single-dimensional to multi-dimensional array accesses were done in a manual manner, as in Polybench all tests are written with flattened array accesses. However, with extra hints from the programmer the GPSME toolkit should be able to treat the single dimensional arrays as multi-dimensional ones.

An interesting observation is that when faced with the same two-dimensional arrays in the 2MM and SYR2K tests, the OpenACC compiler reports more than two times worse performance, as can be observed in Table 2. The reasons for this are not currently clear and will be the subject of some future investigation.

6   Final Results

The results obtained after implementing the proposed enhancements are presented in Fig. 4. The tests which benefited from our enhancements are shown in strong colors, with other tests faded out to indicated that they have not changed since the initial results. The improvements of GPSME over Mint is shown by the hatched bars. These examples rely on triangular loop support, and our improvement has enhanced their performance dramatically.

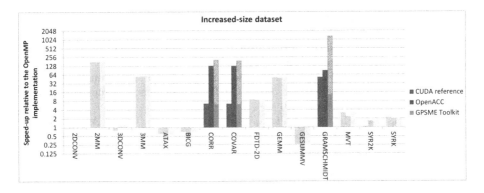

Fig. 4. Speed-up of GPU implementations compared to OpenMP on the enlarged dataset. The faded bars correspond to tests which have not changed since Fig. 2, and for the three tests which have changed the hatched bars show how GPSME has improved over Mint.

We note that the improvements observed from multi-dimensional addressing are not included in this table, as it is not implemented at this point as an automatic transformation.

7   Conclusions

Automatic parallelization through compiler directives is proving to be an effective method of maximising computing resources, and we expect that the coming years will see the approaches achieving the kind of widespread adoption that we currently see with OpenMP. We have shown that both OpenACC and also Mint/GPSME are capable of delivering code with a performance to meet or exceed that provided by the hand-written code supplied with PolyBench, and that the modifications presented in this paper have been enough to push it into the lead on the PolyBench tests.

Future work will revolve around automating some changes which were made manually for the purpose of this paper (such as the enhancements in Sect. 5.2), as well as identifying further opportunities for optimization. It would also be interesting to extend our tests to other paralellization toolkits [2–5] and manual implementations in order to perform a more comprehensive evaluation of the field.

References

1. Owens, J.D., Luekbe, D., Govindaraju, N., Harris, M., Krger, J., Lefohn, A.E., Purcell, T.J.: A survey of general-purpose computation on graphics hardware. Comput. Graph. Forum **26**(1), 80–113 (2007)

2. Amini, M., Creusillet, B., Even, S., Keryell, R., Goubier, O., Guelton, S., McMahon, J.O., Pasquier, F.X., Péan, G., Villalon, P.: Par4All: from convex array regions to heterogeneous computing. In: 2nd International Workshop on Polyhedral Compilation Techniques, Paris, France, Jan 2012

3. Lee, S., Eigenmann, R.: OpenMPC: extended openMP programming and tuning for GPUs. In: Proceedings of the 2010 ACM/IEEE Conference on Supercomputing, November 2010, pp. 1–11 (2010)

4. Meister, B., Vasilache, N., Wohlford, D., Baskaran, M.M., Leung, A., Lethin, R.: R-stream compiler. In: Padua, D. (ed.) Encyclopedia of Parallel Computing, pp. 1756–1765. Springer, Heidelberg (2011)

5. Verdoolaege, S., Juega, J.C., Cohen, A., Gómez, J.I., Tenllado, C., Catthoor, F.: Polyhedral parallel code generation for CUDA. ACM Trans. Archit. Code Optim. **9**(4), 54:1–54:23 (2013)

6. Unat, D., Cai, X., Baden, S.B.: Mint: realizing CUDA performance in 3D Stencil methods with Annotated C. In: Proceedings of the International Conference on Supercomputing, pp. 214–224 (2011)

7. The OpenACC Application Programming Interface, Version 1.0 (2011)

8. OpenMP Application Program Interface, Version 3.1 (2011)

9. Dong, F.: A General Toolkit for "GPUtilisation" in SME Applications. http://www.gp-sme.eu/ (2013). Accessed Oct 2013

10. Lee, S., Vetter, J.S.: Early evaluation of directive-based GPU programming models for productive exascale computing. In: Proceedings of the International Conference on High Performance Computing, Article 23 (2012)

11. Pouchet, L-N.: PolyBench: The Polyhedral Benchmark suite (2011), Version 3.2. http://www.cs.ucla.edu/~pouchet/software/polybench/ (2011)

12. Grauer-Gray, S., Xu, L., Searles, R., Ayalasomayajula, S., Cavazos, J.: Auto-tuning a high-level language targeted to GPU codes. In: Proceedings of Innovative Parallel Computing, pp. 1–10 (2012)

13. Zhou, J., Unat, D., Choi, D.J., Guest, C.C., Cui, Y.: Hands-on performance tuning of 3D finite difference earthquake simulation on GPU fermi chipset. Procedia Comput. Sci. **9**, 976–985 (2012)

14. Fang, J., Varbanescu, A.L., Sips, H.: A comprehensive performance comparison of CUDA and OpenCL. In: Proceedings of the Parallel Processing, pp. 216–225 (2011)

15. Komatsu, K., Sato, K., Arai, Y., Koyama, K., Takizawa, H., Kobayashi, H.: Evaluating performance and portability of OpenCL programs. In: Proceedings of the Automatic Performance Tuning (2010)

16. Che, S., Boyer, M., Meng, J., Tarjan, D., Sheaffer, J.W., Skadron, K.: A performance study of general-purpose applications on graphics processors using cuda. J. Parallel Distrib. Comput. **68**(10), 1370–1380 (2008)

17. Magni, A., Grewe, D., Johnson, N.: Input-aware auto-tuning for directive-based GPU programming. In: Proceedings of the 6th Workshop on General Purpose Processor Using Graphic Processing Units, pp. 66–75 (2013)

18. Reyes, R.N., Lopez, I., Fumero, J.J., de Sande, F.: Directive-based programming for GPUs: a comparative study. In: IEEE 9th International Conference on Embedded Software and Systems (HPCC-ICESS) (2012)

19. Wienke, S., Springer, P., Terboven, C., an Mey, D.: OpenACC — First experiences with real-world applications. In: Kaklamanis, C., Papatheodorou, T., Spirakis, P.G. (eds.) Euro-Par 2012. LNCS, vol. 7484, pp. 859–870. Springer, Heidelberg (2012)
20. Herdman, J.A., Gaudin, W.P., McIntosh-Smith, S., Boulton, M., Beckingsale, D.A., Mallinson, A.C., Jarvis, S.A.: Accelerating hydrocodes with OpenACC, OpeCL and CUDA. In: Proceedings of the High Performance Computing, Networking, Storage and Analysis (SCC), pp. 465–471 (2012)

Real-Time Multiview Human Body Tracking
Using GPU-Accelerated PSO

Boguslaw Rymut[2] and Bogdan Kwolek[1] [✉]

[1] AGH University of Science and Technology,
30 Mickiewicza Av., 30-059 Krakow, Poland
bkw@agh.edu.pl
[2] Rzeszów University of Technology, W. Pola 2, 35-959 Rzeszów, Poland
brymut@prz.edu.pl

Abstract. This paper presents our approach to 3D model-based human motion tracking using a GPU-accelerated particle swarm optimization. The tracking involves configuring the 3D human model in the pose described by each particle and then rasterizing it in each particle's 2D plane. In our implementation, we launch one independent thread for each column of each 2D plane. Such a parallel algorithm exhibits the level of parallelism that allows us to effectively utilize the GPU resources. Owing to such task decomposition the tracking of the full human body can be performed at rates of 15 frames per second. The GPU achieves an average speedup of 7.5 over the CPU. The speedup that achieves the GPU over CPU grows with the number of the particles. For marker-less motion capture system consisting of four calibrated and synchronized cameras, the efficiency comparisons were conducted on four CPU cores and four GTX GPUs on two cards.

Keywords: GPGPU · Real-time computer vision · Human motion capture

1 Introduction

In the early years of computer graphics, the GPU could only be programmed through a graphics rendering interface. Over the years, the GPU has evolved from a highly specialized graphics processor to a versatile and highly programmable architecture that can perform a wide range of data-parallel operations. The GPU architectures benefit from massive fine-grained parallelization, as they are able to execute as many as thousands of threads concurrently. Recently, many research papers reported that general purpose GPUs (GPGPUs) are capable to obtain significant speedups compared to current homogeneous multicore systems in the same price range. These scientific reports initiated a passionate debate on the limits of GPU-supported acceleration for various classes of applications [11]. A comparison of 14 various implementations showed speedups from 0.5× to 15× (GPU over CPU). The experiment was made with Intel Core i7 and NVidia

R. Wyrzykowski et al. (Eds.): PPAM 2013, Part I, LNCS 8384, pp. 458–468, 2014.
DOI: 10.1007/978-3-642-55224-3_43, © Springer-Verlag Berlin Heidelberg 2014

GTX 280. There is a common agreement that in order to achieve satisfactory performance the algorithms to be executed on GPU should be carefully designed.

CPUs are still the most frequently used hardware for image processing, given their versatility and tremendous speed. In general, image processing algorithms are good candidates for GPU implementation, since the parallelization is naturally provided by per-pixel operations. Many research studies confirmed this by showing GPU acceleration of many image processing algorithms [3]. A recent study [13] reports a speedup of 30 times for low-level algorithms and up to 10 times for high-level functions, which contain more overhead and many steps that are not easy to parallelize.

Non intrusive human body tracking is a key issue in user-friendly human-computer communication. This is one of the most challenging problems in computer vision being at the same time one of the most computationally demanding tasks. Particle filters are typically employed to achieve articulated motion tracking. Several improvements of ordinary particle filter were done to achieve fast and reliable articulated motion tracking [4] as well as to obtain the initialization of the tracking [15]. 3D motion tracking can be perceived as dynamic optimization problem. Recently, particle swarm optimization (PSO) [6] has been successfully applied to achieve human motion tracking [7,12]. The motion tracking is achieved by a sequence of static PSO-based optimizations, followed by re-diversification of the particles to cover the possible poses in the next time step.

There are only a few publications that discuss the implementation details of the PSO on GPU. In [16], an approach that restricts the communication of a particle to its two closest neighbors and thus limits the communication between threads was proposed. The authors of [10] compared three different variants of the PSO on GPU, but only parallelized the cost function. In [14], a multi-swarm PSO algorithm was used to achieve a high degree of parallelism. In [7] an approach to PSO-based full body human motion tracking on GPU and using single camera has been proposed. The 3D model with 26 DOF was constructed using cuboids, which were projected into 2D plane and then rendered in parallel. A single thread was responsible for comparing images containing the projected model and the extracted person. The tracking of the full human body was performed with 5 frames per second, whereas the speedup of GTX280 over a CPU was about 15. A common approach to parallelize the PSO consists in executing a local swarm on every processor while optimizing the communication between the swarms. Mussi et al. [12] proposed an approach to articulated human body tracking from multi-view video using PSO running on GPU. Their implementation is far from real-time and roughly requires 7 s per frame. Recently, in [2] a framework for 3D model-based visual tracking using a GPU-accelerated particle filter has been presented. A hand was tracked using both synthetic and real videos. The authors reported a speedup of 9.5 and 14.1 against a CPU for image resolution of 96×72 and 128×96 using 900 and 1296 particles, respectively.

In this work we present an approach that effectively utilizes the advantages of modern graphics card hardware to achieve real-time full body tracking using a 3D human model. The motion tracking was accomplished by a PSO algorithm running on a GPU. The presented approach to 3D articulated human tracking follows the Black Box Optimization paradigm [5], according to which the search processes/particles investigate the hypothesis space of a model state in order to identify the hypothesis that optimally fit a set of observations.

2   GPU Computing

CUDA is a parallel computing platform and programming model invented by NVIDIA. Each function that is executed on the device is called a kernel. A CUDA kernel is executed by an array of threads. Blocks of threads are organized into one, two or three dimensional grid of thread blocks. Blocks are mapped to multiprocessors and each thread is mapped to a single core. A warp is a group of threads within a block that are launched together and usually execute together. When a warp is selected for execution, all active threads execute the same instruction but operate on different data. A unique set of indices is assigned to each thread to determine to which block it belongs and its location inside it.

GPUs offer best performance gains when all processing cores are utilized and memory latency is hidden. In order to achieve this aim, it is common to launch a CUDA kernel with hundreds or thousands of threads to keep the GPU busy. The benefit of having multiple blocks per multiprocessor is that the scheduling hardware is capable to swap out a block that is waiting on a high-latency instruction and replace it with a block that has threads ready to execute. The context switch is very fast because the GPU does not have to store the state, as the CPU does when switching threads between being active and inactive. Thus, it is advantageous to have both high density of arithmetic instructions per memory access as well many more resident threads than GPU cores so that memory latency can be hidden. This permits the GPU to execute arithmetic instructions while certain threads are waiting for access to the global memory.

Memory latency can be hidden by careful design of control flow as well as adequate design of kernels. The kernels can employ not only the global memory that resides off chip, but also they can use shared memory that resides on chip. This memory is shared between all the cores of stream multiprocessor. Its latency is several times shorter than the latency of the global memory. Threads that are executing within the same block can cooperate using it, but threads from different block cannot cooperate via shared memory.

3   Parallel PSO for Object Tracking

Particle Swarm Optimization (PSO) [6] is a bio-inspired meta-heuristic for solving complex optimization problems. The PSO is initialized with a group of random particles (hypothetical solutions) and then searches for optima by updating all particles locations. The particles move through the solution space and undergo evaluation according to some fitness function. Each particle iteratively evaluates the candidate solutions and remembers the personal best location with the best objective value found so far, making this information available to its neighbors.

Particles communicate good positions to each other and adjust their own velocities and positions taking into account such good locations. Additionally each particle utilizes a best value, which can be:

- a global best that is immediately updated when a new best position is found by any particle in the swarm
- neighborhood best where only a specific number of particles is affected if a new best position is found by any particle in the sub-population

Typically, a swarm topology with the global best converges faster since all particles are attracted simultaneously to the best part of the search space. Neighborhood best permits parallel exploration of the search space and decreases the susceptibility of falling into local minima. However, such a topology slows down the convergence speed. Taking into account the faster convergence the topology with the global best has been selected for parallel implementation.

In the ordinary PSO algorithm the update of particle's velocity and position can be expressed by the following equations:

$$v_j^{(i)} \leftarrow w v_j^{(i)} + c_1 r_{1,j}^{(i)}(p_j^{(i)} - x_j^{(i)}) + c_2 r_{2,j}^{(i)}(p_{g,j} - x_j^{(i)}) \tag{1}$$

$$x_j^{(i)} \leftarrow x_j^{(i)} + v_j^{(i)} \tag{2}$$

where w is the positive inertia weight, $v_j^{(i)}$ is the velocity of particle i in dimension j, $r_{1,j}^{(i)}$ and $r_{2,j}^{(i)}$ are uniquely generated random numbers with the uniform distribution in the interval $[0.0, 1.0]$, c_1, c_2 are positive constants, $p^{(i)}$ is the best position that the particle i has found, p_g denotes best position that is found by any particle in the swarm.

The velocity update equation (1) has three main components. The first component, which is often referred to as inertia models the particle's tendency to continue the moving in the same direction. In effect it controls the exploration of the search space. The second component, called cognitive, attracts towards the best position $p^{(i)}$ previously found by the particle. The last component is referred to as social and attracts towards the best position p_g found by any particle. The fitness value that corresponds $p^{(i)}$ is called local best $p_{\text{best}}^{(i)}$, whereas the fitness value corresponding to p_g is referred to as g_{best}. The ordinary PSO algorithm can be expressed by the following pseudo-code:

1. Assign each particle a random position in the problem hyperspace.
2. Evaluate the fitness function for each particle.
3. For each particle i compare the particle's fitness value with its $p_{\text{best}}^{(i)}$.
   If the current value is better than the value $p_{\text{best}}^{(i)}$, then set this value as the $p_{\text{best}}^{(i)}$ and the current particle's position $x^{(i)}$ as $p^{(i)}$.
4. Find the particle that has the best fitness value g_{best}.
5. Update the velocities and positions of all particles according to (1) and (2).
6. Repeat steps 2–5 until a stopping criterion is not satisfied (e.g. maximum number of iterations or a sufficiently good fitness value is not attained).

Our parallel PSO algorithm for object tracking consists of five main phases, namely initialization, evaluation, p_best, g_best, update and motion. At the beginning of each frame, in the initialization stage an initial position $x^{(i)} \leftarrow \mathcal{N}(p_\text{g}, \Sigma)$ is assigned to each particle, given the location p_g that has been estimated in the previous frame. In the evaluation phase the fitness value of each particle is calculated using a cost function. The calculation of the matching score is the most time consuming operation of the tracking algorithm. The calculation of the matching score is discussed in Sect. 4.2, whereas the decomposition of this task into kernels is presented in Sect. 4.3. In the p_best stage the determining of $p_\text{best}^{(i)}$ as well as $p^{(i)}$ takes place. This stage corresponds to operations from the point 3. of the presented above pseudo-code. The operations mentioned above are computed in parallel using available GPU resources, see Fig. 1. Afterwards, the g_best and its corresponding p_g are calculated in a sequential task. Finally, the update stage that corresponds to point 5. in the pseudo-code is done in parallel. That means that in our implementation we employ the parallel synchronous particle swarm optimization. The synchronous PSO algorithm updates all particle velocities and positions at the end of each optimization iteration. In contrast to synchronous PSO the asynchronous algorithm updates particle positions and velocities continuously using currently accessible information.

In order to decompose an algorithm into GPU we should identify data-parallel portions of the program and isolate them as CUDA kernels. In the initialization kernel we generate pseudo-random numbers using the curand library provided by the CUDA$^\text{TM}$ SDK. On the basis of the uniform random numbers we generate normally distributed pseudorandom numbers using Box Mueller transform based on trigonometric functions [1]. The normally distributed random numbers are

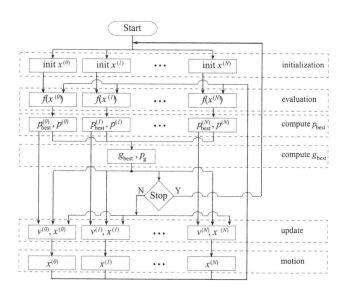

Fig. 1. Decomposition of synchronous particle swarm optimization algorithm on GPU.

generated at the beginning of each frame to re-distribute the particles around the pose in time $t - 1$ and to calculate their velocities. Then the uniform random numbers r_1, r_2 for the optimal pose seeking are generated. This means that for every particle we generate $2 \times D \times K$ uniformly distributed random numbers, where D is dimension and K denotes the maximum number of iterations. They are stored in the memory and then used in the update kernel, see Fig. 1. At this stage the computations are done in $\lceil N/(2 \times W) \rceil$ blocks and W threads on each of them, where W denotes the number of cores per multiprocessor. In the compute p_{best} kernel and the update kernel the number of blocks is equal to $\lceil N/W \rceil$, whereas the number of threads in each block is equal to W. In the update kernel we constrain the velocities of the particles to the assumed maximal velocity values. In the motion stage the model's bone hierarchy is recursively traversed and the internal transformation matrices are updated according to the state vector of the particle.

4   Implementation of Articulated Body Tracking on GPU

At the beginning of this section we detail our approach to 3D model based visual tracking of human motion. Afterwards, we present the cost function. Finally, we discuss the parallelization of the calculations of the cost function.

4.1   3D Model-Based Visual Tracking

The articulated model of the human body has a form of kinematic chain consisting of 11 segments. The 3D model is constructed using truncated cones (frustums) that model the pelvis, torso, head, upper and lower arm and legs.

The model has 26 DOF and its configuration is determined by position and orientation of the pelvis in the global coordinate system and the relative angles between the limbs. Each truncated cone is parameterized by the center of base circle A, center of top circle B, bottom radius $r1$, and top radius $r2$. Given the 3D camera location C and 3D coordinates A and B, the plane passing through the points A, B, C is determined. Since the vectors AB and AC lie in the plane, their cross product, which is perpendicular to the plane of AB and AC, is the normal. The normal is used to determine the angular orientation of the trapezoid to be projected into 2D plane. Each trapezoid of the model is projected into 2D image of each camera via modified Tsai's camera model. The projected image of the trapezoid is obtained by projecting the corners and then a rasterization of the triangles composing the trapezoid. Though projecting all truncated cones we obtain the image representing the 3D model in a given configuration.

In each frame the 3D human pose is reconstructed through matching the projection of the human body model with the current image observations. In most of the approaches to articulated object tracking a background subtraction algorithms are employed to extract the subject undergoing tracking. Additionally, image cues such as edges, ridges and color are often employed to improve the extraction of the person. In the presented approach the human silhouette is extracted via background subtraction. Afterwards, the edges are located within the extracted silhouette. Finally, the edge distance map is extracted [9]. The matching score reflects (i) matching ratio between the extracted silhouette and the projected 3D model and (ii) the normalized distance between the model's projected edges and the closest edges in the image. The objective function of all cameras is the sum of such matching scores. Sample images from the utilized test sequences as well as details of camera setup can be found in [9].

The motion tracking can by attained by dynamic optimization and incorporating the temporal continuity information into the ordinary PSO. Consequently, it can be achieved by a sequence of static PSO-based optimizations, followed by re-diversification of the particles to cover the potential poses that can arise in the next time step. The re-diversification of the particle i can be obtained on the basis of normal distribution concentrated around the best particle location p_g in time $t - 1$, which can be expressed as: $x^{(i)} \leftarrow \mathcal{N}(p_g, \Sigma)$, where $x^{(i)}$ stands for particle's location in time t, Σ denotes the covariance matrix of the Gaussian distribution, whose diagonal elements are proportional to the expected velocity.

4.2   Cost Function

The most computationally demanding operation in 3D model based human motion tracking is calculation of the objective function. In PSO-based approach each particle represents a hypothesis about possible person pose. In the evaluation of the particle's fitness score the projected model is matched with the current image observation. The fitness score depends on the amount of overlapping between the extracted silhouette in the current image and the projected and rasterized 3D model in the hypothesized pose. The amount of overlapping is calculated through checking the overlap degree from the silhouette to the rasterized model as well as from the rasterized model to the silhouette. The larger the overlap is, the larger is the fitness value. The objective function reflects also the normalized distance between the model's projected edges and the closest edges in the image. It is calculated on the basis of the edge distance map [9].

The fitness score for i-th camera's view is calculated on the basis of following expression: $f^{(i)}(x) = 1 - ((f_1^{(i)}(x))^{w_1} \cdot (f_2^{(i)}(x))^{w_2})$, where w denotes weighting coefficients that were determined experimentally. The function $f_1^{(i)}(x)$ reflects the degree of overlap between the extracted body and the projected 3D model into 2D image corresponding to camera i. The function $f_2^{(i)}(x)$ reflects the edge distance map-based fitness in the image from the camera i. The objective function for all cameras is determined according to the following expression: $f(x) = \frac{1}{4} \sum_{i=1}^{4} f^{(i)}(x)$. Since we use synchronous PSO the fitness values are transmitted once in every iteration. The images acquired from the cameras are processed on CPU and then transferred onto the device. They are then utilized in the PSO running on the GPU.

4.3   Parallelization of the Cost Function

In the evaluation phase, see Fig. 1 we employ two kernels. In the first one the 3D models are projected into 2D image of each camera. In the second one we rasterize the models and evaluate the objective functions. In our approach, in every block we rasterize the model in the pose represented by a single particle as well as we calculate its fitness score. Thus, the number of blocks is equal to the number of the particles, see Fig. 2. Each thread is responsible for rasterizing the model in single column and summing the fitness values of the pixels in that column. The number of threads in each block is equal to the image width, whereas the number of running threads in each block is equal to the number of cores per multiprocessor, see Fig. 2.

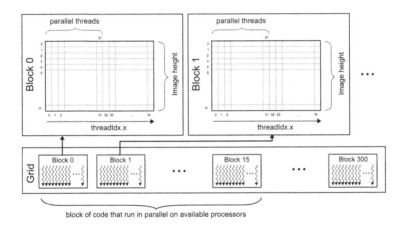

Fig. 2. Parallelization of the cost function.

The cost values of the objective function are summed using parallel reduction. The results from each column of the threaded block are stored in the shared memory. In the next stage, $W/2$ consecutive threads determine the sums of the two adjacent memory cells of the shared memory and then store the results in the shared memory. The next iteration employs $W/4$ threads to add the results of the previous iteration, and so on.

5   Experiments

The experiments were conducted on a PC computer equipped with Intel Xeon X5690 3.46 GHz CPU (6 cores), with 8 GB RAM, and two NVidia GTX 590 graphics cards, each with 16 multiprocessors and 32 cores per multiprocessor. Each card has two GTX GPUs, each equipped with 1536 MB RAM and 48 KB shared memory per multiprocessor.

Table 1. Computation time [ms] for single frame of size 480×270

	# part. (10 it.)	CPU [ms]	GPU [ms]	Speedup
1 camera	100	131.1	24.6	5.3
	300	352.7	44.9	7.9
	1000	1134.7	106.4	10.7
2 cameras	100	132.8	26.8	5.0
	300	352.4	47.4	7.5
	1000	1117.4	113.5	9.9
4 cameras	100	170.3	37.3	4.6
	300	442.3	62.8	7.1
	1000	1391.9	144.7	9.6

Table 1 shows computation time that has been obtained on CPU and GPU for 1, 2, and 4 cameras and PSO executing 10 iterations. For two cameras the computations were conducted on two CPU cores and two GPUs on single card, whereas for four cameras we employed 4 CPU cores and four GPUs. The images acquired from calibrated and synchronized cameras were preprocessed off-line and transferred frame by frame to the GPU. As we can observe, for a system consisting of 2 cameras the speedup that achieves the GPU over the CPU is between 5.0 and 9.9. For 4 cameras the speed up is slightly smaller due to additional transmission overhead between two cards. For MoCap system consisting of 4 cameras and using the PSO algorithm with 300 particles and 10 iterations we can process 16 frames per second. In [9] we demonstrated that for such a PSO configuration the average error on images of size 960×540 is below 75 mm. In this work we employed the images scaled to 480×270 resolution and the average error was about 5 mm larger. Another reason for a slightly larger error is the use of synchronous PSO that achieves worse tracking accuracy in comparison to asynchronous PSO.

The processing times on the CPU were obtained using an implementation presented in [9]. In [8] we showed that a modified PSO algorithm, i.e. annealed particle swarm optimization (APSO) [9], with 300 particles and executing 10 iterations, can be successfully used in 3D gait-based person identification.

6   Conclusions

In this paper we presented an algorithm for articulated human motion tracking on GPU. The tracking has been achieved in real-time using a parallel PSO algorithm. The tracking of full human body can be performed at frame-rates of 16 frames per second using a two high-end graphics cards and images acquired by four cameras. The speedup of the algorithm running on GPU over CPU grows with the number of evaluations of the cost function, i.e. with number of the particles or with the number of iterations. In consequence, on the GPU we can obtain more precise tracking.

Acknowledgment. This work has been supported by the National Science Center (NCN) within the research project N N516 483240.

References

1. Box, G.E.P., Muller, M.E.: A note on the generation of random normal deviates. Ann. Math. Stat. **29**(2), 610–611 (1958)
2. Brown, J., Capson, D.: Framework for 3d model-based visual tracking using a GPU-accelerated particle filter. IEEE Trans. Vis. Comput. Graph. **18**(1), 68–80 (2012)
3. Castano-Diez, D., Moser, D., Schoenegger, A., Pruggnaller, S., Frangakis, A.S.: Performance evaluation of image processing algorithms on the GPU. J. Struct. Biol. **164**(1), 153–160 (2008)
4. Deutscher, J., Blake, A., Reid, I.: Articulated body motion capture by annealed particle filtering. In: IEEE International Conference on Pattern Recognition, pp. 126–133 (2000)
5. Hansen, N., Auger, A., Ros, R., Finck, S., Pošík, P.: Comparing results of 31 algorithms from the black-box optimization benchmarking BBOB-2009. In: Genetic and Evolutionary Computation Conference. GECCO'10, pp. 1689–1696. ACM (2010)
6. Kennedy, J., Eberhart, R.: Particle swarm optimization. In: Proceedings of IEEE International Conference on Neural Networks, pp. 1942–1948. IEEE Press, Piscataway, NJ (1995)
7. Krzeszowski, T., Kwolek, B., Wojciechowski, K.: GPU-accelerated tracking of the motion of 3d articulated figure. In: Bolc, L., Tadeusiewicz, R., Chmielewski, L.J., Wojciechowski, K. (eds.) ICCVG 2010, Part I. LNCS, vol. 6374, pp. 155–162. Springer, Heidelberg (2010)
8. Krzeszowski, T., Michalczuk, A., Kwolek, B., Switonski, A., Josinski, H.: Gait recognition based on marker-less 3D motion capture. In: 10th IEEE International Conference on Advanced Video and Signal Based Surveillance (AVSS), pp. 232–237 (2013)
9. Kwolek, B., Krzeszowski, T., Wojciechowski, K.: Swarm intelligence based searching schemes for articulated 3d body motion tracking. In: Blanc-Talon, J., Kleihorst, R., Philips, W., Popescu, D., Scheunders, P. (eds.) ACIVS 2011. LNCS, vol. 6915, pp. 115–126. Springer, Heidelberg (2011)
10. Laguna-Sanchez, G.A., Olguin-Carbajal, M., Cruz-Cortes, N., Barron-Fernandez, R., Alvarez-Cedillo, J.A.: Comparative study of parallel variants for a particle swarm optimization. J. Appl. Res. Technol. **7**(3), 292–309 (2009)
11. Lee, V.W., Kim, C., Chhugani, J., Deisher, M., Kim, D., Nguyen, A.D., Satish, N., Smelyanskiy, M., Chennupaty, S., Hammarlund, P., Singhal, R., Dubey, P.: Debunking the 100x GPU vs. CPU myth: an evaluation of throughput computing on CPU and GPU. In: Proceedings of the 37th Annual International Symposium on Computer Architecture. ISCA'10, pp. 451–460. ACM, New York, NY, USA (2010)
12. Mussi, L., Ivekovic, S., Cagnoni, S.: Markerless articulated human body tracking from multi-view video with GPU-PSO. In: Tempesti, G., Tyrrell, A.M., Miller, J.F. (eds.) ICES 2010. LNCS, vol. 6274, pp. 97–108. Springer, Heidelberg (2010)
13. Pulli, K., Baksheev, A., Kornyakov, K., Eruhimov, V.: Real-time computer vision with OpenCV. Commun. ACM **55**(6), 61–69 (2012)

14. Solomon, S., Thulasiraman, P., Thulasiram, R.: Collaborative multi-swarm PSO for task matching using graphics processing units. In: Proceedings of the 13th Annual Conference on Genetic and Evolutionary Computation, pp. 1563–1570 (2011)
15. Wu, C., Aghajan, H.: Human pose estimation in vision networks via distributed local processing and nonparametric belief propagation. In: Blanc-Talon, J., Bourennane, S., Philips, W., Popescu, D., Scheunders, P. (eds.) ACIVS 2008. LNCS, vol. 5259, pp. 1006–1017. Springer, Heidelberg (2008)
16. Zhou, Y., Tan, Y.: GPU-based parallel particle swarm optimization. In: IEEE Congress on Evolutionary Computation. CEC'09, pp. 1493–1500 (2009)

Implementation of a Heterogeneous Image Reconstruction System for Clinical Magnetic Resonance

Grzegorz Tomasz Kowalik[1][✉], Jennifer Anne Steeden[1], David Atkinson[2],
Andrew Taylor[1,3], and Vivek Muthurangu[1,3]

[1] Centre for Cardiovascular Imaging, UCL Institute of Cardiovascular Science,
London, UK
kowgrzegorz@gmail.com
[2] Division of Medicine, University College London, Royal Free Campus,
Rowland Hill Street, London, UK
[3] Cardiorespiratory Unit, Great Ormond Street Hospital for Children, London, UK

Abstract. This paper describes development of a novel online, heterogeneous image reconstruction system for Magnetic Resonance data. The system integrates an external computer equipped with a Graphic Processing Unit card into the Magnetic Resonance scanner's image reconstruction pipeline. The system promotes fast online reconstruction for computationally intensive algorithms making them feasible in a busy clinical service. Analysis and improvement of execution time of the complex, iterative reconstruction algorithm as well as networking framework are presented.

The imaging algorithm was broken down into distinctive steps for execution time profiling. Also, steps to achieve overlapping of execution and transmission are described.

The system was successfully used in research and clinical studies requiring high data throughput.

Keywords: GPGPU · CORBA · Distributed system

1 Introduction

Magnetic Resonance Imaging (MRI) is increasingly used in the medical field to make diagnoses and aid patient management. Conventionally, MRI data is acquired in a spatial frequency domain called k-space in a rectilinear manner allowing conversion of k-space data into images with a simple Fast Fourier Transform (FFT). However, the rectilinear filling of k-space is not temporally efficient and can result in temporal blurring when trying to perform *fast MRI* (temporal resolution - TR: < 50 ms). MRI can be sped by using more time efficient k-space filling strategies, called trajectories (e.g. a spiral trajectory). However, reconstruction of data acquired with non-Cartesian trajectories requires additional processing steps, including gridding onto a rectilinear grid prior to FFT, making

R. Wyrzykowski et al. (Eds.): PPAM 2013, Part I, LNCS 8384, pp. 469–479, 2014.
DOI: 10.1007/978-3-642-55224-3_44, © Springer-Verlag Berlin Heidelberg 2014

it more computationally intensive. In addition, the use of time efficient trajectories does not alone provide the necessary speed-up to perform high temporal resolution imaging. Thus, such acquisitions are usually further sped-up using a data reduction technique, called undersampling. Undersampling produces artefacts in the reconstructed images if no additional reconstruction steps are undertaken. However, it is possible to remove these artefacts using information acquired from multiple receiver coils, each of which experiences a different spatial sensitivity. Several reconstruction methods exist for undersampled data, however this work focuses on the use of the SENSE (sensitivity encoding) algorithm [1]. When combined with undersampled arbitrary trajectories, the SENSE algorithm adopts an iterative reconstruction process to produce artefact free images [2]. Consequently, this shifts the bottleneck of MR scanning from data acquisition to reconstruction. As the SENSE reconstruction is highly parallelizable, the use of graphical processing units (GPU) as coprocessors may offer a solution. GPU implementations do exist [3–5] and have been shown to significantly speed-up the reconstruction. However, to be truly effective, such developments must be incorporated into the online scanner reconstruction pipeline. This is vital as it improves the overall efficiency of clinical workflow and enables rapid viewing of the images to check for data integrity prior to finishing the MR exam.

Recently, we have developed a novel online, heterogeneous image reconstructor [6,7] in the form of a distributed system based on client-server architecture, which was designed to address following challenges: (i) use of advanced MR sequences is limited by their reconstruction time and (ii) GPU implementations exist but run in off-line mode. Our motivation was not only to provide fast image reconstruction for computationally intensive reconstructions, but most of all to make them feasible within a busy clinical service. This was achieved by integrating an external computer equipped with a GPU card into a scanner's native image reconstruction system for seamless reconstruction process from a clinician's point of view.

In this article we present a description of development and implementation process of the mentioned heterogeneous MRI reconstruction system.

2   Methods

2.1   Distributed Image Reconstructor

The distributed image reconstruction system, required development of two major components: (i) implementation of networking framework with remote execution based on client-server architecture and (ii) implementation of GPU based MRI reconstruction algorithm for undersampled arbitrary trajectory acquisitions.

Client-server architecture. The distributed client-server architecture (Fig. 1) was based on the Common Object Request Broker Architecture (CORBA) technology. CORBA was chosen to simplify the development of this distributed system, as it ensures parameters passed by one side are transferred unchanged, and are translated into an equivalent format on the target system. The networking layer does not depend on any specific implementation of CORBA. Multiple implementations of CORBA can be used to accommodate different programing languages (e.g. c/c++, Java, Python, MATLAB etc). Nevertheless, CORBA could be replaced with other technology, providing that the necessary interfaces are supplied. The architecture does not force a built reconstruction system to constitute of fixed number of dedicated servers and clients, but it was rather designed to allow flexible arrangement of clients and servers into separate reconstruction systems. This approach is not limited to a single connection - it can contain multiple servers residing on the same/different machines, as well as clients making requests to the same/different servers.

The distributed reconstruction goes through two system states (Fig. 1); system set-up and reconstruction. The system set-up state is maintained by a naming service. This naming service contains a record of servers that registered themself as available to clients. A desired system instantiation is created *ad hoc* by a client searching for, and connecting to, servers providing the required functionality. The system is destroyed by the client disconnecting from the servers, however the servers applications remain awaiting new connections.

The reconstruction state contains four stages (Fig. 1); initialization, data transmission, remote execution and result collection. In the initialization stage a client sends the required parameters for the server to create the necessary data structures. This process is normally done once per reconstruction, since different repetitions usually have the same conditions. After initialisation, the client invokes the remaining reconstruction stages in a desired order until the whole reconstruction task is done.

The system framework (Fig. 1) is divided into three separate layers; Networking layer, Client-reconstruction layer, Server-reconstruction layer. The networking layer provides a set of data transmission and remote execution interfaces for the client-server architecture. The client-reconstruction layer schedules, organises and controls data exchange and remote execution processes, employing interfaces provided by the networking layer. The server-reconstruction layer implements a specific functionality to the interfaces provided by the networking layer. Within this layer, different reconstruction algorithms can be implemented and stored in form of modules. The modules are accessible through a module interface built into the server layer. This approach allows a single server to consist of multiple different processing modules that can be loaded on the client's request.

This layered architecture was adopted to allow disjointed development of each side of communication, as well as flexibility in porting across different hardware and programming technologies. This allows different client side implementations for each MR system to be developed without interfering with the server or networking side of the application. Similarly, technology providing data transmission, remote execution, or the way in which the server manages the reconstruction modules can be changed without affecting the other layers.

Four networking interfaces were declared (Fig. 1), directly relating to each of the reconstruction stages described above. Each interface provides the client with a set of input parameters which specify the type of operation, the means

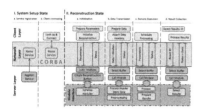

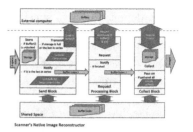

Fig. 1. Client-server architecture for the distributed image reconstruction system.

Fig. 2. Organisation of overlapping data transmission and remote execution for continuous, real-time acquisitions.

to transmit an arbitrary length of data, and return the status of an operation. To allow transmission of data back from a server onto a client, the interface for the result collection stage provides a parameter for the output of an arbitrary length of data.

Real-time Data transmission and remote execution management. The client-server architecture does not force a specific order for the reconstruction stages. However, for real-time applications proper data transmission management can be as important as efficient implementation of the image reconstruction algorithm. Time improvements gained by efficient reconstruction can be counterbalanced by a slow transmission process or wasted on unnecessary synchronizations. For optimal processing of continuous and arbitrary length streams of real-time data, an overlap between data transmission and reconstruction is desired. This can be achieved by buffering of incoming data and assigning different processing threads to each of the communication and execution stages. The optimal situation is when reconstruction time is equal or shorter than data transmission. In this case only one additional storage space is required to allow a constant stream of data between computers. However, in the case of reconstruction being slower than data transmission, it may be beneficial to have more than two buffers.

Figure 2 presents our implementation of a client for an incoming stream of real-time data. The whole process is controlled by three cross-network groups of threads; Send threads, Process threads and Get threads. Each group of threads controls the processing of different aspects of the reconstruction state, enabling overlapping of data transmission and execution. On the client these are represented by three control blocks, which work independently from one another, communicating only by passing messages about the completion of the previous stage. The stream of constantly acquired is divided into sets, which can fit into buffers organised on the external machine. The buffers integrity is protected by a set of locks shared between the control blocks. This mechanism was adopted to prevent overwriting of currently being reconstructed data, with newly incoming data. The number of buffers is an arbitrary parameter that is set during the reconstruction initialization stage.

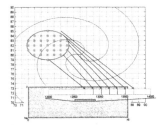

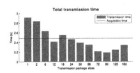

Fig. 3. Construction of a single row of a resampling matrix G^H for a spiral trajectory of three interleaves. The red ring represents a convolution kernel centred on one of trajectory points. The blue squares it encompasses are rectilinear grid points used in calculations. The arrows point places where kernel values are stored within G^H.

Fig. 4. Transmission tests results, showing the decrease in transmission time with increasing transmission package size.

Send threads control the preparation of data for reconstruction. The client separates, labels the data and initiate the sending process. Equivalent threads on the server side store the transferred data in an appropriate format for a reconstruction, within a selected by the client buffer. The incoming data is first pre-stored by the client's send control block to avoid transmission of small chunks of data. The transmission takes place if the storage limits are reached or when the last line of data in a set was received. The send control block is responsible to check the status of the buffer's lock, and only transmit data if the buffer is unlocked. When the buffer is filled-up, the send control block locks the buffer and changes the index of the receiving buffer. This way transmission can continue using a different buffer.

Process threads are responsible for overlooking each side of the remote execution. The client signals readiness for processing by passing an index for the newly filled buffer, to the server. Corresponding threads on the server start the reconstruction and return its status upon completion. If the reconstruction is successful the process control block passes the index of the buffer to the collect control block.

Get threads maintain the process of collecting results. The client sends the index of a result to the server for translation into its specific data storage system. A result is returned if processing for the selected buffer has finished. Next, the client marks the buffer as unlocked and the retrieved data are sent further down the scanner's system for processing, storing or presentation.

This organization of overlapping transmission and remote execution can work smoothly with no interruptions or breaks, providing transmission and reconstruction are faster than the data acquisition.

2.2   Reconstruction Implementation

Reconstruction algorithm. MRI reconstruction algorithms are based on solving a set of linear equations that represent the process of imaging;

$$E\rho = s \tag{1}$$

This formula expresses the relationship between the imaged object (ρ) and the acquired signal (s) as a result of the encoding transformation (E). For the SENSE algorithm the encoding matrix encompasses gridding, FFT and combination with coil sensitivity maps (CSM) [2]. A solution is found using an iterative conjugate-gradient linear solver - this is a very time consuming algorithm that requires significant computational power.

GPUs as vector processors were designed to support matrix operations. As we can express the SENSE reconstruction in the form of matrix multiplications and additions, this makes it a perfect candidate for implementation on a GPU platform.

Implementation. This GPU implementation was based on a previous (original) multi-core CPU implementation. When profiled, the bottleneck of the CPU reconstruction was found to be the gridding, which could take between 70–81 % of the total reconstruction time.

Gridding is the process where the signal is resampled onto a Cartesian grid - this is achieved by convolution with a kernel function. The problem of gridding has been well studied and different ways of implementing it on GPU platform can be found [4]. These are very specific solutions for a GPU platform - they are not trivial in implementation and may require optimizations depending on targeted hardware. This project does not aim to create the fastest gridding implementation, however the following solution is proposed to avoid future reimplementation with new GPU architectures.

The operation of convolution with resampling used for gridding can be expressed in form of a matrix multiplication;

$$G_{(N_c,N_t)}s^t_{(N_t,m)} = s^c_{(N_c,m)} \tag{2}$$

Here gridding of m data sets of N_t samples each acquired on an arbitrary trajectory, is performed by multiplying the acquired data organised into a column-major matrix (s^t) with the gridding matrix (G). The operation produces a new matrix (s^c) of m data sets with N_c samples on a Cartesian grid.

The gridding matrix is created as shown on Fig. 3. The gridding matrix is too big to be stored in dense format. Fortunately, the matrix can be kept in a sparse representation, as only a small fraction of its elements are non-zero values. This approach allows us reuse already existing and optimised code for sparse-dense matrix operations allowing maximisation of GPU's potential.

GPU Implementation specifics. The GPU version of the imaging algorithm was implemented using NVIDIA's CUDA toolkit and which includes libraries; *cusparse* for sparse-dense matrix operations, *cublas* for simple linear vector, matrix operations, reduction and dot product calculations and *cufft* for 2D Fourier Transformations. Some minor operators were not provided with the libraries (e.g. element-wise multiplication) - these were implemented in the form of in-house functions, which were executed on the GPU. Using these libraries significantly reduced the time required for development, and means the implementation benefits from constant optimisation with new releases of the toolkit, as well as their adoption to new GPU architectures.

The entire SENSE algorithm was ported onto the GPU, whilst keeping the number of CPU to GPU communications to an absolute minimum. Incoming real-time data are sent to GPU's memory at the start of the reconstruction and only the results are retrieved upon finishing. The data are stored in continuous allocations, which allow them to be processed by the same kernel. This means that even though a single image may not have enough elements to fully utilise the whole GPU, an operation can be extended to the whole set of images. In this way the data sets can be treated as a dense matrix allowing gridding of all of the sets in a single matrix-matrix multiplication; *cusparseXcsrmm()*. Similarly for 2D Fourier Transformation, a batched transformation can be scheduled with *cufftPlanMany()*. Additionally, the in-house kernels were implemented to take advantage of these data structures. The use of batched functions allowed rapid processing of multiple of data sets. Also, it minimised the number of CPU/GPU synchronisations, since convergence is checked for all of the sets rather than for each individual set.

Implemented System Specifics. The described system was implemented within our clinical environment, on a 1.5 Tesla MR scanner (Avanto, Siemens Medical Solutions, Erlangen, Germany). The scanner's native reconstruction system provides a c/c++ based, multi-threaded programming environment for the implementation of the client side of the system (2x Intel Xeon E5440 2.83 GHz, 16 GB RAM). The scanner was connected with an external computer (2x Intel Xeon E5645, 2.40 GHz, 24 GB RAM) equipped with NVIDIA Tesla C2075 (4 GB RAM, 448 CUDA cores) using a half-duplex ethernet connection. C++ implementation of CORBA technology (omniORB, Apasphere Ltd, Cambridge, UK) was used to implement the networking layer.

3   Tests

Both the data transmission and the GPU reconstruction were tested using a data set of 60 images ($\sim$2.48 s, $\sim$32 MB/s, image size: 128 × 128, TR: $\sim$41 ms). To test for the optimum transmission protocol, the test was run with different transmission package sizes (number of acquisition lines that are sent over network together).

Gridding was the major bottleneck of the native scanner reconstruction and therefore, we initially compared gridding performance alone. We tested two implementations of our new gridding approach and compared them to the native gridding method. The first implementation was on the multi-core CPU in the external machine and tested the speed-up related to realizing gridding as

Table 1. Comparison of the averaged times of each step of the algorithm, the averaged times of an iteration and the total reconstruction times between CPU and GPU.

		CPU [ms]	[%]	GPU [ms]	[%]	CPU/GPU
Per iteration	FFT	212.03	7	69.75	38	3
	Grid	2147.07	70	59.15	33	36
	Matrix combination	610.13	20	26.12	14	23
	Other	78.20	3	20.55	11	4
	Total	3047.42	-	181.51	-	17
Per 60 frames (test data set)	Total	12190.80	-	727.29	-	17

a matrix multiplication and using more modern hardware. The second implementation was on the GPU of the external machine and tested the additional benefit of GPU gridding over the multi-core CPU gridding based on the same algorithm.

In addition, we compared the total native scanner reconstruction with a GPU reconstruction that included all parts of the iterative SENSE algorithm. The different steps of the iterative SENSE reconstruction were timed for comparison between the CPU and GPU implementations.

4 Results

Figure 4 presents results of the data transmission tests, where the transmission time was measured from the beginning of the transmission until the last image of the set was fully transferred on the external computer. The optimal transmission protocol was found for a transmission package size of 12, as this introduced the shortest latency. Importantly, for this protocol the total transmission time (~2.42 s) was faster than the acquisition (~2.48 s).

The native scanner multi-core CPU gridding of 60 images took ~2.14 s. The new multi-core implementation on the external machine was ~8.5x faster than the native scanner gridding (~253 ms to grid a set of 60 images). Thus, in this new implementation gridding was reduced from ~70 % of the total reconstruction time to ~22 %. However, the final reconstruction time would only decrease to ~4.61 s, which is still ~1.9x slower than the acquisition.

The gridding performed on the GPU was ~36x faster than the native scanner gridding and ~4x faster than the multi-CPU gridding run on the external computer. This meant that gridding was no longer the bottleneck of the GPU reconstruction. However, if only the gridding was ported to the GPU, total reconstruction time would still have been greater than acquisition time.

Timing results for each step of the complete GPU reconstruction and the native scanner multi-core CPU reconstruction are presented in Table 1. The total CPU reconstruction time was ~12.2 s for each 60 frame set, which rendered the reconstruction unsuitable for the arbitrary length streams of real-time data.

This is because reconstructing the data would take $\sim$5.0x longer than acquisition. On the other hand, each of the reconstruction steps was sped-up by the GPU implementation and this resulted in the total reconstruction time being $\sim$3.4x faster than the acquisition.

5 Application Example

The system was used in a research project measuring continuous cardiac output during an exercise protocol [7]. This imaging protocol, which uses continuous, real-time reconstruction for undersampled acquisitions with arbitrary trajectories, was shown to benefit the most from the described system. Continuous cardiac output monitoring is clinically important, however impractical with previous reconstruction technology. A spiral MR sequence [8] with 4x undersampling was used to achieve high temporal resolution (TR: $\sim$44 ms) real-time data. Data were acquired over a period of $\sim$10 min producing 13980 images that needed to be reconstructed within clinically acceptable time ($<$ 1 min). The original CPU reconstruction would need over 80 min of additional processing; while the GPU online reconstruction took only $\sim$9 s. This was possible thanks to full overlap between the reconstruction and the data acquisition. The protocol produced raw aortic flow data, which allowed continuous monitoring of changes in response to the exercise.

6 Discussion

The paper describes the challenges underlying translation of advanced MR protocols into clinical workflow. The need for improved computational power for fast and robust reconstructions within the clinical environment were addressed with the development of an external, heterogeneous image reconstructor. The presented, distributed system, based on client-server architecture allowed the creation of a flexible, modular platform that could be implemented on different MR systems and using different reconstruction hardware. Importantly, our implementation was invisible to the end user, which is essential for clinical translation allowing simple transfer of final image data to processing and storage nodes.

For our application of continuous flow quantification, the native CPU reconstruction was longer than the acquisition time. This limits the imaging protocol to short scans and precludes its use in the clinical environment. It has been proven that complex image reconstruction algorithms can be significantly sped-up by porting them onto a GPU platform. In this work the GPU implementation reduced the main bottleneck, the gridding step, from $\sim$70 % to $\sim$33 % of the total reconstruction time. This was significantly better than a novel multi-core CPU reconstruction based on the same gridding approach, demonstrating the importance of using the GPU. However, it was only by speeding up each part of the iterative SENSE reconstruction using the GPU that was it possible to perform reconstruction quicker than acquisition. Nevertheless the GPU implementation must be integrated into the scanner's reconstruction pipeline to truly enable these protocols for clinical use. This introduced a middle step in the form of data transmission that could become the new bottleneck of the reconstruction. The proposed and implemented data management scheme allowed full overlap between all three parts of the reconstruction; acquisition, transmission and execution.

Although, the transmission protocol proved sufficient for the presented applications it may not be enough for more data intensive protocols (e.g. acquisitions with 32 channel receiver coils). For these applications a faster data transmission may be achieved with data compression. This step can be easily introduced to the networking framework, but a suitable compression algorithm needs to be researched to achieve satisfactory compression rates, which are not outweighed by the time needed to run the compression.

Similar works toward offloading of image reconstruction from a scanner had been done [9]. These works concentrated on interventional MRI where resultant images were presented on separate viewing stations. Also, worth of mentioning is a recent work toward an open source platform for implementing and sharing of medical image reconstruction algorithms [10].

In conclusion, we achieved all of the project goals by integrating the GPU based image reconstructor into the scanner's system. We believe it has a potential to revolutionise a type of sequences that are performed on patients and could improve diagnosis and management of patients with cardiovascular disease.

References

1. Pruessmann, K.P., Weiger, M., Scheidegger, M.B., Boesiger, P.: Sense: sensitivity encoding for fast MRI. Magn. Reson. Med. **42**(5), 952–962 (1999)
2. Pruessmann, K.P., Weiger, M., Bornert, P., Boesiger, P.: Advances in sensitivity encoding with arbitrary k-space trajectories. Magn. Reson. Med. **46**(4), 638–651 (2001)
3. Hansen, M.S., Atkinson, D., Sorensen, T.S.: Cartesian sense and k-t sense reconstruction using commodity graphics hardware. Magn. Reson. Med. **59**(3), 463–468 (2008)
4. Sorensen, T.S., Schaeffter, T., Noe, K.O., Hansen, M.S.: Accelerating the nonequispaced fast fourier transform on commodity graphics hardware. IEEE Trans. Med. Imaging **27**(4), 538–547 (2008)
5. Stone, S.S., Haldar, J.P., Tsao, S.C., Hwu, W.M., Sutton, B.P., Liang, Z.P.: Accelerating advanced MRI reconstructions on GPUS. J. Parallel Distrib. Comput. **68**(10), 1307–1318 (2008)
6. Kowalik, G., Steeden, J., Atkinson, D., Muthurangu, V.: A networked GPU reconstructor within the clinical workflow for rapid fat quantification. In: Proceedings of the 19th Annual Meeting of ISMRM, Montreal (2011)
7. Kowalik, G.T., Steeden, J.A., Pandya, B., Odille, F., Atkinson, D., Taylor, A., Muthurangu, V.: Real-time flow with fast GPU reconstruction for continuous assessment of cardiac output. J. Magn. Reson. Imaging **36**(6), 1477–1482 (2012)

8. Steeden, J.A., Atkinson, D., Taylor, A.M., Muthurangu, V.: Assessing vascular response to exercise using a combination of real-time spiral phase contrast MR and noninvasive blood pressure measurements. J. Magn. Reson. Imaging **31**(4), 997–1003 (2010)

9. Roujol, S., de Senneville, B.D., Vahala, E., Sorensen, T.S., Moonen, C., Ries, M.: Online real-time reconstruction of adaptive TSENSE with commodity CPU/GPU hardware. Magn. Reson. Med. **62**(6), 1658–1664 (2009)

10. Hansen, M.S., Sorensen, T.S.: Gadgetron: an open source framework for medical image reconstruction. Magn. Reson. Med. **69**(6), 1768–1776 (2013)

X-Ray Laser Imaging of Biomolecules Using Multiple GPUs

Stefan Engblom[1] and Jing Liu[1,2]($\boxtimes$)

[1] Division of Scientific Computing, Department of Information Technology,
Uppsala University, 751 05 Uppsala, Sweden
{stefan.engblom,jing.liu}@it.uu.se
[2] Laboratory of Molecular Biophysics, Department of Cell and Molecular Biology,
Uppsala University, 751 24 Uppsala, Sweden

Abstract. Extremely bright X-ray lasers are becoming a promising tool for 3D imaging of biomolecules. By hitting a beam of streaming particles with a very short burst of a high energy X-ray and collecting the resulting scattering pattern, the 3D structure of the particles can be deduced. The computational complexity associated with transforming the data thus collected into a 3D intensity map is very high and calls for efficient data-parallel implementations.

We present ongoing work in accelerating this application using multiple GPU nodes. In particular, we look at the scaling properties of the application and give predictions as to the computational viability of this imaging technique.

Keywords: GPU cluster · CUDA/MPI · Single molecule imaging · X-ray laser

1 Introduction

We consider in this paper an emerging computational technology from the field of structural biology. The classical method of determining atomic structures of biological molecules by collecting diffraction patterns from X-ray illuminations of crystals is currently undergoing a rapid development. Modern X-ray free-electron lasers (XFELs) provide extremely intense bursts of X-rays allowing for diffraction patterns to be collected from a *single* protein molecule, a virus particle, or a cell. The idea of using XFELs for the purpose of atomic resolution imaging of non-crystallized samples was first suggested in [6]; for a review, see [2]. In a nutshell, a femtosecond X-ray pulse is faster than the damage processes and the collected diffraction pattern thus belongs to the original object. Notable recent successful applications include the study of photo absorption processes of atoms [11], investigations in nanocrystallography [1], and in imaging of single viruses [9].

The basic idea is simple: target particles are injected into the beam and hit by X-ray pulses. The resulting diffraction pattern is collected by a pixellated

R. Wyrzykowski et al. (Eds.): PPAM 2013, Part I, LNCS 8384, pp. 480–489, 2014.
DOI: 10.1007/978-3-642-55224-3_45, © Springer-Verlag Berlin Heidelberg 2014

detector and represents one 2D projection of the particle. Using a collection of such diffraction patterns from identical particles, a 3D image can in principle be obtained by solving what is remindful of a complicated "puzzle". Solving for the unknown particle rotations when fitting the individual patterns into a consistent 3D structure is a solvable problem, albeit a highly compute intense one. Some different algorithms have been proposed, and the approached considered here is the one successfully employed in [9], and suggested earlier in [4]. The idea is that data is repeatedly matched against tentative rotations of the particles such that after convergence, the final 3D image may be deduced in the sense of a Maximum Likelihood estimator.

With X-ray laser technology developing rapidly, and with diffraction data being delivered at an ever increasing rate, massively parallel approaches for bringing down the run times for analysis become vital. We present in this paper the first results to this effect, namely, a working multi-GPU implementation.

A brief explanation of the imaging algorithm under consideration is found in Sect. 2 and we discuss our data parallel distributed implementation in some detail in Sect. 3. Our results, including a validation, are presented in Sect. 4 and we offer a concluding discussion with some outlooks in Sect. 5.

2   Maximum-Likelihood Imaging with X-Ray Lasers

Denote by $K = (K_k)_{k=1}^{M_{\text{data}}}$ the stream of measured photon counts on the pixel-lated 2D detector. From an increasingly detailed data set an improved estimate of the 3D intensity distribution of the object W can in principle be constructed. In practice the computational challenges due to a small signal to noise ratio are severe and calls for both robust and efficient implementations.

Assuming for now i.i.d. data the Maximum-Likelihood estimator can be formulated as an optimization problem,

$$\hat{W} = \arg_W \max \, M_{\text{data}}^{-1} \sum_{k=1}^{M_{\text{data}}} \log \mathbf{P}(K_k|W), \qquad (2.1)$$

that is, maximizing the *likelihood* of the data given a probabilistic intensity model. This approach is not directly tractable because *(i)* the *rotation* R_k of each sample associated with the measurement K_k is unknown and, *(ii)* the *photon fluency* ϕ_k hitting the sample is in practice also an unknown. A third obstacle has already been mentioned: the measurements are severely contaminated by noise, and care has to be taken in order to arrive at a convergent and robust algorithm.

2.1   The EM Approach

The *Expectation Maximization* algorithm deals with likelihood estimation under hidden data conditions in a constructive way. Briefly, alternating steps of *(i)* assigning probabilities to the hidden states, and *(ii)* likelihood estimates of the parameters of the model can be shown under broad conditions to be a *descent step* of the full likelihood functional [5]. In step *(i)* the model parameters are frozen at the previous step, while in *(ii)* the hidden state probabilities (the *responsibilities* using EM terminology) from step *(i)* are used.

To discuss the EM algorithm in the current context we need to introduce some notation. We discretize the intensity space by introducing the set of points $(q_i)_{i=1}^{M_{\mathrm{pix}}}$ and we also discretize the set of possible rotations by $(R_j)_{j=1}^{M_{\mathrm{rot}}}$. We define W_{ij} as the unknown mass intensity of the sample at position $R_j q_i$. Similarly, ϕ_{jk} is introduced as an estimator of the intensity of the beam that produced data frame k given rotation R_j.

The probabilistic model is that the intensity of the ith pixel in the kth measurement is Gaussian,

$$\mathbf{P}(K_{ik} = \kappa | W_{ij}, R_j, \phi_{jk}) \propto \exp\left(-\frac{(\kappa/\phi_{jk} - W_{ij})^2}{2\sigma^2}\right), \qquad (2.2)$$

with σ a free noise parameter. Up to a constant the log-likelihood is therefore

$$Q_{ijk} = Q_{ijk}(W, \phi) := -\frac{(K_{ik}/\phi_{jk} - W_{ij})^2}{2\sigma^2}, \qquad (2.3)$$

and summing over i produces the joint log-likelihood function,

$$Q_{jk} = Q_{jk}(W, \phi) := \sum_{i=1}^{M_{\mathrm{pix}}} Q_{ijk}(W, \phi). \qquad (2.4)$$

This implies the **"E-step"**

$$
\begin{aligned}
P_{jk}^{(n+1)} = P_{jk}^{(n+1)}(W^{(n)}, \phi^{(n)}) &:= \mathbf{P}(R_j | K_k, W^{(n)}, \phi^{(n)}) \\
&= \frac{w_j T_{jk}(W^{(n)}, \phi^{(n)})}{\sum_{j'=1}^{M_{\mathrm{rot}}} w_{j'} T_{j'k}(W^{(n)}, \phi^{(n)})},
\end{aligned}
\qquad (2.5)
$$

in terms of $T_{jk}(W, \phi) = \exp(Q_{jk}(W, \phi))$, and where w_j are weights normalized such that $\sum_j w_j = 1$. These additional degrees of freedom allow for prior probabilities on the rotations and are necessary when the rotational space is not uniformly discretized. In practice, (2.5) is computed in logarithmic space using an appropriate scaling.

For the case of the current Gaussian model there is no explicit likelihood formula available, but a fix-point iteration step may be used for the **"M-step"**,

$$W_{ij}^{(n+1)} = \frac{\sum_{k=1}^{M_{\mathrm{data}}} P_{jk}^{(n+1)} K_{ik}/\phi_{jk}^{(n)}}{\sum_{k=1}^{M_{\mathrm{data}}} P_{jk}^{(n+1)}}, \qquad (2.6)$$

$$\phi_{jk}^{(n+1)} = \frac{\sum_{i=1}^{M_{\mathrm{pix}}} K_{ik}^2}{\sum_{i=1}^{M_{\mathrm{pix}}} W_{ij}^{(n)} K_{ik}}. \qquad (2.7)$$

This approach of using 'partial steps' can be theoretically justified under certain conditions [5].

2.2   The EMC Algorithm

In principle, the iteration written symbolically as $W^{(n+1)} := (M \circ E)W^{(n)}$ can be used as a method to obtain the required image as W^∞ (E defined by (2.5), M by (2.6)). The problem here is that the problem is severely overdetermined; there are many pairs (i, j) such that $R_j q_i$ is nearly the same point. This problem is of course emphasized when the end result is understood in the sense of a finite resolution pixelized image.

A simple way around this is to add *expansion/compression*-steps before and after the E and M-steps. The purpose of the latter step is to compress (average/smooth) the overdetermined representation into, say, a Cartesian representation with a minimum number of degrees of freedom. *Vice versa*, the expansion step takes us back to the working description.

Consider therefore interpolation weights f such that for a predetermined set of interpolation abscissas $(p_l)_{l=1}^{M_{\mathrm{grid}}}$ and g some smooth function,

$$g(q) \approx \sum_{l=1}^{M_{\mathrm{grid}}} f(p_l - q)g(p_l). \tag{2.8}$$

The expansion operator can now be defined ("**e-step**"),

$$W_{ij} = \sum_{l=1}^{M_{\mathrm{grid}}} f(p_l - R_j q_i)\mathbb{W}_l, \tag{2.9}$$

that is, a map from values on a grid $\mathbb{W}_l := W(p_l)$ into the working description $W_{ij} = W(R_j q_i)$.

In the implementation discussed here, the M- and the c-steps are in fact intertwined in that the normalization is postponed until after W has been determined. Hence we start by computing (compare (2.6))

$$W_{ij}^{(n+1)} = \sum_{k=1}^{M_{\mathrm{data}}} P_{jk}^{(n+1)} K_{ik}/\phi_{jk}^{(n)}, \tag{2.10}$$

with total probability per rotation j given by

$$\nu_j = \sum_{k=1}^{M_{\mathrm{data}}} P_{jk}^{(n+1)}. \tag{2.11}$$

Next the "**c-step**" is determined by averaging,

$$\mathbb{W}_l^{(n+1)} = \frac{\sum_{i=1}^{M_{\mathrm{pix}}} \sum_{j=1}^{M_{\mathrm{rot}}} f(p_l - R_j q_i)W_{ij}^{(n+1)}}{\sum_{i=1}^{M_{\mathrm{pix}}} \sum_{j=1}^{M_{\mathrm{rot}}} f(p_l - R_j q_i)\nu_j}. \tag{2.12}$$

In Algorithm 1 we summarize our view of the EMC algorithm. A key issue to note is the nonlinear character of several of the steps involved, and in particular of the E-step where computations in logarithmic space and a subsequent normalization is made. This precludes the use of highly optimized standard linear algebra packages such as BLAS. On the other hand, as pointed out already in [4], the algorithm can clearly be distributed and is moreover suitable to data-parallel implementations. This is the topic next discussed.

Algorithm 1. The principal logic behind the EMC algorithm.

Input: Initial guess of the 3D intensity distribution $\mathbb{W}^{(0)}$ of the object on the grid $(p_l)_{l=1}^{M_{\mathrm{grid}}}$ and initial estimate of the rotational probabilities $P^{(0)}$.
Output: Improved image $\mathbb{W}$ and probabilities P.

1: **repeat**
2:    $n = 0, 1, \ldots$
3:    $W^{(n)} := e \circ \mathbb{W}^{(n)}$. {Expansion step according to (2.9).}
4:    $P^{(n+1)} := E \circ P^{(n)}$. {Expectation step, (2.5).}
5:    $[W^{(n+1)}, \mathbb{W}^{(n+1)}] := cM \circ W^{(n)}$. {Combined Maximization and compression, (2.10) and (2.12).}
6: **until** change in $\mathbb{W}$ is small enough

---

3 Parallelization Using Multiple GPU Nodes

3.1 Single-Node Data-Parallel Version

Thanks to their high compute density and memory bandwidth, GPUs are an attractive alternative in many applications with a predominantly data parallel character. Our single-node EMC was implemented using CUDA with C/C++-wrappers and the implementation closely follows the logic in Algorithm 1. Briefly, the CPU controls the overall procedure and streams all data to the GPU as well as writes the output. The diffraction patterns are thus first loaded to the CPU, then streamed and copied into GPU memory. At the end of each EMC iteration, the GPU has the option to communicate back various variables to the CPU which is written to disk. On the GPU, the compute intensive steps Eqs. (2.9), (2.5), (2.10) and (2.12) in Algorithm 1 naturally become the kernels.

3.2 Distributed Implementation

Due to the increasing size of diffraction-pattern datasets, we extended the single-node implementation into a distributed version using MPI/CUDA. Related approaches with multi-GPU applications in Computational Fluid Dynamics [3] and NAS-LU benchmarks [7] suggest that good efficiency can be achieved using this combination.

Generally speaking, a good parallel efficiency is obtained when the tasks at each node are roughly of the same computational complexity. There are two immediate ways of distributing data which both target the running time of the most time consuming part (the E-step, see Table 2). Simply put: one either splits the matrix of rotational probabilities P along the rotations or along the images.

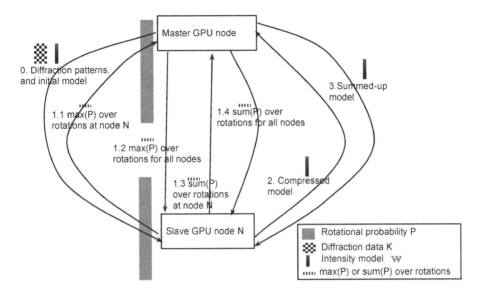

Fig. 1. Distribution of data and communication pattern for our distributed EMC implementation. In the figure, all rectangles representing data have the correct scale with respect to the variable they represent (see Table 1). Step 0 is the initialization phase. For each EMC iteration, intensity model updates are performed via step 1 through 3 among the GPU nodes.

We prefer to split P along rotations, that is, we localize P such that each node has its own segment P_{jk} for j in a certain block of size about M_{rot}/N, where N is the number of nodes, and where $k = 1...M_{\mathrm{data}}$. This approach to distribution is natural when considering the size of typical data (see Table 1) and the associated communication pattern. The only data thus gathered from the slave GPUs are vectors with locally computed maximum values and local sums.

Figure 1 indicates how data flows in our implementation; in the initialization step we distribute the diffraction patterns and the initial model. To normalize the updated rotational probability P at each GPU node, the local maximum and the local sum of P must be transferred to the Master node (steps 1.1–1.4 in Fig. 1). Compression according to (2.12) is first done locally (step 2), while the final reduced average is computed by the Master GPU. The updated intensity model is lastly broadcasted and becomes the initial model for the next iteration (step 3). In this way, each GPU node keeps its own probability estimate, while the reconstruction of the model relies on a global reduction. Algorithm 2 further details our distributed EMC algorithm.

With each GPU bind to one CPU, all communications among GPUs were achieved via OpenMPI using CPU buffers and internal communication within each CPU/GPU-pair. A notable difference to the single GPU implementation is that, due to the global nature of the normalization in steps 1.1–1.4 of Fig. 1, the kernel of the E-step must now be divided into 3 smaller kernels with intertwined global communications.

Algorithm 2. Pseudocode for our distributed version of the EMC algorithm.

Input/output: As in Algorithm 1.

1: Distribute the 3D intensity $\mathbb{W}^{(0)}$ and the diffraction patterns K among the nodes.
2: **repeat**
3:     $n = 0, 1, \ldots$
4:     At each GPU node:
5:     $W^{(n)} := e \circ \mathbb{W}^{(n)}$. {Local expansion step according to (2.9).}
6:     $P^{(n+1)} := E \circ P^{(n)}$. {Local expectation step (2.5), but globally normalized via steps 1.1–1.4 in Fig. 1.}
7:     $[W^{(n+1)}, \mathbb{W}^{(n+1)}] := cM \circ W^{(n)}$. {Local Maximization and compression, (2.10) and (2.12).}
8:     Gather models at the Master GPU and compute the average $\mathbb{W}^{(n+1)}$.
9:     Broadcast the resulting $\mathbb{W}^{(n+1)}$.
10: **until** change in $\mathbb{W}$ is small enough

---

4   Results

In this section we present results from validating our implementation and some first tentative results as to the parallel efficiency obtained on a small multi-GPU computer. All tests ran on a single Linux machine (Red Hat 4.4.6-4) with two Tesla C2075 GPUs and one GeForce GTX 680 GPU. We compiled our code with GCC 4.4.6, CUDA 5.0 and OpenMPI 1.5.4.

In Table 1 we list the sizes of all relevant data in our setup. These runs were configured to reconstruct a $64 \times 64 \times 64$ or a $128 \times 128 \times 128$ 3D intensity model from a collection of 198 diffraction patterns. The setup itself is a reconstruction problem for a mimivirus diffraction dataset [9].

Table 2 shows our initial profiling result from the Nvidia profiler using a single GPU. As can be seen, the E-step dominated the cost of the EMC iterations and took more than 57 % of the total execution time.

4.1   Validation

The goal of the EMC algorithm is to reconstruct a 3D intensity model, which emerges as a function of the estimated rotational probabilities and the raw

Table 1. Sizes of data as used in our experiments. Note that the value of M_{pix} is the result after binning the raw data 1024×1024 into a coarser 64×64 (or 128×128) format

Size	Value
M_{pix}	4,096 (or 16,384)
M_{data}	198
M_{rot}	86,520
M_{grid}	262,144 (or $2,097,152$) $(= M_{\text{pix}}^{3/2})$

Table 2. Profile of the single GPU EMC implementation.

EMC step	Relative time (%)
Expectation step (2.5)	57
Computation of the photon fluency (2.7)	23
Maximization step (2.10)	10
Compression step (2.12)	1.9
Expansion step (2.9)	1.4
Other	6.7

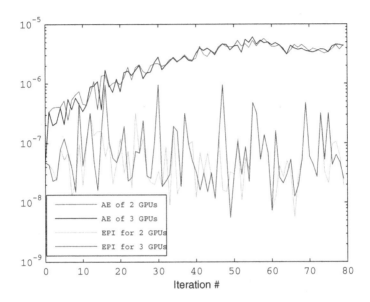

Fig. 2. KS statistics (4.1) as a function of iteration number. Accumulated Error and Error Per Iteration.

diffraction patterns. To validate our code, we relied on the *Kolmogorov-Smirnov statistics* (KS) over these rotational probabilities;

$$D^N = \max_{jk}(|F_{jk}^N - F_{jk}^1|), \tag{4.1}$$

where F_{jk}^N is the cumulative distribution function of the rotational probability P_{jk} for $N \in \{2,3\}$ GPUs, and we thus considered the $N = 1$ case to be the truth.

EMC is an iterative algorithm and in validating our implementation we both considered the error per iteration (EPI) and the accumulated error (AE). Figure 2 displays these for the first 80 iterations for the case $M_{\text{pix}} = 64 \times 64$. The average values are less than about 1.5×10^{-7} (EPI) and 3.3×10^{-6} (AE). Given that single precision was used and that the data consists of about 10^5 floating point numbers we judge that the implementation is validated to within

Table 3. Average execution times per iteration and efficiencies for three different hardware configurations

Set up	$M_{\mathrm{pix}} = 64 \times 64$		$M_{\mathrm{pix}} = 128 \times 128$	
	Time (s)	Efficiency	Time (s)	Efficiency
$1 \times$ C2075	49.9 ± 0.09	$\equiv 1$	179.8 ± 0.13	$\equiv 1$
$2 \times$ C2075	25.2 ± 0.04	0.99	90.5 ± 0.17	0.99
$2 \times$ C2075 $+ 1 \times$ GTX580	16.9 ± 0.03	0.98	62.4 ± 0.17	0.96

the floating point accuracy. These results are very remindful of other similar accuracy comparisons [8].

4.2 Efficiency

We tested our implementation using one to three GPUs and two different problem sizes. The time measured thus included all computations in one EMC iteration and all the necessary data synchronization. As Algorithm 2 describes we globally synchronize for normalizing the rotational probability and for averaging the 3D intensity model. GPUs were therefore idle until all nodes reached these barriers, and hence we cannot reasonably expect a linear speed up. Table 3 shows the average EMC iteration execution time and efficiency for the first 80 iterations.

5   Discussion

With our implementation we obtain a very reasonable and almost linear speedup. Our parallel efficiency is comparable to that obtained in other MPI/CUDA applications [3,10,12].

There are some algorithmic changes that could be made to improve on the efficiency. For instance, we need to synchronize in order to globally broadcast the maximum value of P as used in scaling the result. When this value does not differ substantially from the previous iterations, that value could clearly be reused. Also, in the current implementation we make no effort in hiding the communication time behind the time for computing the photon fluency (2.7), which is independent on the update of the rotational probabilities in (2.5).

A somewhat more technical approach is to prefer to use GPU-to-GPU (or "peer-to-peer"-style) communication to MPI, thus eliminating the need for the CPU to be involved in the communication loop and also the need for the CPU-to-GPU buffer copy.

For future work it is of interest to improve on the resolution of the reconstructed model. By increasing the number of diffraction patterns and use a less coarse binning procedure, it is possible to obtain a considerably higher resolution. For this magnitude of scaling up the input data, which we believe to be realistic in the near future, it is clear that further improvements as to the distribution of data and to the computational efficiency of the individual parts are needed. This is ongoing work.

Acknowledgment. This work was financially supported by the Swedish Research Council, the Röntgen Ångström Cluster, the Knut och Alice Wallenbergs Stiftelse, the European Research Council (JL), and by the Swedish Research Council within the UPMARC Linnaeus center of Excellence (SE, JL).

References

1. Chapman, H.N., et al.: Femtosecond X-ray protein nanocrystallography. Nature **470**(7332), 73–77 (2011). doi:10.1038/nature09750
2. Gaffney, K.J., Chapman, H.N.: Imaging atomic structure and dynamics with ultrafast X-ray scattering. Science **316**(5830), 1444–1448 (2007). doi:10.1126/science. 1135923
3. Jacobsen, D.A., Thibault, J.C., Senocak, I.: An MPI-CUDA implementation for massively parallel incompressible flow computations on multi-GPU clusters. In: 48th AIAA Aerospace Sciences Meeting and Exhibit, vol. 16 (2010). doi:10.2514/6.2010-522
4. Loh, N.D., Elser, V.: Reconstruction algorithm for single-particle diffraction imaging experiments. Phys. Rev. E **80**(2), 026705 (2009). doi:10.1103/PhysRevE.80. 026705
5. Neal, R.M., Hinton, G.E.: A view of the EM algorithm that justifies incremental, sparse, and other variants. In: Jordan, M.I. (ed.) Learning in Graphical Models, pp. 355–368. Kluwer, Dordrecht (1998)
6. Neutze, R., Wouts, R., van der Spoel, D., Hajdu, J.: Potential for biomolecular imaging with femtosecond X-ray pulses. Nature **406**(6797), 752–757 (2000). doi:10. 1038/35021099
7. Pennycook, S.J., Hammond, S.D., Jarvis, S.A., Mudalige, G.R.: Performance analysis of a hybrid MPI/CUDA implementation of the NASLU benchmark. ACM SIGMETRICS Perf. Eval. Rev. **38**(4), 23–29 (2011). doi:10.1145/1964218.1964223
8. Salvadore, F., Bernardini, M., Botti, M.: GPU accelerated flow solver for direct numerical simulation of turbulent flows. J. Comput. Phys. **235**, 129–142 (2013). doi:10.1016/j.jcp.2012.10.012
9. Seibert, M.M., et al.: Single mimivirus particles intercepted and imaged with an X-ray laser. Nature **470**(7332), 78–81 (2011). doi:10.1038/nature09748
10. Wang, Y., Dou, Y., Guo, S., Lei, Y., Zou, D.: CPU-GPU hybrid parallel strategy for cosmological simulations. Concurr. Comput.: Pract. Exper. (2013). doi:10.1002/cpe.3046
11. Young, L., et al.: Femtosecond electronic response of atoms to ultra-intense X-rays. Nature **466**(7302), 56–61 (2010). doi:10.1038/nature09177
12. Zaspel, P., Griebel, M.: Solving incompressible two-phase flows on multi-GPU clusters. Comput. Fluids **80**, 356–364 (2013). doi:10.1016/j.compfluid.2012.01.021

Out-of-Core Solution of Eigenproblems for Macromolecular Simulations

José I. Aliaga[1], Davor Davidović[2], and Enrique S. Quintana-Ortí[1](✉)

[1] Dpto. de Ingeniería y Ciencia de Computadores,
Universidad Jaume I, 12.071 Castellón, Spain
{aliaga,quintana}@uji.es
[2] Centar Za Informatiku I Računarstvo - CIR,
Institut Ruđer Bošković, 10000 Zagreb, Croatia
ddavid@irb.hr

Abstract. We consider the solution of large-scale eigenvalue problems that appear in the motion simulation of complex macromolecules on desktop platforms. To tackle the dimension of the matrices that are involved in these problems, we formulate out-of-core (OOC) variants of the two selected eigensolvers, that basically decouple the performance of the solver from the storage capacity. Furthermore, we contend with the high computational complexity of the solvers by off-loading the arithmetically-intensive parts of the algorithms to a hardware graphics accelerator.

Keywords: Macromolecular motion simulation · Eigenvalue problems · Out-of-core computing · Multicore processors · GPUs

1 Introduction

Coarse-grained models (CGM) combined with normal mode analysis (NMA) has been applied in recent years simulate biological activity at molecular level for extended time scales [1–3]. Concretely, IMOD [4] is a tool chest that exploits the advantage of NMA formulations in internal coordinates (ICs) while extending them to cover multi-scale modeling. Despite the reduction in the degrees of freedom offered by ICs, the diagonalization step remains the major computational bottleneck of this approach, specially for large molecules. In particular, the eigenproblem that has to be solved in this step of CGM-NMA is given by

$$AX = BX\Lambda, \tag{1}$$

where $A \in \mathbb{R}^{n \times n}$ and $B \in \mathbb{R}^{n \times n}$ correspond, respectively, to the Hessian and kinetic matrices that capture the dynamics of the macromolecular complex, $\Lambda \in \mathbb{R}^{s \times s}$ is a diagonal matrix with the s sought-after eigenvalues, and $X \in \mathbb{R}^{n \times s}$ contains the corresponding unknown eigenvectors [5]. Furthermore, when dealing with large macromolecules, A, B are dense symmetric positive definite matrices, $n \geq 10,000$, and typically only the $s \approx 100$ smallest eigenpairs are required.

R. Wyrzykowski et al. (Eds.): PPAM 2013, Part I, LNCS 8384, pp. 490–499, 2014.
DOI: 10.1007/978-3-642-55224-3_46, © Springer-Verlag Berlin Heidelberg 2014

In this paper we address the efficient solution of large-scale generalized symmetric definite eigenproblems arising in the simulation of collective motions of macromolecular complexes using multicore desktop platforms equipped with graphics processing units (GPUs). While there exist other related work [6–8], our paper makes the following original contributions:

- The eigenproblems associated with this particular application involve dense matrices that are, in general, too large to fit into the memory of the GPU and, in some cases, even the main memory of the server. To address this, we consider two specialized algorithms that, by applying out-of-core (OOC) techniques [9], amortize the cost of data transfers with a large number of floating-point arithmetic operations (flops). Besides, to deliver high performance, both "OOC-GPU" algorithms off-load the bulk of their computations to the attached hardware graphics accelerator.
- One of our algorithms is the first OOC-GPU implementation that employs spectral divide-and-conquer (SD&C) based on the polar decomposition proposed recently [10]. We enhance this algorithm with ad-hoc splitting strategies, that aim at reducing the number of SD&C iterations, and are cheap to compute for the biological target application.
- As an alternative algorithm, we revisit an implementation of the two-stage reduction to tridiagonal form [11], where the first stage is also an OOC-GPU code while the subsequent stage operates on a much reduced compact matrix that fits in-core.
- We perform a comparison of these two approaches using several datasets representative of large-scale macromolecular complexes [8].

Overall the major contribution of this paper lies in that it provides a demonstration that complex macromolecular motion simulations can be tackled on desktop servers equipped with GPUs even when the problem data is too large to fit into the memory of the hardware accelerator and, possibly, even the main memory.

The rest of the paper is structured as follows. In Sect. 2 we briefly describe the solution of generalized eigenproblems. In Sect. 3 we review in detail the SD&C method [10], and revisit the two-stage eigensolver, describing our hybrid CPU-GPU approach. Implementations of these eigensolvers are evaluated next, in Sect. 4, using a collection of cases from biological sources. Finally, we close the paper in Sect. 5 with a few concluding remarks.

2   Solution of Symmetric Definite Eigenproblems

All the eigensolvers considered in this work initially compute the Cholesky factorization $B = U^T U$, where $U \in \mathbb{R}^{n \times n}$ is upper triangular [5], to then tackle the *standard* symmetric eigenproblem

$$CY = Y\Lambda \quad \equiv \quad (U^{-T}AU^{-1})(UX) = (UX)\Lambda, \tag{2}$$

where $C \in \mathbb{R}^{n \times n}$ is symmetric and $Y \in \mathbb{R}^{n \times s}$. Thus, the standard eigenproblem (2) shares its eigenvalues with those of (1), while the original eigenvectors

492   J.I. Aliaga et al.

can be recovered from $X := U^{-1}Y$. The initial Cholesky factorization, the construction of $C := U^{-T}AU^{-1}$ in (2), and the solve for X are known to deliver high performance on a large variety of HPC architectures, including multicore processors and GPUs, and their functionality is covered by numerical libraries (e.g., LAPACK, `libflame`, ScaLAPACK, PLAPACK, etc.) including some OOC extensions (SOLAR, POOCLAPACK). Therefore, we will not consider these operations further but, instead, focus on the more challenging solution of the standard eigenproblem (2) on a hybrid CPU-GPU platform when the data matrices are too large to fit into the GPU memory (and, possibly, the main memory).

Among the different solvers for the symmetric eigenproblem, we discard those based on the one-stage reduction to tridiagonal form as well as the Krylov methods [5]. From an OOC viewpoint, the major drawback of these two classes of methods is that they cast a significant part of their computations in terms of the matrix-vector product (MVP). For a matrix of size $n \times n$, this kernel roughly performs $2n^2$ flops on n^2 numbers (i.e., a rate of computation to data of $O(1)$), so that an implementation that operates with OOC data (e.g., a GPU MVP routine where the matrix is on the main memory, or a multicore MVP code with data on disk) is intrinsically limited by data movement and will attain very low performance.

Instead, we will investigate a recent SD&C approach [10], with a much higher computational cost than the one-stage/Krylov-based methods, but which consists mainly of matrix-matrix operations that naturally render it as an appealing candidate for OOC-GPU strategies/platforms. As an alternative, we will also consider a classical eigensolver based on a two-stage reduction to tridiagonal form, which first transforms the matrix C from dense to band form, to then refine this intermediate matrix to tridiagonal form. We have previously described [11] an OOC-GPU practical implementation of this two-stage eigensolver and demonstrated how, by carefully orchestrating the PCI data transfers between host and device, in-core performance is maintained or even increased for the OOC solution of general large-scale eigenproblems on hybrid CPU-GPU platforms.

3   OOC Eigensolvers for GPUs

In this section we review the mathematical methods that underlie our GPU eigensolvers, discuss how to refine the SD&C algorithm to reduce its computational cost for the solution of the eigenproblems arising in macromolecular motion simulation, and offer some practical details about the OOC-GPU implementations using one key numerical kernel that appears in the algorithms.

3.1   The SD&C Algorithm

Numerical Method. For a symmetric matrix $\hat{A} \in \mathbb{R}^{n \times n}$, the following SD&C algorithm [10] starts by computing its polar factor using the QR-based dynamically weighted Halley (QDWH) iterative scheme [12]:

$$\begin{bmatrix} \sqrt{c_j}X_j \\ I_n \end{bmatrix} = \begin{bmatrix} Q_1 \\ Q_2 \end{bmatrix} R \quad \text{(QR factorization)}, \tag{3}$$

$$X_{j+1} := \frac{b_j}{c_j}X_j + \frac{1}{\sqrt{c_j}}\left(a_j - \frac{b_j}{c_j}\right)Q_1 Q_2^T, \quad j \geq 0, \tag{4}$$

where $X_0 := \hat{A}/\alpha$ and I_n denotes the identity matrix. In practice, the scalars α, a_j, b_j, c_j require estimates of the smallest singular value and matrix 2-norm of $\hat{A}$, which are cheap to compute and, upon convergence, the sequence X_j yields the sought-after polar factor U_p.

Assume QDWH has been applied to $\hat{A} := C - \sigma I_n$, with σ a user-defined splitting point for the eigenspectrum of the symmetric matrix C in (2). The subspace iteration [5] is next employed to compute an orthogonal matrix $[V_1, V_2]$, where $V_1 \in \mathbb{R}^{n \times k}$, such that $(U_p + I_n)/2 = V_1 V_1^T$; therefore,

$$\begin{bmatrix} V_1^T \\ V_2^T \end{bmatrix} C [V_1, V_2] = \begin{bmatrix} C_1 & E^T \\ E & C_2 \end{bmatrix}, \tag{5}$$

where C_1 and C_2 contain, respectively, the eigenvalues of C to the left and right of σ, and $\|E\|_F \approx u$, with u the machine unit roundoff [10].

Choosing the Splitting Point. The previous method is designed as a recursive SD&C algorithm: after dividing the spectrum of C into those of C_1 and C_2, the method is applied again to these two subproblems, using appropriate shifts σ_1 and σ_2 to further divide the spectrum. Note that our goal is to compute only a few eigenpairs of the problem, specifically the smallest s. We therefore designed three SD&C strategies, with the common purpose of selecting the appropriate value of σ, that separates the eigenspectrum of C into two subsets C_1 and C_2, with the dimension of the former being equal to (or only slightly larger than) s. Specifically, we designed and evaluated three different SD&C strategies:

SD&C-A. $\sigma = \text{trace}\{A\}/n$, where $\text{trace}\{\cdot\}$ denotes the trace of its argument.
SD&C-B. $\sigma = 4 \, \text{trace}\{A\}/n$.
SD&C-C. In this case, given a macromolecule, we use iMOD to generate the Hessian and kinetic matrix for a problem of much smaller dimension, say $m \approx 1,024$, and choose σ as the $(100 \cdot m/n)$-th largest eigenvalue of this problem.

Note also that just before the application of the subspace extraction, the value $k = \|U_p + I_n\|_F^2/2$ indicates the number of eigenvalues in C_1. Therefore, in case $k < s$, the QDWH iterate has to be recomputed, with a larger value for σ. After the first successful split, the eigenspectrum of C_1 is completely computed using a direct in-core eigensolver based on the reduction to tridiagonal form (and with a negligible cost compared with that of the initial stage).

3.2   Two-Stage Reduction to Tridiagonal Form

The eigensolver based on the two-stage reduction to tridiagonal form performs the major part of the computations in terms of efficient Level-3 BLAS operations, in exchange for a nonnegligible increment in the computational cost when compared with the direct (one-stage) reduction. The two-stage algorithm first computes the decomposition $Q_1^T C Q_1 = \hat{C}$, where $\hat{C} \in \mathbb{R}^{n \times n}$ is a matrix of bandwidth w, and $Q_1 \in \mathbb{R}^{n \times n}$ is orthogonal. In the subsequent stage, $\hat{C}$ is further reduced to a tridiagonal matrix $T \in \mathbb{R}^{n \times n}$ as $Q_2^T \hat{C} Q_2 = T$ with $Q_2 \in \mathbb{R}^{n \times n}$ orthogonal. Finally, the eigenvalues of T (which are also those of the C) and the associated eigenvectors, in $Z \in \mathbb{R}^{n \times s}$, are computed using, e.g., the MR³ solver [13,14], and the eigenvectors are recovered from $Y := Q_1 Q_2 Z$.

Our implementation of this approach is based on the SBR (Successive Band Reduction) toolbox [15] for the two-stage reduction to tridiagonal form, and employs the LAPACK routine for the MR³ method, which in general only adds a negligible cost. We have previously described an OOC-GPU implementation of the reduction to band form [11] that carefully orchestrates computation and communication to deliver performance equal or superior to that of an in-core GPU routine. Furthermore, provided w is carefully chosen, the second stage and the solution of the tridiagonal eigenproblem can proceed with data in core [11].

The OOC-GPU code for the first stage consists basically of three major kernels: QR factorization, one-sided update (for the application of orthogonal transforms from the left), and two-sided update (application from both left and right). These kernels are thus conceptually analogous to some of those appearing in the SD&C algorithm.

3.3   OOC Kernels

We next illustrate the OOC-GPU implementations using (a specialized case of) the QR factorization as a workhorse. Our OOC-GPU algorithm for this operation encodes a left-looking, slab-oriented factorization [9] that transfers data by column blocks (slabs) of width s. Note that, while there exist linear algebra libraries to obtain the QR factorization on GPUs [7], these lack of the specialized kernels that are necessary for our particular operation.

In particular, let us denote the $2n \times n$ matrix that has to be factorized in (3) as D, and consider a partitioning of this matrix into blocks of dimension $s \times s$ each, where $D[i,j]$ denotes the (i,j)-th block and, for simplicity, we assume that n is an integer multiple of s. Here, the parameter s is chosen so that a slab of size $(n+s) \times s$ can fit into the GPU memory. Routine QR_OOC in Listing 1.1 and Fig. 1 (left) describe how to leverage the upper triangular structure of the bottom $n \times n$ half of D during the computation of the QR factorization of this matrix using our OOC-GPU algorithm. For each iteration of the outer loop, the algorithm first updates (part of) the k-th slab of D w.r.t. the transforms that were calculated earlier (as corresponds to a left-looking variant). These transforms are divided into slabs of width s and applied, in the inner loop, to the corresponding fraction of $D[:,k]$ from the left, invoking routine UPDATE_GPU for that purpose. After the update, the algorithm proceeds to factorize the current slab, using routine QR_HYBRID. Note how, at each outer iteration of this loop, one slab of D is transferred from main memory to the GPU, modified there, and the results are sent back to the main memory.

The code for the building kernels UPDATE_GPU and QR_HYBRID is also given in Listing 1.1, and both procedures are illustrated in Fig. 1 (right). In these routines matrices E and F are partitioned into blocks of size $b \times b$, so that $E[i, j]$, $F[i, j]$ stand for the (i, j)-th blocks of the corresponding matrix. For simplicity, we assume now that s is an integer multiple of b. The first routine operates with F (a slab of D of width s) stored in-core (i.e., in the GPU memory), and streams blocks of E, of width b, from the main memory to the GPU, in order to update F with the orthogonal transforms contained in them. The second routine computes a QR factorization of F (stored in-core), using a conventional blocked left-looking procedure with block size b, so that the block factorizations and orthogonal transforms are computed in the CPU, while the updates of the trailing submatrices are performed in the GPU.

```
1   FUNCTION D = QR_OOC( n, s, b, D );
2   r = n/s;
3   FOR k = 1:r
4      COPY D[ 1:r+k, k ] to GPU
5      FOR j = 1:k-1
6         D[ j:r+j, k ] = UPDATE_OOC( n, s, b, D[ j:r+j, j ], D[ j:r+j, k]);
7      END
8      D[ k:r+k, k ] = QR_HYBRID( n, s, b, D[ k:r+k, k ] );
9      COPY D[ k:r+k, k ] to main memory
10  END
11  //----------------------------------------------------------------
12  FUNCTION F = UPDATE_OOC( n, s, b, E, F );
13  r = n/b; t = s/b;
14  FOR k = 1:t
15     COPY E[ k:r+k, k ], containing Qk, to main memory
16     F[ k:r+k, : ] = Qk' * F[ k:r+k, : ]; // Update in GPU
17  END
18  //----------------------------------------------------------------
19  FUNCTION E = QR_HYBRID( n, s, b, E );
20  r = n/b; t = s/b;
21  FOR k = 1:t
22     COPY E[ k:r+k, k ] to main memory
23     E[ k:r+k, k ] = Rk/Qk = QR( E[ k:r+k, k ] );   // Factorize in CPU
24     COPY E[ k:r+k, k ], containing Qk, to GPU
25     E[ k:r+k, k+1:r ] = Qk' * E[ k:r+k, k+1:r ]; // Update in GPU
26  END
```

Listing 1.1. OOC-GPU left-looking slab-based algorithm for the QR factorization QR_OOC and the building kernels UPDATE_OOC and QR_HYBRID.

Optimization of QR_OOC. In our QR_OOC algorithm, only the orthogonal matrix of the resulting QR factorization of D is built/stored while the upper triangular factor is not referenced/kept. Our QR algorithm is a left-looking algorithm that applies all previous transformations to the current slab —in contrast with the traditional right-looking approach that immediately propagates the transforms to the right of the current slab— since left-looking OOC variants in general incur in a smaller number of transfers [9].

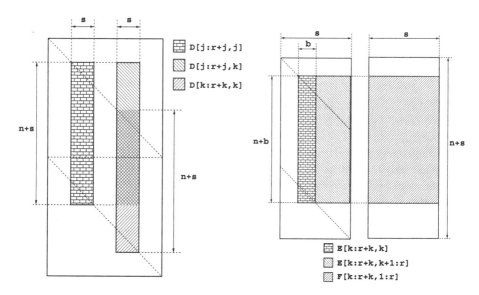

Fig. 1. Accompanying illustrations for the OOC-GPU QR factorization QR_OOC (left), and the building kernels UPDATE_OOC and QR_HYBRID (right).

Furthermore, we leverage the special structure of $D[:,k]$ (Fig. 1, left) to further reduce the number of transfers. Concretely, at each step of the inner loop of routine QR_OOC, $D[j : r + j, k]$, of size $(n + s) \times s$, is stored in the GPU memory. Now, during the next iteration of loop j, $D[j + 1 : r + j + 1, k]$ will be required; see Fig. 1 (left). Thus, the difference between these two slabs corresponds to $D[j, k]$, which does not need to be sent back tó main memory as it belongs to the upper triangular factor; and $D[r + j + 1, k]$, which is stored in main memory and will have to be transferred to the GPU. Therefore, at the end of iteration j, $D[j, k]$ is removed from the GPU memory, as it is not required any longer; and the new block $D[r + j + 1, k]$ is transferred from main memory to the GPU. Applying this approach, at each update step only one block of size $s \times s$ needs to be sent to the GPU, instead of the whole slab of size $(n + s) \times s$.

Optimization of QR_HYBRID. This routine computes the QR factorization of $D[k : r + k, k]$ with the collaboration of both CPU and GPU. This slab is divided into blocks of width b; see Fig. 1 (right). At each iteration of routine QR_HYBRID in Listing 1.1, the orthogonal factor for $E[k : r + k, k]$ is computed on the CPU, and transferred to the GPU; and the submatrix to the right is next updated on the GPU. Thus, the QR factorization computed at each iteration only involves $E[k : r + k, k]$. Following this approach, the special structure of E can be efficiently exploited with little overhead, that depends only on the relation between b and s. (In practice, $b \leq 128$ while s is much larger.)

4   Experimental Results

All the experiments were performed on a server with two Intel Xeon E5520 quad-core processors (total of 8 cores @ 2.27 GHz), 48 Gbytes of RAM, and a Tesla C2050 GPU (2.6 Gbytes of memory, ECC on), using IEEE double-precision arithmetic. The results include the cost of transferring the input data and results between main memory and GPU. The codes were linked to NVIDIA CUBLAS (v5.0) and the BLAS implementation in GotoBLAS2 (v1.13).

For simplicity, we will only consider GPU routines that operate with data residing in the main memory. For matrix decompositions such as the QR factorization and other similar Level-3 BLAS-based kernels, disk latency can be mostly hidden by overlapping it with computation, even in platforms equipped with GPU accelerators [16]. Therefore, we expect these results to carry over to the case where data is stored on disk.

We employed 6 datasets in the experimentation: UTUBSEAM40, UTUBSEAM10, RIBOTIPRE, 1CWP, 1QGT and UTUBSEAM20, leading to eigenproblems of dimension $n = 24,943, 29,622, 30,065, 30,504, 30,785$ and $31,178$, respectively, that in all cases do not fit into the GPU memory.

Our first experiment analyzes the scalability of the OOC-GPU algorithms, measured as the ability of these methods to deliver a constant GFLOPS (billions of flops/second) rate as the problem dimension grows to exceed the capacity of the GPU memory. For this purpose, we employed IMOD to generate matrices of varying dimensions for the UTUBSEAM{10,20,40} benchmarks. Figure 2 shows that the OOC-GPU two-stage and SD&C algorithms are scalable in this sense. At this point, be aware that the much higher GFLOPS ratio of the approach based on the SD&C method do not necessarily imply superior performance since, as we will show in the next experiment, this method also requires a much higher cost than the two-stage alternative.

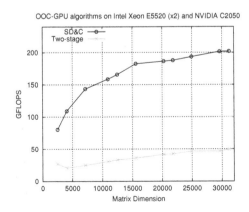

Fig. 2. GFLOPS rate of the OOC-GPU eigensolvers applied to reduced versions of the UTUBSEAM{10,20,40} test cases.

Table 1. Comparison of eigensolvers. Time is reported in seconds in all cases. For the SD&C variants, "#iter" is the number of QDWH iterations and "split" is subproblem size after the first divided-and-conquer step.

Case	Two-stage	SD&C-A			SD&C-B			SD&C-C		
	Time	Time	#iter	split	Time	#iter	split	Time	#iter	split
UTUBSEAM40	1534.3	3087.9	7	8678	2402.8	7	712	2428.2	7	1024
UTUBSEAM10	2536.3	4652.3	7	9006	3871.2	7	936	3877.8	7	1034
RIBOTIPRE	2426.4	5868.2	9	11779	3420.6	6	284	4736.9	7	11448
1CWP	2523.1	5949.8	9	7005	4276.4	7	1412	8264.1	12	16721
1QGT	2622.9	6503.7	10	7362	5525.9	9	1562	9650.2	12	20952
UTUBSEAM20	2780.9	7263.4	10	9288	4521.4	7	815	5937.8	9	2511

Table 1 compares the total execution time of the OOC-GPU two-stage and SD&C algorithms, using the three techniques to choose the splitting parameter σ described in Subsect. 3.1 for the latter. As could be expected, the execution time of the SD&C algorithm strongly varies depending on the properties of the spectrum and the splitting point, and different strategies to select σ greatly affect the convergence speed of the QDWH. For our particular test cases, strategy SD&C-B offers the best results as it combines fast convergence with the decoupling of a subproblem C_1 of reduced size, which renders the cost of the subspace iteration low. However, in all cases, the two-stage approach is clearly superior to the SD&C method.

5   Concluding Remarks

We have presented and evaluated two hybrid CPU-GPU algorithms for the solution of generalized symmetric eigenproblems arising in macromolecular motion simulation, based on the two-stage reduction to tridiagonal form and a new spectral divide-and-conquer approach for the polar decomposition. In both cases, by carefully amortizing the cost of the PCI data transfers with a large number of floating-point arithmetic operations, the implementations attain high performance and, more importantly, offer perfect scalability so that the dimension of the macromolecular problems that can be tackled is not constrained by the capacity of the GPU memory.

Experiments on an desktop platform with two Intel Xeon multicore processors and an NVIDIA "Fermi" GPU, representative of current server technology, illustrate the potential of these methods to address the simulation of biological activity. These results also show the superior performance of the OOC-GPU two-stage approach over the SD&C implementations, despite the former necessarily computes the full eigenspectrum of the problem while the latter can be used, in principle, to obtain only the sought-after part of the spectrum.

As part of future work, we plan to extend these algorithms to operate with data on disk, so that much larger problems can be addressed on desktop platforms with a reduced main memory.

Acknowledgments. D. Davidović's visit to UJI was supported by the COST Action IC0805. The researchers from UJI were supported CICYT TIN2011-23283 and FEDER, the EU FP7 318793 "EXA2GREEN", and P1-1B2011-18 of the Fundació Caixa-Castelló/Bancaixa and UJI. We also thank the Structural Bioinformatics Group, from CSIC, for the datasets.

References

1. Ayton, G.S., Voth, G.A.: Systematic multiscale simulation of membrane protein systems. Curr. Opin. Struct. Biology **19**(2), 138–44 (2009)
2. Bahar, I., Lezon, T.R., Bakan, A., Shrivastava, I.H.: Normal mode analysis of biomolecular structures: functional mechanisms of membrane proteins. Chem. Rev. **110**(3), 1463–97 (2010)
3. Skjaerven, L., Hollup, S.M., Reuter, N.: Normal mode analysis for proteins. J. Mol. Struct. (Theochem) **898**(1–3), 42–48 (2009)
4. Lopez-Blanco, J.R., Garzon, J.I., Chacon, P.: iMOD: multipurpose normal mode analysis in internal coordinates. Bioinformatics **27**(20), 2843–50 (2011)
5. Golub, G.H., Loan, C.F.V.: Matrix Computations, 3rd edn. The Johns Hopkins University Press, Baltimore (1996)
6. Aliaga, J., Bientinesi, P., Davidović, D., Napoli, E.D., Igual, F., Quintana-Ortí, E.S.: Solving dense generalized eigenproblems on multi-threaded architectures. Appl. Math. Comput. **218**(22), 11279–11289 (2012)
7. MAGMA project home page. http://icl.cs.utk.edu/magma/
8. López-Blanco, J.R., Reyes, R., Aliaga, J.I., Badia, R.M., Chacón, P., Quintana, E.S.: Exploring large macromolecular functional motions on clusters of multicore processors. J. Comp. Phys. **246**, 275–288 (2013)
9. Toledo, S.: A Survey of Out-of-core Algorithms in Numerical Linear Algebra. DIMACS Series in Discrete Mathematics and Theoretical Computer Science. American Mathematical Society Press, Providence (1999)
10. Nakatsukasa, Y., Higham, N.J.: Stable and efficient spectral divide and conquer algorithms for the symmetric eigenvalue decomposition and the SVD. Technical Report 2012.52, Manchester Inst. Math. Sci., The University of Manchester (2012)
11. Davidović, D., Quintana-Ortí, E.S.: Applying OOC techniques in the reduction to condensed form for very large symmetric eigenproblems on GPUs. In: 20th Euro. Conf. PDP 2012, pp. 442–449 (2012)
12. Nakatsukasa, Y., Bai, Z., Gygi, F.: Optimizing Halley's iteration for computing the matrix polar decomposition. SIAM J. Matrix Anal. Appl. **31**, 2700–2720 (2010)
13. Dhillon, I.S., Parlett, B.N.: Multiple representations to compute orthogonal eigenvectors of symmetric tridiagonal matrices. Linear Algebra Appl. **387**, 1–28 (2004)
14. Bientinesi, P., Dhillon, I.S., van de Geijn, R.: A parallel eigensolver for dense symmetric matrices based on multiple relatively robust representations. SIAM J. Sci. Comput. **27**(1), 43–66 (2005)
15. Bischof, C.H., Lang, B., Sun, X.: Algorithm 807: the SBR toolbox–software for successive band reduction. ACM Trans. Math. Soft. **26**(4), 602–616 (2000)
16. Quintana-Ortí, G., Igual, F.D., Marqués, M., Quintana-Ortí, E.S., de Geijn, R.A.V.: A run-time system for programming out-of-core matrix algorithms-by-tiles on multithreaded architectures. ACM Trans. Math. Softw. **38**(4), 25:1–25:25 (2012)

Using GPUs for Parallel Stencil Computations in Relativistic Hydrodynamic Simulation

Sebastian Cygert[1], Daniel Kikoła[2], Joanna Porter-Sobieraj[1(✉)], Jan Sikorski[3], and Marcin Słodkowski[3]

[1] Faculty of Mathematics and Information Science,
Warsaw University of Technology, Koszykowa 75, 00-662 Warsaw, Poland
cygerts@gmail.com, j.porter@mini.pw.edu.pl
[2] Department of Physics, Purdue University, 525 Northwestern Ave.,
West Lafayette, IN 47907, USA
dkikola@purdue.edu
[3] Faculty of Physics, Warsaw University of Technology,
Koszykowa 75, 00-662 Warsaw, Poland
slodkow@if.pw.edu.pl, jsikorski@fuw.edu.pl

Abstract. This paper explores the possibilities of using a GPU for complex 3D finite difference computation. We propose a new approach to this topic using surface memory and compare it with 3D stencil computations carried out via shared memory, which is currently considered to be the best approach. The case study was performed for the extensive computation of collisions between heavy nuclei in terms of relativistic hydrodynamics.

Keywords: Finite difference · Riemann solver · MUSTA-FORCE algorithm · Parallel algorithms · CUDA

1 Introduction

Relativistic hydrodynamics is a theory which provides a simple and straightforward solution to many complicated physical problems, for instance in high energy physics, high energy nuclear science and astrophysics [1–4]. Even a complicated, dynamic system can be described with a limited set of relatively simple hyperbolic conservation laws in this framework. All the information regarding the physical process is contained in a single equation of state, which describes the relationship between the thermodynamic properties of a studied system. Assuming the collective fluid system, the knowledge of the details of interactions on the microscopic level is not required. In the case of relativistic hydrodynamics, an accurate representation of relativistic flows and shock waves is crucial for a precise description of many important phenomena, for example jet propagation in nuclear hot matter during heavy nuclei collisions. This requires full 3+1 (the three spatial dimensions + time) dimensional simulations on a larger numerical grid, which are extremely demanding in terms of computing resources.

R. Wyrzykowski et al. (Eds.): PPAM 2013, Part I, LNCS 8384, pp. 500–509, 2014.
DOI: 10.1007/978-3-642-55224-3_47, © Springer-Verlag Berlin Heidelberg 2014

The hydrodynamic simulation is equivalent to solving a set of hyperbolic conservation laws with a given boundary and initial conditions, with additional constraints provided by the equation of state [4,5]. The equations, which describe relativistic hydrodynamic evolution, have a general form of:

$$\frac{\partial \mathbf{U}}{\partial t} + \frac{\partial}{\partial x}(F_x(\mathbf{U})) + \frac{\partial}{\partial y}(F_y(\mathbf{U})) + \frac{\partial}{\partial z}(F_z(\mathbf{U})) = 0, \tag{1}$$

where $\mathbf{U} = [E, M_x, M_y, M_z, R]$ is a vector of conserved quantities in a *laboratory rest frame*, E - energy density, M_x, M_y, M_z - momentum density and R is a charge density; $F_x(\mathbf{U})$, $F_y(\mathbf{U})$ and $F_z(\mathbf{U})$ are fluxes defined as:

$$F_x(\mathbf{U}) = \begin{pmatrix} (E+p)v_x \\ M_x v_x + p \\ M_y v_x \\ M_z v_x \\ R v_x \end{pmatrix} \quad F_y(\mathbf{U}) = \begin{pmatrix} (E+p)v_y \\ M_x v_y \\ M_y v_y + p \\ M_z v_y \\ R v_y \end{pmatrix} \quad F_z(\mathbf{U}) = \begin{pmatrix} (E+p)v_z \\ M_x v_z \\ M_y v_z \\ M_z v_z + p \\ R v_z \end{pmatrix}, \tag{2}$$

and p is a pressure. The equation of state has a general form of:

$$p = f(e, n), \tag{3}$$

where e and n are energy density and charge density in a *rest frame of fluid* (i.e. in a frame where velocity $\mathbf{v}$ is vanishing: $\mathbf{v} = [0, 0, 0]$).

In the numerical applications, all continuous hydrodynamic fields have to be represented as discrete quantities on a numerical grid. In our program, we use a finite-difference scheme on a Cartesian grid for the hydrodynamic simulations. Time evolution in a one-dimensional case for a particular cell i is given by:

$$U_i^{n+1} = U_i^n + \frac{\Delta t}{\Delta x}(F_{i-\frac{1}{2}} - F_{i+\frac{1}{2}}), \tag{4}$$

where U_i^n represents a conserved quantity at the discrete time t_n; Δt and Δx are time and space steps, respectively, and $F_{i-\frac{1}{2}}$, $F_{i+\frac{1}{2}}$, are numerical fluxes through the cell boundaries.

The numerical fluxes are obtained by solving a Riemann problem at each cell boundary [4,5]. We use a second order (in space and time) MUSTA-FORCE [6] approach to obtain $F_{i-\frac{1}{2}}$ and $F_{i+\frac{1}{2}}$. MUSTA-FORCE is a MUlti-STAge predictor-corrector algorithm which uses a relatively simple central scheme called FORCE for solving a Riemann problem in the intermediate steps. The central scheme does not require any assumption to be made about a simulated physical process; therefore it can be used for solving any system of hyperbolic conservation laws. Moreover, this approach provides excellent accuracy and the number of MUSTA steps can be adjusted to obtain a shock wave resolution required for a particular study. However, this approach is more expensive in terms of computing power compared to traditional algorithms. A typical test would use 200–300 cells in each direction (overall up to 27 million cells). Therefore, it is necessary to use parallel processing in order to achieve a reasonable simulation time in the case of (3+1)-dimensional simulations. A promising route towards higher efficiency is computing using graphics processing units, which offer an unprecedented increase in computing power compared to standard CPU simulations.

2   3D Finite Difference Computation on a GPU

2.1   GPU Background

The MUSTA-FORCE algorithm presented in the previous section has been implemented on a GPU with the use of the CUDA parallel programming model [7–9]. The functions executed on a GPU are called *kernels*. Each kernel is executed by launching blocks of threads. Threads from one block run together on the same streaming multiprocessor, each containing streaming processors with on-chip shared memory. All threads have access to common global memory with high access latency. Multiprocessors also contain register memory that guarantees low-latency access that can be completely hidden by the thread scheduler. These registers are partitioned among concurrent threads; when threads perform computations, all the variables are placed by default in registers for as long as they are available. If there is lack of registers, variables awaiting computation are held in local (private for a thread) memory, which is part of the global device memory and which gives the same time penalty for using it. This is called *register spilling* or *register pressure* [8] and slows down the computations. Besides this, a limited amount of shared memory is available for all threads within one block – thus, they can communicate through it. The advantage of using shared memory is that it can be significantly faster than global memory.

Furthermore, CUDA threads are partitioned into *warps*. The threads that comprise a warp are designed to process one common instruction at a time. As a result, when there are conditional sentences in the code and when threads within one warp follow different paths, a warp scheduler needs to issue instructions for all the paths. This unwanted behavior is called branch divergence and therefore, in general, conditional sentences should be used carefully. The warp scheduler selects a warp that is ready to execute its next instruction. This implies that the CUDA program executes efficiently when there are thousands of threads executed in parallel, so the scheduler can hide the memory latencies.

2.2   Related Works

Using graphical processors to speed up computations of 3D finite difference algorithms is a problem that has recently been well addressed in many articles [10–12]. The general approach, called *sliding window*, is to slice the 3D grid into 2D slices and keep the data shared between many threads in shared memory. In Fig. 1 we present a 16×16 data grid and halos that are kept in shared memory. Halos are elements that are required by the border cells to compute their value in the next iteration. Since we use a second order algorithm in space, the width

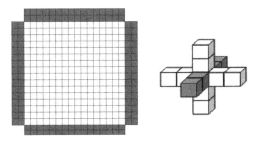

Fig. 1. 16×16 data grid (white) and halos (grey) for the 4th order stencil in shared memory (*left*) and the pattern of data distribution between shared memory (white cells on the xy plane) and registers (grey cells along the z-axis) for a thread for 3D stencil computation (*right*)

of the area of halos is 2. The next important point is also to keep the 2 cells in front and the 2 behind in registers. When the threads are iterating through the z-axis, they shift their registers and load only the next cell.

This approach - although very good - has several drawbacks.

1. *Problems with halos.* When the grid is divided into the blocks there are always some halo elements needed (as seen on the left of Fig. 1) to proceed with the computations. The border threads of the block are used to load these elements. When the size of the grid does not fit perfectly into threadblocks, the last block, of a smaller size, needs to be taken into account during implementation. There is also another case when the size of the last block is smaller than the order of the algorithm - in this case the corresponding threads need to load more cells than normally. This all complicates the code and increases branch divergence, which should be discussed as denoted in Sect. 2.1. Note that the algorithm presented in the appendix of [11] works properly only when all blocks are of the same size.
2. *Data redundancy.* With the presented approach the halo elements will sometimes be read more than once. In such a model, the number of reads per cell is $(n*m+k(n+m))/(n*m)$ [11], where n and m stand for a block's size and k is the order of stencil. For the presented example, with 16×16 data tiles and a 4th order stencil, the read redundancy is equal to 1.5. Furthermore, due to the limited maximum size of shared memory per block (48 KB in contemporary GPUs) in each iteration, the data is read from global to shared memory, read from shared memory and then written back to the global memory, which increases the data redundancy factor by 1. This all means that the overall data redundancy in this example is 2.5.
3. *Relatively heavy usage of registers.* In the finite element method each cell contains a number of variables. When the cell size is multiplied by the number of elements needed to be kept in registers, it may turn out that there are no registers left to proceed with other computations. As denoted in Sect. 2.1, variables not stored in registers are placed in local memory, which is significantly slower.

Despite these drawbacks, the sliding window algorithm is regarded as the best way to approach 3D finite difference algorithms. However, recent advances in CUDA hardware have also opened up the possibility of using another competing method, based on surface memory.

Texture memory was initially designed for graphics usage where thousands of pixels are processed in parallel. Texture memory resides in device memory and is cached in a special cache which is optimized for 2D spatial locality. This means that a read from this memory costs one memory read from global memory on a cache miss, otherwise access to the data is almost instant. However, texture memory allows for only reading operations, and – in devices of compute capability 2.x and higher – surface memory was introduced to handle read/write operations. All this means that texture/surface memory performs well in tasks where there is 2D locality in accessing memory. In particular, solving hydrodynamics equations on a grid suits it perfectly, as well as most 3D finite difference algorithms. Texture memory has already been used for solving fluid dynamics [13–15].

2.3   Implementation Notes

The algorithm presented in Sect. 1 has been implemented on a GPU in two versions - one using shared memory and another using surface memory. The shared memory algorithm is described below.

Algorithm 1. An order-4 stencil computation using shared memory

```
Load data from host to device global memory.
For t_n in 1..N do
    Load two front, a current and one behind cell into registers.
    For z in 3..Z_Dimension-2 do
        Load the second behind cell into register.
        Save a current cell into shared memory.
        For border threads in block do
            Load halos into shared memory.
        Synchronize threads.
        Load neighboring cells from shared memory.
        Synchronize threads.
        Compute cell.
        Write result to global memory.
        Shift registers.
        Synchronize threads.
Send data back to host.
```

It needs to be stressed that the *Compute cell* function is quite complicated. The MUSTA-FORCE algorithm uses many temporary data from interpolated cells. As a result, most variables are sent to local memory because of a lack of registers. Now, due to the operator splitting method we used in integration, and the fact that for most of the interpolated cells the velocity also needs to be computed, this function is called 90 times for a single cell.

Furthermore, in the presented algorithm a cell is a vector of five variables: E, M_x, M_y, M_z and R as mentioned in Sect. 1. Each of these variables is computed separately. When we multiply the size of the cell structure by the number of overall cells – which in our computations reaches millions – it can very clearly be seen that this algorithm is very expensive in terms of both computations and memory usage.

The algorithm using surface memory was built based on the idea of the previous one. One great thing about using surface memory is that implementation becomes extremely simple compared to that with shared memory. Only the way of accessing specific cells is a little more difficult but it can easily be hidden using some functions. The pseudo-code is presented in Algorithm 2.

Algorithm 2. An order-4 stencil computation using surface memory

```
Load data from host to device global memory.
Load data from global memory into surface memory.
For t_n in 1..N do
    For z in 3..Z_Dimension-2 do
        Load neighboring cells from surface memory.
        Synchronize threads.
        Compute cell.
        Write result to surface memory.
        Synchronize threads.
Save data from surface memory to global memory.
Send data back to host.
```

Two surfaces are used in this algorithm; both are used for writing and reading alternately. Because surface memory is part of global memory, in most cases a data grid of almost the same size as in the algorithm using shared memory will fit in the memory. It is very clear that using surface memory simplifies the algorithm. In the main loop the algorithm works using just surface memory, so data redundancy is almost nonexistent and fewer synchronizations are needed. It must be noted that currently surface memory does not support double precision arithmetic. Obviously, if it were necessary, it could be simulated but it would probably negate most of the advantages of using surface memory.

2.4   Results and Analysis

The numerical tests were performed on a PC with an NVIDIA GeForce 610 1 GB graphics card with CC 2.1. The goal of our research was to compare the effectiveness of two implementations: surface and shared memory for solving 3D finite difference algorithms.

The figures present the times for a hydrodynamic simulation for various configurations of input data. The tests were performed for grids of dimensions 60,

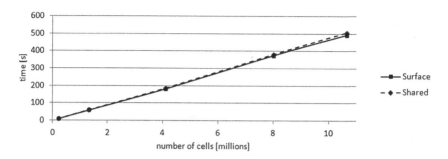

Fig. 2. Execution time for 100 steps of a hydrodynamic simulation using the MUSTA-FORCE algorithm as a function of the total number of cells

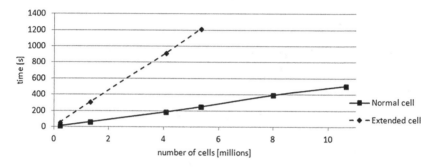

Fig. 3. Execution time for 100 steps of a hydro dynamic simulation using the MUSTA-FORCE algorithm for different sizes of a single cell

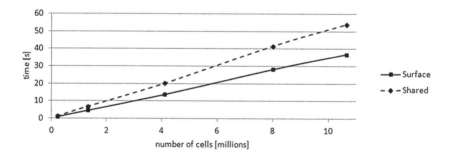

Fig. 4. Execution time for 100 steps of a generalized 3D finite difference algorithm

110, 160, 200, and 220 in each of the three axes and for 100 steps of the algorithm. On the horizontal axis of the graphs the total number of cells in the corresponding grids is presented. The tests were performed using a 16×8 data grid which we found to be the most effective size. 8×8 and 16×16 proved to be only a little slower. Tests on bigger blocksizes (32×16, 32×32) were up to 50 % slower. This was caused by the fact that for bigger blocksizes, a smaller number of blocks could be executed on one streaming multiprocessor in parallel due to the number of register limitations. The maximum grid that fits within the memory limitations for a single cell, including vector $\mathbf{U}$, and thus occupying 20 bytes of memory, was $240 \times 240 \times 240$.

Figure 2 shows that the execution times for both implementations are practically the same with the surface-memory algorithm enjoying a slight advantage. However, we must be aware that some properties of this algorithm affect the results. As was described in Sect. 1, it can be seen that the algorithm is quite difficult, especially because it uses a lot of memory. Due to this fact, we are faced here with quite a big register spill which causes a latency in computations. As a result, more time is spent just loading and saving the variables in local memory instead of doing computations.

To give an overview of how increasing the single cell's size (and thus keeping data in local memory) impacts the time required for simulation, another test was performed. As mentioned in Sect. 1 the algorithm operates on variables in both – the laboratory frame and the fluid rest frame. Initially, it was convenient to keep all this data within a single cell, however while optimizing the code only laboratory frame variables were preserved. Figure 3 presents the timing for both cell sizes using surface memory. The *extended cell* implementation used 46 B instead of 20 B for a single cell. It has to be stressed that in both tests we only keep the laboratory frame variables in surface memory, which means that the *extended cell* does not provide any extra fetching from surface memory. Figure 3 shows that just increasing the size of the cell, without any extra computation cost, causes the algorithm to take about 5 times longer. This confirms the great impact of register pressure on the time needed for computations. Note that the maximum grid size that fit in the memory was just $180 \times 180 \times 180$ and thus the tests could only be performed on a smaller number of cells.

To investigate the impact of register spilling on the timing in our application, we prepared a simplified version of an algorithm for a generalized 3D finite difference method. Instead of using the MUSTA-FORCE algorithm we just compute a simple interpolation between the neighboring cells. The shared memory implementation is about 1.5 times slower than the one with surface memory for a huge number of cells as can be seen in Fig. 4.

To investigate both algorithms in more detail we used Compute Visual Profiler [8] for the simplified algorithm. The most interesting statistic was the number of registers used. The maximum number of registers per thread on the tested graphics card was 63. The surface memory algorithm used 43 of them. The shared memory algorithm used all of them and another 32 bytes were transferred to the local memory. This number fits perfectly our thesis in Sect. 2.2 that the shared memory algorithm is heavily dependent on registers. Another interesting statistic is the control flow divergence [9]. This gives the percentage of thread instructions that were not executed by all threads in the warp, hence causing divergence, which obviously should be as low as possible. In the case of surface memory it is 12.5 % while in the case of shared memory it ranges from 13.5 % to 15.5 %. This is a smaller difference than we expected but surface memory still gives a 15 % relative gain. Another quite important statistic is the occupancy achieved. This ratio provides the actual occupancy of the kernel based on the number of warps executed per cycle. This is the ratio of active warps and active cycles divided by the maximum number of warps that can be executed on a multiprocessor. In the case of surface memory this ratio is equal to 0.45 while in shared memory - 0.32. The rest of the statistics were more or less the same for both implementations.

3   Conclusions

In the paper the possibilities of using surface memory for 3D finite difference algorithms have been examined. To study the performance of this novel idea compared to the shared memory method, which is currently considered to be the best approach for this type of problem, two algorithms using surface memory and one using shared memory were implemented.

The main objective was to investigate the usefulness of surface memory, which seems to be a promising approach for complex 3D finite difference methods. The presented algorithm based on using surface memory provides a number of benefits. There is almost no data redundancy. Only the first and the last iteration use the cells twice, while during the whole computation all the data is kept only in surface memory. It also ensures lower usage of memory per thread which can be considered a great advantage because registers are significantly faster than local memory.

Despite all of these advantages, surface memory turned out to be only slightly faster than the algorithm using shared memory. This is due to the fact that in current graphic cards shared memory is generally the fastest type of memory after the registers. The speed increase gained by using surface memory depends on the characteristics of the algorithm used. For expensive algorithms like MUSTA-FORCE which we used, the register pressure and computation costs cause the time of computations for both algorithms to be comparable. But, as was proved, for algorithms that are not as expensive the usage of surface memory can result in a speed-up of up to 1.5 times.

One important aspect of our algorithm that affects the results is the ratio of computations to shared memory loading. In the case of this algorithm it is very small, the cells are loaded into shared memory, then they are used to compute only one value (in the next timestep), and finally a new batch of cells is loaded into shared memory. If it was possible to perform more computations with one shared memory load, it would decrease the benefits of using surface memory.

Besides these performance related advantages of surface memory, there are also some other benefits. Above all, the implemented code is more general. The code can be easily changed to use any other kind and order of isotropic or anisotropic stencil in any direction.

The studies performed show that surface memory is a promising alternative for shared memory in 3D finite difference algorithms. CUDA technology is relatively new and advances in graphics cards are proceeding very fast. A possible gain in texture cache would definitely increase the benefits of using surface memory. However, an increase in the amount of shared memory would cause the opposite effect. Either way, surface memory proved to be a very competitive alternative for using shared memory in 3D finite difference algorithms an therefore should also be considered in complex stencil computation.

References

1. Adams, J., et al.: (STAR collaboration): experimental and theoretical challenges in the search for the quark-gluon plasma: the STAR Collaboration's critical assessment of the evidence from RHIC collisions. Nucl. Phys. A **757**, 102–183 (2005)
2. Marti, J.M., Muller, E.: Numerical hydrodynamics in special relativity. Living Rev. Relativ. (2003)
3. Duncan, G.C., Hughes, P.A.: Simulations of relativistic extragalactic jets. Astrophys. J. **436**, L119–L122 (1994)
4. Rischke, D.H., Bernhard, S., Maruhn, J.A.: Relativistic hydrodynamics for heavy-ion collisions: general aspects and expansion into vacuum. Nucl. Phys. A **595**, 346–382 (1995)
5. Toro, E.F.: Riemann solvers and numerical methods for fluid dynamics. Springer, Berlin (1997)
6. Toro, E.F.: Multi-stage predictor-corrector fluxes for hyperbolic equations. Isaac Newton Institute for Mathematical Sciences Preprint Series NI03037-NPA, University of Cambridge, UK (2003)
7. CUDA C Best Practices Guide, NVIDIA Corporation (2012)
8. NVIDIA Corporation: Compute Visual Profiler User Guide
9. NVIDIA Corporation: NVIDIA CUDA Programming Guide Version 5.0 (2012)
10. Zumbusch, G.: Vectorized higher order finite difference kernels. In: Manninen, P., Öster, P. (eds.) PARA 2012. LNCS, vol. 7782, pp. 343–357. Springer, Heidelberg (2013)
11. Micikevicius, P.: 3D finite difference computation on GPUs using Cuda. In: Proceedings of the 2nd Workshop on General Purpose Processing on Graphics Processing Units (2009)
12. Michéa, D., Komatitsch, D.: Accelerating a 3D finite-difference wave propagation code using GPU graphics cards. Geophys. J. Int. **182**(1), 389–402 (2010)
13. Brandvik, T., Pullan, G.: Acceleration of a two-dimensional Euler flow solver using commodity graphics hardware. Proc. Inst. Mech. Eng., Part C.: J. Mech. Eng. Sci. **221**(12), 1745–1748 (2007)
14. Elsen, E., LeGresley, P., Darve, E.: Large calculation of the flow over a hypersonic vehicle using a GPU. J. Comput. Phys. **227**(24), 10148–10161 (2008)
15. Phillips, E., Fatica, M.: Implementing the Himeno benchmark with CUDA on GPU clusters. In: IEEE International Parallel & Distributed Processing Symposium, pp. 1–10 (2010)

Special Session
on Multicore Systems

PDNOC: An Efficient Partially Diagonal Network-on-Chip Design

Thomas Canhao Xu[(✉)], Ville Leppänen, Pasi Liljeberg,
Juha Plosila, and Hannu Tenhunen

Department of Information Technology, University of Turku, 20014 Turku, Finland
canxu@utu.fi

Abstract. With the constantly increasing of number of cores in multi-core processors, more emphasis should be paid to the on-chip interconnect. Performance and power consumption of an on-chip interconnect are directly affected by the network topology. The efficiency can also be optimized by proper mapping of applications. Therefore in this paper, we propose a novel Partially Diagonal Network-on-Chip (PDNOC) design that takes advantage of both heterogeneous network topology and congestion-aware application mapping. We analyse the partially diagonal network in terms of area usage, power consumption, routing algorithm and implementation complexity. The key insight that enables the PDNOC is that most communication patterns in real-world applications are hot-spot and bursty. We implement a full system simulation environment using SPLASH-2 benchmarks. Evaluation results shown that, the proposed PDNOC provides up to 25 % improvement in execution time over concentrated mesh, and 3.6x better energy delay product over fully connected diagonal network.

Keywords: Multicore · 3D Chip · Heterogeneous · Network-on-Chip

1 Introduction

In recent years, Chip Multiprocessors (CMPs) have been becoming widely used due to the constraint of chip clock frequency and power consumption. As the number of cores increases, Network-on-Chip (NoC) was proposed as a highly scalable solution to address the communication problems in CMPs [1,2]. The communication methodologies of traditional computer networks are used for on-chip communication. Despite the fact that NoC improves scalability, it also contributes to communication latencies. For example, latency caused by complex router structure and routing algorithm has become a more prominent factor affecting system performance. Moreover, the power consumption and area of routers are becoming more costly overheads. On-chip networks should provide both low latency and high bandwidth that efficiently support various workloads. Several NoC topologies have been proposed, including ring, star, mesh, torus, tree, fat tree, butterfly etc. A topology defines how nodes are connected, which

R. Wyrzykowski et al. (Eds.): PPAM 2013, Part I, LNCS 8384, pp. 513–522, 2014.
DOI: 10.1007/978-3-642-55224-3_48, © Springer-Verlag Berlin Heidelberg 2014

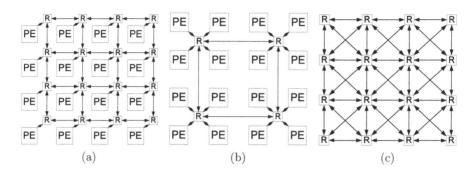

(a)   (b)   (c)

Fig. 1. Three NoC topologies, PEs in (c) are not shown for clarity.

will affect the performance of a network [3]. Mesh (Fig. 1a) has been the most popular topology for on-chip network due to the simple grid-type shape and regular structure. Figure 1a depicts a NoC containing 16 nodes/tiles arranged in mesh topology. Each node consists of a Router R and a Processing Element PE. The PE accommodated a network interface, processor core and related cache. Routers are connected to each other with four cardinal links. Network messages are generated by PEs, serialized and transmitted by routers via links.

The diameter of mesh network increases rapidly with larger networks, degrading system performance and power consumption. Researchers have proposed alternative topologies that can offer high performance or provide reduced area and power consumption. Concentrated mesh (CMesh, Fig. 1b) is proposed to provide low latency for networks with low injection rates [4]. The CMesh network alleviates the scalability problem by sharing the router among multiple network nodes. Concentration degree is defined as the number of nodes sharing a router. By reducing the number of routers, the hop count between nodes are greatly decreased, resulting lower latency in point-to-point communication. Since the router is shared by several nodes, CMesh saturates much earlier than regular mesh under high injection rates [5,6]. A diagonally linked mesh network DMesh is presented in [7] (Fig. 1c). Diagonal links are added to the mesh network, reducing the distance between diagonal nodes and alleviating traffic congestion in the network. It is demonstrated that DMesh improves average latency and saturation traffic load on mesh networks [8]. However, the extra links and high power consumption from complex routers can lead to lower efficiency compared with regular mesh networks. In addition, the overall utilization of DMesh network can be insufficient due to the unbalanced traffic pattern of applications.

In this paper, we propose a Partially Diagonal Network-on-Chip (PDNOC) design focusing on balancing performance and power consumption. We first investigate the traffic pattern of applications. It is discovered that despite the traffic of most applications being self-similar, they appear to contain a common hot-spot and bursty patterns. We explore Partially Diagonal (PD) networks in a mesh. The router design, routing algorithm, number of PD networks and placement of PD networks are analysed. We further propose a congestion-aware mapping algorithm that maps nodes with hot-spot and bursty traffic to PD networks. A full system simulation environment is implemented to verify our design, and compare with other designs. The experiments with 64-core CMP show that the PDNOC provides comparable performance with DMesh while dramatically improved system efficiency.

2 Motivation

To assess the performance of NoCs, different traffic patterns, e.g. synthetic and realistic, can be used. Synthetic traffic patterns, such as uniform random, transpose and hotspot, are abstract models of messages passing in the network. Realistic traffic patterns are traces of real applications running on a NoC-based system. Since NoCs are designed to execute programmes, the most accurate method to evaluate the characteristics of the NoC should be based on realistic application traffic. For this reason, we use the traces of applications to analyse the traffic patterns of the network. The simulation environment and system configuration are described in Sect. 4.

The network injection rate of 64 cores running LU matrix decomposition is illustrated in Fig. 2. Notice that there are several spikes that exceed the limit of the Z-axis. In uniform random traffic pattern, each node should generate around 1.56 % (1/64) traffic. However the traffic trace for LU application shows hotspots and bursty patterns. During the 119.3 M executed cycles, total of 18.99 M messages are transmitted by network nodes. Notice that the time of X-axis is represented as the percentage of total executed cycles, i.e. one percent correspond to 1.19 M-cycles. It is revealed that, 32.4 % of traffic are concentrated on five nodes or PEs (PE0 13.6 %, PE7 6.1 %, PE8 6.1 %, PE37 3.4 % and PE10 3.2 %). The concentrated traffic introduced heavy hot-spot traffic in certain regions of the NoC. Moreover, a small portion of source-destination pairs generated a considerable amount of the traffic. For example, the traffic amount between node 0–38 is 0.791 %, or 33.0x higher than the average traffic volume[1]. Besides hot-

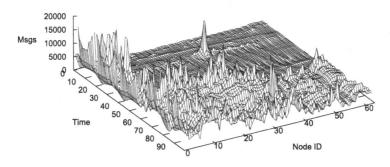

Fig. 2. Network injection rate for 64-core NoC running LU.

---

[1] Considering $64^2 = 4096$ source-destination pairs, each pair should contribute 0.0244 % traffic in uniform random traffic.

spot traffic, we notice the injection rate changes rapidly over time, displaying a bursty pattern. Furthermore, communication peak can be observed, where most of the nodes are actively sending and receiving messages, after around 60 % of the execution time. Despite the fact that we analyse only one application here, we discovered the communication pattern of most applications shows similar behaviour. While fully connected diagonal mesh network can alleviate traffic contention, most of the network resources are wasted due to low overall utilization, i.e. there are only a few hot-spot nodes. Based on this insight, we propose an optimized partially diagonal network design.

3 · The PDNOC Design

PDNOC is designed to be a highly efficient NoC optimized for hot-spot and bursty traffic patterns. Our design leverages several insights: first, the concept of partially diagonal network is introduced and the number and placement of the partially diagonal networks are analysed; second, a task mapping algorithm is proposed for PDNOC to achieve higher performance and efficiency.

3.1   Partially Diagonal Network

The main concept of the partially diagonal network is illustrated in Fig. 3. As is shown in Fig. 3a, node C (*Central*) provides one extra diagonal connections to each of the D (*Diagonal*) nodes. Network congestion should be alleviated in this region due to the extra resources. Furthermore the minimal hop counts of node C for accessing adjacent nodes are reduced as well (Fig. 3b). Based on the fundamental design, we analyse one and two partially diagonal networks in a 8×8 mesh, and the effect of different placements of two networks (Fig. 3c). Comparing with the fully connected diagonal network (Fig. 1c), our design requires less links and smaller routers, which can provide higher efficiency. For example, regular routers in a mesh network have 5 links (4 cardinal and 1 local), each configured with several buffers. The crossbar and buffers can consume over 80 % of router area and power [9]. In [7], the routers of the DMesh design are configured with three sub-routers and two crossbars, each 6×5. There are totally 13 links in each DMesh router, making the router much larger than the conventional mesh router. Meanwhile, extra router delay should be considered with increased number of crossbar and additional sub-routers. Similarly, the routers of the CMesh design with a concentration degree of 4 are implemented with 4 cardinal links and 4 local links. Therefore the size, power consumption and latency are higher than for the mesh router. We analyse our design in terms of routing algorithm and network overhead.

Since there are extra diagonal links, the routing algorithm should be modified accordingly. XY deterministic routing is widely used in the mesh network, in which a flit is first routed to the X direction and last the Y direction. In PDNOC, all regular nodes (nodes without any diagonal connection, i.e. white nodes in Fig. 3a) still apply XY routing. For D nodes, if the destination of a flit is C,

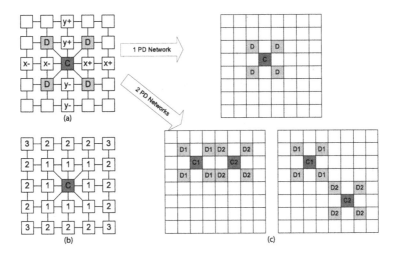

Fig. 3. Partially diagonal network of 5×5 mesh (a), minimal distance from C node (b), different numbers and placements of partially diagonal networks in 8×8 mesh (c) (upper - one network "PDNOC-1", lower left - two adjacent networks "PDNOC-2A", lower right - two interleaving networks "PDNOC-2I").

then it will be transferred via diagonal links, otherwise it will follow XY routing. For C node: if the destination of a flit is directly connected (e.g. nodes with 1 hop in Fig. 3b), then it will be transferred directly; otherwise if the destination is in the $x - /x + /y - /y+$ direction, then XY routing is applied; finally if the destination is in the $x - y+$ direction, then the flit will be sent to $x -$ or the diagonal link of $x - y+$ in a round-robin sequence. Notice that this strategy helps balancing the load between two links. However the routing path is not always minimal. The rest $(x + y+, x + y -$ and $x - y -)$ may be deduced by analogy. Deadlock is avoided since there is no cyclic dependence in the routing path. Livelock is also avoided by using multiple virtual networks.

To calculate the overhead from the routers with diagonal links, we model and simulate the area and power consumption of various router schemes. All routers are modelled under 32 nm processing technology, with 3.0 GHz operating frequency and 1 V voltage. Regular routers with five links consume 0.865 W power when active, and occupy 0.255 mm² area. The D routers have one additional link, which will result in larger crossbar and related buffer area. Moreover the routing logic is more complex. Simulation results show that the area and power consumption of D routers are 0.355 mm² and 1.251 W, 39.2 % and 44.6 % higher than the regular router, respectively. On the other hand, router C is connected with 9 links (8 directions, plus 1 to the PE), requiring the expansion of the crossbar, virtual channels and related buffers. Results revealed that the area and power consumption of C router are 0.749 mm² and 2.616 W respectively, around three times higher than the baseline standard mesh router. Considering 1 partially diagonal network, the overhead of area and power consumption of a 8×8 network are 5.47 % and 5.95 % respectively. Two partially diagonal networks will results higher overhead on area and power consumption, 10.96 % and 11.9 % respectively. It is noteworthy that a full 8×8 diagonal network consumed 2.91x and 2.78x of area and power than the regular mesh network. Therefore the overhead is relatively small in our design.

3.2 Mapping Algorithm

In consideration of the high traffic count between nodes, we further propose a congestion-aware application mapping algorithm for PDNOC. The mapping strategy of applications is a major factor leading to communication congestion. Network latency, resource utilization, execution time and power consumption are affected by application mapping. For example, as aforementioned, the traffic amount between nodes 0–38 is 33x higher than the average traffic volume. However, the Manhattan Distance (MD) [10] between the two nodes is 10 hops in the mesh network with XY routing. System performance can improve provided that the two nodes are mapped close to each other. Considering Fig. 3b, by mapping the hot-spot node to C, and mapping other 8 nodes adjacently to C such that the 8 nodes are selected based on the communication volume with the hot-spot node, the average communication delay among hot-spot node and other 8 nodes are reduced significantly. For example, we discovered node 0 for LU is a hot-spot node, therefore it will be mapped to a C node. Other 8 nodes (38, 14, 29, 15, 8, 28, 36 and 58) are mapped adjacently to node 0 starting with D nodes. Next, 12 nodes with minimum hop count of 2 are selected for mapping according to the same evaluation criteria. If there are 2 partially diagonal networks, then two nodes with highest amount of injection traffic will be selected and mapped to the C nodes at first.

The pseudo code of the mapping algorithm is shown in Algorithm 1. It is noteworthy that two mapping regions can overlap with each other (cf. Fig. 3c). To solve this problem, we first map PD_1 with adjacent nodes of MD=1, once the adjacent nodes are full, the algorithm moves the PD_2 with MD=1 and so forth. It is noteworthy that the result from the mapping algorithm is an optimized mapping, not necessarily optimal.

4 Experimental Evaluation

4.1 Experiment Setup

We evaluate and compare PDNOC with several other NoC designs. The simulation platform is based on a cycle-accurate NoC simulator (GEMS/Simics [11,12]). The source code of GEMS is modified to fit the simulation requirements. We implement a multi-core processor with 64 (8×8) Sun UltraSPARCIII+ cores. Each core is running at 2 GHz, and equipped with private L1 cache (split I + D, 16 KB + 16 KB, 4-way associative, 64-byte line, 3-cycle access delay). The 16MB shared L2 cache is divided into 64 banks/slices, each 512 KB.

Algorithm 1. Mapping algorithm for PDNOC.

Input: The number of partially diagonal networks N_{PD}, network injection
traffic of nodes N_{inj}, node-node communication volume T,
Output: A mapping region R

Sort N_{inj} according to network injection rate
Sort T according to communication volume
Select hot-spot nodes HN_i from N_{inj} according to N_{PD}
foreach HN_i **do**
    foreach *traffic* $T_i \in T$ **do**
        Search node-node communication for HN_i in T_i according to
        communication volume
        Map related free nodes adjacent (MD=1,2,3...) to the HN_i in terms of
        T_i
        Move to the next HN_i if adjacent nodes with smallest MD are full
        Loop until the sub region R_i is full
        Place R_i to R
    end
end

The simulated processor configuration is similar as modern commercial NoCs,
e.g. Tilera TILE [2]. We implement static non-uniform cache architecture [13] for
the memory/cache architecture. MOESI, a two-level distributed directory cache
coherence protocol that has been used widely in modern multi-core processors
[14], is selected for the target system. The routers and links of various designs are
modelled accurately, e.g. each router includes routing computation unit, virtual
channel locator, switch allocator, crossbar switch and related input buffers. We
model delays of links according to their length, and delays of different router
designs according to the router area and complexity of pipeline stages. The
routers and links are running at 3 GHz. Orion2 for on-chip networks is used
to evaluate detailed power characteristics of routers and links [15]. Simulations
are run on Solaris 9 based on UltraSPARCIII+ instruction set in-order issue
structure. Workloads used here are selected from SPLASH-2 [16].

4.2   Performance Analysis

Here we compare the performance and efficiency of PDNOC with a conventional
mesh network (*Mesh* in figures), a fully connected diagonal mesh network [7]
(*DMesh* in figures) and a concentrated mesh network [5] (*CMesh* in figures).
To further analyse the number and placement of the partially diagonal network, we classify the PDNOC into three groups: the first with one partially
diagonal network (*PDNOC-1* in figures), the second with two adjacent partially
diagonal networks (*PDNOC-2A* in figures) and the third with two interleaving
partially diagonal networks (*PDNOC-2I* in figures). The performance is measured in terms of execution time and energy delay product. The energy delay
product is a combination of average network latency and power consumption

of routers and links. Average network latency represents the average number of cycles required for successful transmitting all network messages. The number of cycles of each message is calculated as, from the injection of the message header into the network at the source node, to the reception of the tail flit at the destination node. Systems with a lower energy delay product generally have a better trade-off between the performance and power. To eliminate the effect of arbitrary application mapping in *Mesh*, *DMesh* and *CMesh*, we utilize the same mapping method from *PDNOC-2I* for these topologies. The full system simulation results are illustrated in Fig. 4 and 5, normalized to the *Mesh*.

As Fig. 4 illustrates, the *PDNOC-2I* provides the best overall execution time. Compared with regular mesh network, the average execution time of applications has reduced 9.8 %, 13.5 % and 16.4 % in *PDNOC-1*, *PDNOC-2A* and *PDNOC-2I* respectively. The main reason for these improvements is the additional processing capabilities of routers and links in the areas with hot-spot and bursty traffic. It is noteworthy that the execution times of *PDNOC-2I* are better than that in *DMesh*. The reason being that the additional latencies from the extra sub-routers and crossbar in the more complex *DMesh* routers. We notice that applications with higher overall network injection rate, more hot-spot nodes and more intense bursty traffic, e.g. Cholesky, FFT, LU and Water-Nsq, benefit more from the PDNOC. Increased number of PD network provides improved performance, and the interleaved placement of PD networks performed better than the adjacent placement. On the other hand, *CMesh* suffered from insufficient bandwidth among hot-spot nodes. Results revealed that for all 8 applications, the average execution time of *CMesh* is 11.4 % longer than *Mesh*. The clustered organization of nodes in *CMesh* makes a router shared by 4 nodes. This design is beneficial for network with low loads, due to the reduced delay by eliminating intermediate routers. However the scalability of *CMesh* is much worse than other architectures. Furthermore, intuitively, although hot-spot nodes should be

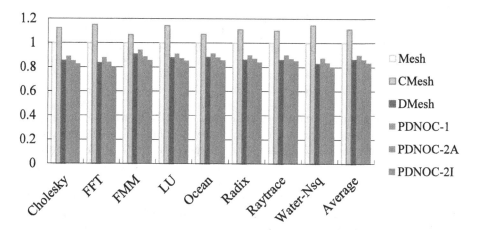

Fig. 4. Normalized execution time with PDNOC designs and other NoC designs.

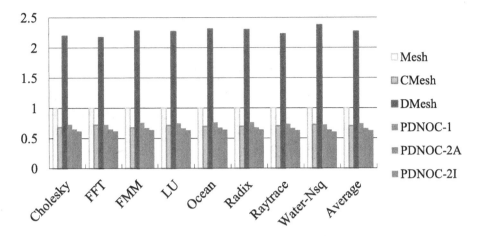

Fig. 5. Normalized energy delay product with PDNOC designs and other NoC designs.

mapped within the same cluster in *CMesh* for better performance, it is questionable whether the router can satisfy the bandwidth requirements.

Considering the energy delay product, *CMesh* is obviously very power-efficient, 29.7 % better than the regular mesh network on average. Notice that only 16 routers are used in *CMesh*, however each router is much larger and consumes more power than other router designs. For *DMesh*, despite the fact that it shows promising performance in terms of execution time, the power consumption is disappointing. The energy delay product of *DMesh* is 2.27x higher than in the *Mesh* for 8 applications. The overhead is mainly from three sub-routers and two crossbars of each *DMesh* router. Taking the 64 *DMesh* routers into account, we calculated that the overall power consumption reaches as high as 160W, implying that it may not be a viable solution for scalable, efficient NoCs. Our partially diagonal network designs demonstrated significantly improved energy delay product compared with *DMesh*. On average, compared with *Mesh*, this metric improved 25.6 %, 33.7 % and 37.1 % for *PDNOC-1*, *PDNOC-2A* and *PDNOC-2I* respectively. This is primarily due to the reduced average network latency. Although PDNOC designs consume more power than *Mesh* and *CMesh*, they provide further gains on network latency. The trade-off from power consumption to performance is worthy in our design.

5   Conclusion

Network-on-Chip (NoC) is expected to be a promising candidate for optimizing the on-chip communication of multicore processors. However, performance and power consumption of an on-chip interconnect are affected by the network topology. We proposed Partially Diagonal NoC (PDNOC) in this paper. The communication traces of several typical applications were analysed. We discovered that the traffic of most applications shows hot-spot and bursty patterns. Based on this insight, we attempted to design an efficient network topology and a congestion-aware mapping algorithm. The partially diagonal network consisted of a central router, several diagonal routers and links. We studied different placements and numbers of the proposed approach. The PDNOC was optimized for high efficiency, that the performance was improved considerably by slightly sacrificing power consumption. Full system evaluation results shown that, on average for 8 applications, the proposed PDNOC with two interleaved networks provided 25 % improvement in execution time over concentrated mesh, and 3.6x better energy delay product over fully connected diagonal network.

References

1. Jantsch, A., et al.: Networks-on-Chip. Kluwer Academic Publishers, Norwell (2003)
2. Bell, S., et al.: Tile64 - processor: a 64-core soc with mesh interconnect. In: IEEE International Solid-State Circuits Conference, ISSCC 2008, Digest of Technical Papers, pp. 88–598 (2008)
3. Agarwal, A.: Limits on interconnection network performance. IEEE Trans. Parallel Distrib. Syst. 2(4), 398–412 (1991)
4. Balfour, J., et al.: Design tradeoffs for tiled cmp on-chip networks. In: Proceedings of the 20th ICS, pp. 187–198. ACM, New York (2006)
5. Das, R., et al.: Design and evaluation of a hierarchical on-chip interconnect for next-generation cmps. In: Proceedings of the 15th HPCA, pp. 175–186, February 2009
6. Kim, J., et al.: Flattened butterfly topology for on-chip networks. Comput. Archit. Lett. 6(2), 37–40 (2007)
7. Wang, C., et al.: Design and evaluation of a high throughput qos-aware and congestion-aware router architecture for network-on-chip. In: Proceedings of the 20th PDP, pp. 457–464 (2012)
8. Wang, C., et al.: Congestion-aware network-on-chip router architecture. In: Proceedings of the 15th CADS, pp. 137–144 (2010)
9. Kim, J.: Low-cost router microarchitecture for on-chip networks. In: 42nd Annual IEEE/ACM International Symposium on Microarchitecture, MICRO-42, pp. 255–266 (2009)
10. Dally, W.J., et al.: Route packets, not wires: on-chip inteconnection networks. In: Proceedings of the 38th DAC, pp. 684–689, June 2001
11. Magnusson, P., et al.: Simics: a full system simulation platform. Computer 35(2), 50–58 (2002)
12. Martin, M.M., et al.: Multifacet's general execution-driven multiprocessor simulator (gems) toolset. Comput. Archit. News 33, 92–99 (2005)
13. Kim, C., et al.: An adaptive, non-uniform cache structure for wire-delay dominated on-chip caches. In: Proceedings of the 10th ASPLOS, pp. 211–222. ACM, New York (2002)
14. Patel, A., et al.: Energy-efficient mesi cache coherence with pro-active snoop filtering for multicore microprocessors. In: Proceeding of the 13th ISLPED, pp. 247–252, August 2008
15. Kahng, A.B., et al.: Orion 2.0: a fast and accurate noc power and area model for early-stage design space exploration. In: Proceedings of the DATE '09, pp. 423–428 (2009)
16. Woo, S.C., et al.: The splash-2 programs: characterization and methodological considerations. In: Proceedings of the 22nd ISCA, pp. 24–36, June 1995

Adaptive Fork-Heuristics for Software Thread-Level Speculation

Zhen Cao$^{(\boxtimes)}$ and Clark Verbrugge

School of Computer Science, McGill University, Montréal, QC H3A 2A7, Canada
zhen.cao@mail.mcgill.ca, clump@cs.mcgill.ca

Abstract. Fork-heuristics play a key role in software Thread-Level Speculation (TLS). Current fork-heuristics either lack real parallel execution environment information to accurately evaluate fork points and/or focus on hardware-TLS implementation which cannot be directly applied to software TLS. This paper proposes adaptive fork-heuristics as well as a feedback-based selection technique to overcome the problems. Adaptive fork-heuristics insert and speculate on all potential fork/join points and purely rely on the runtime system to disable inappropriate ones. Feedback-based selection produces parallelized programs with ideal speedups using log files generated by adaptive heuristics. Experiments of three scientific computing benchmarks on a 64-core machine show that feedback-based selection and adaptive heuristics achieve more than 88 % and 50 % speedups of the manual-parallel version, respectively. For the Barnes-Hut benchmark, feedback-based selection is 49 % faster than the manual-parallel version.

Keywords: Software thread-level speculation · Fork heuristics · Automatic parallelization · Performance tuning

1 Introduction

Thread-level speculation (TLS) is a safety-guaranteed approach to automatic or implicit parallelization. Speculative threads are optimistically launched at *fork points*, executing a code sequence from *join points* well ahead of their parent thread. Safety is preserved in this speculative model by buffering reads and writes of the speculative thread. Once the parent thread reaches the join point the latter may be *joined,* committing speculative writes to main memory and merging its execution state into the parent thread, provided no read conflicts have occurred. In the presence of conflicts the speculative child execution is discarded or rolled back for re-execution by the parent.

The selection of fork/join points plays a key role in the performance of TLS, especially for software implementations as a result of higher overhead than its hardware counterpart. So far there are three sorts of fork-heuristics: static heuristics [5], static profiling heuristics [4,10,11] and dynamic profiling heuristics [9]. The first build mathematical cost-benefit models of speculative execution using

R. Wyrzykowski et al. (Eds.): PPAM 2013, Part I, LNCS 8384, pp. 523–533, 2014.
DOI: 10.1007/978-3-642-55224-3_49, © Springer-Verlag Berlin Heidelberg 2014

compile-time program information, and use the models to predict profitable fork/join points. This approach has the limitation that some model parameters, such as thread dependency probability and iteration count of nested loops, are unknown at compile-time, which in turn limits its effectiveness and application. The second heuristics compile and run the sequential program, collect profiling execution traces, and then use the traces to determine the best fork/join points. The drawback of this approach is lack of real parallel execution environment information, which limits accuracy of the fork point selection decision. The third approach is more promising for real estimation, but is currently based on hardware implementation, which is inappropriate and cannot be directly applied to software TLS.

This paper proposes adaptive fork-heuristics to solve the above problems. Adaptive heuristics are dynamic profiling heuristics for software TLS, which insert all potential fork/join points into the speculative program and rely entirely on the runtime system to determine profitable fork/join points and disable inappropriate ones. Since fork/join points are evaluated during real speculative parallel execution, all necessary information such as the thread conflict ratio and thread execution time is available, enabling accurate estimation of cost-benefit of each thread and thus each pair of fork/join points. On-the-fly fork/join point selection also eliminates the requirement of profiling runs and enables adaptation to different fork/join points for different input data. Our investigation demonstrates feasibility of this approach, as well as providing concrete data on actual performance in a realistic thread-level speculative system.

2   Related Work

The bulk of proposed fork-heuristics are static profiling heuristics. Java runtime parallelizing machine (Jrpm) [4], for instance, first profiles execution of a sequential program with a hardware profiler, and then dynamically speculates on the selected prospective loops after collecting enough profiling data to decide the best loops to parallelize. Du et al. [6] proposed a cost-model-driven compilation framework SPT to select candidate loops for speculative parallelization, which builds control-flow graphs and data-dependence graphs with profiling information of a sequential execution and uses the graphs to evaluate candidate loops based on the cost model. The STAMPede [11] TLS approach selects speculatively parallel loops based on several filter criteria: the loop execution coverage and iteration count are above a threshold and the loop body is neither too large nor too small. The Mitosis [10] compiler/architecture uses arbitrary pairs of basic blocks as fork/join points and models parallel execution based on profiling traces to estimate candidate pairs. The POSH [7] compiler simulates sequential execution on a train input set and models TLS parallel execution to select beneficial fork/join points.

There is also research dedicated to static profiling heuristics. Whaley and Kozyrakis [13] proposed three classes of heuristics for method-level speculation, and found that single-pass heuristics lead to best speedups while simple/complex multi-pass heuristics tend to over/under speculation. Wang et al. [12] constructed a loop-graph and used it for global loop selection to maximize program performance. Liu et al. [8] proposed an online-profiling approach to speculatively parallelize candidate loops. Online static profiling approaches [4,8] have the advantage over offline-profiling that they do not require additional profiling input and can dynamically profile on the real data. However, these still lack parallel execution environment information for accurate estimation of fork/join points. Pure static heuristics are less common, since they lack runtime parameters. Dou and Cintra [5] proposed a thread-tuple cost-model to estimate speedups of candidate loops. As well, some heuristics combine profiling and static approaches, such as SPT [6].

Dynamic profiling heuristics have recently been studied. Luo et al. [9] proposed a dynamic performance tuning technique for selection of candidate loops. It used hardware performance monitors to profile runtime statistics such as instruction fetch penalty and cache miss, and estimated the efficiency of each thread and loop with the statistics. Unfortunately, this approach cannot be directly applied to software TLS without low-level and machine-specific access to hardware performance monitors.

All the above are hardware-TLS heuristics, which tend to focus on finer-granularity parallelism due to hardware resource constraints. In software-TLS, heuristics should focus on coarser-granularity parallelism as software TLS has higher overhead than hardware implementation. So far as we know, the adaptive heuristics we propose are the first heuristics specifically proposed for and validated in software TLS.

HEUSPEC [14] is a software speculation parallel model that dynamically adapts to different value predictors and granularity tasks. While adaptive fork-heuristics target the problem of fork point selection.

3   Adaptive Fork-Heuristics

Adaptive fork-heuristics add potential pairs of fork/join points to the speculative program, evaluate the cost-benefit of each pair during parallel execution and disable unprofitable ones. The design involves three aspects: (1) what the potential fork/join points are, (2) how to estimate the cost-benefit of each pair of fork/join points, and (3) how to disable fork points.

3.1   Potential Fork/Join Points

The potential fork/join points of the design are loop iterations and function (method) calls, since loops usually take the majority of program's execution time and function calls usually represent independent computation tasks. They are also the choice of most other TLS works, known as loop-level speculation and method-level speculation, respectively.

We also apply two optimizations for each loop: "blockize" and "end-barrier." Suppose there are n processors. Blockize splits the loop iterations into n blocks, which in turn avoids creating too many small threads. The exception is loops with a small constant number of iterations, which do not need this optimization. The end-barrier optimization adds a barrierpoint just after the end of the loop. This optimization is beneficial because loops usually have dependency with their continuation, particularly for loop nests, in which case an inner loop thread may cause cascading rollbacks of the outer loop threads.

In this implementation, we add directives of fork/join/barrier points and perform the two optimizations manually. This is a limitation of our prototype—both the adding of fork/join/barrier points annotation and the optimizations can be automated by compiler transformation, enabling full automatic parallelization.

3.2   Cost-Benefit Estimation

The design uses a cost-benefit model to evaluate the profitability of each thread and each pair of fork/join points. The model assumes a constant time $T_{overhead}$ of overhead (thread creation, cache miss, buffering, etc.) for each thread. Although this is an inaccurate approximation since threads with different memory access frequencies have different buffering overhead, we find it works well for our estimation, partly because we are only concerned with whether a thread is profitable, and not how profitable it is.

The runtime $T_{t,run}$ of a speculative thread t comprises two parts: work time $T_{t,work}$ and synchronization/validation/commit/rollback time, which are available through timing. If thread t commits, its cost-benefit is estimated as $\eta_t = T_{t,work}/(T_{t,run} + T_{overhead})$. If it rolls back, its cost-benefit is 0. Given a minimum cost-benefit threshold $\eta_{threshold}$, if $\eta_t < \eta_{threshold}$, then thread t is considered not profitable and should not have been speculated.

If the assumption holds that threads speculated at the same fork point always show similar behaviour (they always commit/rollback and have similar work time / runtime ratio), then we can directly use η_t to estimate the cost-benefit of the fork point. However, the assumption does not hold, even for fork/join points speculating independent loop iterations. The reason is that at the beginning of program execution many threads with dependencies are speculated, causing nondeterministic rollbacks.

We propose 3 independent mechanisms to address this issue: global hint, local hint and interval hint. Global hint uses at least N_{warmup} threads instead of one thread to determine the cost-benefit of a pair of fork/join points. When a thread completes execution, its runtime plus overhead is accumulated to the runtime T_{run} of the pair $T_{run} = T_{run} + T_{t,run} + T_{overhead}$. The exception is when it is cascadingly rolled back, as a cascading rollback does not represent the cost-benefit of a thread. If it commits, its work time is accumulated to the work time T_{work} of the pair $T_{work} = T_{work} + T_{t,work}$. After $N >= N_{warmup}$ threads completes, the cost-benefit of the pair is then estimated as $\eta = T_{work}/T_{run}$. The hint disables the fork point if the cost-benefit is below a threshold. For local hint, if a thread decides not to speculate on a fork point then none of its child threads, grand-child threads, etc. can speculate on the fork point. In other words, a local hint affects the sub-tree of a thread, hence the name. Interval hints directly use the cost-benefit of a thread to decide profitability of its fork point; if a fork-point is disabled, it will try to speculate again after certain amount of time has passed. We find the global hint is the most effective for our benchmarks. It seems to work well on independent loops while the other two might suit more irregular applications. We plan to compare these hints in future work.

3.3   Disabling Fork Points

Each fork point has a globally unique id. The TLS runtime system maintains the attributes of the fork point, which can be accessed given the id. When a thread reaches a fork point, it queries the runtime system with the id whether it can speculate on the fork point. The runtime system then checks a flag variable of fork point attributes and returns the result. When a thread commits/rollbacks, if the adaptive fork-heuristics decide that one fork point is not profitable as discussed in Sect. 3.2, the runtime system then set the flag variable to false to disable the fork point.

If a loop nest has independent outer loops, such as enumerating elements on a matrix, then we can select to speculate on any or all of these loops. Speculating on outer loops enables coarser granularity parallelism but tends to consume more memory than inner ones, while speculating on all loops maximizes parallelism. These decisions have important influence on performance. However, the adaptive fork heuristics will select all speedup loops, even though disabling inner ones may yield further speedups as a result of less thread overhead. Here, we add an option, nest-loop-disabling, to disable an inner loop if its parent nest loop stably commits (its N and η are above the thresholds).

We also propose an optimizing technique called *feedback-based selection* to achieve ideal speedups from the second compilation for our benchmarks. After the program completes execution, it records the cost-benefit of each fork point to a feedback-based selection log file. The next time the TLS compiler compiles the program, it reads the log file and does not insert inappropriate fork/join points as potential candidates. For points that behave differently depending on the input, the programmer can annotate them so that the compiler still insert them even though they are in the log file. The optimization prevents unprofitable fork/join points from hurting performance repeatedly for each compilation.

4   Implementation Framework

We implement the adaptive fork-heuristics into the MUTLS [2,3] software-TLS compiler framework. MUTLS is a language and architecture independent software TLS framework purely based on the LLVM [1] intermediate representation (IR). It can exploit substantial parallelism from both loop- and method-level speculation. It supports compiler directives to annotate fork/join/barrier points. Each annotation also specifies an id. Threads speculated at a fork point will start execution from the join point with the same id, and will be joined when the non-speculative thread reaches that join point. A thread will also stop execution when it reaches a barrier point with the same id.

A sample program as well as its semiautomatically parallelized program annotated with adaptive fork-heuristics and feedback-based selection log file generated by the MUTLS compiler is illustrated in Fig. 1. Given the log file, there are
various criteria to decide inappropriate fork/join points, such as whether a fork
point is disabled, whether the cost-benefit is below a threshold, and whether
the ratio of committed/total threads is below a threshold. Combination of these
criteria is also possible. In the current implementation, we simply decide not to
add a pair of fork/join points as potential ones if the fork point is disabled.

```
void work(int n) {
    ...
    for(i = 0; i < n; i++){
        for(j = 0; j < 8; j++){
            x = f(i, j);
            s[i][j] += x*x;
        }
    }
    ...
}
```

(a) Original Program

```
fork point id 1 selected
cost-benefit 0.9 commit 80 rollback 0
fork point id 2 disabled
cost-benefit 0.3 commit 15 rollback 0
fork point id 3 disabled
cost-benefit 0 commit 0 rollback 10
```

(c) Feedback-based Selection Log File

```
void work(int n) {
    ...
    for(p = 0; p < P; p++){
#pragma tls fork id 1
        for(i = n * p / P; i < n * (p+1) / P; i++){
            for(j = 0; j < 8; j++){
#pragma tls fork id 2
#pragma tls fork id 3
                x = f(i, j);
#pragma tls join id 3
                s[i] += x*x;
            }
#pragma tls join id 2
        }
#pragma tls barrier id 2
    }
#pragma tls join id 1
    }
#pragma tls barrier id 1
    ...
}
```

(b) Program with Adaptive Fork-Heuristics

Fig. 1. Semiautomatic-parallelization of a program

5   Experimental Results

Experiments are performed on a AMD Opteron machine with 64 2.2 GHz processor cores (4×16 core, 8×2 MB L2 cache) and 64 GB memory. We use three
scientific computing benchmarks: Barnes-Hut (bh), molecular dynamics (md)
and Mandelbrot (mb). The performance results are shown in Figs. 2, 3 and 4
and compared in Fig. 5.

For each benchmark, we present the speedups of the manually speculated
program (manual version), the semiautomatically parallelized program using
adaptive fork-heuristics (adaptive version), as well as the optimized program
parallelized by feedback-based selection (feedback version). We also present the
results of the program speculating at every fork point (no hint version) as a baseline. The manual version evenly distributes the computing tasks to N processor
cores, which serves as the reference implementation. We ran the semiautomatic-
parallel programs 10 times each and present the maximum, average (arithmetic

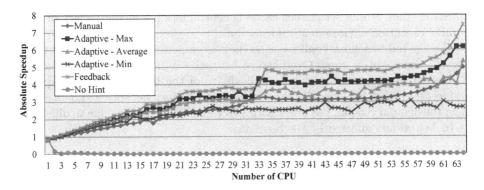

Fig. 2. Speedup results - Barnes Hut

mean) and minimum speedups. The feedback version is parallelized using the log file produced by the maximum speedup run.

We can see that the adaptive fork-heuristics perform excellently on these benchmarks. The feedback and adaptive versions of all benchmarks achieve more than 88 % and 50 % of the manual parallel performance at 64 cores, respectively. The feedback version of Barnes-Hut is even 49 % faster than the manual version. The adaptive version also beats the manual one on average. In addition to the top-level task distribution loops, the heuristics also select some fork points in the task computation functions. However, those fork points result in better speedups not because they contribute to more parallelism, but just because of cache issues. In fact, we find that even just adding pure rollback yields higher speedups than the manual version—even failed speculation acts as pre-fetching. This case demonstrates the power of adaptive fork heuristics that can be directly applied on real execution, which can always avoid inappropriate fork points and try to select as more beneficial ones as possible. In contrast, other fork heuristics use pre-defined fork points that do not guarantee benefits on real execution.

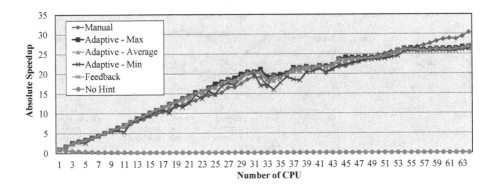

Fig. 3. Speedup results - molecular dynamics

Moreover, they cannot select such cache-beneficial fork points due to lack of real parallel execution environment information.

For the md benchmark, all versions show close performance, with little variance in the adaptive runs. Overhead of heuristics is negligible, which demonstrates satisfactory efficiency and applicability of the heuristics. The adaptive and feedback versions generally show higher performance between 8 and 55 cores than the manual version, but lower with 1 to 7 and 56 to 64 cores, due to different cache behaviours affected by the heuristics.

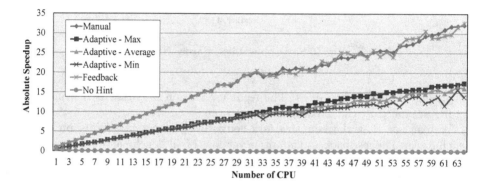

Fig. 4. Speedup results - Mandelbrot

On the other hand, mb is the least efficient benchmark with respect to the heuristics, due to its small innermost loop body. However, the normalized speedups of the adaptive version is relatively stable with the number of cores, and there is not much variance between each run, which guarantees worst-case performance. The nest-loop-disabling option is used for the benchmark, which improves the speedups by 3 times.

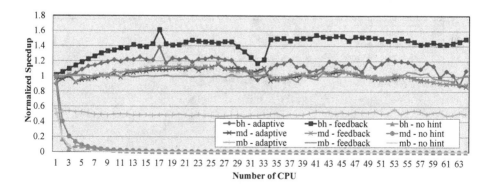

Fig. 5. Performance comparison, speedups normalized to the manual versions

Though the feedback-based selection log file is chosen to be the best speedup one over 10 runs, all the log files select the same fork/join points as the manual version for mb and md. For bh, any of the log files produces a feedback version with significantly higher speedups over the corresponding adaptive ones. Besides, the feedback versions can be further applied feedback-based selection to produce even better feedback versions.

To understand what the best parameters are for the benchmarks, we also experimented with different parameter configurations to see their influence on the performance. The results are illustrated in Fig. 6, with speedups normalized to the average speedups of the adaptive versions. For all experiments, we set $\eta_{threshold} = 0.5$ to indicate overhead should not take more time than useful work. We then set the default parameters $T_{overhead} = 1000000$, $N_{warmup} = 10$ and adjust one of them. The speedup data is computed by geometric means over 10 runs on 64 processor cores.

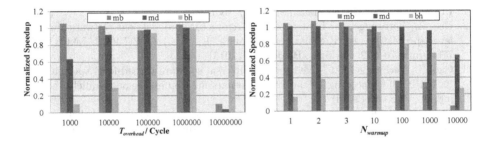

Fig. 6. Parameter results - $T_{overhead}$ and N_{warmup}

It can be seen that the performance is not very sensitive to the parameters of the heuristics, which is encouraging. The benchmarks have relatively stable performance with $T_{overhead}$ between 100000 and 1000000 CPU cycles and N_{warmup} between 3 and 10. In general, the performance degrades outside these ranges, except for mb, which prefers smaller thread overhead and less warmup runs. The significant performance drop of N_{warmup} from 10 to 100 is because a large number of warmup runs prevents the nest-loop-disabling optimization. It is also remarkable that memory-intensive benchmarks (bh) tend to suit larger $T_{overhead}$ than computation-intensive ones (mb, md).

6   Conclusions and Future Work

In this paper we proposed an adaptive fork-heuristics for software TLS, which inserts all potential fork/join points and purely relies on the heuristics and run-time system to disable the inappropriate ones. These adaptive heuristics have ability to utilize the real parallel execution environment information to maximize performance. In addition, we proposed a feedback-based selection technique to achieve ideal speedups.

Experiments on three scientific computing benchmarks on a 64-core machine demonstrate that the adaptive fork-heuristics are both highly effective and efficient. All benchmarks achieve more than 88 % and 50 % speedups of the manual version for the programs parallelized by feedback-based selection and adaptive fork-heuristics, respectively. Moreover, the feedback version of Barnes-Hut are 49 % faster than the manual version due to exploitation of cache efficiency. Experiments also show the encouraging fact that the heuristics are not overly parameter sensitive.

In future work, we will exploit more accurate cost-benefit models and hints that are more stable and even less parameter dependent, and evaluate with more benchmarks. We will also implement the necessary compiler optimizations to enable fully automatic parallelization with adaptive fork-heuristics.

References

1. LLVM (low-level vitrual machine). http://llvm.org
2. Cao, Z., Verbrugge, C.: Language and architecture independent software thread-level speculation. In: Kasahara, H., Kimura, K. (eds.) LCPC 2012. LNCS, vol. 7760, pp. 270–272. Springer, Heidelberg (2013)
3. Cao, Z., Verbrugge, C.: Mixed model universal software thread-level speculation. In: ICPP'13: Proceedings of the 42nd International Conference on Parallel Processing, pp. 651–660 (2013)
4. Chen, M.K., Olukotun, K.: The Jrpm system for dynamically parallelizing Java programs. In: ISCA'03: Proceedings of the 30th Annual International Symposium on Computer Architecture, pp. 434–446, June 2003
5. Dou, J., Cintra, M.: A compiler cost model for speculative parallelization. ACM Trans. Architect. Code Optim. **4**(2), 12 (2007)
6. Du, Z.H., Lim, C.C., Li, X.F., Yang, C., Zhao, Q., Ngai, T.F.: A cost-driven compilation framework for speculative parallelization of sequential programs. In: PLDI'04: Proceedings of the ACM SIGPLAN 2004 Conference on Programming Language Design and Implementation, pp. 71–81, June 2004
7. Liu, W., Tuck, J., Ceze, L., Ahn, W., Strauss, K., Renau, J., Torrellas, J.: POSH: a TLS compiler that exploits program structure. In: PPoPP'06: Proceedings of the 11th ACM SIGPLAN Symposium on Principles and Practice of Parallel Programming, pp. 158–167, Mar 2006
8. Liu, Y., An, H., Liang, B., Wang, L.: An online profile guided optimization approach for speculative parallel threading. In: Choi, L., Paek, Y., Cho, S. (eds.) ACSAC 2007. LNCS, vol. 4697, pp. 28–39. Springer, Heidelberg (2007)
9. Luo, Y., Packirisamy, V., Hsu, W.C., Zhai, A., Mungre, N., Tarkas, A.: Dynamic performance tuning for speculative threads. In: Proceedings of the 36th Annual International Symposium on Computer Architecture (ISCA '09), pp. 462–473 (2009)
10. Quiñones, C.G., Madriles, C., Sánchez, J., Marcuello, P., González, A., Tullsen, D.M.: Mitosis compiler: an infrastructure for speculative threading based on precomputation slices. In: PLDI'05: Proceedings of the 2005 ACM SIGPLAN Conference on Programming Language Design and Implementation, pp. 269–279, June 2005

11. Steffan, J.G., Colohan, C., Zhai, A., Mowry, T.C.: The stampede approach to thread-level speculation. ACM Trans. Comput. Syst. (TOCS) **23**(3), 253–300 (2005)
12. Wang, S., Dai, X., Yellajyosula, K.S., Zhai, A., Yew, P.-C.: Loop selection for thread-level speculation. In: Ayguadé, E., Baumgartner, G., Ramanujam, J., Sadayappan, P. (eds.) LCPC 2005. LNCS, vol. 4339, pp. 289–303. Springer, Heidelberg (2006)
13. Whaley, J., Kozyrakis, C.: Heuristics for profile-driven method-level speculative parallelization. In: ICPP'05: Proceedings of the 2005 International Conference on Parallel Processing, pp. 147–156, June 2005
14. Xu, F., Shen, L., Wang, Z., Guo, H., Su, B., Chen, W.: Heuspec: A software speculation parallel model. In: ICPP'13: Proceedings of the 42nd International Conference on Parallel Processing, pp. 621–630 (2013)

Inexact Sparse Matrix Vector Multiplication in Krylov Subspace Methods: An Application-Oriented Reduction Method

Ahmad Mansour[✉] and Jürgen Götze

Information Processing Lab, TU Dortmund, Otto-Hahn-Str. 4,
44227 Dortmund, Germany
{ahmad.mansour,juergen.goetze}@tu-dortmund.de

Abstract. Iterative solvers based on Krylov subspace method proved to be robust in the presence of well monitored inexact matrix vector products. In this paper, we show that the iterative solver performs well while gradually reducing the number of nonzero elements of the matrix throughout the iterations. We benefit from this robustness in reducing the computational effort and the communication volume when implementing sparse matrix vector multiplication (SMVM) on a Network-on-Chip (NoC).

Keywords: Krylov subspace method · Inexact matrix-vector multiplication · Network-on-chip

1 Introduction

Iterative solvers based on Krylov subspace method are considered to be among the most effective solvers for large linear systems. This method relies on finding an approximate solution in a lower dimension subspace called Krylov subspace. During the run of the iterative solver, matrix vector multiplication (MVM) is performed in each iteration in order to generate and extend the Krylov subspace. If the system matrix is sparse, the computation cost of the iterative solver is mostly spent in sparse matrix vector multiplications (SMVM) [8].

Researchers found that these iterative solvers perform well even with the presence of inexact matrix vector product [2–4,11]. In these papers, the robustness of Krylov subspace methods was only studied from numerical analysis point of view. In this paper, the robustness of these iterative solvers is utilized in reducing the computation cost of SMVM in an NoC. This is done by reducing the number of nonzero elements in the considered matrix while maintaining the convergence of the iterative solver. This can be very beneficial in case of matrices with big differences in the magnitude of their elements as, e.g., matrices from modeling 2D fluid flow in a driven cavity, and matrices representing dynamic analyses in structural engineering. As an example for iterative solvers based on Krylov method, our approach is applied to the Generalized Minimal Residual

R. Wyrzykowski et al. (Eds.): PPAM 2013, Part I, LNCS 8384, pp. 534–544, 2014.
DOI: 10.1007/978-3-642-55224-3_50, © Springer-Verlag Berlin Heidelberg 2014

method (GMRES) [10]. In our previous work [7], a row-based reduction method was used in which matrix elements are compared to a row-dependent threshold. In this paper, a generalized reduction method is introduced, where the threshold is used for the whole matrix elements.

In the following section, the practical background which motivates our investigations is introduced. The theoretical background of the iterative solver is discussed in Sect. 3. In Sect. 4, the reduction mechanism is introduced, which is used to produce the perturbed matrix. The numerical experiments are discussed in Sect. 5. Finally, the paper is concluded in Sect. 6.

2   Practical Background

The computation of SMVM in general purpose processors is inefficient. This is due to the irregular data transfers and the high number of memory accesses compared to the number of floating point operations. Hence, researchers tried to enhance the poor performance of SMVM computations by modifying the existing algorithms or by using accelerators as, e.g., graphics processing units (GPU), and FPGA-based accelerators. In our previous work, an SMVM accelerator is implemented using an FPGA-based NoC [5,12]. In this implementation, only the SMVM is performed on the accelerator while the rest of the iterative algorithm is performed by the host PC. In this case, the matrix has to be loaded in each iteration to perform the matrix vector multiplication.

Matrix elements are read from memory and then injected into a $k \times k$ mesh of nodes, as shown in Fig. 1(a) for $k = 4$. Each node consists of a processing element (PE) and a network interface. The mapping method of matrix and vector elements is shown in Fig. 1(b). The matrix element a_{ij} and the vector element x_j are mapped to PE($mod(j, k \times k + 1)$) for multiplication. Then, the multiplication result b_{ij} is mapped to PE($mod(i, k \times k + 1)$) for accumulation with multiplication results in the i-th row.

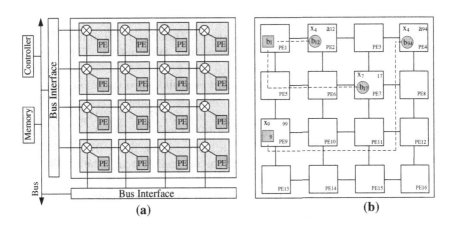

Fig. 1. NoC architecture and data mapping method

For an $N \times N$ sparse matrix with n_z nonzero elements, the number of floating point operations in SMVM is $2n_z - N$. The communication volume (the number of sent messages) cannot be determined from the number of nonzeros as it depends on the sparsity structure of the matrix and the used data mapping method. For example, the multiplication result b_{12} in Fig. 1(b) travels to the next node for accumulation, while b_{94} has to travel five nodes to reach its destination. Regardless of matrix structure and the used data mapping method, the communication volume rises with the number of the nonzero elements. Hence, the objective is reducing the number of nonzeros such that computation and communication costs are reduced while simultaneously taking into consideration the convergence of the iterative solver.

3   Theoretical Background

Iterative solvers have been used in solving large linear systems due to their low storage requirements compared to direct solvers [9]. This is particularly important in case of sparse matrices where the factorization of the matrix for the direct solution leads to an increased number of nonzeros compared to the original matrix. We assume that readers are familiar with Krylov subspace methods. For more details on these methods, refer to Saad [9].

Given a system of linear equations $Ax = b$, the goal is to find a solution (x_*) such that the norm of the residual (r) is less than the targeted tolerance (η), i.e., with $r = b - Ax$, one obtains $\|r\|_2 = \|b - Ax\|_2 < \eta$ for $x = x_*$. For the rest of the paper, the Euclidean norm is denoted by $\|.\|$. In Krylov subspace methods, the iterative solver aims to find an approximate solution in a lower dimension subspace. This subspace of candidates is called *Krylov subspace* $\mathcal{K}_m(A, v_1)$, which is spanned by

$$\{v_1, Av_1, A^2v_1, ..., A^{m-1}v_1\}, \tag{1}$$

where $v_1 = r_0 / \|r_0\|$.

As example for Krylov methods, the Generalized Minimum Residual (GMRES) method is considered in theory and experiments. The details of GMRES can be found in Algorithm 1.

In lines (2-9), Arnoldi iteration is used to build an orthogonal basis (V_m) for the Krylov subspace. After m iterations, one obtains $AV_m = V_{m+1}H_{m+1}$, where $V_m = [v_1, v_2, ..., v_m]$ is an orthogonal matrix, $H_m = V_m^T AV_m$ is an upper Hessenberg matrix and H_{m+1} is the matrix H_m augmented with $h_{k+1,k}e_m^T$, where e_m is the mth canonical vector. As it can be seen from line (10), one solves an m-dimensional system instead of solving the original system. The result of this minimization (y) is used to calculate the solution of the linear system as shown in line (11). The residual in this case can be calculated as follows:

$$r = b - Ax = b - A(x_0 + V_my) = r_0 - AV_my = r_0 - V_{m+1}H_{m+1}y. \tag{2}$$

Algorithm 1. GMRES

1: $r_0 = b - Ax_0, \beta = \|r_0\|, v_1 = r_0/\beta$
2: **for** $i = 1, 2, ...,$ **do**
3:     $w = Av_i$
4:     **for** $k = 1, .., i$ **do**
5:         $h_{k,i} = w^T v_k$
6:         $w = w - h_{k,i}v_k$
7:     **end for**
8:     $h_{i+1,i} = \|w\|$
9:     $v_{i+1} = w/h_{i+1,i}$
10:    Solve $\min_{y} \|\beta e_1 - Hy\|$
11:    If satisfied, Set $x = x_0 + V_i y$ and quit
12: **end for**

The computation and storage cost of GMRES algorithm grows with the number of iterations. In order to overcome this drawback, the algorithm is restarted after a given number of iterations (m) such that the intermediate solution is used to initialize the next restart while old stored data are cleared. This algorithm is denoted by GMRES(m).

In the following section, GMRES with the presence of inexact matrix vector products is discussed as well as the proposed reduction method.

4 Convergence Under Inexact Products

In a Krylov method using inexact matrix vector products, a perturbed matrix is used instead of the original matrix for creating the Krylov subspace. That is, the term $w = Av$ in Algorithm 1 is replaced by $w = \tilde{A}_i v$, where $\tilde{A}_i = (A - \Delta A_i)$ is the perturbed matrix in the ith iteration and ΔA_i is the perturbation matrix. The Arnoldi iteration in the inexact case gives

$$[(A - \Delta A_1)v_1, (A - \Delta A_2)v_2, ..., (A - \Delta A_m)v_m] = V_{m+1}H_{m+1},$$

or

$$AV_m - [\Delta A_1 v_1, \Delta A_2 v_2, ..., \Delta A_m v_m] = V_{m+1}H_{m+1}.$$

In contrast to the residual in the exact case in Eq. (2), the residual in the inexact case can be calculated as

$$\tilde{r} = b - Ax = b - A(x_0 + V_m y) = r_0 - AV_m y = r_0 - V_{m+1}H_{m+1}y + \epsilon = r + \epsilon. \quad (3)$$

where

$$\epsilon = [\Delta A_1 v_1, \Delta A_2 v_2, ..., \Delta A_m v_m]y = \sum_{i=1}^{m} y(i)\Delta A_i v_i, \quad (4)$$

is the gap between the exact and the inexact residuals, which is dependent on the amount of applied perturbation. The norm of the residual gap can be described as follows

$$\|\epsilon\| = \sum_{i=1}^{m} |y(i)| \, \|\Delta A_i v_i\| \leq \sum_{i=1}^{m} |y(i)| \, \|\Delta A_i\|, \tag{5}$$

where (refer to Lemma 5.1 in [11])

$$|y(i)| \leq \frac{\|\tilde{r}_{i-1}\|}{\sigma_{min}(H_m)}. \tag{6}$$

One can see from Eq. (6), that the magnitude of y elements decreases as the residual approaches the tolerance. Therefore, one can increase the norm of the perturbation matrix such that $|y(i)| \, \|\Delta A_i\|$ is always the same and, hence, the residual gap is determined. The residual gap is controlled by bounding the norm of the perturbation matrix to achieve the same convergence as in the exact case. This explains why the norm of the perturbation is allowed to grow throughout iterations.

In this paper the perturbation is expressed by omitting matrix elements which are in magnitude less than some threshold (ε) such that the norm of the perturbation matrix is less than the allowed bound. As the bound on the norm of the perturbation grows, the dropping threshold also increases. That is, the reduction method progressively reduces the number of nonzeros until reaching the solution.

Given a sparse matrix A of size $N \times N$, by dropping elements less than ε_i, the perturbation matrix in the ith iteration can be expressed as follows

$$\Delta A_i = \begin{bmatrix} a_{11} & a_{12} & \cdots & 0 & \cdots & a_{1N} \\ 0 & a_{22} & \cdots & a_{2k} & \cdots & 0 \\ \vdots & \vdots & \ddots & \vdots & \ddots & \vdots \\ 0 & 0 & \cdots & a_{Nk} & \cdots & a_{NN} \end{bmatrix},$$

where the nonzeros of ΔA_i are less than ε_i. Moreover, the positions of the dropped elements in ΔA_i are selected randomly.

To calculate the norm of ΔA_i, the eigenvalues of $\Delta A_i^T \Delta A_i$ need to be determined. The extreme case of ΔA_i is considered by replacing all ignored elements by ε_i. The worst case that $\Delta A_i^T \Delta A_i$ can take is

$$\Delta A_i^T \Delta A_i = N\varepsilon_i^2 \begin{bmatrix} 1 & \cdots & 1 & \cdots & 1 \\ 1 & \cdots & 1 & \cdots & 1 \\ \vdots & & \ddots & & \vdots \\ 1 & \cdots & 1 & \cdots & 1 \end{bmatrix}.$$

As the maximum eigenvalue for this matrix is $\lambda_{max}(\Delta A_i^T \Delta A_i) = (N\varepsilon_i)^2$, then $(\|\Delta A_i\|_2)_{max} = \sqrt{\lambda_{max}} = N\varepsilon_i$, and in the general case: $\|\Delta A_i\|_2 \leq N\varepsilon_i$. By choosing

$$\varepsilon_i = \sigma \frac{\eta}{\|\tilde{r}_{i-1}\|}, \tag{7}$$

where σ is the perturbation parameter and η is the targeted tolerance, the norm of the perturbation matrix becomes

$$\|\Delta A_i\| \leq N\varepsilon_i = N\sigma \frac{\eta}{\|\tilde{r}_{i-1}\|}, \tag{8}$$

and the norm of the residual gap in (5) becomes

$$\|\epsilon\| \leq m \cdot N \cdot \eta \cdot \frac{\sigma}{\sigma_{min}(H_m)}. \tag{9}$$

The problem in determining the value of σ is that $\sigma_{min}(H)$ is not available before the mth iteration, and depends on the amount of perturbation. As in [11], the relation between $\sigma_{min}(H)$ and $\sigma_{min}(A)$ is described as

$$\sigma_{min}(H_m) \geq \sigma_{min}(A) - \|[\Delta A_1 v_1, \Delta A_2 v_2, ..., \Delta A_m v_m]\|, \tag{10}$$

or as in [3],

$$\sigma_{min}(H_m) \geq \sigma_{min}(A)(1 - c), \tag{11}$$

with $0 < c < 1$, depending on the amount of perturbation.

Let us now determine the upper bound of the second term on the right hand side of Inequality (10). As v is a unit vector and ΔA is very sparse, the maximum possible perturbation in w caused by dropping matrix elements less than ε is

$$\Delta w = \Delta A \cdot v = \frac{k_i \varepsilon_i}{\sqrt{N}} \begin{bmatrix} 1 \\ 1 \\ \vdots \\ 1 \end{bmatrix}, \tag{12}$$

where k_i is the maximum number of dropped elements in the matrix rows in the ith iteration. As ΔA is very sparse, then $k \ll N$ and can be replaced by αN, where $\alpha \ll 1$. This gives

$$\Delta w = \Delta A \cdot v = \alpha_i \varepsilon_i \sqrt{N} \begin{bmatrix} 1 \\ 1 \\ \vdots \\ 1 \end{bmatrix}, \tag{13}$$

and

$$[\Delta A_1 v_1, \Delta A_2 v_2, ..., \Delta A_m v_m] = \sqrt{N} \begin{bmatrix} \alpha_1\varepsilon_1 & \alpha_2\varepsilon_2 & \cdots & \alpha_m\varepsilon_m \\ \alpha_1\varepsilon_1 & \alpha_2\varepsilon_2 & \cdots & \alpha_m\varepsilon_m \\ \vdots & \vdots & \ddots & \vdots \\ \alpha_1\varepsilon_1 & \alpha_2\varepsilon_2 & \cdots & \alpha_m\varepsilon_m \end{bmatrix}. \tag{14}$$

The norm of this matrix can be calculated as follows

$$\|[\Delta A_1 v_1, \Delta A_2 v_2, ..., \Delta A_m v_m]\| = \sqrt{N} \sqrt{N \sum_{i=1}^{m} (\alpha_i \varepsilon_i)^2} = N\sigma \sqrt{\sum_{i=1}^{m} (\frac{\alpha_i \eta}{\|\tilde{r}_{i-1}\|})^2}. \tag{15}$$

As $\alpha \ll 1$ and the norm of the residual is always bigger than the tolerance ($\|\tilde{r}_{i-1}\| > \eta$), the term under the square root can be replace by some constant $c \ll 1$.

$$\|[\Delta A_1 v_1, \Delta A_2 v_2, ..., \Delta A_m v_m]\| = N\sigma c, \tag{16}$$

and Inequality (10) becomes

$$\sigma_{min}(H_m) \geq \sigma_{min}(A) - N\sigma c. \tag{17}$$

By choosing $\sigma = \sigma_{min}(A)/N$, the last inequality can be rewritten as

$$\sigma_{min}(H_m) \geq \sigma_{min}(A)(1 - c), \tag{18}$$

or

$$\frac{\sigma_{min}(A)}{\sigma_{min}(H_m)} \leq \frac{1}{1-c} \approx 1, \tag{19}$$

and the norm of the residual gap becomes

$$\|\epsilon\| \leq m \cdot N \cdot \eta \cdot \frac{\sigma_{min}(A)}{N \cdot \sigma_{min}(H_m)} \approx m \cdot \eta. \tag{20}$$

To check whether the selection of σ is correct, one can monitor the effect of the perturbations on the new generated vector by calculating the norm of the perturbation in Eq. (13). The norm can be described as

$$(\|\Delta w_i\|)_{max} = N\alpha_i\varepsilon_i = N\alpha_i \frac{\eta}{\|\tilde{r}_{i-1}\|}\sigma = \alpha_i\sigma_{min}(A)\frac{\eta}{\|\tilde{r}_{i-1}\|}, \tag{21}$$

and the maximum perturbation in all iterations (as $\eta/\|\tilde{r}_{i-1}\| \to 1$) is

$$\|\Delta w\| \leq \alpha\sigma_{min}(A) \leq \alpha(\|w\|)_{min} < (\|w\|)_{min}, \tag{22}$$

where $\|w\| \in [\sigma_{min}(A), \sigma_{max}(A)]$ (remembering that $w = Av$ and $\|v\| = 1$).

From the previous equations it can be seen that with this selection of σ, the residual gap is bounded by $m \cdot \eta$. However, as shown later in Sect. 5, this selection is practically too strict, leading to relatively small reduction rates (the number of ignored nonzeros to the total number of nonzeros). This can obviously be seen from Eq. (22), which allows very small deviations from the exact Krylov subspace. The following experiments use three variants of σ, from which $\sigma_{min}(A)$ seems to be good enough to achieve small residual gaps and big reduction rates.

5   Numerical Experiments

In this section, MATLAB simulations are performed using matrices from Matrix Market [1]. The perturbed matrices are used in the *gmres* MATLAB function instead of the original matrices. For each test matrix, three variants of the perturbation parameter (σ) are considered, which are $\sigma_1 = \sigma_{min}(A)/N$, $\sigma_2 = \sigma_{min}(A)/\sqrt{N}$ and $\sigma_3 = \sigma_{min}(A)$. The true residual (not the GMRES residual) in the following figures needs not to be calculated in the algorithm for both exact and inexact cases. However, it is calculated in the experiments for illustration purpose only.

Figures 2, 3 and 4 show the convergence of residuals and the reduction rates for three cases. In the first case, the non-restarted GMRES (as in Algorithm 1) is considered using ARC130 matrix. The restarted version of GMRES(m) is considered in the second case, in which FS_760_1 matrix is used with restart size $m = 40$. In the last case, the preconditioned GMRES is tested using incomplete LU factorization (ILU) [9]. For this case, CAVITY03 matrix is used with drop tolerance 10^{-3} for the ILU preconditioner.

As it can be seen from the figures, the convergence in the inexact case is identical to that in the exact case even with the presence of growing perturbations. The achieved reduction rate increases by using bigger perturbation parameters. Based on the mathematical proof in Sect. 4, using σ_1 always leads to identical convergence as in the exact case. On the other hand, using σ_2 or σ_3 may lead to bigger residual gaps. In this case, the iterative solver may require more iterations to reach the targeted tolerance.

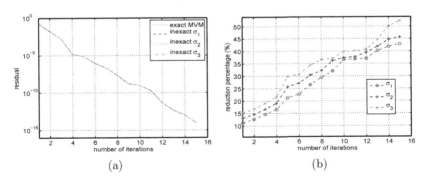

Fig. 2. Convergence and reduction rate for ARC130, 130×130, $m = 130$ and $\eta = 10^{-14}$

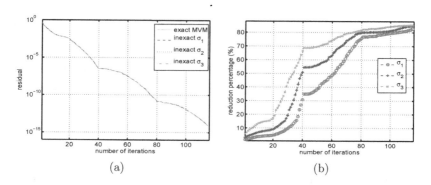

Fig. 3. Convergence and reduction rate for FS_760_1, 760×760, $m = 40$ and $\eta = 10^{-14}$

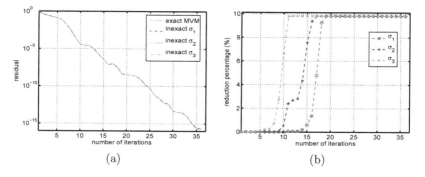

Fig. 4. Convergence and reduction rate for CAVITY03, 317×317, $m = 10$ and $\eta = 10^{-14}$

Going back to our NoC, we compare the effort of performing exact and inexact SMVM in the NoC. An OMNeT++ based NoC simulator [6] is used to compare the performance of SMVM using some test matrices and their reduced versions. The simulator is modular, scalable and parameterizable framework for modeling the NoC-based SMVM accelerator. Benefiting from OMNeT++ built-in support for recording simulation results, the simulator is used to record the number of sent messages in the NoC and the number of floating-point operations performed in each node. Table 1 shows the simulation results for three test matrices using a 4×4 NoC. For each matrix, two reduced versions are used (the reduction percentages are in the brackets).

From these results it can be concluded that using the reduced matrices is beneficial due to the reduction in the number of floating point operations and reduction in communication volume. This leads to higher throughput of the NoC, which is expressed by *Speed-up* in the last column of the table. The reduction in SMVM effort and the speed-up depend on the system architecture, the data mapping method, the sparsity structure of the matrix and the positions of the ignored nonzeros in that matrix. For example, ignoring a_{12} (see Fig. 1(b)) reduces the number of sent messages by one. On the other hand, ignoring a_{94} saves five

Table 1. NoC simulation results

Matrix	Mode	Nonzeros	FLOP	Sent messages	Speed-up
ARC130	Original	1037	1944	11248	-
	Reduced (20 %)	829	1311	10218	1.35
	Reduced (40 %)	622	1009	7508	1.727
FS_760_1	Original	5739	10718	49188	-
	Reduced (40 %)	3487	6214	31846	1.52
	Reduced (60 %)	2316	3872	22622	2
BCSSTM07	Original	7252	14084	78088	-
	Reduced (20 %)	5808	11196	63508	1.23
	Reduced (50 %)	3692	6964	40852	1.85

sent messages. However, it is obvious that the achieved speed-up increases when using higher reduction rates. It has to be noticed that the number of messages in the table includes the number of messages required for injecting the matrix and vector elements and the messages required for collecting the final result.

6   Conclusion and Future Work

In this paper we made use of the robustness of the iterative solvers based on Krylov subspace method in reducing the effort of SMVM in an NoC. It was shown that the reduction mechanism was very beneficial for matrices with big differences in the magnitude of their elements. Besides, the proposed method does not require costly calculations as, e.g., the norm of the perturbation matrix. One of the main challenges in our method, and in other papers, is that the lowest singular value of the matrix needs to be determined. As future work, we aim to find an alternative for the minimum singular value benefiting from the available data during the run of the algorithm.

From hardware point of view, the comparison in performance has to be realized on hardware in order to measure the actual speed-up and the saved power.

References

1. Boisvert, R., Pozo, R., Remington, K., Barrett, R., Dongarra, J.: Matrix market: a web resource for test matrix collections. In: Boisvert, R. (ed.) The Quality of Numerical Software: Assessment and Enhancement, pp. 125–137. Chapman & Hall, London (1997)
2. Bouras, A., Fraysse, V.: Inexact matrix-vector products in Krylov methods for solving linear systems: a relaxation strategy. SIAM J. Matrix Anal. Appl. **26**(3), 660–678 (2005)
3. Giraud, L., Gratton, S., Langou, J.: Convergence in backward error of relaxed GMRES. SIAM J. Sci. Comput. **29**(2), 710–728 (2007)
4. Gratton, S., Toint, P.L., Ilunga, J.T.: Range-space variants and inexact matrix-vector products in Krylov solvers for linear systems arising from inverse problems. Technical Report TR/PA/10/14, CERFACS, Toulouse, France (2011)

5. Jheng, H.Y., Sun, C.C., Ruan, S.J., Götze, J.: FPGA acceleration of sparse matrix-vector multiplication based on network-on-chip. In: 19th European Signal Processing Conference (EUSIPCO), Barcelona, pp. 744–748, August 2011
6. Mansour, A., Götze, J.: An OMNeT++ based network-on-chip simulator for embedded systems. In: IEEE Asia Pacific Conference on Circuits and Systems (APCCAS 2012), Kaohsiung, Taiwan, pp. 364–367, December 2012
7. Mansour, A., Götze, J.: Utilizing robustness of Krylov subspace methods in reducing the effort of sparse matrix vector multiplication. Procedia Comput. Sci. 18, 2406–2409 (2013)
8. Morris, G., Prasanna, V.: Sparse matrix computations on reconfigurable hardware. Computer 40(3), 58–64 (2007)
9. Saad, Y.: Iterative Methods for Sparse Linear Systems. PWS Publishing, Boston (1996)
10. Saad, Y., Schultz, M.H.: GMRES: a generalized minimal residual algorithm for solving nonsymmetric linear systems. SIAM J. Sci. Stat. Comput. 7(3), 856–869 (1986)
11. Simoncini, V., Szyld, D.B.: Theory of inexact Krylov subspace methods and applications to scientific computing. SIAM J. Sci. Comput. 25(2), 454–477 (2003)
12. Sun, C.C., Götze, J., Jheng, H.Y., Ruan, S.J.: Sparse matrix-vector multiplication on network-on-chip. Adv. Radio Sci. 8, 289–294 (2010)

The Regular Expression Matching Algorithm for the Energy Efficient Reconfigurable SoC

Paweł Russek[✉] and Kazimierz Wiatr

AGH University of Science and Technology,
Mickiewicza Av. 30, 30-059 Krakow, Poland
{russek,wiatr}@agh.edu.pl

Abstract. This paper presents an algorithm for a regular expressions pattern matching system. The goal was to achieve better performance and low energy consumption. The proposed scheme is particularly useful when a large set of complex regular expression patterns must be inspected in parallel (e.g. in computer malware and anti-virus systems). The idea of the algorithm derives from a concept of the Bloom filter algorithm. The Bloom filter operation is used to inspect an incoming data and to find static sub-patterns of regular expressions. When the Bloom filter reports a match, a closer inspection is performed. The Bloom filtering is done by a hardware dedicated co-processor. The regular expressions' wildcard matching part is executed by a CPU.

The above concept was implemented and tested on the Xilinx Zynq-7000 All Programmable SoC platform. The results and performance for regular expressions patterns from the ClamAV virus database is given.

Keywords: Regular expressions matching · Energy-efficient systems · Custom architectures · HW–SW Co-Design

1 Motivation

Regular expression matching has become a bottleneck of software based pattern matching systems. Such systems play an important role in data intensive computing. Network Intrusion Detection Systems (NIDS) are one of the examples where regular expression matching is used. In some advanced NID systems deep packet inspection (DPI) is performed. In such systems, received data is assembled from network packets and its content is checked against viruses and another malware software. Known viruses' information is stored in virus databases which are frequently updated. As an example, ClamAV [1], an open source anti-virus toolkit and database, is used to demonstrate the properties of the solution proposed in this paper. As the number of regular expressions increases, matching requires more and more computational effort. ClamAV contains nearly 100,000 regular expression based patterns as of January 2013.

Regular expression pattern matching may consume a lot of CPU run-time in software systems. Basically, it is an example of simple data-intensive computing.

R. Wyrzykowski et al. (Eds.): PPAM 2013, Part I, LNCS 8384, pp. 545–556, 2014.
DOI: 10.1007/978-3-642-55224-3_51, © Springer-Verlag Berlin Heidelberg 2014

In other words, it is not a CPU bound problem, but rather Input/Output (IO) bound problem. For such problems the computer system should support efficient data movement in the system rather than high computing power. The present state-of-the-art, super-scalar, many-core processors fall behind real needs of data-intensive problems. For example, modern processors offer over 200 GFlops of processing power. To accommodate such performance data reuse is necessary. That is why a big cache memory is required for efficient computing. However, when considering a $O(n)$ complex problem, data reuse in many applications is limited. A rough estimation shows that to hold the performance of 200 billions of 64-bit operations per second, data transfer of 1.6 TBps (Trillion Bytes per second) is requested. The comparison of CPU's computing performance to available network traffic speed shows a tremendous gap as well. Above facts put into doubts a policy to use fastest CPUs when tasks are data oriented. Farther, the cost and the energy consumption of powerful processors need to be considered. There is an energy efficiency penalty that has to be paid for flexible performance of super-scalar CISC processors.

A possible approach to solve the above problems is a computer system that is better balanced for data dependent computing. Such systems can be built around low-energy processors. Such processors features lower computing power, but offer an attractive ratio of computing power per energy consumption. Consequently, lower computing power better fits the available IO devices' throughput. In systems for data-intensive computing, lower CPUs' performance is compensated by a higher number of computing nodes. In that policy, by adding to the system an additional node, one adds CPU performance together with an additional IO data throughput that is introduced by the node's IO devices.

Recently, an approach that plays a significant role in data processing is custom processors. They outperform general purpose processors in terms of performance and energy consumption. Today, the field programmable gate array (FPGA) is proven semiconductor technology that is used for implementation of custom computing devices. Examples of data processing in the FPGAs can be found in [2,3].

A recent introduction of the Xilinx's Zynq [4] is the motivation to focus on research capabilities of low-energy processors which are tightly coupled with an FPGA structure. The Zynq-7000 Extensible Processing Platform is a family of silicon devices that combine a complete ARM processor-based SoC with integrated programmable logic.

In this research, the problem of regular expression matching is approached in the manner of HW/SW co-design for embedded SoC. The algorithm was proposed and its performance tested on the Zynq-7000 platform. We applied the bottom-up approach as the algorithm was conceived from scratch with the target platform architecture kept in mind. The most powerful and efficient algorithm that is widely used for string matching is the Bloom filter algorithm. It is suitable for both hardware and software implementations. Although the Bloom filter is very fast it is not suitable for direct regular expression matching. As presented here, adoption of the Bloom algorithm to regular expression matching benefits its performance.

The paper is organized as follows. The next section presents recent and related works chosen from literature. In Sect. 3 an outline of issues which are the background of the work is given. Section 4 presents a decomposition of the problem into hardware and software parts. In Sect. 5 the architecture of the proposed custom hardware accelerator is presented. The results of hardware and software implementation are in Sect. 6. The paper ends with conclusions in Sect. 7.

2 Related Work

Pattern matching is commonly performed by expressing patterns as sets of regular expressions those are converted into automata. Different kinds of automata are used in the literature: deterministic (DFA), non-deterministic (NFA) and alternatively Aho-Corasick automata. The majority of proposals that are met in literature implement DFA or NFA. The NFAs are smaller in size, but the DFAs work faster. In the work of Fang Yu et al. [5], the DFA-based algorithm was run on a PC with a 3.4 Ghz CPU and a 3.7 GB memory. The throughput performance was 0.6–0.7 Gbps for the Linux L7-filter. I was 47.9 to 704 times faster than the NFA-based scanner from a popular software library. On the contrary, the memory usage was 2.6–8.4 times higher. Pasetto et al. [6] have presented Aho-Corasick automata on an Intel Xeon E5472. It could achieve processing rates up to 4.5 Gbps per core on the simpler XML (32 regexs) and SMTP (20 regexs) parsing, and from 0.74 to 1.1 Gbps on the more demanding Snort (500 regexs) and Linux L7 (150 regexs) NIDS sets. The peak performance on eight Intel Xeon cores reaches 46 and 35 Gbps, respectively, for SMTP and XML parsing.

Today, accelerators such as the GPGPUs, FPGAs and also IBM's Cell play an important role in computer systems. As a consequence, many solutions have been proposed in that field. Scarpazza and Russell [7] presented a SIMD solution that delivers up to 14.3 Gbps on one IBM Cell/B.E. chip. Also on Cell, Iorio and van Lunteren [8] proposed the BFSM string matcher for automata achieving 32 Gbps. Lately, the idea of using parallelism offered by the GPUs was considered in [9]. Although the maximum reported throughput was 35 Gbps, the largest of the test cases was limited to around 120 regular expressions. Another GPGPU related work by Naghmouchi et al. [10] focuses on tokenization: a form of regexp matching used to divide a character stream into tokens like telephone number, URLs, etc. The limitation of this approach is the small size of the rule set. Typical, maximal and minimal achieved throughputs after all the introduced optimizations were: 9.4, 44.0 and 7.6 Gbps respectively.

Regarding FPGAs, one of the first papers that proposed to construct an NFA from a regular expression to perform string matching was [11]. This approach was later developed by Sourdis et al. in [12]. The latest achievement can be found in [13]. Thanks to the sharing of common sub-regular expressions the total number of implemented Snort rules was 24,214 characters. The reported throughput was approx. 1 Gbps. Thanks to parallelism in multi-character architecture, Chang et al. [14] was able to increase the maximum system throughput to 4.68 Gbps (2 character design) and 7.27 Gbps (4 character design). Bande Serrano and Palancar [15] proposed an approach that is related to and presented in our work. Unique sub-sequence matchings were used for detecting the possible presence of the string in the data flow. In doing so, they made a reduction of the area cost for processing multiples characters.

Although the original Bloom's algorithm is intended for static patterns, several authors tried its adoption for regular expression matching. A more recent approach was given by [16]. The authors proposed a hashing-table look-up mechanism which uses parallel Bloom filters to enable a large number of fixed-length strings to be scanned in hardware. Based on the Bloom filter, Lockwood et al. [17] proposed a gateway that provides Internet worm and virus protection in networks.

3   Background

3.1   Computing Platform

A hardware processor requires an appropriate host system architecture. The performance of all the components should fit in a well-balanced system. In a general purpose solution the properties of a system cannot be strictly defined because they are different for various algorithms. In practice, the best general purpose solutions are built using state-of-the art components: processors, memory chips, graphics card, storage devices, etc. A policy of choosing the best available on the market component to set up a computer works in practice because usually systems execute different applications.

An essential computing system equipped with a hardware accelerator is depicted in Fig. 1. It consists of a multi/textendash CPU processor, a memory controller connected to a memory block, an IO device and a hardware accelerator. The CPU is tightly coupled with the memory controller that allows for higher

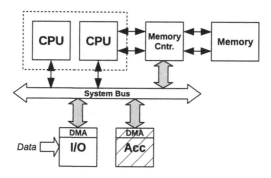

Fig. 1. A simple system with an accelerator.

data throughput and lower latency. In contrary, the IO devices and the accelerators have access to the memory thanks to a system bus (e.g. PCIe, AMBA, etc.). As the memory controller is also connected to the system bus, the IOs and accelerators can also perform memory operations. Preferably, the system bus devices use Direct Memory Access (DMA) transactions to access the main memory. The DMA mechanism allows the CPUs, IO devices and accelerators to work simultaneously. In principle they perform algorithms in a pipeline manner. The functional decomposition of an algorithm is done i.e. separate blocks of data are processed concurrently by IO devices, CPU, accelerator (Acc), and so on. To allow a smooth algorithm execution, the performance of all components must fit. This means that none of the components outperforms other components and none of the components waits for another component to complete its task.

3.2 The Bloom Filter

The fundamental operations that are behind the Bloom filter are both data hashing and memory read. In general, the hashing converts a word into another word of a smaller bit length. In other words, the hashing process converts a $w - bit$ input word into a $h - bit$ output word, where $h < w$. The output word that is generated by the hashing process, is an address in a memory where the stored values are 'true' or 'false'. These indicate the existence or non-existence of the input word in the filter. In the teaching phase the memory is programmed for the given word dictionary. In the check phase, if the location in the memory at the generated address is 'true' a hit occurs. In Bloom's original algorithm, for a single word, the mentioned memory operation is sequentially repeated several times for different hash functions. We denote this number of tries as k. If all the tries are successful, during the check phase, a positive match is generated, i.e., the word exists in the Bloom filter.

There are at least two major disadvantages of the Bloom filter method that need to be considered here. The first, the Bloom algorithm is not a trustworthy tool. The second, in the original algorithm, the search is performed sequentially.

With regard to the first disadvantage, there is a finite probability of an incorrect word match. The Bloom algorithm generates a match signal for the words outside the dictionary. They are further denoted as 'false-positives'. If faultlessness is required, the results ought to be additionally verified by another, perfect algorithm. The nature of the fault indicated in the Bloom filter is a misstatement, i.e., the words that are not in the dictionary can match. Fortunately, the opposite situation, where the algorithm overlooks words is not possible. This means that the words held in the dictionary are always indicated.

The probability of 'false-positives' is given by the formula (1).

$$p_{err} = [1 - (1 - \frac{1}{m})^{(kn)}]^k \approx [1 - e^{-\frac{kn}{m}}]^k \tag{1}$$

where: n is the number of searched patterns, m is the bit size of memory and k is a number of hash operations performed. Additionally, a reasonable assumption is that $m = 2^h$.

In the algorithm presented in this paper we assume that $k = 1$. This assumption simplifies the hardware architecture of the design. As a result, the p_{err} is slightly higher than the optimal value that could be achieved for the adjusted value of k. Moreover, it was also possible to get rid of the hashing operation. We use patterns directly to address the Bloom memory. The algorithm allows for that because the patterns' length can be freely chosen. Their length can be adjusted to the requested value of $log_2(m)$. This phenomenon will be clarified in the next section. In other words, the Bloom filter used in the presented algorithm is reduced to the basic memory look-up operation.

Also the assumption that all patterns are different is taken (in opposite to patterns' hashes). That allows to derive a simplified hit probability (Eq. 2).

$$p_{err} = \frac{n}{m} = \frac{n}{2^p} \tag{2}$$

where: $p = log_2(m)$ is the patterns' length.

The property of the Bloom filtering is that for an input data rate R_{in}, the output data rate is:

$$R_{out} = p_{err} * R_{in} \tag{3}$$

4   Principles of the Proposed Algorithm

In this section the basic idea of the proposed algorithm will be introduced. The main assumption behind the proposed scheme is HW–SW decomposition. To achieve good performance, the part of the algorithm will be executed by the custom processor. The part executed in custom hardware is the Bloom filtering. The CPU is used for regular expression verification (Fig. 2). The streaming architecture was chosen for the custom accelerator data flow as it frequently offers the highest performance.

Obviously, as presented in the previous section the Bloom filter approach cannot be applied directly to detection of regular expressions because it does not support wildcards matching. On the other hand, the performance advantages

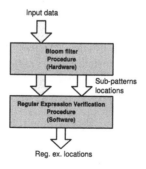

Fig. 2. HW–SW decomposition of the algorithm

of this method cannot be neglected. In order to get rid of the Bloom filter limitations, this operation must be additionally supplemented. Here, the results from the Bloom filtering are further processed by the procedure executed by the CPU. Accordingly, our approach consist of two stages. In the first stage, data is examined against static sub-patterns and then suspected blocks of data and appropriate sub-patterns locations are passed to the second stage for deeper examination. Regular expression matching is implemented in the second stage.

Our algorithm requires regular expression preprocessing before it can be applied. This preprocessing is a very simple process. It requires an extraction of static sub-pattern from the regular expression patterns. The static sub-pattern is the part of the regular expression that does not contain wildcard characters. The bit-length of the sub-pattern can be set to the value that fits $p = \lceil log_2(m) \rceil$.

For the best performance of the software part of the algorithm it is requested that the sub-pattern for the different regular expressions are different. In other words, we want the sub-patterns to distinguish regular expressions as much as possible. For example, for d different regular expressions we want $n = d$ different static patterns of bit-length p. It should be mentioned here that the length of patterns can be bigger than $\lceil log_2(m) \rceil$. In such a case a compression that uses a hash function can be applied. That hash function returns a result of length p. In some cases the use of hashing helps achieve a uniform patterns to regular expressions mapping.

The Bloom filter hardware processor inspects data for characteristic static patterns. Their occurrences in the inspected block of data are returned to the CPU as the locations in the data block. Then, the CPU extracts patterns from locations and checks if a regular expression matches. To start the regular expression matching a proper regular expression must be fetched from the regular expressions database. Its retrieval is sped-up thanks to a hash index. The hash indexes of length h and a hash table of size 2^h is used. The hash table contains positions of the regular expression body in a database. The hashes are used to address the table and the database's position is read from the hash index table. As $h < p$, it happens that more than one regular expression has to be checked for a single hash value. This functionality is provided by the database organization that allows it to store more than one regular expression pattern at its single entry. Merged lists of regexs are created and stored as indexed records in the database.

The scheme of the algorithm's operation is presented in Fig. 3. In a well-balanced system, the performance (i.e. the maximum input data rate) of regular expression verification stage must fit the output data rate of the Bloom filtering stage.

5   The Parallel Architecture of Custom Processor

A custom processor for the Bloom filtering was created in the reconfigurable hardware. Basically, it is implemented as a dual port synchronous RAM memory. Patterns are directly set to the memory's address bus. The size of the bus fits the bit-width of patterns. To achieve the highest throughput, the execution of the incoming data is performed both in a parallel and pipelined manner. The input buffer is matched with a single byte resolution. To fit these requirements with the 32-bit architecture of the host system, the Bloom processor consist of 4 parallel block modules. The modules inspect the input data with a different byte offset (range from 0 to 3). This concept is presented in Fig. 4. The architecture can process 32-bits of data at each clock cycle. The throughput is $4 \times f_{clk}$ bytes per second. The processor can return a single 32-bit word at each cycle. The returned word is a 32-bit value that contains a position of a matching pattern in the input data buffer. As a match can occur in all Bloom blocks simultaneously, it is necessary to include fifos to buffer the output data. The input values are

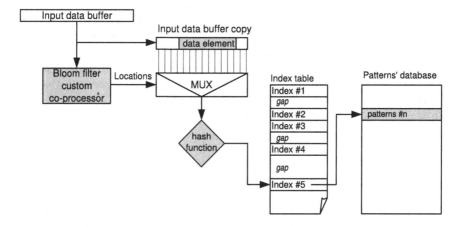

Fig. 3. Principles of the algorithm operation.

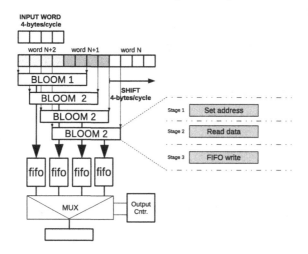

Fig. 4. Parallel and pipeline Bloom filter implementation.

processed in a pipeline. The FPGA's memory consumes one clock cycle to read the data and one clock cycle is necessary to write the data to the fifos. The architecture is programmed with the Bloom filter values during the run-time. Despite the normal operating mode, the matching also implements a teaching mode. In this mode the input data patterns are written to the internal memory.

6 Implementation Results

The presented algorithm was tested on the Zedboard development board [18]. It is built on Xilinx's XC7Z020 System on a Chip that is from the family of the Zynq-7000 All Programmable SoC. The SoC combines the ARM's sub-system and the FPGA sub-system in a single integrated circuit. The ARM sub-system is a dual core Cortex-A9 Processing System (PS). The FPGA sub-system consisting of 85,000 logic cells is the Xilinx's series-7 Programmable Logic (PL) structure. Additionally PL contains 140 blocks of 36-kbit Block RAM (BRAM) memory. Zedboard offers 512 MB of 533 MHz DDR3 memory, which is organized as a 32-bit operating memory. An operating frequency of ARM processor is 800 MHz. All performance tests were performed on the Zedboard under control of the Linux operating system. Linux used Busybox filesystem that was implemented on a ramdisk. The size of BRAM memory that is offered by Zedboard's XC7Z020 is 5.4 Mbit. This allowed us to implement the Bloom filter to $p = 20$. As XC7Z020 is one of the smallest chips in the Zynq-7000 family, additionally, the simulation results of the Bloom filter implemented on XC7Z045 are given in Table 2. The size of the XC7Z045's PL structure allowed us to fit the Bloom filter to $p = 22$. The clock frequency for both FPGA implementations was 100 MHz. For the 32-bit input interface, the peak data throughput of the designs is 400 MB/s.

Table 1. The performance of the CPU and the FPGA for the algorithm's procedures

Processor parameters	Algorithm rate	
	Bloom filtering MB/s	Reg. Ex. Matching patterns/s
Intel Xeon E5645 2.4 GHz CPU 12 MB cache	113	1,582,000
Single core ARM Cortex A-9 800 MHz CPU 512 kB cache	70	365,000
Dual core ARM Cortex A-9 800 MHz $2 \times CPU$ 512 kB cache	140	730,000
Zynq (PL) 100 MHz FPGA	400	-

Table 2. The FPGA Bloom filter implementation results

ZYNQ Bloom filter	Total	Used	Rout (for Rin=320 MB/s)
xc7z020 p = 20	53,200 LUTs	1,580 LUTs (2 %)	0.30 M hits/s (n = 1,000)
	106,400 FFs	960 FFs(1 %)	**0.61 M hits/s** (n = 2,000)
	140 BRAMs	130 BRAMs (92 %)	0.91 M hits/s (n = 3,000)
xc7z045 p = 22	218,600 LUTs	1,993 LUTs (1 %)	0.30 M hits/s (n = 4,000)
	437,200 FFs	966 FFs (1 %)	**0.61 M hits/s (n = 8,000)**
	545 BRAMs	514 BRAMs (94 %)	0.91 M hits/s (n = 12,000)

For reference, the software version of the Bloom filter was implemented as well. The use of a processor's cache was assumed for the Bloom memory. The results of the software performance for $p = 22$ is given in Table 1.

To evaluate the performance of the regular expressions verification procedure, virus bodies from the ClamAV malware database were used. The characteristic static sub-patterns for the ClamAV's viruses were found using an exhaustive search algorithm. The goal was to find the unique sub-patterns for each virus definition. However, that requirement was not met and some of the viruses shared common sub-patterns. It did not obstruct our algorithm to operate properly.

The performance of the regular expression verification procedure was measured for the ARM and Intel CPUs. For this measure, a hash index of size $h = 14$ was established. The average rate of a number of patterns verified in a second is given in Table 1. For the verification procedure, the locations of static sub-patterns in the inspected data buffer is already known from the Bloom filter procedure. A method to obtain the average rate was to measure verification time of input data that contained all viruses' bodies that were concatenated. The measured time was divided by the number of viruses to calculate the average rate of patterns per second. It should be noted that the given rates in Table 1 are pessimistic, the worst-case values. They were measured when all verifications were positive. In practice, only a small fraction of verification would be positive. The algorithm of the verification procedure performs a negative verification faster because the procedure terminates immediately when the first wildcards' match fails.

Table 2 gives an expected output match rate R_{out} of the Bloom filter for $h = 20$ and $h = 22$. According to Eq. 3, the value R_{out} is given for the different numbers of patterns n. The calculations in Table 2 were done for the value $R_{in} = 320MB/s$, as the data to the Bloom filter must be transfered from the operation memory. The data in Table 2 shows that for the XC7Z045 a number of implemented patterns could be 8,000. That value fits the pessimistic (for positive verifications only) Dual Core ARM CPUs performance. The respective value for the XC7Z020 is 2,000 patterns.

We used the Xilinx Power Estimator (XPE) spreadsheet tool [19] to calculate the power consumption of the SoC system. The energy consumption results of the design were as follows:

- single CPU (800 MHz) power consumption: 1.0 W
- dual CPU (800 MHz) power consumption: 1.5 W
- FPGA (100 MHz) power consumption: 1.3 W
- **Total power consumption (2×CPU+FPGA): ~2.8 W**

7   Conclusions and Further Work

The Bloom filter performance of an FPGA custom co-processor outperforms the CPUs' performance for a similar algorithm. The achieved speed-up is six when compared to a single ARM core. The real advantage of the presented solution is that the FPGA custom processor and the CPUs work in parallel so the FPGA does not replace the CPU, but adds additional performance to the system. The energy consumption is probably the most notable measure of the design.

Compared to the solution of Chang et al. [14], our design offers a dest 2.56 Gbps of throughput (7.27 Gbps in [14]). The performance is lower, but the competitive FPGA solutions offer implementation of fewer regular expressions (e.g. 24,214 characters of regular expressions in [13]) Our solution offers a support for 2,000 (8,000 for XC7Z045) of complete regular expressions. Also, the method described in [13] could be adopted to our design. Thanks to the common sub-patterns sharing a number of patterns implemented in the Bloom filter could be reduced. The spared logic could be used to gain additional parallelism that would increase the throughput. On the other hand, the internal system throughput should fit the IO devices capabilities and the value of 2.56 Gbps seems to be a reasonable value when compared to the performance of TCP/IP stack implementations for Ethernet standard.

Further work will be focused on the scaling of the presented single-node solution to a multi-node system. Obviously, a single ARM processor is not capable of handling the size of today's virus database (e.g. over 100,000 body based virus signatures in ClamAV). The presented algorithm can be scaled up using the MPI (Message Passing Interface) programming model for distributed memory systems.

Acknowledgment. This work was supported by: the National Science Centre (NCN) under Grant No. 18.18.120.146 and the National Centre for Research and Development (NCBiR) under Grant No. SP/I/1/77065/10.

References

1. ClamAV: Clam antivirus signature database. http://www.clamav.net. Accessed 1 Feb 2013
2. Jamro, E., Russek, P., Dabrowska-Boruch, A., Wielgosz, M., Wiatr, K.: The implementation of the customized, parallel architecture for a fast word-match program. Comput. Syst. Sci. Eng. **26**(4), 285–292 (2011)
3. Russek, P., Wiatr, K.: The enhancement of a computer system for sorting capabilities using FPGA custom architecture. Comput. Inform. **32**(4), 859–876 (2013)

4. Xilinx.: Zynq-7000 AP SoC Overview. http://www.xilinx.com. Accessed 1 Feb 2013
5. Yu, F., Chen, Z., Diao, Y., Lakshman, T.V., Katz, R.H.: Fast and memory-efficient regular expression matching for deep packet inspection. In: ACM/IEEE Symposium on Architecture for Networking and Communications Systems, ANCS 2006, pp. 93–102 (2006)
6. Pasetto, D., Petrini, F., Agarwal, V.: Tools for very fast regular expression matching. IEEE Comput. **43**(3), 50–58 (2010)
7. Scarpazza, D.P., Russell, G.F.: High-performance regular expression scanning on the Cell/B.E. processor. In: 23rd International Conference on Supercomputing (ICS09) (2009)
8. Iorio, F., Lunteren, J.V.: Fast pattern matching on the cell broadband engine. In: 2008 Workshop on Cell Systems and Applications (WCSA), Affiliated with the 2008 International Symposium on Computer Architecture (ISCA08), Beijing, China, June 2008
9. Cascarano, N., Rolando, P., Risso, F., Sisto, R.: iNFAnt: NFA pattern matching on GPGPU devices. SIGCOMM Comput. Commun. Rev. **40**(5), 20–26 (2010)
10. Naghmouchi, J., Scarpazza, D.P., Berekovic, M.: Small-ruleset regular expression-matching on GPGPUs: quantitative performance analysis and optimization. In: 24th International Conference on Supercomputing (ICS'10), pp. 337–348 (2010)
11. Sidhu, R., Prasanna, V.K.: Fast regular expression matching using FPGAs. In: IEEE Symposium on Field-Programmable Custom Computing Machines (2001)
12. Sourdis, I., Bispo, J.A., Cardoso, J.A.M., Vassiliadis, S.: Regular expression matching in reconfigurable hardware. J. Signal Process. Syst. **51**(1), 99–121 (2008)
13. Lin, C.-H., Huang, C.-T., Jiang, C.-P., Chang, S.-C.: Optimization of regular expression pattern matching circuits on FPGA. In: Proceedings of the Conference on Design, Automation and Test in Europe. Designers Forum, DATE06, European Design and Automation Association, Leuven, Belgium, pp. 12–17 (2006)
14. Chang, Y.-K., Chang, C.-R., Su, C.-C.: The cost effective pre-processing based nfa pattern matching architecture for nids. In: Proceedings of the 2010 24th IEEE International Conference on Advanced Information Networking and Applications, AINA 10, IEEE Computer Society, Washington, DC, USA, pp. 385–391 (2010)
15. Serrano, B.: String alignment pre-detection using unique subsequences for FPGA-based network intrusion detection. Comput. Commun. **35**(6), 720–728 (2012)
16. Dharmapurikar, S., Krishnamurthy, P., Sproull, T.S., Lockwood, J.W.: Deep packet inspection using parallel bloom filters. IEEE Micro **24**(1), 52–61 (2004)
17. Lockwood, J.W., Moscola, J., Kulig, M., Reddick, D., Brooks, T.: Internet worm and virus protection in dynamically reconfigurable hardware. In: Military and Aerospace Programmable Logic Device (MAPLD), p. 10 (2003)
18. Zedboard: ZedBoard Hardware User's Guide. http://www.zedboard.org. Accessed 1 Feb 2013
19. Xilinx: Xilinx Power Estimator. http://www.xilinx.com/power. Accessed 1 Feb 2013

Workshop on Numerical Algorithms on Hybrid Architectures

Performance Evaluation of Sparse Matrix Multiplication Kernels on Intel Xeon Phi

Erik Saule[1]([✉]), Kamer Kaya[1], and Ümit V. Çatalyürek[1,2]

[1] Department of Biomedical Informatics, The Ohio State University, Columbus, USA
[2] Department of Electrical and Computer Engineering, The Ohio State University, Columbus, USA
esaule@uncc.edu, {kamer,umit}@bmi.osu.edu

Abstract. Intel Xeon Phi is a recently released high-performance coprocessor which features 61 cores each supporting 4 hardware threads with 512-bit wide SIMD registers achieving a peak theoretical performance of 1Tflop/s in double precision. Its design differs from classical modern processors; it comes with a large number of cores, the 4-way hyperthreading capability allows many applications to saturate the massive memory bandwidth, and its large SIMD capabilities allow to reach high computation throughput. The core of many scientific applications involves the multiplication of a large, sparse matrix with a single or multiple dense vectors which are not compute-bound but memory-bound. In this paper, we investigate the performance of the Xeon Phi coprocessor for these sparse linear algebra kernels. We highlight the important hardware details and show that Xeon Phi's sparse kernel performance is very promising and even better than that of cutting-edge CPUs and GPUs.

Keywords: Intel Xeon Phi · SpMV · SpMM

1 Introduction

Given a large, sparse, $m \times n$ matrix $\mathbf{A}$, an input vector $\mathbf{x}$, and a cutting edge shared-memory manycore architecture Intel Xeon® Phi, we are interested in analyzing the performance of computing $\mathbf{y} \leftarrow \mathbf{A}\mathbf{x}$ in parallel. The computation, known as the sparse-matrix vector multiplication (SpMV), and with some variants, such as the sparse-matrix matrix multiplication (SpMM), they form the computational core of many applications involving linear systems, eigenvalues, and linear programs, i.e., most large scale scientific applications. For this reason, they have been extremely intriguing in the context of high performance computing (HPC). Efficient shared-memory parallelization of these kernels is

This work was partially supported by the NSF grants CNS-0643969, OCI-0904809 and OCI-0904802. We would like to thank NVIDIA for the K20 cards, Intel for the Xeon Phi prototype, and the Ohio Supercomputing Center for access to Intel hardware.

R. Wyrzykowski et al. (Eds.): PPAM 2013, Part I, LNCS 8384, pp. 559–570, 2014.
DOI: 10.1007/978-3-642-55224-3_52, © Springer-Verlag Berlin Heidelberg 2014

well studied [1–3,9,19], and there exist several techniques such as prefetching, loop transformations, vectorization, register, cache, TLB blocking, and NUMA optimization, which have been extensively investigated to optimize the performance [7,11,12,19]. In addition, company-built support is available for many shared-memory architectures, such as Intel's MKL and NVIDIA's cuSPARSE. Popular 3rd party libraries such as OSKI [18] and pOSKI [8] also exist.

Intel Xeon Phi is a new coprocessor with many cores, hardware threading capabilities, and wide vector registers. Although Intel Xeon Phi has been released recently, performance evaluations already exist in literature [6,15,17]. Eisenlohr et al. investigated the behavior of dense linear algebra factorization on Xeon Phi [6] and Stock et al. proposed an automatic code optimization approach for tensor contraction kernels [17]. For sparse, irregular data, we evaluated the scalability of graph algorithms, coloring and breadth first search (BFS) [15].

Although similar to BFS, SpMV and SpMM are different kernels, in terms of synchronization, memory access, and load balancing. The irregularity and sparsity of SpMV-like kernels create several problems for accelerators. In this paper, we analyze how Xeon Phi performs on SpMV and SpMM. Reference [4] studied the performance of a Conjugate Gradient application which uses SpMV, however this study concerns only a single matrix and is application oriented.

Having 61 cores and hyperthreading capability can help the Intel Xeon Phi to saturate the memory bandwidth during SpMV, which is not the case for many cutting edge processors. Yet, our analysis showed that the memory latency, not the memory bandwidth, is the bottleneck and the reason for not reaching to the peak performance. We observed that the performance of the SpMV kernel highly depends on the nonzero pattern of the matrix and its sparsity: when the nonzeros in a row are aligned and packed in cachelines in the memory, the memory accesses are much faster. We investigate two existing approaches for densifying the computation (namely the reverse Cuthill-McKee ordering *RCM* [5] and dense register blocking). This paper presents concise results and a more detailed version of our work can is available as a technical report [16].

Section 2 presents a brief architectural overview of the Intel Xeon Phi coprocessor. Section 3 describes the sparse-matrix multiplication kernels. In Sects. 4 and 5, we conduct analyze Xeon Phi's performance on these kernels using 22 matrices from UFL Sparse Matrix Collection[1]. Section 6 shows that Xeon Phi's sparse matrix performance is better than that of four modern architectures: two dual Intel Xeon processors, X5680 (Westmere) E5-2670 (Sandy Bridge), and two NVIDIA Tesla® GPUs C2050 and K20. Section 7 concludes the paper.

2   The Intel Xeon Phi Coprocessor

In this work, we use a pre-release KNC card SE10P. There are 61 cores clocked at 1.05 GHz. Each core in the architecture has a 32 kB L1 data cache, a 32 kB L1 instruction cache, and a 512 kB L2 cache. The architecture of a core is based on

---

[1] http://www.cise.ufl.edu/research/sparse/matrices/

the Pentium architecture: though its design has been updated to 64-bit. A core can hold 4 hardware contexts at any time. A core never executes two instructions from the same hardware context consecutively: in other words, if a program only uses one thread, half of the clock cycles are wasted.

Most of the performance of the architecture comes from the vector processing unit (VPU). Each core has 32×512-bit SIMD registers which can be used as a vector of 8×64-bit or 16×32-bit values. The VPU can perform many basic instructions, such as addition or division, and mathematical operations, such as sine and sqrt, allowing to reach 8 double precision operations per cycle (16 single precision). The VPU can also perform both an addition and a multiplication simultaneously using a *Fused Multiply-Add* (FMA) instruction. Therefore, the peak performance of the SE10P card is 1.0248 Tflop/s in double precision (2.0496 Tflop/s in single precision) and half without FMA.

The card has 8 memory controllers; each can execute 5.5 billion transactions/second and has two 32-bit channels. Hence, the controllers can achieve an aggregated total bandwidth of 352 GB/s. The cores' memory interface are 32-bit wide with two channels and the total bandwidth is 8.4 GB/s per core. Thus, the cores can consume 512.4 GB/s at most. However, the bandwidth between the cores and the memory controllers is limited by the ring network which and can theoretically transport at most 220 GB/s. There is a total of 8 GB of memory.

To better understand the performance of Intel Xeon Phi, we designed two simple benchmarks on read and write bandwidth. In both cases, each thread reads or writes large arrays into the memory multiple times. The read-bandwidth benchmark shows four configurations. The first two read the array one byte at a time or four bytes at a time; they are instruction bound and reach respectively 12 GB/s and 60 GB/s. The third benchmark (vect) uses SIMD instructions to process 64 bytes at a time; it reaches 171 GB/s. The last one (vect+pref) adds prefetching instructions and obtains 177 GB/s. Results are presented in Fig. 1(a).

Figure 1(b) shows a similar benchmark for write operations. All three tested configurations use vectorized write instructions to overcome the instruction bound. The first benchmark uses a simple store operation and reaches 65 GB/s. The second configuration disables the Read For Ownership protocol (which forces

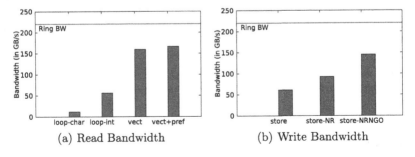

(a) Read Bandwidth   (b) Write Bandwidth

Fig. 1. Benchmarking read and write bandwidth with various instructions. The ring bus theoretical maximal bandwidth is shown.

the processor to bring a cacheline into the cache before being able to write it) by using a No-Read hint (NR) and improves the write bandwidth to 99 GB/s. The last configuration allows the write operations to be committed to the memory in an arbitrary order using the Non Globaly Ordered write instructions (NRNGO); it yields 155 GB/s.

More detailed experiments can be found in our report [16] which shows that when using vect+pref and store-NRNGO, the bandwidth scales sublinearly with the number of cores, indicating a contention on the memory subsystem.

3   Sparse Multiplication Kernels

SpMV is in the form $\mathbf{y} \leftarrow \mathbf{Ax}$ where $\mathbf{A}$ is an $m \times n$ sparse matrix, and $\mathbf{x}$ and $\mathbf{y}$ are $n \times 1$ and $m \times 1$ column vectors. In this kernel, each nonzero is accessed, multiplied with an $\mathbf{x}$-vector entry, and the result is added to a $\mathbf{y}$-vector entry once. That is, there are two reads and one read-and-write per each nonzero accessed. Different from SpMV, in SpMM, $\mathbf{x}$ and $\mathbf{y}$ are $n \times k$ and $m \times k$ dense matrices. Hence, there are k reads and k read-and-writes per each nonzero accessed. Obtaining a good performance for SpMV is difficult on almost any architecture due to the sparsity pattern of $\mathbf{A}$ which yields a non-regular access to the memory. The amount of computation per nonzero is also very small. And most of the operations suffer from bandwidth limitation.

An $m \times n$ sparse matrix $\mathbf{A}$ with τ nonzeros is usually stored in the compressed row storage format CRS which uses three arrays:

- $cids[.]$ is an integer array of size τ that stores the column ids for each nonzero in row-major ordering.
- $rptrs[.]$ is an integer array of size $m+1$. For $0 \leq i < m$, $rptrs[i]$ is the location of the first nonzero of the ith row in the $cids$ array. The first element is $rptrs[0] = 0$, and the last element is $rptrs[m] = \tau$. Hence, all the column indices of row i are stored between $cids[rptrs[i]]$ and $cids[rptrs[i+1]] - 1$.
- $val[.]$ is an array of size τ. $val[i]$ is the value of the ith nonzero.

There exist other sparse matrix representations [14], and the best storage format almost always depends on the pattern of the matrix and the kernel. In this work, we use CRS as it constitutes a solid baseline. Since $\mathbf{A}$ is represented in CRS, it is straightforward to assign a row to a single thread in a parallel execution. Each entry $\mathbf{y}_i$ of the output vector can be computed independently while streaming the matrix row by row. While processing a row i, multiple $\mathbf{x}$ values are read, and the sum of the multiplications is written to $\mathbf{y}_i$. Hence, there are one multiply and one add operation per nonzero, and the total number of floating point operations is 2τ.

4   SpMV on Intel Xeon Phi

For the experiments, we use a set of 22 matrices given in Table 1. The matrices are taken from the UFL Sparse Matrix Collection with one exception *mesh_2048*

Table 1. Properties of the matrices used in the experiments. All matrices are square.

#	name	#row	#nonzero		#	name	#row	#nonzero
1	shallow_water1	81,920	204,800		12	pwtk	217,918	5,871,175
2	2cubes_sphere	101,492	874,378		13	crankseg_2	63,838	7,106,348
3	scircuit	170,998	958,936		14	torso1	116,158	8,516,500
4	mac_econ	206,500	1,273,389		15	atmosmodd	1,270,432	8,814,880
5	cop20k_A	121,192	1,362,087		16	msdoor	415,863	9,794,513
6	cant	62,451	2,034,917		17	F1	343,791	13,590,452
7	pdb1HYS	36,417	2,190,591		18	nd24k	72,000	14,393,817
8	webbase-1M	1,000,005	3,105,536		19	inline_1	503,712	18,659,941
9	hood	220,542	5,057,982		20	mesh_2048	4,194,304	20,963,328
10	bmw3_2	227,362	5,757,996		21	ldoor	952,203	21,723,010
11	pre2	659,033	5,834,044		22	cage14	1,505,785	27,130,349

which corresponds to a 5-point stencil 2048×2048 mesh in 2D. We used the CRS representation, store all the scalar values in double precision, and all the indices via 32-bit integers. In the rest of the paper, the matrices are ordered from 1 to 22 by increasing number of nonzero entries. We repeated each operation 70 times and compute the averages of the last 60 operations. Caches are flushed between each measurement.

4.1   Performance Evaluation

The SpMV kernel is implemented in C++ using OpenMP and processes the rows in parallel. We tested our dataset with multiple scheduling policies when compiled with `icc 13.0` in `-O1` and `-O3` (see [16] for more details). When compiled with `-O1`, the performance obtained varies from 1 to 13GFlop/s. When compiled with `-O3` the performance rises for all matrices and reaches 22GFlop/s on *nd24k*. In total, 5 matrices from our set achieve a performance over 15GFlop/s.

An interesting observation is that the difference on the performance is not constant; it depends on the matrix and not correlated to its size. Analyzing the compiled assembly code for the SpMV inner loop (which computes the dot product between a sparse matrix row and the dense input vector) gives an insight on why the performances differ. When `-O1` is used, the dot product is implemented in a simple way, one element at a time, with 3 memory indirections, one increment, one addition, one multiplication, one test, and one jump per nonzero.

The code generated in `-O3` is much more complex. It uses vectorial operations so as to make 8 operations at once. The compiled code loads 8 consecutive values of the sparse row in a 512-bit SIMD register in a single operation. Then it populates another SIMD register with the values of the input vector. Once populated, the two vectors are multiplied and accumulated with previous results in a single FMA. Populating the SIMD register with the appropriate values from **x** is non trivial since these values are not consecutive in memory. However, Xeon Phi offers an instruction, *vgatherd*, that allows to fetch multiple values at once. The instruction takes an offset vector, a pointer to the beginning of the array, and a destination register. In general, *vgatherd* needs to be called as many times as the number of cachelines the offset vector touches (indicated by a auxiliary

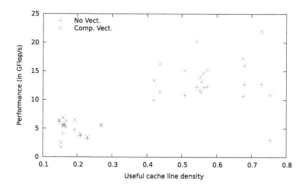

Fig. 2. The improvement of -O3 (Comp. Vect.) is linked to cacheline density.

bit-mask), since it can only simultaneously fetch the elements that are on the same cacheline. So overall, one FMA, two vector loads (one for the nonzero from the matrix and one for the column positions), one increment, one test, and some *vgatherd* are performed for each 8 nonzeros of the matrix.

Figure 2 shows the SpMV performance for each matrix with -O1 and -O3 as a function of the *useful cacheline density* (UCLD), a metric we devised for the analysis. For each row, we computed the ratio of the number of nonzeros on that row to the number of elements in the cachelines of the input vector due to that row. Then we took the average of these values to compute UCLD. For each matrix, there are two points in Fig. 2, one for the performance in -O1 (marked with '+'s) and one for the performance in -O3 (marked with '×'s). These points are horizontally aligned for the same matrix, and their vertical distance represents the improvement on the performance for that matrix. The improvement with vectorization, and in particular with *vgatherd*, is significantly much higher when the UCLD is high.

4.2   Bandwidth Considerations

The nonzeros in the matrix need to be transferred to the core before being processed. Assuming the access to the vectors do not incur any memory transfer, and since each nonzero takes 12 bytes (8 for the value and 4 for the column index) and incurs two floating point operations (multiplication and sum), the flop-to-byte ratio of SpMV is $\frac{2}{12} = \frac{1}{6}$. We saw that the sustained memory bandwidth is about 180 GB/s, which indicates a maximum performance for the SpMV kernel of 30GFlop/s. This is not obtained by our previous experiments.

Assuming only 12 bytes per nonzero need to be transfered to the core gives only a *naive bandwidth* for SpMV: both vectors and the row indices also need to be transferred. For an $n \times n$ matrix with τ nonzeros, the actual minimum amount of memory that need to be transferred in both ways is $2 \times n \times 8 + (n + 1) \times 4 + \tau \times (8 + 4) = 4 + 20 \times n + 12 \times \tau$. Usually, 12τ dominates the equation, but for sparser matrices, $20n$ should not be ignored. The *application bandwidth*,

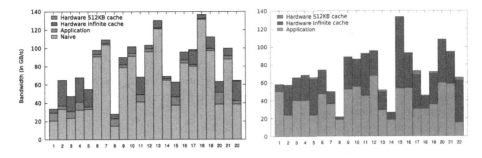

Fig. 3. The achieved bandwidth for SpMV (left) and SpMM (right).

which takes both terms into account, is a common alternative cross-architecture measure of performance on SpMV.

Figure 3 (left) shows that the naive approach which ignores a significant portion of the data for some matrices. The application bandwidth obtained ranges from 22 GB/s to 132 GB/s. Most matrices have a bandwidth below 100 GB/s.

The application bandwidth is computed assuming that every single byte of the problem is transferred exactly once. This assumption is (mostly) true for the matrix and the output vector. However, it does not hold for the input vector for two reasons: first, it is unlikely that each vector element will be used by only a single core's threads, some element will be transferred to multiple cores. Furthermore, a core's cache is only 512 kB, and elements of the input vector may be transfered multiple times to the same core. We analytically computed the number of cachelines accessed by each core assuming that chunks of 64 rows are distributed in a round-robin fashion. We performed the analysis assuming an infinite cache and with a 512 kB cache. We computed the effective memory bandwidth of SpMV and display them as the top two stacks of the bars in Fig. 3 (left). Three observations are striking: first, the difference between the application bandwidth and estimated actual bandwidth is greater than 10 GB/s on 10 instances and more than 20GB/s on three of them. The highest difference is seen on *2cubes_sphere* (#2) where the amount of data transferred is 1.7 times larger than the application bandwidth. Second, there is no significant difference between the assumed infinite cache and 512 kB cache bandwidth. That is, no cache thrashing occurs. Finally, even when we take the actual memory transfers into account, the obtained bandwidth is still way below the architecture peak bandwidth.

4.3   Effect of Matrix Ordering

A widely-used approach to improve SpMV performance is ordering the rows and columns to make the matrix more suitable for the kernel. Such permutations are used in sparse linear algebra for multiple purposes such as improving numerical stability and preconditioning. Here, we employ the reverse Cuthill-McKee algorithm (RCM) [5]. RCM has been widely used for minimizing the maximum distance between the nonzeros and the diagonal of the matrix, i.e., the *bandwidth of the matrix.* We expect that such a densification of the nonzeros can improve both the UCLD of the matrix and reduce the number of times the vector needed to be transfered from the main memory to the core caches.

RCM improves the performance of only 4 matrices by more than 2GFlop/s and most of the matrices benefit less. The performance of 8 matrices degrade. Hence, RCM ordering was not able to significantly improve SpMV on Intel Xeon Phi (see [16] for more details).

4.4 Effect of Register Blocking

One limitation in the original SpMV implementation is that only a single nonzero is processed at a time. Register blocking helps us to process all the nonzeros within a region at once. The region should be small enough that the data associated with it can be stored in the registers so as to minimize memory accesses [7]. Assuming a regular partitioning $\mathbf{A}$ to blocks of size $a \times b$, we use a dense block representation for the blocks containing at least one nonzero. We represent this list of non-empty blocks via CRS. One dimension of the blocks is set to 8 to leverage the Xeon Phi architecture which naturally align on 512 bits, the other dimension varies from 1 to 8. To perform the multiplication, each dense block is loaded into the registers in packs of 8 values allowing to use Fused Multiply-Add operations. Register blocking typically helps the performances by reducing three quantities: (1) the matrix size in memory, (2) the number of instructions to perform the multiplications, (3) the number of load instructions to the vector.

Overall, we could not observe a constant improvement for register blocking on Xeon Phi. The best scheme with 8×1 blocks improved the performance on only 8 instances compared to the original implementation (detailed results in [16]). Register blocking allows to reach a much higher utilization of the hardware (the effective memory bandwidth is over 160GB/s) but this does not compensate the large increase in matrix sizes. Indeed, the matrices we used have a low locality leading to a sharp increase in the size of the matrix when encoded using dense tiles. There is almost no reduction of the load instruction of the vector since the *vgatherd* instruction already reads the input vector per batch.

5 SpMM on Intel Xeon Phi

One idea to obtain more performance is to increase the flop-to-byte ratio by performing more than one SpMV at a time. Many applications can take the advantage of using multiple vectors at once, e.g., graph based recommendation systems [10] or eigensolvers (by the use of the LOBPCG algorithm) [20]. Multiplying several vectors by the same matrix boils down to multiplying a sparse matrix by a dense matrix, which we refer to as SpMM. All the statements above are also valid for existing cutting-edge processors and accelerators. However, with its large SIMD registers, Xeon Phi is expected to perform significantly better.

To implement $\mathbf{Y} \leftarrow \mathbf{A}\mathbf{X}$, we encode the dense $m \times k$ input matrix $\mathbf{X}$ in row-major, so each row is contiguous in memory. To process a row $\mathbf{A}_{i*}$ of $\mathbf{A}$, a temporary array of size k is first initialized to zero. Then for each nonzero in $\mathbf{A}_{i*}$, a row $\mathbf{X}_{j*}$ is streamed to be multiplied by the nonzero and the result is accumulated into the temporary array. We developed three variants of that algorithm: the first variant is generic and relies on compiler vectorization. The second is tuned for values of k which are multiple of 8 and uses FMA to perform the multiplications and additions of 8 at a time. The temporary values are kept in registers by taking the advantage of the large number of SIMD registers available on Xeon Phi. The third variant also uses Non-Globally Ordered write instructions with No-Read hint (NRNGO).

We experimented with $k = 16$. Manual vectorization doubles the performance allowing to reach more than 60GFlop/s in 11 instances. The use of NRNGO write instructions provides significant performance improvements. The achieved performance peaks on the matrix *pwtk* matrix at 128GFlop/s. Figure 3 (right) shows the bandwidth achieved by the best implementation (complete results are in [16]). The application bandwidth is computed assuming the matrix and vector are transferred only once. It surpasses 60 GB/s in only 1 instance. Since there are 16 input vectors, the overhead induced by transferring the values in $\mathbf{X}$ to multiple cores is comparable to the application bandwidth. The impact of having a finite cache is mostly negligable.

6   Against Other Architectures

We compare the performance of Xeon Phi with 4 other architectures including 2 GPU configurations and 2 CPU configurations. We used two CUDA-enabled cards from NVIDIA: an NVIDIA Tesla C2050 (448 CUDA Cores @ 1.15 GHz, 2.6 GB memory @ 1.5 GHz, ECC on, CUDA 4.2) and an NVIDIA Tesla K20 (2,496 CUDA Cores @ 0.71 GHz, 4.8 GB memory @ 2.6 GHz, ECC on, CUDA 5.0). For both GPU configurations, we use the CuSparse library. We also use two Intel CPU systems: the first has a dual Intel Xeon X5680 (Westmere: 6 cores @ 3.33 Ghz, no hyperthreading, 12 MB shared L3 cache). The second has a dual Intel Xeon E5-2670 (Sandy Bridge: 8 cores @ 2.6 GHz, hyperthreading enabled, 20 MB shared L3 cache). The codes for both CPU architectures are compiled with the `icc 13.0` with `-O3` optimization flag. The implementation used is the same as the one used on Xeon Phi except the vector optimizations in SpMM where the instruction sets differ.

Results of the experiments are presented in Fig. 4. We present the configurations as stacked bar charts: K20 on top of C2050 and E5-2670 on top of X5680. Figure 4 (left) shows the SpMV results: E5-2670 appears to be roughly twice faster than X5680. It reaches a performance between 4.5 and 7.6GFlop/s and achieves the highest performance for one instance. For GPU architectures, K20 is faster than the C2050. It performs better for 18/22 instances. It obtains between 4.9 and 13.2GFlop/s and the highest performance on 9 instances. Xeon Phi reaches the highest performance on 12 instances and it is the only architecture which obtains more than 15GFlop/s. Furthermore, it does it for 7 instances.

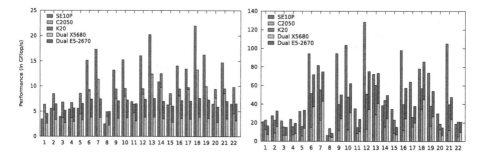

Fig. 4. Architectural comparison between a Intel Xeon Phi coprocessor (Pre-release SE10P), two NVIDIA GPUs (C2050 and K20) and two dual CPU architectures (Intel Xeon X5680 and Intel Xeon E5-2670) for SpMV (left) and SpMM (right).

Figure 4 (right) shows the SpMM results: E5-2670 gets twice the performance of X5680, which is similar to their relative SpMV performances. The K20 is often more than twice faster than C2050, which is much better compared with their relative performances in SpMV. The Xeon Phi coprocessor gets the best performance in 14 instances. Intel Xeon Phi is the only architecture which achieves more than 100GFlop/s. Furthermore, it reaches more than 60GFlop/s on 9 instances. The CPU configurations reach more than 60GFlop/s on 6 instances while the GPU configurations never achieve that performance.

7   Conclusion and Future Work

In this work, we analyze the performance of Intel Xeon Phi coprocessor on SpMV and SpMM. These sparse algebra kernels have been used in many important applications. The analysis gives the first absolute performance results of Xeon Phi. Overall, the performance we obtained is very promising. When compared with cutting-edge processors and accelerators, its SpMV, and especially SpMM, performance are superior thanks to its wide registers and vectorization capabilities.

In particular, we showed that the sparse matrix kernels we investigated are latency bound. Our experiments suggested that having a relatively small 512 kB L2 cache per core is not a problem for Intel Xeon Phi. However, having 61 cores induces a significant data transfer overhead due to accessing similar parts of **x** and **X** from multiple cores, especially in SpMM. We linked the performance of SpMV to the efficacy of the *vgatherd* instruction which allows efficient memory loads. The classical techniques to improve the performance of SpMV appeared to bring little improvements on Xeon Phi. As a future work, we are planning to investigate matrix storage schemes, intra-core locality, and data partitioning to improve the performance of Xeon Phi.

References

1. Bell, N., Garland, M.: Implementing sparse matrix-vector multiplication on throughput-oriented processors. In: Proceedings of the High Performance Computing Networking, Storage and Analysis, SC '09 (2009)
2. Buluç, A., Fineman, J.T., Frigo, M., Gilbert, J.R., Leiserson, C.E.: Parallel sparse matrix-vector and matrix-transpose-vector multiplication using compressed sparse blocks. In: Proceedings of the SPAA '09, pp. 233–244 (2009)
3. Buluç, A., Williams, S., Oliker, L., Demmel, J.: Reduced-bandwidth multithreaded algorithms for sparse matrix-vector multiplication. In: Proceedings of the IPDPS (2011)
4. Cramer, T., Schmidl, D., Klemm, M., an Mey, D.: Openmp programming on intel xeon phi coprocessors: an early performance comparison. In: Proceedings of the Many-core Applications Research Community (MARC) Symposium at RWTH Aachen University, November 2012
5. Cuthill, E., McKee, J.: Reducing the bandwidth of sparse symmetric matrices. In: Proceedings of the ACM National Conference, pp. 157–172 (1969)
6. Eisenlor, J., Hudak, D.E., Tomko, K., Prince, T.C.: Dense linear algebra factorization in OpenMP and Cilk Plus on Intel MIC: development experiences and performance analysis. In: TACC-Intel Highly Parallel Computing Symposium (2012)
7. Im, E.-J., Yelick, K.A.: Optimizing sparse matrix computations for register reuse in SPARSITY. In: Alexandrov, V.N., Dongarra, J., Juliano, B.A., Renner, R.S., Tan, C.J.K. (eds.) ICCS 2001. LNCS, vol. 2073, pp. 127–136. Springer, Heidelberg (2001)
8. Jain, A.: pOSKI: an extensible autotuning framework to perform optimized spmvs on multicore architecture. Master's thesis, UC Berkeley (2008)
9. Krotkiewski, M., Dabrowski, M.: Parallel symmetric sparse matrix-vector product on scalar multi-core CPUs. Parallel Comput. **36**(4), 181–198 (2010)
10. Küçüktunç, O., Kaya, K., Saule, E., Çatalyürek, Ü.V.: Fast recommendation on bibliographic networks. In: Proceedings of the ASONAM'12, August 2012
11. Mellor-Crummey, J., Garvin, J.: Optimizing sparse matrix-vector product computations using unroll and jam. Int. J. High Perform. Comput. Appl. **18**(2), 225–236 (2004)
12. Nishtala, R., Vuduc, R.W., Demmel, J.W., Yelick, K.A.: When cache blocking of sparse matrix vector multiply works and why. Appl. Algebra Eng. Commun. Comput. **18**(3), 297–311 (2007)
13. Potluri, S., Tomko, K., Bureddy, D., Panda, D.K.: Intra-MIC MPI communication using MVAPICH2: early experience. In: TACC-Intel Highly Parallel Computing Symposium 2012 (2012)
14. Saad, Y.: Sparskit: a basic tool kit for sparse matrix computations - version 2 (1994)
15. Saule, E., Çatalyürek, Ü.V.: An early evaluation of the scalability of graph algorithms on the Intel MIC architecture. In: IPDPS Workshop MTAAP (2012)
16. Saule, E., Kaya, K., Çatalyürek, Ü.V.: Performance evaluation of sparse matrix multiplication kernels on intel xeon phi. Technical Report arXiv:1302.1078, ArXiv, Feb. 2013
17. Stock, K., Pouchet, L.-N., Sadayappan, P.: Automatic transformations for effective parallel execution on intel many integrated core. In: TACC-Intel Highly Parallel Computing Symposium (2012)

18. Vuduc, R., Demmel, J., Yelic, K.: OSKI: a library of automatically tuned sparse matrix kernels. In: Proceedings of the SciDAC 2005, J. of Physics: Conference Series (2005)
19. Williams, S., Oliker, L., Vuduc, R., Shalf, J., Yelick, K., Demmel, J.: Optimization of sparse matrix-vector multiplication on emerging multicore platforms. In: Proceedings of the SC '07 (2007)
20. Zhou, Z., Saule, E., Aktulga, H.M., Yang, C., Ng, E.G., Maris, P., Vary, J.P., Çatalyürek, Ü.V.: An out-of-core eigensolver on SSD-equipped clusters. In: Proceedings of the IEEE Cluster, September 2012

Portable HPC Programming
on Intel Many-Integrated-Core Hardware
with MAGMA Port to Xeon Phi

Jack Dongarra[1,2,3], Mark Gates[1], Azzam Haidar[1], Yulu Jia[1], Khairul Kabir[1],
Piotr Luszczek[1(✉)], and Stanimire Tomov[1]

[1] University of Tennessee Knoxville, Knoxville, USA
luszczek@eecs.utk.edu
[2] Oak Ridge National Laboratory, Oak Ridge, USA
[3] University of Manchester, Manchester, USA

Abstract. This paper presents the design and implementation of several fundamental dense linear algebra (DLA) algorithms for multicore with Intel Xeon Phi Coprocessors. In particular, we consider algorithms for solving linear systems. Further, we give an overview of the MAGMA MIC library, an open source, high performance library that incorporates the developments presented, and in general provides to heterogeneous architectures of multicore with coprocessors the DLA functionality of the popular LAPACK library. The LAPACK-compliance simplifies the use of the MAGMA MIC library in applications, while providing them with portably performant DLA. High performance is obtained through use of the high-performance BLAS, hardware-specific tuning, and a hybridization methodology where we split the algorithm into computational tasks of various granularities. Execution of those tasks is properly scheduled over the heterogeneous hardware components by minimizing data movements and mapping algorithmic requirements to the architectural strengths of the various heterogeneous hardware components. Our methodology and programming techniques are incorporated into the MAGMA MIC API, which abstracts the application developer from the specifics of the Xeon Phi architecture and is therefore applicable to algorithms beyond the scope of DLA.

Keywords: Numerical linear algebra · Intel Xeon Phi processor · Many Integrated Cores · Hardware accelerators and coprocessors · Dynamic runtime scheduling using dataflow dependences · Communication and computation overlap

1 Introduction and Background

Solving linear systems of equations and eigenvalue problems is fundamental to scientific computing. The popular LAPACK library [3], and in particular its

---

The authors are listed in an alphabetical order.

R. Wyrzykowski et al. (Eds.): PPAM 2013, Part I, LNCS 8384, pp. 571–581, 2014.
DOI: 10.1007/978-3-642-55224-3_53, © Springer-Verlag Berlin Heidelberg 2014

vendor optimized implementations like Intel's MKL [8] or AMD's ACML [2], have been the libraries of choice to provide these solvers for dense matrices on shared memory systems. This paper considers a redesign of the LAPACK algorithms and their implementation to add efficient support for heterogeneous systems of multicore processors with Intel Xeon Phi coprocessors. This is not the first time that DLA libraries have needed a redesign to be efficient on new architectures – notable examples being the move from LINPACK [7] to LAPACK [3] in the 80's to make algorithms cache friendly, ScaLAPACK [6] in the 90's to support distributed memory systems, and now the PLASMA and MAGMA libraries [1] targeting efficiency on multicore and heterogeneous architectures, respectively.

The development of new high-performance numerical libraries is complex, accounting for the extreme level of parallelism, heterogeneity, and wide variety of accelerators and coprocessors available in current architectures. Challenges vary from new algorithmic designs to choices of programming models, languages, and frameworks that ease development, future maintenance, and portability. This paper addresses these issues while presenting our approach and algorithmic designs in the development of the MAGMA MIC [9] library.

To provide a uniform portability across a variety of coprocessors/accelerators, we developed an API that abstract the application developer from the low level specifics of the architecture. In particular, we use low level vendor libraries, like SCIF for Intel Xeon Phi (see Sect. 3), to define API for memory management and off-loading computations to coprocessors and/or accelerators.

To deal with the extreme level of parallelism and heterogeneity in current architectures, MAGMA MIC uses a hybridization methodology, described in Sect. 4, where we split the algorithms of interest into computational tasks of various granularities, and properly schedule those tasks' execution over the heterogeneous hardware. Thus, we use a Directed Acyclic Graph (DAG) approach to parallelism and scheduling that has been developed and successfully used for dense linear algebra libraries such as PLASMA and MAGMA [1], as well as in general task-based approaches to parallelism, such as runtime systems like StarPU [4] and SMPSs [5].

Besides the use of high-performance low-level libraries, addressed in Sect. 3, obtaining high performance depends on a combination of algorithm and hardware-specific optimizations, discussed in Sect. 4.3. The implication of this on software, in order to maintain its performance portability across hardware, is the need to build in it algorithmic variations that are tunable, e.g., at installation time. This is the basis of autotuning, an example of these advanced optimization techniques.

A performance study is presented in Sect. 5. Besides verifying our approaches and confirming the appeal of the Intel Xeon Phi coprocessors for high-performance DLA, the results open up a number of future work opportunities discussed in our conclusions.

2   Compiler Support for Offload

The primary mode of operation for the Xeon Phi coprocessor is the off-load mode. The device receives work from the host processor and reports back upon completion of the assignment without the host being involved in between these two events. This is very similar to the operation of network off-load engines, specifically, the TCP Off-load Engines (TOEs) that feature an optimized implementation of the TCP stack that handles the majority of the network traffic to lessen the burden of the main processor, which handles other operating system and user application tasks.

This way of using the Xeon Phi device has direct support from the compiler in that it is possible to issue requests to the device and ascertain the completion of tasks directly from the user's C/C++ code. The support for this mode of operation is offered by the Intel compiler through Phi-specific pragma directives: offload, offload_attribute, offload_transfer, and offload_wait.

3   Programming Model: Host-Device with a Server Based on LLAPI

For many scientific applications, the offload model offered by the Intel compiler, described in Sect. 2, is sufficient. This is not the case for a fully equivalent port of MAGMA to the Xeon Phi because of the very rich functionality that MAGMA inherits from both its CUDA and OpenCL ports. We had to use the LLAPI (Low-Level API) based on Symmetric Communication InterFace (SCIF) that offers, as the name suggests, a very low level interface to the host and device hardware. The use of this API is discouraged for most workloads as it tends to be error-prone and offers very little abstraction on top of the hardware interfaces. What motivated us to use it for the port of our library was: (1) the asynchronous events capability that allows low-latency messaging between the host and the device to notify about completion of kernels on Xeon Phi; (2) the possibility of hiding the cost of data transfer between the host and the device which requires the transfer of submatrices to overlap with the computation. The direct access to the DMA (Direct Memory Access) engine allowed us to maximize the bandwidth of data transfers over the PCI Express bus. The only requirement was that the memory regions for transfer be page-aligned and pinned to guarantee their fixed location in the physical memory. Figure 1 shows the interaction between the host and the server running on the Xeon Phi and responding to requests that are remote invocations of numerical kernels on data that have already been transferred to the device.

4   Hybridization Methodology and Optimization Strategies

The hybridization methodology used in MAGMA [10] is an extension of the task-based approach for parallelism and developing DLA on homogeneous multicore systems [1]. In particular,

– The computation is split into BLAS-based tasks of various granularities, with their data dependencies, as shown in Fig. 1b.

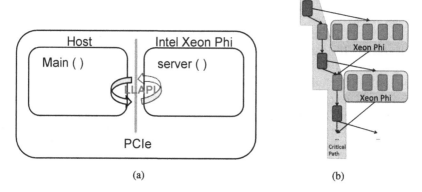

Fig. 1. (a) MAGMA MIC programming model with a LLAPI server mediating requests between the host CPU and the Xeon Phi device. (b) DLA algorithm as a collection of BLAS-based tasks and their dependencies. The algorithm's critical path is, in general, scheduled on the CPUs, and large data-parallel tasks on the Xeon Phi.

- Small, non-parallelizable tasks with significant control-flow are scheduled on the CPUs.
- Large, parallelizable tasks are scheduled on Xeon Phi.

The difference with multicore algorithms is the task splitting, which here is of various granularities to make different tasks suitable for particular architectures, and the scheduling itself. Specific algorithms using this methodology, and covering the main classes of DLA, are described in the subsections below.

4.1   Design and Functionality

The MAGMA interface is similar to LAPACK. For example, compare LAPACK's LU factorization interface vs. MAGMA's:

```
lapackf77_dgetrf(&M,&N, hA,    &lda, ipiv,    &info)
magma_dgetrf_mic(M, N, dA,0, ldda, ipiv,    &info, queue)
```
Here hA is the typical CPU pointer (`double *`) to the matrix of interest in the CPU memory and dA is a pointer in the Xeon Phi memory (`magmaDouble_ptr`). The last argument in every MAGMA call is an Xeon Phi queue, through which the computation will be streamed on the Xeon Phi device (`magma_queue_t`).

To abstract the user from knowing low level directives, main functions, such as BLAS, CPU-Phi data transfers, and memory allocations and deallocations, are redefined in terms of MAGMA data types and functions. This design allows us to more easily port the MAGMA library to other device such as the GPU accelerator using either CUDA or OpenCL and eventually to merge them while maintaining a single source. Also, the MAGMA wrappers provide a complete set of functions for programming hybrid high-performance numerical libraries. Thus, not only users but application developers as well can opt to use the MAGMA wrappers. MAGMA provides the standard four floating point arithmetic precisions – single real, double real, single complex, and double complex. There are routines for the so called one-sided factorizations (LU, QR, and Cholesky), and recently we are developing the two-sided factorizations (Hessenberg, bi-, and tridiagonal reductions), linear system and least squares solvers, matrix inversions, symmetric and nonsymmetric standard eigenvalue problems, SVD, and orthogonal transformation routines.

4.2   LU, QR, and Cholesky Factorizations

The one-sided factorization routines implemented and currently available through MAGMA are:

magma_zgetrf_mic computes an LU factorization of a general M-by-N matrix A
   using partial pivoting with row interchanges;
magma_zgeqrf_mic computes a QR factorization of a general M-by-N matrix A;
magma_zpotrf_mic computes the Cholesky factorization of a complex Hermitian
   positive definite matrix A.

Routines in all standard four floating point precision arithmetics are available, following LAPACK's naming convention. Namely, the first letter of the routine name (after the prefix magma_) indicates the precision – z, c, d, or s for correspondingly double complex, single complex, double real, or single real. The suffix _mic indicates that the input matrix and the output are on the Xeon Phi memory.

4.3   Hybrid Implementation and Optimization Techniques

In order to explain our hybrid methodology and the optimization that we developed, let us give a detailed analysis for the QR decomposition algorithm. While the description below only addresses the QR factorization, it is straightforward to derive with the same ideas the analysis for both the Cholesky and LU factorizations. For that we start briefly by recalling the description of the QR algorithm.

   The QR factorization is a transformation that factorizes an $m \times n$ matrix A into its factors Q and R where Q is a unitary matrix of size $n \times n$ and R is a triangular matrix of size $m \times m$. The QR algorithm can be described as a sequence of steps where, at each step, a QR of a panel is performed based on accumulating a number of Householder transformations in what is called a *"panel factorization"* which are, then, applied all at once by means of high performance Level 3 BLAS operations in what is called the *"trailing matrix update"*. Despite that this approach can exploit the parallelism of the Level 3 BLAS during the trailing matrix update, it has a number of limitations when implemented on massively multithreaded system such as the Intel Xeon Phi coprocessor due to the nature of its operations. On the one hand, the panel factorization relies on Level 2 BLAS operations that cannot be efficiently parallelized on either Xeon Phi or any accelerator such as GPU-based architectures, and thus it can be considered to be close to sequential operations that limit the scalability of the algorithm. On the other hand, this algorithm is referred as the *fork-join approach* since the execution flow will show a sequence of sequential operations (panel factorizations) interleaved with parallel ones (trailing matrix updates). In order to take advantage of the high execution rate of the massively multithreaded system, in particular, the Phi coprocessor we redesigned the standard algorithm in a way to perform the Level 3 BLAS operations (Trailing matrix update) on the Xeon Phi while performing the Level 2 BLAS operations (panel factorization) on the CPU. We also proposed an algorithmic change to remove the fork join bottleneck and to minimize the overhead of the panel factorization by hiding its costs behind the parallel trailing matrix update. This approach can be described as the *"scalable lookahead techniques"*. Our idea is to split of the trailing matrix update into two phases, the update of the lookahead panel (panel of step $i+1$, i.e., dark blue portion of Fig. 2) and the update of the remaining trailing submatrix (clear blue portion of Fig. 2). Thus, during the submatrix update the CPU can receive asynchronously the panel $i+1$ and performs its factorization. As a result, our MAGMA implementation of the QR factorization can be described by a sequence of the three phases described below. Consider a matrix A that can be represented as:

$$A = \begin{pmatrix} A_{11} & A_{12} & A_{13} \\ A_{21} & A_{22} & A_{23} \\ A_{31} & A_{32} & A_{33} \end{pmatrix}, \tag{1}$$

- **Phase 1, the panel factorization:** at a step i, this phase consists of a QR transformation of the panel $A_{i:n,i}$ as in Eq. 2. This operation consists of calling two routines. The DGEQR2 that factorizes the panel and produces nb Householder reflectors (V_{*i}) and an upper triangular matrix R_{ii} of size $nb \times nb$, which is a portion of the final R factor, and the DLARFT that generates the triangular matrix T_{ii} of size $nb \times nb$ used for the trailing matrix update. This phase is performed on the CPU.

$$\begin{pmatrix} A_{11} \\ A_{21} \\ A_{31} \end{pmatrix} \implies \begin{pmatrix} V_{11} \\ V_{21} \\ V_{31} \end{pmatrix}, (R_{1,1}), (T_{1,1}). \tag{2}$$

- **Phase 2, the look ahead panel update:** the transformation that was computed in the panel factorization needs to be applied to the rest of the matrix (trailing matrix, i.e., the blue portion of Fig. 2). This phase consists into updating only the next panel (dark blue portion of Fig. 2) in order to let the CPU start its factorization as soon as possible while the update of the remaining portion of the matrix is performed in phase 3. The idea is to hide the cost of the panel factorization. This operation presented in Eq. 3, is performed on the Phi coprocessor and involves the DLARFB routine which has been redesigned as a sequence of DGEMM's to better take advantage of the Level 3 BLAS operations.

$$\begin{pmatrix} R_{12} \\ \tilde{A}_{22} \\ \tilde{A}_{32} \end{pmatrix} = \left(I - V_{*i} T_{ii}^T V_{*i}^T\right) \begin{pmatrix} A_{12} \\ A_{22} \\ A_{32} \end{pmatrix}. \tag{3}$$

– **Phase 3, the trailing matrix update:** Similarly to phase 2, this phase consists into applying the Householder reflectors generated during the panel factorization of step i according to Eq. 3, but to the remaining portion of the matrix (the trailing submatrix i.e., the clear blue portion of Fig. 2). This operations is also performed on the Phi coprocessor, while in parallel to it, the CPU performs the factorization of the panel $i + 1$ that has been computed in Phase 2.

This hybrid technique of distribution of tasks between CPU-Phi allows us to hide the memory bound operations occurred during the panel factorization (Phase 1) by performing such operation on the CPU in parallel with the trailing submatrix update (Phase 3) on the Phi coprocessor. However, one of the key parameters to performance tuning is the blocking size as the performance and the overlap between the CPU-Phi will be solely guided by it. Figure 2b illustrates the effect of the blocking factor on the performance. It is obvious that, a small nb will reduce the cost of the panel factorization phase 1, but it decreases the efficiency of the Level 3 BLAS kernel of phase 2 and phase 3 and thus resulting a bad performance. As opposed, a large nb will dramatically affect the panel factorization phase 1 which becomes slow and thus the CPU/Phi computation cannot be overlapped, providing a deterioration in the performance as shown in Fig. 2b. As a consequence, the challenging problem is the following: on the one hand, the blocking size nb needs to be large enough to extract high performance from the Level 3 BLAS phase 3 and on the other hand, it has to be small enough to extract efficiency (thanks to the cache speed up) from the Level 2 BLAS phase 1 and overlap CPU/Phi computation. Figure 2b show the performance obtained for different blocking sizes and we can see a trade-off between small and large nb's. Either $nb = 480$ or $nb = 960$ can be considered as a good choice because MKL Phi BLAS is optimized for multiples of 240. Moreover, to extract the maximum performance and allow the maximum overlap between both of the CPU and the Xeon Phi coprocessor, we developed a new variant that can use a variable nb during the steps of the algorithm. The flexibility of our implementation allows an efficient task execution overlap between the CPU host and the Phi coprocessor which enables the algorithm to scale almost perfectly in the Phi coprocessor and provides very good performance close to the practical peak obtained on such system. Our tuned variable implementation is represented by the red curve of Fig. 2b where we can easily observe its advantage over the other variants.

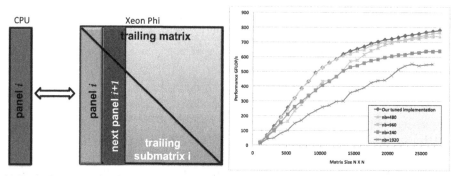

(a) Typical computational pattern for the hybrid one-sided factorizations in MAGMA

(b) Effect of the blocking factor.

Fig. 2. Effect of the blocking factor (Color figure online).

5   Performance Results

This section presents the performance results obtained by our hybrid CPU-Xeon Phi implementation in the context of the development of the state-of-the-art numerical linear algebra libraries.

5.1   Experimental Environment

Our experiments were performed on a system equipped with Intel Xeon-Phi. It is representative of a vast class of servers and workstations commonly used for computationally intensive workloads.

Intel multicore system with dual-socket, 8 core Intel Xeon E5–2670 (Sandy Bridge) processors, each running at 2.6 GHz. Each socket has a 24 MB shared L3 cache, and each core has a private 256 KB L2 and 64 KB L1. The system is equipped with 52 Gbytes of memory. The theoretical peak for this architecture in double precision is 20.8 Gflop/s per core, giving 332 Gflops in total. The system is also equipped with an Intel Xeon Phi cards with 7.7 Gbytes per card running at 1.09 GHz, and giving a double precision theoretical peak of 1046 Gflops.

There are a number of software packages available. On the CPU side we used the MKL (Math Kernel Library) [8] which is a commercial software package from Intel that is a highly optimized numerical library. On the Intel Xeon side, we used the MPSS 2.1.5889-16 as the software stack, icc 13.1.1 20130313 which comes with the composer_xe_2013.3.163 suite as the compiler and the BLAS-3 routine GEMM from MKL 11.00.03.

5.2   Performance Results

Figure 3 reports the performance of the three amigos linear algebra kernels, the Cholesky, QR and LU factorizations with our hybrid implementation and compare it to the performance of the CPU implementation of the MKL libraries.

For our implementation, the blocking factor has been chosen to be flexible in order to achieve the best performance, as a reference it is in the range of 480–960 as described in Sect. 4.3. The graphs show the performance measured using all the cores available on the system (i.e., 60 for the Intel Phi and 16 for the CPU) with respect to the problem size. In order to reflect the time completion, for each algorithm the operation count is assumed to be the same as that of the LAPACK algorithm (i.e., $\frac{1}{3}N^3$, $\frac{2}{3}N^3$, and $\frac{4}{3}N^3$ for the Cholesky factorization, the LU factorization and the QR decomposition respectively)

Figure 3a,b,c provide roughly the same information: our MAGMA algorithm with hybrid techniques delivers higher execution rates than the CPU optimized counterpart. Such comparison is not fair, our goal is not to compare, but it is rather to show the boost that the hybrid CPU+Phi coprocessor implementation provides, versus a CPU implementation. The figures show that the MAGMA hybrid algorithms are capable of completing any of the three amigos algorithms as twice faster as the CPU optimized version for a matrix of size larger than 10000; and more than three times faster when the matrix size is large enough (larger than 20000). The actual curves of Fig. 3 illustrates the efficiency of our

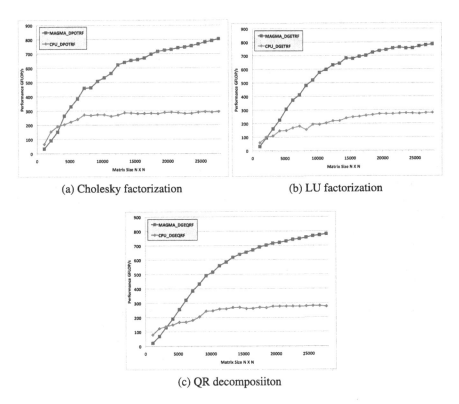

(a) Cholesky factorization

(b) LU factorization

(c) QR decomposiiton

Fig. 3. Comparison of the performance versus the optimized CPU version of the MKL libraries for the three amigos.

hybrid techniques where we note that the performance obtained by our implementation, achieves a very close level to the practical peak of the Intel Xeon Phi coprocessor computed by running the GEMM routine (which is around 850 Gflops). This gain is mostly obtained by two improvements. First the nature of the operations involved in the Phi side which are mostly BLAS Level 3 operations redesigned and redeveloped as a combination of DGEMM's. For more details we denote below the routines executed on the Xeon Phi coprocessor:

– The DSYRK operations for the Cholesky factorization where the DSYRK has been redesigned as a combination of DGEMM's routines,
– The DGEMM for the LU factorization,
– The DLARFB for the QR decomposition where also its has been redesigned as a combination of DGEMM's.

Second, all of the Level 2 BLAS routines that are memory bound and that represent a limit for the performance (i.e., DPOTF2, DGETF2, and DGEQR2 for Cholesky, LU, and QR factorization respectively) are executed on the CPU side while being overlapped with the Phi coprocessor execution as described in Sect. 4.3.

An important remark has to be made here for the Cholesky factorization: the *left-looking* algorithm as implemented in LAPACK is considered as well optimized for memory reuse but at the price of less parallelism and thus is not suitable for massively multicore machines. This variant delivers poor performance when compared to the *right looking* variant that allows more parallelism and thus run at higher speed.

6   Conclusions and Future Work

In this article, we have shown how to extend our hybridization methodology from existing systems to a new hardware platform. The challenge of the porting effort stemmed from the fact that the new coprocessor from Intel, the Xeon Phi, featured programming models and relative execution overheads, that were markedly different from what we have been targeting on GPU-based accelerators. Nevertheless, we believe that the techniques used in this paper adequately adapt our hybrid algorithm to best take advantage of the new heterogeneous hardware. We have derived an implementation schema of the dense linear algebra kernels that also can be applied to either the two-sided factorization used for solving the eigenproblem and the SVD or to the sparse linear algebra algorithms. We plan to further study the implementation of multi-Xeon Phi algorithms in a distributed computing environment. We think that the techniques presented will become more popular and will be integrated into dynamic runtime system technologies. The ultimate goal is that this integration will help to tremendously decrease development time while retaining high-performance.

Acknowledgments. The authors would like to thank the National Science Foundation, the Department of Energy and ISTC for Big Data for supporting this research effort.

References

1. Agullo, E., Demmel, J., Dongarra, J., Hadri, B., Kurzak, J., Langou, J., Ltaief, H., Luszczek, P., Tomov, S.: Numerical linear algebra on emerging architectures: the plasma and magma projects. J. Phys. Conf. Ser. **180**(1), 012037 (2009)
2. AMD. AMD Core Math Library (ACML). http://developer.amd.com/tools/
3. Anderson, E., Bai, Z., Bischof, C., Blackford, S.L., Demmel, J.W., Dongarra, J.J., Du Croz, J., Greenbaum, A., Hammarling, S., McKenney, A., Sorensen, D.C.: LAPACK User's Guide, 3rd edn. SIAM, Philadelphia (1999)
4. Augonnet, C., Thibault, S., Namyst, R., Wacrenier, P.-A.: StarPU: a unified platform for task scheduling on heterogeneous multicore architectures. Concurrency Comput. Pract. Exp. **23**(2), 187–198 (2011)
5. Barcelona Supercomputing Center. SMP Superscalar (SMPSs) User's Manual, Version 2.0. http://www.bsc.es/media/1002.pdf (2008)
6. Blackford, L.S., Choi, J., Cleary, A., D'Azevedo, E., Demmel, J., Dhillon, I., Dongarra, J.J., Hammarling, S., Henry, G., Petitet, A., Stanley, K., Walker, D., Whaley, R.C.: ScaLAPACK Users' Guide. SIAM, Philadelphia (1997). http://www.netlib.org/scalapack/slug/
7. Dongarra, J., Bunch, J., Moler, C., Stewart, G.W.: LINPACK Users' Guide. SIAM, Philadelphia (1979)
8. Intel. Math Kernel Library. http://software.intel.com/en-us/articles/intel-mkl/
9. Software distribution of MAGMA MIC version 1.0. http://icl.cs.utk.edu/magma/software/. Accessed 3 May 2013
10. Tomov, S., Dongarra, J.: Dense linear algebra for hybrid GPU-based systems. In: Kurzak, J., Bader, D.A., Dongarra, J. (eds.) Scientific Computing with Multicore and Accelerators. Chapman and Hall/CRC, London/Boca Raton (2010)

Using Intel Xeon Phi Coprocessor to Accelerate Computations in MPDATA Algorithm

Lukasz Szustak[1]([⊠]), Krzysztof Rojek[1], and Pawel Gepner[2]

[1] Czestochowa University of Technology,
Dabrowskiego 69, 42-201 Czestochowa, Poland
{lszustak,krojek}@icis.pcz.pl
[2] Intel Corporation, Swindon, UK
pawel.gepner@intel.com

Abstract. The multidimensional positive definite advection transport algorithm (MPDATA) belongs to the group of nonoscillatory forward-in-time algorithms, and performs a sequence of stencil computations. MPDATA is one of the major parts of the dynamic core of the EULAG geophysical model.

The Intel Xeon Phi coprocessor is the first product based on the Intel Many Integrated Core (Intel MIC) architecture. In this work, we outline an approach to adaptation of the 3D MPDATA algorithm to the Intel MIC architecture. This approach is based on combination of temporal and space blocking techniques, and allows us to ease memory and communication bounds and better exploit the theoretical floating point efficiency of target computing platforms. In order to utilize computing resources available in Intel Xeon Phi, the proposed approach employs two main levels of parallelism: (i) task parallelism which allows for utilization of more than 200 logical cores, and (ii) data parallelism to use efficiently 512-bit vector processing units.

We discuss performance results obtained on two platforms, including either two Intel Xeon E5-2643 CPUs and Intel Xeon Phi 3120A, or two Intel Xeon E5-2697 v2 CPUs and Intel Xeon Phi7120P. The top-of-the-line Intel Xeon Phi 7120P gives the best performance results for all tests. Notably, this coprocessor executes the MPDATA algorithm 2 times faster than two Intel Xeon E5-2697 v2 CPUs, and 2.86 times faster than two Intel Xeon E5-2643 processors. Both the utilization of Intel Xeon Phi many cores and vectorization play the leading role in performance exploitation.

Keywords: EULAG model · Stencil computation · MPDATA · Intel Xeon Phi · Multi-/manycore programming · OpenMP · Adaptation

1 Introduction

The multidimensional positive definite advection transport algorithm (MPDATA) [7] is one of the two major parts of the dynamic core of the EULAG geophysical

R. Wyrzykowski et al. (Eds.): PPAM 2013, Part I, LNCS 8384, pp. 582–592, 2014.
DOI: 10.1007/978-3-642-55224-3_54, © Springer-Verlag Berlin Heidelberg 2014

model. EULAG (Eulerian/semi-Lagrangian fluid solver) is an established computational model for simulating thermo-fluid flows across a wide range of scales and physical scenarios, including the numerical weather prediction (NWP).

The resent research of EULAG parallelization have been carried out using IBM Blue Gene/Q and CRAY XE6 [4]. Three-dimensional MPI parallelization has been used for running EULAG on these systems with tens of thousands of cores, or even with more than 100K cores. When parallelizing EULAG computation on supercomputers and CPU clusters, the efficiency is declined below 10 %. In this study, we propose to rewrite the EULAG dynamical core and replace standard HPC systems by smaller heterogeneous clusters with accelerators such as GPU [5] and Intel Many Integrated Cores (MIC) [3].

Preliminary studies of porting anelastic numerical models to modern architectures, including hybrid CPU-GPU architectures, were carried out in works [5,10,11]. The results achieved for porting selected parts of EULAG to CPU-GPU architectures revealed potential in running scientific applications on novel hardware architectures.

In this work, we outline an approach to adaptation of the 3D MPDATA algorithm to the Intel MIC architecture. This approach is based on combination of temporal and space blocking techniques, and allows us to ease memory and communication bounds, and better exploit the theoretical floating point efficiency of target computing platforms. We show some of the optimization methods that we found effective, and demonstrate their impact on the performance of both the Intel CPU and MIC architectures. The main focus is using MPDATA to modelling geophysical flows on NWP. The size of computational grid in such problems typically does not exceed 270 thousand grid points ($2048 \times 1024 \times 128$). Here, the starting point is an unoptimized parallel implementation of the MPDATA algorithm. In our work, we use OpenMP standard for multi- and many-core programming.

The content of the paper is organized as follow. In Sect. 2, architecture overview is outlined. The introduction to 3D MPDATA algorithm, including characterization of computation and communication, is presented in Sect. 3. Section 4 introduces the proposed approach to adaptation of MPDATA to Intel MIC Architecture, including block decomposition of 3D MPDATA algorithm, improving efficiency of block decomposition, and parallelization. Preliminary performance results are presented in Sect. 5, while Sect. 6 gives conclusions and future work.

2   Architecture Overview

2.1   Architecture of Intel Many Integrated Cores

The Intel MIC architecture combines many Intel CPU cores onto a single chip [2,3]. The Intel Xeon Phi coprocessor is the first product based on this architecture. The main advantage of these accelerators is that it is built to provide a general-purpose programming environment similar to that provided for Intel

CPUs. This coprocessor is capable of running applications written in industry-standard programming languages such as Fortran, C, and C++.

The Intel Xeon Phi coprocessor includes processing cores, caches, memory controllers, PCIe client logic, and a very high bandwidth, bidirectional ring interconnect [3]. Each coprocessor contains of more than 50 cores clocked at 1 GHz or more. These cores support four-way hyper-threading, which gives more than 200 logical cores. The real number of cores depends on the generation and model of a specific coprocessor. Each core features an in-order, dual-issue x86 pipeline, 32 KB of L1 data cache, and 512 KB of L2 cache that is kept fully coherent by a global-distributed tag directory. As a result, the aggregate size of L2 caches can exceeds 25 MB. The memory controllers and the PCIe client logic provide a direct interface to the GDDR5 memory on the coprocessor and the PCIe bus, respectively. The coprocessor has over 6 GB of onboard memory (maximum 16 GB). The high-speed bidirectional ring connects together all the cores, caches, memory controllers and PCIe client logic of Intel Xeon Phi coprocessors.

An important component of each Intel Xeon Phi processing core is its vector processing unit (VPU) [2], that significantly increases the computing power. Each VPU supports a new 512-bit SIMD instruction set called Intel Initial Many-Core Instructions. The new ability to work with 512-bit vectors enables operating on 16 float or 8 double elements per iteration, instead of a single element.

The Intel Phi coprocessor is delivered in form factor of a PCI express device, and cannot be used as a stand-alone processor. Since the Intel Xeon Phi coprocessor runs Linux operating system, any user can access the coprocessor as a network node, and directly run individual applications in the native mode. These coprocessors also support heterogeneous applications wherein a part of the application is executed on the host (CPU), while another part is executed on the coprocessor (offload mode).

2.2 Target Platforms

A summary of key features of tested platforms is shown in Table 1. In this study, we use two platforms containing a single Intel Xeon Phi coprocessor. The first platform is equipped with two newest CPUs, based on the Ivy Bridge architecture, and the Intel Xeon Phi 3120A card. The second one includes two Sandy Bridge-EP CPUs, and the top-of-the-line Intel Xeon Phi 7120P coprocessor.

It should be noted that values of peak performance shown in Table 1 are given for the double precision arithmetic, with taking into account the usage of SIMD vectorization.

3   Introduction to MPDATA Algorithm

The multidimensional positive definite advection transport algorithm (MPDATA) belongs to the group of nonoscillatory forward-in-time algorithms, and performs a sequence of stencil computations. The full description of the MPDATA algorithm can be found in [6,7].

Table 1. Specification of tested platforms [1]

Product	Code name	# of cores (threads)	SIMD vector [bits]	Freq. [GHz]	Peak DP [GFlop/s]	Cache size [MB]	Memory size [GB]	Memory band. [GB/s]
Intel Xeon E5-2697 v2	Ivy Bridge	2×12 (2×24)	256	2.7	518	2×30	64	2×51.2
Intel Xeon Phi 3120A	Knights Corner	57 (228)	512	1.1	1003	28.5	6	240
Intel Xeon E5-2643	Sandy Bridge-EP	2×4 (2×8)	256	3.3	211	2×10	64	2×51.2
Intel Xeon Phi 7120P	Knights Corner	61 (244)	512	1.238	1208	30.5	16	352

The whole MPDATA computation in each time step are decomposed into a set of 17 stencil sweeps, called further stages. Each stage is responsible for calculating elements of a certain matrix, based on the corresponding stencil. The stages dependent on each other: prior outcomes from stages are usually input data for the subsequent computations. A part of the MPDATA implementation is shown in Fig. 1. It corresponds to the linear version of MPDATA [7].

A single MPDATA time step requires 5 input and 1 output matrices. Other 16 matrices are temporary, and do not play role in the further computational steps. In the basic, unoptimized implementation of the MPDATA algorithm, every stage uses a required set of matrices from the main memory, and writes results to the main memory after computation. This scheme is repeated for all the stages. In consequence, a heavy traffic to the main memory is generated. Moreover, compute units (cores/threads, and VPUs) have to wait for data transfers from the main memory to the cache hierarchy. In order to take full advantage of the novel architecture, the adaptation of MPDATA to the Intel MIC architecture is considered. The new implementation takes into account the memory-bounded character of the algorithm.

4   Adaptation of MPDATA to Intel MIC Architecture

4.1   Block Decomposition of 3D MPDATA Algorithm

Since the MPDATA algorithm includes so many intermediate computation, one of the primary methods for reducing the intensity of memory traffic is to avoid data transfers associated with these computation. For this aim, all the intermediate results must be kept in the cache memory. Such treatment increases the cache reusing. The memory traffic is generated only to transfer the required input and output data. Such an approach is commonly called the temporal blocking [8,9].

In order to implement this approach efficiently, the loop tiling technique is applied. The grid is partitioned into blocks. Every block provides computation for all the 17 stages on the assigned part of the grid. Within a single block,

```
#define fdim(a, b) ( (a>b) ? (a-b):(0.0) )
#define donor(y1, y2, a) ( fdim(a, 0.0) * (y1) - fdim(0.0, a) * (y2) )
//stage 1
for( ... ) // i - dimension
 for( ... ) // j - dimension
 for( ... ) // k - dimension
 f1[i][j][k] = donor(xIn[i-1][j][k],xIn[i][j][k],u1[i][j][k]);
//stage 2
for( ... ) // i - dimension
 for( ... ) // j - dimension
 for( ... ) // k - dimension
 f2[i][j][k] = donor(xIn[i][j-1][k],xIn[i][j][k],u2[i][j][k]);
//stage 2
for( ... ) // i - dimension
 for( ... ) // j - dimension
 for( ... ) // k - dimension
     f3[i][j][k] = donor(xIn[i][j][k-1],xIn[i][j][k],u3[i][j][k]);
//stage 4
for( ... ) // i - dimension
  for( ... ) // j - dimension
    for( ... ) // k - dimension
      x[i][j][k] = xIn[i][j][k]-(f1[i+1][j][k]-f1[i][j][k]+f2[i][j+1][k]-
                   f2[i][j][k]+f3[i][j][k+1]-f3[i][j][k])/h[i][j][k];
/*...*/
```

Fig. 1. Part of MPDATA implementation

each stage computes the adequate chunk of the corresponding matrix. Executing of a sequence of blocks determines the final outcomes for a single MPDATA time step.

The main requirement for this approach is to keep in the cache hierarchy all the data required for MPDATA computation within each block. Therefore, the size $nB \times mB \times lB$ of each block has to be selected in an appropriate way. The idea of block decomposition of the MPDATA algorithm is shown in Fig. 2. This decomposition determines four dimensions of distribution of MPDATA calculation across computing resources: i-, j-, and k-dimensions relate to the grid partitioning, while s-dimension is associated with the order of executing MPDATA stages.

Due to data dependencies between subsequent stages additional computation and communication within each data block are required. These overheads are needed on the borders between adjacent blocks. In the proposed method, the computation associated with all the 17 stages, and executed within each block are extended by adequate halo areas. Adding halo allows to avoid data dependency between blocks within a single MPDATA time step.

The sizes of halo areas are determined in all the four dimensions (i, j, k and s), according to data dependencies between MPDATA stages. Thus, each of 5 input,

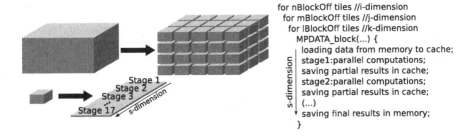

```
for nBlockOff tiles //i-dimension
  for mBlockOff tiles //j-dimension
    for lBlockOff tiles //k-dimension
      MPDATA_block(...) {
        loading data from memory to cache;
        stage1:parallel computations;
        saving partial results in cache;
        stage2:parallel computations;
        saving partial results in cache;
        (...)
        saving final results in memory;
      }
```

Fig. 2. Idea of block decomposition of MPDATA computation

Table 2. Sizes of halo areas for MPDATA algorithm

Halo areas	Input					Temporary																Output
	u1	u2	u3	h	x	S1	S2	S3	S4	S5	S6	S7	S8	S9	S10	S11	S12	S13	S14	S15	S16	S17
iL	2	2	2	2	3	2	2	2	2	1	1	1	1	1	1	1	1	1	0	0	0	0
iR	3	2	2	2	3	3	2	2	2	2	1	1	1	1	1	1	1	1	1	0	0	0
jL	2	2	2	2	3	2	2	2	2	1	1	1	1	1	1	1	1	1	0	0	0	0
jR	2	3	2	2	3	2	3	2	2	1	2	1	1	1	1	1	1	1	0	1	0	0
kL	2	2	2	2	3	2	2	2	2	1	1	1	1	1	1	1	1	1	0	0	0	0
kR	2	2	3	2	3	2	2	3	2	1	1	2	1	1	1	1	1	1	0	0	1	0

one output, and 16 temporary matrices, is partitioned into chunks of size $nB \times mB \times lB$, which further is expanded by adequate halo areas with sizes iL, iR, and jL, jR as well as kL, kR. Table 2 presents the sizes of halo areas in i-, j-, and k-dimensions for chunks of all the matrices.

This approach allows us to avoid data transfers for intermediate computation at the cost of extra computation associated with halo areas in chunks of temporary matrices, as well as extra communication between the main and cache memories, corresponding to halo areas in chunks of the input matrices. Another advantage of this approach is reducing the main memory consumption because all the intermediate results are stored in the cache memory only. In the case of coprocessors, it plays an important role because the size of main memory is fixed, and significantly smaller than for traditional CPU solutions.

The requirement of expanding halo areas is one of the major difficulties when applying the proposed approach, taking into account data dependencies between MPDATA stages. It requires to develop a dedicated task scheduling for the MPDATA block decomposition.

4.2 Improving Efficiency of Block Decomposition

Although the block decomposition of MPDATA allows for reducing the memory traffic, it still does not guarantee a satisfying utilization of used platforms. The main difficulty here results from extra computation and communication, which have impact on the performance degradation. In particular, there are three groups of extra computation and communication, corresponding to i-, j-, and k-dimensions. Some of them can be reduced or even avoided by applying the following rules:

1. The additional computation and communication in k-dimension can be avoided if $lB = l$, and the size $nB \times mB \times lB$ of block is small enough to save in cache all the required data. This rule is especially useful when the value of l is relatively small, as it is in the case of NWP, where l is in range [64, 128].
2. The overheads associated with j-dimension is avoided by leaving partial results in the cache memory. It becomes possible when extra computation are repeated by adjacent blocks. In this case, some results of intermediate computation have to reside in cache for executing the next block. This rule requires to develop a flexible management of computation for all the stages, as well as an adequate mapping of partial results onto the cache space. In consequence, all the chunks are still expanded by their halo areas (Table 2), but only some portions of these chunks are computed within the current block. It means that this approach does not increase the cache consumption. The idea of improving the efficiency of block decomposition is shown in Fig. 3.
3. In order to reduce additional calculations in i-dimension, the size nB should be as large as possible to save in the cache hierarchy all the data required to compute a single block.

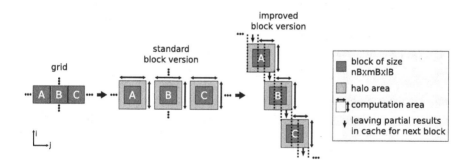

Fig. 3. Idea of leaving partial results in cache memory

4.3 Parallelization

In order to utilize computing resources available in the Intel Xeon Phi coprocessor, the proposed approach employs two main levels of parallelism:

– task parallelism which allows for utilization of more than 200 logical cores;
– data parallelism to use efficiently 512-bit vector processing units.

Different MPDATA blocks are processed sequentially, following the order proposed for the CPU block decomposition in the previous subsection (Fig. 3). For a fixed MPDATA block, a sequence of stages is executed, taking into account the adequate sizes of halo areas. All computation executed within every stage are distributed across available threads. Assigned chunk to each stage is partitioned into sub-chunk of size $nB^* \times mB^* \times lB$, where partitioning takes place along i and j dimensions. As a result, a task is assigned to each thread, as a part of MPDATA block. Due to the data dependencies of MPDATA, appropriate synchronizations between MPDATA stages are necessary.

Another level of parallelization is vectorization applied within each thread, so the resulting SIMDification is performed within k-dimension. In consequence, the value of size lB has to be adjusted to the vector size.

Because of intra-cache communication between tasks, the overall system performance depends strongly on a chosen task placement onto available threads. Therefore, the physical core affinity plays a significant role in optimizing the system performance. In this work, the affinity is adjusted manually, to force communication between tasks placed onto the closest adjacent cores. This increases the sustained intra-cache bandwidth, as well as reduces cache misses, and the latency of access to the cache memory.

5   Preliminary Performance Results

In this section we present preliminary performance results obtained for the double precision 3D MPDATA algorithm on the platforms introduced in Sect. 2. In all the tests, we use the ICC compiler as a part of Intel Parallel Studio 2013, with the same optimization flags. The best configurations for our approach is chosen in an empirical way, individually for each platform. Moreover, we use Intel Xeon Phi in the native mode.

Currently, only the first four stages are implemented and tested. These four stages correspond to the linear version of the MPDATA algorithm. Since all the input matrices are required to provide the correctness of calculation, the overall performance for this part of MPDATA is strongly limited by the memory traffic between the main memory and cache memory.

Figure 4 presents the normalized execution time of the 3D MPDATA algorithm, for 500 time steps and the grid of size $1022 \times 512 \times 63$. The achieved performance results correspond to the following setups: (a) comparison of the block and improved block versions; (b) advantages of using vectorization; (c) performance for different numbers of threads per core; (d) comparison of Intel Xeon CPU and Intel Xeon Phi (best configurations with SIMD).

Figure 4a presents a performance gain for the improved version of block decomposition. The proposed method of reducing extra computation allows us to speedup MPDATA block version from 2 to 4 times, depending on the platform used and size of the grid.

The advantages of using vectorization is observed for all the platforms. In particular, for Intel Xeon Phi 7120P, it allows us to accelerate computation more than 3 times, using all the available threads/cores (Fig. 4b).

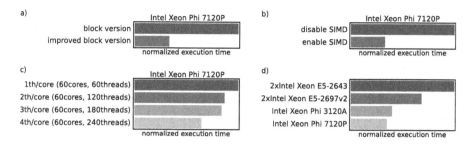

Fig. 4. Preliminary performance results: (a) comparison of block and improved block versions; (b) advantages of using vectorization; (c) performance for different numbers of threads per core; (d) comparison of Intel Xeon CPU and Intel Xeon Phi (best configurations with SIMD)

Figure. 4c shows the performance obtained for different numbers of threads per core, using Intel Xeon Phi 7120P. The best efficiency of computation is achieved when running 4 threads per each core.

The performance comparison of all the platforms is shown in Fig. 4d. For each platform, we use all the available cores with vectorization enabled. As expected, the best performance result is obtained using Intel Xeon Phi 7120P. This coprocessor executes the MPTADA algorithm 2 times faster than two Intel Xeon E5-2697 v2 CPU, totally containing 24 cores. The both models of the Intel Xeon Phi coprocessor give similar performance results.

6   Conclusions and Future Work

Using the Intel Xeon Phi coprocessor to accelerate computations in the 3D MPDATA algorithm is a promising direction for developing the parallel implementation of this algorithm. Rewriting the EULAG code, and replacing conventional HPC systems with heterogeneous clusters using accelerators such as Intel MIC is a perspective way to improve the efficiency of using this model in practical simulations.

The main challenge of the proposed parallelization is to take advantage of many- and multi-core, vectorization, and cache reusing. For this aim, we propose the block version of the 3D MPDATA algorithm, based on combination of temporal and space blocking techniques. Such an approach gives us the possibility to ease memory bounds by increasing the efficient cache reusing, and reducing the memory traffic associated with intermediate computations. Furthermore, the proposed method of reducing extra computation allows us to accelerate the MPDATA block version up to 4 times, depending on the platform used and size of the grid.

In all the performed tests, the Intel Xeon Phi 7120P coprocessor gives the best performance results. This coprocessor executes the MPTADA algorithm 2 times faster than two Intel Xeon E5-2697 v2 CPUs, totally containing 24 cores, and 2.86 times faster than two Intel Xeon E5-2643. Both the manycore and vectorization features of the Intel MIC architecture play the leading role in the performance exploitation. The other important features are the number of threads per core, as well as an adequate thread placement onto physical cores. All these features have a significant impact on the sustained performance.

At this point of our research, only the first four stages of the MPDATA algorithm are implemented, and tested. They correspond to the linear part of MPDATA. The performance achieved for this part of MPDATA is still limited by memory traffic, mostly because all the input data of the whole MPDATA algorithm are required to provide the correctness of computation for the linear part. As a result, the tested part of MPDATA does not extract the full potential of applying this coprocessor to implement MPDATA computation. Moreover, since the remaining part is unleashed from the memory-cache communication, it gives the opportunity for increasing the efficiency of computation. Implementing and optimizing this part of MPDATA will be studied in future works.

The achieved performance results provide the basis for further research on optimizing the distribution of MPDATA computation across all the computing resources of the Intel MIC architecture, taking into consideration features of its on-board memory, cache hierarchy, computing cores, and vector units. Additionally, the proposed approach requires to develop a flexible data and task scheduler, supported by adequate performance models. Another direction of future work is adaptation to heterogeneous clusters with Intel MICs, with a further development and optimization of code.

Acknowledgments. This work was supported in part by the Polish National Science Centre under grant no. UMO-2011/03/B/ST6/03500.

We gratefully acknowledge the help and support provided by Jamie Wilcox from Intel EMEA Technical Marketing HPC Lab.

References

1. Intel Architectures Comparison. http://ark.intel.com/pl/compare/75799,75797, 64587,75283
2. Intel: Intel Xeon Phi Coprocessor System Software Developers Guide. Intel Corporation (2013)
3. Colfax International: Parallel Programming and Optimization with Intel Xeon Phi Coprocessors. Handbook on the Development and Optimization of Parallel Applications for Intel Xeon Processors and Intel Xeon Phi Coprocessors. Colfax International (2013)
4. Piotrowski, Z., Wyszogrodzki, A., Smolarkiewicz, P.: Towards petascale simulation of atmospheric circulations with soundproof equations. Acta Geophys. **59**, 1294–1311 (2011)
5. Rojek, K., Szustak, L.: Parallelization of EULAG model on multicore architectures with GPU accelerators. In: Wyrzykowski, R., Dongarra, J., Karczewski, K., Waśniewski, J. (eds.) PPAM 2011, Part II. LNCS, vol. 7204, pp. 391–400. Springer, Heidelberg (2012)

6. Rojek, K., Szustak, L., Wyrzykowski, R.: Performance analysis for stencil-based 3D MPDATA algorithm on GPU architecture. In: Wyrzykowski, R., Dongarra, J., Karczewski, K., Waśniewski, J. (eds.) PPAM 2013, Part I. LNCS, vol. 8384, pp. 145–154. Springer, Heidelberg (2014)
7. Smolarkiewicz, P.: Multidimensional positive definite advection transport algorithm: an overview. Int. J. Numer. Meth. Fluids **50**, 1123–1144 (2006)
8. Treibig, J., Wellein, G., Hager, G.: Efficient multicore-aware parallelization strategies for iterative stencil computations. J. Comput. Sci. **2**, 130–137 (2011)
9. Wittmann, M., Hager, G., Treibig, J., Wellein, G.: Leveraging shared caches for parallel temporal blocking of stencil codes on multicore processors and clusters. Parallel Process. Lett. **20**(4), 359–376 (2010)
10. Wyrzykowski, R., Rojek, K., Szustak, L.: Model-driven adaptation of double-precision matrix multiplication to the cell processor architecture. Parallel Comput. **38**, 260–276 (2012)
11. Wyrzykowski, R., Rojek, K., Szustak, L.: Using blue gene/P and GPUs to accelerate computations in the EULAG model. In: Lirkov, I., Margenov, S., Waśniewski, J. (eds.) LSSC 2011. LNCS, vol. 7116, pp. 670–677. Springer, Heidelberg (2012)

Accelerating a Massively Parallel Numerical Simulation in Electromagnetism Using a Cluster of GPUs

Cédric Augonnet$^{(\boxtimes)}$, David Goudin, Agnès Pujols, and Muriel Sesques

CEA/CESTA, 15 Avenue des Sablières, CS 600001, 33116 Le Barp, France
cedric.augonnet@cea.fr

Abstract. We have accelerated a legacy massively parallel code solving 3D Maxwell's equations on a hybrid cluster enhanced with GPUs. To minimize the impact on our existing code, we combine its original Full-MPI approach with task parallelism to design an efficient accelerated LL^t solver that efficiently shares the same GPUs between different processes and relies on an optimized communication patterns. On 180 nodes of the Tera100 cluster, our GPU-accelerated LL^t decomposition reaches 80 TFlop/s on a problem with 247980 unknowns, whereas the sustained machine's CPU and GPU peaks are respectively 13 and 153 TFlop/s.

Keywords: GPU · Dense linear algebra · Cluster computing · Application · Electromagnetism

1 Context

Accelerators are a promising way to build powerful energy-efficient machines. Transitioning to these architectures is however a significant challenge, especially for legacy codes. In this paper, we consider a massively parallel 3D electromagnetic Full-MPI code that requires a tremendous amount of processing resources, and show how we have modified it to exploit a large cluster enhanced with GPUs. This optimized production code was written in FORTRAN since the 90s, so we had to adopt a suitable porting methodology, based on pragmatic constraints. We have followed a gradual porting methodology with a limited impact on our code, and which is flexible enough to be adapted to other architectures in the future (e.g. Intel Xeon Phi processors).

1.1 Physical Problem

The design of stealthy objects requires the computation of the Radar Cross Section (or RCS) of complex 3D targets with complex coatings. The RCS is defined as the ratio between reflected and incident energy in a specific direction. This implies numerically solving Maxwell's equations with the harmonic hypothesis into penetrable bodies and the unbounded surrounding free space. Objects

R. Wyrzykowski et al. (Eds.): PPAM 2013, Part I, LNCS 8384, pp. 593–602, 2014.
DOI: 10.1007/978-3-642-55224-3_55, © Springer-Verlag Berlin Heidelberg 2014

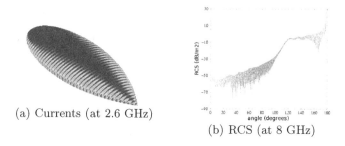

(a) Currents (at 2.6 GHz)

(b) RCS (at 8 GHz)

Fig. 1. An example of RCS computation on NASA almond object

can be composed with both conducting and dielectric bodies. The problem consists in numerically solving Maxwell's equations. The electric and magnetic fields at any point in space can be expressed in terms of surface integral of the charges and currents induced on the surface of the body, as shown on Fig. 1(a). This problem is discretized using the finite element method, so that we can approximate the electric current at the surface of the domain by solving a complex symmetric (non hermitian) dense linear system $AX = B$. On Fig. 1(b), we depict the ratio of energy reflected in each direction due to the currents we have computed at the surface of the objects. We here consider stealth objects with really low RCS, so double precision is required to achieve a sufficient accuracy. Besides, a direct method is used because iterative solvers would not be suitable with many right-hande sides.

With the time harmonic hypothesis, the mesh size is constraint by the wavelength of the illuminating wave, and typically for linear finite elements, ten discretization points per wavelength are required. Consequently the size of the discretized problems grows as a cubic (square) function of the frequency; for large objects or high frequency illuminating waves the size of the associated linear systems can increase up to millions of unknowns. Due to the use of a dense method, the size the impedance matrix A is in $\mathcal{O}(n^2)$ and its factorization is in $\mathcal{O}(n^3)$ operations. millions of unknowns with dense linear systems. Thanks to the recent advances of the parallel architectures based on accelerators (GPU, MIC), the on-going research on linear algebra (especially direct methods for sparse and dense linear systems), large 3D simulations become realistic.

1.2 Parallelization and Performance Analysis of Legacy Code

All the experiments in this paper were carried out on the *Hybrid* partition of the TERA100 cluster of the CEA/DAM. It consists of 180 nodes interconnected by an INFINIBAND QDR network. Each node contains two Xeon E5620 quad-core processors clocked at 2.40 GHz and 24 GB of main memory along with two NVIDIA M2090 FERMI cards with 6 GB of ECC-enabled embedded memory.

This legacy application is a FORTRAN-based massively parallel code written using a *Full-MPI* paradigm. Given the size and the complexity of the code, a full rewrite of the entire application is currently not a realistic option.

The integral formulation leads to the steps presented in Table 1. The number of unknowns is denoted as n, the number of right-hand sides is denoted as n_{rhs}. As an illustration, the actual time measurements on the existing code are also given for a problem with 247980 unknowns, on the CPUs of 64 nodes of the Hybrid partition of the TERA100 cluster. It is worth noting that about 458 GB are required to store the assembled matrix for this problem. The factorization step clearly dominates the overall computation time in Table 1, and thus constitutes the first portion of the code that we must accelerate using GPUs, in Sect. 4. Resolution time grows linearly with the number of incidence angles. While the cost for a single scan angle is limited, the overall scanning may thus incur a significant cost. Section 5 therefore optimizes and accelerates the resolution phase, especially in the context of a large number of scan angles. Result exploitation, and the assembly of the impedance matrix and of the right-hand sides is also a non negligible step, however the corresponding code is currently not suitable to be adapted to GPU computing. Its acceleration, for example by the means of code annotations (e.g. OpenACC, HMPP), is a future work.

Table 1. Steps of the legacy code and their complexity

Step	Description	Complexity	Duration (relative cost)
Assembly	Compute A and B	$\mathcal{O}(n^2)$	197.97 s (3.4 %)
Factorisation	Find L such as $A = LL^t$	$\mathcal{O}(n^3)$	5542.61 s (95.6 %)
Resolution	Find X such as $LL^tX = B$	$\mathcal{O}(n^2 \times n_{rhs})$	1.80 s (0.03 %)
I/O	Interpret and save results	-	55.8 s (1.0%)

2   Related Work

3D Maxwell equations are commonly solved using Fast Multipole Method (FMM) [10], but our code achieves a complete physic model that would be particularly difficult with FMM. Methods like Multigrid [4] or Discontinuous Galerkin [6] are also well suited for accelerators but they have not been tested in our code for all the physics we have to deal with.

Dense symmetric (non hermitian) complex matrices are not common practice, so there is no such variant of Cholesky decomposition readily available in libraries such as Scalapack. Dense linear algebra is a problem mostly studied on a single machine equipped with accelerators [1,7], but there are also a few studies which consider hybrid clusters. Parsec (formerly known as DAGUE) [5] and libflame [8] provide kernels on top of hybrid clusters. They also use tiled algorithms and optimize collective communications. We differ from them as we consider a Full-MPI legacy code, so that we have developed specific techniques to efficiently share each GPU between multiple processes. On the other hand, for efficiency purposes, most runtime systems [3,5,8] assume that we can use a hybrid programming model with one process per node or per GPU. This would require to rework our entire code, while our approach is to gradually adapt

portions of our code using a Full-MPI approach. S_GPU is a library that virtu-
alizes GPUs shared among multiple MPI processes [9]. It offers support for load
balancing and data management, but S_GPU does not automate out-of-card
algorithms.

3 Full-MPI vs. Hybrid Approach

The original code was written using a Full-MPI approach, which means there is
one process per processing unit. This approach does not take advantage of the
memory hierarchy available in machines nowadays. Even though MPI implementations may internally take advantage of this hierarchy, costly synchronizations
and memory transfers are still required within a MPI node. A Full-MPI approach
may also increase the size of the buffers used to exchange data between processes,
which limits the scalability in terms of memory consumption. Hybrid programming models allow to reduce these costs by only creating a few MPI process per
node (typically one per NUMA node or per GPU), and to use a multi-threaded
approach within the process (e.g. using OpenMP or pthreads). This allows using
shared memory, and thereby to reduce memory footprint and to avoid superfluous data transfers within a machine. Having fewer processes per GPU device also
avoids contention and the overhead of blocking operations. However, a hybrid
model usually implies porting efforts to use thread-safe data structures along
with multi-threaded algorithms. OpenMP language extensions could help to
write a multi-threaded assembly phase, but the data structures in our legacy
code are not thread safe: therefore, we did not modify the assembly phase yet
and concentrated on accelerating the factorization and resolution phases.

Table 2. Performance projections depending on the parallelization strategy

	Legacy code	GPU-accelerated projections	
	1 process/core	1 process/GPU	1 process/core
Assembly	200 s	$200 \times 4 = 800$ s	200 s
Factorization	5500 s	$5500/8 = 690$ s	$5500/7 = 785$ s
Total	5700 s	1490 s ($\times 3.8$)	985 s ($\times 5.6$)

Table 2 shows performance projections under different scenarios for the same
test case detailed before with 247980 unknowns. The first column evaluates the
legacy Full-MPI code with a process per core. The second and the third columns
respectively assume we have a Full-MPI code with a process per GPU, or per
core. Using 2 cores out of 8 approximately results in a slowdown of 4 for assembly
phase. We typically measure a speedup of 8 by using GPUs instead of CPUs for
this code. In the third column, we however approximate the slowdown due to
multiplexing each GPU between multiple cores by only assuming a speedup
of 7. One would intuitively try to optimize factorization as much as possible as

this accounts for most of processing time. However, due to Amdhal's law, these estimations show that without a hybrid assembly phase, only using one MPI process per GPU would result in a significant slowdown of the entire application.

Due to the porting efforts that would have been required to port the assembly phase with a hybrid paradigm, we therefore ported our entire application using a Full-MPI strategy which enables an incremental porting strategy. In Sect. 4.3, we show how to effectively design an algorithm that fully exploits GPUs which are shared between multiple processes.

4 Factorization Phase

The factorization phase of the algorithm constitutes the most time consuming part of our application. It consists in factorizing the impedance matrix A with a LL^t decomposition where L is a lower triangular matrix.

4.1 Tiled Cholesky Algorithm

The legacy LL^t decomposition in our code was written similarly to the SCALA-PACK, and adopted a 2D-cyclic data distribution which ensures a good load balancing. However, this approach is based on a fork-join parallelism, which is known not to be scalable enough for manycore platforms. Instead, we have used a tiled LL^t decomposition algorithm, which is well suited for multicore platforms equipped with accelerators [2].

We obtain an efficient load balancing thanks to the existing 2D-cyclic data distribution, however the data layout within a MPI node was restructured in order to have matrices divided into contiguous tiles. Tiled algorithms indeed refer to blocked dense linear algebra algorithms where the matrix is organized in blocks of data stored contiguously. Besides enabling much more parallelism than available in the original algorithms, this data organization significantly improves cache efficiency by allowing a better data locality or avoiding false sharing. Transferring contiguous blocks over MPI or between accelerators and the host is also much more efficient. Blocks can also be stored independently, so that it is possible to only allocate a subset of blocks on the accelerators when the entire matrix does not fit into embedded memory.

Task Parallelism. We have implemented a Full-MPI distributed version of the tiled LL^t algorithm, which naturally translates into a graph of tasks. Each task is an asynchronous piece of computation accessing a few tiles. Edges account for data dependencies, and are translated into MPI transfers over MPI or copies between host and the accelerators.

This is indeed a convenient and portable paradigm to encapsulate computation. It maintains a separation of concerns between the actual execution of the graph over the parallel machine (load balancing, efficient data transfers, etc.) and the design of efficient numerical kernels on the available architectures. Porting a code already written with tasks on a new type of accelerator requires to re-implement a few kernels (e.g. on top of an optimized library such as CUBLAS), and to provide the mechanisms to transfer data between main memory and the accelerator.

We have for instance implemented an optimized hybrid kernel which factorizes diagonal blocks by combining calls to the MKL and CUBLAS libraries, regardless of how the overall graph would be scheduled over the parallel machine.

Granularity Considerations. Block sizes in the tiled algorithm were chosen with regards to the performance obtained by the matrix multiplication kernel which dominates computation time. While MKL's ZGEMM performs well on (40×40) blocks, reaching 5 GFlop/s per core on TERA100, its CUBLAS counterpart only performs well for larger matrix sizes, typically above (128×128) to obtain up to 210 GFlop/s per device.

A large blocking size ensures a good efficiency of the different kernels, but a large granularity reduces the amount of parallelism and requires to allocate large buffers to store temporary blocks. We have thus empirically selected an appropriate block size by considering trade-offs between the efficiency of kernels, the amount of parallelism, and limitations of the memory footprint.

4.2   Data Management

A static schedule is derived from a 2D-cyclic data distribution. Data dependencies are enforced over the cluster by exchanging coherent data over MPI and between the host and the accelerators.

Within each node, we have implemented a light-weight runtime layer that facilitates transfers between the host and the accelerators. It provides simple mechanisms such as synchronization methods and functions to lazily copy an up-to-date version of a piece of data. This is achieved by keeping track of data replicates to determine whether a piece of data is valid on the host and/or on the accelerators, and by performing data transfers from a valid source if the local copy is out-of-date. Maintaining such a cache avoids numerous data transfers between the host and the accelerator. This is a portable approach because this thin layer can easily be adapted to new architectures by reimplementing a few core functionality using vendor's specific API.

Our runtime layer also exploits asynchronous data transfers to overlap them with computation, and to reduce the impact of kernel submission latency. Latency indeed becomes critical when the accelerator is shared with other MPI processes, as they could maintain the processing unit busy for a long time before it becomes available. There also exists much more evolved general-purpose runtime systems that provide similar features along with task scheduling within a single MPI node [1,5,8], but the use of such full-fledged environment and their potential overhead is not necessary here as we consider a very regular algorithm with a static mapping.

We are also severely constrained by the amount of memory required to store the impedance matrix. Not to be limited to the memory available on the GPU devices, our runtime layer transparently keeps a list of pre-allocated blocks on the devices, and keeps reference counts to evict unused blocks and reuse them when the algorithm needs to access new blocks of data. Maintaining such a pre-allocated set of blocks also avoids having to regularly allocate and deallocate data, which would induce costly synchronization as described in Sect. 4.3.

4.3 Efficient Multiplexing of CUDA Devices Between Multiple MPI Processes

It is well known that asynchronous task and data management is essential to properly exploit GPUs. When a GPU is shared by multiple processes, relying on blocking operations results in waiting for all other pending computation on other processes to end before actually executing any operation. Such a high latency severely impacts performance as it delays most computation and avoids to overlap computation with data transfers. However, ensuring that CUDA does not explicitly or implicitly performs any blocking operation is a delicate problem.

Using multiple CUDA streams allows programmers to submit independent non-blocking operations on the device. However, presumably non-blocking operations become blocking API calls if there are too many pending operations. CUDA also inserts implicit dependencies when different streams need to use the same resource: for example, two independent asynchronous data transfers would be serialized on a C1060 TESLA device, even if they are in different streams and in different directions (i.e. from or to the device).

Synchronization should be performed at stream-level with CUDA events or `cudaStreamSynchronize()` rather than synchronizing at context scope with `cudaThreadSynchronize()`. One should also rely on the non-blocking `cudaStreamWaitEvent` method to express dependencies between two CUDA streams. Modifications of the address space (e.g. with `cudaFree`) should be avoided because they force CUDA to block the device and to flush all pending computation.

4.4 Optimizing Communication Patterns

Figure 2 shows a trace of the *panel update* phase of the tiled LL^t algorithm. Note that we here depict a small problem with a small amount of parallelism

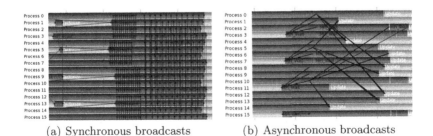

(a) Synchronous broadcasts   (b) Asynchronous broadcasts

Fig. 2. Trace showing the optimization of collective communications

to visually emphasize data transfers. On Fig. 2(a), collective data transfers are performed using `MPI_Bcast` which is blocking. Since a set of independent tiles are broadcasted at each step, blocking MPI calls introduce a significant performance bottleneck because a large amount of time is wasted broadcasting each tile one by one. Furthermore, this prevents the host from submitting any work to the devices during the entire data diffusion phase. We thus observed that our initial implementation on Fig. 2(a) does not scale with the number of tiles to be broadcasted, and therefore with the number of processors because a large number of tiles is required to create enough parallelism. underlined in a accelerator-based environment: GPUs may accelerate computation by an order of magnitude, but the network is unchanged and thus becomes relatively slower.

We therefore replaced all synchronous broadcast operations by an asynchronous *flat tree* algorithm, in which the root node sends point-to-point asynchronous messages to all destination nodes individually. This allows to transfer multiple tiles at the same time, and to wait for transfers completion only once needed. flat-tree broadcast algorithm is usually not as efficient as tree-based hierarchical algorithms for a single transfer. However, the tiled LL^t algorithm successively broadcasts multiple blocks to different sub-set of the processors, so that a set of simultaneous asynchronous broadcasts based on a flat-tree algorithm turns out to be an efficient approach in our case, as shown on Fig. 2(b).

4.5   Performance Evaluation

Each node of the machine described in Sect. 1.2 has 8 CPU cores and 2 GPUs. The asymptotic speed of the ZGEMM$_{NT}$ kernel, which we will refer as the *sustained peak* on Fig. 3(a), is 9 GFlop/s per core and 425 GFlop/s per GPU . It is worth noting that the accelerated Factorization phase only involves CPUs to perform part of the diagonal block LL^t decomposition. The legacy code takes 7417 s to factorize the matrix, which results in an overall execution time of 7705 s for the entire application. Our accelerated version only takes 777 s to factorize the matrix, and the entire application lasts 1088 s. We thus obtain a speedup of ×9.54 for the matrix decomposition on Fig. 3(b), which results in a speedup of

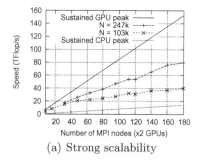
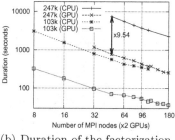

(a) Strong scalability   (b) Duration of the factorization

Fig. 3. Performance evaluation on the factorization step

7.08 for the overall application. We measure 80 TFlop/s on 180 nodes, whereas the sustained peak 180 nodes would be 13.0 TFlop/s (resp. 153.0) using CPUs (resp. GPUs).

5 Solve Phase

The number of observation angles may grow considerably when computing the RCS around an entire object. Rather than solving two symmetric linear systems per angle (one resolution per polarisation), we can solve a single system with multiple right-hand sides. This reduces parallelization overhead because more computation is made during each solve phase. This is also much more efficient on accelerators which are especially suited for vector processing. sides at the same time. However, to avoid superfluous memory consumption, the number of systems to be solved simultaneously must remain limited to fit into memory along with the factorized matrix.

Similarly to the factorization phase, we have implemented a GPU-accelerated resolution algorithm with a tiled algorithm. We also used the data management runtime layer introduced in Sect. 4.2, so that the factorized impedance matrix need not be transferred back to the host between factorization and resolution phases. To preserve parallelism and avoid communication overhead, collective operations were also implemented by the means of asynchronous broadcasts based on point-to-point transfers described in Sect. 4.4.

We measured that the resulting GPU-accelerated implementation is typically three times faster than our CPU-based legacy algorithm, which leads to dramatic improvements for a large angular scanning with thousands of incidence angles.

6 Conclusion

We have ported a real legacy application on a hybrid cluster and obtained significant speedup. We have described the *pragmatic* methodology we followed to determine which parts of the application should and could be accelerated, and which parallelization strategy to adopt. Our scalable Full-MPI LL^t solver show that it is possible to efficiently share GPUs between multiple MPI processes. This conservative approach enabled a gradual porting strategy with a limited impact on the rest of our Full-MPI legacy code. We have shown that task parallelism combined with a runtime system is a convenient paradigm from performance, portability and programmability points of view. Besides, we have shown how to design CUDA applications that are really asynchronous, which is essential to design any scalable algorithms.

In the future, we will use multicore CPUs in conjunction with GPUs, either within a hybrid factorization, or to use idle CPUs to perform other computation such as matrix assembly or post-processing. Matrix assembly will be accelerated using an annotation-based approach such as OpenACC to minimize the impact on existing code. We will extend task paradigm to exploit dynamic scheduling, possibly by the means of a runtime system, and to adapt it on other architectures such as Intel Xeon Phi. We will consider numerical improvements and use our approach on other solvers, for example based on the ACA method [11].

References

1. Agullo, E., Augonnet, C., Dongarra, J., Faverge, M., Ltaief, H., Thibault, S., Tomov, S.: QR factorization on a multicore node enhanced with multiple GPU accelerators. In: International Parallel and Distributed Processing Symposium, pp. 932–943. IEEE (2011)
2. Agullo, E., Demmel, J., Dongarra, J., Hadri, B., Kurzak, J., Langou, J., Ltaief, H., Luszczek, P., Tomov, S.: Numerical linear algebra on emerging architectures: the PLASMA and MAGMA projects. J. Phys. Conf. Ser. **180**, 12–37 (2009). IOP Publishing
3. Augonnet, C., Thibault, S., Namyst, R., Wacrenier, P.A.: Starpu: a unified platform for task scheduling on heterogeneous multicore architectures. Concurr. Comput. Pract. Exp. **23**(2), 187–198 (2011)
4. Baker, A.H., Falgout, R.D., Kolev, T.V., Yang, U.M.: Scaling hypres multigrid solvers to 100,000 cores. High-Performance Scientific Computing, pp. 261–279. Springer, London (2012)
5. Bosilca, G., Bouteiller, A., Herault, T., Lemarinier, P., Saengpatsa, N.O., Tomov, S., Dongarra, J.J.: Performance portability of a GPU enabled factorization with the DAGuE framework. In: IEEE Cluster, pp. 395–402 (2011)
6. Hesthaven, J.S., Warburton, T.: Nodal high-order methods on unstructured grids: I. time-domain solution of Maxwell's equations. J. Comput. Phys. **181**(1), 186–221 (2002)
7. Humphrey, J.R., Price, D.K., Spagnoli, K.E., Paolini, A.L., Kelmelis, E.J.: CULA: hybrid GPU accelerated linear algebra routines. In: SPIE Defense, Security, and Sensing, pp. 770502–770502. International Society for Optics and Photonics (2010)
8. Igual, F.D., Chan, E., Quintana-Ortí, E.S., Quintana-Ortí, G., Van De Geijn, R.A., Van Zee, F.G.: The flame approach: from dense linear algebra algorithms to high-performance multi-accelerator implementations. J. Parallel Distrib. Comput. **72**(9), 1134–1143 (2012)
9. Ospici, M., Komatitsch, D., Mehaut, J.F., Deutsch, T., et al.: SGPU 2: a runtime system for using of large applications on clusters of hybrid nodes. In: Second Workshop on Hybrid Multi-Core Computing, held in Conjunction with HiPC (2011)
10. Song, J., Lu, C.C., Chew, W.C.: Multilevel fast multipole algorithm for electromagnetic scattering by large complex objects. IEEE Trans. Antennas Propag. **45**(10), 1488–1493 (1997)
11. Zhao, K., Vouvakis, M.N., Lee, J.F.: The adaptive cross approximation algorithm for accelerated method of moments computations of EMC problems. IEEE Trans. Electromagn. Compat. **47**(4), 763–773 (2005)

Multidimensional Monte Carlo Integration on Clusters with Hybrid GPU-Accelerated Nodes

Dominik Szałkowski[(✉)] and Przemysław Stpiczyński

Institute of Mathematics, Maria Curie–Skłodowska University,
Pl. Marii Curie-Skłodowskiej 1, 20-031 Lublin, Poland
{dominisz,przem}@hektor.umcs.lublin.pl

Abstract. The aim of this paper is to show that the multidimensional Monte Carlo integration can be efficiently implemented on clusters with hybrid GPU-accelerated nodes using recently developed parallel versions of LCG and LFG pseudorandom number generators. We explain how to utilize multiple GPUs and all available cores of CPUs within a single node and how to extend computations on all available nodes of a cluster using MPI. The results of experiments performed on a Tesla-based GPU cluster are also presented and discussed.

Keywords: Multidimensional integration · Monte Carlo methods · Parallelized pseudorandom number generators · GPU clusters

1 Introduction

Recently, GPU clusters have become a very attractive computer architecture for achieving high performance execution of scientific applications at low costs [1–3], especially for linear algebra computations [4]. The detailed description of NVIDIA CUDA and Fermi architectures can be found in [5] and [6]. GPU clusters comprise three principal components: host nodes, GPUs and interconnection network, which can be treated as a single system. GPUs are connected to host nodes using PCIe bus (PCIe ×16 slots are required for connecting NVIDIA Tesla cards). In order to achieve really high performance system, interconnection network should be of high throughput and low latency. It should also be scalable. Thus, QDR or FDR Infiniband 4× interconnect is highly desirable. Message Passing Interface [7] and CUDA SDK [6] are usually used in software development process for GPU clusters.

Many problems in physics involve computing of multidimensional integrals. Very often such problems have to be solved numerically because their analytical solutions are known only in a few cases. Monte Carlo methods are a broad class of computational algorithms that rely on repeated random sampling to obtain numerical results approximating exact analytical solutions. Such methods are very attractive for solving multidimensional integration problems [8].

It is clear that the process of generation pseudorandom numbers is the most important part of the Monte Carlo integration. Linear congruential generator

R. Wyrzykowski et al. (Eds.): PPAM 2013, Part I, LNCS 8384, pp. 603–612, 2014.
DOI: 10.1007/978-3-642-55224-3_56, © Springer-Verlag Berlin Heidelberg 2014

(LCG) and Lagged Fibonacci generator (LFG) are rather simple generators (sometimes even better than other "high quality" generators [9]) and their parallel versions have been introduced in the SPRNG Library [10,11]. This library has been developed under the assumption that a parallel generator should produce a totally reproducible stream of pseudorandom numbers without any interprocessor communication [12] using cycle division or parameterizing techniques [13,14]. Parallel generators should also be portable between serial and parallel platforms and they should be tested for possible correlations and "high quality" properties [15]. Our approach for developing parallel pseudorandom number generators is quite different [16]. We parallelize recurrence relations for LCG and LFG and statistical properties of our parallel generators are exactly the same as for corresponding sequential ones, thus there is no need to perform special statistical (and rather expensive) tests.

In this paper we show how to implement our parallel versions of LCG and LFG to utilize multiple GPUs and all available cores of CPUs within a single node and how to extend computations on all available nodes of a cluster using MPI. The results of experiments performed on a Tesla-based GPU cluster are also presented and discussed.

2   Parallel Pseudorandom Numbers Generators

Let us consider the following two well-known pseudorandom number generators which can be used in Monte Carlo integration algorithms:

1. **Linear Congruential Generator (LCG):** $x_{i+1} \equiv (ax_i + c)(\mathrm{mod}\, m)$, where x_i is a sequence of pseudorandom values, $m > 0$ is the *modulus*, a, $0 < a < m$ is the *multiplier*, c, $0 \leq c < m$ is the *increment*, x_0, $0 \leq x_0 < m$ is the *seed* or *start value*,
2. **Lagged Fibonacci Generator (LFG):** $x_i \equiv (x_{i-p_1} \diamond x_{i-p_2})(\mathrm{mod}\, m)$, where $0 < p_1 < p_2$, $\diamond \in \{+, -, *, \mathrm{xor}\}$ (for example $p_1 = 5$, $p_2 = 17$, which was the standard parallel generator in the Thinking Machines Connection Machine Scientific Subroutine Library).

Usually, $m = 2^M$, and $M = 32$ or $M = 64$, thus the generators produce numbers from $\mathbb{Z}_m = \{0, 1, \ldots, m - 1\}$. It allows the modulus operations to be computed by merely truncating all but the rightmost 32 or 64 bits, respectively. Thus, when we use unsigned int or unsigned long int data types, we can neglect "$(\mathrm{mod}\, m)$". Note that the integers x_k are between 0 and $m - 1$. They can be converted to real values $r_k \in [0, 1)$ by $r_k = x_k/m$. It is clear that LCG and LFG can be considered as special cases of linear recurrence systems [17]. The LCG generator can be defined as

$$\begin{cases} x_0 = d \\ x_{i+1} = ax_i + c, \quad i = 0, \ldots, n - 2, \end{cases} \tag{1}$$

and similarly for the LFG generator we have

$$\begin{cases} x_i = d_i & i = 0, \ldots, p_2 - 1 \\ x_i = x_{i-p_1} + x_{i-p_2}, & i = p_2, \ldots, n - 1. \end{cases} \tag{2}$$

Let us assume that $n = rs$, where $r, s > 1$. Then (1) can be rewritten in the following block form:

$$\begin{bmatrix} A & & & \\ B & A & & \\ & \ddots & \ddots & \\ & & B & A \end{bmatrix} \begin{bmatrix} \mathbf{x}_0 \\ \mathbf{x}_1 \\ \vdots \\ \mathbf{x}_{r-1} \end{bmatrix} = \begin{bmatrix} \mathbf{f}_0 \\ \mathbf{f} \\ \vdots \\ \mathbf{f} \end{bmatrix}, \tag{3}$$

where $\mathbf{x}_i = (x_{is}, \ldots, x_{(i+1)s-1})^T \in \mathbb{Z}_m^s$ and $\mathbf{f}_0 = (d, c, \ldots, c)^T \in \mathbb{Z}_m^s$, and $\mathbf{f} = (c, \ldots, c)^T \in \mathbb{Z}_m^s$, and the matrices A and B are defined as follows

$$A = \begin{bmatrix} 1 & & & \\ -a & 1 & & \\ & \ddots & \ddots & \\ & & -a & 1 \end{bmatrix} \in \mathbb{Z}_m^{s \times s}, \qquad B = \begin{bmatrix} 0 & \cdots & 0 & -a \\ \vdots & \ddots & 0 & 0 \\ \vdots & & \ddots & \vdots \\ 0 & \cdots & \cdots & 0 \end{bmatrix} \in \mathbb{Z}_m^{s \times s}.$$

From (3) we have $A\mathbf{x}_0 = \mathbf{f}_0$ and $B\mathbf{x}_{i-1} + A\mathbf{x}_i = \mathbf{f}$, $i = 1, \ldots, r-1$. When we set $\mathbf{t} = A^{-1}\mathbf{f}$ and $\mathbf{y} = A^{-1}(a\mathbf{e}_0)$, where $\mathbf{e}_0 = (1, 0, \ldots, 0)^T \in \mathbb{Z}_m^s$, then we get the main formula for the parallel version LCG:

$$\begin{cases} \mathbf{x}_0 = A^{-1}\mathbf{f}_0 \\ \mathbf{x}_i = \mathbf{t} + x_{is-1}\mathbf{y}, & i = 1, \ldots, r-1. \end{cases} \tag{4}$$

The equation (4) has a lot of potential parallelism. The algorithm comprises the following steps. First (Step 1) we have to find $\mathbf{x}_0$, $\mathbf{y}$, $\mathbf{t}$ (using three separate tasks). Then (Step 2) we find the last entry of each vector $\mathbf{x}_i$, $i = 1, \ldots, r-1$. Finally (Step 3), we find in parallel $s - 1$ entries of the vectors $\mathbf{x}_1, \ldots, \mathbf{x}_{r-1}$.

Similarly we can consider the following algorithm for finding a pseudorandom sequence using (2). Let $n = rs$, $r, s > p_2$. To find a sequence $x_0, \ldots, x_{n-1}$, we have to solve the following system of linear equations

$$\begin{bmatrix} A_0 & & & \\ B & A & & \\ & \ddots & \ddots & \\ & & B & A \end{bmatrix} \begin{bmatrix} \mathbf{x}_0 \\ \mathbf{x}_1 \\ \vdots \\ \mathbf{x}_{r-1} \end{bmatrix} = \begin{bmatrix} \mathbf{f} \\ 0 \\ \vdots \\ 0 \end{bmatrix}, \tag{5}$$

where the matrices A_0, A and B are of the forms showed in Fig. 1. The vectors are defined as $\mathbf{f} = (d_0, \ldots, d_{p_2-1}, 0, \ldots, 0)^T \in \mathbb{Z}_m^s$, $\mathbf{x}_i = (x_{is}, \ldots, x_{(i+1)s-1})^T \in \mathbb{Z}_m^s$. Thus we have

$$\begin{cases} A_0\mathbf{x}_0 = \mathbf{f} \\ B\mathbf{x}_{i-1} + A\mathbf{x}_i = 0, & i = 1, \ldots, r-1. \end{cases} \tag{6}$$

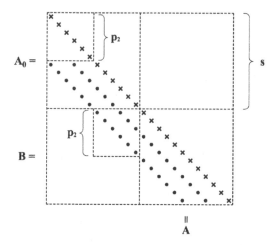

Fig. 1. The matrix of the system (5) for $n = 24$, $s = 12$, $r = 2$, $p_1 = 3$, $p_2 = 6$. Symbols "$\bullet$" and "$\times$" denote -1 and 1, respectively.

Let $\mathbf{y}_k$ satisfies the system of linear equations $A\mathbf{y}_k = \mathbf{e}_k$, $k = 0, \ldots, p_2 - 1$, where $\mathbf{e}_k$ is the k-th unit vector from $\mathbb{Z}_m^s$, and $\mathbf{y} \equiv \mathbf{y}_0 = (1, y_1, \ldots, y_{s-1})^T$. It is easy to verify that

$$\mathbf{y}_k = (\underbrace{0, \ldots, 0}_{k}, 1, y_1, \ldots, y_{s-1-k})^T.$$

Thus it is sufficient to find only the vector $\mathbf{y}$. Finally using (6) we get the main formula for the parallel version LFG:

$$\begin{cases} \mathbf{x}_0 = A_0^{-1}\mathbf{f} \\ \mathbf{x}_i = \sum_{k=0}^{p_2-1} x_{is-p_2+k}\mathbf{y}_k + \sum_{k=0}^{p_1-1} x_{is-p_1+k}\mathbf{y}_k, \qquad i = 1, \ldots, r-1. \end{cases} \tag{7}$$

Note that (7) is the generalization of (4). Thus one can develop a similar parallel *divide-and-conquer* algorithm. During the first step we have to find the vectors $\mathbf{x}_0$ and $\mathbf{y}_0$, then (Step 2) using (7) we find p_2 last entries of $\mathbf{x}_1, \ldots, \mathbf{x}_{r-1}$. Finally (Step 3) we use (7) to find $s - p_2$ first entries of these vectors (in parallel). Note that Step 2 requires interprocessor communication. To implement the parallel algorithms efficiently, let us observe that we have to find the following matrix

$$Z = [\mathbf{x}_0, \ldots, \mathbf{x}_{r-1}] \in \mathbb{Z}_m^{s \times r}, \tag{8}$$

where all vectors $\mathbf{x}_i$ are defined by (4) or (7). The details of our single-GPU implementations can be found in [16].

3   Multidimensional Monte Carlo Integration

The idea of Monte Carlo methods for multidimensional integration can be found in [18]. For a given dimension $d \geq 1$, let $I^d = [0, 1]^d$ be the d-dimensional

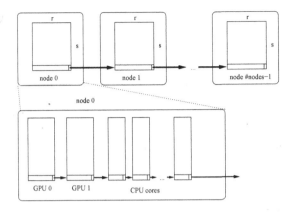

Fig. 2. Implementation of the algorithm

unit cube and let $f(\mathbf{v})$ be a bounded Lebesgue-integrable function on I^d. The approximation for the Lebesgue integral of f over I^d can be found using

$$\int_{I^d} f(\mathbf{v})d\mathbf{v} \approx \frac{1}{N} \sum_{i=1}^{N} f(\mathbf{v}_i), \tag{9}$$

where $\mathbf{v}_1, \ldots, \mathbf{v}_N$ are random points from I^d. The strong law of large numbers guarantees that the numerical integration using (9) converges almost surely and from the central limit theorem it follows that the expected error is $O(N^{-1/2})$ (see [18] for more details).

The idea of Monte Carlo algorithm for multidimensional integration can be rather simple. It comprises three main steps:

1. generate N pseudorandom points $\mathbf{v}_i \in I^d$, $i = 1, \ldots, N$, using Nd pseudorandom real numbers from $[0, 1)$,
2. calculate the values of $f(\mathbf{v}_i)$, $i = 1, \ldots, N$,
3. perform the global reduction (summation) according to (9).

In order to implement the multidimensional Monte Carlo integration using the parallel versions of LCG and LFG for clusters with hybrid GPU-accelerated nodes, let us assume the following:

1. each cluster node is responsible for generating n real random numbers and n/d random points from I^d,
2. the global number of random points N is equal to pn/d, where p denotes the number of cluster nodes,
3. each cluster node computes the matrix Z_j, $j = 0, \ldots, p-1$, of the form (8),
4. matrices Z_j are computed using available GPUs and CPU cores, thus Z_j are divided into blocks of columns (see Fig. 2),
5. the ratio of the number of columns in GPU blocks to the number of columns in CPU blocks follows from the performance ratio of GPUs to CPU cores,

6. GPU blocks are computed according to (4) or (7) using CUDA kernels, while CPU blocks are computed using (1) or (2),
7. GPU blocks are stored in the GPU's global memory, while CPU blocks are computed "on the fly", so there is no need to store them in RAM,
8. as soon as the last stripe of rows of Z_j is calculated (Step 2), the required number (LCG)/p_2 numbers (LFG) is/are sent to the next node,
9. after computing entire GPU blocks, GPUs calculate values of integrand function and perform the reduction,
10. CPU cores calculate values of integrand function and the sum of them "on the fly",
11. OpenMP is used to utilize CPU cores and multiple GPUs within a single node,
12. we use two levels of communication: two GPU cards inside of a single node communicate using peer-to-peer memory access with `cudaMemcpyPeer()` routine omitting CPU, then all nodes perform the standard `MPI_reduce()` function (the next level of communication).

Our approach is equivalent to the process of generating a single sequence of random numbers with known statistical properties. Thus there is no need to use parametrization techniques that need testing and sometimes can lead to obtain sequences with unwanted correlations between numbers [10,11].

4   Results of Experiments

The considered algorithm for the multidimensional integration (9) using the parallel versions of LCG and LFG respectively, has been tested on a GPU cluster of 32 nodes, each with two Intel Xeon X5650 (6 cores each with hyper-threading, 2.67 GHz, 48 GB RAM) and two NVIDIA Tesla M2050 (448 cores, 3 GB GDDR5 RAM with ECC off), connected using 40 Gbit/s Infiniband, running under Linux with NVIDIA CUDA Toolkit ver. 5.0 and Intel Cluster Studio ver. 2012. We have used the set of the following test functions [19] ($\mathbf{c}, \mathbf{w} \in \mathbb{R}^d$ are fixed coefficients):

1. Oscillatory: $f_1(\mathbf{v}) = \cos(\mathbf{c} \cdot \mathbf{v} + 2\pi w_1)$,
2. Product peak: $f_2(\mathbf{v}) = \prod_{i=1}^{d} \frac{1}{(v_i - w_i)^2 + c_i^{-2}}$,
3. Corner peak: $f_3(\mathbf{v}) = \frac{1}{(1 + \mathbf{c} \cdot \mathbf{v})^{d+1}}$,
4. Gaussian: $f_4(\mathbf{v}) = \exp(-\mathbf{c}^2(\mathbf{v} - \mathbf{w})^2)$,
5. C^0-continuous: $f_5(\mathbf{v}) = \exp(-\mathbf{c} \cdot |\mathbf{v} - \mathbf{w})|$,
6. Discontinuous: $f_6(\mathbf{v}) = \begin{cases} 0 & \text{for } v_1 > w_1 \vee v_2 > w_2, \\ \exp(\mathbf{c} \cdot \mathbf{v}) & \text{otherwise.} \end{cases}$

Figure 3 shows the speedup of the integration using LCG and LFG using two GPUs vs one CPU core. We can observe that in case of LCG the speedup is from 2 up to 8, depending on the type of integrand functions. For some functions, the dimension d also influences the speedup. The best speedup can be observed for "Product peak" and "Discontinuous". In case of LFG, the values of p_1 and p_2 are

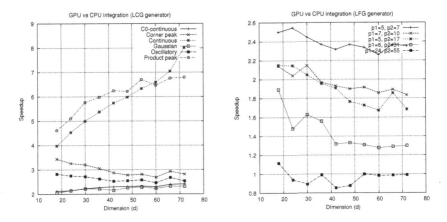

Fig. 3. Speedup of the integration using LCG (left) and LFG (right) using two GPUs vs one CPU core

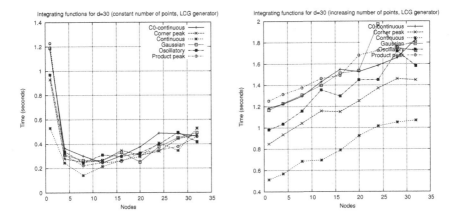

Fig. 4. Time of the integration for $d = 30$ and with constant number of global points (left) and with increasing number of global points (right)

crucial for the efficiency of the algorithm. Bigger values of these parameters result in substantial decrease of the performance. It is a straightforward conclusion from the analysis of the formula (7), where the number of operations grows when p_1 and p_2 grow. The results from Fig. 3 have been used to determine the number of columns in GPU and CPU blocks.

Let $p_{GPU} = 2$ be the number of GPU devices and $p_{CPU} = 22$ be the number of CPU cores. We use $n_{GPU} \approx 377 \cdot 10^6$ random numbers per each GPU (this is to fill its 3 GB global memory) and for each CPU core we use $n_{CPU} = n_{GPU} \cdot r$ random numbers, where r is the performance ratio of GPUs to CPU cores (see Fig. 3). Thus, the global number of generated random numbers is $n = p_{GPU} \cdot n_{GPU} + p_{CPU} \cdot n_{CPU}$. Depending on the function family, dimension and the type

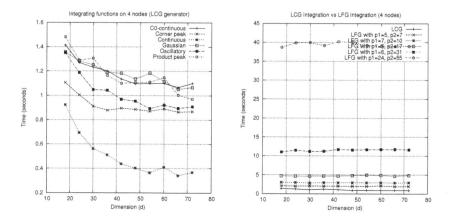

Fig. 5. Time of the integration using four nodes with fixed number of global random numbers and increasing dimensions (left). Comparison of the integration time using LCG and LFG (various p_1 and p_2, right)

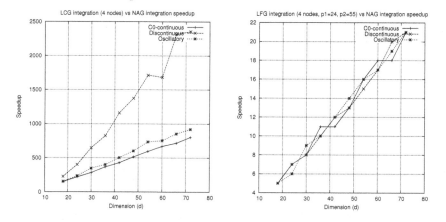

Fig. 6. Speedup of the integration using LCG (left) and LFG (right) using four nodes vs D01GBF NAG routine using one node

of generator, we use n from $7 \cdot 10^9$ to $67 \cdot 10^9$. Then the global number of points is $N = n/d$, where d is the dimension.

The scalability of the algorithm in case of the fixed number of global points (we use n random numbers) is shown in Fig. 4 (left). We can observe that the performance grows (the time decreases) to a certain number of nodes. From this threshold value, the performance decreases slightly. It is due to the ratio of the computation to communication cost. For Monte Carlo methods it is very important to use sufficiently large number of points to achieve desired accuracy. Thus it is better to increase the number of global points when additional cluster nodes are used (we add n random numbers for each new node). Figure 4 (right) shows that in such a case the execution time grows slightly.

Table 1. Accuracy of the LCG integration and D01GBF for "Oscillatory"

Dimension	LCG integration	LCG accuracy	NAG integration	NAG accuracy
24	-0.48923137	2.48E-5	-0.48917838	1.58E-4
30	-0.69255621	2.71E-5	-0.69255412	1.90E-4
36	0.12922529	3.11E-5	0.12918500	2.45E-4
42	0.61335932	3.14E-5	0.61327319	2.86E-4
48	0.20421457	3.39E-5	0.20427389	3.24E-4
54	-0.49101977	3.82E-5	-0.49094998	3.86E-4
60	-0.34648007	4.00E-5	-0.34634743	4.26E-4
66	0.02768727	4.20E-5	0.02758309	4.49E-4
72	0.42734610	4.40E-5	0.42736322	5.27E-4

It should be noticed that for the increasing dimension and the fixed number of global random numbers, the number of global points from I^d decreases resulting in shorter execution time (Fig. 5, left). The comparison of the time of the algorithm using LCG and LFG respectively is shown in Fig. 5 (right). Again we can see the influence of the values of p_1 and p_2.

Figure 6 compares our algorithm with D01GBF routine from the well-known NAG Library. D01GBF uses an adaptive Monte Carlo method based on the algorithm described in [20]. Although we have used the version of NAG intended for multicore shared-memory machines, we have not observed any change of the performance of D01GBF in case of the use of multiple cores of Xeon processors. We have observed that our algorithm is much faster than the NAG routine. Moreover, our algorithm produces more accurate results (see Table 1).

5   Conclusions

We have showed that the multidimensional Monte Carlo integration can be efficiently implemented on clusters with hybrid GPU-accelerated nodes using recently developed parallel versions of LCG and LFG pseudorandom number generators. We have explained how to utilize multiple GPUs, all cores of CPUs and all available nodes of a cluster using MPI. The results of experiments performed on a Tesla-based GPU cluster have shown the our implementation is much more efficient than the corresponding NAG routine. Although, we use the NVIDIA CUDA Fermi architecture, the implementation does not rely heavily on it. We use basic CUDA interface, thus the use of advanced hardware features is hidden. However we can expect that the use of the new Kepler architecture would be profitable but it will be the subject of our further research.

References

1. Kindratenko, V.V., Enos, J., Shi, G., Showerman, M.T., Arnold, G.W., Stone, J.E., Phillips, J.C., mei Hwu, W.: GPU clusters for high-performance computing. In: Proceedings of the 2009 IEEE International Conference on Cluster Computing, pp. 1–8, New Orleans, LA, USA. IEEE, 31 Aug–4 Sept 2009

2. Bueno, J., Planas, J., Duran, A., Badia, R., Martorell, X., Ayguade, E., Labarta, J.: Productive programming of GPU clusters with OmpSs. In: 26th International Conference on Parallel Distributed Processing Symposium (IPDPS), 2012, pp. 557–568. IEEE (2012)
3. Göddeke, D., Strzodka, R., Mohd-Yusof, J., McCormick, P.S., Buijssen, S.H.M., Grajewski, M., Turek, S.: Exploring weak scalability for FEM calculations on a GPU-enhanced cluster. Parallel Comput. 33(10–11), 685–699 (2007)
4. Fatica, M.: Accelerating linpack with CUDA on heterogenous clusters. In: Proceedings of 2nd Workshop on General Purpose Processing on Graphics Processing Units, GPGPU 2009, pp. 46–51, Washington, DC, USA, 8 Mar 2009
5. NVIDIA Corporation: NVIDIA next generation CUDA compute architecture: Fermi. http://www.nvidia.com/ (2009)
6. NVIDIA Corporation: CUDA Programming Guide. NVIDIA Corporation. http://www.nvidia.com/ (2012)
7. Pacheco, P.: Parallel Programming with MPI. Morgan Kaufmann, San Francisco (1996)
8. Bull, J.M., Freeman, T.L.: Parallel globally adaptive quadrature on the KSR-1. Adv. Comput. Math. 2, 357–373 (1994)
9. Ferrenberg, A., Landau, D., Wang, Y.J.: Monte Carlo simulations: hidden errors from good random number generators. Phys. Rev. Lett. 69, 3382–3384 (1992)
10. Mascagni, M., Srinivasan, A.: Algorithm 806: SPRNG: a scalable library for pseudorandom number generation. ACM Trans. Math. Softw. 26(3), 436–461 (2000)
11. Mascagni, M., Srinivasan, A.: Corrigendum: Algorithm 806: SPRNG: a scalable library for pseudorandom number generation. ACM Trans. Math. Softw. 26(4), 618–619 (2000)
12. Mascagni, M.: Parallel linear congruential generators with prime moduli. Parallel Comput. 24(5–6), 923–936 (1998)
13. Mascagni, M., Chi, H.: Parallel linear congruential generators with Sophie-Germain moduli. Parallel Comput. 30(11), 1217–1231 (2004)
14. Mascagni, M., Srinivasan, A.: Parameterizing parallel multiplicative lagged-Fibonacci generators. Parallel Comput. 30(5–6), 899–916 (2004)
15. Srinivasan, A., Mascagni, M., Ceperley, D.: Testing parallel random number generators. Parallel Comput. 29(1), 69–94 (2003)
16. Stpiczyński, P., Szałkowski, D., Potiopa, J.: Parallel GPU-accelerated recursion-based generators of pseudorandom numbers. In: Proceedings of the Federated Conference on Computer Science and Information Systems, pp. 571–578, Wroclaw, Poland. IEEE Computer Society Press, 9–12 Sept 2012
17. Stpiczyński, P.: Solving linear recurrence systems on hybrid GPU accelerated manycore systems. In: Proceedings of the Federated Conference on Computer Science and Information Systems, Szczecin, Poland, pp. 465–470. IEEE Computer Society Press, 18–21 Sept 2011
18. Niederreiter, H.: Quasi-Monte Carlo methods and pseudo-random numbers. Bull. Am. Math. Soc. 84, 957–1041 (1978)
19. Hahn, T.: CUBA–a library for multidimensional integration. Comput. Phys. Commun. 168, 78–95 (2005)
20. Lautrup, B.: An adaptive multi-dimensional integration procedure. In: Proceedings of the 2nd Colloquium on Advanced Methods in Theoretical Physics, Marseille

Efficient Execution of Erasure Codes on AMD APU Architecture

Roman Wyrzykowski, Marcin Woźniak$^{(\boxtimes)}$, and Lukasz Kuczyński

Institute of Computer and Information Sciences,
Czestochowa University of Technology,
Dabrowskiego 73, 42-201 Czestochowa, Poland
{roman,marcell,lkucz}@icis.pcz.pl

Abstract. Erasure codes such as Reed-Solomon codes can improve the availability of distributed storage in comparison with replication systems. In previous studies we investigated implementation of these codes on multi/many-core architectures, such as Cell/B.E. and GPUs. In particular, it was shown that bandwidth of PCIe bus is a bottleneck for the implementation on GPUs.

In this paper, we focus on investigation how to map systematically the Reed-Solomon erasure codes onto the AMD Accelerated Processing Unit (APU), a new heterogeneous multi/many-core architecture. This architecture combines CPU and GPU in a single chip, eliminating costly transfers between them through the PCI bus. Moreover, APU processors combine some features of Cell/B.E. processors and many-core GPUs, allowing for both vectorization and SIMT processing simultaneously.

Based on the previous works, the method for the systematic mapping of computation kernels of Reed-Solomon and Cauchy Reed-Solomon algorithms onto the AMD APU architecture is proposed. This method takes into account properties of the architecture on all the levels of its parallel processing hierarchy.

Keywords: Erasure codes · Reed-Solomon codes · Multicore architectures · GPU · APU · OpenCl

1 Introduction

There is a rapid increase in sensitive data, such as biomedical records or financial data. Protecting such data while in transit as well as while at rest is crucial [7]. An example are distributed data storage systems in grids [18] and clouds [1], that have different security concerns than traditional file systems, since data are now spread across multiple hosts. Failure of a single host could lead to loss of sensitive data, and compromise the whole system. Consequently, suitable techniques, e.g., cryptographic algorithms and data replication, should be applied to fulfill such important requirements as confidentiality, integrity, and availability [1,18].

R. Wyrzykowski et al. (Eds.): PPAM 2013, Part I, LNCS 8384, pp. 613–621, 2014.
DOI: 10.1007/978-3-642-55224-3_57, © Springer-Verlag Berlin Heidelberg 2014

A classic concept of building fault-tolerant systems consists of replicating data on several servers. Erasure codes can improve the availability of distributed storage by splitting up the data into n blocks, encoding them redundantly using m blocks, and distributing the blocks over various servers [4]. As was shown in [16], the use of erasure codes reduces "mean time of failures by many orders of magnitude compared to replication systems with similar storage and bandwidth requirements". This approach can be also used to build large-scale storage installation that requires fault-protection beyond RAID-5 in the case of multiple disk failures [1,13].

There are many ways of generating erasure codes. A standard approach is the use of the Reed-Solomon (or RS) codes [11]. These codes were applied, among others, to provide a reliable data access in the well known persistent data store, OceanStore [9]. The main disadvantage of this approach is a relatively large computational cost because all operations are implemented over the Galois field $GF(2^w)$ arithmetic, which is traditionally not supported by microprocessors, where

$$2^w \geq n + m. \tag{1}$$

All operations, such as addition, multiplication or division, have to be performed in the Galois field $GF(2^w)$, which in the case of multiplication or division leads to a significant growth of computational cost. The computational cost of RS codes grows with the value of n [15] (and m, as well).

The development of high-performance multicore architectures opens a way to take advantages of RS erasure codes, since performance delivered by this architectures is no longer an obstacle to utilization of RS codes in practical data storage systems [1,5,10,19]. In our previous studies [10,18–20], we investigated implementation of these codes on multi/many-core architectures, such as Cell/B.E. and GPUs. In particular, it was shown that bandwidth of PCIe bus is a bottleneck for the implementation on GPUs.

Therefore, the APU (Accelerated Processing Unit) [2], where RAM memory access from the GPU does not require the PCIe transfer, appears to be an interesting alternative to alleviate such a bottleneck. In this work, we focus on investigating how to map systematically the RS erasure codes on the AMD APUs, a new heterogeneous multi/many-core architecture, which combines CPU and GPU in a single chip.

The material of this paper is organized as follows. Details of the Reed-Solomon encoding/decoding algorithm are introduced in Sect. 2. In Sect. 3, we discuss the previous research using heterogeneous architectures: Cell/B.E. hybrid processor and NVIDIA Tesla M2070Q GPU. Sect. 4 provides overview of the AMD APU architecture as well as discusses the proposed methodology for mapping the RS codes on the APU. The performance results achieved on A8-3870 APU are presented in Sect. 5. The last Sect. 6 gives conclusions and future work.

2   Reed-Solomon Codes and Linear Algebra Algorithms

Applying EC codes to increase reliability of distributed data management systems can be described in the following way [11]. A file F of size $|F|$ is partitioned into n blocks (stripes) of size B words each, where:

$$B = |F|/n .\tag{2}$$

Each block is stored on one of n data devices $D_0, D_1, \ldots, D_{n-1}$. Also, there are m checksum devices $C_0, C_1, \ldots, C_{m-1}$. Their contents are derived from contents of data devices, using an encoding algorithm. This algorithm has to allow for restoring the original file from any n (or a bit more) of $n + m$ storage devices $D_0, D_1, \ldots, D_{n-1}, C_0, C_1, \ldots, C_{m-1}$, even if m of these devices failed, in the worst case.

The application of the Reed-Solomon erasure codes includes [11,12] two stages: (i) encoding, and (ii) decoding. At the encoding stage, an input data vector $\mathbf{d}_n = [d_0, d_1, \ldots, d_{n-1}]^T$, containing n words each of size w bits, is multiplied by a special matrix

$$\mathbf{F}_{(n+m)\times n} = \begin{bmatrix} \mathbf{I}_{n\times n} \\ \mathbf{F}^*_{m\times n} \end{bmatrix}\tag{3}$$

with elements defined over $GF(2^w)$. As a result of the encoding procedure, we obtain an $(n + m)$ column vector

$$\mathbf{e}_{n+m} = \mathbf{F}_{(n+m)\times n}\mathbf{d}_n = \begin{bmatrix} \mathbf{d}_n \\ \mathbf{c}_m \end{bmatrix} ,\tag{4}$$

where:

$$\mathbf{c}_m = \mathbf{F}^*_{m\times n}\mathbf{d}_n .\tag{5}$$

Therefore, the encoding stage can be reduced to performing many times the matrix-vector multiplication (5), where all operations are carried out over $GF(2^w)$.

The decoding stage consists in deleting those rows of the matrix $\mathbf{F}_{(n+m)\times n}$ that correspond to failed nodes. The reconstruction of failed elements (words) of the vector $\mathbf{d}_n$ is based on applying the following expression:

$$\mathbf{d}_n = \phi^{-1}_{n\times n} \times \mathbf{e}^*_n ,\tag{6}$$

where the column $\mathbf{e}^*_n$ consists of entries of the original vector $\mathbf{e}_{n+m}$ located in nodes that did not fail. The decoding procedure ends with determining those entries of the checksum vector $\mathbf{c}_m$ that correspond to failed nodes.

In our investigations we focus on mapping only the first stage, namely encoding, since our main objective is to determine opportunities given by applying innovative multicore architectures to accelerate computations required to implement the classic version of the Reed-Solomon codes. It follows from the above assumption that there is no need to consider the decoding phase, since it differs from the encoding phase only by an additional procedure of computing the inverse matrix of size $n \times n$, which is performed just once in order to reconstruct a given file after the failure. Moreover, the complexity of the inversion procedure is small for relatively small values of m, which are of our primary interest. So, for sufficiently large files, efficiency aspects of entire computations depend in practice on the efficiency of operation (5).

3   Previous Research Using Cell/B.E. and Tesla M2070Q GPU

The computational power of Cell/B.E. [3], coupled with its security features, make it a suitable platform to implement algorithms aimed at improving data confidentiality, integrity, and availability [10,18], such as Reed-Solomon codes. Also, basic features of multicore GPUs [6,17] such as utilization of a large number of relatively simple processing units operating in the SIMD fashion, as well as hardware supported multithreading, enable the efficient implementation of this type of computations.

Previous investigations [10,18–20] confirmed the advantage of using Cell/B.E. for the efficient implementation of the classic Reed-Solomon codes, as well as the Cauchy modification of Reed-Solomon Codes. The implementation for Cell/B.E. covered all three levels of processor parallelism: eight SPE cores, vector processing, and two pipelines. In the case of massively parallel NVIDIA GPU, the main factor limiting the encoding efficiency was the PCIe bus bandwidth, in spite of using the stream processing, which allows for overlapping GPU computations with data transfers between the GPU and CPU. The achieved encoding bandwidth was up to 9.5 GB/s in the case of the Cell processor, and 3.5 GB/s for the NVIDIA Tesla GPU.

4   Mapping Reed-Solomon Algorithm on AMD APU Architecture

4.1   AMD APU Architecture

Accelerated Processing Unit is combination of CPU and GPU in a single chip. An architecture overview of APU is shown in Fig. 1. In addition to the CPU and GPU, the AMD APU architecture also consists of High Performance Bus and Memory Controller, as well as Unified Video Decoder and Platform Interfaces providing communication with external components. The access to the RAM memory from both the CPU and GPU is realized by the High Performance Bus and Memory Controller with the bandwidth of 29.8 GB/s. Such a combination of CPU, GPU and RAM enables for the elimination of time-consuming transfers over PCIe when performing memory access. Furthermore, the multi-core Radeon GPU used in this architecture allows for running thousands of threads, and vectorization, as well as provides a relatively low power consumption (TDP at 100 W), and very low hardware costs. Both CPU and GPU cores can be used through the OpenCL programming framework [8].

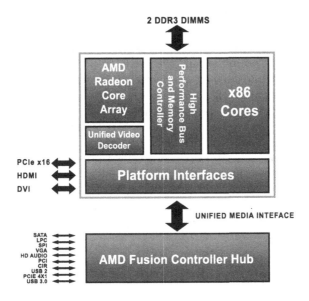

Fig. 1. AMD APU Architecture

In the presented study, the AMD A8-3870 (Liano) processor with the Radeon HD 6550D GPU is used. This GPU consists of 5 SIMD engines with 20 texture units each, which gives 400 GPU cores. For 600 MHz clock frequency, its peak performance is 480Gflops in single precision.

4.2 Mapping Details

The basic algorithm for mapping the Reed-Solomon encoding on the APU GPU is based on the vectorization algorithm proposed in [19] for the Cell/B.E. implementation. The general form of the algorithm is as follows:

$$\mathbf{c} = [0, 0, \ldots, 0]$$
$$\textbf{for } i = 0, 1, \ldots, m - 1 \textbf{ do } \{$$
$$\quad \textbf{for } j = 0, 1, \ldots, n - 1 \textbf{ do}$$
$$\quad\quad \mathbf{c}_i := \mathbf{c}_i \oplus \mathbf{f}_{i,j}^* \odot \mathbf{d}_j \qquad (7)$$
$$\}$$

Here $\oplus$ and $\odot$ symbols denote respectively the addition and multiplication operations carried out over $GF(2^w)$.

As mentioned before, the basic operation of the Reed-Solomon algorithm is a matrix-vector multiplication over the $GF(2^w)$ arithmetic. For the efficient implementation, the multiplication operation of the form $c = f*d$ is implemented using table lookups [12], based on the following formula:

$$c = gfilog(gflog(f) + gflog(d)). \qquad (8)$$

Here $gflog$ and $gfilog$ denote respectively logarithms and antilogarithms tables of 256 elements each, defined over $GF(2^w)$. Values of these tables are stored in the fast local memory. The addition is implemented as a bitwise XOR operation.

The important factor influencing the encoding performance is the structure of F-matrix. Taking into account the fact that the first row and the first column have the same values ($\mathbf{f}_{0,j}^* = \mathbf{f}_{i,0}^* = 1$), it possible to reduce the number of instructions performed by each GPU thread. Also, the situation in which an encoded data item has a value of 0 requires a different treatment. In the proposed approach, the conditional instruction is eliminated by introducing some additional calculations so that every single GPU thread will perform exactly the same instructions shown below:

$k = get\_global\_id(0);$

$c[0][k] = d[0][k];$

for $j = 1, ..., n - 1$ **do** $c[0][k]$ $^\wedge= d[j][k];$

for $i = 1, ..., m - 1$ **do** {

$\quad c[i][k] = d[0][k];$

$\quad$ **for** $j = 1, ..., n - 1$ **do** {

$\quad\quad tmp = d[j][k]$ && **true**;

$\quad\quad c[i][k]$ $^\wedge= tmp * gfilog[gflog[f^*[i][j] + gflog[d[j][k]]];$ $\quad$ (9)

$\quad$ }

}

In the case of $d[j][k] == 0$, the right side of expression (9) takes the value of 0. In any other case, the value of expression (9) is determined using the lookup table.

As one of the main assumptions in our development is keeping the full compatibility with the open-source *Jerasure* library [14], values of $gflog$ and $gfilog$ tables, as well as F-matrix, are taken from this library.

5   Performance Results

The performance experiments are carried out for the platform based on the AMD A8-3870 APU with the Radeon HD 6550D GPU, 8 GB RAM. To program this platform we use the OpenCL v1.2 programming standard and gcc v4.7.2 compiler. As in the previous studies, two variants of encoding are tested. They assume the use of either 4 data nodes and 4 checksums nodes, or 8 data nodes and 4 checksums nodes. In both cases, the maximum encoding bandwidth of 1.5 GB/s is achieved, while on CPU with the *Jerasure* library the encoding bandwidth is about 0.25 GB/s only (Fig. 2).

Since the APU processor allows for the vector processing in each single GPU thread, a vectorized version of the RS algorithm is also examined. It is based

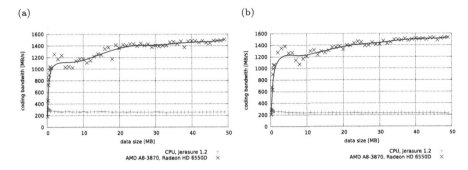

Fig. 2. Bandwidth achieved on AMD A8-3870 (Radeon 6550D) for different variants of implementing the Reed-Solomon encoding, depending on size of encoded file: (a) first variant of encoding with $n, m = 4$; (b) second variant of encoding with $n = 8$, $m = 4$

on vectorizing accesses to lookup tables [18,19]. However, in the case of APU, where *gflog* and *gfilog* tables are placed in the fast local memory, the vectorized version of lookup table introduces too many additional operations, which results in decreasing the performance.

Table 1 presents the performance comparison for the RS algorithm implementation on AMD A8-3870 APU and NVIDIA Tesla M2070Q GPU, where results for the GPU are taken from our previous work [19].

The bandwidth of 1.5 GB/s on the AMD A8 APU seems to be not much in comparison with 3.5 GB/s on the Tesla GPU processor. However, taking into account the power consumption, the APU architecture is a better choice. In fact, because the Tesla GPU processor can not work separately, the TDP factor in this case should also include the CPU power consumption. So based on the ratio of encoding bandwidth to TDP of each system, we conclude about a clear advantage of APU over GPU. At the same time, the deciding factor in favor of the AMD A8 APU architecture is its much lower hardware cost.

Table 1. Reed-Solomon performance comparison

	Tesla M2070Q GPU	AMD A8-3870 APU
Peak performance (Gflops)	1030	480 56
TDP (W)	225 + 100	100
RAM bandwidth (GB/s)	(over PCIev2.0 x16) 8	29,8
$\frac{Peak\ performance}{TDP}$ $\left(\frac{Gflops}{Watt}\right)$	2,68	4,80
Encoding bandwidth (GB/s)	3,5	1,5
$\frac{Encoding\ bandwidth}{TDP}$ $\left(\frac{GB/s}{Watt}\right)$	0,011	0,015
Hardware cost	$2100	$100

6   Conclusions

The preliminary performance results achieved on AMD A8 APU show the possibility of using the APU architecture for the efficient implementation of the Reed-Solomon codes. The proposed approach allows us to encode data on the fly in case of distributed data storage systems based on the 10GbEth network. Low hardware costs and possibility to eliminate time-consuming CPU-GPU transfers over the PCIe bus makes APUs a perspective architecture for many demanding applications.

References

1. Gomez, L.B., Nicolae, B., Maruyama, N., Cappello, F., Matsuoka, S.: Scalable Reed-Solomon-based reliable local storage for HPC applications on IaaS clouds. In: Kaklamanis, C., Papatheodorou, T., Spirakis, P.G. (eds.) Euro-Par 2012. LNCS, vol. 7484, pp. 313–324. Springer, Heidelberg (2012)
2. Branover, A., Foley, D., Steinman, W.: Amd fusion APU: llano. IEEE Micro **32**(2), 28–37 (2012)
3. Chen, T., Raghavan, R., Dale, J., Iwata, E.: Cell broadband engine architecture and its first implementation: a performance view. IBM J. Res. Dev. **51**, 559–572 (2007)
4. Collins, R., Plank, J.: Assessing the performance of erasure codes in the wide-area. In: Proceedings of 2005 International Conference on Dependable Systems and Networks - DSN'05, pp. 182–187. IEEE Computer Society (2005)
5. Curry, M.L., Skjellum, A., Lee Ward, H., Brightwell, R.: Gibraltar: a Reed-Solomon coding library for storage applications on programmable graphics processors. Concurr. Comput. Pract. Exp. **23**(18), 2477–2495 (2011)
6. Fatahalian, K., Houston, M.: A closer look at gpus. Commun. ACM **51**(10), 50–57 (2008)
7. Kher, V., Kim, Y.: Securing distributed storage: challenges, techniques, and systems. In: Proceedings of the 2005 ACM Workshop on Storage Security and Survivability, StorageSS '05, pp. 9–25. ACM, New York (2005)
8. Khronos Group: The opencl specification version 1, 2 (2012)
9. Kubiatowicz, J., Bindel, D., Chen, Y., Czerwinski, S., Eaton, P., Geels, D., Gummadi, R., Rhea, S., Weatherspoon, H., Weimer, W., Wells, C., Zhao, B.: OceanStore: an architecture for global-scale persistent storage. In: Proceedings of 9th International Conference on Architectural Support for Programming Languages and Operating Systems - ASPLOS 2000, pp. 190–201 (2000)
10. L. Kuczynski and R. Wyrzykowski. Efficient Data Management in PC Meta-Clusters. The Publishing Office of Czestochowa University of Technology, 2011.
11. Plank, J.: A tutorial on Reed-Solomon coding for fault-tolerance in raid-like systems. Softw. Pract. Exp. **27**, 995–1012 (1997)
12. Plank, J., Ding, Y.: Note: correction to the 1997 tutorial on Reed-Solomon coding. Softw. Pract. Exp. **35**, 189–194 (2005)
13. Plank, J., Luo, J., Schuman, C., Xu, L., Wilcox-O'Hearn, Z.: A performance evaluation and examination of open-source erasure coding libraries for storage. In: FAST-09: 7th USENIX Conference on File and Storage Technologies, pp. 253–265 (2009)

14. Plank, J., Simmerman, S., Schuman, C.: Jerasure: a library in C/C++ facilitating erasure coding for storage applications. https://www.cs.utk.edu/plank/plank/papers/CS-08-627.pdf
15. Plank, J., Thomason, M.: A practical analysis of low-density parity-check erasure codes for wide-area storage applications. In: Proceeding of 2004 International Conference on Dependable Systems and Networks - DSN'04, pp. 115–124. IEEE Computer Society (2004)
16. Weatherspoon, H., Kubiatowicz, J.D.: Erasure coding vs. replication: a quantitative comparison. In: Druschel, P., Kaashoek, M.F., Rowstron, A. (eds.) IPTPS 2002. LNCS, vol. 2429, p. 328. Springer, Heidelberg (2002)
17. Wozniak, M., Olas, T., Wyrzykowski, R.: Parallel implementation of conjugate gradient method on graphics processors. In: Wyrzykowski, R., Dongarra, J., Karczewski, K., Wasniewski, J. (eds.) PPAM 2009, Part I. LNCS, vol. 6067, pp. 125–135. Springer, Heidelberg (2010)
18. Wyrzykowski, R., Kuczynski, L.: Towards secure data management system for grid environment based on the cell broadband engine. In: Wyrzykowski, R., Dongarra, J., Karczewski, K., Wasniewski, J. (eds.) PPAM 2007. LNCS, vol. 4967, pp. 825–834. Springer, Heidelberg (2008)
19. Wyrzykowski, R., Kuczynski, L., Wozniak, M.: Towards efficient execution of erasure codes on multicore architectures. In: Jónasson, K. (ed.) PARA 2010, Part II. LNCS, vol. 7134, pp. 357–367. Springer, Heidelberg (2012)
20. Wyrzykowski, R., Kuczynski, L., Wozniak, M.: Systematic mapping of Reed-Solomon erasure codes on heterogeneous multicore architectures. In: High-Performance Computing on Complex Environments, 25 pp. Wiley, New york (2014)

AVX Acceleration of DD Arithmetic Between a Sparse Matrix and Vector

Toshiaki Hishinuma[1(✉)], Akihiro Fujii[1],
Teruo Tanaka[1], and Hidehiko Hasegawa[2]

[1] Major of Informatics, Kogakuin University,
Tokyo, Japan
[2] Faculty of Lib., Info. and Media Sci., University of Tsukuba,
Tsukuba, Japan
em13015@ns.kogakuin.ac.jp

Abstract. High precision arithmetic can improve the convergence of Krylov subspace methods; however, it is very costly. One system of high precision arithmetic is double-double (DD) arithmetic, which uses more than 20 double precision operations for one DD operation. We accelerated DD arithmetic using AVX SIMD instructions. The performances of vector operations in 4 threads are 51–59 % of peak performance in a cache and bounded by the memory access speed out of the cache. For SpMV, we used a double precision sparse matrix A and DD vector x to reduce memory access and achieved performances of 17–41 % of peak performance using padding in execution. We also achieved performances that were 9–33 % of peak performance for a transposed SpMV. For these cases, the performances were not bounded by memory access.

Keywords: Double-double arithmetic · AVX · SpMV · High precision

1 Introduction

In many cases, the kernel of a numerical simulation is the solution of a large and sparse system of linear equations. Well-known algorithms for this solution are the Krylov subspace methods, but these methods diverge, stagnate, and increase iterations because of rounding errors. High precision arithmetic may be able to improve the convergence of these methods [1]; however, it is very costly. One system of high precision arithmetic is double-double (DD) arithmetic [2], which does not need any special hardware and runs on general-purpose processors but uses more than 20 double precision operations for one DD operation.

In this study, we accelerate DD arithmetic using AVX (Intel Advanced Vector Extensions) [3]. The targeted operations of Krylov subspace methods are the vector operation, double precision sparse matrix and DD vector product (SpMV), and transposed double precision sparse matrix and DD vector product.

R. Wyrzykowski et al. (Eds.): PPAM 2013, Part I, LNCS 8384, pp. 622–631, 2014.
DOI: 10.1007/978-3-642-55224-3_58, © Springer-Verlag Berlin Heidelberg 2014

2 SIMD Instruction and DD Arithmetic

2.1 Test Bed

The CPU is a 4-core 8-thread Intel Core i7 2600 K 3.4 GHz (Sandy Bridge), which can use AVX. It has an 8 MB L3 cache and 16×256 bit SIMD registers. ALU of Sandy Bridge operates an FP adder and multiplier in parallel. AVX calculates four double precision variables at once. The peak performance of this CPU is 108.8 GFLOPS (3.4×4 (cores) $\times 2$ (adder and multiplier) $\times 4$ (AVX)).

Memory is a 16 GB DDR3–1333 dual channel and memory bandwidth is 21.2 GB/s (10.6 GB/s $\times 2$ (dual channel)).

OS is Fedora 16 and the compiler is an Intel C/C++ compiler 12.0.3. Compiler options -O3, -xAVX, -openmp, and -fp-model precise are used for enabling C code optimization, AVX instructions, OpenMP based multi-threading, and value-safe optimization.

2.2 DD Arithmetic

DD arithmetic consists of combinations of double precision values only and uses two double precision variables to implement one quadruple precision variable [2]. It is based on the error-free floating-point arithmetic algorithms by Dekker [4] and Knuth [5]. A DD addition consists of 11 double precision addition instructions, and a DD multiplication consists of 15 double precision addition instructions and 9 double precision multiplication instructions.

An IEEE 754 quadruple precision variable consists of a 1 bit sign part, 15 bit exponent part, and 112 bit significant part. A DD precision variable consists of a 1 bit sign part, 11 bit exponent part, and 104 (52×2) bit significant part. The exponent part of a DD precision variable is 4 bits shorter and the significant part is 8 bits shorter than the exponent and significant parts of an IEEE 754 quadruple precision variable, respectively.

The simplest way to use IEEE 754 quadruple precision is with Fortran REAL*16. We compared Fortran REAL*16 using an Intel Fortran compiler 12.0.3 (ifort) and DD arithmetic without any SIMD instructions. The compiler option in ifort was -O3. Fortran REAL*16 in ifort was implemented only by integer operations. We computed $y = \alpha \times x + y$, where x and y are quadruple precision vectors and α is a quadruple precision variable. Two vectors x and y whose sizes are 10^5 can be stored in the cache. The elapsed time of Fortran REAL*16 was 3 ms and that of DD Arithmetic was 0.76 ms in 1 thread, which means that DD arithmetic was 3.9 times faster than Fortran REAL*16.

The exponent and significant parts of DD variables were shorter than those of quadruple variables but DD arithmetic was faster than quadruple precision arithmetic in Fortran.

Table 1. Double-double vector operations

	Operation	Load	Store	Complexity (add + sub:mult)
axpy	$y = \alpha x + y$	2	1	35 (26:9)
axpyz	$z = \alpha x + y$	2	1	35 (26:9)
xpay	$y = x + \alpha y$	2	1	35 (26:9)
dot	val = $x \cdot y$	2	0	35 (26:9)
nrm2	val = $\|x\|$	1	0	31 (24:7)
scale	$x = \alpha x$	1	1	24 (15:9)

NOTE: α and val are DD variables, x, y and z are DD vectors.

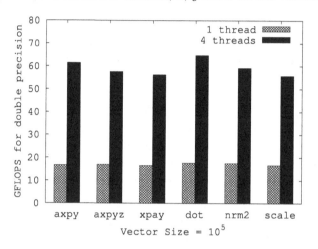

Fig. 1. Performances of vector operations on AVX

3   Double-Double Vector Operations

3.1   Vector Operations

Table 1 lists DD vector operations. "Load" means the number of elements of DD vector referred in a kernel, "Store" means the number of elements of DD vector moved to memory in a kernel, and "Complexity" means the number of double precision operations in a kernel. We computed "GFLOPS for double precision" using (Complexity × N)/elapsed-time.

We used an average at least 70 experiments and used the static scheduling in OpenMP. Figure 1 shows the performances of DD vector operations in 1 thread and 4 threads on AVX when N is 10^5. In this case, all variables were stored in the cache.

The performances of vector operations in 1 thread are from 16.6 to 17.7 GFLOPS and 61–65 % of peak performance of one core, while those in 4 threads are from 55.7 to 64.7 GFLOPS and 51–59 % of peak performance. The performances of vector operations in 4 threads are 3.4–3.7 times higher than those of 1 thread. Multi-threading worked well for vector operations in the cache on AVX.

The performance of dot is the highest, i.e., 64.7 GFLOPS in 4 threads and 59 % of peak performance. All vector operations were expressed by the same instructions and only in the case of dot, store was eliminated from inside the loop by the compiler. The performance of scale is the lowest, i.e., 55.7 GFLOPS and 51 % of peak performance. Scale consisted of one load and one store. This ratio was considered to be low performance.

The peak performance on the Intel core i7 2600 K is on the premise that FP adder and multiplier are performed in parallel. However, DD arithmetic has a different number of double precision addition and multiplication instructions. For example, axpy consists of 26 double precision addition and 9 double precision multiplication instructions in a kernel. Theoretically, ALU calculates 26 double precision addition and 26 double precision multiplication instructions, i.e., a total of 52 flops. However, axpy can calculate a maximum of 35 flops. Therefore, the peak performance of axpy diminishes 67 % (35/52) of the peak performance of hardware. We defined diminished peak performance as corrected peak performance. The corrected peak performance of axpy, axpyz, xpay, dot was 73.2, that of nrm2 was 70.4, and that of scale was 87.2 GFLOPS. The performances of vector operations in 1 thread were 91–97 % of corrected peak performances of one core, and those in 4 threads were 65–88 % of corrected peak performances.

3.2   Memory Access

The 8 MB L3 cache can store two vectors when N is less than 2.6×10^5. Figure 2 shows the performances of axpy in 1 thread and 4 threads on AVX when the vector size changes from 10^3 to 8.0×10^5.

In the cache, when N is 10^5, the performance of axpy in 1 thread is 17 GFLOPS, which is 62 % of peak performance of one core and 93 % of corrected peak performance of one core. The performance of axpy in 4 threads is 61.4

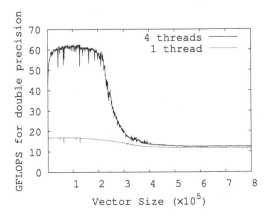

Fig. 2. Performances of axpy on AVX

GFLOPS, which is 56 % of peak performance and 84 % of corrected peak performance, meaning that the performance of axpy in 4 threads is 3.6 times higher than that in 1 thread. Out of the cache, when N is 8.0×10^5, the performance of axpy diminishes to 12 GFLOPS in 1 thread and 4 threads. The speed of data moved from/to memory was $(8.0 \times 10^8 \times 16$ (bytes) $\times 3)$/elapsed-time (2.2 ms) $= 17.5$ GB/s. It is 83 % of the peak memory access speed of 21.2 GB/s. In the DDR3-1333 dual channel, the maximum theoretical performance of axpy is 15.4 GFLOPS, i.e., $21.2/(16 \times 3 \times 35$ (flops)). Out of the cache, we considered performance to be bounded by the memory access speed and multi-threading was not effective.

4   Sparse Matrix Vector Product in DD Arithmetic

4.1   Product of the Double Precision Matrix and DD Vector

Out of the cache, DD arithmetic for vectors is bounded by the memory access speed; therefore, we need to reduce memory access to accelerate computation. In many cases, for an iterative solver library, input matrix A is given by double precision and iteratively used. To reduce memory access and accelerate the sparse matrix and vector product, we used the double precision sparse matrix A and DD precision vector x product (SpMV). This allowed the size of value in the sparse matrix to half, compared to store in DD values.

The complexity of the DD matrix and DD vector product is 35 flops. SpMV consists of 25 double precision addition and 8 double precision multiplication instructions and its complexity is 33 flops. We computed GFLOPS for SpMV using $(33 \times$ the number of non-zero elements (nnz))/elapsed-time. The corrected peak performance of SpMV is 72 GFLOPS in 4 threads.

To store the double precision sparse matrix A, Compressed Row Storage (CRS) [6] is used. The CRS format is expressed by the following three arrays:col_ind, row_ptr, and val; one for matrix value (val), and the other two for integers (col_ind and row_ptr). The val array stores the values of the non-zero elements of matrix A, as they are traversed row-wise. The col_ind array stores the column indexes of the elements in the val array, i.e., if val[k] = a_{ij} then col_ind[k] = j. The row_ptr array stores the locations in the val array that start a row, i.e., if val[k] = a_{ij} then row_ptr[i] $\leq$ k $<$ row_ptr[i + 1].

The memory requirements of SpMV in its kernel is 50 bytes, consisting of 8 bytes for matrix A, 16 bytes for vectors x and y, and 4 bytes for vector col_ind. However, the DD matrix and DD vector product needs 58 bytes in a kernel. The value of byte per flops is 1.5 for SpMV and 1.7 for the DD matrix and DD vector product. SpMV can reduce the required memory to 88 % of that of the DD matrix and DD vector product.

4.2   Fraction Processing on AVX

AVX must calculate four double precision instructions at once. Processing for the remainder, which has one, two, or three elements, occurs at most once for a vector. However, in SpMV, the remainder occurs for each row. We call the processing of the remainder as fraction processing.

Four methods exist for fraction processing. Table 2 lists these methods and the performance of fraction processing for a band matrix in 4 threads, where N is 10^4 and the bandwidth is 63 and 1023. There were three elements in the remainder in each row. This was the worst case. In SpMV, we used an average of 500 experiments and guided scheduling in OpenMP.

Table 2. Performance of fraction processing in ms (GFLOPS) (N = 10^4, 4 threads)

	Bandwidth 63	1023
Padding in execution	49 (42.2)	71 (47.4)
Padding in creation CRS	47 (44.3)	71 (47.4)
Using SSE2 and normal instruction (without padding)	53 (39.0)	81 (41.1)
Using normal instruction (without padding)	48 (41.1)	71 (47.1)

"Padding" means that unnecessary elements are added to the remainder and the remainder is eliminated. "Padding in execution" assigns zero to the operand of AVX instructions at the execution. "Padding in creation" assigns zero to val and col_ind of CRS when creating CRS, but it enlarges the size of vector val and col_ind. "Using SSE2 (Streaming SIMD Extensions 2) and normal instruction" means the following code, where r represents the number of elements in the remainder:

if (r $\geqq$ 2)
process two elements with SSE2 instruction; r = r − 2;
if (r == 1)
process one element with normal instruction.

"Using normal instruction" means the following code:

for (; r < 0 ; r = r − 1)
process one element with normal instruction.

The performance of "using SSE2 and normal instruction" was the least (from 39.0 to 41.1 GFLOPS), because the ymm register is the same hardware as the xmm register. When switching from AVX to SSE2, there is some processing to save the register's values; therefore, frequently switching AVX and SSE2 is not recommended.

The performance of "padding in creation CRS" was the best (from 44.3 to 47.4 GFLOPS). However, the difference between "padding in execution" and "using normal instruction" was from 1 % to 7 %. We chose the "padding in execution" and "using normal instruction", which did not need an extra cost for the creation of the CRS matrix or any extra storage for a matrix.

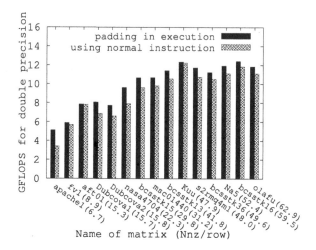

Fig. 3. Performances of SpMV on AVX (1 thread)

We used a set of 15 sparse matrices that were taken from the University of Florida Sparse Matrix Collection [7]. Figure 3 shows the performances of SpMV. It arranges by nnz/row in 1 thread. The performances of SpMV using "padding in execution" are from 5.1 to 12.4 GFLOPS, 19–46 % of peak performance of one core, and 28–69 % of corrected peak performance of one core. The performances of SpMV "using normal instruction" are from 3.4 to 12.2 GFLOPS, 12–45 % of peak performance of one core, and 19–68 % of corrected peak performance of one core. The performances of "padding in execution" are 1.0–1.2 times higher than those of "using normal instruction", except for apache1.

In this result, "padding in execution" is the best overall condition because it can calculate any remaining numbers at once. However, "using normal instruction" needs a loop of processing fraction, which is the number of the remainder.

4.3   Multi-threading and Memory Access

Figure 4 shows the performances of SpMV in 1 thread and 4 threads "using padding in execution". The performances of SpMV in 1 thread are from 5.1 to 12.4 GFLOPS, 19–46 % of peak performance of one core, and 28–69 % of corrected peak performance of one core. The performances of SpMV in 4 threads are from 18.3 to 44.5 GFLOPS, 17–41 % of peak performance, and 26–62 % of corrected peak performance, meaning that the performances of SpMV in 4 threads are 3.3–3.6 times higher than those in 1 thread. The performances of SpMV show that matrices with more nnz/row show higher performances.

For the evaluation of memory access, we changed the size of the band matrix from 10^3 to 4.0×10^5 when matrix bandwidth was 32. The 8 MB L3 cache can store one double precision band matrix and two DD vectors when the matrix size is less than 1.9×10^4. In the cache, when N is 10^4, performance of SpMV in 1 thread is 12.7 GFLOPS, while that in 4 threads is 41.2 GFLOPS, meaning

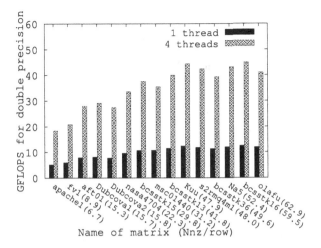

Fig. 4. Performances of SpMV on AVX (padding in execution)

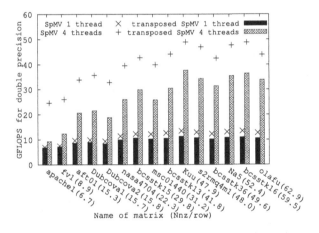

Fig. 5. Performances of the transposed SpMV on AVX (padding in execution)

that the performance of SpMV in 4 threads is 3.2 times higher than that in 1 thread. Out of the cache, when N is 4.0×10^5, the performance of SpMV in 1 thread is 11.8 GFLOPS, while that in 4 threads is 38.1 GFLOPS, meaning that the performance of SpMV in 4 threads is 3.2 times higher than that in 1 thread. The performance of SpMV in 4 threads in the cache is 1.1 times higher than out of the cache. As the difference in performance between in and out of the cache is small, we concluded that SpMV is not bounded by the memory access speed.

4.4 Product of the Transposed Sparse Matrix and Vector

A transposed SpMV is also necessary for a Krylov subspace method. However, the performance and memory access patterns of the transposed SpMV are different from those of the original SpMV with the same storage format. An effective storage format for a transposed SpMV is available, but storage space needs twice because of the transposed matrix. We evaluated a double precision transposed sparse matrix A^T and DD vector product x $(y = A^T x)$.

The difference between $A^T x$ and Ax is the cache hit ratio of DD vector x and y. When a sparse matrix has a complicated structure, SpMV has a diminishing cache hit ratio of loading x, while a transposed SpMV has a diminishing cache hit ratio of loading and storing y.

Figure 5 shows the performances of $A^T x$ and Ax in 1 thread and 4 threads using padding in execution. The performances of the transposed SpMV in 1 thread are from 6.8 to 11.3 GFLOPS and those in 4 threads are from 9.3 to 36.3 GFLOPS. The performances of the transposed SpMV in 4 threads are 1.3–3.3 times higher than those in 1 thread. The performances of SpMV are 1.3–1.7 times higher than those of the transposed SpMV in 4 threads, except for apache1 and aft01 that have a few the number of nnz/row. The difference in performances between SpMV and transposed SpMV were small.

5   Conclusion

We accelerated DD arithmetic using AVX SIMD instructions. The peak performance is on the premise that FP adder and multiplier were performed in parallel. However, DD arithmetic has a different number of double precision addition and multiplication instructions. It does not perform FP adder and multiplier in parallel and does not reach peak performance of hardware. We defined diminished peak performance as corrected peak performance.

In the cache, the performances of DD vector operations in 4 threads were 51–59 % of peak performance and 65–88 % of corrected peak performance. The performances of DD vector operations in 4 threads were 3.4–3.7 times higher than those in 1 thread. Multi-threading worked well, but performances were bounded by memory access speed out of the cache. In the theoretical memory access speed, the maximum performance of axpy was 15.4 GFLOPS. We concluded that performance was bounded by the memory access speed and multi-threading was not effective.

For SpMV, we used the double precision sparse matrix A and DD precision vector x product to reduce the required memory access to 88 % of that of the DD matrix and DD vector product. To use the AVX, processing of the remainder was necessary for each row. We chose "padding in execution" and "using normal instruction", which did not need any extra cost for the creation of the CRS matrix or any extra storage for a matrix. "Padding in execution" was 1.0–1.2 times faster than "using normal instruction". "Padding in execution" was the best overall condition.

In the cache, when the size of the bandmatrix was 10^4 and bandwidth was 32, using "padding in execution", the performance of SpMV in 1 thread was 47 % of peak performance of one core and 70 % of the corrected peak performance of one core. The performance of SpMV in 4 threads was 38 % of peak performance and

57 % of corrected peak performance, meaning that the performances of SpMV in 4 threads were 3.2 times higher than those in 1 thread. Out of the cache, when the size of the bandmatrix was 4.0×10^5, the performance of SpMV in 1 thread was 44 % of peak performance of one core and 66 % of corrected peak performance of one core. The performance of SpMV in 4 threads was 35 % of peak performance and 53 % of corrected peak performance, meaning that the performance of SpMV in 4 threads was 3.2 times higher than that in 1 thread. For these cases, the performance was not bounded by memory access speed. The performances of SpMV were 1.3–1.7 times higher than that of the transposed SpMV in 4 threads. The difference in performance between SpMV and the transposed SpMV were small, except in some matrices which had a few nnz/row. The ratio of corrected peak performance were good except for vector operations out of the cache. ALU and multi-threading worked well. AVX acceleration of DD arithmetic was effective. The problem of acceleration was in the different number of addition and multiplication instructions of DD arithmetic. In the future, we will improve the number of addition and multiplication instructions.

Acknowledgement. The authors would like to thank the reviewers for their helpful comments.

References

1. Hasegawa, H.: Utilizing the quadruple-precision floating-point arithmetic operation for the Krylov subspace methods. In: The 8th SIAM Conference on Applied Linear, Algebra (2003)
2. Bailey, D.H.: QD (C++ / Fortran-90 double-double and quad-double package), http://crd-legacy.lbl.gov/dhbailey/mpdist/
3. Intel: Intrinsics Guide, http://software.intel.com/en-us/articles/intel-intrinsics-.guide
4. Dekker, T.: A floating-point technique for extending the available precision. Numer. Math. **18**, 224–242 (1971)
5. Knuth, D.E.: The Art of Computer Programming: Seminumerical Algorithms, vol. 2. Addison-Wesley, Reading (1969)
6. Barrett, R., et al.: Templates for the Solution of Linear Systems: Building Blocks for Iterative Methods, pp. 57–65. SIAM, Philadelphia (1994)
7. The University of Florida Sparse Matrix Collection, http://www.cise.ufl.edu/research/sparse/matrices/

Using Quadruple Precision Arithmetic to Accelerate Krylov Subspace Methods on GPUs

Daichi Mukunoki[1,3]([✉]) and Daisuke Takahashi[2]

[1] Graduate School of Systems and Information Engineering,
University of Tsukuba, Tsukuba, Japan
mukunoki@hpcs.cs.tsukuba.ac.jp
[2] Faculty of Engineering, Information and Systems, University of Tsukuba,
1-1-1 Tennodai, Tsukuba, Ibaraki 305–8573, Japan
daisuke@cs.tsukuba.ac.jp
[3] Japan Society for the Promotion of Science, 5-3-1 Kojimachi,
Chiyoda-ku, Tokyo 102-0083, Japan

Abstract. The convergence of the Krylov subspace methods is affected by round-off errors. The number of iterations until convergence may be decreased by reducing round-off errors through the use of quadruple precision arithmetic instead of double precision. We implemented the CG and BiCGStab methods using quadruple precision arithmetic and compared the performance with the standard double precision implementations on an NVIDIA Tesla K20X GPU. Our results show that in some cases our implementations using quadruple precision arithmetic outperform the double precision versions. We will show that quadruple precision arithmetic is not costly for the CG and BiCGStab methods on GPUs and the use of quadruple precision arithmetic may be a more effective alternative to the use of preconditioning.

Keywords: Krylov subspace method · CG method · BiCGStab method · Quadruple precision · GPU

1 Introduction

The convergence of the Krylov subspace methods, which are iterative methods for solving linear systems, is significantly affected by round-off errors. Thus, there are cases where reducing round-off errors with multiple precision arithmetic, such as quadruple precision, causes the algorithm to converge more quickly [8].

Although multiple precision arithmetic operations generally require a large amount of computation time, we have shown that dense matrix-vector multiplication using software implemented quadruple precision arithmetic is memory-bound on GPUs. This is due to the low Bytes/Flop of GPUs, and thus the execution time is only about twice that of double precision operation [10]. This shows that the use of quadruple precision arithmetic is not always costly on modern processors such as GPUs.

R. Wyrzykowski et al. (Eds.): PPAM 2013, Part I, LNCS 8384, pp. 632–642, 2014.
DOI: 10.1007/978-3-642-55224-3_59, © Springer-Verlag Berlin Heidelberg 2014

In this paper, we will describe the implementation and performance of the Conjugate Gradient (CG) and Bi-Conjugate Gradient Stabilized (BiCGStab) methods, which are Krylov subspace methods, using quadruple precision arithmetic on an NVIDIA Tesla K20X GPU. Then, we will compare the performance with the standard double precision implementations on the GPU. Since the Krylov subspace methods are generally regarded as being memory-intensive, using quadruple precision arithmetic operations on the GPU will only double the execution time of 1 iteration of double precision versions. Even if the use of quadruple precision arithmetic increases the execution time of 1 iteration, the time until convergence may be reduced if increasing the precision can reduce the number of iterations enough to compensate. Therefore, we expect that the implementations using quadruple precision arithmetic may outperform the standard double precision versions.

This paper is organized as follows: In Sect. 2 we will introduce related work. In Sect. 3 we will explain the CG and BiCGStab methods using quadruple precision arithmetic. In Sect. 4 we will show our implementation for GPUs. In Sect. 5 we will compare the performance of our quadruple precision implementation with the double precision versions. In Sect. 6 we will discuss the performance and effectiveness of quadruple precision arithmetic. Finally, we will conclude the paper in Sect. 7.

2 Related Work

Hasegawa [8] compared the performance of an unpreconditioned BiCG method using quadruple precision arithmetic to the preconditioned method using only double precision arithmetic on various architectures. He did not show cases where implementations using quadruple precision arithmetic outperformed those using only double precision arithmetic, but he expected that the use of quadruple precision arithmetic may be an effective alternative to preconditioning which has low parallelism on parallel architectures.

Furuichi et al. [6] implemented the Generalized Conjugate Residual (GCR) methods using quadruple precision arithmetic on the NEC SX-9 supercomputer. They applied quadruple precision arithmetic to the preconditioned methods save for the preconditioning operations, which were performed using double precision operations. As a result, they improved the convergence without significantly increasing the execution time. Saito et al. [15] also showed convergence improvement of the GCR methods on the Scilab toolbox they developed by using quadruple precision arithmetic for certain parts of the algorithm.

Such studies showed that the use of quadruple precision arithmetic improves the convergence and is useful for solving problems which cannot be solved using standard double precision solvers. However, we expect that the use of quadruple precision arithmetic can also be used to accelerate double precision solvers even when quadruple precision arithmetic is not necessary. In addition, although Krylov subspace methods have been implemented on GPUs [7,11], there is no research on the implementation and performance of methods using quadruple-precision arithmetic on GPUs.

$$r_0 = b - Ax_0$$
$$\tilde{r} = r_0$$
for : $k = 1, 2, \dots$ do
    $\rho_{k-1} = \langle \tilde{r}, r_{k-1} \rangle$
    if $\rho_{k-1} = 0$ method fails
    if $k = 1$ then
        $p_k = r_{k-1}$
    else
        $\beta_{k-1} = (\rho_{k-1}/\rho_{k-2})(\alpha_{k-1}/\omega_{k-1})$
        $p_k = r_{k-1} + \beta_{k-1}(p_{k-1} - \omega_{k-1}v_{k-1})$
    end if
    solve $p_k = M\hat{p}$
    $v_k = A\hat{p}$
    $\alpha_k = \rho_{k-1}/\langle \tilde{r}, v_k \rangle$
    $s = r_{k-1} - \alpha_k v_k$
    if $||s||/||r_0|| < \epsilon$ then
        $x_k = x_{k-1} + \alpha_k \hat{p}$
        break
    end if
    solve $s = M\hat{s}$
    $t = A\hat{s}$
    $\omega = \langle t, s \rangle/\langle t, t \rangle$
    $x_k = x_{k-1} + \alpha_k \hat{p} + \omega_k \hat{s}$
    $r_k = s - \omega_k t$
    if $||r_k||/||r_0|| < \epsilon$ break
    if $\omega = 0$ break
end for

$$r_0 = b - Ax_0$$
for : $k = 1, 2, \dots$ do
    solve $Mz_{k-1} = r_{k-1}$
    $\rho_{k-1} = \langle r_{k-1}, z_{k-1} \rangle$
    if $k = 1$ then
        $p_1 = z_0$
    else
        $\beta_{k-1} = \rho_{k-1}/\rho_{k-2}$
        $p_k = z_{k-1} + \beta_{k-1}p_{k-1}$
    end if
    $q_k = Ap_k$
    $\alpha_k = \rho_{k-1}/\langle p_k, q_k \rangle$
    $x_k = x_{k-1} + \alpha_k p_k$
    $r_k = r_{k-1} - \alpha_k q_k$
    if $||r_k||/||r_0|| < \epsilon$ break
end for

Fig. 1. Preconditioned CG method

Fig. 2. Preconditioned BiCGStab method

3   CG and BiCGStab Methods Using Quadruple Precision Arithmetic

The CG method (Fig. 1) and the BiCGStab method (Fig. 2) [2] are Krylov subspace methods which are often used to solve large sparse linear systems $Ax = b$. The CG method is applied when the coefficient matrix A is a symmetric positive definite matrix, and the BiCGStab method can be used when the coefficient matrix A is asymmetric. The convergence of the Krylov subspace methods depends on the spectral properties of the coefficient matrix. To improve the spectral properties, preconditioners which approximate the coefficient matrix are often used. In the algorithms shown in Figs. 1 and 2 use the preconditioning matrix M. By setting $M = I$, the algorithms become the same as the unpreconditioned ones.

For this paper, we implemented both the unpreconditioned and preconditioned methods. For the unpreconditioned methods, we used quadruple precision arithmetic everywhere except for the norm computation for checking convergence, where we used double precision. Double precision is enough for the norm computation because the operation is unrelated to the convergence. The input matrix A and vector b are stored in the double precision format. The vector x and all other floating-point data are stored in the quadruple precision format.

For the preconditioned methods, we use an incomplete-LU preconditioner while preserving the non-zero pattern of the coefficient matrix A, as known as ILU(0), one of the most popular preconditioners for Krylov subspace methods. The ILU(0) performs incomplete-LU factorization which approximates $A \approx M = LU$, where L and U are the lower and upper triangular matrices, respectively. Thus, the system is solved as $M^{-1}Ax = M^{-1}b$ using sparse triangular solvers with a forward substitution with L and a backward substitution with U. In general, a preconditioning process has serial processing portions, and these portions may be the most time-consuming portion in the Krylov subspace methods on parallel architectures. In fact, Naumov [11] reports that the most time-consuming portion of the ILU(0) preconditioned BiCGStab method is triangular solvers on GPUs. On the other hand, preconditioning can be performed using lower precision arithmetic since the objective of preconditioning is computing an approximation of the coefficient matrix. Therefore, even when using quadruple precision arithmetic, we used a double precision ILU(0) preconditioner.

4 Implementation

In our research, we implemented both the double and quadruple precision versions to compare the performance. We used CUDA which is a C/C++ based programming environment for GPU computing. Our target architecture is the NVIDIA Kepler architecture GPUs of compute capability 3.5.

4.1 CG and BiCGStab Methods

We used the GPU to perform vector operations, and the CPU to perform scalar operations. We implemented SpMV ($y = Ax$), DOT ($r = \langle x, y \rangle$) and some GPU kernel functions which perform scalar multiplication and vector addition similar to AXPY ($y = \alpha x + y$). For some double precision kernels, vendor provided libraries such as CUBLAS [12] and cuSPARSE [13] are available. However, in order to measure the performance impact of the different precisions accurately, we implemented all vector operation subroutines that require both double and quadruple precision versions from scratch. By doing this we can ensure that the algorithms of both quadruple and double precision versions are completely the same including the number of threads for GPU kernel functions.

Among the kernels we implemented, SpMV is generally the most time-consuming operation. We used the Compressed Row Storage (CRS) format which is one of the most widely used storage formats for storing sparse matrices. In the CRS format, a sparse matrix is stored into the data array by scanning the matrix in the row direction using two arrays to represent the position of the non-zero elements: an index array, which represents the column number of the non-zero elements in the data array, and a pointer array, which points to the first non-zero element of each row. Our SpMV implementation is based on the CRS-vector method [3] which assigns 32 threads to calculate a single row. Reguly and Giles [14] improved the method by selecting the optimal number of threads for the calculation of a single row from among 1, 2, 4, 8, 16 and 32 in proportion to the average number of non-zero elements per row. We also used this approach and optimized the implementation for the Kepler architecture. The Kepler architecture supports some new features that we expect will improve the performance of SpMV, such as the 48 KB read-only data cache and shuffle instructions. We used these features to improve the performance of SpMV.

For the preconditioned methods, we used the double precision ILU(0) preconditioning subroutines provided by the cuSPARSE library. The usage of the subroutines is shown in Naumov's report [11]. In the iterative portion, the cuSPARSE subroutine `cusparseDcsrsv_solve()` is executed two and four times on the CG and BiCGStab methods, respectively. The subroutine solves a sparse lower or upper triangular system with either a forward substitution or a backward substitution.

4.2   Quadruple Precision Arithmetic

Quadruple precision arithmetic operations are not natively supported by GPUs. We implemented double-double (DD) operations [1] which are often used to perform quadruple precision floating-point arithmetic operations in software. In the DD operations, a quadruple precision floating-point value $a^{(q)}$ is represented as $a^{(q)} = a_{hi}^{(d)} + a_{lo}^{(d)}$ using two double precision floating-point values, $a_{hi}^{(d)}$ and $a_{lo}^{(d)}$ ($|a_{lo}^{(d)}| \leq 0.5 ulp(a_{hi}^{(d)})$). The total significand precision of the quadruple precision format on the DD operations is twice that of the IEEE double precision format but the exponent precision is the same. Therefore, the significand and exponent are less than that of the "binary128" 128-bit floating-point format as defined in IEEE 754–2008 [9]. The quadruple precision arithmetic operations are computed using only double precision floating-point arithmetic operations. Fundamentally, the algorithm computes two-digit numbers much the same way as humans do on paper. Such techniques were proposed in 1971 by Dekker [5].

We implemented the DD addition and multiplication as CUDA device functions. Implementation details are the same as in our previous work [10]. One DD value is stored using a "double2" type value which is a vector type consisting of two double precision values defined in CUDA. For scalar value computations on the CPU side, the DD operations are performed using the QD library [1].

5   Performance Evaluation

5.1   Evaluation Method

We evaluated the execution time of the iterative portion of the CG and BiCGStab methods. We used an NVIDIA Tesla K20X Kepler architecture GPU (6 GB

Table 1. Properties of the test matrices

	bmwcra_1	pdb1HYS	Lin	SiO
# of rows	148,770	36,417	256,000	33,401
# of nonzeros	10,641,602	4,344,765	1,766,400	1,317,655
Structure	symmetric	symmetric	symmetric	symmetric
Positive definite	yes	yes	no	no
Application	structural problem	weighted undirected graph	structural problem	theoretical/quantum chemistry problem

(a) bmwcra_1   (b) pdb1HYS   (c) Lin   (d) SiO

Fig. 3. Nonzero patterns of the test matrices

GDDR5 memory with ECC-enabled) and a CUDA 5.0 environment. The host is an Intel Xeon E5–2609 (2.40 GHz) with 16 GB DDR3 memory, and CentOS 6.4 (kernel: 2.6.32–358.2.1.el6.x86_64). The programs were compiled with nvcc 5.0 (-O3 -arch sm_35) and gcc 4.4.6 (-O3). The "-arch sm_35" compiler flag for nvcc is required in order to use the features of the Kepler architecture.

In the next section, we will show four cases where the use of quadruple precision arithmetic is effective. The test matrices we used and their nonzero patterns are shown in Table 1 and Fig. 3, respectively. The four matrices are real square matrices selected from The University of Florida Sparse Matrix Collection [4]. We used the CG and BiCGStab methods for the positive-definite and non positive-definite matrices, respectively. The following conditions are used on all the methods: right-hand side vector $b = (1, 1, ..., 1)^T$, $x_0 = 0$, the stopping criterion $\epsilon = 10^{-12}$, and the maximum number of iterations is 20,000.

In this paper, we call the unpreconditioned CG method using double precision arithmetic "DP-CG"and that using quadruple precision arithmetic "QP-CG". For the implementations using the double precision ILU(0) preconditioner, we add the suffix "+DP-ILU(0)". This is also the same for the BiCGStab methods. For example, the preconditioned BiCGStab method using quadruple precision arithmetic is "QP-BiCGStab+DP-ILU(0)".

5.2   Result

The results for the matrices "bmwcra_1"and "pdb1HYS" on the CG methods are shown in Tables 2 and 3, respectively. For the unpreconditioned method, using quadruple precision arithmetic reduces the total time until convergence to approximately 89 % of the double precision version on "bmwcra_1"and to

Table 2. Results for matrix "bmwcra_1"

Implementation	# of iteration	Execution time [s]		$\frac{||b-Ax||_2}{||b||_2}$
		1 iteration	Total	
DP-CG	18442	1.31E-03	24.1	6.26E-08
QP-CG	10077	2.14E-03	21.6	2.06E-09
DP-CG+DP-ILU(0)	2191	0.0195	42.6	2.31E-08
QP-CG+DP-ILU(0)	1387	0.0202	28.0	2.06E-09

Table 3. Results for matrix "pdb1HYS"

Implementation	# of iteration	Execution time [s]		$\frac{||b-Ax||_2}{||b||_2}$
		1 iteration	Total	
DP-CG	9083	5.73E-04	5.20	4.21E-04
QP-CG	4428	7.83E-04	3.47	3.06E-05
DP-CG+DP-ILU(0)	1593	0.216	343.70	1.76E-04
QP-CG+DP-ILU(0)	1062	0.216	229.40	3.10E-05

approximately 67 % on "pdb1HYS". The total time is also reduced when using the preconditioned method, to approximately 66 % on "bmwcra_1"and approximately 67 % on "pdb1HYS". However for both the problems, the total time for the preconditioned method is longer than the unpreconditioned method on both the double and quadruple precision implementations.

The results for the matrices "Lin"and "SiO" on the BiCGStab methods are shown in Tables 4 and 5, respectively. The total time until convergence is reduced to approximately 76 % of the double precision version on "Lin"with the preconditioned method and to approximately 42 % on "SiO"with the unpreconditioned method by using quadruple precision arithmetic. On the other hand on "Lin", although the DP-BiCGStab calculated the solution in 9.89 s, the method broke down with $\rho = 0$ on the 7619th iteration, but the QP-BiCGStab got the solution faster without breaking down. On "SiO", quadruple precision arithmetic is useful with the unpreconditioned method, but not on with the preconditioned method except for improving the accuracy of the solution.

Note that, in all four cases the accuracy of the solution is improved on both the unpreconditioned and preconditioned methods by using quadruple precision arithmetic.

6   Discussion

6.1   Computation Cost for Quadruple Precision Arithmetic

In the four cases shown in the previous section, the execution time of 1 iteration of the unpreconditioned methods using quadruple precision arithmetic is approximately 1.4 – 1.6 times more than the double precision versions. We investigated

Table 4. Results for matrix "Lin"

Implementation	# of iteration	Execution time [s]		$\frac{\|b-Ax\|_2}{\|b\|_2}$
		1 iteration	Total	
DP-BiCGStab	7619*	1.30E-03	9.89	8.17E-10
QP-BiCGStab	4606	2.09E-03	9.62	7.49E-13
DP-BiCGStab+DP-ILU(0)	1820	0.0118	21.49	9.00E-12
QP-BiCGStab+DP-ILU(0)	1294	0.0127	16.39	4.79E-13

* broke down at the iteration

Table 5. Results for matrix "SiO"

Implementation	# of iteration	Execution time [s]		$\frac{\|b-Ax\|_2}{\|b\|_2}$
		1 iteration	Total	
DP-BiCGStab	1524	8.39E-04	1.28	2.37E-11
QP-BiCGStab	444	1.21E-03	0.54	8.99E-13
DP-BiCGStab+DP-ILU(0)	187	0.485	90.64	3.46E-12
QP-BiCGStab+DP-ILU(0)	209	0.485	101.40	6.81E-13

the computation cost for quadruple precision arithmetic on the main vector operations of the CG and BiCGStab methods. Table 6 shows the performance of the double and quadruple precision SpMV, DOT and AXPY for the four problems. For quadruple precision operations, we used "DDFlops" which means DD-type floating point operations per second, instead of the standard "Flops". We found the performance ratio, DP:QP, to be ≈ 1.5–1.9:1 on SpMV, 1.3–1.7:1 on DOT, 1.3–1.9:1 on AXPY.

We can infer that these operations are memory-bound on the GPU on both double and quadruple precision as evidenced by the Byte/Flop and Bytes/DDFlop of the GPU and the operations. For example, quadruple precision SpMV is approximately $(8+4) \times$ NNZ [Bytes]$/(2 \times$ NNZ) [DDFlop] = 6.0 [Bytes/DDFlop], where NNZ $\gg$ N and the input matrix is double precision. For the Tesla K20X GPU, the theoretical peak double precision performance is 1.31 TFlops and the actual bandwidth is approximately 170 GB/s with ECC-enabled. The theoretical peak performance of the quadruple precision operations using DD operations is just 1/20 of double precision on multiply-add operations [10]: 1.31[TFlops]/20 = 65.5[GDDFlops]. Thus, the GPU has a Bytes/DDFlop ratio of: 170/65.5 ≈ 2.6.

Therefore, in theory quadruple precision arithmetic operations can be performed in about twice the time of double precision on SpMV, DOT and AXPY, however the performance ratio of DP:QP may vary because the execution efficiency may vary depending on problem size and precision. However, in practice, the execution time for quadruple precision operations may be less than twice that of the double precision operations due to precision-independent costs such as the kernel launch cost and the overhead incurred while handling the SpMV index arrays. Hence, we can conclude that the use of quadruple precision arithmetic on the CG and BiCGStab methods may be not as costly on GPUs.

Table 6. Performance of double and quadruple precision SpMV, DOT and AXPY

Problem	DP [GFlops]			QP [GDDFlops]		
	SpMV	DOT	AXPY	SpMV	DOT	AXPY
bmwcra_1	20.87	5.49	12.45	12.55	3.58	6.84
pdb1HYS	23.45	1.87	11.82	15.92	1.49	8.60
Lin	12.55	7.70	13.49	8.42	4.55	7.17
SiO	14.06	1.73	11.08	7.52	1.38	8.24

6.2 Effectiveness of Quadruple Precision Arithmetic

As we previously discussed, quadruple precision arithmetic operations can be performed in about twice the time of double precision for SpMV, DOT and AXPY on GPUs. Therefore, when the use of quadruple precision operations can reduce the number of iterations to about half that of the double precision version, the time until convergence may be shortened.

However, on the preconditioned methods on GPUs, the preconditioner takes an extremely long time. In general, the preconditioning process includes serial processing portions. The ILU(0) preconditioner also includes serial processing portions in the forward and a backward substitution step. In the four cases presented in this paper, the preconditioning process occupies approximately 83.5 – 99.8 % of the execution time of 1 iteration. As a result, the execution time of 1 iteration is almost the same on both the double and quadruple precision versions when using a double precision preconditioner. In such cases, when quadruple precision arithmetic decreases the number of iterations, the implementations using quadruple precision arithmetic outperform the double precision versions. In addition, preconditioning is generally used when the problem does not converge when using unpreconditioned methods, but as an example of "Lin", there are cases where the use of quadruple precision arithmetic may be preferable to preconditioning. Hence, the use of quadruple precision arithmetic may be a more effective alternative to the use of preconditioning on GPUs.

On the other hand, we found cases where the number of iterations increases when using quadruple precision arithmetic, for example on "SiO" with the preconditioned method. We need to do further research in order to determine under which conditions the number of iterations decreases when using quadruple precision arithmetic.

7 Conclusion

We have implemented the CG and BiCGStab methods using quadruple precision arithmetic on GPUs and evaluated the performance on a Tesla K20X GPU. Our results show that there are cases where the time until convergence can be reduced by using quadruple precision arithmetic instead of double precision even when quadruple precision arithmetic is not necessary. We have shown the following:

1. unpreconditioned CG and BiCGStab methods using quadruple precision arithmetic outperform those using only double precision arithmetic
2. unpreconditioned CG and BiCGStab methods using quadruple precision arithmetic outperform the methods using only double precision arithmetic with the double precision ILU(0) preconditioner
3. CG and BiCGStab methods using quadruple precision arithmetic with the double precision ILU(0) preconditioner outperform the methods using only double precision arithmetic with the same preconditioner

We conclude that quadruple precision arithmetic may be not costly on the CG and BiCGStab methods on GPUs and there are cases where the use of quadruple precision arithmetic is an effective alternative to the use of preconditioning on GPUs. In addition, we need to do further research to determine in which cases quadruple precision arithmetic is effective.

Acknowledgment. This research was supported by JST, CREST.

References

1. Bailey, D.H.: QD (C++/Fortran-90 double-double and quad-double package). http://crd.lbl.gov/~dhbailey/mpdist/
2. Barrett, R., Berry, M., Chan, T.F., Demmel, J., Donato, J., Dongarra, J., Eijkhout, V., Pozo, R., Romine, C., der Vorst, H.V.: Templates for the Solution of Linear Systems: Building Blocks for Iterative Methods, 2nd edn. SIAM, Philadelphia (1994)
3. Bell, N., Garland, M.: Efficient sparse matrix-vector multiplication on CUDA. NVIDIA Technical Report NVR-2008-004 (2008)
4. Davis, T., Hu, Y.: The University of Florida Sparse Matrix Collection. http://www.cise.ufl.edu/research/sparse/matrices/
5. Dekker, T.J.: A floating-point technique for extending the available precision. Numer. Math. **18**, 224–242 (1971)
6. Furuichi, M., May, D., Tackley, P.: Development of a stokes flow solver robust to large viscosity jumps using a schur complement approach with mixed precision arithmetic. J. Comput. Phys. **230**(24), 8835–8851 (2011)
7. Gravvanis, G., Filelis-Papadopoulos, C., Giannoutakis, K.: Solving finite difference linear systems on GPUs: CUDA based parallel explicit preconditioned biconjugate conjugate gradient type methods. J. Supercomput. **61**(3), 590–604 (2012)
8. Hasegawa, H.: Utilizing the quadruple-precision floating-point arithmetic operation for the Krylov Subspace Methods. In: Proceedings of the SIAM Conference on Applied Linear Algebra (LA03) (2003)
9. IEEE Computer Society: IEEE Standard for Floating-Point Arithmetic. IEEE Std 754-2008, pp. 1–58 (2008)
10. Mukunoki, D., Takahashi, D.: Implementation and evaluation of triple precision BLAS subroutines on GPUs. In: Proceedings of the 2012 IEEE 26th International Parallel and Distributed Processing Symposium Workshops & PhD Forum (IPDPSW 2012), The 13th Workshop on Parallel and Distributed Scientific and Engineering, Computing (PDSEC-12), pp. 1378–1386 (2012)
11. Naumov, M.: Incomplete-LU and Cholesky Preconditioned Iterative Methods Using CUSPARSE and CUBLAS. NVIDIA White Paper, WP – 06720 – 001_v5.5 (2013)

12. NVIDIA Corporation: CUBLAS Library. https://developer.nvidia.com/cublas
13. NVIDIA Corporation: cuSPARSE Library. https://developer.nvidia.com/cusparse
14. Reguly, I., Giles, M.: Efficient sparse matrix-vector multiplication on cache-based GPUs. In: Proceedings of the Innovative Parallel Computing: Foundations and Applications of GPU, Manycore, and Heterogeneous Systems (InPar 2012), pp. 1–12 (2012)
15. Saito, T., Ishiwata, E., Hasegawa, H.: Analysis of the GCR method with mixed precision arithmetic using QuPAT. J. Comput. Sci. **3**(3), 87–91 (2012)

Effectiveness of Sparse Data Structure for Double-Double and Quad-Double Arithmetics

Tsubasa Saito[1], Satoko Kikkawa[2],
Emiko Ishiwata[3], and Hidehiko Hasegawa[4](✉)

[1] Tokyo Metropolitan Ome Sogo High School, Tokyo, Japan
[2] Canon Inc., Tokyo, Japan
[3] Tokyo University of Science, Tokyo, Japan
[4] University of Tsukuba, Tsukuba, Japan
hasegawa@slis.tsukuba.ac.jp

Abstract. Double-double and Quad-double arithmetics are effective tools to reduce the round-off errors in floating-point arithmetic. However, the dense data structure for high-precision numbers in MuPAT/Scilab requires large amounts of memory and a great deal of the computation time. We implemented sparse data types `ddsp` and `qdsp` for double-double and quad-double numbers. We showed that sparse data structure for high-precision arithmetic is practically useful for solving a system of ill-conditioned linear equation to improve the convergence and obtain the accurate result in smaller computation time.

Keywords: Ill-conditioned matrix problem · Sparse matrix · Multiple precisions

1 Introduction

In floating-point arithmetic, we cannot avoid the computation errors. Therefore, for example, it is known that the iterative method for solving a system of ill-conditioned linear equation may not converge when double-precision arithmetic is used. Double-double and Quad-double arithmetics facilitate the use of high-precision arithmetic on ordinary double-precision arithmetic environment. We have developed MuPAT [1,2], which is "Multiple Precision Arithmetic Toolbox" on Scilab [3] (cf. [4]), and have shown the effectiveness of double-double and quad-double arithmetics for ill-conditioned problems [5]. MuPAT has only dense data structures. Because of the large amount of memory and much more computation time, double-double and quad-double arithmetics cannot be applied for large matrices.

We developed sparse data structures for quadruple and octuple-precision arithmetics as a part of MuPAT. This implementation enables the users to treat large matrices with lower memory consumption and small computation time. We

---

Tsubasa Saito and Satoko Kikkawa were at Tokyo University of Science while conducting this research.

R. Wyrzykowski et al. (Eds.): PPAM 2013, Part I, LNCS 8384, pp. 643–651, 2014.
DOI: 10.1007/978-3-642-55224-3_60, © Springer-Verlag Berlin Heidelberg 2014

defined new data types for a sparse matrix which have double-double and quad-double numbers, and made it possible to use a combination of double, double-double, and quad-double arithmetics for both dense and sparse data structures.

We compared the memory consumption and the computation time of the matrix computations with sparse and dense data structures for double-double and quad-double arithmetics. We also showed that high-precision sparse data structure is practically useful for ill-conditioned matrices by applying double, double-double and quad-double arithmetics for the Biconjugate Gradient (BiCG) method.

2   Double-Double and Quad-Double

Double-double and Quad-double arithmetics were proposed for quasi-quadruple-precision and quasi-octuple-precision arithmetics by Hida et al. [6]. A double-double number is represented by two, and a quad-double number is represented by four, double-precision numbers. A double-double number $x_{(dd)}$ and a quad-double number $y_{(qd)}$ are represented by an unevaluated sum of double-precision numbers $x_0, x_1, y_0, y_1, y_2, y_3$ as follows:

$$x_{(dd)} = x_0 + x_1, \quad y_{(qd)} = y_0 + y_1 + y_2 + y_3,$$

where $x_0, x_1, y_0, y_1, y_2, y_3$ satisfy the following inequalities:

$$|x_1| \leq \frac{1}{2}\mathrm{ulp}(x_0), \quad |y_{i+1}| \leq \frac{1}{2}\mathrm{ulp}(y_i), \quad i = 0, 1, 2,$$

where ulp stands for "units in the last place". A double-double(quad-double) number has 31(63) significant decimal digits. They can be computed by using only double-precision arithmetic operations (see [6,7] for details).

In Scilab, double-precision numbers are defined by the data type named constant. Scalars, vectors and matrices are treated in the same way as constant. In MuPAT, double-double and quad-double numbers are defined as data types named dd and qd, consisting of two or four constant data. We can use constant, dd and qd types at the same time, with the same operators $(+, -, *, /)$ and the same functions such as abs, sin and norm.

3   Sparse Data Structure for MuPAT

MuPAT has only the dense data structures of the three data types constant, dd, and qd. Sparse data structure is important to reduce the memory consumption and the computation time. Especially using double-double and quad-double arithmetics, sparse data structures are more important because they require twice or four times memories compared with double-precision arithmetic, and also require much more computation time. We developed the sparse data structures for double-double and quad-double arithmetics with considering the following points;

- The same arithmetic operators $(+, -, *)$ can be used among these data types.
- Operations for sparse and dense data in different precision numbers are available at the same time.

The users can compare problems in different precisions with lower memory consumption and small computation time.

3.1   Sparse Data Structure for Double Precision Number

Sparse data structure stores non-zero entries with its row and column indices. In Scilab, the following matrix

$$a = \begin{pmatrix} 0 & 0 & 9 & 0 \\ 0 & 0 & 7 & 1 \\ 1 & 0 & 0 & 0 \\ 0 & 0 & 0 & 8 \end{pmatrix}$$

can be represented by a sparse data type sparse as follows.

```
a   =
(    4,      4) sparse matrix
(    1,      3)        9.
(    2,      3)        7.
(    2,      4)        1.
(    3,      1)        1.
(    4,      4)        8.
```

The first line (4, 4) means the size of the matrix. The row and column indices and values of the matrix are stored after line 2. The entries are stored row-by-row. The same arithmetic operators $(+, -, *)$ for constant can be used for sparse, and mixed operations between constant and sparse are also allowed. The results of these binary operations become constant or sparse depending on the operations.

3.2   Sparse Data Structures for Double-Double and Quad-Double Numbers

To treat high-precision arithmetic for sparse data structure, we defined two new sparse data types; one is ddsp for double-double numbers, and the other is qdsp for quad-double numbers. These data types are based on CCS (Compressed Column Storage) format, which contains some vectors; row index, column pointer and values. ddsp has two and qdsp has four value vectors to represent a double-double number and a quad-double number respectively. By these definitions of data types, MuPAT has six data types: constant, dd, qd for dense data and sparse, ddsp, qdsp for sparse data of double, double-double and quad-double numbers respectively.

3.3   Definition of Matrix Operators

Now we have three sparse data types sparse, ddsp, and qdsp. To enable the use of the same matrix operators $(+, -, *)$ for these data types, operator overloading was applied to perform arithmetic operations among every existing data types constant, dd, and qd, and sparse data types sparse, ddsp, and qdsp.

In many cases, the sparsity cannot be kept after sparse matrix operations. Especially for sparse matrix multiplication, the result tends to have many non-zero entries and become a dense matrix. Therefore, we should allocate memory space dynamically.

3.4   Functions for Sparse Matrix

Some functions for sparse are extended to ddsp and qdsp. For example, full for changing a sparse data type into a dense data type, and nnz for returning the number of non-zero entries, and so on can be used in the same syntax among sparse, ddsp, and qdsp. A' for transposition of A and insertion and extraction of matrix elements can be performed in the same syntax for all data types.

4   The Memory Consumption and the Computation Time

To confirm the effectiveness of a sparse matrix computation with using ddsp and qdsp, we compared the memory consumption and the computation time between sparse and dense data structures. All experiments were carried out on Intel Core i5 1.7 GHz, 4 GB memory and Scilab version 5.3.3 running on Mac OS X Lion. An explicit parallelization, for example OpenMP, was not applied.

4.1   Memory Consumption

We prepared some 1000×1000 random sparse matrices with different sparsity for constant, dd, qd, sparse, ddsp, and qdsp. The sparsity patterns are random. Table 1 shows the sparsity and the memory consumption of each matrix. Sparsity is defined as the percentage of non-zero entries of the matrix. If the sparsity is less than 66 % for double-double number or 80 % for quad-double number, the memory consumptions of the sparse data structures are smaller than that of the dense data structures.

4.2   Matrix Operations

Using the matrices in Table 1, we measured the following matrix operations.

- Matrix vector product $A\boldsymbol{x}$, $B\boldsymbol{x}$, $C\boldsymbol{x}$
- Matrix addition $A + B$, $B + C$, $C + A$
- Matrix multiplication AB, BC, CA

We executed each operation repeatedly 100 times. Tables 2 and 3 show the results.

Table 1. Memory consumption

Matrix		Memory (MB)					
Sparsity		constant	sparse	dd	ddsp	qd	qdsp
A	1 %	8.00	0.12	16.00	0.25	32.00	0.41
B	5 %	8.00	0.60	16.00	1.21	32.00	2.01
C	10 %	8.00	1.21	16.00	2.41	32.00	4.01
D	66 %	8.00	7.92	16.00	15.85	32.00	26.40
E	80 %	8.00	9.60	16.00	19.21	32.00	32.01

Table 2. Results of matrix operations (Memory)

	Sparsity (%)	Memory (MB)	
		ddsp	qdsp
$A + B$	6	1.43	2.39
$C + A$	11	2.63	4.39
$B + C$	15	3.49	5.82
AB	40	9.51	15.98
CA	63	15.17	25.17
BC	99	23.86	39.73

Table 3. Results of matrix operations (Time)

	Time (s)					
	dd	ddsp	dd/ddsp	qd	qdsp	qd/qdsp
$A\boldsymbol{x}$	4.10	0.03	141.5	20.73	0.15	134.6
$B\boldsymbol{x}$	4.13	0.14	30.2	20.76	0.74	28.0
$C\boldsymbol{x}$	4.10	0.32	13.0	20.81	1.49	14.0
$A + B$	6.36	0.64	10.0	15.97	1.16	13.7
$C + A$	6.40	1.25	5.1	15.66	2.14	7.3
$B + C$	6.35	1.69	3.7	15.85	2.90	5.5
AB	2245.59	5.21	430.9	14909.50	12.27	1214.7
CA	2288.42	8.10	282.6	14964.51	20.84	718.1
BC	2282.17	16.54	138.0	14954.12	71.61	208.8

Matrix Vector Product. The computation time of matrix vector product for ddsp is 141.5 times smaller than that of dd when the sparsity of the matrix is 1 % ($A\boldsymbol{x}$) and 13.0 times faster when the sparsity is 10 % ($C\boldsymbol{x}$). The computation time for qdsp is 134.6 times smaller than that of qd when the sparsity is 1 % and 14.0 times faster when the sparsity is 10 %. The speedup values increase as the matrix sparsity decreases.

Matrix Addition. The computation time of matrix addition for ddsp is 10.0 times and 3.7 times smaller than that of dd when the sparsity is 6 % ($A + B$)

Table 4. Properties of test matrices

Matrix	Dimension	Non-zero	Sparsity (%)	Condition number
west0497	497	1,721	0.70	4.62×10^{11}
gre_1107	1,107	5,664	0.46	3.19×10^{7}
tols2000	2,000	5,184	0.13	5.99×10^{6}
sherman3	5,005	20,033	0.08	3.49×10^{18}

and 15 % $(B+C)$ respectively. The computation time for qdsp is 13.7 times and 5.5 times smaller than that of qd when the sparsity is 6 % and 15 % respectively.

Matrix Multiplication. The sparsity may be increased in matrix multiplication. In case of quad-double arithmetic, the computation result of BC (Sparsity 99 %) by using a dense data type qd requires 32 MB memory. On the other hand, the result by using a sparse data type qdsp requires 40 MB memory. When the sparsity of the result is more than 66 % for double-double number or 80 % for quad-double number, the memory usage of sparse data types ddsp and qdsp are larger than that of dense data types dd and qd. However, the computation times using qd and qdsp are 14954.1 s and 71.6 s respectively. The computation time for qdsp is 208.8 times smaller than that of qd. In case of AB, AB keeps low sparsity, and the computation time of ddsp is 430.9 times and qdsp is 1214.7 times smaller than that of dd and qd respectively.

5   Using High-Precision Arithmetic with Sparse Data Structure for Ill-Conditioned Problems

We show the effectiveness of sparse data structure for high-precision arithmetic on Scilab by applying the Biconjugate Gradient (BiCG) method for ill-conditioned matrices. Theoretically, the BiCG method, which is one of the Krylov subspace method, converges after at most n iterations, where n is the dimension of the matrix [8]. However, in floating-point arithmetic, the norm of the residual may diverge and oscillates, and then the iteration process may not converge. Sometimes it may require more than n iterations.

The iteration was started with $x_0 = 0$ and the right-hand side vector b was given by substituting the solution $x^* = (1, 1, ..., 1)^\top$ into $b = Ax^*$. Stopping criterion was $\|r_k\|_2 \leq 10^{-12}\|r_0\|_2$. The initial shadow residual was $r_0^* = r_0$. Iteration process was terminated at 10^4 iterations if it did not converge.

We took up four ill-conditioned test sparse matrices from [9]. These matrices are constructed double-precision numbers, then lower components of double-double and quad-double numbers are filled with zero. Condition numbers were obtained using the Scilab function cond in double-precision. Table 4 shows the list of test matrices.

Table 5 shows the results for double (D), double-double (DD), and quad-double (QD). "Iterations" denotes the number of iterations required for convergence, "Residual" denotes the relative residual norm $\|r\|_2/\|r_0\|_2$ and "Error"

Table 5. Computation results

	Matrix	Iterations	Residual	Error	Time (s)		
					constant	sparse	c/s
D	west0497	†	1.02e+02	3.70e+05	39.1	2.8	13.8
	gre_1107	†	6.97e+03	1.69e+04	278.7	4.1	68.7
	tols2000	†	8.06e+02	2.34e+06	998.6	4.7	211.7
	sherman3	†	1.73e-03	6.24e-01	6749.1	11.7	577.6

	Matrix	Iterations	Residual	Error	Time (s)		
					dd	ddsp	dd/ddsp
DD	west0497	†	2.18e-01	7.73e+02	303.7	15.0	20.2
	gre_1107	†	2.40e-01	9.08e-01	1828.9	21.2	86.2
	tols2000	1586	9.29e-13	3.55e-09	938.3	4.1	228.7
	sherman3	7696	9.98e-13	1.05e-13	31227.4	45.8	681.9

	Matrix	Iterations	Residual	Error	Time (s)		
					qd	qdsp	qd/qdsp
QD	west0497	2676	6.09e-13	3.50e-08	306.6	7.0	43.84
	gre_1107	3401	8.59e-13	3.05e-11	2136.2	17.6	121.3
	tols2000	1080	6.77e-13	1.96e-09	2342.8	7.1	328.5
	sherman3	4884	9.35e-13	1.73e-13	−	91.1	

† : More than 10^4 iterations,   − : Out of Memory,   c/s : constant/sparse

denotes the relative error norm $\|x - x^*\|_\infty / \|x^*\|_\infty$. constant/sparse is abbreviated to "c/s". The values of "Residual" and "Error" were obtained by using sparse data structures.

Using double-precision arithmetic, the BiCG method did not converge for all matrices. Especially, for west0497 and gre_1107, the BiCG method converged by only using quad-double arithmetic. For sherman3, even if the BiCG method converged by using double-double arithmetic, the number of iteration became more than n. Using quad-double arithmetic, the convergence improved, and the number of iteration decreased and became less than n. High-precision arithmetic produces great improvement and enables us to obtain the accurate result that cannot be obtained by double-precision arithmetic.

However, using dense data types dd and qd, iteration process requires a great deal of the computation time. Sparse data types ddsp and qdsp can save the computation time. For sherman3, the computation time of ddsp is 680 times smaller than that of dd, in the best case. Thus, high-precision sparse data structure provides more accurate results with practicable computation time. In case of dense data structure, sherman3 could not be stored by a quad-double number because of Out of Memory error. High-precision sparse data structure is also important in terms of the memory consumption.

An improvement of the accuracy by high-precision arithmetic depends on the problems and the methods. Although double-double and quad-double arithmetics do not perform well for all problems, high-precision sparse data structures surely increase the number of problems which can be solved accurately.

6 Conclusion

We developed the sparse data structures for quadruple-precision and octuple-precision arithmetics in MuPAT/Scilab, and showed that high-precision sparse data structure is practicable for solving a system of ill-conditioned linear equation.

MuPAT covers all arithmetic operators for double, double-double, and quad-double numbers for both dense and sparse data structures. Using MuPAT, six data types are available at the same time and operations for mixed-precision and mixed data structure are also available by the same operators and functions. To use double-double and quad-double arithmetics with lower memory consumption and smaller computation time, only a modification to definition of numbers is needed.

The memory consumption of the sparse data types is smaller for a matrix whose sparsity is less than 66 % for double-double number or 80 % for quad-double number when comparing with dense data types. In matrix vector product, the computation time of sparse data structures for a double-double number and a quad-double number are 141.5 times and 134.6 times smaller than that of dense data structures respectively, when the sparsity of the matrix is 1 %. In matrix addition, the computation time of sparse data structure for a double-double number and a quad-double number are 10.0 times and 13.7 times smaller than that of dense data structures respectively, when the sparsity of the result is 6 %. In matrix multiplication, even if the result becomes a dense matrix whose sparsity is more than 99 %, needing more memory than using dense data structures, the computation time can be reduced.

As a case study, we investigated the convergency of the BiCG method for ill-conditioned matrices (cf. [5] for the results of the GCR method with double-double arithmetic). Double-double and Quad-double arithmetics are crucial to improvement of the accuracy. However, dense data structures for double-double and quad-double numbers and arithmetics require large amounts of memory and considerably long computation time. For some situations, a matrix cannot be stored by a quad-double number because it requires four times as large memory as a double-precision number. High-precision sparse data structure facilitates the pragmatic problems of these restriction. Using sparse data structure, the computation time became 20–680 times smaller than using dense data structure. Sparse data structure for double-double and quad-double arithmetics is a practicable way to improve the convergence for ill-conditioned matrices, and to increase the number of problems that can be solved.

Parallelization of sparse data structures for a double-double number and a quad-double number is another big issue. As our future works, we will discuss elsewhere.

Acknowledgement. The authors would like to thank the reviewers for their helpful comments.

References

1. MuPAT, http://www.mi.kagu.tus.ac.jp/qupat.html
2. Kikkawa, S., Saito, T., Ishiwata, E., Hasegawa, H.: Development and acceleration of multiple precision arithmetic toolbox MuPAT for Scilab. JSIAM Lett. **5**, 9–12 (2013)
3. Scilab, http://www.scilab.org/
4. Baboulin, M., Buttari, A., Dongarra, J., Kurzak, J., Langou, J., Langou, J., Luszczek, P., Tomov, S.: Accelerating scientific computations with mixed precision algorithms. Comput. Phys. Comm. **180**, 2526–2533 (2009)
5. Saito, T., Ishiwata, E., Hasegawa, H.: Analysis of the GCR method with mixed precision arithmetic using QuPAT. J. Comput. Sci. **3**, 87–91 (2012)
6. Hida, Y., Li, X.S., Bailey, D.H.: Quad-double arithmetic: algorithms, implementation, and application, Technical Report LBNL-46996, Lawrence Berkeley National Laboratory, Berkeley, CA 94720 (2000)
7. Dekker, T.J.: A floating-point technique for extending the available precision. Numer. Math. **18**, 224–242 (1971)
8. Barrett, R., et al.: Templates for the solution of linear systems: building blocks for iterative methods, 2nd edn. SIAM, Philadelphia (1994)
9. The University of Florida Sparse Matrix Collection, http://www.cise.ufl.edu/research/sparse/matrices/

Efficient Heuristic Adaptive Quadrature on GPUs: Design and Evaluation

Daniel Thuerck[1]([✉]), Sven Widmer[2], Arjan Kuijper[1,3], and Michael Goesele[2]

[1] TU Darmstadt, Darmstadt, Germany
daniel.thuerck@gris.informatik.tu-darmstadt.de
[2] Graduate School of Computational Engineering,
TU Darmstadt, Darmstadt, Germany
[3] Fraunhofer IGD, Darmstadt, Germany

Abstract. Numerical integration is a common sub-problem in many applications. It can be solved easily in CPU-based applications using adaptive quadrature such as the adaptive Simpson's rule. These algorithms rely, however, on error estimation yielding a significant computational overhead. In addition, they require recursive function evaluations, which are not well suited for parallel computation on graphics processing units (GPUs) due to warp divergence issues. In this paper, we introduce heuristic forward quadrature as an alternative that is not only more efficient than traditional methods, but also better suited for accelerated massively-parallel calculation on GPUs. Additionally, we will give an error estimate for our method and demonstrate performance results for 1D and 2D integral applications which show that the algorithm leverages quadrature for the efficient implementation on GPUs.

Keywords: Numerical integration · GPGPU · Numerical algorithms · Heuristics · Interval estimation

1 Introduction

General purpose programming on graphics processing units (GPGPU) has been popularized with the advent of CUDA [8], OpenCL and related techniques and is currently one of the state-of-the-art approaches both inside and outside the computer science domain. GPGPU is often used to numerically solve ordinary or partial differential equations (ODEs, PDEs), e.g. in flow simulations, image processing [9], or the economic sciences (option pricing via the Black-Scholes equation) [5]. Especially when solving PDEs in financial mathematics, integration is required at some point. If there is no analytical solution available, we need to rely on numerical integration (also known as *quadrature*). Adaptive methods such as the adaptive Simpson's method or the Gauss-Kronrod algorithm are used with a given error tolerance to ensure exact values.

The principle of standard adaptive quadrature algorithms is shown in Algorithm 1. The integration is first performed on the whole interval. Then, the

R. Wyrzykowski et al. (Eds.): PPAM 2013, Part I, LNCS 8384, pp. 652–662, 2014.
DOI: 10.1007/978-3-642-55224-3_61, © Springer-Verlag Berlin Heidelberg 2014

Algorithm 1. Principle of adaptive quadrature algorithms.

---

function ADAPTIVEQUADRATURE(f, a, b, ε)

    $q \overset{\approx}{\leftarrow} \int_a^b f(x) dx$

    $\delta \leftarrow |q - \int_a^b f(x) dx|$ ▷ Using some given error estimator.

    if $\delta > \varepsilon$ **then**

        $q \leftarrow$ ADAPTIVEQUADRATURE$(f, a, a + (b - a)/2, \varepsilon)$ + ADAPTIVEQUADRATURE$(f, a + (b - a)/2, b, \varepsilon)$

---

error is estimated by a given heuristic (in Simpson's case, the result is compared with a second integration using standard Simpson's rule to check the difference against a user-defined error tolerance). If the threshold is exceeded, the interval is subdivided in at least two parts and the method is called recursively on the subintervals.

While the method is able to guarantee a given error threshold, it requires a significant computational overhead: Each time the interval is subdivided, the last result is discarded. Additionally, the error estimates are complex and computationally intensive as they need to compute a better approximation than the current algorithm's level. Given those two characteristics, the algorithm is not well suited for GPUs. First, interval subdivision often yields branching which severely affects the performance of threads in the same warp. Second, recursive kernels are not yet possible in consumer cards (but will become available with Nvidia HyperQ). In CUDA, recursion is limited to device functions while it is not available at all in OpenGL (and OpenGL ES). Although transformation in a non-recursive algorithms is possible, it is quite complex [7]. In this paper, we propose an alternative method that results in a smaller number of function evaluations (and thus in a significant improvement of performance) and is especially well suited for GPUs.

Our contributions are as follows: We propose *heuristic adaptive forward* quadrature as an alternative quadrature method. We motivate and investigate a special heuristic, give an error estimate and its proof, and show that the number of function evaluations is significantly smaller than with today's standard routine. Furthermore, we show a GPU implementation and analyze its performance using an application from the image processing domain.

2 Related Work

Adaptive quadrature is a well-investigated topic. Research has, however, been discontinued in recent years. A good overview can be found in Gander and Gautschi [6]. The precision of the traditional, recursive algorithm depends largely on the error metric. Typically, an integrand is integrated with two different methods, one more precise than the other, and it is tested whether a given error threshold is exceeded [6]. Further investigation on those estimators has been conducted by Shapiro [12] and Berntsen et al. [3,4]. Both methods rely on the

Algorithm 2. Principle of heuristic adaptive forward quadrature.

---

function HEURISTICQUADRATURE(f, a, b, h_*, h^*)

    $p \leftarrow a$

    $q \leftarrow 0$

    while $p < b$ **do**

        $h \leftarrow$ ESTIMATEINTERVALLENGTH(f, p, h_*, h^*)

        $q \leftarrow q + h\frac{f(p)+f(p+h)}{2}$

        $p \leftarrow p + h$

---

traditional method and improve the performance by clever interval subdivision and error estimators that combine global and local precision.

As stated above, these traditional approaches use recursion intensively and are not well suited for GPUs. Quadrature implementations on GPUs rely mostly on non-adaptive integration [15], which is embarrassingly parallel. Another approach for multidimensional quadrature was proposed by Arumugam et al. [2]. Integration is done in two phases – interval division and integration – and relies on recursion. The recursive subdivision part is, however, implemented on the CPU using a hybrid CPU/GPU architecture. Anson et al. [1] presented a similar method on a reconfigurable FPGA architecture. Existing CPU libraries as QUADPACK [10] implement only the Simpson and Gauss-Kronrod methods, which are badly suited for the GPU. Currently, there is no published method for adaptive integration on GPUs available. In contrast to these libraries, we leverage adaptive quadrature for efficient use on commodity GPUs by introducing a new algorithm and a suitable implementation.

3   Non-recursive Adaptive Quadrature

As mentioned in the introduction, there are two ways of adaptive quadrature: One can either estimate the quadrature error *a posteriori* and subdivide intervals thereafter or estimate the interval length *a priori*. Our method concentrates on the latter. An overview over this algorithm is given in Algorithm 2. In essence, we apply the trapezoid rule for every interval. In usual quadrature, the result is discarded if it exceeds the error threshold and the interval is subdivided, effectively returning to its begin. This method could be called *forward-backward*, while our method is of the *forward* kind: One after another, the heuristic selects intervals and arbitrary algorithms can be applied to them, never discarding the result. Of course the success depends largely on the interval selection heuristic. Before we propose a particular heuristic, let us shortly review advantages and disadvantages of this approach.

The clear advantage is, as mentioned, that all intervals contribute to the final result and the number of function evaluations is minimized. As the evaluation of an error estimator is unnecessary, we save computational power and do need to evaluate conditionals. On the other hand, the error of our method is clearly dependent on the heuristic. Additionally, there is no possibility to implement a hard error threshold: the heuristic defines the interval lengths and such (indirectly) the error. Using careful design, the error can, however, be bounded and practical results are promising as shown in Sect. 4.

3.1   The ∂^2 Heuristic

Every heuristic has the goal of providing small interval sizes in regions where the integrand is curved while using larger intervals on near-linear parts. Ideally (in terms of error), every interval would only contain a linear subset of the integrand's graph. To predict how a function develops in a given interval, we can use its first and second derivatives. For the error estimation (see Sect. 4) we assume that the integrand is C^2-continuous on $[a, b]$. The second derivative yields information about the curvature of the integrand and thus about the development of the rate of value change. If the second derivative is small or even zero, the curve's steepness will remain almost constant and a greater interval length can be used. Alternatively, if the second derivative grows, the curvature of the integrand increases and we need to consider smaller intervals.

A first approximation of this heuristic (given a minimal interval size h_* and maximal interval size h^*) is

$$h_i = h_* + (\alpha - f''(p_i))h_{i-1}(h^* - h_*). \tag{1}$$

p_i is the last integration point (with interval length h_i). α is a given constant. Remember that we use a forward method: each interval length depends on the length of the last interval to model the integrand's change. Unfortunately, Eq. 1 shows bad behavior when the second derivative of the integrand is small, e.g. in the case of the sine function. To improve the estimation here, we replace $\alpha - f''(p_i)$ by $\beta = \max(|f''(p_i)|, |f'(p_i) - f'(p_{i-1})|)$. The second max term captures the behavior of functions where the curvature is small but nonetheless, the rate of change (hence, the first derivative) is huge such as piecewise linear functions. Note that although the later error proof needs the continuity assumption, in practice the algorithm can be applied to non-continuous functions, too.

A linear involvement of the curvature is, however, not very useful since for only small changes, no interval length change is necessary. Following this intuition, we introduce weighting by $e^{-\beta}$. This yields the final heuristic

$$h_i = h_* + e^{-\max(|f''(p_i)|, |f'(p_i) - f'(p_{i-1})|)}(h^* - h_*). \tag{2}$$

A visualization on how this heuristics works in an application is given in Fig. 1.

Often, derivatives are not directly given and thus not available for calculation. In this case, we need to approximate first and second derivative using simple central differences (Algorithm 3).

With this heuristic, we tested several functions on the unit interval and compared the error and the number of integrand evaluations required to MATLAB's quad function, an improved adaptive Simpson's method with $\epsilon = 10e - 6$. As Table 1 shows, this approach is much more efficient for most functions while loosing only very little precision.

Algorithm 3. ∂^2 heuristic algorithm implemented using simple central difference approximations.

```
function FAQ(f, a, b, h_*, h*)
    q ← 0
    h ← h_*
    p ← a
    lastVal ← f(p)
    lastHill ← (f(p + 0.1h) − lastVal)/(0.1h)    ▷ Capture extreme behaviour at graph start.
    while p < b do
        p ← p + h
        thisVal ← f(p)
        thisHill ← |thisVal − lastVal|/h                        ▷ Approximate f'.
        q ← q + h · (thisVal + lastVal)/2
        d ← max(|(thisHill − lastHill)/thisHill|, |thisHill − lastHill|)    ▷ Approximate f''.
        h ← h_* + e^{−d}(h* − h_*)
        if p + h > b then
            h ← b − p
        lastVal ← thisVal
        lastHill ← thisHill
```

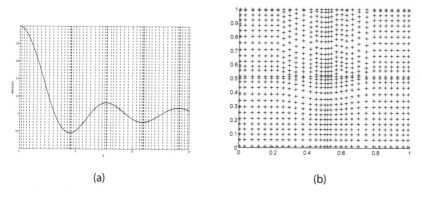

(a)   (b)

Fig. 1. Interval selection for ∂^2-heuristic and integrand $\frac{sin(x)}{x}$ visualized (a) and in 3D for a Gaussian Kernel $e^{-|\mathbf{x}-0.5|_1/0.25^2}$ view from above (b) with $h_* = 0.01$, $h^* = 0.05$.

4   Error Estimation

After motivating and introducing our heuristic in the previous sections, we still need to discuss how the method's error can be estimated. Note that while other quadrature methods offer the possibility to define a hard error threshold, we are unable to do so. Instead, we need to estimate the methods's error a priori. As for non-adaptive methods, a theorem on how the error behaves can be derived. Using the trapezoidal rule error estimation and the fact that our integrand f is C^2-continuous we arrive at the following error bound theorem:

Theorem 1. *Let I be the result of adaptive quadrature with heuristic (2) applied to the function f and bounds a, b with minimum and maximum interval sizes*

Table 1. Error and performance results for our forward quadrature method in comparison to adaptive Simpson quadrature.

Function	Relative error	Absolute error	Percentage of function evaluations compared to Matlab quad
$\sin(x)$	0.01	−0.01	48 %
$\sinh(x)$	0.01	0.02	48 %
$x^8 + x^4 + x^3 + x + 1$	0.0	0.18	37 %
$\exp(-x)$	0.01	0.01	65 %
$\sin(x)/x$	0.0	0.0	64 %
$\log(x)$	−0.15	−0.09	21 %
$\sqrt{(x)}$	0.0	−0.01	27 %

h_*, h^* and let further f be a twice continuously differentiable function on $\mathbb{R}$. Then the error $\Delta I = |I - \int_a^b f(x)dx|$ is bounded by

$$\Delta I \leq \sup_{\eta \in [a,b]} \frac{\tilde{h}_i^{\,2}(b-a)}{24} f''(\eta)$$

where $\tilde{h}_i$ is defined as given in the proof.

Proof. By the given algorithm, it is obvious that we can express the interval lengths as a sequence $(h_i)_{i \in \mathbb{R}^+}$ with

$$h_i = h_* + e^{-\frac{|f'(a+\sum_{j=0}^{i-1} h_j) - f'(a+\sum_{j=0}^{i-2} h_j)|}{f'(a+\sum_{j=0}^{i-2} h_j)}} (h^* - h_*) \tag{3}$$

where $a + \sum_{j=0}^{i-2} h_j$ is the second-to-previous quadrature anchor point and $a + \sum_{j=0}^{i-1} h_j$ the previous. The quotient

$$\frac{|f'(a + \sum_{j=0}^{i-1} h_j) - f'(a + \sum_{j=0}^{i-2} h_j)|}{f'(a + \sum_{j=0}^{i-2} h_j)} \tag{4}$$

in the algorithm is essentially an approximation of the second derivative f'', which we can use for error estimation. As f'' is continuous, we can use the mean value theorem, such that there is an $\xi \in [a, b]$ that

$$f''(\xi) = \frac{f'(b) - f'(a)}{b - a} \tag{5}$$

or

$$f''(\xi)\frac{b-a}{f'(a)} = \frac{f'(b) - f'(a)}{f'(a)} \tag{6}$$

whose last part is – when applied to each interval h_i (so that $h_i = b - a$) – exactly the approximated quotient mentioned earlier. Hence we can provide an upper bound for the interval length with a being the starting point of interval h_i

$$h_i \leq \tilde{h}_i = \sup_{\xi \in [a-h_{i-1},a]} \left(h_* + e^{-f''(\xi)\frac{h_{i-1}}{f'(a-h_{i-1})}} (h^* - h_*) \right). \tag{7}$$

Let now $|\mathcal{H}| = \frac{b-a}{\tilde{h}_i}$, then we can estimate the quadrature error by using the estimate for the rectangle rule (as integration in a given interval is done by rectangle rule in the algorithm), being x_i the calculation point in interval h_i:

$$\Delta I \leq \sum_{i=1}^{|\mathcal{H}|} \frac{\tilde{h}_i^3}{24} f''(x_i) \leq \sup_{\eta \in [a,b]} |\mathcal{H}| \frac{\tilde{h}_i^3}{24} f''(\eta). \tag{8}$$

4.1   Extension on Two-Dimensional Integrands

So far, the integrand was implicitly a function $f : [a, b] \to \mathbb{R}$. This is, however, not always the case. Often, an integral needs to be evaluated on an area $\Omega \subset \mathbb{R}^2$. As our integrand f is differentiable, it is also continuous. Let now $[a, b]$ be compact, then we can use Fubini's theorem to formulate the approximation for $f : \mathbb{R}^2 \to \mathbb{R}$ on a rectangle $\Omega = I \times J$:

$$\int_{\Omega} f(x,y)\, dxdy = \int_J \left(\int_I f(x,y)dx \right) dy \tag{9}$$

$$\approx \int_J \left(\sum_{i=1}^{|\mathcal{H}_x|} |h_{x,i}| f(p_i, y) \right) dy \tag{10}$$

$$\approx \sum_{i=1}^{|\mathcal{H}_x|} |h_{x,i}| \left(\sum_{k=1}^{|\mathcal{H}_{y,i}|} |h_{y,k}| f(p_i, p_{i,k}) \right) \tag{11}$$

where $\mathcal{H}_d$ is the set of intervals chosen by the heuristic in direction d. Effectively, we get a grid on Ω for quadrature. An example grid is shown in Fig. 1(b). By repeating this method, we can extend the algorithm to n dimensions.

5   Implementation and Performance on GPUs

As the proposed quadrature algorithm is well suited for GPUs, we implemented it using CUDA Version 5.0 [8]. The performance is evaluated using two different applications, one for the one dimensional and the other for the two dimensional case. An extension to n dimensions is straightforward. We compare the performance against a multi-threaded CPU version with our heuristic as well as a GPU and multi-threaded CPU implementation of the quadrature by the standard adaptive Simpson's rule. To the authors' best knowledge, there are no implementations for quadrature on CUDA that can be considered as industry standard and baseline. The well-known QUADPACK [10] is limited to one-dimensional functions and despite its popularity, there is no massively-parallel implementation available. Hence, programmers usually create their own implementations of popular methods such as the Simpson rule or the Gauss-Kronrod method. Although the last one is considered state-of-the-art, it is computationally more expensive than using the Simson rule which is why we compare our performance to the Simpson rule.

All experiments were performed on a PC running Ubuntu 12.04 with the latest Nvidia drivers (version 304.88). The system is equipped with an Intel Core i7-3930K hexacore CPU 64 GB of RAM, a Nvidia Geforce GTX 620 (primary device) and a Nvidia Geforce GTX 680 card with two gigabyte of RAM as a headless compute device.

5.1   One-Dimensional Case

As an application for the one-dimensional case, we use the Perona and Malik [9] diffusion. This filter models the physical process of diffusion described by the PDE

$$I_t = \nabla \cdot (c(|\nabla I|)\nabla I) \tag{12}$$

where I is the image intensity function of a given grayscale image on a region Ω. The function c is called the *diffusivity*. In contrast to linear diffusion which is equivalent to convolving the image with a Gaussian the diffusion strength varies in nonlinear diffusion over the image domain. For the diffusivities proposed by Perona and Malik, quadrature is not necessary. However, Thuerck and Kuijper [13,14] presented a diffusivity which leads to a well-posed process but has no analytical integral. To implement this model, we need numerical integration and can apply our algorithm.

The CUDA implementation of the proposed quadrature algorithm is straightforward. For each pixel, finite differences with its neighbors are calculated and used to calculate the diffusivity in this point by quadrature. Essentially, Algorithm 3 can be implemented in CUDA as a device function directly as there is no further inherent parallelism. Figure 2 shows that both GPU implementations outperform the CPU equivalents by more than one order of magnitude. We can observe a performance increase of about 20 percent over the GPU based Simpson algorithm as well.

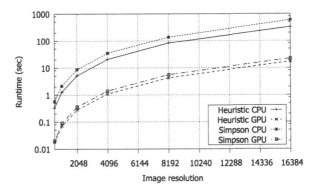

Fig. 2. Performance of the modified Perona and Malik diffusion by Thuerck and Kuijper [13] using the adaptive Simpson's rule.

5.2 Two-Dimensional Case

Application cases of 2D integration are quite prominent in fluid simulations and financial mathematics. Especially in the field of option pricing using the Black-Scholes [5] equation, quadrature of a Gaussian kernel is required when the influence of other options is included in the calculation.

To evaluate the key aspects we developed a simplified prototype for performance evaluation. As input data, we take an image of a given size. The CUDA kernel then reads the intensity of each pixel in the whole image and performs a quadrature of an Gaussian bell in the region $[0, 0]$ to $[i, i]$ where i is the intensity to ensure that the threads operate on different data. The difference to a Black-Scholes implementation is thus only a constant for runtime purposes.

The algorithm is split up into three phases. First, one CUDA thread calculates the samples for the first dimension as needed by Fubini's theorem. Afterwards, each thread can use the 1D implementation concurrently to execute integration in the second dimension, which results in a grid as shown in Fig. 1(b). Upon finish, each thread writes its result to shared memory and, after parallel reduction, the final result is written to global memory.

By executing the integration in each of the dimensions in sequence, we reduce the complexity to $\mathcal{O}(\text{dimensions})$ rather than $\mathcal{O}(\text{x-samples})$. Hence, we observe a much higher speed up than in the 1D case. While in this case, the speed up is only a result of the reduced number of integration steps, the 2D case generates enough workload to satisfy the GPU and a better performance is achieved by doing a large number of integration steps in parallel in addition to each 1D integration being less complex. However, there is room for improvement: A carefully designed 2D heuristic could improve thread utilization and enable us to execute integration in two/dimensions at least partly concurrent. Nevertheless, the performance evaluation shows that the given heuristic algorithm effectively enables us to use the GPU for quadrature, which usual methods cannot do.

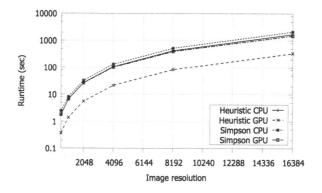

Fig. 3. Performance of the two dimensional case.

In the two dimensional case the adaptive Simpson's rule's performance is similar to both CPU implementations, as seen in Fig. 3. Our presented algorithm outperforms all three comparison implementations with one order of magnitude.

Naturally, the speed-up in the 2D case is dramatically higher than in the 1D case. This is due to the quadratic number of integrations in the 2D case compared to the linear number of integrations in the 1D case.

6   Conclusion and Future Work

In this paper, we showed that heuristic adaptive integration can speed up CPU as well as GPU implementations while keeping the accuracy constant when choosing suitable bounds h_*, h^*. In the GPU case, the performance evaluation resulted in a strong recommendation to use the heuristic on the GPU as sensible speed ups cannot be achieved with traditional Simpson quadrature. The prototypical implementations confirm this fact. Furthermore, the presented algorithm can be extended to n dimensions using Fubini's theorem.

Most of our future plans are already mentioned above. Currently, when integrating in n dimensions, we first create the sampling coordinates and interval sizes for $n - 1$ dimensions and use the nth dimension for the actual integral calculation. The sampling process in each dimension can be parallelized. However, as the result of every dimensions depend on the previous dimensions, the algorithm scales linear with the number of dimensions for fixed grid sizes. Therefore, we wish to improve this behaviour in the future. Second, we would like to fully implement a Black-Scholes option pricing kernel to determine the speed-up in a non-artificial application. Although a reference implementation from NVIDIA exists [11], it is restricted to 1D quadrature. Lastly, we consider developing suited heuristics for the multidimensional case so there is no need to revert to the one-dimensional heuristic via Fubini's theorem.

References

1. Tse, A.H.T., Chow, G.C.T., Jin, Q., Thomas, D.B., Luk, W.: Optimising performance of quadrature methods with reduced precision. In: Choy, O.C.S., Cheung, R.C.C., Athanas, P., Sano, K. (eds.) ARC 2012. LNCS, vol. 7199, pp. 251–263. Springer, Heidelberg (2012)
2. Arumugam, K., Godunov, A., Ranjan, D., Terzic, B., Zubair, M.: An efficient deterministic parallel algorithm for adaptive multidimensional numerical integration on GPUs. http://on-demand.gputechconf.com/gtc/2013/poster/pdf/P0237_KameshArumugam.pdf (2013)
3. Berntsen, J.: Practical error estimation in adaptive multidimensional quadrature routines. J. Comput. Appl. Math. **25**(3), 327–340 (1989)
4. Berntsen, J., Espelid, T.O., Sørevik, T.: On the subdivision strategy in adaptive quadrature algorithms. J. Comput. Appl. Math. **35**(1), 119–132 (1991)
5. Black, F., Scholes, M.: Taxes and the pricing of options. J. Finan. **31**(2), 319–332 (1976)

6. Gander, W., Gautschi, W.: Adaptive quadrature revisited. BIT Numer. Math. **40**(1), 84–101 (2000)

7. McKeeman, W.M., Tesler, L.: Algorithm 182: nonrecursive adaptive integration. Commun. ACM **6**(6), 315 (1963). http://doi.acm.org/10.1145/366604.366640

8. NVIDIA: CUDA Compute Unified Device Architecture. www.nvidia.com/object/cuda_home_new.html

9. Perona, P., Malik, J.: Scale-space and edge detection using anisotropic diffusion. IEEE Trans. Pattern Anal. Mach. Intell. **12**(7), 629–639 (1990)

10. Piessens, R., Doncker-Kapenga, D., Überhuber, C., Kahaner, D., et al.: Quadpack, A Subroutine Package for Automatic Integration, 301 p. Springer, Heidelberg (1983)

11. Podlozhnyuk, V.: Black-scholes option pricing. Part of CUDA SDK documentation (2007)

12. Shapiro, H.D.: Increasing robustness in global adaptive quadrature through interval selection heuristics. ACM Trans. Math. Softw. (TOMS) **10**(2), 117–139 (1984)

13. Thuerck, D., Kuijper, A.: Cosine-driven non-linear denoising. In: Kamel, M., Campilho, A. (eds.) ICIAR 2013. LNCS, vol. 7950, pp. 245–254. Springer, Heidelberg (2013)

14. Thuerck, D., Kuijper, A.: Lazy nonlinear diffusion parameter estimation. In: Petrosino, A. (ed.) ICIAP 2013, Part I. LNCS, vol. 8156, pp. 211–220. Springer, Heidelberg (2013)

15. Windisch, A., Alkofer, R., Haase, G., Liebmann, M.: Examining the analytic structure of greens functions: Massive parallel complex integration using GPUs. Comput. Phys. Commun. **184**, 101–116 (2012)

An Efficient Representation on GPU for Transition Rate Matrices for Markov Chains

Jarosław Bylina$^{(\boxtimes)}$, Beata Bylina, and Marek Karwacki

Institute of Mathematics, Marie Curie-Skłodowska University, Lublin, Poland
{jaroslaw.bylina,beata.bylina}@umcs.pl, marek.karwacki@gmail.com

Abstract. The authors present a novel modification of the HYB format — known from the CUSP library. The new format is suitable for sparse Markovian transition rate matrices and enables processing two times bigger matrices on single GPU, also improving computation performance at the same time. Particularly, the SpMV operation — that is the multiplication of a sparse matrix by a vector — is analyzed for this format on one GPU and two GPUs. Numerical experiments for transition rate matrices of Markov chains from [18] show that the proposed format allows to process matrices of sizes about 3.6×10^7 rows with the use of single GPU (3 GB RAM). When the plain HYB format is used the matrices of these sizes do not fit in one GPUs memory. Moreover, the use of the modified HYB format can give the speedup even up to 13 times in comparison to multi-threaded CPU (12 cores).

Keywords: GPU · SpMV · Markov chains · Transition rate matrix · CUSP · Sparse matrices

1 Introduction

Markov chains are a tool for modeling various computer appliances and networks, as well as other natural processes and systems. Recently, they have been used to model wireless networks [2,7,8] and they often appear in computational biology [3].

An efficient usage of Markov chains for modeling such systems is quite difficult because of computational problems. There are two aspects of these problems — the first is the size of the matrix (and the memory requirements) and the second is a usually long time of computations (needed to find solutions of Markov chain).

To speed up computations associated with Markov chains we can use accelerators — such as GPUs [3,5,6,12]. GPUs have possibilities to accelerate the scientific and engineering computations because they are equipped with a big number of quite independent processing units.

Any Markov chain can be described in terms of linear algebra with the use of a square matrix — which is a matrix of probabilities or a matrix of transition rates. We will be interested in the latter.

A transition rate matrix Q describing a Markov chain which models a system or a phenomenon has some particular traits. It is a square singular matrix with

R. Wyrzykowski et al. (Eds.): PPAM 2013, Part I, LNCS 8384, pp. 663–672, 2014.
DOI: 10.1007/978-3-642-55224-3_62, © Springer-Verlag Berlin Heidelberg 2014

a weak row-dominant diagonal. All the diagonal elements are negative and all the off-diagonal ones are not negative. The sum of every row is zero. Usually, the matrix is a huge one and very sparse.

Sparse matrices are stored in special data structures and special algorithms are used to process these structures optimally. We can find descriptions of many such storage schemes in the literature (e.g. [1, 4, 11]). The sparse storage schemes designed for CPU works also for GPU, but because of different properties of these architectures, different schemes are preferred.

Some sparse storage schemes for GPU and the performance of the matrix-vector multiplication were described thoroughly in [1]. In [4] the authors investigated the following storage formats for Markovian transition rate matrices: COO, CSR, ELL and HYB from the CUSP library. After some tests for huge Markov chain transition rate matrices the HYB format appeared to be the best. GPUs are not intended to make computations on large sets of data — as large as a typical matrix Q. So, in [6], details of an implementation of the HYB format on two GPUs were presented to fit bigger matrices in the GPU memory and to make the operations on bigger matrices possible.

In [12] other sparse storage formats were presented which are used in of the probabilistic model checking algorithms.

Currently, the performance of algorithms strongly depends not only on the arithmetic operations, but also — and what is even more important nowadays — on the number of references to the memory. That is because the arithmetic operations are cheap (that is: fast) in comparison with memory reading/writing. Moreover, the computer memory is hierarchically structured in layers of different speeds and sizes. Thus, one of the aims of designing new algorithms is to reorganize data to be able to use smaller but faster type of memory.

The contribution of this paper is HYBIV: a novel modification of an existing sparse storage format — namely HYB from the CUSP library [14]. The HYBIV format is adapted to sparse transition rate matrices. Besides, in the HYBIV format the amount of data read from the slower memory is less than in the HYB format.

The main motivation for this modification was the desire to put the biggest possible transition rate matrix in the GPU memory — and (if only possible) to speed up the computation at the same time.

Additionally — to the HYBIV format — the paper presents the sparse matrix-vector multiplication (SpMV) for the modified format (HYBIV) for one GPU and two GPUs. The numerical experiment was carried out for two groups of transition rate matrices (for NCD and MUTEX models from [18]) — to compare the HYB format from the CUSP library with the new HYBIV format for one GPU and for two GPUs. A comparative analysis was also done with the CSR format from the MKL library (on CPU).

The structure of the article is following. Section 2 presents the problem of sparse matrix storage on GPUs and the HYB format from the CUSP library [14]. Section 3 contains a description of the proposed modification of the HYB format (that is, the HYBIV format) which takes advantage of some properties of transition rate matrices. Section 3 describes the numerical experiments. Section 4 shows some conclusions.

2   Sparse Matrices on GPU

One of the main directions of numerical algorithms development on GPUs is construction of data structures and algorithms which uses all of the advantages of GPUs. It is quite hard to build an efficient algorithm fulfilling this condition from the scratch. That is why it is good to use existing libraries. Some examples of them include: CUBLAS [13], CUSP [14], Thrust [19], MAGMA [17].

The sparse matrix-vector multiplication (SpMV) $y \leftarrow Ax$ — where A is a sparse matrix and x is a dense vector — is the most commonly used operation in sparse matrix computations. Performance of many scientific and engineering applications highly depends on operation SpMV. Among others, this operation is used in various algorithms that find both the stationary and transient states of Markov chain [5,6,11].

The matrix-vector multiplication for dense linear algebra is easily implemented with the use of many threads of GPU. However, for sparse matrix computations, implementing it on a GPU is a big problem because of the irregularity in accessing the elements which are unordered. That is why it is very important to prepare and use an adequate format to store sparse matrices on GPU.

The CUSP library is written in C++ and it is an open library of generic parallel algorithms for sparse matrix operations computed on CUDA devices. CUSP includes a highly efficient interface to manipulate sparse matrices. Both the sparse storage formats and the algorithms on them are defined. Among the algorithms, the sparse matrix-vector multiplication is defined as well.

The HYB format in the CUSP library is a hybrid format between the ELL format (known from ELLPACK [15]) and the COO format (well-known coordinate list sparse format). So the matrix in the HYB format is stored as two parts — one in the ELL format (two two-dimensional arrays) and another in COO format (three one-dimensional arrays):

- *ell_data* stores values of non-zero elements of the original matrix — as a two-dimensional rectangular array of the size $N \times M$, where N is the number of rows and M is a mode of the number of non-zero elements in a row (highest in their histogram); rows which are originally shorter than M are complemented with zeros;
- *ell_indices* stores column indices of the respective non-zero elements stored in *ell_data*;
- *coo_data* stores values of non-zeros that do not fit in *ell_data*;
- *coo_col* stores column indices of the respective non-zero elements stored in *coo_data*;
- *coo_row* stores row indices of the respective non-zero elements stored in *coo_data*.

Figure 1 shows a square matrix of size 5. Figure 2 shows the same matrix stored in the HYB format from the CUSP library.

2.1   HYBIV: The HYB with Indexed Values Format

The motivation for the new format was:

– very large matrices,
– quite limited set of different values in Markov chain transition rate matrix —
  it is caused by the fact that Markov chains often consist of repetitive patterns
  of behaviors and the transitions implied by them have the same values,
– ease of modification of the existing format and existing functions (the matrix-
  vector multiplication, among others) in the CUSP library.

The HYBIV format is an extension of the HYB format from the CUSP
library. To limit the amount of memory needed to store the matrix:

– all the different values of the elements of the matrix Q are stored in a separate
  array — named *data* (every value only once);
– both the arrays containing original values of the matrix Q (as floating point
  numbers) are replaced by arrays containing indices of the real values of the
  element of the matrix — in the array *data*; that is:

• the array *ell_data* is replaced by an array *ell_iv*,
• the array *coo_data* is replaced by an array *coo_iv*.

In the GPU implementation we put the *data* vector in the constant memory
or in the texture memory (it depends on its size) and the values in *data* are
stored in double precision. These types of memory are intended only for reading
and they provide low latency. Moreover, they are accessible from all running
threads simultaneously.

The arrays *ell_iv*, *ell_indices*, *coo_iv*, *coo_col*, *coo_row* are stored in the global
memory of the GPU and their values are normally 4-byte integers. However, for

$$A = \begin{bmatrix} 14 & 0 & 0 & 0 & 11 \\ 0 & 21 & 0 & 19 & 0 \\ 0 & 0 & 0 & 15 & 0 \\ 11 & 13 & 0 & 12 & 11 \\ 0 & 0 & 0 & 0 & 18 \end{bmatrix}$$

Fig. 1. A not compressed square matrix

$$ell\_data = \begin{bmatrix} 14 & 11 \\ 12 & 19 \\ 15 & * \\ 11 & 13 \\ 18 & * \end{bmatrix} \quad ell\_indices = \begin{bmatrix} 0 & 4 \\ 1 & 3 \\ 3 & * \\ 0 & 1 \\ 4 & * \end{bmatrix}$$

$$coo\_data = \begin{bmatrix} 12 & 11 \end{bmatrix} \quad coo\_col = \begin{bmatrix} 3 & 4 \end{bmatrix} \quad coo\_row = \begin{bmatrix} 3 & 3 \end{bmatrix}$$

Fig. 2. A square matrix stored in the HYB format

a smaller array *data*, the indices in the arrays *ell_iv* and *coo_iv* are stored as 2-byte integers, what makes the HYBIV format even more memory-saving.

Figure 3 presents matrix from Fig. 1, but in the HYBIV format.

$$data = \begin{bmatrix} 14 & 11 & 21 & 19 & 15 & 13 & 12 & 18 \end{bmatrix}$$

$$ell\_iv = \begin{bmatrix} 0 & 1 \\ 2 & 3 \\ 4 & * \\ 1 & 5 \\ 7 & * \end{bmatrix} \quad ell\_indices = \begin{bmatrix} 0 & 4 \\ 1 & 3 \\ 3 & * \\ 0 & 1 \\ 4 & * \end{bmatrix}$$

$$coo\_iv = \begin{bmatrix} 6 & 1 \end{bmatrix} \quad coo\_col = \begin{bmatrix} 3 & 4 \end{bmatrix} \quad coo\_row = \begin{bmatrix} 3 & 3 \end{bmatrix}$$

Fig. 3. A square matrix stored in the HYBIV format

2.2 HYBIV on Many GPUs

The HYBIV format is also suitable for data distribution between many GPUs. The method of the partitioning is quite analogous to the one presented in [6] (however, that one concerned HYB). Authors also considered column-wise 1D and 2D blocks partitioning, but both schemes are not suitable for ELLPACK format, because right side of ELLPACK matrix is less dense then left, which leads to significantly uneven distribution.

Namely, the original $n \times n$ matrix Q^T (we store transposed transition rate matrices because they are more suitable for our algorithms — but the principles are identical) is partitioned among many GPUs, with each GPU storing contiguous block of complete rows of the matrix (row-wise 1D partitioning) — every row in only one GPU.

For our purposes — the sparse matrix-vector multiplication $y \leftarrow Q^T x$ — we need also a whole $n \times 1$ vector x in each of z GPUs. Since each GPU performs computations with the use of a $\frac{n}{z} \times n$ matrix and the $n \times 1$ vector x, therefore after the multiplication each GPU holds a partial y vector of the size $\frac{n}{z} \times 1$.

3 Numerical Experiments

In this section we tested the memory requirements, the time and the performance of the SpMV operation using:

- the CUSP library with the HYB storage format (on one GPU and two GPUs),
- the CUSP library with our modification of HYB — the HYBIV storage format (on one GPU and two GPUs),
- and MKL [16] with the CSR storage format (on CPU).

Our intention was to investigate and compare memory requirements on one GPU and on two GPUs for the original HYB format and for the modification — HYBIV format. We also wanted to compare the SpMV operation for the HYB and HYBIV formats on one and two GPUs (with the CUSP library) and for the CSR format on CPU (with the MKL library). We also wanted to check, what the speed up of GPU over CPU in this computational problem is — that is why we used MKL as a CPU representative.

We tested the SpMV operation on CPU and GPU under Linux with gcc and NVIDIA nvcc compilers. The input matrices are generated (by the authors) from two classical Markovian models [18]: NCD (*nearly completely decomposable*) [10] and MUTEX (*mutual exclusion*) [9].

Table 1 shows the specification of the hardware and the software used in the experiment.

Table 1. Hardware and software used in the experiment

CPU	2× Intel Xeon X5650 2.67 GHz (2 × 6 cores)
Host memory	48 GB DDR3 1333 MHz
GPU	2× Tesla M2050 (515 Gflops DP, 3 GB memory)
OS	Debian GNU Linux 6.0
Libraries	CUDA Toolkit 4.0, CUSP 0.2

3.1 Size and GPU Memory Requirements

Tables 2 and 3 present details about test matrices, where:

- n is the number of rows,
- nz is the number of non-zero elements,
- nz/n represents the matrix density,
- uv is the number of unique values of matrix' elements,
- **HYB** and **HYBIV** are memory usages by the respective format in the algorithm run on one GPU,
- **HYB2** and **HYBIV2** are memory usages by the respective format in the algorithm run on two GPU (left on the first GPU, right on the second GPU).

Memory usage was checked by the function `cudaMemGetInfo`.

HYBIV format on larger matrices required almost twice less memory than HYB. Memory usage per GPU was smaller in HYB2 and HYBIV2 than in HYB and HYBIV but it was higher than half because vector x was stored on each GPU. Performance of HYBIV was a little better than HYB, because in HYBIV less data is stored in slow global memory. Using 2 GPUs we almost doubled the performance in comparison to single GPU. On smaller matrices HYBIV and splitting data across two GPUs did not bring benefits.

Table 2. Properties of the matrices for the MUTEX model and their memory usage (in MB) on GPU

n	nz	nz/n	uv	**HYB**	**HYB2** gpu1 + gpu2	**HYBIV**	**HYBIV2** gpu1 + gpu2
3 797	47 381	12.5	1 872	**64**	65 + 65	65	65 + 65
63 019	1 049 483	16.6	29 772	74	71 + 70	**68**	68 + 67
4 194 050	96 458 530	23.0	348 710	1166	726 + 694	**630**	450 + 418
8 386 560	201 242 020	24.0	398 635	–	1438 + 1374	**1246**	862 + 798

Table 3. Properties of the matrices for the NCD model and their memory usage (in MB) on GPU

n	nz	nz/n	uv	**HYB**	**HYB2** gpu1 + gpu2	**HYBIV**	**HYBIV2** gpu1 + gpu2
2 667 126	18 480 126	6.93	2 249	276	240 + 220	**179**	187 + 166
15 390 826	107 124 226	6.96	4 048	1295	1090 + 972	**737**	782 + 664
21 084 251	146 835 251	6.96	4 497	1751	1470 + 1309	**987**	1048 + 887
36 361 101	253 442 301	6.97	5 397	–	2490 + 2212	**1657**	1854 + 1712

3.2 Time

All the processing times are reported in seconds. The time is measured with an MKL function `dsecnd`.

Tables 4 and 5 show the time in seconds for double precision SpMV using the HYB and HYBIV storage formats on one GPU and two GPUs and the CSR storage format from the MKL library on CPU. The bolded values denote the fastest computation times.

Table 4. SpMV run-time (in seconds) on CPU (CSR, SpMV from MKL); on one GPU (HYB and HYBIV); on two GPUs (HYB2 and HYBIV2) — MUTEX

n	**CPU-CSR-MKL**	**HYB**	**HYB2**	**HYBIV**	**HYBIV2**
3 797	0.0052	**0.0004**	0.0021	**0.0004**	0.0020
63 019	0.0063	0.0005	0.0005	0.0005	**0.0004**
4 194 050	0.1673	0.0840	0.0425	0.0813	**0.0413**
8 386 560	0.4777	–	0.0854	0.1662	**0.0839**

For bigger Markov matrices, the best format for storing them (counting the performance time and the memory) is HYBIV in its 2-GPU form (that is, HYBIV2). The HYB storage scheme does not perform very well in many potentially suitable cases. The reason is that the HYB SpMV granularity is not sufficient (one thread per row) and the matrix needs to be large enough to utilize the GPU. For comparison in CSR kernel one row is processed by a warp (32 threads).

Table 5. SpMV run-time (in seconds) on CPU (CSR, SpMV from MKL); on one GPU (HYB and HYBIV); on two GPUs (HYB2 and HYBIV2) — NCD

n	CPU-CSR-MKL	HYB	HYB2	HYBIV	HYBIV2
2 667 126	0.0234	0.0059	0.0032	0.0051	**0.0030**
15 390 826	0.1077	0.0369	0.0196	0.0330	**0.0174**
21 084 251	0.1790	0.0512	0.0267	0.0452	**0.0237**
36 361 101	0.2445	–	0.0475	0.0803	**0.0426**

Table 6. SpMV run-time (in seconds) on one GPU (HYBIV) for the NCD model — comparison of *data* storage in constant memory and in texture memory

n	HYBIVconst	HYBIVtex	$\frac{\text{HYBIVtex}}{\text{HYBIVconst}}$
2 667 126	0.0051	0.0052	1.01
15 390 826	0.0330	0.0351	1.06
21 084 251	0.0452	0.0506	1.12
36 361 101	0.0803	0.0914	1.14

Table 6 shows the processing time of the SpMV operation (in seconds) in the HYBIV format on one GPU for two cases:

– *data* stored in the constant memory,
– *data* stored in the texture memory.

This comparison is done only for the NCD model — because the array *data* for the MUTEX model was too big to fit in the constant memory (too many distinct values in matrix Q).

3.3   Performance

Figure 4 shows performance results for double precision SpMV on one GPU and on two GPUs using the CUSP library with the HYB and HYBIV storage formats,

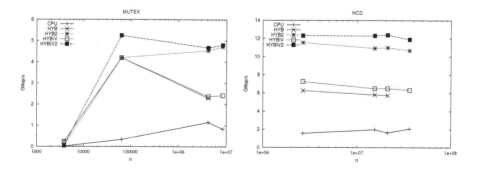

Fig. 4. Performance of SpMV for MUTEX (left) and for NCD (right)

as well as the MKL library with the CSR storage scheme on CPU. In all examined examples we find that SpMV on one GPU gives much better performance than on CPU and SpMV on two GPUs gives two times better performance than on one GPU.

4   Conclusion and Future Work

In this paper we proposed HYBIV — a new modification of an existing sparse storage scheme, namely HYB — to enable storing bigger transition rate matrices. The original format HYB is implemented in the CUSP library and the new format is a plain extension to the old one. Thus, the new format can be implemented very easily, allowing computations with two times bigger matrices.

Large matrices needed almost twice less memory in HYBIV than in HYB. In very sparse matrices with less unique values it is also possible to store data values in constant memory (as in the NCD model), which is, as shown, faster than texture memory. Moreover, for both the models there exist matrices which do not fit in the GPU memory with the use of the HYB format, but they can be placed in the GPU memory when we use HYBIV — without any degradation of the performance.

The sparse matrix-vector multiplication (SpMV) is faster in the HYBIV format than in HYB. Moreover, on two GPUs it is almost twice as fast as on one GPU — irrespective of the matrix and the format. In comparison to CPU computations it can be up to 13 times faster on one GPU and 26 times faster on two GPUs.

For very sparse matrices (NCD) two GPUs perform the best irrespective of storage scheme. For the MUTEX model the results are not so clear. The structure and the size of the matrix affect the performance.

Generally, bigger matrices achieve better performance.

Our format not only requires less memory, but it also speeds up the computations and can be used for more than one GPU. The numerical experiments done for transition rate matrices from two classical Markovian models confirms usefulness of the proposed HYBIV storage scheme — at least for one GPU and two GPUs.

Furthermore, the HYBIV format can be applied not only to matrices arising from Markov chains, but also to other sparse problems — for example, to the finite element method (FEM).

Acknowledgments. This work was partially supported within the project N N516 479640 of the Ministry of Science and Higher Education (MNiSW) of the Polish Republic "Modele dynamiki transmisji, sterowania zatłoczeniem i jakością usług w Internecie".

References

1. Bell, N., Garland, M.: Efficient Sparse Matrix-Vector Multiplication on CUDA. NVIDIA Technical Report No. NVR-2008-004 (2008)
2. Bianchi, G.: Performance analysis of the IEEE 802.11 distributed coordination function. IEEE J. Sel. Areas Commun. **18**, 535–547 (2000)
3. Bustamam, A., Burrage, K., Hamilton, N.A.: Fast parallel Markov clustering in bioinformatics using massively parallel computing on GPU with CUDA and ELLPACK-R sparse format. IEEE/ACM Trans. Comput. Biol. Bioinform. **9**, 679–692 (2012)
4. Bylina, B., Bylina, J., Karwacki, M.: Computational aspects of GPU-accelerated sparse matrix-vector multiplication for solving Markov models. Theor. Appl. Inform. **23**, 127–145 (2011)
5. Bylina, B., Karwacki, M., Bylina, J.: A CPU-GPU hybrid approach to the uniformization method for solving Markovian models – a case study of a wireless network. In: Kwiecień, A., Gaj, P., Stera, P. (eds.) CN 2012. CCIS, vol. 291, pp. 401–410. Springer, Heidelberg (2012)
6. Bylina, B., Karwacki, M., Bylina, J.: Multi-GPU implementation of the uniformization method for solving Markov models. In: Proceedings of Federated Conference on Computer Science and Information Systems (FedCSIS), pp. 533–537 (2012)
7. Bylina, J., Bylina, B.: A Markovian queuing model of a WLAN node. In: Kwiecień, A., Gaj, P., Stera, P. (eds.) CN 2011. CCIS, vol. 160, pp. 80–86. Springer, Heidelberg (2011)
8. Bylina, J., Bylina, B., Karwacki, M.: A Markovian model of a network of two wireless devices. In: Kwiecień, A., Gaj, P., Stera, P. (eds.) CN 2012. CCIS, vol. 291, pp. 411–420. Springer, Heidelberg (2012)
9. Fernandes, P., Plateau, B., Stewart, W.J.: Efficient descriptor-vector multiplication in stochastic automata networks. J. ACM **45**, 381–414 (1998)
10. Philippe, B., Saad, Y., Stewart, W.J.: Numerical methods in Markov chain modelling. Oper. Res. **40**, 1156–1179 (1992)
11. Stewart, W.J.: Introduction to the Numerical Solution of Markov Chains. Princeton University Press, Princeton (1994)
12. Wijs, A.J., Bošnački, D.: Improving GPU sparse matrix-vector multiplication for probabilistic model checking. In: Donaldson, A., Parker, D. (eds.) SPIN 2012. LNCS, vol. 7385, pp. 98–116. Springer, Heidelberg (2012)
13. CUBLAS. https://developer.nvidia.com/cublas
14. CUSP. http://code.google.com/p/cusp-library/
15. ELLPACK. http://www.cs.purdue.edu/ellpack/
16. Intel Math Kernel Library. http://software.intel.com/en-us/articles/intel-mkl/
17. MAGMA. http://icl.cs.utk.edu/magma/index.html
18. Stewart, W.J.: MARCA_Models: a collection of Markov chain models. http://www4.ncsu.edu/~billy/MARCA_Models/MARCA_Models.html
19. Thrust. http://code.google.com/p/thrust/

Eigen-G: GPU-Based Eigenvalue Solver for Real-Symmetric Dense Matrices

Toshiyuki Imamura[1,3]($\boxtimes$), Susumu Yamada[2,3], and Masahiko Machida[2,3]

[1] RIKEN Advanced Institute for Computational Science,
7-1-26 Minatojima-minami-machi, Chuo-ku, Kobe-shi, Hyogo 650-0047, Japan
imamura.toshiyuki@riken.jp
[2] CCSE, Japan Atomic Energy Agency, Kashiwa-shi, Chiba 277-8587, Japan
[3] CREST, Japan Science and Technology Agency, Tokyo, Japan

Abstract. This paper reports the performance of Eigen-G, which is a GPU-based eigenvalue solver for real-symmetric matrices. We confirmed that Eigen-G outperforms state-of-the-art GPU-based eigensolvers such as magma_dsyevd and magma_dsyevd_2stage implemented in the MAGMA version 1.4.0. Applying the best-tuned CUDA BLAS libraries and the GPU-CPU hybrid DGEMM yields an even better performance improvement. We observe an approximately 2.3 times speedup over magma_dsyevd on a Tesla K20c.

Keywords: Eigenvalue solver · GPGPU · CPU-GPU collaborative model

1 Introduction

The theoretical computational performance of a flagship GPU card such as an NVIDIA GeForce TITAN or an AMD Radeon HD8000 series card has exceeded 1 TFLOPS for not only single-precision but also double-precision floating point arithmetic. What is more, GPU cluster systems such as Titan and Tianhe-1A take first place in the 40-th and 36-th editions of the top 500 list of the world's most powerful supercomputers, respectively. This is evidence that GPGPU technology has already been popularized.

In the field of numerical linear algebra, GPGPU has big advantages due to the hardware features of GPUs, especially *higher computing power* and *wider memory bandwidth*, as well as *low power consumption*. Particularly, the severe problem that matrix–vector multiplication is adversely affected by the poor memory bandwidth of CPUs can be resolved by the utilization of a GPU. Furthermore, the computation of matrix–matrix products, which is often used for performance benchmarking of systems, can be performed very efficiently on GPUs. In fact, the performance of a matrix–matrix product routine is a couple of times faster than the total performance of a single multicore processor.

The main purpose of the present study is to develop the GPU-based eigenvalue solver Eigen-G in order to take advantage of these points, namely, the wider

R. Wyrzykowski et al. (Eds.): PPAM 2013, Part I, LNCS 8384, pp. 673–682, 2014.
DOI: 10.1007/978-3-642-55224-3_63, © Springer-Verlag Berlin Heidelberg 2014

Block Householder tridiagonalization

$\{\boxed{A} \leftarrow A\}$ // host to device
for $j = n, \ldots, 1$ **step** $-m$
  (0) $U \leftarrow \emptyset, V \leftarrow \emptyset, W \leftarrow \boxed{A}_{(:,j-m+1:j)}$ // device to host
  for $k = 0, \ldots, m-1$
    (1) Compute a Householder reflector:
      $(\beta, u^{(k)}) := H(W_{(:,j-k)})$
      $\{\boxed{u^{(k)}} \leftarrow \hat{u}^{(k)}\}$ // host to device
    (2) Matrix-Vector multiplication
      $\boxed{\hat{v}^{(k)}} \leftarrow \boxed{A}_{1:j-k-1,1:j-k-1} \boxed{u^{(k)}}$
      $\{\hat{v}^{(k)} \leftarrow \boxed{\hat{v}^{(k)}}\}$ // device to host
    (3) $\tilde{v}^{(k)} \leftarrow \hat{v}^{(k)} - (UV^T + VU^T)u^{(k)}$
    (4) $v^{(k)} \leftarrow \beta(\tilde{v}^{(k)} - su^{(k)}), s = \frac{1}{2}\beta(u^{(k)}, \tilde{v}^{(k)})$
      $\{\boxed{v^{(k)}} \leftarrow \tilde{v}^{(k)}\}$ // host to device
      $\boxed{U} \leftarrow [\boxed{U}, \boxed{u^{(k)}}], \boxed{V} \leftarrow [\boxed{V}, \boxed{v^{(k)}}],$
    (5) $W \leftarrow W - (u^{(k)}v^{(k)T} + v^{(k)}u^{(k)T})$ for [Row,Col]=[:, $j - k : j$]
  endfor
  $A_{(:,j-m+1:j)} \leftarrow W$
  (6) $2m$ rank-update
    $\boxed{A} \leftarrow \boxed{A} - (\boxed{U}\boxed{V}^T + \boxed{V}\boxed{U}^T)$ for [Row,Col]=[1 : $j - m$, 1 : $j - m$]
endfor

⬅ Matrices enclosed by boxes are on GPU.
⬇

Block Householder backtransformation

$\{\boxed{X} \leftarrow X_{\bullet,1:M}\}$ // host to device
for $j = 1, \ldots, n$ **step** m
  (0) $U \leftarrow A_{(\bullet,j:j+m-1)}$
  (1) Compute G:
    $G \leftarrow \text{diag}^{-1}(\beta_j, \ldots, \beta_{j+m-1}) + \text{tril}(\hat{G}, -1). \hat{G} \leftarrow U^T U$
  (2) Compute W and Y:
    $W \leftarrow UC(= UG^{-1})$
    $\{\boxed{U} \leftarrow U$ and $\boxed{W} \leftarrow W\}$ // host to device
    $\boxed{Y}, Y_{m+1:n} \leftarrow [\boxed{U^T}\boxed{X}], UX_{\bullet,m+1:n}]$
  (3) Update X:
    $[\boxed{X}, X_{\bullet,m+1:n}] \leftarrow [\boxed{X}, X_{\bullet,m+1:n}] - [\boxed{W}\boxed{Y}], WY_{\bullet,m+1:n}]$
endfor
$\{X_{\bullet,1:M} \leftarrow \boxed{X}\}$ // device to host

Cuppen's Divide and Conquer

1. Divide a tridiagonal matrix into two parts with a rank-one perturbation as
$$T = T_1 \oplus T_2 + \epsilon vv^T.$$

2. Solve SEV for the subproblems $[D_1, Q_1] = \text{eig}(T_1)$ and $[D_2, Q_2] = \text{eig}(T_2)$.

3. Rewrite the original SEV as
$$D + \epsilon uu^T,$$
where $D = D_1 \oplus D_2$ and $u = (Q_1^T \oplus Q_2^T)v$.

4. Deflate the eigenmodes if necessary from the updated SEV.

5. Solve the SEV via solving a secular equation
$$1 + \epsilon \sum_k v_k^2/(d_k - \lambda_i) = 0$$
$$q_i = \text{normalize}((D - \lambda_i I)^{-1}u) \quad \text{for each } i.$$

6. Compute $\Lambda = \text{diag}(\lambda_1, \lambda_2, \ldots)$ and $Q = (Q_1 \oplus Q_2)[q_1, q_2, \ldots]$.

7. Return $[\Lambda, Q]$.

Fig. 1. Details of the standard three-step algorithm for SEV

memory bandwidth and higher performance. The code for Eigen-G has been extensively optimized for CUDA GPGPU. We adopt the collaborative model to execute GPU kernel codes and introduce a supervisor thread to establish concurrent processing on both a GPU and a CPU. Furthermore, we select the best-tuned CUDA BLAS in order to improve the performance of Eigen-G.

The rest of this paper is organized as follows. In Sect. 2, we briefly present two eigenvalue solvers, MAGMA and Eigen-G. In Sect. 3, details of the GPU implementation are described. In Sect. 4, preliminary benchmark results on three types of GPUs are described. In Sect. 5, summaries are presented and future research is discussed.

2   Eigenvalue Solver on a Single GPU

The standard algorithm of the standard eigenvalue computation (SEV) for a real symmetric dense matrix is divided into three steps. The first step and the second step are *the Householder tridiagonalization* and the eigenvalue computation for the tridiagonal format, respectively. In the present study, *Cuppen's divide and conquer algorithm* is employed. The last step is *back-transformation* of the computed eigenvectors to the original matrix. Figure 1 presents the details of the steps enhanced for GPGPU.

2.1   MAGMA

MAGMA [1] is an enhancement of LAPACK for multiple GPUs and multicore processors. MAGMA is implemented by introducing the effective implementation of native BLAS codes and task scheduling onto heterogeneous/hybrid architectures. The eigenvalue driver routine of the MAGMA library is `magma_dsyevd`, which consists of three routines corresponding to the three steps of the above-mentioned algorithm, `magma_dsytrd` [2], `magma_dstedx` [3], and `magma_dormtr`. These subroutines call GPU kernel functions of MAGMABLAS such as `magmablas_dgemm` and `magmablas_dsymv`. Furthermore, the MAGMA library is designed to reduce the cost of data transfer by a DAG-based tool.

2.2   Eigen-G

Eigen-G is a newly developed eigensolver for GPGPU. The numerical algorithm adopted in Eigen-G is similar to that of `magma_dsyevd`, a three-step scheme comparable to Householder tridiagonalization, Cuppen's divide and conquer, and back-transformation. We modified the source code of Eigen-s [4,5], which was developed for peta-scale supercomputer systems. Details of the design and the CUDA implementation are described in the next section.

3   CUDA Implementation

3.1   Offload Execution Model

In the present study, we *offload* particular linear algebraic operations by CUDA BLAS calls, DGEMM and DSYMV, which are the most time-consuming parts in Eigen-G. There are several offload execution models for GPGPU. We consider the following typical models (Fig. 2);

1. *offload model* is a simple GPU-call mechanism. In this model, CPU cores wait until the GPU kernel completes. This model uses only one-side resource.
2. In the *collaborative model*, the supervisor thread on the CPU is dedicated to handling the GPU at the expense of a single core of the CPU, and the other threads run on the rest of cores. Thus, both CPU and GPU work simultaneously. Furthermore, asynchronous data transfer is performed behind the computation.

In order to utilize both the CPU and GPU efficiently, we mainly adopt the collaborative model to implement Eigen-G.d

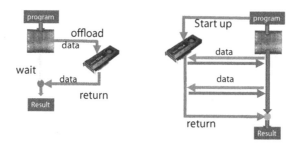

Fig. 2. Typical execution model for GPGPU: offload model (left) and collaborative model (right)

3.2 Thunking Mode

The simplest way to offload the CUDA BLAS routines is to introduce a *thunking mode wrapper*. Generally, thunking mode wrappers remap the internal data layout automatically, and they allow the API to be used without any changes to the application. In the case of the CUDA programming model, an extra array in the device memory is allocated and data transfer between host and device memory is issued automatically by the thunking wrapper.

3.3 Best-Tuned CUDA BLAS Libraries

In the present study, it is obvious that the performance of Eigen-G depends strongly on the performance of CUDA BLAS libraries. The most successful CUDA BLAS libraries are CUBLAS, MAGMABLAS [1], ASPEN.K2 [6], and etc. We must select the best-tuned CUDA BLAS routines from CUBLAS, MAGMA-BLAS, and ASPEN.K2 in terms of the performance of DGEMM and DSYMV. Table 1 shows the performance of DGEMM and DSYMV on a Tesla K20c, a GeForce GTX580, and a Tesla C2050. The results suggest that we should select CUBLAS and ASPEN.K2 for DGEMM and DSYMV, respectively.

3.4 GPU+CPU Hybrid DGEMM

The DGEMM call can be divided using a block style such as $AX = A[X_1, X_2] = [AX_1, AX_2]$. In this representation, we consider a hybrid task assignment so that AX_1 and AX_2 are mapped onto a CPU and a GPU, respectively.

We assume a simple cost model of DGEMM as follows:

$$T = t_{\text{other}} + \max\left\{t_{\text{GPU}} \cdot \frac{M}{N}, \; t_{\text{CPU}} \cdot \frac{N - M}{N}\right\}. \tag{1}$$

Here, M refers to the number of vectors computed on the GPU, and $t_{\text{CPU}} = 2N^3/\text{FLOPS}_{\text{CPU}}$ and $t_{\text{GPU}} = 2N^3/\text{FLOPS}_{\text{GPU}}$ hold. The optimal solution is obtained when $t_{\text{GPU}} \cdot M/N = t_{\text{CPU}} \cdot (N - M)/N$ holds. Thus,

$$\frac{M^{\text{opt}}}{N} = \frac{t_{\text{CPU}}}{t_{\text{CPU}} + t_{\text{GPU}}} = \frac{(\text{FLOPS}_{\text{CPU}})^{-1}}{(\text{FLOPS}_{\text{CPU}})^{-1} + (\text{FLOPS}_{\text{GPU}})^{-1}} \tag{2}$$

Table 1. Performance (GFLOPS) of the DGEMM-NN kernel in the case of square matrices ($N = M = L$) (top) and the DSYMV-U kernel (bottom) on a Tesla K20c, a GeForce GTX580, and a Tesla C2050. In addition, results with a CPU (Corei7-3930K) is presented.

		Matrix dimension								
		1088	2112	3136	4160	5184	6208	7232	8256	9280
K20c	MAGMABLAS 1.4	554.3	593.3	597.2	590.8	587.6	584.0	585.8	586.5	587.0
	CUBLAS 5.5	901.5	1016	1015	1035	1039	1044	1044	1045	1046
GTX580	MAGMABLAS 1.4	178.2	186.3	187.8	188.4	188.6	188.8	188.9	—	—
	CUBLAS 5.0	189.4	192.0	192.4	192.6	192.6	192.6	192.6	—	—
C2050	MAGMABLAS 1.4	283.2	292.7	293.8	294.9	295.1	295.1	295.3	295.5	295.7
	CUBLAS 5.0	304.6	312.7	314.9	315.0	308.3	306.6	304.4	289.3	300.8
CPU	MKL 11.0	98.34	28.95	143.8	147.5	149.4	150.0	150.2	151.7	151.4

		Matrix dimension								
		1088	2112	3136	4160	5184	6208	7232	8256	9280
	CUBLAS 5.5	9.0	13.0	14.8	15.9	16.6	17.0	17.4	17.7	17.8
K20c	MAGMABLAS 1.4	10.4	21.9	31.6	34.6	37.0	40.0	42.3	44.0	45.2
	ASPEN.K2	24.8	36.6	44.4	50.0	53.2	54.7	56.2	56.7	57.0
	CUBLAS 5.0	12.6	17.9	21.1	21.9	22.8	23.7	24.1	23.9	24.5
GTX580	MAGMABLAS 1.4	17.4	31.0	36.4	39.4	43.3	45.5	46.8	46.3	48.9
	ASPEN.K2	29.3	39.5	50.3	56.9	61.5	63.7	66.2	67.3	68.8
	CUBLAS 5.0	9.1	13.4	14.5	14.8	15.7	16.3	16.3	15.8	16.6
C2050	MAGMABLAS 1.4	13.0	23.3	26.9	28.9	31.2	32.8	33.5	32.3	34.9
	ASPEN.K2	18.1	26.2	31.9	35.2	37.2	38.4	39.5	39.7	40.7
CPU	MKL 11.0	0.46	14.8	7.68	7.80	7.71	7.81	7.52	7.03	7.25

minimizes the computational time as

$$T^{\text{opt}} = t_{\text{other}} + \frac{2N^3}{\text{FLOPS}_{\text{CPU}} + \text{FLOPS}_{\text{GPU}}}. \qquad (3)$$

This implies that we can utilize both the CPU and GPU almost perfectly by adopting the optimal task assignment. We benchmark a DGEMM code and use the average FLOPS values; FLOPS_(Core i7-3930(K), Core i7-860, K20c, GTX580, C2050) = (100G, 30G, 1000G, 180G, 290G).

3.5   Other Issues of a CUDA Implementation

Householder Tridiagonalization is designed to reduce the load of data transfer in eigensolvers. As described previously, we introduce a supervisor thread dedicated to handling a GPU. The supervisor thread watches data transfers and exclusive execution control of the GPU. This enables us to take advantage of both the CPU and GPU concurrently by concealing data transfer between host and device behind the computation.

Let us explain in more detail about the data transfer. The matrix and vector data enclosed by boxes in Fig. 1 are stored on the device memory. Matrix data A is transferred from host memory to device memory in advance of entering the main iteration. The panel data W is transferred from the device to the host at step (0). After the calculation of u in step (1), the data u is transferred back to the device. Then DSYMV at step (2) is offloaded to the GPU, and the result v is moved to the host memory. Following steps (3) and (4), v is transferred back to the device behind the computation steps (5) and (1). At step (6), DGEMM is offloaded to the GPU.

The Divide and Conquer Method exhibits very natural parallelism in the internal processes and can also be applied to subproblems recursively, as in line 2 in the bottom left of Fig. 1. The computationally most dominant part of the algorithm is the matrix–matrix multiplication $(Q = (Q_1 \oplus Q_2)[q_1, q_2 \dots])$ at step 6. Here, we will make a GPGPU modification such that we offload the DGEMM call to a GPU card. In the general implementation of LAPACK, a large working array is reused flexibly in internal subroutines. Although a similar working array can be introduced on the GPU and mirrored on the host side, the GPU has a severely limited memory capacity. Therefore, we adopt the thunking call of DGEMM. A small modification is made in `dlaed3()` by replacing the DGEMM calls with CUDA-BLAS calls.

Householder Back-Transformation is the inverse one-sided procedure of the Householder transformation matrices, such as $(I - \beta_k u_k u_k^T) \cdots (I - \beta_2 u_2 u_2^T)(I - \beta_1 u_1 u_1^T)$. The eigenvectors are transformed by multiplying this matrix from the left. Since two consecutive operations of the Householder transformation are written as

$$I - UCU^T = (I - \beta_2 u_2 u_2^T)(I - \beta_1 u_1 u_1^T) \tag{4}$$

($U = [u_1, u_2]$), the compact WY representation is applied successively. In the general case, we compute C and update X with $U = [u_1, \dots, u_k]$ by the following implicit scheme (the notation `tril` is borrowed from the Matlab script language):

$$C = (\text{diag}^{-1}(\beta_1, \dots, \beta_k) - \text{tril}(U^T U, -1))^{-1} \tag{5}$$
$$= (D^{-1} - S)^{-1} = (I - DS)^{-1} D. \tag{6}$$

These calculations comprise only the multiple matrix–matrix multiplications. Therefore, as described in the previous subsection, GPU+CPU hybrid DGEMM performs well and we expect to take advantage of both the CPU and GPU.

4 Preliminary Benchmark Results for Single-GPU Environments

4.1 Configuration of Hardware and Software

We benchmarked Eigen-G on three [Intel-CPU]+[NVIDIA-GPU] platforms: (i) a Core i7-3930 (3.2 GHz, 6 cores, AVX) + a Tesla K20c (1.17 TFLOPS in the

DP mode, 2496 CUDA cores), (ii) a Core i7-3930K (3.2 GHz, 6 cores, AVX) + a GeForce GTX580 (790 GFLOPS in the DP mode, 512 CUDA cores), and (iii) a Core i7-860 (2.8 GHz, 4 cores, SSE3) + a Tesla C2050 (515 GFLOPS in the DP mode, 448 CUDA cores).

On the host side (denoted by CPU, hereinafter), thread parallelization is carried out by OpenMP and 6 or 4 threads perform on a CPU where one thread is assigned to the supervisor thread. On the other hand, GPU codes are written in CUDA 5.0 and 5.5. Appropriate CUDA BLAS libraries already mentioned above run on a GPU. The BLAS and LAPACK libraries installed on the host computers are Intel MKL 11.0, and the latest MAGMA library 1.4.0 is employed on GPUs.

4.2 Results and Discussions

Table 2 presents the elapsed time and the time breakdown for each step in the eigenvalue computation on a Tesla K20c, a GeForce GTX580 and a Tesla C2050. In the second column, 'tridi', 'D&C', and 'back' refer to the Householder tridiagonalization, divide and conquer, and the back-transformation, respectively. We summarize the benchmark results as follows.

1. We see a 1.25 to 3 times speedup over LAPACK.
2. Eigen-G outperforms MAGMA version 1.4.0 in terms of the total time and the first step of the Householder tridiagonalization, seen especially in x2.3 on a Tesla K20c when the matrix dimension is 8256.
3. The results on the second and the third steps look quite similar on a Tesla K20c and a GeForce GTX580. However, MAGMA performs 20 % faster on a Tesla C2050.

Figure 3 presents an additional performance result of the brand new MAGMA 2-stage solver (magma_dsyevd_2stage) on a Tesla K20c and a GeForce GTX580. The 2-stage algorithm is known as one of the prominent algorithms to resolve the severe problem of the Householder transformation mentioned in the introduction. The remarkable point is that Eigen-G and the 2-stage solver perform comparably on K20c. Since it is known that the 2-stage algorithm has a big overhead in the 2-stage back-transformation (it is equivalent to twice the cost of the 1-stage back-transformation algorithm), Eigen-G performs comparably while still adopting the 1-stage algorithm. This result suggests that the 1-stage algorithm is still powerful when we compute all the eigenpairs.

Figure 4 shows the hybrid performance of the back-transformation by varying M/N, which is the ratio of the number of vectors handled by the GPU to the total number. As shown in Table 1 and the theoretical peak performance, the optimal performance is seen at a high M/N ratio on a Tesla GPU, whereas the GeForce GPUs perform at the same speed as the CPUs. Taking the relation (2) into consideration, a great performance improvement is achieved in the hybrid computation. The relation (3) also suggests that this is especially true when

Table 2. Elapsed time and time-breakdown (in units of seconds) of Eigen-G, MAGMA (magma_dsyevd), and LAPACK (dsyevd) for each step on an NVIDIA Tesla K20c (top), an NVIDIA GeForce GTX580 (middle), and an NVIDIA Tesla C2050 (bottom).

			Matrix dimension							
			1088	2112	3136	4160	5184	6208	7232	8256
K20c	Eiegn-G $M/N = 29/32$	tridi	0.10	0.35	0.82	1.56	2.67	4.16	6.22	8.83
		D&C	0.04	0.15	0.34	0.64	0.74	1.07	1.48	1.98
		back	0.02	0.06	0.15	0.28	0.47	0.72	1.07	1.52
		total	0.15	0.56	1.32	2.48	3.88	5.96	8.77	12.33
	MAGMA 1stage	tridi	0.04	0.58	1.63	3.51	6.50	10.79	16.68	24.49
		D&C	0.04	0.13	0.28	0.48	0.77	1.13	1.57	2.07
		back	0.02	0.07	0.13	0.28	0.50	0.78	1.20	1.65
		total	0.10	0.79	2.07	4.31	7.82	12.78	19.57	28.36
	LAPACK	tridi	0.04	0.28	1.14	2.96	6.16	10.69	17.30	26.68
		D&C	0.09	0.37	1.84	1.67	2.82	4.40	6.36	8.88
		back	0.03	0.17	0.54	1.20	2.28	3.91	6.03	9.10
		total	0.16	0.84	3.56	5.88	11.35	19.13	29.86	44.89
GTX580	Eiegn-G $M/N = 20/32$	tridi	0.09	0.35	0.88	1.72	2.97	4.74	7.06	10.09
		D&C	0.04	0.17	0.41	0.79	1.24	1.92	2.82	3.95
		back	0.03	0.11	0.30	0.64	1.17	1.91	2.94	4.31
		total	0.15	0.63	1.59	3.16	5.38	8.57	12.82	18.35
	MAGMA 1stage	tridi	0.04	0.48	1.35	3.00	5.32	8.80	13.58	20.77
		D&C	0.04	0.16	0.39	0.77	1.31	2.05	3.04	4.27
		back	0.03	0.14	0.42	0.89	1.67	2.80	4.34	6.37
		total	0.11	0.79	2.19	4.69	8.36	13.73	21.07	31.56
	LAPACK	tridi	0.04	0.28	1.02	2.62	5.97	9.45	15.39	24.33
		D&C	0.08	0.35	0.99	1.57	2.82	4.04	5.97	8.37
		back	0.03	0.17	0.53	1.10	2.29	3.56	5.70	8.64
		total	0.14	0.82	2.58	5.34	11.17	17.16	27.22	41.55
C2050	Eiegn-G $M/N = 29/32$	tridi	0.12	0.50	1.24	2.44	4.25	6.82	10.32	15.03
		D&C	0.07	0.33	0.84	1.68	1.85	2.80	4.00	5.52
		back	0.03	0.14	0.37	0.78	1.41	2.30	3.59	5.29
		total	0.23	0.98	2.45	4.91	7.52	11.93	17.92	25.85
	MAGMA 1stage	tridi	0.08	0.69	1.97	4.35	7.83	13.03	20.23	31.75
		D&C	0.07	0.25	0.56	1.02	1.62	2.43	3.44	4.64
		back	0.02	0.13	0.32	0.65	1.17	1.92	3.05	4.45
		total	0.18	1.07	2.86	6.03	10.63	17.40	26.74	40.87
	LAPACK	tridi	0.07	0.99	3.72	10.23	21.93	35.52	57.06	95.20
		D&C	0.12	0.56	1.49	2.93	5.80	8.29	12.60	18.39
		back	0.07	0.54	1.63	3.54	7.33	11.32	17.94	27.59
		total	0.23	2.10	6.87	16.74	35.13	55.21	87.72	141.34

the performance difference between a CPU and a GPU is small. In fact, the performances on a K20c with the hybrid DGEMM mode and the GPU-oly mode (M/N) are 780 and 710 GFLOPS, respectively. On the other hand, they perform 270 GFLOPS and 175 GFLOPS on a GTX580, respectively.

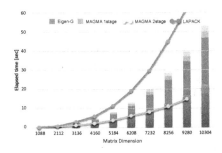
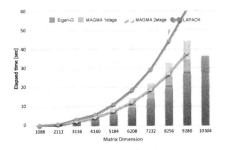

Fig. 3. Elapsed time of Eigen-G, MAGMA 1stage algorithm (`magma_dsyevd`), MAGMA 2-stage algorithm (`magma_dsyevd_2stage`), and LAPACK (`dsyevd`) for each step on an NVIDIA Tesla K20c (left) and an NVIDIA GeForce GTX580 (right).

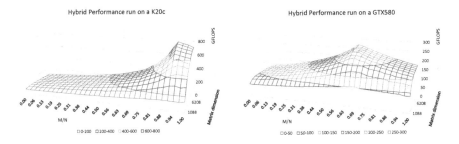

Fig. 4. Effect of the GPU-CPU hybrid scheduling of DGEMM in the backtransformation, K20c (left) and GTX580 (right).

5 Conclusion

In this paper, we have presented our developed GPU-based eigenvalue solver Eigen-G. We adopt the collaborative offload model. Concurrent processing on both the CPU and GPU is realized by introducing a supervisor thread which is dedicated to handling the GPU. In the code optimization, we selected the best-tuned BLAS implementation from CUBLAS, MAGMABLAS, and ASPEN.K2. The GPU-CPU hybrid DGEMM also performs in the present code. The approach yields a good performance improvement, especially for Householder tridiagonalization.

As a more advanced implementation for multiple GPUs has already been reported [7], the present study on a single GPU is limited to small eigenvalue problems. However, since the tuning techniques in the present study are applicable to multiple GPUs, it is indispensable that Eigen-G be developed for multiple GPUs in the near future.

Acknowledgment. This research was supported in part by the Ministry of Education, Scientific Research on Priority Areas, 21013014.

References

1. Agullo, E., Demmel, J., et al.: Numerical linear algebra on emerging architectures: the PLASMA and MAGMA projects. J. Phys. Conf. Ser. **180**(1), 012037 (2009)
2. Tomov, S., Nath, R., Dongarra, J.: Accelerating the reduction to upper Hessenberg, tridiagonal, and bidiagonal forms through hybrid GPU-based computing. Parallel Comput. **36**(12), 645–654 (2010)
3. Vömel, C., Tomov, S., Dongarra, J.: Divide and conquer on hybrid GPU-accelerated multicore systems. SIAM J. Sci. Comput. **34**, 70–82 (2012)
4. Yamada, S., Imamura, T., Kano, T., Machida, M.: High-performance computing for exact numerical approaches to quantum many body problems on the earth simulator. In: Proceedings of the 2006 ACM/IEEE Conference on Supercomputing SC06 (2006)
5. Imamura, T., Yamada, S., Machida, M.: Development of a high performance eigensolver on the peta-scale next generation supercomputer system. Prog. Nucl. Sci. Technol. AESJ **2**, 643–650 (2011)
6. Imamura, T.: ASPEN-K2: automatic-tuning and stabilization for the performance of CUDA BLAS level 2 kernels. In: 15th SIAM Conference on Parallel Processing for Scientific Computing (2012)
7. Yamazaki, I., Dong, T., et al.: Tridiagonalization of a dense symmetric matrix on multiple GPUs and its application to symmetric eigenvalue problems. Concurrency Comput. Pract. Exper. (2013). doi:10.1002/cpe.3152

A Square Block Format for Symmetric Band Matrices

Fred G. Gustavson[1,2], José R. Herrero[3](✉), and Enric Morancho[3]

[1] IBM T.J. Watson Research Center, New York, USA
fg2935@hotmail.com
[2] Umeå University, Umeå, Sweden
[3] Computer Architecture Department, Universitat Politècnica de Catalunya,
BarcelonaTech, Barcelona, Spain
{josepr,enricm}@ac.upc.edu

Abstract. This contribution describes a Square Block, SB, format for
storing a banded symmetric matrix. This is possible by rearranging "in
place" LAPACK Band Layout to become a SB layout: store submatrices
as a set of square blocks. The new format reduces storage space, provides
higher locality of memory accesses, results in regular access patterns, and
exposes parallelism.

Keywords: Upper Square Block Band Format · Banded Cholesky factorization · New data storage format · Locality · Parallelism

1 Introduction

A banded matrix A can be stored as a dense matrix (Fig. 1a). However, this
implies the storage of null elements outside of the band. LAPACK [1] specifies
a format for storing a band matrix using a rectangular array AB. The elements
outside of the band are not stored. We consider the case where the matrix is
symmetric. Thus, we only need to store either the lower or the upper part. In
this paper we consider the former case, i.e. uplo = 'L' (Fig. 1b).

In Dense storage, value $A_{i,j}$ is referenced in the code as $A(i,j)$. Thus, the
j^{th} diagonal element is stored in $A(j,j)$. In LAPACK Lower Band storage, uplo
= 'L', the j-th column of A is stored in the j-th column of AB such that the
diagonal element $A(j,j)$ is stored in $AB(1,j)$. Consequently, in LAPACK lower
band codes $A_{i,j}$ is referenced as $AB_{1+i-j,j}$. This means that the correspondence
is written in the code as $AB(1+i-j,j) = A(i,j)$ for $j <= i <= min(n,j+kd)$.
This makes the code less readable than it could be. Figure 2 highlights the details
of the storage of a *panel*, a set of contiguous columns.

However, when kd, the *half bandwidth*, is small, the difference between dense
and LAPACK storage requirements can be very large, clearly in favor of LAPACK

This work was supported by the Spanish Ministry of Science and Technology
(TIN2012-34557) and the Generalitat de Catalunya, Dep. d'Innovació, Universitats
i Empresa (2009 SGR980).

R. Wyrzykowski et al. (Eds.): PPAM 2013, Part I, LNCS 8384, pp. 683–689, 2014.
DOI: 10.1007/978-3-642-55224-3_64, © Springer-Verlag Berlin Heidelberg 2014

1a. Dense Storage   1b. LAPACK Storage

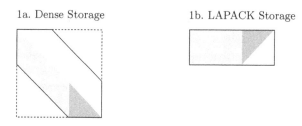

Fig. 1. Band matrix storage in dense (left) and LAPACK (right) formats

2a. Dense Storage   2b. LAPACK Storage

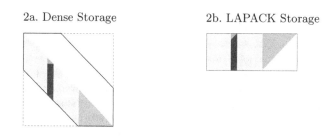

Fig. 2. Storage of a panel within the band in dense (left) and LAPACK (right) formats

storage. For a symmetric matrix of dimension n and half bandwidth kd, dense storage requires the storage of n^2 values. Using LAPACK storage the banded matrix is stored in a rectangle containing $kd+1$ by n values. The rectangle holds a parallelogram P of dimension $kd+1$ by $n-kd$; and an isosceles triangle T of side equal to kd. The rightmost white isosceles triangle seen in Fig. 1b within the rectangle corresponds to data allocated but not used. Clearly, this storage scheme incurs in space overhead, wasting about half the storage allocated to T.

1.1 Goals

Our goals include improving programmability and reducing storage requirements when operating on banded symmetric matrices. At the same time, we improve data locality and make parallelization more efficient. To do so, we are willing to rearrange the data so that:

- Space requirements are close to the optimum
- No further data copies or transformations are necessary at computation time
- Data management is more efficient
- Parallelization is easier and more efficient.

1.2 Related Work

Improved Programmability. In [2] the authors describe a minor data format change for storing a symmetric band matrix AB using the same array space specified by LAPACK [1]. In LAPACK lower band codes $AB_{i,j}$ is referenced in its code as $AB_{i-j+1,j}$. This makes the code less readable than it could be as one would like to reference the (i, j) element of a matrix AB as $AB_{i,j}$. Furthermore, the layout of lower AB in the LAPACK's user Guide, page 142 of [1] shows the user a rectangular matrix with the diagonal of AB residing in the first row. Clearly, a layout description where the diagonal of AB resides on the main diagonal of AB, see again page 142 of [1], is more suggestive and other things being equal is preferable. In [2] the authors improve Programmability of LAPACK Lower Band Cholesky by changing the Leading Dimension of AB from $LDAB$ to $LDAB - 1$ in the array declaration of AB. This tells the compiler that the distance in the 2nd dimension is one less. As a result one can write $AB(i, j)$ to access value $A(i, j)$. Also, in the layout description of AB the diagonal of AB is depicted as laying on the diagonal of AB.

Improved Data Locality and Parallelization. In [3] the authors propose Lower Blocked Column Packed and Upper Square Blocked Packed Formats, also known as Lower and Upper Block Packed Format (BPF) respectively. Both versions of BPF are alternatives to the Packed storage of a matrix used traditionally to conserve storage when that matrix has special properties. Two examples are symmetric and triangular matrices. By using BPF we may partition a symmetric matrix where each submatrix block is held contiguously in memory. This gives another way to pack a symmetric matrix and it avoids the data copies, that are inevitable when Level-3 BLAS are applied to matrices held in standard Column Major (CM) or Row Major (RM) format as well as in standard packed format.

3a. Lower Blocked Packed Format   3b. Upper Blocked Packed Format

```
0                                   0  2 | 4  6 | 8 10 |12 14
1  9                                3  5 | 7    | 9 11 |13 15
2 10 |16                              16 18 |20 22 |24 26
3 11 |17 23                              19 |21 23 |25 27
4 12 |18 24 |28                                28 30 |32 34
5 13 |19 25 |29 33                                31 |33 35
6 14 |20 26 |30 34 |36                               36 38
7 15 |21 27 |31 35 |37 39                               39
```

Fig. 3. Lower and Upper Blocked Packed Formats of a triangular matrix

We define *lower* and *upper* BPF via an example in Fig. 3 with varying length rectangles of width $nb = 2$ and SB of order $nb = 2$ superimposed. Figure 3 gives the memory addresses of the array that holds the matrix elements of BPF. The rectangles making up the array of Fig. 3 are in standard Fortran format and hence BPF supports calls to level-3 BLAS. The rectangles in Fig. 3a are *not* further divided into SB as these SB are *not* contiguous. Figure 3 is a collection of $N = \lceil n/nb \rceil$ rectangular matrices concatenated together. The rectangles in Fig. 3b are the transposes of the rectangles in Fig. 3a and vice versa. Figure 3b rectangles have a *major* advantage over the rectangles of Fig. 3a: the ith rectangle consists of $N - i$ order nb SB. This gives two dimensional contiguous granularity for _GEMM calls using upper BPF which lower BPF *cannot* possess. Lower BPF is *not* a preferred format over upper BPF as it does not give rise to contiguous SB. Another advantage of using upper BPF is one may at factor stage i call _GEMM $(N - i - 1)(i - 1)$ times where each call is a parallel SB _GEMM update. This approach was used by LAPACK multicore Cholesky implementations [4,5] among others. This implies that a BPF layout supports both traditional and multicore LAPACK implementations. Upper BPF is the preferred format. For further details see [3] and the references therein.

2   Upper Square Block Band Format

We could store a band matrix using Upper BPF (see Fig. 4). If we did so, we would be unnecessarily storing elements marked with * in the figure.

$$
\begin{array}{cc|cc|cc|cc}
0 & 2 & 4 & 6 & * & * & * & * \\
* & 3 & 5 & 7 & 9 & * & * & * \\
\cline{1-2}
 & & 16 & 18 & 20 & 22 & * & * \\
 & & * & 19 & 21 & 23 & 25 & * \\
\cline{3-4}
 & & & & 28 & 30 & 32 & 34 \\
 & & & & * & 31 & 33 & 35 \\
\cline{5-6}
 & & & & & & 36 & 38 \\
 & & & & & & * & 39 \\
\end{array}
$$

Fig. 4. A band matrix stored in Upper Blocked Packed Format

Trying to reduce the unused space we can avoid storing those SB which only store null elements outside of the band. With this we could reduce the storage considerably. In the example in Fig. 4 we could avoid the storage of the top rightmost SB. However, we want to reduce further the storage of the part of the matrix which stores the parallelogram P. Let us observe in Fig. 4 the blocks which keep the boundaries of the band. Blocks which keep the main diagonal store a lower triangle L which is not used. For instance, for the blocks of size $nb = 2$ in Fig. 5 we can observe that memory address 1 is not used. Similarly, blocks which include the outermost diagonal store an upper triangle U which is not used at all. We can only observe that memory address 9 is the only one being used within the block that contains it. From this observations we conclude that we could save space by storing in address 1 the value originally hold in address 9. The same could be done with the value in memory address 25 which could be stored in the unused space in address 17. If we did so, the blocks containing the values in memory addresses 9 and 25 would not be needed. In general, in each block row holding a slab of P we could avoid storing the lower triangle originally stored in the rightmost non-null block by storing it in the unused space corresponding to the lower triangle of the leftmost block in that slab. We refer to this new format as Upper Square Block Band Format (USBBF). The final triangular part T can just be stored in Upper BPF, which is compatible with USBBF.

$$
\begin{array}{cc|cc|cc|cc}
0 & 2 & 4 & 6 & * & * & * & * \\
1 & 3 & 5 & 7 & 9 & * & * & * \\
\cline{3-8}
 & & 16 & 18 & 20 & 22 & * & * \\
 & & 17 & 19 & 21 & 23 & 25 & * \\
\cline{5-8}
 & & & & 28 & 30 & 32 & 34 \\
 & & & & * & 31 & 33 & 35 \\
\cline{7-8}
 & & & & & & 36 & 38 \\
 & & & & & & * & 39 \\
\end{array}
$$

Fig. 5. A band matrix stored in Upper Blocked Packed Format

2.1 Data Transformation Process

Figure 6[1] shows graphically the transformation process from a panel within the band (P part) stored in LAPACK format into a slab in USBBF. The panel needs to be transposed and the bordering triangles joined in a single block.

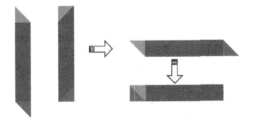

Fig. 6. From LAPACK Lower Band Format into Upper Square Block Band Format

It is possible to perform these data transformations fast in-place based on the work published in [6] and [7]. The process implies partitioning the matrix into submatrices and transposing them. This is achieved with a series of Shuffle/Unshuffle, and Transposition operations described in [7].

2.2 Final Layout

The new layout for uplo = 'L' consists of two geometric figures; a parallelogram P and a lower isosceles triangle T of side equal to kd. P and T must be stored in compatible formats:

---

[1] Readers can get a color version of the figures via email to the authors.

– P is stored in Upper Square Block Band Format (USBBF)
– T is stored in Upper Square Block Packed Format (Upper BPF).

The final layout stores P and T as shown in Fig. 7.

Fig. 7. Final Layout of a Band Matrix transformed to USBBF.

We must note that kd is arbitrary while nb is not. This means the boundary between the band ending and the blocked T beginning is not necessarily on a multiple of nb as Fig. 7 suggests. This issue can be handled in general. However, for clarity of presentation we make the simplifying assumption that $kd + 1$ is a multiple of nb. Then Fig. 7 is accurate. This also eases the programming effort.

The parallelogram is partitioned into slabs of width nb. Each slab of P is also a parallelogram P_i of size $kd + 1$ by nb. P_i consists of two isosceles triangles of sizes nb and $nb - 1$ and a rectangle R_i of size $kd + 1 - nb$ by nb. Now the two triangles concatenate to form a SB of order nb. Hence, P_i also consists of just a SB and R_i. By transposing R_i in-place R_i becomes $\lfloor (kd + 1)/nb - 1 \rfloor$ SB's plus a leftover rectangular block. We note that transposing AB gives an uplo = 'U' LAPACK implementation starting from the uplo = 'L' implementation. Thus, to get our SB formulation we follow this procedure. Triangle T now becomes an upper isosceles triangle. We also map T into upper blocked packed format [3] so it becomes "compatible" with the transposed parallelogram P.

The band in P can be stored with minimal storage. Using full format to store the final triangle T as in LAPACK requires that LDA $\geq KD + 1$. Clearly, this wastes about half the storage allocated by Fortran or C to T. On the other hand, for each SB, LDA $= nb$. This means *minimal* storage is wasted for large KD when T is stored in Upper BPF. Therefore, this implies space savings w.r.t. LAPACK band storage.

3   Ongoing Work

We are currently implementing an optimized parallel band Cholesky factorization based on USBBF. As we have shown in this paper, the new format stores submatrices as a set of square blocks. This provides higher locality of memory accesses, results in regular access patterns, and exposes parallelism. Consequently, this allows for efficient execution of kernels working on square blocks in parallel:

- No further data copies or transformations are necessary at computation time;
- Data off-loading is more efficient;
- Can use regular BLAS or LAPACK codes, or Specialized kernels [8,9];
- Can be parallelized more easily with Dynamic Task Scheduling based on a Task Dependency Graph [10].

4   Conclusions

The new Upper Square Block Band Format (USBBF) stores submatrices as a set of square blocks. This reduces storage space, provides higher locality of memory accesses, results in regular access patterns, and exposes parallelism. The data transformation can be done very efficiently in-place and in parallel.

References

1. Anderson, E., Bai, Z., Bischof, C., Blackford, L.S., Demmel, J., Dongarra, J.J., Du Croz, J., Greenbaum, A., Hammarling, S., McKenney, A., Sorensen, D.: LAPACK Users' Guide, 3rd edn. Society for Industrial and Applied Mathematics, Philadelphia (1999)
2. Gustavson, F.G., Quintana-Ortí, E.S., Quintana-Ortí, G., Remón, A., Waśniewski, J.: Clearer, simpler and more efficient LAPACK routines for symmetric positive definite band factorization. To appear in: PARA'08. IMM-Technical Report-2008-19. Technical University of Denmark, DTU Informatics, Building 321 (2008)
3. Gustavson, F.G., Waśniewski, J., Dongarra, J.J., Herrero, J.R., Langou, J.: Level-3 Cholesky factorization routines improve performance of many Cholesky algorithms. ACM Trans. Math. Softw. **39**(2), 9:1–9:10 (2013)
4. Kurzak, J., Buttari, A., Dongarra, J.: Solving systems of linear equations on the cell processor using Cholesky factorization. IEEE Trans. Parallel Distrib. Syst. **19**(9), 1175–1186 (2008)
5. Quintana-Ortí, G., Quintana-Ortí, E.S., Remón, A., Geijn, R.A.: An algorithm-by-blocks for supermatrix band Cholesky factorization. In: Palma, J., Amestoy, P.R., Daydé, M., Mattoso, M., Lopes, J. (eds.) VECPAR 2008. LNCS, vol. 5336, pp. 228–239. Springer, Heidelberg (2008)
6. Gustavson, F.G., Karlsson, L., Kågström, B.: Parallel and cache-efficient in-place matrix storage format conversion. ACM TOMS **38**(3), 17:1–17:32 (2012)
7. Gustavson, F.G., Walker, D.W.: Algorithms for in-place matrix transposition. In: Wyrzykowski, R., Dongarra, J., Karczewski, K., Waśniewski, J. (eds.) PPAM 2013, Part II. LNCS, vol. 8385, pp. 105–117. Springer, Heidelberg (2014)
8. Herrero, J.R., Navarro, J.J.: Compiler-optimized kernels: an efficient alternative to hand-coded inner kernels. In: Gavrilova, M.L., Gervasi, O., Kumar, V., Tan, C.J.K., Taniar, D., Laganá, A., Mun, Y., Choo, H. (eds.) ICCSA 2006. LNCS, vol. 3984, pp. 762–771. Springer, Heidelberg (2006)
9. Herrero, J.R.: New data structures for matrices and specialized inner kernels: low overhead for high performance. In: Wyrzykowski, R., Dongarra, J., Karczewski, K., Wasniewski, J. (eds.) PPAM 2007. LNCS, vol. 4967, pp. 659–667. Springer, Heidelberg (2008)
10. Herrero, J.R.: Exposing inner kernels and block storage for fast parallel dense linear algebra codes. To appear in: PARA'08

Workshop on Models, Algorithms, and Methodologies for Hierarchical Parallelism in New HPC Systems

Transparent Application Acceleration by Intelligent Scheduling of Shared Library Calls on Heterogeneous Systems

João Colaço, Adrian Matoga, Aleksandar Ilic, Nuno Roma,
Pedro Tomás, and Ricardo Chaves[⊠]

INESC-ID / IST, Rua Alves Redol, 9, 1000-029 Lisboa, Portugal
`ricardo.chaves@inesc-id.pt`

Abstract. Transparent application acceleration in heterogeneous systems can be performed by automatically intercepting shared libraries calls and by efficiently orchestrating the execution across all processing devices. To fully exploit the available computing power, the intercepted calls must be replaced with faster accelerator-based implementations and intelligent scheduling algorithms must be incorporated. When compared with previous approaches, the framework herein proposed does not only transparently intercepts and redirects the library calls, but it also incorporates state-of-art scheduling algorithms, for both divisible and indivisible applications. When compared with highly optimized implementations for multi-core CPUs (e.g., MKL and FFTW), the obtained experimental results demonstrate that, by applying appropriate light-weight scheduling and load-balancing mechanisms, performance speedups as high as 7.86 (matrix multiplication) and 4.6 (FFT) can be achieved.

Keywords: Transparent acceleration · Heterogeneous computing · Automatic scheduling · Load balancing

1 Introduction

Over the past decade, Graphics Processing Units (GPUs) have evolved into general-purpose computing devices, offering a substantial performance boost for highly parallel applications. Together with other types of accelerators, including high-end Field-Programmable Gate Arrays (FPGAs) or parallel coprocessors (such as Intel Xeon Phi), GPUs have become widely used in the High-Performance Computing (HPC) domain, and their share is still growing.

However, the adoption of these powerful accelerators in heterogeneous environments often requires a substantial amount of work by skilled programmers to identify the parallelizable kernels and optimize them for specific architectures. In contrast, many application developers usually use different software packages or computing environments (such as Matlab, Octave, R, etc.) to perform the computations, mainly focusing on the algorithm correctness and efficiency, and not on optimizing the code for any specific device architecture. In fact, highly

R. Wyrzykowski et al. (Eds.): PPAM 2013, Part I, LNCS 8384, pp. 693–703, 2014.
DOI: 10.1007/978-3-642-55224-3_65, © Springer-Verlag Berlin Heidelberg 2014

optimized computational libraries are often adopted to attain high performance on different devices, such as BLAS, LAPACK and FFTW (for general-purpose CPUs), or cuBLAS, CULA and cuFFT (for GPUs). However, additional speed-ups can be achieved by effectively selecting the most efficient library implementation on a per device basis or even by dividing the work load and simultaneously executing it across multiple devices.

Beisel et al. [1], proposed an interposition scheme to intercept calls to shared libraries and delegate them to one of the existent library implementations, allowing to transparently accelerate applications in heterogeneous systems without changing the original code. To effectively select which library is called at each time, a static scheduling algorithm was used, based on pre-defined performance models. However, this approach does not take into account the performance differences in software implementations, hardware devices or even real-time system usage. Furthermore, to fully exploit the computational power of modern heterogeneous systems, it is absolutely crucial to provide the means for efficient cross-device execution.

In this paper, the idea of transparently accelerating existing applications by replacing the kernels that are implemented as library functions is further investigated. The original idea is extended and further improved by adding intelligence to the system. Two approaches are considered: (a) when the problem cannot be divided due to data/control dependencies, the system selects the best available accelerator; (b) when the problem can be divided into multiple parallel computations, the system automatically assigns each device with different computation portions, thus achieving a collaborative execution. To perform this task, a dynamic load balancing algorithm is adopted that relies on partial estimations of performance models for multiple devices in the heterogeneous system, which are built and updated in real-time. By relying on the proposed framework, it is possible to exploit the full computing capacity of the system, without requiring any special intervention on the system or on the original code by the end-user.

The remaining of this paper is organized as follows. The next section presents the architecture of the proposed framework to transparently accelerate existing applications. Section 3 describes the adaptive scheduling algorithms for both indivisible and divisible load problems. Section 4 presents the experimental results obtained with the adopted transparent acceleration approach and discusses the achieved speed-ups for two case-study applications. The last section concludes and addresses some future work directions.

2   Framework Architecture

The architecture of the developed framework, illustrated in Fig. 1, is based on the model proposed in [1]. Accordingly, the LD_PRELOAD environment variable is used to specify the *wrapper library* which transparently redirects the function execution to one or more of the available *plugins*. An one-to-one mapping between plugins and the underlying devices is assumed.

In the execution environment illustrated in Fig. 1, an application natively using the BLAS library is accelerated using more efficient implementations, such

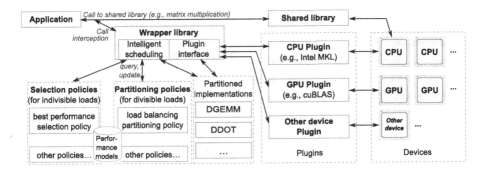

Fig. 1. Architecture of the proposed framework.

as MKL for CPU or cuBLAS for GPU (see *Plugins* module). Other similar examples could also be devised using other widely used libraries, such as the FFTW/cuFFT or the LAPACK/CULA libraries.

2.1 Selection and Partitioning Policies

Upon a call to a library function, the wrapper library first determines the problem size based on the function arguments and classifies the requested computation as divisible or indivisible in the *Intelligent scheduling* module. If the considered function has an implementation that allows work load partitioning across multiple devices, the configured divisible load *partitioning policy* is queried in order to find a balanced distribution of the loads given to each device. If the call is recognized as indivisible, the fastest plugin for the detected problem size is selected using the configured *selection policy.*

Subsequently, the appropriate function is executed by the selected plugin(s) via *Plugin Interface.* For indivisible work loads, the plugin usually only calls an equivalent function of the accelerated library, such as MKL or cuBLAS, translating the arguments and results where appropriate. Divisible work loads require partitioning of the computation according to the previously computed distribution. The partitioning depends on the actual function that was called and must be implemented separately for individual functions. All available plugins are called with a different portion of the computation and the results are combined accordingly. After the execution, the wrapper requests the used policy to update its *performance model* according to the actual measured performance.

Section 3 presents two policies based on run-time updated performance models. However, the proposed framework is not limited to these two strategies, as it allows integrating different selection and partitioning policies.

2.2 Library Generation

The adopted library generation procedure is described in more detail in [6] along with the developed application profiling framework that transparently collects

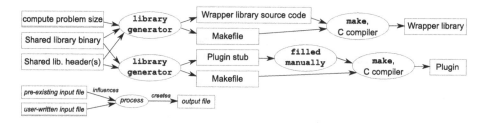

Fig. 2. Generation procedure of the wrapper library and plugins.

extensive profile information (e.g., processor performance counter values and the estimation of the amounts of data to be transferred to and from the accelerator). The approach differs from [1] in what concerns the *micro-generators* concept [3]. Each micro-generator is defined as a class that generates pieces of code related to a specific feature, which are inserted at several distinct scopes in the generated source, such as: global declarations, global definitions, library initialization, library finalization, wrapper function prefix, and wrapper function postfix. Multiple micro-generators can be combined and their options adjusted by the framework user, in order to produce a library dedicated and optimized for a specific purpose. A special kind of micro-generator was considered to allow the programmer to specify external definition files containing portions of code to be inserted at different scopes of the generated library, without modifying the generator itself. This feature is used to insert the code required to estimate the problem size.

The adopted procedure to generate the described wrapper library is presented in Fig. 2. The library generator tool finds and parses the library binary and header files installed in the system and produces the wrapper library incorporating the basic logic that manages the available scheduling policies and plugins. The generated plugin stubs can be used to add support for new accelerators to the system.

Once the above configuration is complete, the programmers and the application users can automatically benefit from the heterogeneous computing resources, without even being aware of their existence.

3   Adaptive Multi-Device Task Scheduling

A selection policy and a partitioning policy are also herein proposed, which use performance models for the available plugins, built at run time. An individual performance model is constructed for each function of each plugin. Each model is stored as an ordered map which associates the problem size N with the performance s expressed as the ratio between the size of computation assigned to a plugin and the time taken to perform computation and communication. The definition of the problem size depends on the function. For example, for the multiplication of two square matrices, the problem size N may be defined as the matrix dimension M or the number of scalar multiplications involved, i.e. M^3.

3.1   Best Performance Selection Policy

In the most general case, each function call is considered as an indivisible work that must be executed by a single plugin. The goal is therefore to choose the fastest plugin for a particular call, based on the problem size. Assuming that the performance characteristics of the available plugins are not known a-priori, the proposed algorithm dynamically builds the performance models during run time. If, for a given problem size and model, the fastest plugin is not yet known, all plugins are simultaneously executed and their performance is recorded. As soon as the collected information is sufficient to determine the best choice for a particular call, only the fastest plugin is executed, and the model is updated with its last achieved performance. This way, the scheduling scheme is not only self-learning, but is also adaptable to the changing characteristics of the system, such as concurrent application execution that might compete for shared resources.

Given the problem size of N, the detailed procedure to determine the fastest plugin for a particular call uses the two existing neighboring points (left and right) in the performance model (n_L and n_R), such that $n_L \leq N \leq n_R$:

1. INITIALIZE maximum performance variables for n_L and n_R : $s_L := 0$, $s_R := 0$.
2. LOOP: for the performance model of <u>each</u> plugin i, $1 \leq i \leq p$:
   (a) Find n_L and n_R defining the narrowest range, such that $n_L \leq N \leq n_R$;
   (b) If n_L is found and performance at n_L is greater than s_L, then $best_L := i$;
   (c) If n_R is found and performance at n_R is greater than s_R, then $best_R := i$;
3. IF both $best_L$ and $best_R$ are assigned and hold the same value, this value determines which plugin should be used to execute the function.
   ELSE, the optimal selection is not known and all plugins execute the function simultaneously. The execution time for each of them is measured and the obtained speed values used to update the models.

Additional overheads are imposed by the need to pass the function arguments to the several threads in which the plugins are concurrently run, as well as to allocate temporary buffers for the results (in order to isolate the plugins from each other) and to copy the results back to the application buffer upon the fastest plugin completes its execution.

While the overheads mentioned above may be significant, generally they only play a role during the first few executions of a given function within the application run time, and further calls will choose the optimal implementation with a minimum overhead. The actual costs related to this approach will be precisely measured and discussed in the experimental evaluation section of this paper.

3.2   Load Balancing Policy

Whenever the intercepted function call allows work load partitioning across multiple devices, a divisible load partitioning policy is used to partition and balance the load given to each plugin. The problem that arises here is how to partition the problem size N across several heterogeneous devices i ($1 \leq i \leq p$), such that the overall cross-device execution (computation and communication) is finished in the shortest possible time. In detail, each device i should process a certain number of independent parcels n_i of the problem size, such that load balancing is achieved. In contrast to other usual approaches in heterogeneous systems, where the speed s_i of each device is described with constants, a more realistic Functional Performance Modeling (FPM) principle [5] is used. In this model the performance of each device i is modeled as a continuous function of the assigned fraction of the problem size n_i, and defined within the interval $[0, N]$.

The shortest parallel processing time is attained when all devices finish their execution and communicate the results back at the same time (load balancing condition), such that:

$$\frac{n_1}{s_1(n_1)} = \frac{n_2}{s_2(n_2)} = \cdots = \frac{n_p}{s_p(n_p)}; \quad \sum_{i=1}^{p} n_i = N \tag{1}$$

Since n_i must be integers, Eq. (1) is solved by using a two step algorithm. The first step starts by defining the upper and lower bounds of the solution search space, converging towards the optimal distribution by bisecting the angle between these two boundaries, assigning such bisection to one of the search limits in each iteration. Then, it approximates the load distributions by rounding down to the nearest integers. With such preliminary distribution, the algorithm proceeds to the second step (*refinement*), which iteratively redistributes the remaining load to the processing devices according to the devices speed s_i, until assigning the total problem size N to all processors. This results in an algorithm with $\mathcal{O}(p \log N)$ complexity, whose formal proof can be found in [5].

However, the process of building the full performance models for each application and device in the system (i.e., the model for a full range of problem sizes) might be very time consuming. Hence, the adaptive load balancing approach proposed in [2] is adopted, which builds the partial estimations of the full FPMs during the application run-time. For each device, the partial FPMs are built by applying piecewise linear approximations on a set of points, which are obtained from previous application runs according to the number of performed loads and the time taken to process them [2]. Since the performance models are not known *a-priori*, the adaptive load balancing starts by assigning each device with an equal amount of loads to process. Hence, in the first iteration, the total problem size N is evenly partitioned, such that each device is assigned with $n_i = N/p$. This part of the load balancing is referred to as the *initial phase*. The subsequent *iterative phase*, used to obtain the new load distributions, consists of two steps: (i) update of the partially built FPMs; and (ii) determination of the new load distributions, by applying the previously described algorithm to the newly approximated FPMs [2].

4   Experimental Results

Computationally intensive mathematical applications are particularly suitable to evaluate the proposed framework. For such purpose, GNU Octave was naturally selected since it relies on several libraries to execute different types of mathematical operations. From all the Octave required libraries we selected two of the most commonly used: BLAS, for double-precision dense matrix multiplication (DGEMM), and double-precision complex to complex Fast-Fourier Transforms (FFTs). The following subsections describe the experimental setup and present the obtained application acceleration when using these two example libraries.

4.1 Experimental Setup

For the purpose of evaluating the proposed framework, a machine with a Quad-core Intel i7-950 CPU (3.07 GHz), with 12 GB DDR3-1033, and two NVIDIA GTX 580 GPUs were used. Given the limited performance of the native GNU Octave BLAS library (cBLAS), the conducted analyses uses the highly optimized MKL multi-threaded library for the BLAS baseline and the FFTW library [4] as the reference for the FFT performance.

To improve Octave performance, the mechanisms provided by the proposed framework were used to automatically create the plug-in wrappers [6] for the accelerated libraries, namely MKL, cUBLAS, FFTW and cuFFT [7,8]. It should be noticed that while all the considered libraries adapt the algorithms to the work load and underlying architecture, for the FFT plugins, the micro-generation is more challenging. In detail, the computation of the FFT is split into two phases, *planning* and *execution*, which were incorporated into the framework.

4.2 Performance Characterization

In order to evaluate the efficiency of the proposed **Best Performance Selection** policy (BPS), we executed multiple calls and with different problem sizes for the DGEMM and FFT libraries within Octave. The obtained experimental results for the DGEMM function, depicted in Fig. 3, show the execution time obtained when using the BPS policy. As can be observed, BPS was capable of selecting in all cases the plugin with the best performance. In the specific DGEMM case, the MKL library delivers better performance for matrix sizes of up to 350×350, after which the cUBLAS library becomes faster. This is mainly due to the fact that for smaller problem sizes the data transfer overheads are too large and the GPU computing potential is under-used. Naturally, for other libraries the selection point will be different, according to the characteristics of the algorithm. For example, for the considered 1D FFT problem sizes (65 k to 4.2 M) the best performance is always achieved with cuFFT library (see Fig. 4).

For the specific cases of divisible problems, additional speed-ups can still be achieved by relying on the **Load Balancing** (LB) policy. When applying the LB policy to the DGEMM function, we consider the basic column based matrix partitioning algorithm, where each plugin computes a different set of columns of the resulting matrix. In this scope, the source matrix B is divided into column-sets according to the partial estimations of the full performance model of the available plugins (which are constructed in run-time). Figure 5 presents the speed-up values regarding the highly optimized, multi-threaded MKL implementation,

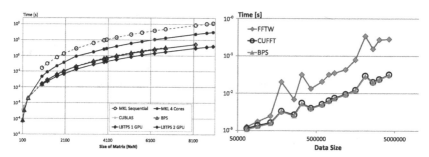

Fig. 3. DGEMM execution time.   **Fig. 4.** 1D FFT execution time.

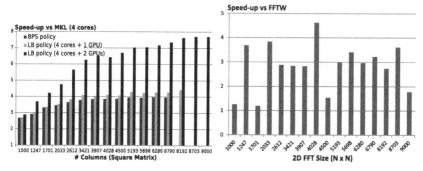

Fig. 5. DGEMM speed-up.   **Fig. 6.** 2D FFT speed-up

running on all four cores. It can be concluded that the adaptive LB policy not only delivers a performance which greatly surpasses a single device, but it also allows executing the DGEMM function on problem sizes that do not fit on the GPU memory (see right side of Fig. 5). Figure 6 presents the speed-ups obtained when collaboratively performing 2D FFT across all four CPU cores and a single GPU. The results herein reported reflect the performance of parallel 2D FFT CPU + GPU implementation which performs 1D FFT on different matrix dimensions, each of them followed by a matrix transposition (performed with Eigen linear algebra library). Experimental results for the 2D FFT, show that despite the matrix transposition overhead, speed-up of up to 4.6 can be achieved, regarding the original FFTW 2D implementation.

A temporal diagram of DGEMM execution is presented in Fig. 7, where not only it can be observed the achieved load balancing between the devices, but also the insignificant overhead of the function call interception, as discussed in the following subsection.

Overhead. To evaluate the scheduling overhead that is introduced by the proposed framework, the amount of time required by the several steps of the implemented algorithm were properly characterized, as presented in Table 1.

The first component corresponds to the function call *interception, redirection and return*. As can be observed, it represents a rather insignificant amount of

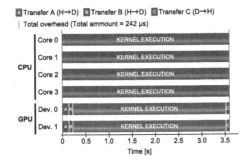

Fig. 7. Temporal diagram of BLAS dgemm execution after load-balancing for a 8703×8703 matrix multiplication case.

Table 1. Framework overheads, considering C function calls of a given work size, using D devices.

Overhead	Amount	Time
Library interception, Redirection and return	C	0.16 μs
BPS: Model update	1	0.34 μs
Thread dispatch	D	36 μs
Fastest selection	$C - 1$	3.16 μs
LBTPS: Model update	C	0.42 μs
Partition distribution	C	25.05 μs
Thread dispatch	$C \times D$	36 μs
GPU INIT: cuFFT	1	1.3 s
cuBLAS	1	0.273 s

time, independent of the problem size. This overhead must be considered once per function call.

Then, when the *Best Performance* selection policy (BPS) is used and it is not possible to find the fastest plugin for a given problem size, all implementations are run at the same time. This implies an added overhead to dispatch a thread to each of the D devices and a consequent model update. As soon as the information regarding a given problem size is collected and stored in the model, the only overhead for consequent calls in the scheduler will be the plugin selection. Therefore, in a best case scenario, the total overhead of BPS is just 3.327 μs. In the case of the *Load-Balancing* policy (LB) the incurred overhead for each function call is related to partition distribution and model update. Although the model update overhead is stable, the partition distribution depends on the information already gathered in the model (hence, the presented values represent average values).

Finally, since several implementations run in parallel, there is an average of 36 μs per plugin overhead for spawning threads. This parcel occurs only once with BPS (for each device), while LB imposes this overhead at every execution.

The last two entries in this table represent the *library initialization* phases. They are independent of the scheduling framework and have to be considered only once, before the actual scheduling takes place. In the considered experimental procedure, both cuFFT and cuBLAS imply large overheads. To mask these CUDA libraries initialization times, they can be executed asynchronously. As such, provided that the target application does not use them at start-up time, this overhead component can be completely hidden. As a consequence, these initialization times were ignored in the presented benchmark tests.

5   Conclusions

A new application acceleration framework based on a transparent redirection of shared-library function calls to the several existing devices in a heterogeneous system was proposed in this paper. The adoption of dynamically constructed performance models allowed this framework to reduce the overheads and to quickly adapt to the user application behaviour, without the need to modify the program source code. By recording the actual performance of the available devices for different library functions and problem sizes, intelligent scheduling mechanisms were implemented in order to allow divisible work loads to be partitioned across all devices in the system, achieving the best balance and attaining the maximum performance. Indivisible work loads are redirected to the fastest single-device implementation that is available, based on the corresponding problem size. The framework was evaluated by comparing the attained performance with state-of-the art single-device implementations, i.e. Intel MKL and FFTW. The obtained results have shown that speedups as high as 7.86 for matrix multiplication and 4.6 for FFT can be obtained, with negligible overheads imposed by the proposed call interception and scheduling mechanisms.

Acknowledgments. This work was partially supported by national funds through Fundação para a Ciência e a Tecnologia (FCT) under projects , "Threads" (ref. PTDC/ EEA-ELC/117329/2010), "P2HCS" (ref. PTDC/EEI-ELC/3152/2012), "HELIX" (ref. PTDC/EEA-ELC/113999/2009), "TAGS" (PTDC/EIA-EIA/112283/2009) and project PEst-OE/EEI/LA0021/2013.

References

1. Beisel, T., Niekamp, M., Plessl, C.: Using shared library interposing for transparent acceleration in systems with heterogeneous hardware accelerators. In: Proceedings of the ASAP (2010)
2. Clarke, D., Lastovetsky, A., Rychkov, V.: Dynamic load balancing of parallel computational iterative routines on highly heterogeneous HPC platforms. Parallel Proc. Lett. **21**(02), 195–217 (2011)
3. Fetzer, C., Xiao, Z.: A flexible generator architecture for improving software dependability. In: Proceedings of the ISSRE. pp. 102–113 (2002)
4. Frigo, M., Johnson, S.G.: The design and implementation of FFTW3. Proc. IEEE **93**(2), 216–231 (2005)

5. Lastovetsky, A., Reddy, R.: Data partitioning with a functional performance model of heterogeneous processors. Int. J. High Perform. Comput. Appl. **21**(1), 76–90 (2007)
6. Matoga, A., Chaves, R., Tomás, P., Roma, N.: A flexible shared library profiler for early estimation of performance gains in heterogeneous systems. In: Proceedings of the HPCS (2013)
7. NVIDIA: CUBLAS Llibrary : User Manual (2012). http://docs.nvidia.com/cuda/pdf/CUDA_CUBLAS_Users_Guide.pdf
8. NVIDIA: CUFFT Llibrary : User Manual (2012). http://docs.nvidia.com/cuda/pdf/CUDA_CUFFT_Users_Guide.pdf

A Study on Adaptive Algorithms for Numerical Quadrature on Heterogeneous GPU and Multicore Based Systems

Giuliano Laccetti[1], Marco Lapegna[1(✉)], Valeria Mele[1], and Diego Romano[2]

[1] Department of Mathematics and Applications, University of Naples Federico II,
Complesso Universitario Monte S. Angelo, Via Cintia, Naples, Italy
{giuliano.laccetti,marco.lapegna,valeria.mele}@unina.it
[2] ICAR-CNR, Via P. Castellino 111, Naples, Italy
diego.romano@na.icar.cnr.it

Abstract. In this work, a parallel adaptive algorithm for the computation of a multidimensional integral on heterogeneous GPU and multicore based systems is described. Two different strategies have been combined together in the algorithm: a first procedure is responsible for the load balancing among the threads on the multicore CPU and a second one is responsible for an efficient execution on the GPU of the computational kernel. The performance is analyzed and experimental results on a system with a quad-core CPUs and two GPUs have been achieved.

Keywords: Hierarchical parallelism · Hybrid algorithms · Adaptive algorithms · Multidimensional integration

1 Introduction

Modern HPC systems are today characterized by hybrid computing nodes, where traditional multicore CPUs live together with special purpose hardware such as Graphical Processing Units (GPUs) used as floating point accelerator. These components have very different features and require different algorithmic development methodologies, so that, in order to efficiently use such emerging hybrid hardware, the development of algorithms and scientific software implies a suitable combination of several methodologies to deal with the various forms of parallelism corresponding to each device.

The aim of our work is to study a special class of algorithms for numerical quadrature for such hybrid computing nodes. More precisely we deal with the numerical computation of multidimensional integrals:

$$I(f) = \int_U f(\underline{t})\,d\underline{t} = \int_U f(t_1,...,t_d)\,dt_1 \cdots dt_d, \tag{1}$$

where $U = [a_1, b_1] \times \cdots \times [a_d, b_d]$ is a d-dimensional hyperrectangular region. In the last thirty years, several efficient routines have been developed for the

R. Wyrzykowski et al. (Eds.): PPAM 2013, Part I, LNCS 8384, pp. 704–713, 2014.
DOI: 10.1007/978-3-642-55224-3_66, © Springer-Verlag Berlin Heidelberg 2014

solution of this problem on traditional CPUs. Most of them (see for example [3,15]) are based on adaptive algorithms, that allow high accuracy with a reasonable computational cost.

2 Parallelization of Adaptive Algorithms on Hybrid Nodes

Given a family of hyperrectangular subdomains $s(k)$ $(k = 1, .., K)$ of a partition S of U, a basic multidimensional quadrature rule $r(k)$ and an absolute error estimate procedure $e(k)$ defined on $s(k)$, an adaptive algorithm for the computation of (1) is an iterative procedure that, at each iteration j, evaluates an approximation $Q^{(j)}$ of $I(f)$ and an estimate $|E^{(j)}|$ of the error $|Q^{(j)} - I(f)|$:

$$Q^{(j)} = \sum_{s(k) \in S} r(k) \simeq I(f) \qquad |E^{(j)}| = \sum_{s(k) \in S} e(k) \simeq |Q^{(j)} - I(f)|$$

To achieve this, the algorithm computes a sequence $Q^{(j)}$ of composite quadrature rules approaching $I(f)$ and a sequence $|E^{(j)}|$ of approximations of the error $|Q^{(j)} - I(f)|$ approaching 0, until a stopping criterion is satisfied. For our purposes we remark that the basic quadrature rules $r(k)$ are based on a summation such as:

$$r(k) = \sum_{i=1}^{n} A_i \, f(\underline{t}_i) \tag{2}$$

For dimension up to dimension $d = 15$ there are several methods to compute the basic rules $r(k)$ and the absolute errors $e(k)$ in standard regions $s(k)$ [2,6].

Since the convergence rate of this procedure depends on the behaviour of the integrand function (presence of peaks, oscillations, etc.), in order to reduce as soon as possible the error, at the iteration j, the subdomain $\hat{s} \in S$ with maximum error estimate $\hat{e}$ is split in two parts $s(\lambda)$ and $s(\mu)$ that take the place of $\hat{s}$ in the partition S, that is $S = S - \{\hat{s}\} \cup \{s(\lambda) , s(\mu)\}$. In a similar way the approximations $Q^{(j)}$ and $E^{(j)}$ are updated, evaluating the (2) in the new subdomains.

Algorithm 1.

Initialize $\mathcal{H}$, $Q^{(0)}$ and $E^{(0)}$
while (stopping criterion not satisfied) **do** iteration j
    1) select $\hat{s} \in \mathcal{H}$ such that $\hat{e} = \max_{k=1,..,K} e(k)$
    2) divide $\hat{s}$ in two parts $s(\lambda)$ and $s(\mu)$
    3) compute $r(\lambda)$, $e(\lambda)$, $r(\mu)$ and $e(\mu)$
    4) sort the subdomains according to their errors
    5) update $\mathcal{H}$, $Q^{(j)}$ and $E^{(j)}$
endwhile

To implement an adaptive algorithm for numerical quadrature, it is necessary to store all the subdomains $s(k)$ of the partition $\mathcal{S}$ in a suitable data structure, where the subdomain with maximum error estimate $\hat{e}$ can be found with a small computational cost. This can be achieved by storing the data related to the subdomains $s(k)$ in a partially ordered binary tree $\mathcal{H}$ called *heap*, where the subdomain with the largest error estimate is in the root. The computational cost to sort a heap is $\log_2 K$, where K is the number of subdomains in $\mathcal{H}$. A framework for a sequential global adaptive algorithm for the computation of multidimensional integrals is therefore the Algorithm 1 [14]. There are several approaches to introduce parallelism in adaptive algorithms [14]. The main strategies are the following:

- Integrand Level Parallelism: the degree of parallelism is given by the number of integrand functions that have to be eventually computed at the same time. Since the integrals are distinct, this is an pleasingly form of parallelism [18], and is well suited to computer systems that do not require frequent communications and/or synchronizations between tasks, such as geographically distributed systems;
- Subdivision Level Parallelism: the degree of parallelism is given by the number of subdomains that are subdivided at the same time, so that it is possible to process several subdomains at each iteration. This is a high form of parallelism suited for a SPMD programming model such as that one used for clusters or MPP systems;
- Subregion Level Parallelism: in this case only one subdomain is divided in several parts concurrently processed , and the degree of parallelism is given by the number of these parts. This is a more tight form of parallelism with respect to the previous level.
- Integration Formula Level Parallelism: the degree of parallelism is given by the number of integrand functions required by the integration rule (2). This is a low level form of parallelism that does not require MIMD based computing systems, because the function evaluations all have the same expression. So it is well suited to SIMD or GPU accelerated systems.
- Integrand Function Level: the degree of parallelism is given by the simultaneous calculation of different tasks of the integrand function, so it depends strongly by its analytical form.

For our aims, consider then an environment represented by a computing node (e.g. a cluster node or a blade in a server) with a node main memory, one o more host multicore CPUs and one or more floating point accelerator devices such as the NVIDIA's GPUs or the Intel Xeon Phy. Furthermore the acceleration device has a private memory and cannot access directly the node main memory, so that the data have to be moved from the host memory to the device memory and viceversa. From the above, the best strategy to develop a hybrid algorithm for this environment is then to use a combination of the Subdomain Level Parallelism for the subdomains management on the host multicore CPU, and an Integration Formula Level Parallelism to evaluate the basic rule (2) on the GPU device.

2.1   The Host Algorithm

To introduce a Subdomain Level Parallelism in Algorithm 1, consider a multicore based computing environment, where N threads T_i ($i = 0, .., N - 1$) are in execution, one on each core, sharing the node main memory.

In a such environment it is then possible to process N subregions concurrently by different threads. This can be achieved by storing the data related to the subdomains $s(k)$ in a shared heap $\mathcal{H}$ stored in the node main memory. But, in this centralized approach, where all threads access a single shared heap with a global synchronization, all the basic operations on the heap must be carried out in a critical section, so that the synchronization cost depends on the number of threads N, with a strong scalability degradation [10].

In order to avoid global critical sections, we give up the idea of a single centralized heap, and we split the heap $\mathcal{H}$ in N separate heaps $\mathcal{H}_i$, one for each thread, each of them accessing its private data structure without synchronizations with other threads. However also this approach has a side effect: since of the N items $\hat{s}_i$ with the largest error, resident in the heap roots of $\mathcal{H}_i$ are not those that globally have the highest priority, some threads can process unimportant items with a slow numerical convergence. At this regard note that the sequence of items with large error is unpredictable, so that it is impossible to distribute the subdomains $\hat{s}_i$ with large errors uniformly among the heaps $\mathcal{H}_i$ before the computation.

In order to combine fast convergence with high efficiency, in our approach, at each iteration j, the threads compare the maximum error $\hat{e}_i$ in the roots of $\mathcal{H}_i$ and, if the critical items are not equally distributed among the heaps, they attempt to reorganize the subdomains in a more suitable way.

To this aim we propose a loosely coordinated approach, where the N threads are logically organized according to a 2-dimensional periodical mesh $\mathcal{M}_2$. This structure is a grid of $\Lambda_0 \times \Lambda_1 = N$ threads, arranged along the points of a 2-dimensional space with integer non negative coordinates in which a shared memory between each couple of connected nodes is established. The shared memories are used as buffer to exchange data between two threads according to a producer-consumer protocol. In addition, the corresponding threads on the opposite faces of the mesh are connected too, so that the mesh is periodical.

In a 2-dimensional periodical mesh, each thread T_i has 4 neighbors: 2 for each direction. In the horizontal direction ($dir = 0$), we define $T_{i-}^{(0)}$ and $T_{i+}^{(0)}$ respectively the leftmost and the rightmost thread of T_i in $\mathcal{M}_2$. Analogously in the vertical direction ($dir = 1$) we define $T_{i-}^{(1)}$ and $T_{i+}^{(1)}$ the lowermost and the uppermost threads of T_i.

We then define $\mathcal{H}^*$ a *loosely coordinated heap* as a collection of heap $\mathcal{H}_i$ $i = 0, .., N - 1$, where the roots are connected among them according to the $\mathcal{M}_2$ topology.

With the described threads organization, at the iteration j, each thread T_i attempts to share its item $\hat{s}_i \in \mathcal{H}_i$, with largest error $\hat{e}_i$, only with the neighbor thread $T_{i+}^{(dir)}$ alternatively in the two directions horizontal and vertical. More precisely, in a fixed direction dir, let $\hat{e}_i$ $\hat{e}_{i+}$ and $\hat{e}_{i-}$ be respectively the errors of the subdomains in the heap root of $\mathcal{H}_i$, $\mathcal{H}_{i+}^{(dir)}$ and $\mathcal{H}_{i-}^{(dir)}$. If $\hat{e}_i > \hat{e}_{i+}$ then the item $\hat{s}_i \in \mathcal{H}_i$ with largest error $\hat{e}_i$ is moved forward to the heap $\mathcal{H}_{i+}$ along the direction dir, using a producer-consumer protocol on the shared space. In the same way if $\hat{e}_{i-} > \hat{e}_i$ the item $\hat{s}_{i-} \in \mathcal{H}_{i-}$ with largest error $\hat{e}_{i-}$ is moved to the heap $\mathcal{H}_i$. In this way, the critical items with large error are shared among the heaps with a faster convergence.

Furthermore, it should be noted that in this proposed data redistribution, at each iteration j, there are not global synchronizations among threads T_i and each of them exchanges data only with the two threads $T_{i+}^{(dir)}$ and $T_{i-}^{(dir)}$ with $dir = mod(j, 2)$, so that the cost of threads synchronization is constant and it does not depend on the number of threads N, so that the resulting algorithm can be considered scalable [10].

2.2   The Device Algorithm

Modern GPUs are designed to efficiently deal with problems in the field of computer graphics. In this field, it is typically necessary to perform the exact same operations on all pixels in the image where you want to recreate the same effect. For this reason, modern GPUs provide a SIMD type parallelism where hundreds of single computing elements work synchronously on different data, under the control of a single Control Unit. On the other hand, each computing element is designed as simple as possible in order to keep its production cost low, so that the power of the single elements is much lower in comparison to those of the traditional CPUs. These characteristics mean that only some algorithms are suitable for an efficient implementation on these devices. More precisely only a fine grained parallelism on many data is able to unleash the computing power of these devices.

From this point of view, the computation of (2) is well suited for an execution on a GPU because of the large value of the number of nodes n, so that the n products $A_i f(\underline{t}_i)$ are evaluated concurrently by the GPU computing elements according to the Integration Formula Level Parallelism.

It should be noted, however, that the use of these environments involves a heavy overhead. For example in CUDA (the computing platform and programming model created by NVIDIA for its GPUs), the computing elements cannot directly access the data stored in the node memory, so that it is necessary to allocate space on the memory graphics card and to transfer data in it. This transfer is a tremendous bottleneck for the computation: just think that the NVIDIA Tesla C1060 has a peak performance $p^* = 933$ Gflops (single precision) and a memory bandwidth of only $m^* = 102$ GByte/s (i.e. 25.5 Gwords/s, about 3 % of the peak performance). For such a reason, a key parameter for the development of efficient algorithms for such computing device is the ratio $\Theta = p^*/m^*$, which gives a measure of the number of floating point operations required for each data transferred, in order to support the peak performance. For the NVIDIA Tesla C1060 we have $\Theta \simeq 35$.

To this aim we observe that the integrand formula (2) requires the transfer from the host memory to the device memory of 2 d-dimensional array (the center of the region and the length of its edges) and it is based on n independent function evaluations where $d^3 < n < d^4$ (see for example [11]), large enough to support the parameter Θ.

In any case it is important to observe that in a sum-based formula (2), after the parallel evaluation of the n products $A_i\, f(\underline{t}_i)$, it is necessary to collect these values together, by summing pairs of partial sums in parallel. Each step of this pair-wise summation cuts the number of partial sums in half and ultimately produces the final sum after $\log_2 n$ steps. This procedure that computes a single value from a set of data by using an associative operation (e.g. sum or maximum) is called *reduction*, and its optimization is a key problem in the development of algorithms for the GPUs, due to a decreasing number of active threads in the cascade scheme required to calculate a single value from data produced by several processing units. For such a reason we use the optimization strategies described in [13] to compute (2).

Algorithm 2.

---

initialize $\mathcal{H}_i$, $Q_i^{(0)}$ and $E_i^{(0)}$
while (local stopping criterion not satisfied) **do** iteration j
   define $dir = mod(j, 2)$
   if $\hat{e}_i > \hat{e}_{i+}$ **then**
      remove $(\hat{s}_i)$ from $\mathcal{H}_i$
      produce $(\hat{s}_i)$ for $T_{i+}^{(dir)}$
   endif
   if $\hat{e}_{i-} > \hat{e}_i$
      consume $(\hat{s}_{i-})$ produced by $T_{i-}^{(dir)}$
      insert $(\hat{s}_{i-})$ in $\mathcal{H}_i$
   endif
      1) select $\hat{s}_i \in \mathcal{H}_i$ such that $\hat{e}_i = \max_{k=1,..,K} e(k)$
      2) divide $\hat{s}_i$ in two parts $s_i(\lambda)$ and $s_i(\mu)$
      3) compute $r_i(\lambda)$, $e_i(\lambda)$, $r_i(\mu)$ and $e_i(\mu)$ on the GPU device
      4) sort the subdomains according to their errors
      5) update $Q_i^{(j)}$ and $E_i^{(j)}$
endwhile

We conclude this section reporting, in Algorithm 2, the description of the hybrid algorithm obtained by integrating the two described methods. More precisely, using the programming model SPMD, we describe the subdomains management based on the parallelization at Subdivision Level between the threads T_i, and at the same time we remark the section of the algorithm with the evaluation of the quadrature formula in step 3) executed in SIMD mode on the GPU using a Formula Level Parallelism.

3   Test Results

In this section we present the experimental results achieved on a system composed by a quad-core CPU, an Intel Core I7 950 operating at 3.07 Ghz, and two

NVIDIA's C1060 GPUs (Tesla). Each NVIDIA C1060 GPU has 240 streaming processor cores operating at 1.3 Ghz with a peak performance of 933 Gflops in single precision arithmetic (78 Gflops in double precision arithmetic). The host main memory is 12 GBytes large and the bandwidth between the host memory and the device memory is 102 GByte/s.

In this computational environment we implemented our hybrid Algorithm 2 in double precision using C language, with the CUDA library for the implementation of the step 3) on the GPU, and POSIX thread library and semaphores for the redistribution of the subdomains among the threads in the host algorithm. For the experiments we used a standard procedure based on the well known Genz package [12]. This package is composed by six different families of functions, each of them characterized by some issues making the problem (1) hard to integrate numerically (peaks, oscillations, singularities...). Each family is composed by 10 different functions where the parameters α_i and β_i change and average test results are computed (execution time, error,). Here we report the results for the following three families:

$$f^{(1)}(\underline{x}) = cos(2\pi\beta_1 + \textstyle\sum_{i=1}^{d} \alpha_i x_i) \quad \text{Oscillating functions}$$
$$f^{(2)}(\underline{x}) = (1 + \textstyle\sum_{i=1}^{d} \alpha_i x_i)^{-d-1} \quad \text{Corner peak functions} \quad (3)$$
$$f^{(3)}(\underline{x}) = \exp(-\textstyle\sum_{i=1}^{d} \alpha_i |x_i - \beta_i|) \quad C^{(0)} \text{ functions}$$

on the domain $U = [0,1]^d$ with dimension $d = 10$. We selected these functions because their different analytical features. However, for other functions in the Genzs package we achieved similar results. We remark that in our algorithm we use the Genz and Malik quadrature rule with $\phi = 1,245$ function evaluations so that at each iteration $2\phi = 2,490$ function evaluations are computed in the two new subdomains s_λ and s_μ.

A first set of experiments is aimed to study the parallelization at the subdivision level implementing only the host algorithm. In these experiments we measured

- the Scaled Speed-up SS_N and the Scaled Efficiency SE_N [10] with N=1, 2, 3 and 4 threads.
- The minimum (MinErr) and the maximum (MaxErr) relative error $|I(f) - Q(f)|/|I(f)|$ on the 10 functions of each family.

To compute SS_N we set $F = 10 \times 10^6$ function evaluations in each threads, so that the total number of function evaluations is $FVAL = N \times 10 \times 10^6$ when the number of threads increases. The local stopping criterion is based on a maximum allowed number of iterations in each thread $Maxit = F/2\phi = 4016$.

Table 1 refers to the experiments executed only on the CPU and it reports the Scaled Speed-up for the three families of functions by using 1, 2, 3 and 4 threads . We observe a good scalability when the number of threads increases. As already remarked, the evaluation of the multidimensional integration rules are tasks with a favorable ratio of floating point computation on data movement so that the data can be easily stored in the core caches and reused in the next iterations with an extensive use of cached data.

Table 1. Scaled Speed-up and Scaled Efficiency for the three families of functions $f^{(1)}$, $f^{(2)}$ and $f^{(3)}$ with 1, 2, 3 and 4 cores. The workload in each processing unit is $F = 10 \times 10^6$ when the number of core increases. The average execution times with 1 core for the three families of functions are: $Time(f^{(1)}) = 0.27\,s$, $Time(f^{(2)}) = 0.22\,s$ and $Time(f^{(3)}) = 0.28\,s$.

	$N = 1$	$N = 2$	$N = 3$	$N = 4$
Family $f^{(1)}$				
SS_N	1	1.9	2.8	3.4
SE_N	1	0.95	0.93	0.85
Family $f^{(2)}$				
SS_N	1	1.9	2.8	3.6
SE_N	1	0.95	0.93	0.90
Family $f^{(3)}$				
SS_N	1	1.9	2.7	3.4
SE_N	1	0.95	0.90	0.85

Table 2. Execution time and number of function evaluations per second with $N = 4$ threads without the use of GPU, when the number of node n in the basic rule changes. The total number of function evaluations is $F = 4 \times 10 \times 10^6$.

	$n = 1245$	$n = 2585$	$n = 9385$	$n = 37384$
Family $f^{(1)}$				
exec. time	0.079	0.071	0.064	0.055
$FVAL$/time	506×10^6	563×10^6	625×10^6	727×10^6
Family $f^{(2)}$				
exec. time	0.059	0.054	0.048	0.041
$FVAL$/time	677×10^6	740×10^6	833×10^6	975×10^6
Family $f^{(3)}$				
exec. time	0.082	0.074	0.066	0.057
$FVAL$/time	487×10^6	540×10^6	606×10^6	701×10^6

A second set of experiments is aimed to investigate the performance gain using a GPU device as a floating point accelerator. In this case the quadrature formula (2) has been implemented in the CUDA programming environment for a scheduling on the GPU, as described in previous section. More precisely, for our experiments, we have been used several quadrature formulas belonging to the family of Genz and Malik [11] with number of nodes $n = 1245, 2585, 9385, 37389$ respectively. This is because the utilization of the GPU involves a high overhead due to the data transfer between the host memory and device memory, which is balanced only by an intensive use of its computational capabilities.

In Tables 2 and 3 are reported the performance results of the hybrid algorithm by using only the quad-core CPU and by using also the GPU devices as a floating point accelerator respectively. As a performance measure, we used the number of function evaluations per second. Also in this case the local stopping criterion is based on the maximum function evaluations in each thread $F = 10 \times 10^6$, so

Table 3. Execution time and number of function evaluations per second with $N = 4$ threads with the use of GPU, when the number of node n in the basic rule changes. The total number of function evaluations is $F = 4 \times 10 \times 10^6$.

	$n = 1245$	$n = 2585$	$n = 9385$	$n = 37384$
Family $f^{(1)}$				
exec. time	0.080	0.053	0.031	0.018
$FVAL$/time	500×10^6	754×10^6	1290×10^6	2222×10^6
Family $f^{(2)}$				
exec. time	0.065	0.049	0.028	0.015
$FVAL$/time	615×10^6	816×10^6	1428×10^6	2666×10^6
Family $f^{(3)}$				
exec. time	0.085	0.056	0.033	0.020
$FVAL$/time	470×10^6	714×10^6	1212×10^6	2000×10^6

that the total number of function evaluations is $FVAL = N \times 10 \times 10^6$ ($N = 4$ is the number of threads) for all test.

From these Tables it is evident that a basic rule with a small number of function evaluations ($n = 1245$ and $n = 2585$) is unable to exploit the computing power of the GPU used in these experiments. More precisely, we can observe that the performance gain obtained with the use of the GPU is wasted because of the overhead related to the memory device allocation and the data transfer, without significant benefit for the performance. Only with a large number of nodes in the basic rule ($n = 9385$ and $n = 37384$) we report a significant performance gain. Compared with the value in Table 2, the performance gain reported in Table 3 is about $3\times$.

4   Conclusions

We presented a hybrid multicore CPU/GPU approach that can exceed $3\times$ the performance of traditional quadrature adaptive algorithms running just on current homogeneous multicore CPUs. In any case we report a significant performance gain only with a large number of function evaluations of the basic rule ($n > 10^4$), because the overhead introduced by the memory device management. In any case our approach demonstrates the utility of graphics accelerators for multidimensional quadrature problems in a large number of dimensions. Furthermore we remark that our approach can be combined with other hybrid strategies for multidimensional quadrature, such as that described in [16] or [8], as well as for other on going works [1,4,5,7,9,17].

References

1. Antonelli, L., Carracciuolo, L., Ceccarelli, M., D'Amore, L., Murli, A.: Total variation regularization for edge preserving 3D SPECT imaging in high performance computing environments. In: Sloot, P.M.A., Tan, C.J.K., Dongarra, J., Hoekstra, A.G. (eds.) ICCS-ComputSci 2002, Part II. LNCS, vol. 2330, pp. 171–180. Springer, Heidelberg (2002)

2. Berntsen, J.: Practical error estimation in adaptive multidimensional quadrature routines. J. Comput. Appl. Math. **25**, 327–340 (1989)

3. Berntsen, J., Espelid, T., Genz, A.: Algorithm 698: DCUHRE - an adaptive multidimensional integration routine for a vector of integrals. ACM Trans. Math. Softw. **17**, 452–456 (1991)

4. Boccia, V., D'Amore, L., Guarracino, M.R., Laccetti, G.: A grid enabled PSE for medical imaging: experiences on MediGrid. In: Proceedings of the IEEE Symposium on Computer Based Medical Systems, pp. 529–536 (2005)

5. Carracciuolo, L., D'Amore, L., Murli, A.: Towards a parallel component for imaging in PETSc programming environment: a case study in 3-D echocardiography. Parallel Comput. **32**, 67–83 (2006)

6. Cools, R., Rabinowitz, P.: Monomial cubature rules since "Stroud": a compilation. J. Comput. Appl. Math. **48**, 309–326 (1993)

7. D'Amore, L., Casaburi, D., Galletti, A., Marcellino, L., Murli, A.: Integration of emerging computer technologies for an efficient image sequences analysis. Integr. Comput. Aided Eng. **18**, 365–378 (2011)

8. D'Alessio, A., Lapegna, M.: A scalable parallel algorithm for the adaptive multidimensional quadrature. In: Sincovec, R., et al. (eds.) SIAM Conference on Parallel Processing for the Scientific Computing, pp. 933–936. SIAM (1993)

9. D'Amore, L., Murli, A.: Image sequence inpainting: towards numerical software for detection and removal of local missing data via motion estimation. J. Comput. Appl. Math. **198**, 396–413 (2007)

10. Dongarra, J., Foster, I., Fox, G., Gropp, W., Kennedy, K., Torczon, L., White, A.: Sourcebook of Parallel Computing. Morgan Kaufmann, San Francisco (2003)

11. Genz, A., Malik, A.: An embedded family of fully symmetric numerical integration rules. SIAM J. Numer. Anal. **20**, 580–588 (1983)

12. Genz, A.: Testing multiple integration software. In: Ford, B., Rault, J.C., Thommaset, F. (eds.) Tools, Methods and Language for Scientific and Engineering Computation. North Holland, New York (1984)

13. Harris, M.: Optimizing parallel reduction in CUDA. Technical report, presentation packaged with CUDA Toolkit, NVIDIA Corporation (2007)

14. Krommer, A.R., Ueberhuber, C.W.: Numerical Integration on Advanced Computer Systems. LNCS, vol. 848. Springer, Heidelberg (1994)

15. Laccetti, G., Lapegna, M.: PAMIHR. A parallel FORTRAN program for multidimensional quadrature on distributed memory architectures. In: Amestoy, P.R., Berger, P., Daydé, M., Duff, I.S., Frayssé, V., Giraud, L., Ruiz, D. (eds.) Euro-Par 1999. LNCS, vol. 1685, pp. 1144–1148. Springer, Heidelberg (1999)

16. Laccetti, G., Lapegna, M., Mele, V., Romano, D., Murli, A.: A double adaptive algorithm for multidimensional integration on multicore based HPC systems. Int. J. Parallel Prog. **40**, 397–409 (2012)

17. Maddalena, L., Petrosino, A., Laccetti, G.: A fusion-based approach to digital movie restoration. Pattern Recogn. **43**, 1485–1495 (2009)

18. Murli, A., D'Amore, L., Laccetti, G., Gregoretti, F., Oliva, G.: A multi-grained distributed implementation of the parallel Block Conjugate Gradient algorithm. Concurrency Comput. Pract. Experience **22**, 2053–2072 (2010)

Improving Parallel I/O Performance Using Multithreaded Two-Phase I/O with Processor Affinity Management

Yuichi Tsujita[1,4]([⊠]), Kazumi Yoshinaga[1,4], Atsushi Hori[1,4], Mikiko Sato[2,4], Mitaro Namiki[2,4], and Yutaka Ishikawa[1,3]

[1] RIKEN Advanced Institute for Computational Science, Kobe, Japan
yuichi.tsujita@riken.jp
[2] Tokyo University of Agriculture and Technology, Tokyo, Japan
[3] The University of Tokyo, Tokyo, Japan
[4] JST CREST, Tokyo, Japan

Abstract. I/O has been one of the performance bottlenecks in parallel computing. Using a parallel I/O API such as MPI-IO is one effective approach to improve parallel computing performance. The most popular MPI-IO implementation, ROMIO, utilizes two-phase I/O technique for collective I/O for non-contiguous access patterns. Furthermore, such two-phase I/O is frequently used in application oriented parallel I/O libraries such as HDF5 through an MPI-IO interface layer. Therefore performance improvement in the two-phase I/O may have a big impact in improving I/O performance in parallel computing. We report enhancements of the two-phase I/O by using Pthreads in order to improve I/O performance in this paper. The enhancements include overlapping scheme between file I/O and data exchanges by multithreaded operations and the processor affinity for threads dedicated for file I/O and data exchanges. We show performance advantages of the optimized two-phase I/O with an appropriate processor affinity management relative to the original two-phase I/O in parallel I/O throughput evaluation of HDF5.

Keywords: MPI-IO · HDF5 · Two-phase I/O · Multithreaded I/O · Processor affinity management

1 Introduction

MPI [10] is currently the de facto standard communication interface in parallel computing. With the increase in the number of MPI processes in recent parallel computing, the size of data which are read or written by MPI processes is increasing dramatically. I/O for huge-scale data is one of the performance bottlenecks in parallel computing. MPI-IO in the MPI standard [10] provides a variety of parallel I/O API in order to achieve scalable I/O operations. A well-known MPI-IO library, ROMIO [14], has the two-phase I/O performance optimization scheme [15] (hereinafter, TP-IO), which is used in collective I/O for handling

R. Wyrzykowski et al. (Eds.): PPAM 2013, Part I, LNCS 8384, pp. 714–723, 2014.
DOI: 10.1007/978-3-642-55224-3_67, © Springer-Verlag Berlin Heidelberg 2014

non-contiguous access patterns. TP-IO consists of repetitions of contiguous file I/O and communications for the purpose of I/O performance improvement. Since I/Os and communications are carried out in a sequential manner, room remains to overlap I/O with communications regarding performance improvement. We have already implemented such a scheme in TP-IO by using Pthreads [5] in a multithreaded manner, and this multithreaded TP-IO outperformed the original one [17].

In scientific applications, application-oriented I/O libraries such as HDF5 [16] have been widely used. HDF5's parallel I/O interface is built upon an MPI-IO interface layer, and collective I/O for non-contiguous access patterns by using derived data types is frequently used. Therefore performance improvement in TP-IO may lead to performance improvement in HDF5.

In this paper, we describe enhancements of the TP-IO to have the management of CPU core bindings for threads which participate in the multithreaded TP-IO. The multithreaded TP-IO without the processor affinity management outperforms the original TP-IO with about 4.5 % minimization in I/O time measurements. Furthermore the processor affinity management in the multithreaded TP-IO minimizes I/O time about 2.9 % compared with the multithreaded TP-IO without the processor affinity management. Consequently the processor affinity management indicates effectiveness in I/O performance improvement through the I/O time measurements.

2   Multithreaded Two-Phase I/O

The TP-IO consists of I/O and communication phases in order to improve parallel I/O performance. A typical collective read scheme in the original TP-IO is illustrated in Fig. 1. The entire data file including data gaps is divided evenly between the MPI processes which take part in file I/O. File I/O and communications are carried out using a temporary buffer referred to as the collective buffer (hereinafter, CB).

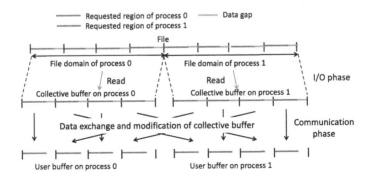

Fig. 1. Collective read in TP-IO

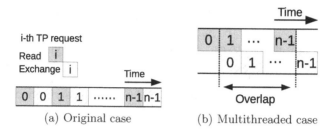

(a) Original case   (b) Multithreaded case

Fig. 2. TP-IO read scheme consisting of file I/O and communication

In general, an assigned data space is larger than the CB. As a result, a combination of file I/O and communications is repeated until the entire assigned data space has been accessed as shown in Fig. 2(a). Each I/O request numbered from 0 includes information about not only an assigned CB but also other parameters, such as offset and size of the assigned data region. In contrast, our multithreaded implementation shown in Fig. 2(b) can overlap an I/O phase with a communication phase. It is noted that this figure depicts a very ideal case. In general, communication and I/O costs are not equal, thus overlapping effect may decrease compared with the case in Fig. 2(b).

A functional diagram of the multithreaded TP-IO is shown in Fig. 3. An I/O thread is invoked by a main thread in each process using `pthread_create` when `MPI_File_open` is called. The main and I/O threads are deployed on a specified CPU core through an `MPI_Info` object with the help of processor affinity management by using a CPU core affinity API for Pthreads such as `pthread_attr_setaffinity_np`. We also have prepared our own function set to extract process information from a /proc file system to know a current CPU core ID.

Once an MPI-IO function such as `MPI_File_write_all` is called, associated I/O requests are firstly enqueued in a read queue. The I/O thread periodically checks the status of the queue, and dequeues one request and then executes file I/O according to the request. After the file I/O has completed, the I/O thread

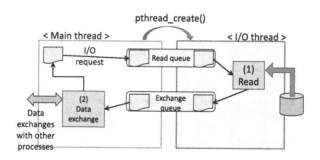

Fig. 3. Multithreaded TP-IO by using Pthreads

enqueues the I/O request in an exchange queue. The main thread checks the exchange queue periodically and dequeues a request; this is followed by data exchanges with other MPI processes according to the request. This sequence is repeated until all the requests have been carried out. Every MPI process can buffer multiple I/O requests in order to minimize idle times due to congestion in communications and I/O operations. The maximum number of I/O requests in queues can be specified through an MPI_Info object. Good CPU core bindings based on communication and I/O patterns also improve I/O performance too. For example, a main thread might be placed on a CPU socket which manages network interfaces in a communication intensive case.

3   Performance Evaluation

A performance evaluation was carried out on a PC cluster system of the Information Technology Center, the University of Tokyo (hereinafter, T2K-Todai). Its node specification is shown in Table 1. We used 32 PC nodes of the T2K-Todai, which were dedicated only for our system software development. Each node has four quad-core AMD Opteron processors, and thus we had 16 cores per node. Every processor had four HyperTransport links, one for 8 GiB DDR2-667 memory accesses and the other three for connections to the other processors on the same node. Network connections between nodes were established with two 1 Gbps Ethernet links for control and two Myrinet 10 Gbps links [12] for MPI communications. In addition to the nodes, we utilized a Lustre file system [8] dedicated to the 32 nodes, where the file system consisted of 1 MDS and 4 OSTs, via one Myrinet 10 Gbps link.

We used an MPICH2 library [11], version 1.4.1p1, in order to implement our proposed scheme in ROMIO. For the evaluation of the various CPU core bindings, we deployed one MPI process per node to maintain CPU core resource availability on each node, and thus 32 MPI processes executed I/O operations. An I/O thread was invoked by a main thread, and then both threads were deployed on the specified CPU core on the same node. Here MPI communications between nodes were established with IP over Myrinet for the modified MPICH2 library. I/O operations were performed on the Lustre file system with 1 MiB striping among the four OSTs.

In this paper, we focus on collective I/O for non-contiguous patterns which include data of other MPI processes in addition to own data on each MPI process

Table 1. Node specification of T2K-Todai

CPU	AMD Opteron 8356 Barcelona (2.3 GHz, 4 cores, L2 cache: 512 KiB/core, L3 cache: 2 MiB/CPU) × 4
Memory	32 GiB (8 GiB × 4)
Interconnect	Myrinet 10 Gbps × 2, 1 Gbps Ethernet × 2
OS	Linux kernel 2.6.18-53 with glibc version 2.5
Parallel file system	Lustre version 1.8.1

```
...
dataset1 = H5Dopen2(fid1, "Data1", H5P_DEFAULT);
file_dataspace = H5Dget_space(dataset1);
ret = H5Sselect_hyperslab(file_dataspace, H5S_SELECT_SET, start, stride,
            count, NULL);
mem_dataspace = H5Screate_simple(2, count, NULL);
xfer_plist = H5Pcreate(H5P_DATASET_XFER);
ret = H5Pset_dxpl_mpio(xfer_plist, H5FD_MPIO_COLLECTIVE);

io_time = MPI_Wtime();
ret = H5Dread(dataset1, H5T_NATIVE_INT, mem_dataspace, file_dataspace,
          xfer_plist, &(data_array1[0][0]));
io_time = MPI_Wtime() - io_time;

H5Sclose(file_dataspace);
H5Sclose(mem_dataspace);
H5Pclose(xfer_plist);
H5Dclose(dataset1);
...
```

Fig. 4. Pseudo code using an HDF5 collective read API for parallel I/O performance evaluation

because this kind of access pattern is commonly used in applications and TP-IO is used in such access pattern.

We used the most recent version of HDF5 (version 1.8.10); however, a benchmark program named IOR [6] which included HDF5 benchmark codes (version 2.10.3) did not support some newly implemented HDF5 functions. Therefore, we used a performance evaluation program in the HDF5 release with some modifications. Figure 4 shows a pseudo code of part of the I/O performance evaluation program. We evaluated collective I/O by using HDF5's H5Dread for non-contiguous I/O accesses. We invalidated the file system cache by remounting the Lustre file system prior to every I/O performance evaluation. In order to measure I/O times and give some hints such as CB size or the number of I/O requests to an MPI-IO layer, we slightly modified it to manage them through an MPI_Info object. Since our multithreaded TP-IO supported only the read operation at the time of the evaluation, we measured only collective read performance. In this evaluation, two-dimensional data with 24,320 integers in both row and column (2.2 GiB in total) was read by all 32 MPI processes on the Lustre file system. In order to have TP-IO, data was split evenly along with a column index, where non-contiguous access patterns were generated for every process. In this paper, we show mean values for every measured time.

Figure 5 shows I/O times of HDF5 collective read operations in terms of total CB size. Total CB size was calculated as a product of the number of I/O requests and CB size, which gives the total utilized memory size for all the CBs in queues. For example, we indicated 64 MiB in total when we had a 16 MiB CB with 4 I/O requests. Note that a shorter time is better in terms of I/O performance. "original" in Fig. 5(a) stands for parallel I/O using the original ROMIO. "ior-2," "ior-4," and "ior-8" denote multithreaded collective I/O with at most 2, 4, and 8 I/O requests in queues.

Figure 5(a) shows I/O times without processor affinity management. We can see that the "ior-8" case outperformed other cases with more than 32 MiB in total

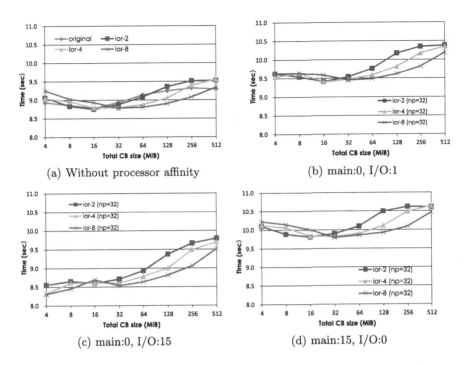

Fig. 5. Collective read times for approximately 2.2 GiB with 32 MPI processes by using an HDF5 collective read API on a Lustre file system, where "main" and "I/O" in (b), (c), and (d) denote assigned CPU core IDs for main and I/O threads, respectively

CB size. Compared with the "original" case, we have achieved 4.5 % improvement with 32 MiB in total CB size.

On the other hand, Fig. 5(b)–(d) shows I/O times of multithreaded TP-IO with the CPU core binding by using our processor affinity management function. Note that the CPU core IDs from 0 to 3 were given to a CPU socket 0, and so forth in the T2K Todai case. A main thread was placed on the CPU core 0 in both Fig. 5(b) and (c) while deploying an I/O thread on the CPU cores 1 and 15, respectively. Figure 5(d) shows the combination opposite to the (c) case. The "ior-8" case also outperformed other cases with more than 32 MiB in total CB size.

Overall of the evaluation results in Fig. 5, only the results in Fig. 5(c) minimized I/O times relative to the associated results in Fig. 5(a) regarding the number of I/O requests. For instance the "ior-8" case in Fig. 5(c) minimized about 2.9 % relative to the same case in Fig. 5(a) with 32 MiB in total CB size. While the I/O times of the same case in both the Fig. 5(b) and (d) were longer than those in the Fig. 5(a).

We interpret this performance difference as coming from the CPU core binding effect. From the point of view of inter-node MPI communications, it might be better to place a main thread as close as possible to network interface cards in a communication-intensive case. In our evaluation environment, the first CPU socket, which had CPU cores numbered from 0 to 3, managed the Myrinet interface card. Thus performance results in Fig. 5(b) and (c) were expected to be better than those in Fig. 5(d).

Figure 5(b), which deployed the main and I/O threads on the same CPU socket, was expected to be better than Fig. 5(c) in terms of cache effectiveness. However the CPU cache was not effective in our evaluation since each thread had multiple CBs larger than the CPU cache size. Furthermore the both threads used computing resources of the same CPU socket. As a result, we had I/O performance degradation. In contrast, the threads were separated into two CPU sockets in Fig. 5(c), and, as a result, this case could minimize resource utilization per CPU somehow and so I/O times were minimized in Fig. 5(c).

In order to examine the behavior of the TP-IO in the optimized HDF5, we checked communication and I/O times inside the TP-IO scheme as shown in Fig. 6. We show only the 8 I/O request case in the figure because this case outperformed others. In this figure, "comm" and "read" stand for communications and read operations inside TP-IO, respectively. "calculated" denotes multithreaded TP-IO times calculated by $max(t_{comm}, t_{read}) \cdot (1 - S_{CB}/S_{data}) + (t_{comm} + t_{read}) \cdot S_{CB}/S_{data}$, where t_{comm} and t_{read} denote mean communication and read times, respectively, and S_{CB} and S_{data} stand for the sizes of CB and amount of accessed data per process, respectively. In addition, the number of TP-IO cycles is also shown in every figure for reference.

The case of Fig. 6(c) shows shorter communication times than the case of Fig. 6(a) when total CB size is smaller than 32 MiB, while the times in the cases of Fig. 6(b) and (d) are longer than the case of Fig. 6(a). Read times in Fig. 6(c) are almost the same with the times in Fig. 6(a), while the times in Fig. 6(b) and (d) are longer than the times in Fig. 6(a). As a result, calculated times in Fig. 6(b), and (d) are longer than those in Fig. 6(a), while the times in Fig. 6(b) are shorter than the times in Fig. 6(a).

The more the number of MPI processes we have, the more communication times we may have. Data exchanges in the current TP-IO including our multi-threaded TP-IO are done by nonblocking point-to-point communications among MPI processes. Besides, small CB size leads to an increase in the number of TP-IO cycles. Thus it means an increase in communication times proportional to the number of TP-IO cycles and squared the number of MPI processes.

As a consequence of this performance evaluation, we note that it is better to deploy main and I/O threads on different CPU sockets. Furthermore placing a main thread on a CPU core which is close to a network interface card is preferable when communication times are comparable with or higher than file I/O times.

4   Related Work

Parallel netCDF (PnetCDF) [7] is also a well-known application-oriented parallel I/O library, which is also built on an MPI-IO layer as same as the HDF5 library is. PnetCDF also frequently utilize derived data types for non-contiguous access

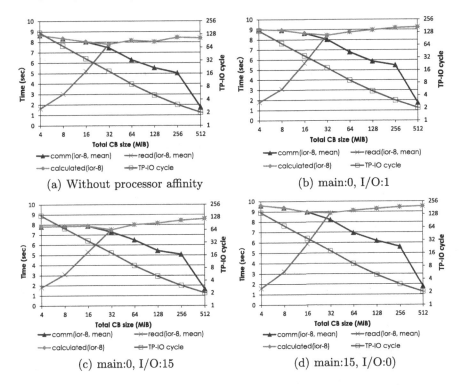

(a) Without processor affinity

(b) main:0, I/O:1

(c) main:0, I/O:15

(d) main:15, I/O:0)

Fig. 6. Times for communication and file reading by main and I/O threads, respectively, during I/O operations by `H5Dread` in addition to calculated times of the I/O operations and the number of TP-IO cycles in terms of total CB size

patterns, thus TP-IO is used in collective I/O for such access patterns. Our multithreaded TP-IO is also expected to improve parallel I/O performance of PnetCDF.

Both [4] and [9] showed excellent overlapping strategies by using a multithreaded scheme. However, they overlapped computations with file I/O. Therefore their approaches differ from ours in terms of technical target and strategy.

View-based I/O [3] is addressing to optimize TP-IO by eliminating the extra communication cost of generating both file and memory access patterns. Every client MPI process exchanges information to adjust assigned data region dynamically in the original TP-IO. In contrast, the view-based I/O sends file view data to every I/O aggregators prior to I/O operations. Thus the view-based I/O avoids communication costs to exchange information as the original TP-IO does.

In the extended TP-IO evaluated here, every process communicates once just before its I/O accesses in order to achieve higher performance. Blas et al. [2] utilized a GPFS file system [1] to realize background writing and read-ahead techniques. In the proposed TP-IO, a multithreaded method is used to overlap. In addition, optimization of MPI-IO implementations on a GPFS provided overlapped manner of I/O operations against data communications by using a double buffering scheme [13]. Although our proposal is similar to this idea, this idea is tightly coupled with the nature of GPFS, and thus it is file system dependent. While our proposal is independent of underlying file systems, making it usable on any kind of file system for which POSIX I/O and Pthreads are available.

5   Concluding Remarks

We have reported our performance optimization approach using processor affinity management in the TP-IO through performance evaluation for parallel I/O of HDF5. We evaluated non-contiguous accesses by using an HDF5 collective read API on a PC cluster system with a Lustre file system.

The multithreaded implementation without the processor affinity management minimized I/O times about 4.5 % relative to the times of the original TP-IO implementation at 32 MiB in total CB size, for instance. Moreover, the multi-threaded TP-IO with the processor affinity management minimized I/O times about 2.9 % relative to the times of the multithreaded one without processor affinity management at the same total CB size. The improvement was realized by a good CPU core binding. Main and I/O threads were deployed on CPU cores 0 and 15, respectively, where the core 0 was closer to a Myrinet interface card compared with the core 15, in the evaluation. It is also noted that the performance was degraded when we had inappropriate CPU core bindings, where a main thread was on the CPU core 15, which was far from the Myrinet interface card or main and I/O threads were on the same CPU socket. The same effect was also observed in further examinations for communication and I/O operation times inside ROMIO.

Consequently it is remarked that placing the main thread on a CPU core which is close to the Myrinet interface card is preferable when communication times were comparable with or higher than file I/O times like the performance evaluation. Furthermore the I/O thread should be placed on a different CPU socket apart from a CPU socket on which a main thread is running.

I/O performance evaluation with different CPU core binding patterns that we have not measured is our future work for further analysis. Besides, a more detailed evaluation by using, e.g., system resource monitoring such as CPU and I/O utilization, also remains as future work.

Acknowledgment. This research work is partially supported by JST CREST. The authors would like to thank the Information Technology Center, the University of Tokyo for their assistance in using the T2K-Todai cluster system.

References

1. General Parallel File System. http://www-03.ibm.com/systems/software/gpfs/
2. Blas, J.G., Isaila, F., Carretero, J., Singh, D., Garcia-Carballeira, F.: Implementation and evaluation of file write-back and prefetching for MPI-IO over GPFS. Int. J. High Perform. Comput. Appl. **24**, 78–92 (2010)

3. Blas, J.G., Isaila, F., Singh, D.E., Carretero, J.: View-based collective I/O for MPI-IO. In: CCGRID, pp. 409–416 (2008)
4. Dickens, P., Thakur, R.: Improving collective I/O performance using threads. In: Proceedings of the Joint International Parallel Processing Symposium and IEEE Symposium on Parallel and Distributed Processing, pp. 38–45 (1999)
5. Institute of Electrical, Electronic Engineers: Information Technology – Portable Operating Systems Interface – Part 1: System Application Program Interface (API) – Amendment 2: Threads Extensions [C Languages] (1995)
6. IOR. http://sourceforge.net/projects/ior-sio/
7. Li, J., Liao, W.K., Choudhary, A., Ross, R., Thakur, R., Gropp, W., Latham, R., Siegel, A., Gallagher, B., Zingale, M.: Parallel netCDF: a high-performance scientific I/O interface. In: Proceedings of the 2003 ACM/IEEE Conference on Supercomputing. SC '03, p. 39. ACM, Nov 2003
8. Lustre. http://wiki.lustre.org/index.php/Main_Page
9. Ma, X., Winslett, M., Lee, J., Yu, S.: Improving MPI-IO output performance with active buffering plus threads. In: Proceedings of the 17th International Parallel and Distributed Processing Symposium (IPDPS'03), p. 68b. IEEE Computer Society, Apr 2003
10. MPI Forum. http://www.mpi-forum.org/
11. MPICH. http://www.mpich.org/
12. Myricom Inc. http://www.myricom.com/
13. Prost, J.P., Treumann, R., Hedges, R., Jia, B., Koniges, A.: MPI-IO/GPFS, an optimized implementation of MPI-IO on top of GPFS. In: SC '01: Proceedings of the 2001 ACM/IEEE Conference on Supercomputing, p. 58. IEEE Computer Society (2001)
14. Thakur, R., Gropp, W., Lusk, E.: On implementing MPI-IO portably and with high performance. In: Proceedings of the Sixth Workshop on Input/Output in Parallel and Distributed Systems, pp. 23–32 (1999)
15. Thakur, R., Gropp, W., Lusk, E.: Optimizing noncontiguous accesses in MPI-IO. Parallel Comput. **28**(1), 83–105 (2002)
16. The National Center for Supercomputing Applications. http://hdf.ncsa.uiuc.edu/HDF5/
17. Tsujita, Y., Muguruma, H., Yoshinaga, K., Hori, A., Namiki, M., Ishikawa, Y.: Improving collective I/O performance using pipelined two-phase I/O. In: Proceedings of the 2012 Symposium on High Performance Computing. HPC '12, pp. 7:1–7:8. Society for Modeling and Simulation International, CD-ROM, Mar 2012

Storage Management Systems for Organizationally Distributed Environments PLGrid PLUS Case Study

Renata Słota[1](✉), Lukasz Dutka[2], Michał Wrzeszcz[2], Bartosz Kryza[2],
Darin Nikolow[1], Dariusz Król[2], and Jacek Kitowski[1,2]

[1] Department of Computer Science, Faculty of Computer Science,
Electronics and Telecommunications, AGH University of Science and Technology,
Al. Mickiewicza 30, 30–059 Krakow, Poland
[2] ACC Cyfronet AGH, AGH University of Science and Technology,
Ul. Nawojki 11, 30–950 Krakow, Poland
{rena,dutka,wrzeszcz,bkryza,darin,dkrol,kito}@agh.edu.pl

Abstract. With the increasing amount of data the research community is facing problems with methods of effectively accessing, storing, and processing data in large scale and geographically distributed environments. This paper addresses major data management issues, in particular use cases and scenarios (on the basis of Polish research community organized around the PLGrid PLUS Project) and discusses architectures of data storage management systems available in both PL-Grid and other similar federated environments. On that basis, a concept of a new meta storage system, named VeilFS, is presented. The proposed system unifies file access methods for geographically distributed large scale systems and hides complexity of data access and management in such environments. However, it should be emphasized that the main purpose of this article is identification and discussion about users' requirements and existing solutions. The VeilFS system will be described in detail in the future.

Keywords: Storage system · Data management · Organizationally distributed environment · Grid · Cloud

1 Introduction

Modern scientific research is guided by several paradigms including theoretical methods, experimental science, simulation and most recently an emerging 4th paradigm. The 4th paradigm relates to scientific research based on processing and analysis of large amounts of data, which is believed to be the most important research paradigm in the coming years [10]. Applications ranging from natural science research, bioinformatics, mathematics, economy as well as social sciences require more and more computational and data processing and storage capabilities to handle increasingly complex algorithms and applications. Such data intensive applications [4] are highly heterogeneous in terms of their

R. Wyrzykowski et al. (Eds.): PPAM 2013, Part I, LNCS 8384, pp. 724–733, 2014.
DOI: 10.1007/978-3-642-55224-3_68, © Springer-Verlag Berlin Heidelberg 2014

architecture (e.g. some use simple files others use complex centralized databases), data complexity (e.g. some use raw measurements while other process complex data structured), time constraints (e.g. real time processing vs batch processing) and several other aspects.

In the near future the research community will be faced with major problems concerning access, store, search and sharing data generated either by sensors or through simulations and needed as input to complex processing jobs and workflows. This is often referred to in the literature as the 'Big Data' revolution [15], often defined not only through overall data volume, but also such aspects as variety of data and the processing speed required in order to actually use the potential of access to such amount of information. the Polish research community organized around the PL-Grid infrastructure [13] in the frame of the PLGrid PLUS project [17]. It is worth to emphasize that the topic of data management in Grid environment is investigated by our team for many years [8,20]. The rest of the paper is organized as follows. Section 2 presents PLGrid PLUS project use-cases of user data management. The main data storage management systems used in the PL-Grid infrastructure are described in Sect. 3. Section 4 suggests other possible solutions, in the form of state of the art, to enable provision of storage resources in grid environments. Section 5 summarizes the use cases that need to be supported by a new system solution in our opinion and presents our vision on this issue. Finally, Sect. 6 concludes the paper.

2 Data Access Requirements of Domain Grids – Use Cases

PLGrid PLUS is a continuation of the PL-Grid project which provided basic Grid infrastructure spanning 5 major supercomputing centers in Poland. The follow-up project is focused on integrated domain specific research communities and helping them use the distributed high performance computing resources to improve their everyday research activities. These 13 domain specific Grids include such diverse scientific communities as for example Ecology, High Energy Physics, Bioinformatics, Nanotechnology, Material Science, Acoustics and Astronomy.

Table 1 presents the results of research on users' requirements concerning typical data use cases. The results were obtained from a questionnaire distributed among the representatives of each of the domains and highlights the current problems with typical data management use cases.

The Importance column reflects the number of specific scientific domains adopted in the PLGrid PLUS project for which this use case is significant and the Difficulty column shows whether currently most users find it easy or not to perform such a use case.

Currently, the existing data management approach in PL-Grid shows up some inconveniences for both the users and the administrators, due to: heterogeneous solutions between different computer centres, various long term storage policies in computer centres, difficult block access to files managed by Grid storage and

Table 1. PLGrid PLUS user data management use cases.

Use case	Importance	Difficulty	Comment
Archivization	High	High	Archivization is only available outside of the Pl-Grid infrastructure through PLATON infrastructure [2]
Temporary files access	High	Low	Temporary files are stored locally at sites using NFS or Lustre, however different sites have different data deletion policy and different paths
Permanent files access	High	Average	Files can be accessed through DPM [24], LFC [5], Unicore [6]. Block access is possible through RFIO protocol. However DPM still requires specific commands for data management, which makes many users fall back to manual transfers
Data staging	High	Average	Users can specify necessary job input files in the job description file, provided these files are already registered in the permanent Grid storage (e.g. LFC)
Data transfer from/to Grid	High	Easy	Users can use simple SSH/SCP commands
Data transfer between sites	Average	Average	Users can manually transfer files between sites, either using Grid middleware (e.g. LFC) or manually over SSH based protocols.
Relational database access	Average	Easy	Users can create a custom database on a central MySQL server, this however poses certain performance and scalability issues
Metadata	Low	Average	Users can only search using the files logical names, however no mechanism for structured metadata descriptions is available

lack of advanced metadata support. One of the major problems faced by these users is the lack of uniform and transparent methods of data management, which result in non-optimal usage of storage and computing resources by manually managed data transfers using SSH based protocols for both file sharing and staging before job execution.

In the next sections we present our considerations on systems capable of providing storage resources in the context of the presented use cases and scenarios.

3   Data Storage in PL-Grid Infrastructure

Variety of storage management systems are used by the PL-Grid infrastructure. The reason for such high heterogeneity of storage originates from the following facts:

- users need storage resources which have different characteristics depending on the nature of applications,
- the local sites providing computational resources make autonomic decisions about the storage hardware and software, which should be used, depending on the requirements of their key users,
- the sites adopt some of the spare storage resources which already exist at the given location.

The storage hardware is not used directly by the PL-Grid users but via some storage software layers which arrange the storage resources into file systems. The mostly used ones are presented below.

3.1   Lustre

Lustre [7,14] is a parallel distributed filesystem for computational clusters.

The Lustre filesystem within the PL-Grid infrastructure is typically used for a high performance scratch filesystem. There are different Lustre instances on the different sites, which means that the data stored on this filesystem can only be shared within the local cluster of the given site. In order to keep these data from deleting they should be copied to a permanent storage outside of the local cluster, e.g., the home directory or LFC (described below). Cyfronet are shown in Table 2. The measured Lustre instance consists of disk array volumes attached via FC 8 Gbps interfaces to a set of servers which further provides the Lustre filesystem to the worker nodes of the cluster via InfiniBand interfaces. Sequential reads and writes have been tested using the Linux tool **dd** to measure the performance.

3.2   QStorMan

QstorMan [22,23], developed in PL-Grid project [12], is aimed at delivering storage QoS and resource usage optimization for applications which use the Lustre filesystem. QStorMan fulfills these goals by continuous monitoring of the Lustre nodes and dynamically forwarding data access requests to the most appropriate storage resources. The forwarding is done by defining storage pools containing selected storage resources and forcing the Lustre system to use the best storage pool for the given request taking into account the provided storage QoS requirements. QStorMan supports two types of usage [21]: via its API in the cases when small changes to the application's source code can be made or via switching the system libraries to use the QStorMan code for legacy applications. The usage of QStorMan has been proved to improve the data access efficiency of PL-Grid data-intensive applications.

3.3   LFC

LFC (LCG File Catalog) [5] is a storage software for meta-data management which provides common filesystem functionality for distributed storage resources.

Table 2. Performance of Lustre and LFC filesystems in Cyfronet's PL-Grid site

Test no.	Filesize [MB]	Lustre transfer rate		LFC transfer rate	
		Read [MB/s]	Write [MB/s]	Read [MB/s]	Write [MB/s]
1	128	20.7	15.8	40.1	110
2	512	18.9	26.4	99.7	109
3	1024	17.9	23.7	60.8	58.4
4	2048	25.6	35.5	59.5	27.4

LFC supports user initiated file replication for better data protection and availability. The common way of using LFC is via the command line utilities [26].
Another less popular way is by using the GFAL API [9] to access LFC directly
from the application's source code. Finally, a FUSE-based implementation of a
filesystem called GFAL-FS [9] can be used to provide access to the data in the
same manner as to a regular Unix-like filesystem. Unfortunately this is the least
efficient (compared to the previous two methods) way. Only read-only mode
is currently supported by GFAL-FS. Typical performance data for the LFC
filesystem using the storage resources installed at Cyfronet are shown in Table 2.
Sequential reads and writes have been tested using the LCG utility `lcg-cp` to
measure the performance.

4   Alternative Proposed Solutions - State of the Art and Discussion

Besides the tools presented in the previous section a number of data management systems have been developed for organizationally distributed environments.
These systems can be categorized based on supported use cases. In this section, a
few data management systems oriented on different use cases are discussed. The
standard POSIX filesystem interface is arguably a preferable interface to any
data management system for most applications. Hence, many efforts have been
undertaken to develop various tools, which would allow to abstract any specific
interface with the POSIX interface. Parrot [25] is a tool for attaching existing
programs to remote data management systems, which expose other access protocols, e.g. HTTP, FTP or XRootD, through the filesystem interface. It utilizes
the *ptrace* debugging interface to trap system calls of the program and replace
them with remote I/O operations. As a result, remote data can be accessed in the
same way as local files. However, using ptrace can generate significant overhead
[25], which can be unacceptable for HPC applications.

Another integrating tool, developed for Grid environments is iRODS [11,18],
which operates at a much higher level. It is a service for distributed storage
resources integration with metadata support and rule-oriented management. It
is often referred to as an adaptive middleware, since its data management behaviour can be adjusted to administrators/users needs using rules. It has an extend-
able modular architecture, which can be divided as follows:

- *A metadata catalog* called iCAT, which handles metadata information about actual data stored in the system, e.g. filename, size, or location. In addition, user defined metadata can be stored as well. Then, the user can search for data, which have been tagged. An administrator can query the metadata catalog directly using an SQL-like language to provide an aggregated information about the system. However, iCAT constitutes a central point of information about the system, thus it can be treated as a single point of failure.
- *Rule Engine* is responsible for tracking users actions and executing predefined rules. Each rule is a chain of activities provided by low-level modules, e.g. data replication or checksum calculation, built in or supported by the users or administrators to provide required functionality. Rules are triggered by various system events, e.g. putting/getting data or authentication.
- *Data Servers* are resources for actual data storage. In a basic setup, iRODS can store data using designated folders on any number of servers. In addition, by using plugin mechanism, iRODS can be integrated with external data management systems, e.g. GridFTP-enabled systems, SRM-compatible systems, or even Amazon S3 service. Hence, iRODS can provide a coherent view of user data stored in different systems.
- *User interface* for exposing the service to external clients. iRODS provides multiple user interfaces, starting with iCommands, which are counterparts of common unix commands, e.g. `ls`, `cp`, `cd` etc., web-based browser and Explorer for Windows. In addition, iRODS provides a FUSE-based file system, which can be mounted on any FUSE-compatible Unix system [1] and utilized with the POSIX interface.

The iRODS system has a built-in support for federalization, i.e. connecting organizationally distributed installations of iRODS. In such a case, users from one iRODS installation (referred to as Zone) can access data located in another Zone. Special user accounts can be created in a remote Zone with a pointer to the home Zone of the user, hence during the authentication process the user is challenged with the home Zone. Each iRODS Zone is a separated entity that can be managed in a different way and can include different storage resources. data access and federalization. However, it also has some drawbacks. iCAT, which is involved in most requests, is implemented as a relational database, hence it can be considered as a bottleneck of the whole system. Moreover, iRODS does not provide location transparency of data stored across multiple federated iRODS installations. In such a case, the user has to manage the location of data among different installations on his own. which requires fast and scalable storage systems. Hadoop Distributed File System (HDFS) [19] is a distributed file system designed to support the map-reduce framework called Hadoop. HDFS intends to store large data sets reliable, which have to be streamed at high bandwidth to user computation processes. Similarly to other distributed file systems, HDFS stores metadata and actual data separately. Hence, storage resources for actual data can be scaled easily just by adding more servers. HDFS takes care of distribution and replication data among available storage resources. On the other hand, the metadata server can be considered as a single point of failure, which constraints scalability of the system and decreases its fault-tolerance. Due to increasing popularity of the map-reduce paradigm, other tools are also developed. Tachyon [3] is a relatively new project, which provides a high performance for map-reduce applications by using memory aggressively.

In the PLGrid PLUS project heterogeneity is a very important issue so Cloud storage has been also investigated. It is designed to deliver many online storage services, whereas traditional storage systems are primarily designed for high performance computing and transaction processing. It places great importance on data security, reliability, and efficiency. Moreover, Cloud storage systems also support mass data management for providing public service support functions, and maintaining data in the background [27]. For our research, we have chosen OpenStack Object Store, known as Swift [16]. Swift is able to provide common file names within Grid and Cloud infrastructure of PLGrid PLUS project which is a very important feature for the users. We have compared it to LFS which is also able to provide common file names. Even at very small testbed (4 nodes) we have achieved transfer rates similar to LFC deployed on the PL-Grid infrastructure. We have verified that Swift can be used to stream files directly to process memory. However, Swift file sharing mechanism that is base on an API access key sharing or a session token, is more difficult to use for most of PL-Grid users than LFC file sharing mechanism based on Unix permissions. Moreover, the users have a lot of data stored at LFC so although, for some cases Swift would be better that LFC, it would be difficult to replace LFC with Swift. Results of this analysis confirm that choice of one type of storage for the PLGrid PLUS project is inconvenient.

5   Distributed Data Sources Veil - VeilFS

The proposed VeilFS system (see Fig. 1) unifies access to files stored at heterogeneous data storage management systems that belong to geographically distributed organizations. It is currently being developed by ACC Cyfronet AGH under PLGrid PLUS project. VeilFS addresses the main users' requirements described in Sect. 2. It provides a user space file system (FUSE) that wraps several types of storage that are used for archivization and storing of temporary and permanent files. To make use of VeilFS as simple as possible, the users do not have to choose type of storage they want to use. The files' locations are chosen by the system. The users operate only on logical names of files. node accesses files through FUSE. FUSE is connected with user account so many FUSE systems may exist at one worker node if processes of many users are running on it. Each FUSE cooperates with VeilFS which indicates location of files on the basis of their logical names. Information about users' files is stored in the distributed database so any user has coherent view of his/her data regardless the location where he connects to the system. Moreover, the user is always able to access his/her data - if the file is stored in other computing center, it is downloaded if needed. In the database, user defined metadata may be also stored.

To avoid creation of single points of failure and bottlenecks, no function is assigned to machines used by VeilFS. FUSE may send request to any machine

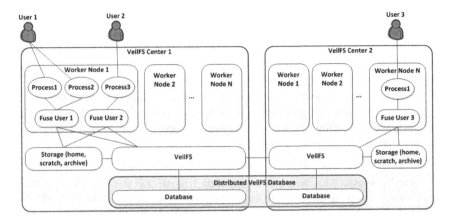

Fig. 1. VeilFS scheme

and it will be redirected to appropriate one. If one of the machines fails, the other one takes its functionality. The data in the database may be replicated to many physical machines and various directories (home, scratch, archive) may use redundant storage. All connections with VeilFS are encrypted, GSI authentication is used. All of that makes the proposed system secure.

Rules defined by administrators are used by VeilFS to select location for created files and to control the system in the center. On the basis of these rules files are migrated (e.g., least used files are archived) and user's activity is monitored (e.g., quote is controlled). VeilFS instances in computing centers are independent. However, the organization-wide rules are used to coordinate their cooperation. For instance, if user processes in one center often use a file that is stored in other center, this data may be permanently migrated.

6   Conclusions

Data processing and managing in large scale environments is the major problem which the research community will face in the near future. Our users' requirements analysis has shown that access to files is too complicated for many of them. Variety of used storage solutions confuses users. Users expect that data access will be simple using one tool - preferably based on standard POSIX filesystem interface. In short, the actual users expect access to the data in large scale computational environment in the same way as they do on their own personal computers. The fact of the distribution of the large scale computational environments they work with in computer centres should not bring more barriers, but be the opportunity for intra-community data sharing and collaboration. exist. However, they are still full of complexity and barriers for their users if we come to globally distributed environments. There is no transparency in data center selection in multiple data centres scenarios.

The proposed meta-filesystem - VeilFS unifies data access despite the geographical distribution of computational resources or heterogeneity of the actual storages used by the computing centers. The decision where files will be stored is made by the system - the user may only provide advisory information. Furthermore, the system migrates files when their usage profile changes which is completely transparent to the user. We believe that managing real file location by the system (the user operates only on logical names) is the right approach because the users are not always aware of the specificity of storage management systems so their choices may not be optimal.

Acknowledgments. This research is supported partly by the European Regional Development Fund program no. POIG.02.03.00-00-096/10 as part of the PLGrid PLUS project and AGH-UST grants no. 11.11.230.015 and 15.11.230.097.

References

1. FUSE: Filesystem in Userspace. http://fuse.sourceforge.net/ (2013). Accessed 21 April 2013 (Online)
2. PLATON Storage Service U4. http://www.storage.pionier.net.pl/ (2013). Accessed 21 April 2013 (Online)
3. Tachyon Project. http://tachyon-project.org/ (2013). Accessed 21 April 2013
4. Atkinson, M., et al.: Data-intensive research workshop report. Technical Report, e-Science Institute. http://research.nesc.ac.uk/files/DIRWS.pdf (2010)
5. Baud, J.P.B., Caey, J., Lemaitre, S., Nicholson, C., Smith, D., Stewart, G.: LCG data management: from EDG to EGEE. In: UK e-Science All Hands Meeting, Nottingham, UK (2005)
6. Benedyczak, K., Rekawek, T., Rybicki, J., Schuller, B.: UNICORE data management: recent advancements. In: Romberg, M., Bala, P., Mller-Pfefferkorn, R., Mallmann, D. (eds.) UNICORE Summit 2011 Proceedings, Torun, Poland, 7–8 July 2011. IAS Series, vol. 9, pp. 24–27, Forschungszentrum Jülich (2011)
7. Braam, P.J., Schwan, P.: Lustre: the intergalactic file system. In: Ottawa Linux Symposium, June 2002
8. Dutka, Ł., Kitowski, J.: Application of component-expert technology for selection of data-handlers in CrossGrid. In: Kranzlmüller, D., Kacsuk, P., Dongarra, J., Volkert, J. (eds.) PVM/MPI 2002. LNCS, vol. 2474, pp. 25–32. Springer, Heidelberg (2002)
9. Grid File Access Library 2.0 official page. https://svnweb.cern.ch/trac/lcgutil/wiki/gfal2 (2013). Accessed 14 April 2013
10. Hey, A., Tansley, S., Tolle, K.: The Fourth Paradigm: Data-Intensive Scientific Discovery. Microsoft Research, Redmond (2009)
11. Hunich, D., Muller-Pfefferkorn, R.: Managing large datasets with iRODS: a performance analysis. In: Proceedings of the 2010 International Multiconference on Computer Science and Information Technology (IMCSIT), pp. 647–654 (2010)
12. Kitowski, J., et al.: Polish computational research space for international scientific collaborations. In: Wyrzykowski, R., Dongarra, J., Karczewski, K., Waśniewski, J. (eds.) PPAM 2011, Part I. LNCS, vol. 7203, pp. 317–326. Springer, Heidelberg (2012)

13. Kitowski, J., Turała, M., Wiatr, K., Dutka, Ł.: Pl-grid: foundations and perspectives of national computing infrastructure. In: Bubak, M., Szepieniec, T., Wiatr, K. (eds.) PL-Grid 2011. LNCS, vol. 7136, pp. 1–14. Springer, Heidelberg (2012)

14. Lustre. http://www.whamcloud.com/lustre/ (2013). Accessed 10 January 2013

15. Mills, S., Lucas, S., Irakliotis, L., Rappa, M., Carlson, T., Perlowitz, B.: DEMYSTIFYING BIG DATA: a practical guide to transforming the business of Government. Technical report. http://www.ibm.com/software/data/demystifying-big-data/ (2012)

16. OpenStack Object Storage ("Swift"). https://wiki.openstack.org/wiki/Swift (2013). Accessed 14 April 2013

17. PLGrid Plus project. http://www.plgrid.pl/en#section-1t (2013). Accessed 14 April 2013

18. Roblitz, T.: Towards implementing virtual data infrastructures a case study with iRODS. Comput. Sci. **13**(4), 21–33 (2012). http://journals.agh.edu.pl/csci/article/view/43

19. Shafer, J., Rixner, S., Cox, A.L.: The Hadoop distributed filesystem: balancing portability and performance. In: ISPASS, pp. 122–133, March 2010

20. Słota, R., Nikolow, D., Skitał, Ł., Kitowski, J.: Implementation of replication methods in the grid environment. In: Sloot, P.M.A., Hoekstra, A.G., Priol, T., Reinefeld, A., Bubak, M. (eds.) EGC 2005. LNCS, vol. 3470, pp. 474–484. Springer, Heidelberg (2005)

21. Słota, R.: Storage QOS provisioning for execution programming of data-intensive applications. Sci. Program. **20**(1), 69–80 (2012)

22. Słota, R., Król, D., Skałkowski, K., Orzechowski, M., Nikolow, D., Kryza, B., Wrzeszcz, M., Kitowski, J.: A toolkit for storage QOS provisioning for data-intensive applications. Comput. Sci. **13**(1), 63–73 (2012). http://journals.agh.edu.pl/csci/article/view/26

23. Słota, R., Nikolow, D., Kitowski, J., Król, D., Kryza, B.: FiVO/QStorMan semantic toolkit for supporting data-intensive applications in distributed environments. Comput. Inform. **31**(5), 1003–1024 (2012)

24. Stewart, G.A., Cameron, D., Cowan, G.A., McCance, G.: Storage and data management in EGEE. In: Proceedings of the fifth Australasian Symposium on ACSW frontiers, ACSW'07, Australia, vol. 68, pp. 69–77. Australian Computer Society Inc, Darlinghurst (2007)

25. Thain, D., Livny, M.: Parrot: an application environment for data-intensive computing. J. Parallel Distrib. Comput. Pract. **6**(3), 9–18 (2005)

26. Worldwide LHC Computing Grid. http://wlcg.web.cern.ch/ (2013). Accessed 10 April 2013

27. Zhou, K., Wang, H., Li, C.: Cloud storage technology and its application. ZTE Commun. **16**(4), 24–27 (2010)

The High Performance Internet of Things: Using GVirtuS to Share High-End GPUs with ARM Based Cluster Computing Nodes

Giuliano Laccetti[1], Raffaele Montella[2(✉)], Carlo Palmieri[2],
and Valentina Pelliccia[2]

[1] Department of Mathematics and Applications, University of Naples Federico II,
Complesso Universitario Monte S. Angelo, Via Cintia, Naples, Italy
`giuliano.laccetti@unina.it`
[2] Department of Science and Technologies, Centro Direzionale di Napoli,
Parthenope University of Napoli, Isola C4, 80143 Naples, Italy
`{raffaele.montella,carlo.palmeiri,valentina.pelliccia}@uniparthenope.it`

Abstract. The availability of computing resources and the need for high quality services are rapidly evolving the vision about the acceleration of knowledge development, improvement and dissemination. The Internet of Things is growing up. The high performance cloud computing is behind the scene powering the next big thing. In this paper, using the GVirtuS, general purpose virtualization service, we demonstrate the feasibility of accelerate inexpensive ARM based computing nodes with high-end GPUs hosted on x86_64 machines. We draw the vision of a possible next generation of low-cost, off the shelf, computing clusters we call Neowulf characterized by high heterogenic parallelism and expected as low electric power demanding and head producing.

Keywords: Hyerarchical parallelism · Hybrid algorithms · Adaptive algorithms · Multidimensional integration

1 Introduction

The Cloud Computing is an internet-based model in which virtualized and standard resource are provided as a service over the Internet. It provides a minimal management effort or service provider interaction and users interact with a virtual and dynamically scalable set of resources that can manage depending on their needs. Cloud Computing providers differ for the service provisioned and for the kind of the cloud architecture. The main consolidated service models are: Software as a Service (SaaS), Platform as a Service (PaaS) and Infrastructure as a Service (IaaS).

The High Performance Computing (HPC) is one of the leading edge disciplines in information technology with a wide range of demanding applications in science [12,13], engineering, economy, medicine [1] and creative arts [7].

R. Wyrzykowski et al. (Eds.): PPAM 2013, Part I, LNCS 8384, pp. 734–744, 2014.
DOI: 10.1007/978-3-642-55224-3_69, © Springer-Verlag Berlin Heidelberg 2014

The High Performance Cloud Computing (HPCC) model might offer a solution applying the elasticity concept of cloud computing to HPC resources, resulting in an IaaS delivery model. The cloud computing approach promises increased flexibility and efficiency in terms of cost, energy consumption and environmental friendliness [11] changing the point of view on performance contract systems [3].

Researchers and developers have become interested in harnessing this power for general-purpose computing, an effort known collectively as GPGPU (for General-Purpose computing on the GPU). Especially in the field of parallel computing applications, virtual clusters instanced on cloud infrastructures suffers from the poorness of message passing performances between virtual machine instances running on the same real machine and also from the impossibility to access hardware specific accelerating devices as GPUs. Recently, scientific computing has experienced on general-purpose graphics processing units to accelerate data parallel computing tasks. Presently, virtualization allows a transparent use of accelerators as CUDA based GPUs, as virtual/real machines and guest/host real machines communication issues rise serious limitations to the overall potential performance of a cloud computing infrastructure based on elastically allocated resources using split-driver based components as GVirtuS [6].

The Internet of Things (IoT) services are build on the top of other services as a sort of construction game thanks to well documented public interfaces strongly leveraging on different web services technologies. IoT generally refers to uniquely identifiable objects and their virtual representations in an Internet-like structure. It is interesting consider a large number of this low power, low-performance processors teamed up to build a data center with similar processing power than regular CPUs, but smaller energy consumption. ARM processors, designed for the embedded mobile market, operate at about 1 GHz frequencies and consume just 0.25 W. There is already a significant trend towards using ARM processors in data servers and cloud computing environments. Those workloads are limited by the I/O and memory systems, not by the CPU performance. Recently, ARM processors are also taking significant steps towards increased double precision (DP) floating point (FP) performance, making them competitive with state-of-the-art server performance. The ARM Cortex-A15, targeted as the computing unit in the Barcelona Supercomputing Center Mont Blanc project, will increase super-scalar issue to two arithmetic instructions per cycle, and has a fully pipelined FMA unit, delivering 4 GFLOPS at 1 GHz, on potentially the same 0.25 W budget, achieving 16 GFLOPS/W. The new ARMv8 instruction set, which will be implemented in future generations of ARM cores, features a 64-bit address space, and adds DP to the NEON SIMD ISA1, allowing for 8 ops/cycle on an A15 pipeline: 8 GFLOPS at 1 GHz, for 32 GFLOPS/W.

In this paper we present our preliminary results in accelerating inexpensive HPC clusters, known as Beowulf clusters, made by off the shelf computing components using of low power ARM based computing nodes grouped in sub-clusters leveraging on one or more high-end GPGPU devices hosted on accelerator nodes. We perform some really promising experiments setting up a controlled testing environment imitating the core of a more complex architecture.

The rest of this paper is organized as follows: in the section two we draw out our vision of the next generation of really hybrid HPC clusters accelerated by Internet of Things based components and high-end GPUs; the third section deals with design and technical issues of the hybrid GPU/x86_64/ARM software architecture using GVirtuS as transparent bridge between the ARM living applications and the GPUs. The section number four is on implementation details, while in the one number five some tests and preliminary results are described and discussed. Finally, the last section, the sixth, is about the usual conclusions and future directions on those promising issues.

2   Vision and Contextualization

In the world of supercomputing the two top charts, Top 500 and Green 500, show, we have two trends: the number of core increases thanks the use of dedicated accelerators (GPUs, CPU array boards) and the compute/cost efficiency is increasing its important in the technology development, so, in the future the two charts will merge in just one considering the environmental (and economical) footprint of a HPC iron giant as a primary requirement. For many applications as operational computations [10] or for the cloud hosting providers the energy saving is no more a freak item but a mandatory issue. In the recent past a good amount of the world spread computing power has been achieved using the low/medium costs off the shelf Beowulf commodity clusters. A Beowulf is a cluster of machines interconnected by a high performance network employing the message-passing model for parallel computation. The key advantages of this approach are high performance for low price, system scalability and rapid adjustment to new technological advances. The latter point is the key for the next step of the Beowulf evolution in the vision described in this paper. As the now days CPU computing power increases, the need for electric power rises needing more cooling. The availability of Internet of Things derived ARM CPUs in their high performance incarnation (64 bit, multicore) lead the HPC world to ARM based clusters powered with on chip or on board GPUs. The idea we show here is dedicated to the low-end / middle-end in house solutions designing what could be defined as Neowulf the next generation of Beowulf clusters (Fig. 1).

The computing nodes of a regular old-style cluster behave as input/output nodes for ARM based inexpensive sub-clusters. In this way the amount of heat producers decrease while the high computing power demanding applications have to be refactored in order to fit this new heterogenic approach. Tanks to the software component we show in this paper, these devices are seen by each of the ARM based sub-cluster computing nodes as directly connected to them in a transparent way. This vision permits to gain more computing power reducing the expensive, power hungry and heat producer x86_64 based computing nodes, increase the parallelism at the sub-cluster level and, last but not the least, unchain the high-end GPGPU power to ARM based computing nodes.

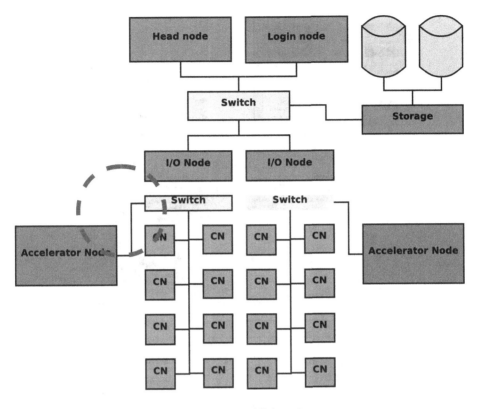

Fig. 1. The "Neowulf" big picture.

3   Design and Technical Issues

We use the GVirtuS framework model in order to design of our split driver implementation classically parted in front-end, communicator and back-end.

The front-end is a kernel module that uses the driver APIs supported by the platform. The interposer library provides the familiar driver API abstraction to the guest application. It collects the request parameters from the application and passes them to the back-end driver, converting the driver API call into a corresponding frontend driver call. When a callback is received from the frontend driver, it delivers the response messages to the application. In GVirtuS the front-end runs on the virtual machine instance and its implemented as a stub library.

The communicator maps the request parameters from the shared ring and converts them into driver calls to the underlying wrapper library. Once the driver call returns, the backend passes the response on the shared ring and notices the guest domains. The wrapper library converts the request parameters from the backend into actual driver API calls to be invoked on the hardware. It also relays the response messages back to the backend. The driver API is the vendor provided API for the device. The back-end is a component serving frontend

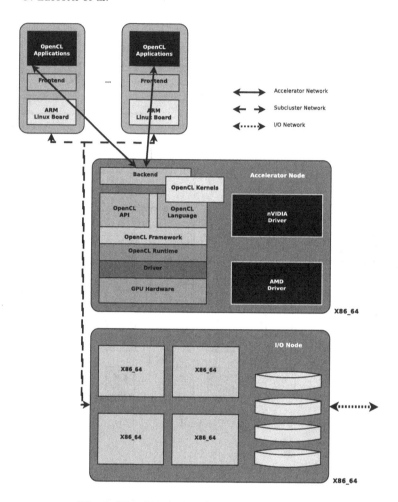

Fig. 2. The GVirtuS on ARM block diagram.

requests through the direct access to the driver of the physical device. This
component is implemented as a server application waiting for connections and
responding to the requests submitted by frontends. In an environment requiring shared resource the back-end must offer a form of resource multiplexing.
Another source of complexity is the need to manage multithreading at the guest
application level (Fig. 2).

3.1 GVirtuS on ARM

The GVirtuS porting on arm idea raised from different application fields such as
High Performance Internet of Things (IPIoT) and HPC. In HPC infrastructures
the ARM processors are used as computing nodes often provided by tiny GPU
on chip or integrated on the CPU board. We developed the idea to share one or

more regular high-end GPU devices hosted on a small number of x86 machines with a good amount of low power/low cost ARM based computing sub-clusters better fitting into the HPC world.

From the architectural point of view this is a big challenge because involving word size, endianness and programming models. For our prototype we used the 32 bits ARMV6K processor supporting both big and little endian so we had to set the little endian mode in order to make data transfer between the ARM and the x86 full compliant. Due to the prototypal nature of the system all has been set to work using 32 bits. The solution is the full recompilation of the framework with a specific reconfiguration of the ARM based system. As we will migrate on 64 bits ARMs this point will be revise.

In a previous work we used GVirtuS as nVidia CUDA virtualization tool achieving good results in terms of performances and system transparency [5]. In order to fit the GPGPU/x86_64/ARM application into our generic virtualization system we mapped the back-end on the x86_64 machine directly connected to the GPU based accelerator device and the front-end on the ARM board(s) using the GVirtuS tcp/ip based communicator.

We chose to design and implement a GVirtuS plugin implementing OpenCL. This have been strongly motivated by several issues:

1. Since the CUDA version 4 the library design appears to be made not fitting with the split driver approach on which leverages GVirtuS and other similar products [];
2. The OpenCL is intrinsically open and all interfaces are public and well documented and, above all, work with nVidia devices, but is not limited to a particular vendor or architecture as GVirtuS itself;
3. OpenCL applications can be compiled directly on the ARM board without any installation of ad hoc libraries.

3.2   GVirtuS - OpenCL Plugin

OpenCL (Open Computing Language) is an open standard and royalty-free allowing to perform multi/single core generale purpose programming on highly heterogeneous systems. OpenCL allows developers to write their code once and run on CPUs and GPUs and different accelerator boards as mic based Intel Phi. In order to access a GPU in a virtual environment has been developed a wrapper for libOpencl.so. The virtualized library has the same interface of the original one and the independence from the communicator is guaranteed. The compatibility between the virtualized interface and libOpenCL.so allow the users to get a transparent virtualization system to run OpenCL applications. It is possible to run any of OpenCL applications without writing or recompile anything. Each GVirtuS OpenGL plugin components participate as follows:

Front-end side: For each OpenCL routine a stub method has been implemented with the same interface of the original one. All the stubs method have a common implementation consisting in the next five steps:

- Create a connection between back-end and front-end and flush all the buffers;
- Each parameters will be sent to the back-end through the input buffer;
- Request the execution of a routine using its name as parameter;
- Get and Use the exit code only if the execution is successful;
- Return the exit code the same one as the OpenCL routine.

Back-end side: Back-end has a stub method for each OpenCL routine in order to handle the frontend requests. All the handlers method have a common implementation consisting in the next five steps:

- Deserialize all the parameters from the input buffer;
- Execute the OpenCL routine and store the exit result;
- Insert the output parameters in a new buffer;
- Create an object Result containing the previous created buffer and the exit code;
- Exit and deliver the result to the frontend.

There are tree main input parameters types available:

- Host Pointer: back-end and front-end have different addressing space so a valid pointer on the front-end is invalid on the back-end and vice-versa. Aligning the addressed region makes the address translation.
- Device Pointer: the memory address is sent to the back-end or front-end. There is no need for translation because both, be and fe, refer to the device addressing space.
- Variables: It is really simple to add a scalar variable as a parameter.

In order to make the implementation effective and high performance, but with a good trade off in development straightforwardness we deeply used an OOP coding approach.

4   Implementation

The implementation, in C++ for all components, on the back-end side is related to an x86-based multi-core hardware platform with multiple accelerators attached via PCIe devices, running Linux as both host and guest operating system. In the font-end we used the same core running in a similar, but ARM based, Linux environment.

4.1   OpenCLFrontend

The OpenCLFrontend class establishes connections with the back-end and executes the OpenCL routine through the compiled library libGvirtus-frontend. The constructor method creates an object of the class Frontend from the libGvirtus-frontend library using the method GetFrontend using a factory/instance design pattern. All the stubs methods have a common schema. Every stub follows the same interface of the handled OpenCL routine. The first step is to get the unique instance of the GVirtus Frontend class. This task is accomplished by the constructor method. The Prepare method reset the input buffer that will contain the parameters to send to the back-end. After that all the parameters are inserted in to the input buffer. The execute method forward the request for the routine using the name of the routine as parameter. If the method is successfully executed so we can get the output parameters. At last the method GetExitCode returns the exit code of the routine executed by the backend. The clGetDeviceIDs routine can be used to obtain the list of available devices on a platform. This simple explicative schema is common to all the stubs coded.

4.2 OpenCLBackend

The main task of GVirtuS back-end is to start a communication in server mode and waiting then accepting new incoming connections. It handles the loading of plugins previously installed. GVirtuS back-end invokes the GetHandler method in order to create a new instance of OpenclHandler class containing all the methods needed in order to serve the requests of OpenCL routine execution. In this class its possible to find all the methods to handle the execution of OpenCL routines. In the OpenclHandler class there is a table, mpsHandlers, associating function pointers to the name of the routines, so any routine can be handled in the right way. As in the front-end there is a stub method for each OpenCL method, in the back-end there is a function managing the execution of each method.

5 Evaluation

We set a prototypal hardware environment in order to evaluate the performance on ARM acceleration using external x86_64 GPUs, the GVirtuS overhead and the result reliability of a software testing suite. That evaluation process has two specific goals: (1) check the software stack accountability; (2) gather results on performance test. The OpenCL SDK provides a software suite which each component performs computations in bot CPU and GPU modes checking the result coherence and showing the brute performance results. All tests available on the standard OpenCL SDL have been successfully run using the GVrtuS-OpenCL SDK. We used a Raspberry Pi Mod.B rev.2 ARM 11 equipped with Wheezy Raspbian Linux as computing node and a Genensis GE-i940 Tesla powered by an i7-940 2.93 GHz fsb, Quad Core HT 8 Mb cache with one nVIDIA Qudro FX5800 4 Gb as GP device and two nVIDIA Tesla C1060 4 Gb as GPGPU device as accelerator node. For those tests no I/O node has been provided and the setup is related on a single node sub-cluster. In this context the GVirtuS fron-end was run on the ARM computing nodes while the back-end has been executed on the acceleration node. We used the OpenCL version of the testing software known as MatrixMul, DotProduct and Histogram (Fig. 3). ScalarProd computes k scalar products of two real vectors of length m. Notice that an OpenCL thread on

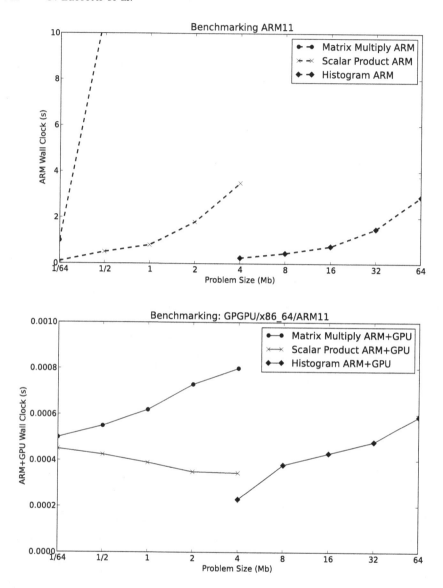

Fig. 3. ARM CPU without (up) and with (down) GPU acceleration.

the GPU executes each product so no synchronization is required. MatrixMul computes a matrix multiplication. The matrices are m n and n p, respectively. It partitions the input matrices in blocks and associates a OpenCL thread to each block. As in the previous case, there is no need of synchronization. Histogram returns the histogram of a set of m uniformly distributed real random numbers in 64 bins. The set is distributed among the OpenCL threads each computing a local histogram. The final result is obtained through synchronization and reduction

Table 1. Performance tests results.

Test	Input size (MB)	Relative (%)
MatrixMul	8	0.04 %
DotProduct	16	0.27 %
Histogram	64	0.65 %

techniques. The Table 1 is a synthesis of the obtained results considering the regular ARMV6K as the reference:

During the DotProduct testing process we change the problem dimension from 2^{20} to 2^{22}. The ARM performance are varying with the same problem dimension trend. The wall clock remains almost constant when is used the GPU acceleration. This demonstrates that the GVirtuS-OpenCL is fine working and the performances are not affected by the communication time. In the MatrixMul test the problem dimension has been varied in this steps $2^6 \times 2^9$, $2^9 \times 2^{12}$ and $2^{10} \times 2^{11}$. The performance results are pretty similar to the previous case with the GPU version having wall clock times almost unchanged. The Histogram has been used varying the problem size to 2^4, 2^5 and 2^6. The results are trivially the same.

6   Conclusions and Future Directions

In this paper has been presented our preliminary results about the design and the implementation of an OpenCL wrapper library as GVirtuS framework plugin. The most challenging result achieved by our work is the implementation of a base tool unchaining the development of really distributed and heterogenic hardware architectures and software applications. The experiments we performed validate our promising vision. The incredible performance results we achieved, the wall clock using acceleration is less than the 1 % compared with the non-accelerated ARM board, have been affected by the computing power of the ARM side: they need for more investigation and developments. The next step will be setup a sub-cluster made by high performance ARM based boards provided by multicore ARM 64 bit CPUs and high bandwidth network interfaces. We expect some improvements from the ARM side, but even a better scalability because a more performing communication. In this scenario some other actors will get playing as the use of MPICH [2] for ARM to ARM and ARM to x86_64 message passing, the OpenMP for intra ARM board parallelism and, above all, one or more GPU devices hosted on the accelerator node have to be multiplexed by several ARM processes. As long range future directions we planned a complete reverse of the point of view has been planned: using GVirtuS components in order to abstract and virtualize the ARM HPC sub-cluster acting as an accelerator board for x86_64 machines and applications on instruments shared on the cloud [4,8].

References

1. Boccia, V., D'Amore, L., Guarracino, M.R., Laccetti, G.: A grid enabled PSE for medical imaging: experiences on MedIGrid. In: Proceedings - IEEE Symposium on Computer-Based Medical Systems, pp. 529–536 (2005)
2. Gregoretti, F., Laccetti, G., Murli, A., Oliva, G., Scafuri, U.: MGF: a grid-enabled MPI library. Future Gener. Comput. Syst. **24**(2), 158–165 (2008)
3. Caruso, P., Laccetti, G., Lapegna, M.: A performance contract system in a grid enabling, component based programming environment. In: Sloot, P.M.A., Hoekstra, A.G., Priol, T., Reinefeld, A., Bubak, M. (eds.) EGC 2005. LNCS, vol. 3470, pp. 982–992. Springer, Heidelberg (2005)
4. Di Lauro, R., Lucarelli, F., Montella, R.: SIaaS-sensing instrument as a service using cloud computing to turn physical instrument into ubiquitous service. IEEE 10th International Symposium on Parallel and Distributed Processing with Applications (ISPA), 2012, pp. 861–862. IEEE (2012)
5. Giunta, G., Montella, R., Laccetti, G., Isaila, F., Blas F.J.G.: A GPU accelerated high performance cloud computing infrastructure for grid computing based virtual environmental laboratory. In: Dr. Constantinescu, Z. (ed.) Advances in Grid Computing. ISBN: 978-953-307-301-9, InTech (2011)
6. Giunta, G., Montella, R., Agrillo, G., Coviello, G.: A GPGPU transparent virtualization component for high performance computing clouds. In: D'Ambra, P., Guarracino, M., Talia, D. (eds.) Euro-Par 2010, Part I. LNCS, vol. 6271, pp. 379–391. Springer, Heidelberg (2010)
7. Maddalena, L., Petrosino, A., Laccetti, G.: A fusion-based approach to digital movie restoration. Pattern Recogn. **42**(7), 1485–1495 (2009)
8. Montella, R., Agrillo, G., Mastrangelo, D., Menna, M.: A globus toolkit 4 based instrument service for environmental data acquisition and distribution. In: Proceedings of the 3rd International Workshop on Use of P2P, Grid and Agents for the Development of Content Networks, pp. 21–28. ACM (2008)
9. Montella, R., Coviello, G., Giunta, G., Laccetti, G., Isaila, F., Blas, J.G.: A general-purpose virtualization service for HPC on cloud computing: an application to GPUs. In: Wyrzykowski, R., Dongarra, J., Karczewski, K., Waśniewski, J. (eds.) PPAM 2011, Part I. LNCS, vol. 7203, pp. 740–749. Springer, Heidelberg (2012)
10. Montella, R., Giunta, G., Laccetti, G.: Multidimensional environmental data resource brokering on computational grids and scientific clouds. In: Furht, B., Escalante, A. (eds.) Handbook of Cloud Computing, pp. 475–492. Springer, New York (2010)
11. Montella, R., Foster, I.: Using hybrid grid/cloud computing technologies for environmental data elastic storage, processing, and provisioning. In: Furht, B., Escalante, A. (eds.) Handbook of Cloud Computing, pp. 595–618. Springer, New York (2010)
12. Murli, A., Boccia, V., Carracciuolo, L., D'Amore, L., Laccetti, G., Lapegna, M.: Monitoring and migration of a PETSc-based parallel application for medical imaging in a grid computing PSE. In: Gaffeny, P.W., Pool, J.C.T. (eds.) Grid-Based Problem Solving Environments. IFIP, vol. 239, pp. 421–432. Springer, Boston (2007)
13. Pham, Q., Malik, T., Foster, I., Di Lauro, R., Montella, R.: SOLE: linking research papers with science objects. In: Groth, P., Frew, J. (eds.) IPAW 2012. LNCS, vol. 7525, pp. 203–208. Springer, Heidelberg (2012)

Workshop on Power and Energy Aspects of Computation

Monitoring Performance and Power for Application Characterization with the Cache-Aware Roofline Model

Diogo Antão, Luís Taniça, Aleksandar Ilic, Frederico Pratas,
Pedro Tomás, and Leonel Sousa[✉]

INESC-ID / Instituto Superior Técnico, Universidade de Lisboa, Lisbon, Portugal
las@inesc-id.pt

Abstract. Accurate on-the-fly characterization of application behaviour requires assessing a set of execution-related parameters at runtime, including performance, power and energy consumption. These parameters can be obtained by relying on hardware measurement facilities built-in modern multi-core architectures, such as performance and energy counters. However, current operating systems (OSs) do not provide the means to directly obtain these characterization data. Thus, the user needs to rely on complex custom-built libraries with limited capabilities, which might introduce significant execution and measurement overheads. In this work, we propose two different techniques for efficient performance, power and energy monitoring for systems with modern multi-core CPUs. Here we propose two monitoring tools that allow capturing the run-time behaviour of a wide range of applications at different system levels: (i) at the user-space level, and (ii) at kernel-level, by using the OS scheduler to directly capture this information. Although the importance of the proposed monitoring facilities is patent for many purposes, we focus herein on their employment for application characterization with the recently proposed Cache-aware Roofline model.

Keywords: Power and performance monitoring · Application characterization · Power and performance counters

1 Introduction

Modern computing systems are complex heterogeneous platforms capable of sustaining high computing power. While in the past designers have been able to improve processing performance by applying power hungry techniques, e.g., by increasing the pipeline depth and, therefore, the overall working frequency, such techniques have become unbearable due to the well known power wall. To overcome this issue, processor manufacturers turned to multi-core designs, typically by replicating a number of identical cores on a single die to increase performance, where each core includes a set of private coherent caches and dedicated execution engines, and in some cases hardware support for multiple threads. Although

R. Wyrzykowski et al. (Eds.): PPAM 2013, Part I, LNCS 8384, pp. 747–760, 2014.
DOI: 10.1007/978-3-642-55224-3_70, © Springer-Verlag Berlin Heidelberg 2014

these solutions are able to provide extra processing power, they also introduce additional complexity into the design, making it harder for application designers to fully exploit the available processing power. In particular, all cores share the access to a common higher level memory organization, typically containing the last level cache and the main memory, and the contention caused by multiple cores competing for the shared resources can drastically affect the execution efficiency. In addition to the these issues, current trends show us that future processors and applications will have to consider novel techniques to improve power and energy efficiency, potentially leading to extra architectural complexities.

In order to characterize and understand the behaviour of such complex computational systems, we require accurate real-time monitoring tools. These allow, for example, to identify application and architectural efficiency bottlenecks for real-case scenarios, thus giving both the programmer and the computer architect hints on potential optimization targets. While many profiling tools have been developed in the latest years, e.g., PAPI [1] and OProfile [2], it is not always easy to convert the acquired data into insightful information. This is particularly true for modern processors, which comprise very complex architectures, including deep memory hierarchy organizations, and for which several architectural events must be analysed.

Taking into account the complexity of modern processor architectures and the effects of having different applications running concurrently in multiple cores, Ilic et al. [3], proposed a Cache-aware Roofline model to unveil architectural details that are fundamental in nowadays application and architectural optimization. The Cache-aware Roofline Model [3] is a single-plot model that shows the practical limitations and capabilities for modern multi-core general-purpose architectures. It shows the attainable performance of a computer architecture as an upper-bound, by relating the peak floating-point performance (Flops/s), the operational intensity (Flops/byte), and the peak memory bandwidth for each cache level in the memory hierarchy (Bytes/s), all in the same plot. The model considers data traffic across both on-chip and off-chip memory domains, as it is perceived by the core [3].

In this paper we propose two monitoring methods that combine the advantages of the recently proposed Cache-aware Roofline Model [3], with real-time accurate monitoring facilities in a way that allows application developers to easily relate the application behaviour with the architecture characteristics, thus fostering new application optimizations. The two different monitoring tools proposed herein rely on the Hardware Performance Measurement Counters (HPMCs) and are able to extract in real-time important power and performance characteristics of the running application. While the first of these tools (SpyMon) aims at providing a simple environment that can be used from the user-space, it is complemented by a second tool (KerMon) which allows accurate application profiling by measuring application execution at the kernel level. Both tools are designed to be lightweight, easily adjusted for the user needs, and do not require changing the application code. It thus eases the monitoring of complex parallel applications that spawn multiple threads, e.g., OpenMP or OpenCL applications running on the CPU.

The results reported in this paper illustrate the differences between the two proposed monitoring tools, and show the importance of using such monitoring techniques for the understanding and characterization of application execution on modern general-purpose processors. Overall both monitoring methods are able to provide insightful information about the behaviour of the applications and how its execution is affected by the processor architectural limitations.

2   Performance Monitoring

Most modern processors contain HPMCs that can be configured to count micro-architectural events such as clock cycles, retired instructions, branch miss-predictions and cache misses. To count these events, a small set of Model-Specific Registers (MSRs) is provided by each architecture, which limits the total number of events that can be simultaneously measured. For example, on the Intel Sandy Bridge and Haswell architectures, the HPMC facility provides three MSRs types:

- the IA32_PERFEVTSELx MSRs, which allow selecting and configuring an event to be monitored by a corresponding IA32_PMCx MSR HPMC;
- the IA32_PMCx, which contains the actual counter value;
- the MSR_OFFCORE_RSPx MSR, which allows selecting and configuring other events such as the amount of DRAM traffic.

To setup an event, one needs to determine the adequate configuration word for that event, which needs to be written into one IA32_PERFEVTSELx MSR. In order to read the counter value it is necessary to read the corresponding IA32_PMCx MSR. Optionally, it is possible to set an initial value into the counter by writing the desired value into the corresponding IA32_PMCx MSR, which is particularly useful to avoid register overflows. If the configured event is an uncore event, it is also necessary to determine the adequate uncore configuration word and write it into the corresponding MSR_OFFCORE_RSPx MSR [4].

On the Intel architectures, HPMC does not provide energy or power consumption measurements. In order to assess this information, the Running Average Power Limit (RAPL) interface must be used. This allows simultaneously obtaining real-time energy consumption readings for several different domains, such as cores, package, uncore or DRAM.

The configuration and the reading procedures on these monitoring interfaces are not trivial and require special permissions to be accessed. In order to overcome the referred constraints from the user-space perspective, an MSRDriver library was developed, which provides low level routines for reading and writing to MSRs, allowing us to configure, reset and read the HPMCs. However, for the kernel-space monitoring tool (**KerMon**) it is possible to directly read and write the MSRs directly. In such a case the MSRDriver tool is not required.

2.1 Related Work

Many tools can be found in the literature that allow monitoring applications using the above referred HPMCs, e.g., PAPI [5], OProfile [2], PerfCtr [6], Permon2 [7], Intel PCM [8] and LIKWID [9].

Here we present a set of two different tools: **SpyMon**, which targets application monitoring from the user-space, and **KerMon**, which monitors processor events directly from the kernel-space. The main difference between most of the previously described tools and **SpyMon** is the fact that the latter follows a core-oriented topology, instead of process-oriented. Thus, it allows monitoring any given application thread, running at any given time, not worrying on which process may have launch them. This allows reducing overheads on the filtering process, keeping the tool simple and headed to its purpose. Furthermore, **SpyMon** allows for real-time reconfiguration of the HPMCs to be read. Although LIKWID is also one of the few tools which is core-oriented, that tool was designed as a wrapper to the user application and, so far, does not take power consumption into account. Moreover, it does not allow plotting the information according to the Cache-Aware Roofline Model.

Some of the above referred tools, do not access the HPMCs directly. They invoke a Linux kernel subsystem called "Perf Events" (originally "Performance Counters for Linux") that provides support for performance events monitoring on the user-space. It is available from Linux 2.6.31 [10] and it is still the only available framework native to the standard Linux kernel to support reading the HPMCs. Since this subsystem is available, user-space tools started to use it in detriment of other drivers, due to the latter requiring kernel or module compilation and installation. PAPI is an example of such a tool that nowadays uses "Perf Events" instead of another drivers such as the PerfCtr or Permon2 [5].

The "Perf Events" subsystem, although residing in the kernel, is designed to be invoked by user-space applications. It cannot be called in kernel space without several modifications. Furthermore, on restricted contexts, e.g. an interrupt handler or the task scheduler (where the functions must be fast, simple and cannot sleep), it is impossible to use "Perf Events" due to its inner complexity.

KerMon uses an approach different than **SpyMon**. It allows accurate profiling by measuring application execution at the kernel level. This allows transparent monitoring of an application even when it is scheduled to a different core. It thus has a wider range of scenarios where it can be used, providing a novel solution to the existing state-of-the-art.

3   User-Space Application Monitoring (SpyMon)

The proposed **SpyMon** tool is developed for monitoring CPU performance and power consumption. It is designed to be lightweight, simple and adjustable for the user needs. In contrast to most of the state-of-the-art tools, **SpyMon** follows a core-oriented approach, monitoring the behaviour of each logical CPU and therefore being able to capture the information of all running applications.

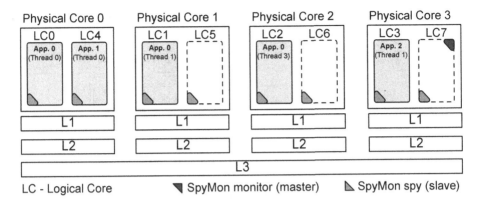

Fig. 1. Spacial perception of **SpyMon** while monitoring 5 threads from 3 applications.

The proposed tool is composed of a "monitor" and several "spies", following a master-slave methodology. The monitor (master) is the main process of the tool and is responsible for all the required initializations, for data analysis and processing and for controlling the flow of the whole tool. The spies (slaves) are lightweight processes bonded to a pre-defined logical core and have the sole purpose of configuring and fetching the performance counter readings.

The typical **SpyMon** configuration is to launch a spy to monitor each of the logical cores and to pin the monitor to the last logical core, as shown in Fig. 1. The communication is made by means of signals and communication pipes and it is minimized in order to reduce the cache pollution, thus reducing the **SpyMon** interference with the running applications. The pipes are bidirectional, allowing the master to change, at run-time, the events to be monitored in a logical core (e.g., because a running application changed state, or because it changed logical core). Moreover, other configurations are possible, e.g., if the user requires the monitoring of an application pinned to a specific logical cores, a single spy configuration can be adopted.

According to the user needs, the tool can also be configured regarding the performance events to monitor (e.g., retired instructions, L1 misses and number of loads), the time between samples and the number of samples, thus defining the execution time of the tool. The number of events is limited by the number of HPMCs and the number of events available on the architecture. The performance events configuration of each logical core is stored in temporary files during the tool execution. This allows the user to reconfigure the counters of a specific logical core at run-time.

SpyMon follows the steps shown in Fig. 2. When the application is launched, the master initializes the required drivers (step 1), namely the MSRDriver and RAPL (even though other drivers could be used, as long as they give the tool access to the special purpose registers). Then, the master creates and binds a process (spy) for each logical core to be monitored (step 2), where each spy is

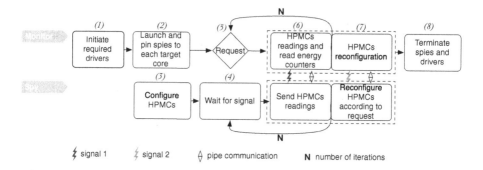

Fig. 2. **SpyMon** execution diagram.

responsible for configuring the HPMCs of its logical core (step 3). The configuration is sent by the master through a communication pipe. When the configuration is done, each spy gets into a sleep state until its action is required by
the monitor (step 4). The monitor has two possible requests (step 5): it may ask
for the counter readings (step 6) or for a reconfiguration of the HPMC events
to monitor in a specific (or several) logical core (step 7). In step 6, the master
signals each spy with signal 1. Each spy reads the raw values of the HPMCs and
sends them to the master using the established pipe. Then, the monitor reads
the energy HPMCs and derives power consumption, thus completing a sample.
In step 7, the monitor signals the corresponding spy (or spies) with signal 2 and
sends the required configuration through the pipe. The default request is the
one represented in step 6. However, before sending a request, the master checks
the time stamp of the performance events configuration files and, if any of them
has been updated since the last configuration, step 7 is executed. This procedure
is repeated periodically for each sample. During the time between samples, all
SpyMon processes are suspended in order to reduce the interference of the tools
with the running applications.

4   Kernel-Space Application Monitoring (KerMon)

KerMon follows a different approach than **SpyMon**; it uses a process-oriented
approach by monitoring applications directly from the kernel scheduler. Thus, it
allows getting accurate power and performance readings for all individual threads
spawn by a monitored application, at the cost of requiring kernel modification
by means of a patch. Naturally, the performance readings may still be affected
by other running applications, since some HPMCs are not specific to a logical
core (e.g., uncore events and energy consumption are shared between multiple
logical cores). In other cases, it is possible to get individualized readings, but
these are still influenced by other running threads (e.g., L1 and L2 cache misses
from logical core 0 are likely to be influenced by the thread executing on logical
core 4, as depicted in Fig. 1).

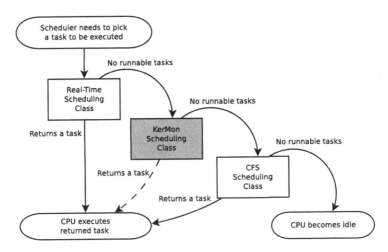

Fig. 3. Algorithm to pick a task to be executed. The dashed elements are implemented by the **KerMon** patch.

The Linux task scheduler is currently implemented in a manner that provides a set of classes that can be expanded to design different scheduling levels, organized by priority. The levels that co-exist in standard Linux distributions are, from highest to lowest priority, Real-Time scheduler and Completely Fair Scheduler (CFS). According to the task scheduler algorithm, each running process is assigned with a specific scheduling level and is scheduled by distributing processor run time to all processes in a fair manner, but taking into account the scheduling levels priority.

Thus, when the task scheduler is invoked, it will iterate over the scheduling classes until one of the classes returns a task to be executed, as shown in Fig. 3. That implies that the class order in this iterative process reflects its priority. Since the real-time class is the first in the task scheduler order, no CFS task can be executed while there is a real-time task in a runnable state [11].

Taking advantage of the existing scheduling algorithm, a **KerMon** class was introduced with a priority level between real-time and CFS (see Fig. 3). This allows reducing the interference of Linux system applications (e.g. daemons and terminals) on applications being monitored and provides implicit isolation for accurate benchmarking.

The new **KerMon** class is based on modifying a copy of the CFS class to enable power and performance measurement whenever a task is scheduled in/out or a schedule tick occurs. Since within the Operating System (OS) task scheduler interrupts and preemption are disabled, the implementation must be fast, lightweight and cannot go into sleep mode. Due to those strict restrictions, **KerMon** must rely on raw low-level Central Processing Unit (CPU) interfaces. Thus, for configuring and reading event counters, it uses the HPMC interface; to obtain a measure of time, it simply accesses the Time-Stamp Counter (TSC)

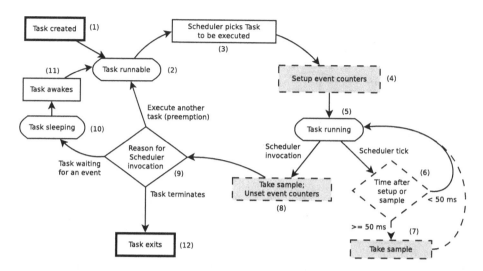

Fig. 4. KerMon task lifecycle. The dashed elements are implemented by the **KerMon** patch and the grey elements are performance monitoring actions.

register, reading the number of cycles since the system boot; energy consumption is obtained through the RAPL interface.

To monitor an application, one needs to use the standard Linux system call `sched_setscheduler` in order to change the application's scheduler class into **KerMon**. Then, the new system call *set_events* must be invoked to instruct **KerMon** which events need be monitored in this application. The **KerMon** scheduling class then works as depicted in the task life-cycle in Fig. 4. When a task is created (1) and becomes runnable (2), it is put on a logical core runnable list. Thus, when the scheduler picks this task among the runnable ones (3), **KerMon** configures and starts the event counters by writing to the appropriate HPMCs (4), just before the logical processor is assigned to the task execution (5). The **KerMon** will then read the HPMCs, forming a monitoring sample and storing it in a task associated buffer, in any of the following occasions:

(7) When a scheduler tick occurs[1] (6) and at least 50 ms of continuous unsampled runtime has occurred.

(8) When the scheduler is called for preemption purposes, either because the running time has expired or because the task goes into sleep mode while waiting for some event.

When the task is scheduled out, depending on the reason why the scheduler is invoked (9), the task can: go back to the runnable state (2) if it was preempted; sleep waiting for an event to occur (10); or exit, ending its life cycle.

An user-space application can then retrieve the samples of a monitored application through the new system call *read_event_log*, that fetches the samples from

---

[1] A scheduler tick is a periodic interruption to update the execution statistics and to check if is necessary to preempt the running task.

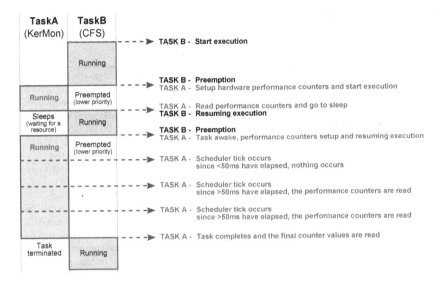

Fig. 5. Example of a timeline of two tasks sharing a CPU, where task **A** executes on **KerMon** while **B** executes on CFS. For simplification, in this figure, the tick period is 25 ms.

the buffer associated with the application. The samples can be post-processed for any purpose such as benchmarking. By having one buffer per monitored application, it is possible to monitor multiple applications in a simultaneous and independent manner. Moreover, it is possible to define different events for various applications.

Figure 5 presents an example timeline of two tasks executing on the same logical processor, where task **A** executes on **KerMon** scheduling class, while task **B** executes on the default CFS scheduling class. As it can be observed, applications scheduled on the **KerMon** class have higher priority than those on the CFS class. Thus, task **A** forces task **B** to be preempted whenever it goes to the runnable state. Also, monitoring of task **A** is triggered whenever it is preempted or whenever a tick occurs and more than 50 ms have elapsed since the last sampling event.

5   Experimental Results

As stated before, in order to assess the potential of the proposed methods, we show the outcome of combining them with the Cache-aware Roofline Model [3]. For that we have executed a group of floating-point applications, selected from the SPEC CPU 2006 benchmark set, on an Intel i7 3770 K processor[2].

---

[2] The Intel i7-3770 K is an Ivy Bridge based micro-architecture with 4 cores. It operates at 3.5 GHz and its memory organization comprises 3 cache levels of 32 KB, 256 KB and 8192 KB, respectively. The DRAM memory controllers support up to two channels (8 B) of DDR3 operating at 2×933 MHz.

Table 1. Sets of monitoring events used to characterize the execution.

Event Set	Even	Description
0	IVY_FP_COMP_OPS_EXE_SSE_FP_SCALAR_SINGLE	Number of SSE single-precision FP scalar μops executed
	IVY_FP_COMP_OPS_EXE_SSE_PACKED_SINGLE	Number of SSE single-precision FP packed μops executed
	IVY_SIMD_FP_256_PACKED_SINGLE	Number of AVX 256-bit packed single-precision FP instructions executed
	IVY_SIMD_FP_256_PACKED_DOUBLE	Number of AVX 256-bit packed double-precision FP instructions executed
1	IVY_FP_COMP_OPS_EXE_SSE_SCALAR_DOUBLE	Number of SSE double-precision FP scalar μops executed
	IVY_FP_COMP_OPS_EXE_SSE_FP_PACKED_DOUBLE	Number of SSE double-precision FP packed μops executed
	IVY_MEM_UOP_RETIRED_ALL_LOADS	Qualify any retired memory μops that are loads
	IVY_MEM_UOP_RETIRED_ALL_STORES	Qualify any retired memory μops that are stores
2	IVY_L1D_REPLACEMENT	Number of lines brought into the L1 data cache
	IVY_LLC_REFERENCE	Last level cache references
	IVY_L2_RQSTS_CODE_RD_MISS	Number of instruction fetches that missed the L2 cache
	IVY_OFF_CORE_MISSES_0	Number of L3 Misses

SSE - Streaming SIMD Extensions; FP - floating-point; AVX - Advanced Vector Extensions; μops - micro-operations

Moreover, since the Intel i7 3770 K processor only supports simultaneous access to 4 HPMCs, our experimental test set required repeating the execution of every application for each set of 4 events as specified in Table 1. The monitoring results obtained for the different events' sets were combined *a posteriori* by using as reference the number of instructions executed in each time interval (the number of instructions was obtained from the fixed counter IINSTRUCTION:RETIRED [4]).

Figure 6 shows the performance results for each SPEC CPU 2006 benchmark application as obtained with the two proposed monitoring methods. The results are plotted as points representing the overall (average) performance results, superimposed to the roofline model derived in [3]. Moreover, the points are plotted against different lines that represent the maximum achievable performance of the Intel 3770 K processor for double-precision MUL and ADD arithmetic instructions of different vectorization widths: scalar, Streaming SIMD Extensions (SSE) and Advanced Vector Extensions (AVX). Aside from *calculix*, which we analyse further below, all other applications show a similar performance behaviour in both cases. Moreover, we also show the overall power and energy results in Fig. 7 obtained with both monitoring methods. As for performance, also for

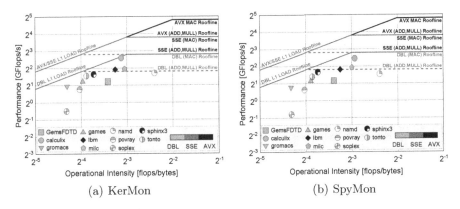

(a) KerMon   (b) SpyMon

Fig. 6. Application roofline model plot for i7 3770 K, showing the floating-point SPEC 2006 benchmarks; the application colour characterization was made according to average classification (double, SSE or AVX)

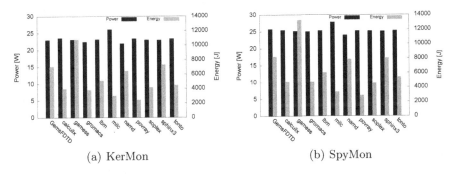

(a) KerMon   (b) SpyMon

Fig. 7. Average power and energy consumption for the execution of the different benchmarks

power and energy we observe the same trends for both methods, although Spy-Mon incurs in a small constant overhead due to the active monitoring threads.

In order to explore the potential of the features introduced in the proposed tools we show a more thorough analysis for two applications, *calculix* and *tonto*. A more fine-grained analysis of the application behaviour is important because, in many cases, the same application may show different behaviours during the execution. Events that may affect the execution over time are: fluctuations of the pressure on the memory subsystem, different types of instructions may be executed simultaneously creating bottlenecks in different points of the microarchitecture, overflow of HPMCs, system changes, execution uncertainties and other non-deterministic behaviours, just to name a few. The outcome of these effects is clearly depicted for *calculix* that gives a different result for each monitoring method. In fact, just by observing the results depicted in Fig. 6 would lead us to classify *calculix* as a memory-bound (KerMon) or as a compute-bound (SpyMon). Nevertheless, these effects are also valid for other applications, for

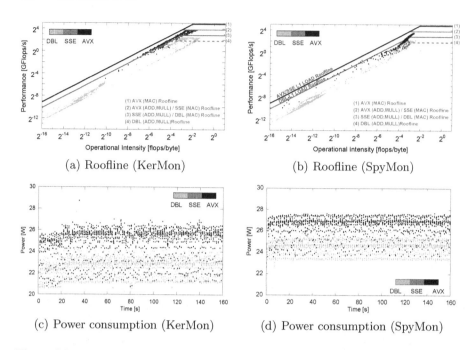

(a) Roofline (KerMon)   (b) Roofline (SpyMon)

(c) Power consumption (KerMon)   (d) Power consumption (SpyMon)

Fig. 8. Monitoring performance and power of Calculix; for simplification purposes, the L2, L3 and DRAM load rooflines for AVX, SSE and DBL instructions are not represented in plots (a) and (b)

which the single observation of the average point may be misleading, thus requiring a more detailed analysis. However, when the execution is broken into smaller samples (50 ms) one can perform a more accurate analysis. This is illustrated for *calculix* in Fig. 8(a, b), where we can observe that the application comprises two very distinct behaviours, which are now consistent across both methods: (*i*) on the bottom-left of the roofline plot a significant number of samples appears in the memory-bound region, and (*ii*) on the top-right side of the roofline we observe two patterns of points, one for AVX instructions, and a second for simple double-precision instructions, both more towards the compute-bound region. In particular, in (*ii*) we can observe that the program sections where AVX instructions are more predominant tend to be more affected by the memory execution than the program sections dominated by scalar double-precision instructions. Moreover, Fig. 8(c, d) shows the power results sampled over time.

The characterization of *tonto* is depicted in Fig. 9(a). In this case we also observe two distinct regions in the roofline, namely corresponding to scalar instructions and to SSE instructions. In contrast to the *calculix* case, here the application shows a memory-bound behaviour, although near the compute bound region.

The results obtained for the power consumption over time are also depicted in Fig. 9(b). In this case, the results show that the distinct regions observed on

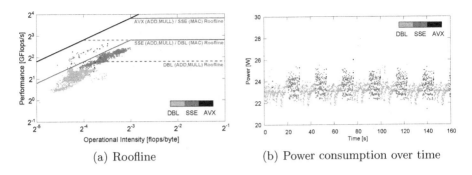

(a) Roofline   (b) Power consumption over time

Fig. 9. Monitoring performance and power of tonto using KerMon

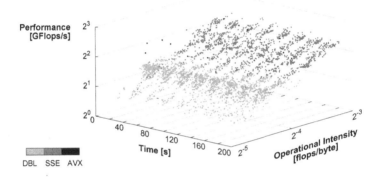

Fig. 10. Temporal representation of the roofline for tonto (KerMon)

the roofline plot, are also clearly differentiated in time, showing a heterogeneous execution pattern. We can conclude from this analysis that *tonto* uses different parts of the architecture more intensively in very distinct time instants. In order to better illustrate this behaviour, we combined the results shown in Fig. 9(a) with the execution over time and have created a temporal representation of the application Roofline. The plot in Fig. 10 eases the visualization of the different program sections over time.

6   Conclusions

In this paper we propose two new tools for application monitoring and characterization, which extract runtime information at different OS levels, namely: SpyMon at user-space level and KerMon at kernel-space level (OS scheduler). These tools combine the accuracy of hardware measurement facilities, which are integrated in modern multi-core architectures, with the Cache-Aware Roofline Model. This allows run-time characterization of application execution and allows extracting important guidelines for application optimization.

The experimental results presented in this paper show that both KerMon and SpyMon obtain similar performance characterization results. However, SpyMon, which performs core-oriented characterization, may show an increased power-usage if set to monitor cores which are not being used by any of the running applications. On the other hand, since KerMon requires changing the OS scheduler, it is harder to install in a system and requires root access. Despite these differences, in overall, both monitoring methods allow a user/programmer to get a clear picture of the behaviour of the application and how its execution is affected by the processor architectural limitations.

Acknowledgments. This work was partially supported by national funds through Fundação para a Ciência e a Tecnologia (FCT) under projects P2HCS (ref. PTDC/EEI-ELC/3152/2012), Threads (ref. PTDC/EEA-ELC/117329/2010), and project PEst-OE/EEI/LA0021/2013.

References

1. Browne, S., Dongarra, J., Garner, N., Ho, G., Mucci, P.: A portable programming interface for performance evaluation on modern processors. Int. J. High Perform. Comput. Appl. **14**(3), 189–204 (2000)
2. OProfile: About oprofile. http://oprofile.sourceforge.net/about/ (2012)
3. Ilic, A., Pratas, F., Sousa, L.: Cache-aware roofline model: upgrading the loft. IEEE Comput. Architect. Lett. **99**, 1 (2013)
4. Intel: Intel 64 and ia-32 architectures software developer's manual: volume 3b, pp. 120–251. http://download.intel.com/products/processor/manual/253669.pdf (2012)
5. PAPI: Papi: Supported platforms: Currently supported. http://icl.cs.utk.edu/papi/custom/index.html?lid=62&slid=96
6. Curtis-Maury, M., Nikolopoulos, D., Antonopoulos, C.: Pacman: A performance counters manager for intel hyperthreaded processors. In: 3rd International Conference on Quantitative Evaluation of Systems, 2006. QEST 2006, pp. 141–144 (2006)
7. Eranian, S.: Perfmon2: a flexible performance monitoring interface for linux, Citeseer (2006)
8. Intel: Intel performance counter monitor - a better way to measure cpu utilization. http://software.intel.com/en-us/articles/intel-performance-counter-monitor-a-better-way-to-measure-cpu-utilization (2012)
9. Treibig, J., Hager, G., Wellein, G.: Likwid: a lightweight performance-oriented tool suite for x86 multicore environments. In: 2010 39th International Conference on Parallel Processing Workshops (ICPPW), pp. 207–216. IEEE (2010)
10. LWN.net: Perfcounters added to the mainline. http://lwn.net/Articles/339361 (2009)
11. Molnar, I.: Goals, design and implementation of the completely fair scheduler. https://www.kernel.org/doc/Documentation/scheduler/sched-design-CFS.txt

Energy and Deadline Constrained Robust Stochastic Static Resource Allocation

Mark A. Oxley[1(✉)], Sudeep Pasricha[1,2], Howard Jay Siegel[1,2],
and Anthony A. Maciejewski[1]

[1] Department of Electrical and Computer Engineering, Fort Collins, CO 80523, USA
[2] Department of Computer Science, Colorado State University,
Fort Collins, CO 80523, USA
{mark.oxley,sudeep,hj,aam}@colostate.edu

Abstract. In this paper, we study the problem of energy and deadline constrained static resource allocation where a collection of independent tasks ("bag-of-tasks") is assigned to a heterogeneous computing system. Computing systems often operate in environments where task execution times vary (e.g., due to data dependent execution times), therefore we model the execution time of tasks stochastically. This research focuses on the design of energy-constrained resource allocation heuristics that maximize robustness against the uncertainties in task execution times. We design and evaluate a new resource allocation heuristic based on Tabu Search that employs dynamic voltage and frequency scaling (DVFS) and exploits heterogeneity by incorporating novel local search techniques.

Keywords: Heterogeneous computing · Static resource allocation · Power-aware computing · DVFS · Robustness

1 Introduction

The electricity used by data centers has increased by 56 % worldwide between the years 2005 and 2010 [10]. This increase highlights the need for energy-aware resource management for these data centers. Many of these data centers form heterogeneous computing environments that utilize a mixture of different machines to execute workloads with diverse computational requirements. In this work, we assume that nodes are heterogeneous in both performance and power consumption, where the performance of individual nodes in the cluster are inconsistent in the following sense: if machine A is faster than machine B for a given task, machine A may not be faster for all tasks.

Resource allocation decisions often rely on estimated values for task execution times where actual values may differ from available estimates (e.g., due to data dependent execution times). To account for these variations, we model task execution times stochastically using random variables.

This research addresses the problem of *statically* allocating a workload of independent (non-communicating) tasks to a high-performance heterogeneous

R. Wyrzykowski et al. (Eds.): PPAM 2013, Part I, LNCS 8384, pp. 761–771, 2014.
DOI: 10.1007/978-3-642-55224-3_71, © Springer-Verlag Berlin Heidelberg 2014

computing cluster. We aim to develop and analyze resource management techniques that maximize *robustness*, the probability that our workload finishes by a common deadline, while maintaining a specified probability that the energy consumed is within a given energy budget.

We make three main contributions in this work. Our first contribution is the design of three new intelligent local search techniques that maximize robustness within an energy constraint for a heterogeneous computing environment. The second contribution is an analysis of the impact of each local search technique's ability to improve a solution and the benefits associated with combining the local search techniques. The third contribution is the design and analysis of two global search techniques and the effects of combining global and local search into a new Tabu Search heuristic.

2   Related Work

Robust resource allocation has been previously studied (e.g., [2,3,13,14]). We adopt the system model from [2], however that work addresses the problem of minimizing energy consumption with a robustness constraint, where in our work we attempt to maximize robustness under an energy constraint and change the robustness calculation to be more generally applicable. The other studies listed do not consider energy when optimizing for robustness. Our work uses energy-aware resource allocation techniques and DVFS to balance the tradeoff between performance and energy consumption to maximize robustness under an energy constraint.

The use of DVFS to balance performance and energy consumption has been studied widely (e.g., [5,11,15]). In [5], the focus is on minimizing energy consumption, rather than maximizing performance, and considers deterministic rather than stochastic execution times as we do in our work. DVFS is used to tradeoff between energy consumption and makespan in [11], where the workload consists of precedence-constrained applications represented by a directed acyclic graph (DAG). Scheduling is performed on a task-by-task basis using a novel objective function that considers both energy consumption and the estimated finishing time of a task, using DVFS to exploit slack times from precedence constraints. Our work considers stochastic execution times, robustness, the probability of meeting an energy budget, and the overhead energy consumption of compute nodes. Using DVFS to schedule tasks with uncertain execution times is addressed in [15], however the system model is homogeneous and the work does not consider an energy constraint.

3   System Model

3.1   Compute Nodes

Our compute cluster model, adopted from [2], consists of N heterogeneous compute nodes where the performance and power consumption of each node may vary substantially. Nodes are also heterogeneous in their processor configuration. Each compute node i consists of n_i multicore processors, and each multicore processor j in compute node i has n_{ij} cores. We assume that all cores and multicore processors in a given compute node are homogeneous. We use the triple ijk to denote core k on multicore processor j in compute node i.

We assume the processors are DVFS-enabled to use the Advanced Configuration and Power Interface (**ACPI**) performance states (**P-states**) [7] that allow the processor to change operating voltage and frequency. We assume each core has a set of five P-states, denoted P, available: P_0, P_1, P_2, P_3, and P_4. Each core in the system can operate in an individual P-state, and our resource allocation techniques are designed such that P-states do not switch during task execution. Lower-numbered P-states consume more power, but provide better performance. The power consumption of a core within compute node i in P-state P_π is $\rho_i^{(\pi)}$.

3.2 Workload

The workload in this environment consists of a collection of T independent tasks to be completed before a given system deadline δ. Such a collection of independent tasks is known as a *bag-of-tasks* [6], the primary workload in various types of distributed computing systems [9]. We assume the task execution times follow a Gaussian distribution, and the means and variances are known *a priori* so that we can perform a static (i.e., off-line) mapping. The use of Gaussian distributions allows us to sum task execution times using a closed-form equation rather than having to perform convolution, however our proposed resource management techniques are applicable for task execution times that follow any distribution with a calculable mean (expected value). In an actual system, the means and variances of the execution time distributions can be obtained by historical, experimental, or analytical techniques [13]. We work closely with Oak Ridge National Labs, and in their environment, as well as others, similar types of tasks are executed frequently allowing for the collection of historical information of the execution times of tasks on machines.

For a given resource allocation, let the set T_{ijk} denote tasks in T that have been assigned to core ijk and let $t_{ijk}^x \in T_{ijk}$. Let $\mathbf{PS}(t_{ijk}^x)$ denote the assigned P-state for task t_{ijk}^x. We denote the mean execution time associated with task t_{ijk}^x in P-state π as $\mu(t_{ijk}^x, \pi)$, and the associated variance as $V(t_{ijk}^x, \pi)$.

3.3 Energy Constraint Calculation

The energy required to execute tasks in a single core can be calculated as the product of the power consumption of the core in each P-state and the amount of time the core spends in each P-state. Let T_{ijk}^π denote the subset of tasks assigned to core ijk in P-state π. The energy consumed is calculated as the product of execution time (a random variable) and average power (a deterministic value). Multiplying a random variable with a known value has the effect of multiplying the expected value by that value and the variance by the square of that value. For core ijk, the expected value (S_{ijk}^{π}) and variance (Var_{ijk}^{π}) are

$$S_{ijk}^{\pi} = \sum_{t_{ijk}^{x} \in T_{ijk}^{\pi}} \rho_i^{(\pi)} \mu(t_{ijk}^{x}, \pi) \tag{1}$$

and

$$Var_{ijk}^{\pi} = \sum_{t_{ijk}^{x} \in T_{ijk}^{\pi}} (\rho_i^{(\pi)})^2 V(t_{ijk}^{x}, \pi). \tag{2}$$

We also account for the overhead power consumption at the processor level (e.g., for interference in shared caches) and the node level (e.g., for disk drives, fans, and memory). Overhead power provides a constant power that is not affected by DVFS, thus the greater execution times resulting from low-power P-states can result in greater energy consumption than when just executing the task as fast as possible in P-state P_0. We assume multicore processors can deactivate when all of their cores have finished their assigned workload, and compute nodes are able to shut down when all of their multicore processors are deactivated. We assume deactivated processors and deactivated compute nodes consume negligible power. For our static resource allocation problem with independent tasks, nodes do not have idle time as nodes are active when processing tasks then deactivate when finished with their assigned workload.

Let F_{ij} be the maximum expected completion time among cores in multicore processor j on node i, and σ_{ij}^2 be the associated variance. Also, let F_i be the maximum expected completion time among multiprocessors in node i, and σ_i^2 be the associated variance. Let ω_i^{MP} be the power overhead for each multicore processor in compute node i, and ω_i^{node} be the power overhead for compute node i. We calculate the expected energy required to process the entire workload across all compute nodes, denoted ζ, as

$$\zeta = \sum_{i=1}^{N} \left(F_i * \omega_i^{node} + \sum_{j=1}^{n_i} (F_{ij} * \omega_i^{MP}) + \sum_{j=1}^{n_i} \sum_{k=1}^{n_{ij}} \sum_{\forall \pi \in P} S_{ijk}^{\pi} \right). \tag{3}$$

The variance of the energy required to process the entire workload, denoted γ, is calculated similarly to ζ, with the exceptions of using the associated variances $(\sigma_i^2, \sigma_{ij}^2, \text{ and } Var_{ijk}^{\pi})$ instead of the expected values $(F_i, F_{ij}, \text{ and } S_{ijk}^{\pi})$ and squaring the constants.

The distribution of the energy used to process the workload can be expressed as $\mathcal{N}(\zeta, \gamma)$. Given an energy budget of Δ, we can compute the probability that $\mathcal{N}(\zeta, \gamma)$ is less than Δ (i.e., $\mathbb{P}(\mathcal{N}(\zeta, \gamma) \leq \Delta)$) by converting $\mathcal{N}(\zeta, \gamma)$ to its associated cumulative density function (**cdf**). We denote this probability as ϕ, and we require resource allocations to meet an energy constraint of $\phi \geq \eta$. The values for Δ and η are set by the system administrator.

3.4   Robustness Calculation

We define a "robust" resource allocation as one that can mitigate the impact of uncertain task execution times on our performance objective of finishing all tasks by deadline δ. The calculation of the completion time of core k when using a stochastic model for task execution times is performed by taking the sum of the random variables for each task assigned to core k. The sum of two independent normally distributed random variables α and β produces a normally distributed random variable with its mean being the sum of the means of α and β and the variance being the sum of the variances of α and β.

For core ijk, the expected finishing time ($\boldsymbol{F_{ijk}}$) and variance ($\boldsymbol{\sigma_{ijk}^2}$) are

$$F_{ijk} = \sum_{\forall t_{ijk}^x \in T_{ijk}} \mu(t_{ijk}^x, \mathrm{PS}(t_{ijk}^x)) \tag{4}$$

and

$$\sigma_{ijk}^2 = \sum_{\forall t_{ijk}^x \in T_{ijk}^x} V(t_{ijk}^x, \mathrm{PS}(t_{ijk}^x)). \tag{5}$$

Thus, the completion time distribution of ijk can be expressed as $\mathcal{N}(F_{ijk}, \sigma_{ijk}^2)$. Given deadline δ, we can compute the probability that $\mathcal{N}(F_{ijk}, \sigma_{ijk}^2)$ is less than δ (i.e., $\mathbb{P}(\mathcal{N}(F_{ijk}, \sigma_{ijk}^2) \leq \delta)$) by converting $\mathcal{N}(F_{ijk}, \sigma_{ijk}^2)$ to its associated cdf. We define the overall *system robustness*, denoted Ψ, as the minimum probability across all cores: $\Psi = \min_{\forall i \in N} \left(\min_{\forall j \in n_i} \left(\min_{\forall k \in n_{ij}} \mathbb{P}(\mathcal{N}(F_{ijk}, \sigma_{ijk}^2) \leq \delta) \right) \right)$.

3.5   Combining Performance Metrics

In this study, we use a penalized objective function to incorporate the energy constraint into our optimization techniques. The idea behind a penalized objective function is to reduce the objective function value of infeasible solutions based on the solution's distance from feasibility, but still allow infeasible solutions to be considered when searching for an optimal feasible solution. A *solution* is a complete mapping of tasks to machines. Recall that ϕ is the probability that the energy consumption of a resource allocation is less than Δ and a solution is considered feasible if $\phi \geq \eta$. The distance from feasibility for a solution, denoted $\boldsymbol{d}$, is the difference of ϕ from η: $d = \eta - \phi$.

When $d > 0$, it indicates that $\phi < \eta$, and we penalize the objective function by subtracting a weighted value of the distance from feasibility. Our penalized objective function, denoted ψ, is: $\psi = \Psi - c * d$, where c indicates a constant used to control how strongly the constraint will be enforced. A high value of the coefficient c can restrict exploration, however c must be large enough that a feasible solution is found. A value of $d \leq 0$ indicates the energy constraint has been met and we do not need to penalize the solution, and our objective function becomes: $\psi = \Psi$.

4   Heuristics

4.1   Overview

In this section we describe two heuristics: Min-Min Balance and our Tabu Search. Due to lack of space, we omit descriptions for genetic and memetic algorithms, although these are considered in the experiments we conduct.

4.2   Min-Min Balance

Inspired by the greedy iterative maximization heuristic proposed by [14], we create the Min-Min Balance heuristic. Min-Min Balance starts with an initial solution created using the well-known Min-Min Completion Time (Min-Min CT) heuristic [4,8,12] and tries to improve the solution using greedy modifications. The *min-robustness core*, denoted $core_{minR}$, refers to the core that has the least probability of finishing its assigned workload by the deadline, that is, the core that determines the robustness metric of the solution. The *max-robustness core*, denoted $core_{maxR}$, refers to the core that has the greatest probability of finishing its workload by the deadline. Min-Min Balance starts by generating an initial mapping using Min-Min CT with all tasks assigned in P_4 (the lowest energy P-state) to ensure the energy constraint is met. The solution is then modified using two steps that reassign tasks and P-states, keeping moves that result in a greater robustness value without violating the energy constraint. The first step iteratively reassigns arbitrary tasks from $core_{minR}$ to $core_{maxR}$ with the goal of increasing robustness by balancing the workload until no tasks from $core_{minR}$ can be reassigned without improving the solution. After the first step, the second step attempts to raise robustness by changing the P-states of tasks on $core_{minR}$ to lower-numbered (i.e., better performing) P-states until no tasks on $core_{minR}$ can change P-states without improving the solution.

4.3   Tabu Search

Overview: The distinguishing feature of Tabu Search is its exploitation of memory, generally through the use of a *Tabu List*. We use a Tabu List to store regions of the search space that have been searched and should not be searched again. Our implementation of Tabu Search combines intelligent local search ("short hops") with the global search ("long hops") in an attempt to find a globally optimal solution.

Local search is performed using three short-hop operators: (1) *task swap*, (2) *task reassignment*, and (3) *P-state reassignment*. One short-hop consists of one iteration of all three operators. Long-hops are performed when local search terminates, with the purpose of jumping to a new neighborhood in the search space, while avoiding areas already searched. After each long-hop, short-hops are again performed to locally search the region near the long-hop solution. The Tabu List stores unmodified long-hops (i.e., starting solutions) that indicate neighborhoods that have been searched before, and may not be searched again. A new solution generated by a long-hop must differ from any solution in the Tabu List by 25 %, otherwise a new long-hop is generated. The heuristic terminates after a given number of long-hops are performed.

To evaluate potential task to node allocations we make use of a mean execution time rank matrix (**MET rank matrix**) that contains the rank of each heterogeneous node for each task, based on mean execution times. That is, for a given task, the nodes are ranked by how fast the nodes can execute the task (e.g., if node i can execute task t faster than node j, node i is given a better rank for task t). Let the rank of task t on node i be $\mathbf{rank}(t_i)$. When comparing the rank of any two tasks A and B on node i, task A is ranked lower (better) than task B if rank(A_i) is less than rank(B_i).

Long-hops: The initial solution (first long-hop) in our heuristic is generated using a Min-Min CT allocation with all tasks running on cores in P-state P_4 (to help ensure the energy constraint is met). Subsequent long-hop solutions are generated by first unmapping an arbitrary 25 % of tasks, and then reassigning the tasks randomly (i.e., to random cores with random P-states) or heuristically. We try both approaches: one that randomly generates new solutions (*random long-hops*) and one that uses Min-Min CT in P_4 to generate new solutions (*Min-Min long-hops*). The three short-hop operators are now described in more detail.

Task Swap: The goal of the *task swap* short-hop operator is to swap tasks that are assigned to poorly *ranked* nodes to better ranked nodes, a move that can potentially improve both robustness and energy consumption. We first choose an arbitrary core k and create a task list consisting of all tasks assigned to core ijk, recalling that the notation for such a task list is T_{ijk}. T_{ijk} is sorted in descending order by the rank of each task for node i (e.g., the worst-ranked task is first). We select the first task in the task list, denoted $task_A$, and find the best-ranked node for the task, denoted $node_{best}$. Within $node_{best}$, an arbitrary core z is chosen. The task from core z that has the best rank for node i, $task_B$, is selected for swap. The core assignments for $task_A$ and $task_B$ are swapped, and the best P-state combination (according to the objective function ψ) is found and assigned to the tasks. If the solution improves (higher ψ), the swap is kept and *task swap* ends. Otherwise, the swap is not kept, $task_A$ is removed from T_{ijk}, and *task swap* repeats until the solution improves or T_{ijk} is empty.

Task Reassignment: It can also be useful to transfer tasks from one core to another instead of swapping task assignments, thus we implement a *task reassignment* operator to transfer tasks. The goal of *task reassignment* is to improve robustness (Ψ) by removing tasks from the min-robustness core. We start by choosing the min-robustness core (core k) and create a task list consisting of all tasks assigned to the min-robustness core (T_{ijk}). T_{ijk} is sorted in descending order by the rank of each task for node i. We select the first task in the task list ($task_A$), and find the best-ranked node for the task ($node_{best}$). Within $node_{best}$, the max-robustness core is selected as the target core (core z), and $task_A$ is assigned to core z. The best P-state (according to ψ) is found to run core z in when executing $task_A$. If the solution improves (higher ψ), the new assignment is kept and *task reassignment* ends. Otherwise, the new assignment is not kept, task A is removed from T_{ijk}, and *task reassignment* repeats until the solution improves or T_{ijk} is empty.

P-state Reassignment: Depending on whether or not the energy constraint has been met, *P-state reassignment* increases or decreases the P-states of tasks to either reduce energy consumption or improve robustness. If the energy constraint is satisfied (i.e., $\phi \geq \eta$), the min-robustness core (core k) is chosen, and a task list is generated consisting of all tasks assigned to the min-robustness core (T_{ijk}). A task is chosen arbitrarily from the task list ($task_A$), and the P-state of the task is decreased by 1 if not already currently assigned to execute in P_0. If the energy constraint has not been met, the max-robustness core (core k) is chosen, and a task list is generated consisting of all tasks assigned to the max-robustness core (T_{ijk}). A task is chosen arbitrarily from the task list ($task_A$), and the P-state of the task is increased by 1 if not currently assigned to execute in P_4. If the solution improves, the new P-state is kept and *P-state reassignment* ends. Otherwise, the new P-state is not kept, task A is removed from T_{ijk}, and *P-state reassignment* repeats until the solution improves or T_{ijk} is empty.

5   Results

The cluster we simulate consists of 250 compute nodes (N), where the number of multicore processors in a node can vary from one to four (n_i), and the number of cores in a multicore processor can vary from two to sixteen (n_{ij}), giving a total of 613 multicore processors and 5,430 cores. We conducted 100 distinct simulation trials, with the means and variances of the task execution times varying among trials. These trials simulate numerous diverse heterogeneous workload/system environments. The figures presented in this section show data collected over the 100 simulation trials by displaying the mean values and 95 % confidence interval error bars around the mean. The workload of each trial consisted of 40,000 tasks. The mean and variance values of the task execution times for each P-state in each node were generated using the Coefficient of Variation (COV) method [1] and a scaling procedure.

Power consumption values of cores for each node in each P-state ($\rho_i^{(\pi)}$) were generated by sampling a normal distribution for P-states P_4 and P_0 and interpolating the intermediate states using a quadratic curve. Power overhead values (ω_i^{node} and ω_i^{MP}) were generated by sampling a uniform distribution with bounds chosen such that the total average overhead power comprised approximately 30 % of the total power consumed by the system. We have simulated two different-sized heterogeneous platforms to demonstrate the efficacy of our heuristics. Our primary contribution is not to determine a universally applicable set of parameters, but rather show how parameters for resource allocation heuristics can be calibrated for any given platform through our simulation experiments.

To show the benefits and weaknesses of each local search operator in our Tabu Search heuristic, we tested each operator's performance separately when improving a Min-Min CT P_4 solution over time. Figure 1 shows a comparison of each operator separately, as well as all operators working together when improving the Min-Min CT P_4 solution, with repeated applications of each operator over the execution time specified on the x-axis. We observed that the improvement when applying the *task reassignment* and *P-state reassignment* operators saturates quickly, whereas the *task swap* operator continues to improve the solution slowly. *Task reassignment* and *P-state reassignment* take greedy approaches by only considering the min-robustness core for task and P-state reassignments, giving strong initial performance but causing the operators to reach local maxima quickly. The *task swap* operator is able to swap assignments on all cores,

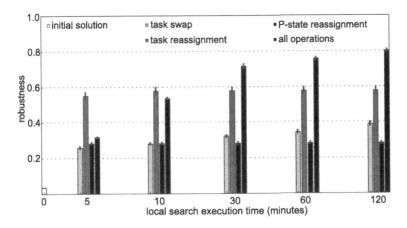

Fig. 1. Local search operator comparison. The figure shows how each operator can improve the robustness of a Min-Min CT P_4 solution over time.

taking a less greedy approach that allows a broader search. Local search performs the best when all operators are used, thus all operators are used in our Tabu Search heuristic when comparing with other heuristics.

Figure 2a shows a comparison of Tabu Search when using random long-hops versus Min-Min long-hops. Random long-hops can potentially search more of the solution space at the expense of generating poor initial solutions, whereas Min-Min long-hops can generate good solutions but limit the search space. Results are shown using varying numbers of long-hop and short-hop iterations to try and understand the tradeoffs between global search and local search, with the number of iterations of each chosen such that the total combined heuristic execution times were the same for the different iteration values. In our environment, we can see that using Min-Min long-hops in our Tabu Search provided significantly better robustness values than using random long-hops, as random long-hops generate poor initial solutions. One conclusion from Fig. 2a is that our local search is not improved by global hops for our problem domain when the long-hops are performed randomly, however, global hops become beneficial when performed intelligently using Min-Min.

Figure 2b shows the results obtained from the Min-Min Balance and Tabu Search heuristics, as well as genetic and memetic algorithms (GA and MA) adapted for this environment. The GA is similar to that from [2], and we created the MA by executing the three local search operators from Tabu on offspring chromosomes before they are added to the general population of a GA. The numbers inside the bars indicate the heuristic execution times. The superior performance of Tabu Search and the MA compared to the GA and Min-Min Balance within the computation times given further demonstrate the significance of the local search operations.

We experimented with the heuristics on a smaller system consisting of 25 heterogeneous nodes, 178 total cores, and 4,000 tasks. For that system we observed

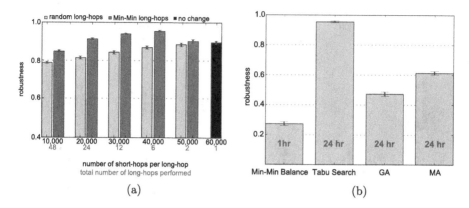

Fig. 2. Comparison of (a) random long-hops and Min-Min long-hops and (b) robustness of Min-Min Balance, Tabu Search, GA, MA. All solutions meet the energy constraint.

similar relative performance among the heuristics (as in Fig. 2b), with the Tabu Search and MA outperforming the GA and Min-Min Balance when Tabu Search, GA, and MA are given 24 h to execute. Though we observed similar relative performances for the two different-sized systems, we leave scalability analysis for future work.

6   Conclusions

In this paper, we study the problem of statically allocating a bag-of-tasks to a heterogeneous computing system. We proposed new local search techniques that use knowledge of the problem and model to maximize robustness under an energy constraint. We also evaluated and compared Tabu Search short-hop and long-hop techniques, and then combined the local and global search techniques to create a new Tabu Search heuristic. These local search techniques can be adapted to other models and problem domains to enhance Tabu Search, GAs, and other search heuristics with the performance of our Tabu Search heuristic and MA demonstrating the significance of the local search operators in the resource allocation problem domain.

Acknowledgments. The authors thank T. Hansen, M. Amini Salehi, J. Potter, and G. Pfister for their valuable comments. This research was supported by NSF grants CNS-0905399 and CCF-1302693, and by the CSU Abell Endowment. This research used the CSU ISTeC Cray System supported by NSF grant CNS-0923386.

References

1. Ali, S., Siegel, H.J., Maheswaran, M., Hensgen, D.: Representing task and machine heterogeneities for heterogeneous computing systems. Tamkang J. Sci. Eng., Special 50th Anniversary Issue **3**(3), 195–207 (2000)

2. Apodaca, J., Young, D., Briceño, L., Smith, J., Pasricha, S., Maciejewski, A.A., Siegel, H.J., Bahirat, S., Khemka, B., Ramirez, A., Zou, Y.: Stochastically robust static resource allocation for energy minimization with a makespan constraint in a heterogeneous computing environment. In: International Conference on Computer Systems and Applications (AICCSA '11), pp. 22–31, December 2011
3. Boloni, L., Marinescu, D.: Robust scheduling of metaprograms. J. Sched. 5(5), 395–412 (2002)
4. Braun, T.D., Siegel, H.J., Beck, N., Bölöni, L., Freund, R.F., Hensgen, D., Maheswaran, M., Reuther, A.I., Robertson, J.P., Theys, M.D., Yao, B.: A comparison of eleven static heuristics for mapping a class of independent tasks onto heterogeneous distributed computing systems. J. Parallel Distrib. Comput. 61(6), 810–837 (2001)
5. Chang, P.C., Wu, I.W., Shann, J.J., Chung, C.P.: ETAHM: an energy-aware task allocation algorithm for heterogeneous multiprocessor. In: Design Automation Conference (DAC '08), pp. 776–779 (2008)
6. Cirne, W., Brasileiro, F., Sauv, J., Andrade, N., Paranhos, D., Santos-neto, E., Medeiros, R., Gr, F.C.: Grid computing for bag of tasks applications. In: IFIP Conference on E-Commerce, E-Business and E-Government, September 2003
7. Hewlett-Packard, Intel, Microsoft, Phoenix Technologies, and Toshiba: Advanced Configuration and Power Interface Specification, rev. 5.0, http://www.acpi.info (2011)
8. Ibarra, O.H., Kim, C.E.: Heuristic algorithms for scheduling independent tasks on nonidentical processors. J. Assoc. Comput. Mach. 24(2), 280–289 (1977)
9. Iosup, A., Sonmez, O., Anoep, S., Epema, D.: The performance of bags-of-tasks in large-scale distributed systems. In: International Symposium on High Performance Distributed Computing (HPDC '08), pp. 97–108, June 2008
10. Koomey, J.: Growth in data center electricity use 2005 to 2010. Technical report, Analytics Press, http://www.analyticspress.com/datacenters.html (2011)
11. Lee, Y.C., Zomaya, A.Y.: Minimizing energy consumption for precedence-constrained applications using dynamic voltage scaling. In: IEEE/ACM International Symposium on Cluster Computing and the Grid (CCGRID '09), pp. 92–99, May 2009
12. Maheswaran, M., Ali, S., Siegel, H.J., Hensgen, D., Freund, R.F.: Dynamic mapping of a class of independent tasks onto heterogeneous computing systems. J. Parallel Distrib. Comput. 59(2), 107–131 (1999)
13. Shestak, V., Smith, J., Maciejewski, A.A., Siegel, H.J.: Stochastic robustness metric and its use for static resource allocations. J. Parallel Distrib. Comput. 68(8), 1157–1173 (2008)
14. Sugavanam, P., Siegel, H.J., Maciejewski, A.A., Oltikar, M., Mehta, A., Pichel, R., Horiuchi, A., Shestak, V., Al-Otaibi, M., Krishnamurthy, Y., Ali, S., Zhang, J., Aydin, M., Lee, P., Guru, K., Raskey, M., Pippin, A.: Robust static allocation of resources for independent tasks under makespan and dollar cost constraints. J. Parallel Distrib. Comput. 67(4), 400–416 (2007)
15. Xian, C., Lu, Y.H., Li, Z.: Dynamic voltage scaling for multitasking real-time systems with uncertain execution time. IEEE Trans. Comput. Aided Des. Integr. Circuits Syst. 27(8), 1467–1478 (2008)

Performance and Energy Analysis of the Iterative Solution of Sparse Linear Systems on Multicore and Manycore Architectures

José I. Aliaga[1], Hartwig Anzt[2(✉)], Maribel Castillo[1], Juan C. Fernández[1], Germán León[1], Joaquín Pérez[1], and Enrique S. Quintana-Ortí[1]

[1] Dpto. de Ingeniería y Ciencia de Computadores, Universidad Jaume I, 12.071 Castellón, Spain
{aliaga,castillo,jfernand,leon,al001566,quintana}@uji.es
[2] Innovative Computing Lab (ICL), University of Tennessee, Knoxville, USA
hanzt@icl.utk.edu

Abstract. In this paper we investigate the performance-energy balance of a variety of concurrent architectures, from general-purpose and digital signal multicore systems to graphics processors (GPUs), representative of current technology. This analysis employs the conjugate gradient method, an important algorithm for the iterative solution of linear systems that is basically composed of the sparse matrix-vector product and other (minor) vector kernels. To allow a fair comparison, we leverage simple implementations of the numerical methods and underlying kernels, and rely only on those optimizations applied by the target compiler.

Keywords: Energy efficiency · High-performance computing · Sparse linear algebra · Multicore processors · Low-power processors · GPUs

1 Introduction

Competing for the world's first exascale system, many high performance computing (HPC) initiatives have identified the power wall as a key challenge that will have to be confronted, resulting in an unmistakable call for power-efficient systems [5,9]. At the other end of the spectrum, energy-efficient components are essential for extended battery life of mobile appliances like smart phones and tablets, and hardware companies devote considerable effort to integrating sophisticated energy-saving mechanisms into embedded devices. These two trends seem to be converging, though, and a small number of recent HPC research prototypes aim at delivering high performance-power ratios by adopting technology originally designed for the mobile market [1,2].

Although most manufacturers advertise the power-efficiency of their products by providing theoretical energy specifications, an equitable comparison between different hardware architectures remains difficult. The reason is not only that

R. Wyrzykowski et al. (Eds.): PPAM 2013, Part I, LNCS 8384, pp. 772–782, 2014.
DOI: 10.1007/978-3-642-55224-3_72, © Springer-Verlag Berlin Heidelberg 2014

distinct devices are often designed for one particular type of computation, but also that they are tailored for either performance or power efficiency. New energy-related metrics have been recently proposed to analyze the balance between these two key figures [7], but the situation becomes increasingly difficult once the different levels of optimization applied to an algorithm enter the picture. In particular, extensive software optimization that results in significant performance and power improvements for one specific hardware are also likely to hamper the portability of the code to other architectures.

In this paper we provide a map of the energy-performance landscape of a variety of general-purpose and specialized hardware architectures using the conjugate gradient (CG) method, a key algorithm for the numerical solution of symmetric positive definite (SPD) sparse linear systems [12]. The cornerstone of this iterative method is the sparse matrix-vector product, which is also a crucial operation for many other numerical methods [4]; the significance of the results carries beyond the scope of this inspection. To avoid promoting one particular hardware architecture, we develop and evaluate simple implementations of the CG method and the sparse matrix-vector product, either using the CSR or the ELLPACK sparse matrix formats [12]. A motivation for not applying optimizations to the code, other than those intrinsic to the compilation process, stems from the fact that many complex numerical codes running on large HPC facilities —e.g., for CFD applications, weather or economic simulations— are used out-of-the-box without hands-on optimization.

The rest of the paper is structured as follows. In Sect. 2 we briefly introduce the mathematical formulation of the CG method as well as the implementation and basic storage formats for the sparse matrix-vector operation. In Sect. 3 we present the benchmark matrices. In Sect. 4 we review the main characteristics of the different architectures: four general-purpose multicore processors (Intel Xeon E5504 and E5-2620, and AMD Opteron 6128 and 6276); a low-power multicore digital signal processor (Texas Instruments C6678); two low-power multicore processors (ARM Cortex A9 and Intel Atom D510); and three GPUs with different capabilities (NVIDIA Quadro M1000, Tesla C2050 and Kepler K20). Section 5 contains the main contribution of this paper, namely, the practical evaluation of these architectures, from the point of view of both performance and energy efficiency. Finally, in Sect. 6 we offer a few concluding remarks.

2   Solving Sparse SPD Linear Systems

2.1   Krylov-Based Iterative Solvers

The CG method [12] is among the best known Krylov subspace methods for the solution of linear systems $Ax = b$, where $A \in \mathbb{R}^{n \times n}$ is SPD, $b \in \mathbb{R}^n$ contains the independent terms, and $x \in \mathbb{R}^n$ is the solution. The method is mathematically formulated in Fig. 1, where the user-defined parameters *maxres* and *maxiter* set upper bounds, respectively, on the relative residual for the computed approximation to the solution x_k, and the maximum number of iterations.

$$x_0 := 0 \text{ // or any other initial guess}$$
$$r_0 := b - Ax_0, \qquad d_0 := r_0$$
$$\beta_0 := r_0^T r_0, \qquad \tau_0 := \| r_0 \|_2 = \sqrt{\beta_0}, \qquad k := 0$$
while $(k < maxiter)$ & $(\tau_k > maxres)$
$$\quad z_k := Ad_k$$
$$\quad \rho_k := \beta_k / d_k^T z_k$$
$$\quad x_{k+1} := x_k + \rho_k d_k$$
$$\quad r_{k+1} := r_k - \rho_k z_k$$
$$\quad \beta_{k+1} := r_{k+1}^T r_{k+1}$$
$$\quad \alpha_k := \beta_{k+1} / \beta_k$$
$$\quad d_{k+1} := r_{k+1} + \alpha_k d_k$$
$$\quad \tau_{k+1} := \| r_{k+1} \|_2 = \sqrt{\beta_{k+1}}, \qquad k := k + 1$$
end

Fig. 1. Mathematical formulation of the CG method.

In practical applications, the computational cost of the CG method is dominated by the matrix-vector multiplication $z_k := Ad_k$. Given a sparse matrix A with n_z nonzero entries, this operation roughly requires $2n_z$ floating-point arithmetic operations (flops). Additionally, the loop body contains several vector operations (for the updates of x_{k+1}, r_{k+1}, d_{k+1}, and the computation of ρ_k and β_{k+1}) that cost $O(n)$ flops each.

2.2   The Sparse Matrix-Vector Product

The sparse matrix-vector product is ubiquitous in scientific computing, being a key operation for the iterative solution of linear systems and eigenproblems [12] as well as the PageRank algorithm [11], among others [4]. For sparse matrices, the irregular memory access pattern of this operation and the limited memory bandwidth of current general-purpose architectures has resulted in a considerable number of efforts proposing specialized matrix storage layouts as well as optimized implementations for a variety of architectures; see, e.g., [6,10,14].

In our implementation of the sparse matrix-vector product for the CG method, matrix A is stored either in the compressed sparse row (CSR) format, for multicore architectures, or the ELLPACK format for the GPUs.

In general, ELLPACK incurs some storage overhead, but the aligned structure allows for a more efficient hardware usage when targeting streaming processors like GPUs [8,13]. Figures 2 and 3 sketch, respectively, the two sparse matrix formats and the corresponding actual algorithms that we utilize for the sparse matrix-vector product. In both routines, n refers to the matrix size; the matrix is stowed using arrays values, colind and, in the case of routine spmv_csr, also array rowptr (see Fig. 2); the input and output vectors of the product $y := Ax$ are, respectively, x and y; finally, nzr refers to the number of entries per row

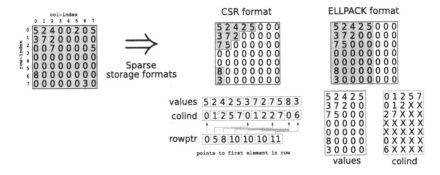

Fig. 2. Dense and sparse matrix storage formats. The memory demand corresponds to the grey areas.

```
1     void spmv_csr ( int n,
2                      int * rowptr, int * colind, float * values,
3                      float * x, float * y ) {
4         int i, j;    float tmp;
5
6         #pragma omp parallel for private ( j, tmp )
7         for ( i = 0; i < n; i++ ) {
8             tmp = 0.0;
9             for ( j = rowptr [ i ]; j < rowptr [ i+1 ]; j++ )
10                tmp += values [ j ] * x [ colind[ j ] ];
11            y[i] = tmp;
12        }
13    }
14    // --------------------------------------------------------------
15    __global__
16    void spmv_ell ( int n, int nzr,
17                      int * colind, float * values,
18                      float * x,    float * y ) {
19        int i, j, k; float tmp;
20
21        i = blockDim.x * blockIdx.x + threadIdx.x;
22        if ( i < n ){
23            tmp = 0.0;
24            for ( j = 0; j < nzr; j++ ) {
25                k = n * j + i;
26                if ( values[ k ] != 0 )
27                    tmp += values [ k ] * x [ colind [ k ] ];
28            }
29            y[i] = tmp;
30        }
31    }
```

Fig. 3. Sparse matrix-vector product using the CSR and ELLPACK formats (spmv_csr and spmv_ell, respectively).

in the ELLPACK-based routine. Although other formats exist that are more appealing for specific architectures or sparsity patterns —e.g., jagged diagonal storage format [12]— we choose CSR and ELLPACK here for their simplicity. Also, no attempt is made to exploit the symmetric structure of matrix A.

Table 1. Description and properties of the test matrices.

Source	Acronym	Matrix	#nonzeros (n_z)	Size (n)	n_z/n
Laplace	A159	A159	27,986,067	4,019,679	6.94
UFMC	AUDI	AUDIKW_1	77,651,847	943,645	82.28
	BMW	BMWCRA1	10,641,602	148,770	71.53
	CRANK	CRANKSEG_2	14,148,858	63,838	221.63
	F1	F1	26,837,113	343,791	78.06
	INLINE	INLINE_1	38,816,170	503,712	77.06
	LDOOR	LDOOR	42,493,817	952,203	44.62

2.3 Vector Operations

For multicore architectures, we employed the legacy implementation of BLAS from *netlib*[1] for the level-1 (vector) operations. No attempt was made to parallelize the vector operations of the CG method on these architectures. For GPUs we instead used the implementation in NVIDIA's CUBLAS. We consider this a fair comparison since (i) in general, the time cost of the vector operations is significantly lower than that of the sparse matrix-vector product; and (ii) due to their reduced cost, there is very little opportunity to benefit from a concurrent execution of vector operations on multicore processors.

3 Matrix Benchmarks

We have selected six SPD matrices from the University of Florida Matrix Collection (UFMC)[2], corresponding to finite element discretizations of several structural problems arising in mechanics, and an additional case derived from a finite difference discretization of the 3D Laplace problem; see Table 1. In the linear systems, vector b was initialized so that all the entries of the solution x were equal to 1, and the CG iteration was started with the initial guess $x_0 \equiv 0$. All tests operated with single-precision (SP) floating-point arithmetic. While the use of double precision (DP) arithmetic is mandatory for the solution of sparse linear systems of equations, in [3] we show how the use of mixed SP-DP combined with iterative refinement improves execution time and energy consumption.

4 Hardware Setup and Compilers

Table 2 lists the main hardware features of the systems and compilers involved in the experimentation. We used the GNU C compiler (`gcc`) in as many platforms as possible, with similar versions, and the optimization flag `-O3` in all cases. For each multi-core processor we also report the different frequencies that were evaluated in the experimentation.

---

[1] http://www.netlib.org
[2] http://www.cise.ufl.edu/research/sparse/matrices.

Table 2. Architectures with their corresponding idle power characteristics. For the GPU systems (FER, KEP and QDR), the idle power includes the accelerator.

Acron.	Architecture	Total #cores	Frequency (GHz) – Idle power (W)	RAM size, type	Compiler
AIL	AMD Opteron 6276 (Interlagos)	8	1.4–167.29, 1.6–167.66 1.8–167.31, 2.1–167.17 2.3–168.90	64GB, DDR3 1.3GHz	gcc 4.4.6
AMC	AMD Opteron 6128 (Magny-Cours)	8	0.8–107.48, 1.0–109.75, 1.2–114.27, 1.5–121.15, 2.0–130.07	48GB, DDR3 1.3GHz	gcc 4.4.6
IAT	Intel Atom D510	2	0.8–11.82, 1.06–11.59, 1.33-11.51, 1.6–11.64	1GB, DDR2 533MHz	gcc 4.5.2
INH	Intel Xeon E5504 (Nehalem)	8	2.0–280.6, 2.33–281.48, 2.83–282.17	32GB, DDR3 800MHz	gcc 4.1.2
ISB	Intel E5-2620 (Sandy-Bridge)	6	1.2–93.35, 1.4–93.51, 1.6–93.69, 1.8–93.72, 2.0–93.5	32GB, DDR3 1.3GHz	gcc 4.1.2
ARM	ARM Cortex A9	4	0.62–11.7, 1.3–12.2	2GB, DDR3L	gcc 4.5.2
FER	Intel Xeon E5520	8	1.6–222.0, 2.27–226.0	24 GB,	gcc 4.4.6
	NVIDIA Tesla C2050 (Fermi)	448	1.15	3 GB, GDDR5	nvcc 4.2
KEP	Intel Xeon i7-3930K	6	1.2–106.30, 3.2–106.50	24 GB,	gcc 4.4.6
	NVIDIA Tesla K20 (Kepler)	2,496	0.7	5 GB, GDDR5	nvcc 4.2
QDR	ARM Cortex A9	4	0.102–11.2, 1.3–12.2	2 GB, DDR3L	gcc 4.5.2
	NVIDIA Quadro 1000 M	96	1.4	2 GB, DDR3	nvcc 4.2
TIC	Texas Instruments C6678	8	1.0–18.0	512 MB, DDR3	cl6x 7.4.1

In order to measure power, we leveraged a WATTSUP?PRO wattmeter, connected to the line from the electric socket to the power supply unit (PSU), with an accuracy of ±1.5 % and a rate of 1 sample/sec, and results were collected on a separate server. All tests were executed for a minimum of 3 min (in case the solver converged in a shorter time, the same process was repeated till the minimum time was reached), after a warm up period of 5 min. Since the platforms where the processors are embedded contain other devices (e.g., disks, network interface cards, and on IAT even the LCD display), on each platform we measured the average power while idle for 30 s, and then subtracted the corresponding value (see Table 2) from all the samples obtained from the wattmeter to get a baseline reading of the system components at idle. We expect this setup renders a fair comparison of the energy-efficiency of the different architectures, as we only evaluate the power that is drawn to do the actual work.

5 Experimental Results

Tables 3 and 4 and Fig. 4 collect the results of the experimental study. For each platform and benchmark case, we evaluated the performance, net energy (i.e., energy after subtracting the cost of idle power), and total energy for different combinations of the number of cores and processor frequency. For brevity, we only report the results that correspond to the best case from the perspectives of time and net energy and the corresponding pairs (net energy, total energy) and

Table 3. Optimal configuration, time and energy for general-purpose architectures. c denotes the #cores, f the frequency (in MHz), T the time per iteration (in seconds), and E_{net} and E_{tot} the net and total energy, respectively, per iteration (in Joules).

	Matrix	\multicolumn{5}{c}{Optimized w.r.t time}	\multicolumn{5}{c}{Optimized w.r.t net energy}
		c	f	T	E_{net}	E_{tot}	c	f	T	E_{net}	E_{tot}
AIL	A159	8	2300	7.52e−02	2.06e+01	3.33e+01	8	2100	7.65e−02	1.24e+01	2.52e+01
	AUDI	8	2300	1.94e−01	5.16e+01	8.44e+01	8	2100	2.80e−01	4.29e+01	8.97e+01
	BMW	8	2300	1.46e−02	3.92e+00	6.38e+00	8	2100	1.92e−02	3.05e+00	6.27e+00
	CRANK	8	2300	2.47e−02	6.66e+00	1.08e+01	8	2100	3.56e−02	5.52e+00	1.15e+01
	F1	6	2300	4.99e−02	1.13e+01	1.98e+01	6	2100	6.84e−02	9.82e+00	2.12e+01
	INLINE	8	2300	6.59e−02	1.76e+01	2.87e+01	8	2100	9.15e−02	1.43e+01	2.96e+01
	LDOOR	8	2300	7.50e−02	1.99e+01	3.26e+01	8	2100	9.15e−02	1.43e+01	2.96e+01
AMC	A159	8	2000	8.57e−02	9.92e+00	2.20e+01	8	2000	8.57e−02	9.92e+00	2.20e+01
	AUDI	8	2000	3.03e−01	2.78e+01	7.05e+01	8	2000	3.03e−01	2.78e+01	7.05e+01
	BMW	8	2000	2.11e−02	2.39e+00	5.36e+00	8	2000	2.11e−02	2.39e+00	5.36e+00
	CRANK	8	2000	3.57e−02	3.85e+00	8.87e+00	8	2000	3.57e−02	3.85e+00	8.87e+00
	F1	6	2000	6.98e−02	7.65e+00	1.75e+01	6	2000	6.98e−02	7.65e+00	1.75e+01
	INLINE	8	2000	9.67e−02	1.06e+01	2.42e+01	8	2000	9.67e−02	1.06e+01	2.42e+01
	LDOOR	8	2000	9.88e−02	1.13e+01	2.53e+01	8	2000	9.88e−02	1.13e+01	2.53e+01
IAT	A159	2	1600	4.59e−01	2.53e+00	7.87e+00	1	800	1.46e+00	1.79e+00	1.90e+01
	BMW	2	1600	1.03e−01	5.40e−01	1.74e+00	1	800	3.73e−01	3.73e−01	4.78e+00
	CRANK	2	1600	1.48e−01	7.15e−01	2.44e+00	1	800	4.82e−01	4.63e−01	6.16e+00
	F1	2	1600	3.38e−01	1.69e+00	5.62e+00	2	800	6.36e−01	1.20e+00	8.71e+00
	INLINE	2	1600	3.81e−01	1.99e+00	6.43e+00	1	800	1.34e+00	1.37e+00	1.72e+01
	LDOOR	2	1600	5.04e−01	2.60e+00	8.47e+00	1	800	1.73e+00	1.85e+00	2.23e+01
INH	A159	4	2830	1.04e−01	8.86e+00	3.81e+01	4	2000	1.06e−01	7.40e+00	3.72e+01
	AUDI	6	2830	1.94e−01	1.79e+01	7.27e+01	4	2000	2.10e−01	1.39e+01	7.29e+01
	BMW	8	2333	2.19e−02	2.02e+00	8.18e+00	4	2000	2.23e−02	1.59e+00	7.85e+00
	CRANK	6	2830	2.96e−02	2.79e+00	1.11e+01	4	2000	3.20e−02	2.17e+00	1.12e+01
	F1	6	2333	6.25e−02	5.21e+00	2.28e+01	4	2000	6.66e−02	4.55e+00	2.32e+01
	INLINE	6	2830	8.13e−02	7.72e+00	3.07e+01	4	2000	8.40e−02	5.85e+00	2.94e+01
	LDOOR	4	2830	1.04e−01	9.01e+00	3.85e+01	4	2000	1.06e−01	7.60e+00	3.74e+01
ISB	A159	6	2000	4.61e−02	2.36e+00	6.67e+00	4	1200	9.47e−02	1.68e+00	1.05e+01
	AUDI	6	2000	1.48e−01	6.39e+00	2.02e+01	6	1200	2.42e−01	4.32e+00	2.69e+01
	BMW	6	2000	1.13e−02	5.56e−01	1.61e+00	4	1400	2.32e−02	4.71e−01	2.64e+00
	CRANK	6	2000	1.93e−02	8.70e−01	2.67e+00	6	1200	3.17e−02	6.20e−01	3.58e+00
	F1	6	2000	3.11e−02	1.51e+00	4.41e+00	6	1200	5.08e−02	1.27e+00	6.02e+00
	INLINE	6	2000	4.84e−02	2.26e+00	6.79e+00	6	1200	7.91e−02	1.61e+00	8.99e+00
	LDOOR	6	2000	5.27e−02	2.62e+00	7.54e+00	6	1200	8.59e−02	1.94e+00	9.96e+00

(time, total energy), respectively. As the tables and figure convey a considerable amount of information, we limit the following analysis to some central aspects.

Optimization with Respect to Time. If run time (or, analogously, performance) is the figure of merit, the Tesla K20 (KEP), followed by the Tesla C2050 (FER), are the architectures of choice (see the top-left plot of Fig. 4). For all test cases, they outperform all other systems by almost an order of magnitude. However, their performance strongly depends on the matrix structure, and the distinct test cases exhibit flop rates that differ by a factor of 6.32. This sensitivity to the matrix sparsity pattern is also shared by the Quadro 1000 M (QDR), which outperforms the C6678 DSP (TIC) only for those cases where the use of the ELLPACK format does not incur significant overhead on GPUs. While TIC shows some variation on performance as well as the performance-per-watt ratio

Table 4. Optimal configuration, time and energy for specialized architectures. c denotes the #cores, f the frequency (in MHz), T the time per iteration (in seconds), and E_{net} and E_{tot} the net and total energy, respectively, per iteration (in Joules).

		Optimized w.r.t time					Optimized w.r.t net energy				
	Matrix	c	f	T	E_{net}	E_{tot}	c	f	T	E_{net}	E_{tot}
ARM	A159	4	1300	8.20e−01	2.54e+00	1.25e+01	1	620	2.12e+00	9.75e−01	2.58e+01
	AUDI	4	1300	7.90e−01	2.86e+00	1.25e+01	2	1300	8.20e−01	1.69e+00	1.17e+01
	BMW	4	1300	1.10e−01	3.86e−01	1.73e+00	1	620	3.20e−01	1.41e−01	3.88e+00
	CRANK	4	620	1.30e−01	5.28e−01	2.05e+00	1	620	4.20e−01	2.02e−01	5.12e+00
	F1	4	1300	2.90e−01	1.04e+00	4.58e+00	1	620	9.10e−01	4.00e−01	1.10e+01
	INLINE	4	1300	3.80e−01	1.38e+00	6.01e+00	1	620	1.15e+00	5.64e−01	1.40e+01
	LDOOR	4	1300	5.00e−01	1.92e+00	8.02e+00	1	620	1.41e+00	6.49e−01	1.71e+01
FER	A159	1	2270	5.43e−03	9.58e−01	2.18e+00	1	1600	5.43e−03	9.15e−01	2.12e+00
	AUDI	1	2270	2.95e−02	5.28e+00	1.20e+01	1	1600	2.95e−02	5.08e+00	1.16e+01
	BMW	1	2270	4.05e−03	6.97e−01	1.61e+00	1	1600	4.05e−03	6.58e−01	1.56e+00
	CRANK	1	2270	1.69e−02	2.85e+00	6.68e+00	1	1600	1.70e−02	2.69e+00	6.46e+00
	F1	1	2270	1.49e−02	2.65e+00	6.02e+00	1	1600	1.49e−02	2.51e+00	5.83e+00
	LDOOR	1	2270	7.73e−03	1.46e+00	3.21e+00	1	1600	7.74e−03	1.39e+00	3.11e+00
KEP	A159	1	3200	3.93e−03	5.85e−01	1.00e+00	1	1200	3.96e−03	4.14e−01	8.35e−01
	AUDI	1	3200	1.75e−02	2.66e+00	4.52e+00	1	1200	1.75e−02	1.91e+00	3.77e+00
	BMW	1	3200	2.86e−03	4.09e−01	7.14e−01	1	1200	2.89e−03	2.87e−01	5.94e−01
	CRANK	1	3200	1.08e−02	1.49e+00	2.64e+00	1	1200	1.09e−02	1.04e+00	2.19e+00
	F1	1	3200	8.20e−03	1.22e+00	2.09e+00	1	1200	8.23e−03	8.57e−01	1.73e+00
	INLINE	1	3200	1.81e−02	2.51e+00	4.43e+00	1	1200	1.81e−02	1.73e+00	3.65e+00
	LDOOR	1	3200	5.42e−03	8.41e−01	1.42e+00	1	1200	5.45e−03	6.07e−01	1.19e+00
QDR	A159	1	1300	3.42e−02	9.39e−01	1.36e+00	1	1300	3.42e−02	9.39e−01	1.36e+00
	BMW	1	1300	3.39e−02	8.47e−01	1.26e+00	1	1300	3.39e−02	8.47e−01	1.26e+00
	CRANK	1	1300	1.49e−01	3.63e+00	5.44e+00	1	102	1.57e−01	3.54e+00	5.30e+00
	F1	1	1300	1.19e−01	3.15e+00	4.60e+00	1	102	1.27e−01	3.03e+00	4.46e+00
	LDOOR	1	1300	4.69e−02	1.31e+00	1.88e+00	1	1300	4.69e−02	1.31e+00	1.88e+00
TIC	A159	8	1000	2.43e−01	6.66e−01	5.04e+00	1	1000	5.28e−01	4.86e−01	9.99e+00
	BMW	8	1000	1.90e−02	7.04e−02	4.13e−01	8	1000	1.90e−02	7.04e−02	4.13e−01
	CRANK	8	1000	2.64e−02	9.15e−02	5.66e−01	4	1000	3.60e−02	9.00e−02	7.38e−01
	F1	8	1000	9.10e−02	2.68e−01	1.91e+00	8	1000	9.10e−02	2.68e−01	1.91e+00

depending on the matrix structure, its energy efficiency is unmatched by any other architecture. The two other low-power processors, ARM and Atom (IAT), show better ratios than the conventional general-purpose processors (Interlagos AIL, Magny-Cours AMC, Nehalem INH, and Sandy-Bridge ISB), but only for some matrix cases higher energy-efficiency than the GPUs. For FER and QDR, it is interesting to notice that, although they yield very different performance, their performance-per-watt ratios are almost equal.

Optimization with Respect to the Net Energy Efficiency. The first general observation is that, in many cases, reducing the operating voltage and frequency pays off. Interestingly, this is not always the case for the GPU-accelerated architectures (see, e.g., the results for QDR in Table 4): For those test cases where the vector and matrix-vector operations are fast, a host that operates at low frequency becomes a bottleneck from the point of view of performance, and the associated energy overhead blurs the power savings. However, for all GPU implementations, rescaling frequency (and voltage) of the host system has only

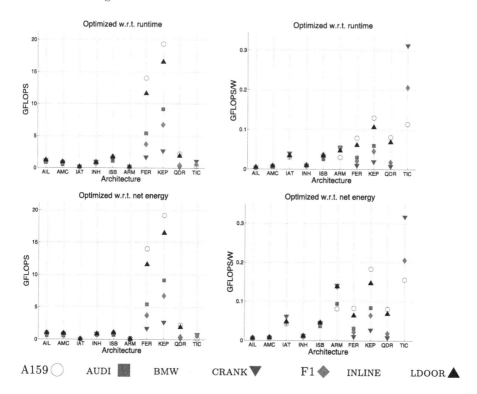

Fig. 4. Comparison of performance (left) and performance/watt (right) when optimizing with respect to run time (top) or net energy (bottom), measured respectively in terms of GFLOPS and GFLOPS/W (1 GFLOPS=10^9 flops/s).

a negligible impact on performance and energy. This is very different for the low-power processors, where rescaling voltage and frequency can improve the energy efficiency by a wide margin; see example values of the results for ARM and IAT in Tables 3 and 4. These architectures provide not only higher optimization potential, but also exhibit superior energy efficiency compared with the conventional general-purpose CPUs or GPUs. When optimizing for net energy, TIC is again the absolute winner, followed by ARM, with the latter still exhibiting higher performance-per-watt rates than the GPUs or the general-purpose CPUs, see Fig. 4. IAT is, in terms of energy efficiency, competitive with the FER and QDR GPU-based systems. Unfortunately, the factors gained when improving energy efficiency translate into the related run time as the reduced watt-per-iteration values for TIC, ARM, and IAT come at the price of significantly higher execution times (see Tables 3 and 4). Finally, while differences of that scale are not attained by any of the general-purpose CPUs, optimizing with respect to net energy is not necessarily equivalent to optimization for the total energy consumption. For instance, the net energy consumption for the AUDI matrix on INH can be reduced by decreasing frequency (and voltage), at the cost of an increase of the total energy (see Table 3).

6 Summary and Future Work

Alternative directions are taken today by hardware manufacturers to lower the resource cost of numerical operations. In order to provide a broad overview about the potential of different approaches, we have analyzed the performance and energy efficiency of a large variety of architectures. We observed that the flops-per-watt rate of manycore systems, like the graphics processors from NVIDIA, can be matched by low-power devices such as the Intel Atom, the ARM A9 or a digital signal processor from Texas Instruments. While GPUs traditionally achieve a high power efficiency through outstanding performance, the low-power architectures provide it with less cores, a lower power dissipation and/or smaller memories. This reduces the suitability of these inexpensive architectures for general-purpose computing, but makes them appealing candidates for mobile and embedded appliances (their original target) as well as specific applications. Despite the fact that conventional general-purpose processors attained neither the performance nor the performance-per-watt rates of the GPUs, they were the only architectures able to process all test cases due to their superior memory capacity. Future research should increase the scope of the study complexity by adding problems not directly related to linear algebra, e.g., sorting algorithms or image processing.

Acknowledgements. This work was supported by the CICYT project TIN2011-23283 and FEDER, and by EU FET grant "EXA2GREEN" 318793.

References

1. CRESTA: collaborative research into Exascale systemware, tools and applications. http://cresta-project.eu
2. The Mont Blanc project. http://montblanc-project.eu
3. Anzt, H., Heuveline, V., Aliaga, J., Castillo, M., Fernández, J., Mayo, R., Quintana-Ortí, E.S.: Analysis and optimization of power consumption in the iterative solution of sparse linear systems on multi-core and many-core platforms. In: Green Computing Conference and Workshops (IGCC), pp. 1–6 (2011)
4. Asanovic, K., et al.: The landscape of parallel computing research: a view from Berkeley. Technical Report UCB/EECS-2006-183, University of California at Berkeley, Electrical Engineering and Computer Sciences (2006)
5. Ashby, S., et al.: The opportunities and challenges of Exascale computing. Summary Report of the Advanced Scientific Computing Advisory Committee (ASCAC) Subcommittee, November 2010
6. Barrett, R., Berry, M., Chan, T.F., Demmel, J., Donato, J., Dongarra, J., Eijkhout, V., Pozo, R., Romine, C., der Vorst, H.V.: Templates for the Solution of Linear Systems: Building Blocks for Iterative Methods, 2nd edn. SIAM, Philadelphia (1994)
7. Bekas, C., Curioni, A.: A new energy aware performance metric. Comput. Sci. Res. Dev. **25**, 187–195 (2010). doi:10.1007/s00450-010-0119-z
8. Bell, N., Garland, M.: Efficient sparse matrix-vector multiplication on CUDA. NVIDIA Technical Report NVR-2008-004, NVIDIA Corporation, December 2008

9. Bergman, K., et al.: Exascale computing study: Technology challenges in achieving exascale systems. DARPA IPTO ExaScale Computing Study (2008)

10. Buluç, A., Williams, S., Oliker, L., Demmel, J.: Reduced-bandwidth multi-threaded algorithms for sparse matrix-vector multiplication. In Proceedings of the IPDPS, pp. 721–733 (2011)

11. Langville, A., Meyer, C.: Google's PageRank and Beyond: The Science of Search Engine Rankings. Princeton University Press, Princeton (2009)

12. Saad, Y.: Iterative Methods for Sparse Linear Systems. Society for Industrial and Applied Mathematics, Philadelphia (2003)

13. Vázquez, F., Fernández, J.J., Garzón, E.M.: A new approach for sparse matrix vector product on nvidia gpus. Concurrency Comput. Pract. Experience **23**(8), 815–826 (2011)

14. Williams, S., Bell, N., Choi, J., Garland, M., Oliker, L., Vuduc, R.: Sparse matrix vector multiplication on multicore and accelerator systems. In: Kurzak, J., Bader, D.A., Dongarra, J. (eds.) Scientific Computing with Multicore Processors and Accelerators. CRC Press, Boca Raton (2010)

Measuring the Sensitivity of Graph Metrics to Missing Data

Anita Zakrzewska[✉] and David A. Bader

Georgia Institute of Technology, Atlanta, GA, USA
azakrzewska3@gatech.edu

Abstract. The increasing energy consumption of high performance computing has resulted in rising operational and environmental costs. Therefore, reducing the energy consumption of computation is an emerging area of interest. We study the approach of data sampling to reduce the energy costs of sparse graph algorithms. The resulting error levels for several graph metrics are measured to analyze the trade-off between energy consumption reduction and error. The three types of graphs studied, real graphs, synthetic random graphs, and synthetic small-world graphs, each show distinct behavior. Across all graphs, the error cost is initially relatively low. For example, four of the five real graphs studied needed less than a third of total energy to retain a degree centrality rank correlation coefficient of 0.85 when random vertices were removed. However, the error incurred for further energy reduction grows at an increasing rate, providing diminishing returns.

Keywords: Graphs · Graph algorithms · Sensitivity analysis · Missing data · Energy consumption · Power

1 Introduction

Power consumption has become a critical issue in computing. This is a concern both for supercomputers, where massive energy use poses a financial and an environmental cost, and for embedded in-the-field processing systems, which have a limited energy supply or battery lifetime. Achieving maximum computational capabilities on embedded systems while limiting power use is an important task.

We address energy reduction for irregular, sparse graph algorithms through data sampling or removal. Sparse networks often represent relationships, communication, or information flow. For example, a graph may represent an online social network, network traffic, biological networks, or financial transactions. Often such graphs are constructed from a massive, and constant, stream of data, which leads to large graphs and energy-expensive computations. However, in cases where an approximate solution suffices, it is not always necessary to store and use the entire graph. For example, when calculating distances, approximate results for shortest paths may be acceptable for a given application. If the goal

R. Wyrzykowski et al. (Eds.): PPAM 2013, Part I, LNCS 8384, pp. 783–792, 2014.
DOI: 10.1007/978-3-642-55224-3_73, © Springer-Verlag Berlin Heidelberg 2014

is to find the most influential, or important, vertices, it is only necessary to calculate top scores correctly since low-scoring vertices are of no interest. Since approximations are often satisfactory for real-world graph metrics, a certain degree of error in the underlying graph data, such as missing or incorrect edges and vertices, may be tolerated. Real-time streams of data may also amass too much information to be stored or lead to over-saturation, in which case certain vertices and edges may need to be removed over time.

Vertex and edge removal can also be performed intentionally with the goal of reducing energy consumption. Sampling results in a smaller graph, with fewer memory accesses, fewer compute operations, and a shorter overall running time, all of which contribute to less energy use. However, in order for this to be a feasible approach, it is necessary to determine the resulting level of error. We investigate the sensitivity of several graph metrics to missing vertices and edges, which can be used to set tolerable error level thresholds.

Previous work has compared the sensitivity of scale free and random networks to vertex removal [1]. Sampling and contraction methods have been used to reduce the size of internet topology graphs [13]. Graph analytic sensitivity to noisy data has been addressed by Borgatti *et al.* [5]. However, that work only considers vertex centrality measures on Erdős-Rényi random graphs [9], whose structure differs from that of real networks. Because the authors focus on errors in the data due to noise instead of conscious data sampling for power reduction, many of the errors analyzed, such as false positive edges, are not as applicable to the goal of energy reduction. Kossinets [12] studies the effects of missing data in social networks by analyzing a bipartite scientific collaboration network of authors and papers as well as bipartite random graphs. Our work differs because we focus on filtering methods for the purpose of decreasing the size of the graph and therefore the energy needed to compute various analytics.

2   Energy Model

The energy consumption of an algorithm can be modeled in terms of the energy per memory operation, energy per arithmetic operation, and constant energy that must be expended until the computation terminates, as given in Eq. (1), where W is the number of memory operations, ϵ_{flop} is the fixed energy cost of a compute operation, Q is the number of arithmetic operations, ϵ_{mem} is the fixed energy cost of a memory operation, T is the duration of the algorithm, and π_0 is the fixed constant energy cost, which may be idle energy or leakage [7,11].

$$E = W\epsilon_{flop} + Q\epsilon_{mem} + T\pi_0 \qquad (1)$$

Because sparse graph algorithms tend to exhibit a low arithmetic intensity and are memory bound,we focus on the number of memory operations and energy per memory operation. Many real-world graphs have a low diameter and irregular structure with little or no locality in the data access pattern. Sparse graph algorithms tend to exhibit low data reuse and focus on traversing the graph structure [15]. Therefore, focusing on memory cost is an appropriate proxy for the energy consumption of sparse graph algorithms.

Dynamic power management is a technique used to reduce power consumption in which system components are switched to a low-performance, or idle, state when load demands are low [4]. Memory power reduction can be achieved by dynamically adjusting memory voltage and frequency based on bandwidth utilization [8]. We describe three possible situations in which energy considerations can cause analytics to be run on incomplete graphs. From the algorithmic perspective, all have the same result. A subset of the vertices and edges of a graph are not used when calculating a graph analytic.

1. The system may choose not to access a subset of the graph in memory. This reduces the number of memory accesses.
2. Portions of memory may be turned to a low power mode to conserve energy, resulting in some data being unavailable. In-the-field embedded systems, for example, may do this after having detected low energy supplies.
3. The system may have insufficient storage for the entire graph and so a subset of the graph must be removed or never stored in the first place.

3   Methodology

Our approach to measuring sensitivity to missing data is as follows. We start with a true, base graph G and compute the value of a metric, called the true metric value. For each sampling level k, the graph is sampled in several ways and a subset of the vertices and edges is removed, creating the sampled graph $G_{k,sampled}$. The metric is recomputed on $G_{k,sampled}$, which gives the observed metric value. We then compare the true metric value to the observed metric value, resulting in a metric error. The energy required is calculated as the ratio of energy needed for $G_{k,sampled}$ to the energy needed for G. The relationship between the average metric error and energy required for each sampling level can then be examined. This process is repeated for all sampling methods described in Sect. 3.2.

3.1   Datasets

Testing is performed on both real and synthetic networks, listed in Table 1. The real graphs come from the 10th DIMACS Implementation Challenge [2] and include citation networks, collaboration networks, a graph of users of the Pretty-Good-Privacy algorithm for secure information interchange, and a graph of the structure of the Internet from 2006. The synthetic graphs used were produced by an RMAT generator [6]. These include both Erdős-Rényi random graphs and small-world graphs that have many properties of real-world social networks, such as a power law degree distribution and low diameter [3,10,14]. We used parameters $\alpha = 0.25, \beta = 0.25, \gamma = 0.25, \delta = 0.25$ for the Erdős-Rényi random graph and $\alpha = 0.55, \beta = 0.1, \gamma = 0.1, \delta = 0.25$ for the small-world graph.

Table 1. Graph instances used in testing

Name	Vertices	Edges
citationCiteseer	268,495	1,156,647
coAuthorsCiteseer	227,320	814,134
coAuthorsDBLP	299,067	977,676
as-22july06	22,963	48,436
PGPgiantcompo	10,680	24,316
SmallWorld EF 8	32,768	237,523
SmallWorld EF 16	32,768	456,626
SmallWorld EF 32	32,768	861,878
Random EF 8	32,768	262,085
Random EF 16	32,768	524,031
Random EF 32	32,768	1,047,549

3.2   Graph Sampling Methods

The four approaches used to sample data are listed below for a graph with n vertices. For each one, we consider values of $p = 0.01, 0.05, 0.1, 0.15, \ldots, 0.8, 0.85$.

- **RandEdge:** Edges in the graph are chosen to be removed with equal probability p so that the error is distributed evenly across the network.
- **RandVertex:** Each vertex in the graph is chosen to be removed with equal probability p. When a vertex is removed, all of its incident edges are removed as well.
- **HighDegVertex:** The highest degree vertices and incident edges are removed. The top $p * n$ vertices are selected.
- **LowDegVertex:** The $p * n$ lowest degree vertices and their edges are removed.

3.3   Metrics Evaluated

We evaluate the graph connectivity, clustering coefficients, and degree centrality. The degree centrality of a vertex measures the number of edges incident on it and is the most basic centrality measure. We evaluate the sensitivity of degree centrality by measuring how much the rank of vertices' degree centrality changes when data is removed. For a given sampling level, the degree centrality rank is calculated for each vertex present in both the original and sampled graphs, resulting in two vectors. The Spearman correlation coefficient of these two rank vectors is then measured.

A connected component of a graph is a set of vertices linked by paths of edges. As vertices and edges are removed, the connected components of a graph may disconnect. While the number of components may increase, measuring the error in the number of connected components offers little information about how the structure of the graph has changed. A component splitting in half is a very different scenario from a few vertices disconnecting. We define the error in connectivity as the proportion of pairs of vertices that were in the same component in the original graph and remain in the same component in the sampled graph. Only vertices with nonzero degree in the sampled graph are considered. Let $c(G, u, v)$ be an indicator function whose value is one when vertices u and v are in the same component in graph G and zero otherwise. The connectivity retained is then given by Eq. (2). Using the Shiloach-Vishkin algorithm [16], the estimated cost of computing connected components is $mlog(n)$ where m is the number of edges and n vertices.

$$ConnectivityRetained = \frac{\sum_{v,u \in G_{sampled}} c(G_{sampled}, u, v)}{\sum_{v,u \in G_{sampled}} c(G, u, v)} \tag{2}$$

The clustering coefficient measures the density of triangles in a graph and is one measure of the degree to which the graph is clustered. The local clustering coefficient of v is the ratio of closed triplets to open triplets of v and the global clustering coefficient is the ratio of total triangles to total triplets in the graph. High clustering coefficients suggest a small-world graph [17]. The global clustering coefficient can be used to characterize the entire graph, while local coefficients can reveal entities that engage in the most or least clustered activity. We measure the absolute and relative error in global clustering coefficient. To measure the sensitivity of local clustering coefficients, we calculate the Spearman correlation coefficient of the per-vertex rank in local clustering coefficient. Each vertex compares its list of adjacent vertices with the adjacency list of each of its neighbors, searching for intersections. Thus, the adjacency list of vertex v is accessed $d_v + 1$ times, once for itself, and once for each neighbor. Thus, the energy cost is given by $E = \epsilon_{mem} * \sum_v d_v^2$.

4   Results

The proportion of connectivity retained for each graph against energy required is plotted in Fig. 1. For each dataset, the energy value on the x-axis is the ratio of energy needed for $G_{sampled}$ to the energy needed for G. For all sampling types, the connected components of synthetic graphs are far more robust to missing data than those of real ones, which can be explained by the regular structure of RMAT graphs. Of the synthetic graphs, random graphs are the most robust with almost no error, while small-world graphs behave more similarly to real data. Removing low degree vertices causes the least amount of error across datasets. Removing high degree vertices, random vertices, and random edges provides diminishing returns as can be seen by the change in slope of the curves in Fig. 1.

The clustering coefficient of a graph also affects the sensitivity of its connected components. Among datasets studied, networks with a high global clustering coefficient require a higher proportion of energy to retain their connectivity structure. Figure 2 plots the clustering coefficient against the proportion of energy necessary to retain a connectivity of 0.85 and 0.95. For random edge, random vertex, and high degree vertex removal, the connected components of highly clustered graphs are least robust.

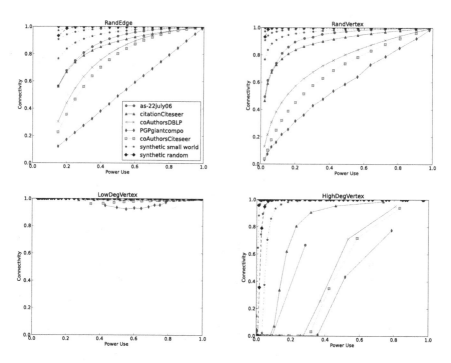

Fig. 1. The connectivity retained, or proportion of pairs of vertices that remain in the same connected component after sampling

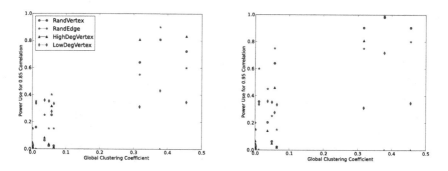

Fig. 2. Graph global clustering coefficient versus percentage of energy needed for 0.85 and 0.95 connectivity

Figure 3 plots the local clustering rank correlation coefficient against energy required. Real graphs are least sensitive to low and high degree vertex removal and most sensitive to random edge and vertex removal. In order to achieve a correlation coefficient of at least 0.85, the real graphs, in order listed in Table 1, need a 0.28, 0.47, 0.45, 1.0, and 0.13 proportion of energy with high degree vertex removal and 0.47, 0.48, 0.48, 0.2, and 0.86 with low degree vertex removal.

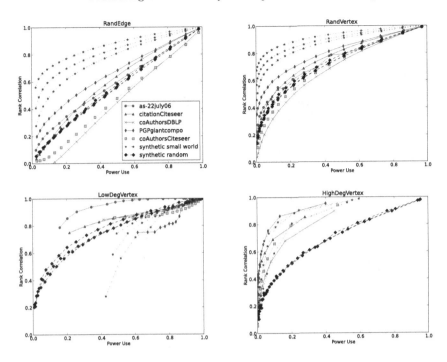

Fig. 3. Clustering coefficient rank correlation

With random edge and vertex removal, the real graphs need from 0.81 to 0.9 and from 0.62 to 0.73 energy, respectively. These relatively narrow bands show that random missing data may produce more consistent results, but at the cost of more energy usage. Random graphs are least sensitive to missing low degree vertices, requiring 0.61 to 0.7 energy for 0.85 correlation and most to missing random edges, requiring 0.81 to 0.9. Unlike random or real graphs, synthetic small-world graphs showed the most sensitivity when low degree vertices are removed. Despite these differences between the three network categories, Fig. 3 shows similar behavior for all types. As with connectivity error, data removal provides diminishing returns across the graphs studied. As the removal rate increases and the energy use decreases, the rate at which the clustering rank correlation coefficient falls increases. This suggests that significant energy savings could be achieved at relatively low error levels. It is interesting to note that for all graphs, this behavior is least prominent with random edge removal, where the curves are closer to linear.

Figure 4 plots the degree centrality rank correlation coefficient against energy used. A clear distinction can be seen between the sensitivity of real graphs, synthetic small-world graphs, and synthetic random graphs. Synthetic small-world graphs are most robust to all sampling methods, random graphs are the least robust, and the behavior of real graphs is in between the two. The robustness of small-world graphs compared to random ones can be explained by their skewed

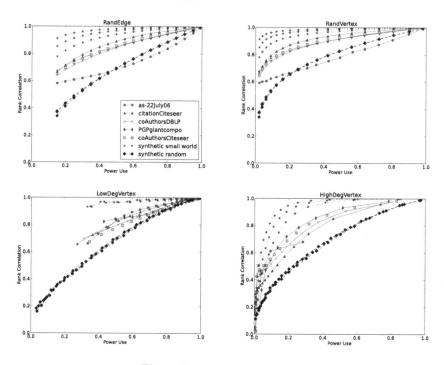

Fig. 4. Degree rank correlation

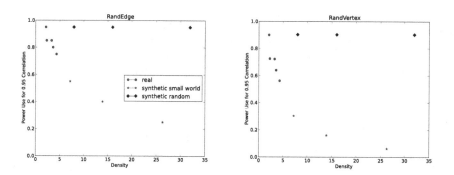

Fig. 5. Graph density versus energy needed to retain degree rank correlation of 0.95 using random edge removal (left) and random vertex removal (right)

degree distribution. Since there is little variation in vertex degree centrality in random graphs, this metric is very sensitive to missing data. Small-world graphs exhibit a large variation in degree centrality and so more data must be removed to change the metric. However, the results cannot be explained solely by a skewed degree distribution. The top 1 % of vertices contain a greater proportion of network edges in the real graphs than in the synthetic small-world graphs. Thus, the real graphs may have the most skewed degree distribution, but their degree centrality is not most robust. Although the real datasets come from a variety of sources, their degree centrality values are affected similarly by missing data. As with connected components and local clustering coefficients, the gradient of the curves across datasets increases from right to left, suggesting that significant energy savings can be initially achieved with relatively low error, but that the error cost grows as more data is removed. Four of the five real graphs use less than a third of total energy to retain a degree centrality rank correlation coefficient of 0.85 with random vertex removal and less than 0.55 energy with random edge or high degree vertex removal.

The density of a graph, its ratio of edges to vertices, does affect its sensitivity to random edge and vertex sampling. In Fig. 5, density is plotted against the energy required to retain a degree centrality rank correlation coefficient of 0.95. Red stars denote synthetic small-world graphs and blue circles denote real graphs. For both of these categories, as the graph density increases, the proportion of energy needed decreases. This trend does not hold for random graphs, which are represented in the scatter plots with black diamonds.

5   Conclusion

We have investigated an approach for reducing the energy consumption of sparse graph algorithms with edge and vertex sampling. Such data removal will naturally result in errors which may or may not be tolerable, depending on the metric and application. We have examined the sensitivity of clustering coefficients, degree centrality, and connected components to various sampling strategies and analyzed the trade-off between energy reduction and error. Synthetic random graphs, synthetic small-world graphs, and real small-world graphs each tended to react distinctly. The structure of the graph is important in predicting the sensitivity to missing data and in choosing the best sampling technique and conclusions drawn from synthetic graphs may not be applicable to real data. Although the real networks came from a variety of sources, they tended to exhibit similar behavior that was distinct from that of either type of synthetic graph. Structural features such as the degree of clustering and density also have an effect on a network's robustness. It is interesting to note that in most cases, a similar pattern exists in the trade-off between energy savings and metric error. The gradient of the curve increases as energy use decreases, showing that the error cost of power saving is initially low, but grows at an increasing rate. This pattern suggests that significant energy savings might be achieved with relatively low error levels.

Acknowledgement. The work depicted in this paper was partially sponsored by Defense Advanced Research Projects Agency (DARPA) under agreement #HR0011-13-2-0001. The content, views and conclusions presented in this document do not necessarily reflect the position or the policy of DARPA or the U.S. Government, no official endorsement should be inferred. Distribution Statement A: "Approved for public release; distribution is unlimited."

References

1. Albert, R., Jeong, H., Barabási, A.L.: Error and attack tolerance of complex networks. Nature **406**(6794), 378–382 (2000)
2. Bader, D.A., Meyerhenke, H., Sanders, P., Wagner, D.: Graph partitioning and graph clustering. In: Proceedings of the 10th DIMACS Implementation Challenge Workshop. AMS (2013)
3. Barabási, A., Albert, R.: Emergence of scaling in random networks. Science **286**(5439), 509–512 (1999)
4. Benini, L., Bogliolo, A., De Micheli, G.: A survey of design techniques for system-level dynamic power management. IEEE Trans. Very Large Scale Integr. (VLSI) Syst. **8**(3), 299–316 (2000)
5. Borgatti, S.P., Carley, K.M., Krackhardt, D.: On the robustness of centrality measures under conditions of imperfect data. Soc. Netw. **28**(2), 124–136 (2006)
6. Chakrabarti, D., Zhan, Y., Faloutsos, C.: R-MAT: a recursive model for graph mining. In: SIAM International Conference on Data Mining (2004)
7. Choi, J., Bedard, D., Fowler, R., Vuduc, R.: A roofline model of energy. In: Proceedings of the IEEE International Parallel and Distributed Processing Symposium (IPDPS) (2013)
8. David, H., Fallin, C., Gorbatov, E., Hanebutte, U.R., Mutlu, O.: Memory power management via dynamic voltage/frequency scaling. In: Proceedings of the 8th ACM International Conference on Autonomic Computing, pp. 31–40. ACM (2011)
9. Erdős, P., Rényi, A.: On the evolution of random graphs. Publ. Math. Inst. Hungar. Acad. Sci. **5**, 17–61 (1960)
10. Faloutsos, M., Faloutsos, P., Faloutsos, C.: On power-law relationships of the internet topology. In: Proceedings of the Conference on Applications, Technologies, Architectures, and Protocols for Computer Communication, SIGCOMM '99, pp. 251–262. ACM (1999)
11. Korthikanti, V.A., Agha, G.: Towards optimizing energy costs of algorithms for shared memory architectures. In: Proceedings of the 22nd ACM Symposium on Parallelism in Algorithms and Architectures, SPAA '10, pp. 157–165. ACM (2010)
12. Kossinets, G.: Effects of missing data in social networks. Soc. Netw. **28**(3), 247–268 (2006)
13. Krishnamurthy, V., Faloutsos, M., Chrobak, M., Lao, L., Cui, J.-H., Percus, A.G.: Reducing large internet topologies for faster simulations. In: Boutaba, R., Almeroth, K.C., Puigjaner, R., Shen, S., Black, J.P. (eds.) NETWORKING 2005. LNCS, vol. 3462, pp. 328–341. Springer, Heidelberg (2005)
14. Leskovec, J., Kleinberg, J., Faloutsos, C.: Graphs over time: densification laws, shrinking diameters and possible explanations. In: Proceedings of the Eleventh ACM SIGKDD International Conference on Knowledge Discovery in Data Mining, pp. 177–187. ACM (2005)
15. Lumsdaine, A., Gregor, D., Hendrickson, B., Berry, J.: Challenges in parallel graph processing. Parallel Process. Lett. **17**(1), 5–20 (2007)
16. Shiloach, Y., Vishkin, U.: An o(logn) parallel connectivity algorithm. J. Algorithms **3**, 57–67 (1982)
17. Watts, D., Strogatz, S.: Collective dynamics of small world networks. Nature **393**, 440–442 (1998)

The Energy/Frequency Convexity Rule: Modeling and Experimental Validation on Mobile Devices

Karel De Vogeleer[1]([✉]), Gerard Memmi[1], Pierre Jouvelot[2], and Fabien Coelho[2]

[1] TELECOM ParisTech – INFRES – CNRS LTCI - UMR, 5141 Paris, France
[2] MINES ParisTech – CRI, Fontainebleau, France
{karel.devogeleer,gerard.memmi}@telecom-paristech.fr,
{pierre.jouvelot,fabien.coelho}@mines-paristech.fr

Abstract. This paper provides both theoretical and experimental evidence for the existence of an Energy/Frequency Convexity Rule, which relates energy consumption and CPU frequency on mobile devices. We monitored a typical smartphone running a specific computing-intensive kernel of multiple nested loops written in C using a high-resolution power gauge. Data gathered during a week-long acquisition campaign suggest that energy consumed per input element is strongly correlated with CPU frequency, and, more interestingly, the curve exhibits a clear minimum over a 0.2 GHz to 1.6 GHz window. We provide and motivate an analytical model for this behavior, which fits well with the data. Our work should be of clear interest to researchers focusing on energy usage and minimization for mobile devices, and provide new insights for optimization opportunities.

Keywords: Energy consumption and modeling · DVFS · Power consumption · Execution time modeling · Smartphone · Bit-reverse algorithm

1 Introduction

The service uptime of battery-powered devices, e.g., smartphones, is a sensitive issue for nearly any user [9]. Even though battery capacity and performance are hoped to increase steadily over time, improving the energy efficiency of current battery-powered systems is essential because users expect right now communication devices to provide data access every time, everywhere to everyone. Understanding the energy consumption of the different features of (battery-powered) computer systems is thus a key issue. Providing models for energy consumption can pave the way to energy optimization, by design and at run time.

The power consumption of Central Processing Units (CPUs) and external memory systems is application and user behavior dependent [2]. Moreover, for cache-intensive and CPU-bound applications, or for specific Dynamic Voltage

R. Wyrzykowski et al. (Eds.): PPAM 2013, Part I, LNCS 8384, pp. 793–803, 2014.
DOI: 10.1007/978-3-642-55224-3_74, © Springer-Verlag Berlin Heidelberg 2014

and Frequency Scaling (DVFS) settings, the CPU energy consumption may dominate the external memory consumption [15]. For example, Aaron and Carroll [2] showed that, for an embedded system running `equake`, `vpr`, and `gzip` from the SPEC CPU2000 benchmark suite, the CPU energy consumption exceeds the RAM memory consumption, whereas `crafty` and `mcf` from the same suite showed to be straining more energy from the device RAM memory.

Providing an accurate model of energy consumption for embedded and, more generally, energy-limited devices such as mobile phones is of key import to both users and system designers. To reach that goal, our paper provides both theoretical and first experimental evidence for the existence of an Energy/Frequency Convexity Rule, that relates energy consumption and CPU frequency on mobile devices. This convexity property seems to ensure the existence of an optimal frequency where energy usage is minimal.

This existence claim is based on both theoretical and practical evidence. More specifically, we monitored a Samsung Galaxy SII smartphone running Gold-Rader's Bit Reverse algorithm [7], a small kernel based on multiple nested loops written in C, with a high-resolution power gauge from Monsoon Solutions Inc. Data gathered during a week-long acquisition campaign suggest that energy consumed per input element is strongly correlated with CPU frequency and, more interestingly, that the corresponding curve exhibits a clear minimum over a 0.2 GHz to 1.6 GHz window. We also provide and motivate an analytical model of this behavior, which fits well with the data. Our work should be of clear interest to researchers focusing on energy usage and minimization on mobile devices, and provide new insights for optimization opportunities.

The paper is organized as follows. Section 2 introduces the notions of energy and power, and how these can be decomposed in different components on electronic devices. Section 3 describes the power measurement protocol and methodology driving our experiments, and the C benchmark we used. Section 4 introduces our CPU energy consumption model, and shows that it fits well with the data. Section 5 outlines the Energy/Frequency Rule derived from our experiment and modeling. Related work is surveyed in Sect. 6. We conclude and discuss future work in Sect. 7.

2   Power Usage in Computer Systems

The total power P_{total} consumed by a computer system, including a CPU, may be separated into two components: $P_{total} = P_{system} + P_{CPU}$, where P_{CPU} is consumed by the CPU itself and P_{system} by the rest of system. In a battery-powered hand-held computer device P_{system} may include the power needed to light the LCD display, to enable and maintain I/O devices (including memory), to keep sensors online (GPS, gyro-sensors etc.), and others.

The power consumption P_{CPU} of the CPU we focus on here can be divided into two parts: $P_{CPU} = P_{dynamic} + P_{leak}$, where $P_{dynamic}$ is the power consumed by the CPU during the switching activities of transistors during computation. P_{leak} is power originating from leakage effects inherent to silicon-based transistors, and is in essence not useful for the CPU's purposes. $P_{dynamic}$ may be split into the power P_{short} lost when transistors briefly *short-circuit* during gate state changes and P_{charge}, needed to charge the gates' capacitors: $P_{dynamic} = P_{short} + P_{charge}$. In the literature P_{charge} is usually [17] defined as $\alpha\,CfV^2$, where α is a proportional constant indicating the percentage of the system that is active or switching, C the capacitance of the system, f the frequency at which the system is switching and V the voltage swing across C.

P_{short} originates during the toggling of a logic gate. During this switching, the transistors inside the gate may conduct simultaneously for a very short time, creating a direct path between V_{CC} and the ground. Even though this peak current happens over a very small time interval, given current high clock frequencies and large amount of logic gates, the short-circuit current may be non-negligible. Quantifying P_{short} is gate specific but it may be approximated by deeming it proportional to P_{charge}. Thus the power $P_{dynamic}$ stemming from the switching activities and the short-circuit currents in a CPU is thus $P_{charge} + (\eta-1)P_{charge}$, i.e., $\eta \cdot \alpha C_L f V^2$, where η is a scaling factor representing the effects of short-circuit power.

P_{leak} originates from leakage currents that flow between differently doped parts of a metal-oxide semiconductor field-effect transistor (MOSFET), the basic building block of CPUs. The energy in these currents are lost and do not contribute to the information that is held by the transistor. Some leakage currents are induced during the *on* or *off*-state of the transistor, or both. Six distinct sources of leakage are identified [12]. Despite the presence of multiple sources of leakage in MOSFET transistors, the sub-threshold leakage current, gate leakage, and band-to-band tunneling (BTBT) dominate the others for sub-100 nm technologies [1]. Leakage current models, e.g., as incorporated in the BSIM [12] micro models, are accurate yet complex since they depend on multiple variables. Moreover, P_{leak} fluctuates constantly as it also depends on the temperature of the system. Consequently P_{leak} cannot be considered a static part of the system's power consumption. Given the different sources of power consumption in a MOSFET based CPU, the portal power can be rewritten as $P_{total} = P_{system} + P_{leak} + P_{dynamic}$.

The relationship between the *power* $P(t)$ (Watts or Joules/s) and the *energy* $E(\Delta t)$ (Joules) consumed by an electrical system over a time period Δt is given by

$$E(\Delta t) = \int_0^{\Delta t} P(t)\ dt = \int_0^{\Delta t} I(t) \cdot V(t)\ dt, \tag{1}$$

where $I(t)$ is the current supplied to the system, and $V(t)$ the voltage drop over the system. Often $V(t)$ is constant over time, hence $dP(t)/dt$ only depends on $I(t)$. If both *current* and *voltage* are constant over time, the energy integral becomes the product of *voltage*, *current* and *time*, or alternatively *power* and *time*.

3 Power Measurement Protocol on Mobile Devices

A Samsung Galaxy S2 is used in our testbed sporting the Samsung Exynos 4 Systems-on-Chip (SoC) 45 nm dual-core. The Galaxy S2 has a 32 KB L1 data and instruction cache, and a 1 MB L2 cache. The mobile device runs Android 4.0.3 on the Siyah kernel adopting Linux 3.0.31. The frequency scaling governor in Linux was set to operate in *userspace* mode to prevent frequency and voltage scaling on-the-fly. The second CPU core was disabled during measurements. The smartphone is booted in *clockwork recovery* mode to minimize noisy side-effects of the Operating System (OS) and other frameworks.

During the experiments, the phone's battery was replaced by a power supply (Monsoon Power Monitor) that measures the power consumption at 5 kHz with an accuracy of 1 mW. The power of the system and the temperature of the CPU were simultaneously logged. The kernel was patched to print a temperature sample to the kernel debug output at a rate of 2 Hz.

The *bit-reverse* algorithm is used as benchmark kernel. This is an important operation since it is part of the ubiquitous Fast Fourier Transformation (FFT) algorithm, and rearranges deterministically elements in an array. The bit-reversal kernel is CPU intensive, induces cache effects, and is economically pertinent. The Gold-Rader implementation of the bit-reverse algorithm, often considered the reference implementation [7], is given below:

```
void bitreverse_gold_rader (int N, complex *data) {
    int n = N, nm1 = n-1; int i = 0, j = 0;
    for (; i < nm1; i++) {
        int k = n >> 1;
        if (i < j) {
            complex temp = data[i]; data[i] = data[j]; data[j] = temp;
        }
        while (k <= j) {j -= k; k >>= 1;}
        j += k;
    }
}
```

The input of the bit-reversal algorithm is an array with a size of 2^N; the elements are pairs of 32 bit integers, representing complex numbers. Note that array sizes up to 2^9 fit in the L1 cache, while sizes over 2^{18} are too big to fit in the L2 cache.

During the measurements, N is set between 6 and 20 in steps of 2, while varying the CPU frequency from 0.2 GHz to 1.6 GHz in steps of 0.1.

To minimize overhead, 128 copies of the kernel are run sequentially. For time measurement purposes, this benchmark is repeated 32 times for at least 3 s each time (this may require multiple runs of the 128 copies). For the power and temperature measurements, the benchmark is repeated in an infinite loop until 32 samples can be gathered. The benchmark is compiled with GCC 4.6, included in Google's NDK, generating ARMv5 thumb code.

Data was fitted using R and the nls() function employing the *port* algorithm.

4   Modeling Energy Consumption

Energy is the product of time by power. We look at each of these factors in turn here.

4.1   Execution Time

Since applications run over an OS, we need to take account for it when estimating computing time. Indeed, an OS needs a specific amount of time, or clock cycles, to perform (periodical) tasks, e.g., interrupt handling, process scheduling, processing kernel events, managing memory etc. When the processor is not spending time in kernel mode, the processor is available for user-space programs, e.g., our benchmark. From a heuristic point of view, it can be assumed that the OS kernel needs a fixed amount of clock cycles cc_k per time unit to complete its tasks. Thus, we propose to model the amount of clock cycles to complete a benchmark sequence of instructions cc_b as $t(f^\beta - cc_k)$, where cc_k are the number of clock cycles spent in the OS, t the total time needed to complete the program, f the system's clock frequency and β an architecture-dependent scaling constant, to be fitted later on with the data. The definition of cc_b is rewritten to isolate the execution time:

$$t = \frac{cc_b}{f^\beta - cc_k}.$$
(2)

Note that t tends to zero for $f \to \infty$ and there is a vertical asymptote at $\sqrt[\beta]{cc_k}$.

Table 1 shows the fitting errors of Eq. 2 on the execution time measurement data, averaged over all tested input sizes. The fitting exhibits a vertical asymptote around 115 MHz. This may indicate the minimum amount of clock cycles required by the OS of the phone to operate. The measurement data for input sizes 2^6 up to 2^{16} are well described by Eq. 2. However, sizes 2^{18} and 2^{20}, too large to fit within cache L2, seem to operate under different laws. Therefore, from now on, the attention is focused on data that fit in the cache of the CPU.

4.2   Power Consumption

If dynamic power modeling is rather easy (see Sect. 2), the case for leakage is more involved, and warrant a longer presentation. In particular, leakage power is

Table 1. Average absolute execution time (t), power (P), and energy (E) fitting errors (%) of Eq. 2, 4 and 5 respectively, on the measured data given different CPU frequencies (f) at a $37\,^\circ\mathrm{C}$ core temperature.

f (GHz)	0.2	0.3	0.4	0.5	0.6	0.7	0.8	0.9	1.0	1.1	1.2	1.3	1.4	1.5	1.6
error t	1.18	2.71	1.55	0.55	0.56	4.21	0.62	1.63	4.71	3.68	1.86	0.44	5.87	0.75	2.61
error P	2.94	1.00	0.20	0.99	1.31	1.46	1.24	0.49	0.02	0.70	0.86	0.82	0.03	7.40	0.58
error E	18.39	0.83	0.92	2.93	3.31	1.34	2.73	2.37	4.69	4.68	3.01	1.46	5.83	8.02	3.27

heavily temperature-dependent [12]. For example, our CPU at 1.3 GHz shows an inflated power consumption of around 5 % between a CPU temperature of 36 °C and 46 °C. You et al. [18] shows similar results for a 0.1 μm processor; a temperature increase from 30 °C to 40 °C leads to a 3 % power increase. On the other hand, the power P_{charge} required for a given computation does not change with regards to the CPU temperature. The Berkeley Short-channel IGFET Model (BSIM) [12] shows that the leakage current micro models depend on a multitude of variables. The temperature itself appears several times in the sub-threshold and BTBT leakage models; the gate leakage however is not temperature dependent. Mukhopadhyay et al. [13] showed via simulation that for 25 nm technology the sub-threshold leakage current is dominant over the BTBT leakage current, but the latter cannot be neglected. Under normal conditions, the temperature of the CPU's silicon varies continuously depending on the load of the CPU and the system's ambient temperature. Therefore, to have a fair comparison of energy consumption between different code pieces one needs to compare the measurements at a reference temperature.

Finding a temperature scaling factor for the leakage current is however not a straightforward task. Nevertheless, approximative scaling factors have been analytically obtained or experimentally defined via simulations (mainly SPICE) [5, 11, 16]. After analysis on our data, we discovered that none of the cited approximations would fit well. This is because the rationale on which these approximations are based assume conditions, which are not entirely realistic, to simplify the leakage current micro models. Most previous research works focus solely on the sub-threshold leakage effect, neglecting other leakage effects. This may be appropriate for technologies larger than the 45 nm technology we use.

Skadron et al. [14] studied the temperature dependence of I_{leak} as well. Skadron et al. deducted a relationship between the leakage power P_{leak} and dynamic power P_{dynamic} based on International Technology Roadmap for semiconductors (ITRS) measurement traces (variables indexed with 0 are reference values):

$$R_T = \frac{P_{\text{leak}}}{P_{\text{dynamic}}} = \frac{R_0}{V_0 T_0^2} e^{\frac{B}{T_0}} V T^2 e^{\frac{-B}{T}}. \tag{3}$$

If the temperature T is stable across different operating voltages, then the value of R_T is a function of V multiplied by a constant γ, which includes the temperature dependent variables and other constants. Total power P_{total} is thus:

$$\begin{aligned} P_{\text{total}} &= P_{\text{system}} + P_{\text{leak}} + P_{\text{dynamic}} \\ &= P_{\text{system}} + \gamma V P_{\text{dynamic}} + P_{\text{dynamic}} \\ &= P_{\text{system}} + (1 + \gamma V) \cdot \eta \alpha C f V^2. \end{aligned} \tag{4}$$

This formulation of P_{total} incorporates three parameters: P_{system}, γ, and $\eta \alpha C$. The values of these variables can be obtained via fitting power traces on Eq. 4. V and f are linked via the DVFS process inherent to the Linux kernel and the hardware technicalities. Experimental values for our CPU are found inside the Siyah kernel; they are shown in Table 2. The power fitting errors are shown in

Table 2. Frequency and voltage relationship for the Dynamic Voltage and Frequency Scaling (DVFS) process in the default Siyah kernel.

f (MHz)	200	300	400	500	600	700	800	900	1000	1100	1200	1300	1400	1500	1600
V (mV)	920	950	950	950	975	1000	1025	1075	1125	1175	1225	1250	1275	1325	1350

Table 1 for a 37 °C CPU temperature. The fitting errors are on the average not larger than 3 % except for the measurement point at 1.5 GHz. This measurement point was obtained at different independent occasions but appears, for obscure reasons, to disobey persistently the model in Eq. 4.

4.3   Energy Consumption

Typical compute-intensive programs incur approximately a constant load on the CPU and system, barring user interactions. Moreover, if the time to complete one program is also much smaller than the sampling rate of the power gauge, then $P(t)$ in Eq. 1 is constant. Hence, it suffices to sample the power P_{bench} of a benchmark at a given CPU temperature and multiply this value by the execution time of the benchmark t_{bench} to get an energy estimate. As a result the definition of time in Eq. 2, and power in Eq. 4, can be used to model the energy of one benchmark kernel. The energy consumed by the CPU E_{CPU} is given by

$$E_{\text{CPU}} = E_{\text{leak}} + E_{\text{dynamic}}$$
$$= P_{\text{bench}} \cdot t_{\text{bench}}$$
$$= \left((1 + \gamma V) \cdot \eta \alpha C f V^2 \right) \cdot \frac{cc_{\text{b}}}{f^\beta - cc_{\text{k}}}. \tag{5}$$

Constants γ, $\eta \alpha C$, cc_{b}, and cc_{k} in this formulation were evaluated before via fitting the power and time traces.

5   The Energy/Frequency Convexity Rule

Using the testbed and models described above, Fig. 1 shows the measured and modeled energy E_{CPU} for our benchmark kernel over the different frequencies; data have been normalized over the benchmark input size. Table 1 shows the average absolute energy error between our fitted model and the measured data. The average fitting error stays below 6 % except for measurement points 1.5 GHz and 200 MHz. The large fitting error in the 200 MHz case stems from the large execution time that amplifies the power measurement fitting error (see Table 1). It can also be seen that, for larger benchmark input sizes, on the average more energy is required. This is the result of higher level cache utilization.

Figure 1 exhibits a clear convex curve, with a minimum at Frequency f_{opt}, suggesting the existence of an Energy/Frequency Convexity Rule for compute-intensive programs. Why is the energy consumption curve convex? The energy

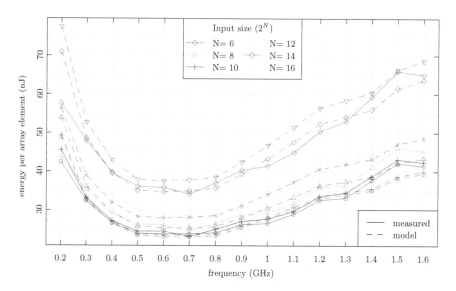

Fig. 1. Energy required by the CPU at 37 °C to complete our benchmark kernel given an input size. The dashed lines denote the theoretical curve as per Eq. 5.

consumption of the benchmark kernel scales approximately linearly with the number of instructions. The time Δt it takes to execute an instruction sequence increases more than linearly with decreasing operating frequency. P_{leak} is independent of the type of computation, and E_{leak} builds up linearly with time: $E_{\text{leak}} = P_{\text{leak}}\Delta t$. P_{leak} becomes increasingly important in the part where the CPU frequency f is smaller than f_{opt}. For the part where $f > f_{\text{opt}}$, the inflated E_{CPU} can be attributed to the increasing supply voltage V_{CC} affecting P_{dynamic} quadratically. Furthermore, had P_{system} been incorporated in the picture, then f_{opt} would have moved to a higher frequency because the additional consumed energy of the system could have been minimized by a faster completion of the computations on the CPU.

Our proposal for the existence of an Energy/Frequency Convexity Rule can be further supported using our previous models. Indeed, we can model the relationship in Table 2 between the *frequency* (GHz) and *voltage* (V) in the Linux kernel with a linear approximation: $V = m_1 f + m_2$. Now the derivative of E_{CPU} defined in Eq. 5 over f or V can be computed. The energy curve shows a global minimum $E_{\text{CPU,min}}$ for f_{opt} when its derivative is equal to zero ($\partial E_{\text{CPU}}/\partial f = 0$) and its second derivative is positive.

Given that E_{CPU} only shows one minimum, f_{opt} is the global minimum if the following equality holds:

$$\frac{(1 + \gamma V)V f^\beta \beta}{f^\beta - cc_{\text{k}}} = fm_1(3\gamma V + 2) + (1 + \gamma V)V. \tag{6}$$

Four parameters appear in this formulation that affect the optimal frequency f_{opt}: β and cc_{k}, which are related to the execution time of the benchmark, m_1

the slope between V and f, and γ related to the leakage current ratio. Simulations show that, if β or cc_k decreases, $E_{CPU,min}$ will shift to a higher frequency, and, if m_1 or γ decreases, f_{opt} will decrease as well. γ is temperature dependent; if the temperature increases, γ will increase accordingly. Hence, $E_{CPU,min}$ and f_{opt} increase with temperature as well. For the presented measurements Eq. 5 shows a minimum on the average around 700 MHz. This holds for all input sizes of the benchmark between 2^6 and 2^{16}. As a result, this implies that there exists an operating frequency, which is neither the maximum nor the minimum operating frequency, at which the CPU would execute a code sequence on the top of the OS in the most energy efficient way.

6   Related Work

The convex property of the energy consumption curve has been hinted at before in the literature. A series of papers, approaching the problem from an architectural point of view, have shown a convex energy consumption curve with respect to DVFS [4,10,15]. The authors put forward some motivation, but do not provide an analytical framework. Other studies, e.g., Hager et al. [8] and Freeh et al. [6], discuss what the consequences are of said behavior and how to exploit them, from a high-level point of view. A detailed explanation or an analytical derivation, as presented here, is however not highlighted.

7   Conclusion

We provide an analytical model to describe the energy consumption of a code sequence running on top of the OS of a mobile device. The energy model is parameterized over five parameters abstracting the specifics of the Dynamic Voltage and Frequency Scaling (DVFS) process, the execution time related parameters, and the power specifications of the CPU. Measurement traces from a mobile device were used to validate the appropriately fitted model. It is shown that the model is on the average more than 6 % accurate. The importance of power samples obtained at a reference temperature is also pointed out.

It is also shown that the analytical energy model is convex (representing what we call the Energy/Frequency Convexity Rule) and yields a minimum energy consumption of a code sequence for a given CPU operation frequency. This minimum is a function of the temperature, execution time related parameters, and technical parameters related to the hardware. A more in depth analysis of the Energy/Frequency Convexity Rule can be found in our technical report [3].

Future work includes checking the validity of our model and its parameters over a wide range of compute-intensive benchmarks. Also, extending the presented model to better handle memory access operations, in particular the impact of caches, is deserved. Finally a generalization of the model to encompass the impact of other programs running in parallel with benchmarks or system power effects would be useful.

References

1. Agarwal, A., Mukhopadhyay, S., Kim, C., Raychowdhury, A., Roy, K.: Leakage power analysis and reduction: models, estimation and tools. IEEE Proc. Comput. Digital Tech. **152**(3), 353–368 (2005)
2. Carroll, A., Heiser, G.: An analysis of power consumption in a smartphone. In: Proceedings of the USENIX Conference on USENIX, Berkeley (2010)
3. De Vogeleer, K., Memmi, G., Jouvelot, P., Coelho, F.: Energy consumption modeling and experimental validation on mobile devices. Technical Report 2013D008, TELECOM ParisTech, December 2013
4. Fan, X., Ellis, C.S., Lebeck, A.R.: The synergy between power-aware memory systems and processor voltage scaling. In: Falsafi, B., VijayKumar, T.N. (eds.) PACS 2003. LNCS, vol. 3164, pp. 164–179. Springer, Heidelberg (2005)
5. Ferre, A., Figueras, J.: Characterization of leakage power in cmos technologies. In: 1998 IEEE International Conference on Electronics, Circuits and Systems, vol. 2, pp. 185–188 (1998)
6. Freeh, V.W., Lowenthal, D.K., Pan, F., Kappiah, N., Springer, R., Rountree, B.L., Femal, M.E.: Analyzing the energy-time trade-off in high-performance computing applications. IEEE Trans. Parallel Distrib. Syst. **18**(6), 835–848 (2007)
7. Gold, B., Rader, C.M.: Digital Processing of Signals. McGraw-Hill, New York (1969)
8. Hager, G., Treibig, J., Habich, J., Wellein, G.: Exploring performance and power properties of modern multicore chips via simple machine models. CoRR abs/1208.2908 (2012)
9. Ickin, S., Wac, K., Fiedler, M., Janowski, L., Hong, J.H., Dey, A.: Factors influencing quality of experience of commonly used mobile applications. IEEE Commun. Mag. **50**(4), 48–56 (2012)
10. Le Sueur, E., Heiser, G.: Dynamic voltage and frequency scaling: the laws of diminishing returns. In: Proceedings of the 2010 International Conference on Power Aware Computing and Systems, HotPower'10, Berkeley, pp. 1–8 (2010)
11. Liao, W., He, L., Lepak, K.M.: Temperature and supply voltage aware performance and power modeling at microarchitecture level. Trans. Comp. Aided Des. Integ. Cir. Sys. **24**(7), 1042–1053 (2006)
12. Liu, W., Jin, X., Kao, K., Hu, C.: BSIM 4.1.0 MOSFET model-user's manual. Technical Report UCB/ERL M00/48, EECS Department, University of California, Berkeley (2000)
13. Mukhopadhyay, S., Raychowdhury, A., Roy, K.: Accurate estimation of total leakage current in scaled cmos logic circuits based on compact current modeling. In: Proceedings of the Design Automation Conference 2003, pp. 169–174, June 2003
14. Skadron, K., Stan, M.R., Sankaranarayanan, K., Huang, W., Velusamy, S., Tarjan, D.: Temperature-aware microarchitecture: modeling and implementation. ACM Trans. Archit. Code Optim. **1**(1), 94–125 (2004)
15. Snowdon, D.C., Ruocco, S., Heiser, G.: Power management and dynamic voltage scaling: Myths and facts. In: 2005 WS Power Aware Real-time Comput. New Jersey, September 2005
16. Su, H., Liu, F., Devgan, A., Acar, E., Nassif, S.: Full chip leakage estimation considering power supply and temperature variations. In: Proceedings of the 2003 International Symposium on Low power Electronics and Design, ISLPED '03, pp. 78–83. ACM, New York (2003)

17. Weste, N.H.E., Eshraghian, K.: Principles of CMOS VLSI Design: A Systems Perspective. Addison-Wesley Longman Publishing Co. Inc., Boston (1985)
18. You, Y.-P., Lee, Ch., Lee, J.-K.: Compiler analysis and supports for leakage power reduction on microprocessors. In: Pugh, B., Tseng, C.-W. (eds.) LCPC 2002. LNCS, vol. 2481, pp. 45–60. Springer, Heidelberg (2005)

Author Index

Printed in the United States
By Bookmasters